The Calcium Channel: Structure, Function and Implications

Bayer AG Centenary Symposium
Stresa/Italy, May 11–14, 1988

Editors: M. Morad
W. Nayler
S. Kazda
M. Schramm

Springer-Verlag
Berlin Heidelberg New York
London Paris Tokyo

Martin Morad
Professor of Physiology and Medicine
University of Pennsylvania, Department of Physiology
Philadelphia, PA 19104–6085, USA

Winifred G. Nayler
Principal Research Investigator
University of Melbourne
Department of Medicine, Austin Hospital
Heidelberg, 3084, Victoria, Australia

Stanislav Kazda
Associate Professor of Pharmacology and Toxicology
Bayer AG, Institute of Pharmacology
Aprather Weg 18a, D-5600 Wuppertal 1, FRG

Matthias Schramm
Associate Professor of Physiology
Bayer AG, Institute of Pharmacology
Aprather Weg 18a, D-5600 Wuppertal 1, FRG

The editors are indebted to Miss Annelore Schmidt
for her valuable assistance

ISBN-13:978-3-540-50061-2   e-ISBN-13:978-3-642-73914-9
DOI: 10.1007/978-3-642-73914-9

This work is subject to copyright. All rights are reserved, whether the whole or part of the material is concerned, specifically the rights of translation, reprinting, reuse of illustrations, recitation, broadcasting, reproduction on microfilms or in other ways, and storage in data banks. Duplication of this publication or parts thereof is only permitted under the provisions of the German Copyright Law of September 9, 1965, in its version of June 24, 1985, and a copyright fee must always be paid. Violations fall under the prosecution act of the German Copyright Law.

© Springer-Verlag Berlin Heidelberg 1988

The use of general descriptive names, trade names, trade marks, etc. in this publication, even if the former are not especially identified, is not to be taken as a sign that such names, as understood by the Trade Marks and Merchandise Marks Act, may accordingly be used freely by anyone.

Product Liability: The publisher can give no guarantee for information about drug dosage and application thereof contained in this book. In every individual case the respective user must check its accuracy by consulting other pharmaceutical literature.

Contents

History and Philosophy of Bayer Pharmaceutical Research
W.-D. Busse 1

The Effect of Lime and Potash on the Frog Heart:
An Imaginative Reconstruction of a Paper Presented by Sydney Ringer
to the Physiological Society on 9th December 1882
R. A. Chapman 9

Homology of Calcium-Modulated Proteins:
Their Evolutionary and Functional Relationships
R. H. Kretsinger, N. D. Moncrief, M. Goodman, and J. Czelusniak 16

1 The Function of the Calcium Channel

Kinetics, and β-Adrenergic Modulation of Cardiac Ca^{2+} Channels
W. Trautwein and D. Pelzer 39

Dihydropyridines, G Proteins, and Calcium Channels
A. M. Brown, L. Birnbaumer, and A. Yatani 54

Electrophysiology of Dihydropyridine Ca Agonists
M. Bechem and M. Schramm 63

Proton-Induced Transformation of Ca^{2+}-Channel:
Possible Mechanism and Physiological Role
M. Morad and G. Callewaert 71

Membrane Potential and Dihydropyridine Block of Calcium Channels
in the Heart: Influence of Drug Ionization on Blocking Activity
R. S. Kass and J. P. Arena 81

Dihydropyrine-Sensitive and Insensitive Ca^{2+} Channels in Normal and Transformed Fibroblasts

C. Chen, M. J. Corbley, T. M. Roberts and P. Hess 92

Protein Phosphorylation and the Inactivation of Dihydropyridine-Sensitive Calcium Channels in Mammalian Pituitary Tumor Cells

D. Kalman and D. L. Armstrong 103

Sodium Currents Through Neuronal Calcium Channels: Kinetics and Sensitivity to Calcium Antagonists

E. Carbone and H. D. Lux 115

Block of Sodium Currents Through a Neuronal Calcium Channel by External Calcium and Magnesium Ions

H. D. Lux, E. Carbone, and H. Zucker 128

The Voltage Sensor of Skeletal Muscle Excitation-Contraction Coupling: A Comparison with Ca^{2+} Channels

G. Pizarro, G. Brum, M. Fill, R. Fitts, M. Rodriguez, I. Uribe, and E. Rios 138

2 The Structure of the Calcium Channel

Biochemistry and Molecular Pharmacology of Ca^{2+} Channels and Ca^{2+}-Channel Blockers

J. Barhanin, M. Fosset, M. Hosey, C. Mourre, D. Pauron, J. Qar, G. Romey, A. Schmid, S. Vandaele, and M. Lazdunski 159

The Structure of the Ca^{2+} Channel: Photoaffinity Labeling and Tissue Distribution

H. Glossmann, J. Striessnig, L. Hymel, G. Zernig, H.-G. Knaus, and H. Schindler 168

Site-Specific Phosphorylation of the Skeletal Muscle Receptor for Calcium-Channel Blockers by cAMP-Dependent Protein Kinase

A. Röhrkasten, H. E. Meyer, T. Schneider, W. Nastainczyk, M. Sieber, H. Jahn, S. Regulla, P. Ruth, V. Flockerzi, and F. Hofmann 193

Molecular Properties of Dihydropyridine-Sensitive Calcium Channels from Skeletal Muscle

M. J. Seagar, M. Takahashi, and W. A. Catterall 200

Molecular Characterization of the 1,4-Dihydropyridine Receptor in Skeletal Muscle

P. L. VAGHY, E. MCKENNA, and A. SCHWARTZ 211

Reconstitution of Solubilized and Purified Dihydropyridine Receptor from Skeletal Muscle Microsomes as Two Single Calcium-Channel Conductances with Different Functional Properties

D. PELZER, A. CAVALIE, V. FLOCKERZI, F. HOFMANN, and W. TRAUTWEIN 217

1,4-Dihydropyridines as Modulators of Voltage-Dependent Calcium-Channel Acitivity

D. W. CHESTER and L. G. HERBETTE 231

Dihydropyridine Pharmacology of the Reconstituted Calcium Channel of Skeletal Muscle

H. VALDIVIA and R. CORONADO 252

Expression of mRNA Encoding Rat Brain Ca^{2+} Channels in Xenopus Oocytes

H. A. LESTER, T. P. SNUTCH, J. P. LEONARD, J. NARGEOT, and N. DAVIDSON . . . 272

3 Calcium Channels in the Cardiovascular and Endocrine System

Ca^{2+} Pathways Mediating Agonist-Activated Contraction of Vascular Smooth Muscle and EDRF Release from Endothelium

N. J. LODGE and C. VAN BREEMEN 283

Calcium Channels and the Heart

J. S. DILLON, X. H. GU, and W. G. NAYLER 293

Role of Calcium Channels in Cardiac Arrythmias

E. CARMELIET 310

General Anaesthetics and Antiarrhythmics Antagonize the Ca^{2+} Current in Single Frog Cardiac Cells

P. SCAMPS, K. MONGO, A. UNDROVINAS, and G. VASSORT 317

The Calcium Channel and Vascular Injury

S. KAZDA 326

Endocrine Effects of Calcium Channel Agonists and Antagonists
in a Pituitary Cell System
P. M. Hinkle, R. N. Day, and J. J. Enyeart 335

L-Type Calcium Channels and Adrenomedullary Secretion
C. R. Artalejo, M. G. López, C. F. Castillo, M. A. Moro,
and A. G. García 347

Regulation of Signal Transduction by G Proteins in Exocrine Pancreas Cells
I. Schulz, S. Schnefel, and R. Schäfer 363

Calcium Channel Blockers and Renin Secretion
P. C. Churchill 372

Action of Parathyroid Hormone on Calcium Transport in Rat Brain
Synaptosomes is Independent of cAMP
C. L. Fraser and P. Sarnacki 394

Calcium and the Mediation of Tubuloglomerular Feedback Signals
P. D. Bell, M. Franco, and M. Higdon 404

Inhibition of Parathyroid Hormone Secretion by Calcium:
The Role of Calcium Channels
L. A. Fitzpatrick and H. Chin 418

4 Neuropharmacology of Calcium-Channels

The Neuropharmacology of Ca^{2+} Channels
R. J. Miller, D. A. Ewald, S. N. Murphy, T. M. Perney, I. J. Reynolds,
S. A. Thayer, and M. W. Walker 433

Kinetic Characteristics of Different Calcium Channels
in the Neuronal Membrane
P. G. Kostyuk, Y. M. Shuba, A. N. Savchenko, and V. I. Teslenko 442

Increased Calcium Currents in Rat Hippocampal Neurons During Aging
P. W. Landfield 465

Regulation of Brain 1,4-Dihydropyridine Receptors by Drug Treatment
V. Ramkumar and E. E. El-Fakahany 478

Nimodipine and Neural Plasticity in the Peripheral Nervous System
of Adult and Aged Rats

W. H. GISPEN, T. SCHUURMAN, and J. TRABER 491

Anticonvulsant Properties of Dihydropyridine Calcium Antagonists

F. B. MEYER, R. E. ANDERSON, and T. M. SUNDT JR. 503

Interaction Between Certain Antipsychotic Drugs and Dihydropyridine
Receptor Sites

R. QUIRION, D. BLOOM, and N. P. V. NAIR 520

Effects of Organic Calcium-Entry Blockers on Stimulus-Induced Changes in
Extracellular Calcium Concentration in Area CA 1 of Rat Hippocampal Slices

U. HEINEMANN, P. IGELMUND, R. S. G. JONES, G. KÖHR, and H. WALTHER 528

Alcohol, Neurodegenerative Disorders and
Calcium-Channel Antagonist Receptors

D. A. GREENBERG, R. O. MESSING, S. S. MARKS, C. L. CARPENTER,
and D. L. WATSON 541

5 Endogenous Ligands and Antibodies

Endogenous Ligands for the Calcium Channel: Myths and Realities

D. J. TRIGGLE 549

Endogenous 1,4-Dihydropyridine-Displacing Substances Acting
on L-Type Ca^{2+} Channels:
Isolation and Characterization of Fractions from Brain and Stomach

R. A. JANIS, D. E. JOHNSON, A. V. SHRIKHANDE, R. T. MCCARTHY,
A. D. HOWARD, R. GREGUSKI, and A. SCRIABINE 564

Endothelium-Derived Novel Vasoconstrictor Peptide Endothelin:
A Possible Endogenous Agonist for Voltage-Dependent Ca^{2+} Channels

M. YANAGISAWA, H. KURIHARA, S. KIMURA, K. GOTO, and T. MASAKI 575

Calcium-Channel Antibodies:
Subunit-Specific Antibodies as Probes for Structure and Function

K. P. CAMPBELL, A. T. LEUNG, A. H. SHARP, T. IMAGAWA, and S. D. KAHL 586

Endogenous Ligands for Voltage-Sensitive Calcium Channels in Extracts
of Rat and Bovine Brain

B. J. EBERSOLE and P. B. MOLINOFF 601

An Endogenous Purified Peptide Modulates Ca^{2+} Channels in Neurons and Cardiac Myocytes
I. HANBAUER, E. SANNA, G. CALLEWAERT, and M. MORAD 611

Antibodies Against the ADP/ATP Carrier Interact with the Calcium Channel and Induce Cytotoxicity by Enhancement of Calcium Permeability
H.-P. SCHULTHEISS, I. JANDA, U. KÜHL, G. ULRICH, and M. MORAD 619

Calcium, Aging and Disease

Calcium, Aging and Diseases
T. FUJITA 635

Subject Index 640

Contributors

R. E. Anderson
Mayo Graduate School, Rochester, Minnesota 55905, USA

J. P. Arena
Department of Physiology, University of Rochester School of Medicine and Dentistry, Rochester, NY 14642, USA

D. L. Armstrong
Laboratory of Cellular and Molecular Pharmacology, National Institute of Environmental Health Sciences, P. O. Box 12233, Research Triangle Park, NC 27709, USA

C. R. Artalejo
Departamento de Farmacologia, Facultad de Medicina, Universidad Autónoma de Madrid, Arzobispo Morcillo, 4, 28029 Madrid, Spain

J. Barhanin
Centre de Biochimie du Centre National de la Recherche Scientifique, Parc Valrose, 06034 Nice Cedex, France

M. Bechem
Institut für Pharmakologie, Bayer AG, Aprather Weg 18a, 5600 Wuppertal 1, FRG

P. D. Bell
Dept. of Physiology and Biophysics, Nephrology Research and Training Center, Room 701 Sparks Center, University of Alabama at Birmingham, Birmingham, Alabama 35294, USA

D. Bloom
Douglas Hospital Research Centre and Department of Psychiatry, Faculty of Medicine, McGill University, 6875 Boulevard LaSalle, Verdun, Québec, Canada H4H 1R3

L. BIRNBAUMER
Departments of Physiology and Molecular Biophysics, and Cell Biology, Baylor College of Medicine, One Baylor Plaza, Houston, TX 77030, USA

C. VAN BREEMEN
Department of Pharmacology, University of Miami School of Medicine, Miami, FL 33101, USA

A. M. BROWN
Department of Physiology and Molecular Biophysics, Baylor College of Medicine, One Baylor Plaza, Houston, TX 77030, USA

G. BRUM
Departamento di Biofísica, Facultad de Medicina, Montevideo, Uruguay

W.-D. BUSSE
Bayer AG, Leitung Fachbereich Pharma-Forschung, Aprather Weg 18a, D-5600 Wuppertal 1, FRG

G. CALLEWAERT
Department of Physiology, University of Pennsylvania, Philadelphia, PA 19104, USA

K. P. CAMPBELL
Department of Physiology and Biophysics, The University of Iowa College of Medicine, Iowa City, IA 52242, USA

E. CARBONE
Dipartimento di Anatomia e Fisiologia Umana, Corso Raffaello 30, 10125 Torino, Italy

E. CARMELIET
Laboratory of Physiology, University of Leuven, Campus Gasthuisberg, Herestraat, 3000 Leuven, Belgium

C. L. CARPENTER
Department of Neurology, University of California, and Ernest Gallo Clinic and Research Center, San Francisco General Hospital, San Francisco, CA 94110, USA

C. F. CASTILLO
Departamento de Farmacología, Facultad de Medicina, Universidad Autónoma de Madrid, Arzobispo Morcillo, 4, 28029 Madrid, Spain

W. A. CATTERALL
Department of Pharmacology, University of Washington, Seattle, WA 98195, USA

A. CAVALIE

II. Physiologisches Institut, Medizinische Fakultät, Universität des Saarlandes, 6650 Homburg/Saar, FRG

R. A. CHAPMAN

The Laboratory of Cellular Cardiology, School of Veterinary Science, Park Row, The University of Bristol, Bristol BS1 5LS, United Kingdom

C. CHEN

Department of Cellular and Molecular Physiology and Program in Neuroscience, Harvard Medical School, Boston, MA 02115, USA

D. W. CHESTER

Department of Medicine and Biomolecular Structure Analysis Center, University of Connecticut Health Center, Farmington, CT 06032, USA

H. CHIN

Laboratory of Biochemical Genetics, National Heart, Lung and Blood Institute, National Institutes of Health, Bethesda, MD 20892, USA

P. C. CHURCHILL

Department of Physiology, Wayne State University School of Medicine, 540 East Canfield, Detroit, Michigan 48201, USA

J. CORBLEY

Dana-Farber Cancer Institute and Departments of Biological Chemistry and Molecular Pharmacology, Harvard Medical School, Boston, MA 02115, USA

R. CORONADO

Department of Physiology and Molecular Biophysics, Baylor College of Medicine, Houston, TX 77030, USA

J. CZELUSNIAK

Department of Anatomy and Cell Biology, Wayne State University School of Medicine, Detroit, MI 48201, USA

N. DAVIDSON

Dept. of Biological Sciences, University of Illinois at Chicago, Chicago, IL 60680, USA

R. N. DAY

Department of Pharmacology and the Cancer Center, University of Rochester School of Medicine and Dentistry, Rochester, NY 14642, USA

J. S. DILLON

Department of Medicine, University of Melbourne, Austin Hospital, Heidelberg, Victoria, 3084, Australia

B. J. EBERSOLE
Department of Pharmacology, University of Pennsylvania School of Medicine,
Philadelphia, PA 19104-6084, USA

E. E. EL-FAKAHANY
Department of Pharmacology and Toxicology, University of Maryland School of
Pharmacy, 20 N. Pine Street, Baltimore, MD 21201, USA

J. J. ENYEART
Department of Pharmacology and the Cancer Center, University of Rochester
School of Medicine and Dentistry, Rochester, NY 14642, USA

D. A. EWALD
Department of Pharmacological and Physiological Sciences, University of Chicago,
Chicago, IL 60637, USA

M. FILL
Department of Physiology and Molecular Biophysics, Baylor College, Houston,
TX, USA

R. FITTS
Department of Biology, Marquette University, Milwaukee, Wisconsin, USA

L. A. FITZPATRICK
Department of Medicine, Division of Endocrinology, University of Texas Health
Science Center, 7703 Floyd Curl Drive, San Antonio, TX 78284, USA

V. FLOCKERZI
Institut für Physiologische Chemie, Medizinische Fakultät, Universität des
Saarlandes, 6650 Homburg/Saar, FRG

M. FOSSET
Centre de Biochimie du Centre National de la Recherche Scientifique,
Parc Valrose, 06034 Nice Cedex, France

M. FRANCO
Instituto de Cardiología, I. Chavez, México City, México

C. L. FRASER
Department of Medicine, Divisions of Nephrology and Geriatrics, Veterans
Administration Medical Center, 4150 Clement Street, San Francisco, CA 94121,
USA

T. FUJITA
Third Division, Department of Medicine, Kobe University, School of Medicine,
7-5-2 Kusunoki-cho, Chuo-ku, Kobe 650, Japan

A. G. García
Departamento de Farmacologia, Facultad de Medicina, Universidad Autónoma de Madrid, Arzobispo Morcillo, 4, 28029 Mardrid, Spain

W. H. Gispen
Rudolf Magnus Institute for Pharmacology, University of Utrecht, Padualaan 8, 3584 CH Utrecht, The Netherlands

H. Glossmann
Institut für Biochemische Pharmakologie, Universität Innsbruck, 6020 Innsbruck, Austria

M. Goodman
Department of Anatomy and Cell Biology, Wayne State University School of Medicine, Detroit, MI 48201, USA

K. Goto
Institute of Basic Medical Sciences, University of Tsukuba, Tsukuba, Ibaraki 305, Japan

D. A. Greenberg
San Francisco General Hospital, Building 1, Room 101, San Francisco, CA 94110, USA

R. Greguski
Institute for Preclinical Pharmacology, Miles Inc., 400 Morgan Lane, West Haven, CT 06516, USA

X. H. Gu
Department of Medicine, University of Melbourne, Austin Hospital, Heidelberg, Victoria, 3084, Australia

I. Hanbauer
HE-B, National Heart, Lung, and Blood Institute, National Institutes of Health, Bethesda, MD 20892, USA

U. Heinemann
Institut für normale und pathologische Physiologie, Universität zu Köln, Robert-Koch-Str. 39, 5000 Köln 41, FRG

L. G. Herbette
Department of Medicine and Biomolecular Structure Analysis Center, University of Connecticut Health Center, Farmington, CT 06032, USA

P. Hess
Department of Cellular and Molecular Physiology, Harvard Medical School, 25 Shattuck Street, Boston, MA 02115, USA

M. Higdon
   Nephrology Research and Training Center, Departments of Physiology and
   Biophysics and of Medicine, University of Alabama at Birmingham, Birmingham,
   AL 35294, USA

P. M. Hinkle
   Box Pharmacology, University of Rochester Medical Center,
   601 Elmwood Avenue, Rochester, NY 14642, USA

F. Hofmann
   Institut für Physiologische Chemie, Medizinische Fakultät, Universität des
   Saarlandes, 6650 Homburg/Saar, FRG

M. M. Hosey
   Dept. of Biological Chemistry & Structure, University of Health Sciences/The
   Chicago Medical School, 3333 Green Bay Road, North Chicago, IL 60064, USA

A. D. Howard
   Institute for Preclinical Pharmacology, Miles Inc., 400 Morgan Lane, West Haven,
   CT 06516, USA

L. Hymel
   Institut für Biophysik, Universität Linz, 4020 Linz, Austria

P. Igelmund
   Institut für normale und pathologische Physiologie, Universität zu Köln,
   Robert-Koch-Str. 39, 5000 Köln 41, FRG

T. Imagawa
   Department of Physiology and Biophysics, University of Iowa College of Medicine,
   Iowa City, Iowa 52242, USA

H. Jahn
   Institut für Physiologische Chemie, Medizinische Fakultät, Universität des
   Saarlandes, 6650 Homburg/Saar, FRG

I. Janda
   Department of Internal Medicine, Klinikum Großhadern, Universität München,
   8000 München 70, FRG

R. A. Janis
   Institute for Preclinical Pharmacology, Miles Inc., 400 Morgan Lane, West Haven,
   CT 06516, USA

D. E. Johnson
   Institute for Preclinical Pharmacology, Miles Inc., 400 Morgan Lane, West Haven,
   CCT 06516, USA

R. S. G. Jones
Institut für normale und pathologische Physiologie, Universität zu Köln,
Robert-Koch-Str. 39, 5000 Köln 41, FRG

S. D. Kahl
Department of Physiology and Biophysics, University of Iowa College of Medicine,
Iowa City, Iowa 52242, USA

D. Kalman
Department of Biology, University of California, 405 Hilgard Avenue,
Los Angeles, CA 90024, USA

R. S. Kass
Department of Physiology, University of Rochester School of Medicine and
Dentistry, Rochester, NY 14642, USA

S. Kazda
Institut für Pharmakologie, Bayer AG, Aprather Weg 18a, 5600 Wuppertal 1,
FRG

S. Kimura
Institute of Basic Medical Sciences, University of Tsukuba, Tsukuba, Ibaraki 305,
Japan

H.-G. Knaus
Institut für Biochemische Pharmakologie, Universität Innsbruck, 6020 Innsbruck,
Austria

G. Köhr
Institut für normale und pathologische Physiologie, Universität zu Köln,
Robert-Koch-Str. 39, 5000 Köln 41, FRG

P. G. Kostyuk
Bogomoletz Institute of Physiology, Ukrainian Academy of Sciences, Kiev, USSR

H. Kretsinger
Department of Biology, University of Virginia, Charlottesville, VA 22901, USA

U. Kühl
Department of Internal Medicine, Klinikum Großhadern, Universität München,
8000 München 70, FRG

H. Kurihara
Institute of Basic Medical Sciences, University of Tsukuba, Tsukuba, Ibaraki,
Japan

P. W. LANDFIELD
Department of Physiology and Pharmacology, Bowman Gray School of Medicine of Wake Forest University, 300 South Hawthorne Road, Winston-Salem, N. C. 27103, USA

M. LAZDUNSKI
Centre de Biochimie du Centre National de la Recherche Scientifique, Parc Valrose, 06034 Nice Cedex, France

J. P. LEONARD
Dept. of Biological Sciences, University of Illinois at Chicago, Chicago, IL 60680, USA

H. A. LESTER
156-29 Caltech, Pasadena, CA 91125, USA

A. T. LEUNG
Department of Physiology and Biophysics, The University of Iowa College of Medicine, Iowa City, IA 52242, USA

N. J. LODGE
Department of Pharmacology, University of Miami School of Medicine, Miami, FL 33101, USA

M. G. LÓPEZ
Departamento de Farmacología Facultad de Medicina, Universidad Autónoma de Madrid, Arzobispo Morcillo, 4, 28029 Madrid, Spain

H. D. LUX
Max-Planck-Institut für Psychiatrie, Abteilung Neurophysiologie, Am Klopferspitz 18 A, 8033 Planegg, FRG

S. S. MARKS
Department of Neurology, University of California, and Ernest Gallo Clinic and Research Center, San Francisco General Hospital, San Francisco, CA 94110, USA

T. MASAKI
Institute of Basic Medical Sciences, University of Tsukuba, Tsukuba, Ibaraki 305, Japan

R. T. MCCARTHY
Institute for Preclinical Pharmacology, Miles Inc., 400 Morgan Lane, West Haven, CT 06516, USA

E. MCKENNA
Department of Pharmacology and Cell Biophysics, University of Cincinnati College of Medicine, Cincinnati, Ohio 45267, USA

R. O. Messing
Department of Neurology, University of California, and Ernest Gallo Clinic and Research Center, San Francisco General Hospital, San Francisco, CA 94110, USA

F. B. Meyer
Department of Neurosurgery, Mayo Clinic, 200 First Street Southwest, Rochester, Minnesota 55905, USA

H. E. Meyer
Institut für Physiologische Chemie, Abteilung Biochemie Supramolekularer Systeme, Ruhr-Universität Bochum, 4630 Bochum, FRG

R. J. Miller
Dept. of Pharmacological and Physiological Sciences, University of Chicago, 47 E. 58th Street, Chicago, IL 60637, USA

P. B. Molinoff
Department of Pharmacology, University of Pennsylvania School of Medicine, Philadelphia, PA 19104–6084, USA

N. D. Moncrief
Department of Biology, University of Virginia, Charlottesville, VA 22901, USA

K. Mongo
Laboratoire de Physiologie Céllulaire Cardiaque, INSERM U-241, Université Paris-Sud, Bat. 443, 91405 Orsay, France

M. Morad
Department of Physiology, University of Pennsylvania, Philadelphia, PA 19104, USA

M. A. Moro
Departamento de Farmacologia, Facultad de Medicina, Universidad Autónoma e Madrid, Arzobispo Morcillo, 4, 28029 Madrid, Spain

C. Mourre
Centre de Biochimie du Centre de la Recherche Scientifique, Parc Valrose, 6034 Nice Cedex, France

S. N. Murphy
Department of Pharmacological and Physiological Sciences, University of Chicago, Chicago, IL 60637, USA

N. P. V. Nair
Douglas Hospital Research Centre and Department of Psychiatry, Faculty of Medicine, McGill University, 6875 Boulevard LaSalle, Verdun, Québec, Canada H4H 1R3

J. NARGEOT
Centre de biochimie macromoleculaire, C.N.R.S., 34033 Montpellier, France

W.G. NAYLER
Department of Medicine, University of Melbourne, Austin Hospital, Heidelberg, Victoria, 3084, Australia

W. NASTAINCZYK
Institut für Physiologische Chemie, Medizinische Fakultät, Universität des Saarlandes, 6650 Homburg/Saar, FRG

D. PAURON
Centre de Biochimie du Centre National de la Recherche Scientifique, arc Valrose, 06034 Nice Cedex, France

D. PELZER, II
Physiologisches Institut der Universität des Saarlandes, 6650 Homburg/Saar, FRG

T.M. PERNEY
Department of Pharmacological and Physiological Sciences, University of Chicago, Chicago, IL 60637, USA

G. PIZARRO
Department of Physiology, Rush University, 1750 W. Harrison St., Chicago, L 60612, USA

J. QAR
Marine Science Station of Aqaba, P.O.B., Jordan

R. QUIRION
Douglas Hospital Research Centre, 6875 Boulevard LaSalle, Verdu, Québec, Canada H4H 1R3

V. RAMKUMAR
Department of Medicine, Duke University Medical Center, Box 3444, Durham, NC 27710, USA

S. REGULLA
Institut für Physiologische Chemie, Medizinische Fakultät, Universität des Saarlandes, 6650 Homburg/Saar, FRG

I.J. REYNOLDS
Department of Pharmacological and Physiological Sciences, University of Chicago, Chicago, IL 60637, USA

Contributors

M. ROBERTS
Dana-Farber Cancer Institute and Departments of Pathology, Harvard Medical School, Boston, MA 02115, USA

G. ROMEY
Centre de Biochimie du Centre National de la Recherche Scientifique, Parc Valrose, 06034 Nice Cedex, France

P. RUTH
Institut für Physiologische Chemie, Medizinische Fakultät, Universität des Saarlandes, 6650 Homburg/Saar, FRG

E. Ríos
Department of Physiology, Rush University, 1750 W. Harrison St., Chicago, IL 60612, USA

M. RODRÍGUEZ
Department of Physiology, Rush University, 1750 W. Harrison St., Chicago, IL 60612, USA

A. RÖHRKASTEN
Institut für Physiologische Chemie, Medizinische Fakultät, Universität des Saarlandes, 6650 Homburg/Saar, FRG

E. SANNA
HE-B, National Heart, Lung, and Blood Institute, National Institutes of Health, Bethesda, MD 20892, USA

P. SARNACKI
University of California at San Francisco, San Francisco, California, USA

A.N. SAVCHENKO
Bogomoletz Institute of Physiology, Ukrainian Academy of Sciences, Kiev, USSR

P. SCAMPS
Laboratoire de Physiologie Céllulaire Cardiaque, INSERM U-241, Université Paris-Sud, Bat. 443, 91405 Orsay, France

R. SCHÄFER
Max-Planck-Institut für Biophysik, Kennedy-Allee 70, 6000 Frankfurt 70, FRG

H. SCHINDLER
Institut für Biophysik, Universität Linz, 4020 Linz, Austria

A. SCHMID
Centre de Biochimie du Centre National de la Recherche Scientifique, Parc Valrose, 06034 Nice Cedex, France

S. SCHNEFEL
Max-Planck-Institut für Biophysik, Kennedy-Allee 70, 6000 Frankfurt 70, FRG

T. SCHNEIDER
Institut für Physiologische Chemie, Medizinische Fakultät, Universität des Saarlandes, 6650 Homburg/Saar, FRG

M. SCHRAMM
Institut für Pharmakologie, Bayer AG, Aprather Weg 18a, 5600 Wuppertal 1, FRG

H.-P. SCHULTHEISS
Dept. of Internal Medicine, Klinikum Großhadern, Universität München, 8000 München, FRG

I. SCHULZ
Max-Planck-Institut für Biophysik, Kennedy-Allee 70, 6000 Frankfurt 70, FRG

T. SCHUURMAN
Institute of Molecular Biology and Medical Biotechnology, University of Utrecht, Padualaan 8, 3584 CH Utrecht, The Netherlands

A. SCHWARTZ
Department of Pharmacology and Cell Biophysics, University of Cincinnati College of Medicine, Cincinnati, Ohio 45267, USA

A. SCRIABINE
Institute for Preclinical Pharmacology, Miles Inc., 400 Morgan Lane, West Haven, CT 06516, USA

M.J. SEAGAR
Laboratoire de Biochimie, CNRS UA 1179-INSERM U 172, Faculté de Medecine Nord, 13326 Marseille Cédex 15, France

A.H. SHARP
Department of Physiology and Biophysics, The University of Iowa College of Medicine, Iowa City, IA 52242, USA

A.V. SHRIKHANDE
Institute for Preclinical Pharmacology, Miles Inc., 400 Morgan Lane, West Haven, CT 06516, USA

Y.M. SHUBA
Bogomoletz Institute of Physiology, Ukrainian Academy of Sciences, Kiev, USSR

M. SIEBER
Institut für Physiologische Chemie, Medizinische Fakultät,
Universität des Saarlandes, 6650 Homburg/Saar, FRG

T. P. SNUTCH
Divisions of Biology and of Chemistry and Chemical Engineering, California
Institute of Technology, Pasadeny, CA 91125, USA

J. STRIESSNIG
Institut für Biochemische Pharmakologie, Universität Innsbruck, 6020 Innsbruck,
Austria

T. M. SUNDT JR.
Cerebrovascular Research Laboratories, Department of Neurosurgery,
Mayo Clinic, 200 First Street Southwest, Rochester, MN 55905, USA

M. TAKAHASHI
Mitsubishi-Kasei Institute of Life Sciences, Machida-Shi, Tokyo, Japan

V. L. TESLENKO
Institute for Theoretical Physics, Ukrainian Academy of Sciences, Kiev, USSR

S. A. THAYER
Department of Pharmacological and Physiological Sciences, University of Chicago,
Chicago, IL 60637, USA

J. TRABER
Neurobiology Department, Troponwerke, Neurather Ring 1, 5000 Köln, FRG

W. TRAUTWEIN
Physiologisches Institut, Medizinische Fakultät, Universität des Saarlandes, 6650
Homburg/Saar, FRG

D. J. TRIGGLE
126 Cooke, School of Pharmacy, SUNY/Buffalo, New York, 14260, USA

G. ULRICH
Department of Internal Medicine, Klinikum Großhadern, Universität München,
8000 München 70, FRG

A. UNDROVINAS
Laboratoire de Physiologie Céllulaire Cardiaque, INSERM U-241, Université
Paris-Sud, Bat. 443, 91405 Orsay, France

I. URIBE
Department of Physiology, Rush University, 1750 W. Harrison St., Chicago,
IL 60612, USA

P. L. VAGHY
Department of Pharmacology and Cell Biophysics, University of Cincinnati College of Medicine, Cincinnati, Ohio 45267, USA

H. VALDIVIA
Department of Physiology and Molecular Biophysics, Baylor College of Medicine, Houston, TX 77030, USA

S. VANDAELE
Centre de Biochimie du Centre National de la Recherche Scientifique, Parc Valrose, 06034 Nice Cedex, France

G. VASSORT
Laboratoire de Physiolgie Céllulaire Cardiaque, INSERM U-241, Université Paris-Sud, Bat. 443, 91405 Orsay, France

M. W. WALKER
Department of Pharmacological and Physiological Sciences, University of Chicago, Chicago, IL 60637 USA

H. WALTHER
Institut für normale und pathologische Physiologie, Universität zu Köln, Robert-Koch-Str. 39, 5000 Köln 41, FRG

D. L. WATSON
Department of Neurology, University of California, and Ernest Gallo Clinic and Research Center, San Francisco General Hospital, San Francisco, CA 94110, USA

M. YANAGISAWA
Institute of Basic Medical Sciences, University of Tsukuba, Tsukuba, Ibaraki 305, Japan

N. YATANI
Department of Physiology and Molecular Biophysics, Baylor College of Medicine, One Baylor Plaza, Houston, TX 77030, USA

G. ZERNIG
Institut für Biochemische Pharmakologie, Universität Innsbruck, 6020 Innsbruck, Austria

H. ZUCKER
Abteilung Neurophysiologie, Max-Planck-Institut für Psychiatrie, 80033 Planegg, FRG

Moments to remember

① Kazda – a man with 3 hands!
② Schwartz, promises to say no more!
③ "Caged-Ca^{2+}" Kaplan and the grey eminence Carmeliet – 2 am conference!
④ Kostyuk, typical Russian tourist.
⑤ Morad, showing off newly-acquired baton!
⑥ Hille, a perfect chairman!
⑦ Pelzer, Kass, Kaplan, Beam, Hanbauer, Heinemann, engaged in a serious scientific discussion. Is that the disgenic mouth?
⑧ Moe! Where is Church?
⑨ Schultheiss (Herr und Frau) – "If you ask, you better listen!"
⑩ Bean vs. Molinoff: Harvard 1: Penn O.
⑪ Seagar & friend, batting for Catterall!
⑫ Reuter – happy to see that Ca^{2+} channels are still center stage
⑬ Group picture – a family gathering.
⑭ Trautwein, provocative as ever!
⑮ Scriabine – the mystery of the endogenous factor.
⑯ Lester, a new man – "expressed"!
⑰ Endotheline-Yanagisawa and Van Breemen with MM urging other climbers.
⑱ Franz Hofmann. Whatever you say, Prof. "W".
⑲ Lux. "You must stop now – next paper!"
⑳ Schramm – Big Brother is watching you.
㉑ The presidents – the odd couple.
㉒ Janis – drinking endogenous-free sherry!
㉓ Richard Miller – "I always need 10 minutes more!"
㉔ Carbone – caught looking.
㉕ Fitzpatrick – the "parathyroid lady"!
㉖, ㉗ Lazdunski & Schultz, the best two dancers!
㉘, ㉙ Vassort & Frazer, from Paris to San Francisco!
㉚ Dr. & Dr. Casteels – always Flemish!
㉛ Hess, telling Coronado about the open Ca^{2+} channel (see Figure).
㉜ Rios, the man from Montevideo.
㉝ Landfield – "I want to go shopping!"
㉞ Nayler – In response to Sydney Ringer – "Sir, I am a scientist."
㉟ Bechem – man of purpose!
㊱ Almers – a German capitan or a Japanese school boy!
㊲ Fox to Kostyuk – look, I did not say that!
㊳ Campbell with DHP-receptor (Tanabe) in hand but selling antibodies: Churchill, Hosey, Beam are not buying!
㊴ Buzz-Brown: G proteins, the best game in Texas!
㊵ Glossman, puffing on the photoaffinity pipe.
㊶ Dr. & Dr. Kazda, polka Italian style

History and Philosophy of Bayer Pharmaceutical Research

W.-D. Busse

Fachbereich Forschung Pharma, Bayer AG, Aprather Weg 18a, D-5600 Wuppertal, FRG

Research has a long tradition at Bayer. This year, 1988, marks 100 years of our pharmaceutical activities. In 1888 the first Bayer drug – phenacetin, an antipyretic – was synthesized, starting from a by-product of dye manufacture. This finding led to the establishment of a Pharmaceutical Department within Friedrich Bayer and Co. in Elberfeld, with all its associated facilities (Figs. 1, 2). The beginning of the company itself, however, dates back to 1863. In that year, Friedrich Bayer started the production of aniline dyestuffs in his private house in Barmen-Rittershausen. During the 125 years of the company's life and the 100 years of pharmaceuticals, many developments and products have been discovered by Bayer researchers and launched by Bayer. But we need consider only the landmarks of these pharmaceutical achievements to illustrate how man's research works for men.

Fig. 1. Original head offices of the company at Barmen-Rittershausen in 1888

Fig. 2. The Pharma Research Center at Wuppertal-Elberfeld in 1988

Fig. 3. Felix Hoffmann (1868–1946), who synthesized acetylsalicylic acid in 1897

Fig. 4. The first stable form of acetylsalicylic acid was developed in 1899. The Aspirin bottle from 1900 contains such powder

Felix Hoffmann, born in 1868, was searching for a drug to help his father, who suffered from rheumatism. He developed a pure and stable form of acetylsalicylic acid, better known under the trade name aspirin, which was launched in 1899 by Bayer. It became the analgesic of choice for millions of people, and even today 80 years later, it has not lost its importance in the pharmaceutical field (Figs. 3, 4). Moreover, its indications have been extended, after the elucidation of its mechanism of action in the late 1970s, and more recently by the results of clinical trials which demonstrated its efficacy in the prevention of heart infarction.

Worldwide recognition of Bayer was firmly established after the substantial achievements in the treatment of tropical diseases. Bayer 205 or germanin, active against sleeping sickness and launched in 1923, was developed by the Bayer chemists Oscar Dressel and Richard Kothe from an idea of Wilhelm Röhl. This achievement of curing sleeping sickness was regarded as so significant that a South American farmer and composer wrote the Bayer 205 tango. Thus, germanin is probably the only drug for which a piece of music has been composed.

The antimalarials Atebrin, Plasmochin, and Resochin came soon afterwards. Resochin is today still the most widely used antimalarial drug. These achievements in tropical medicine have made the Bayer cross a symbol for achievement, trust, safety, and quality.

There is no better example of therapeutic success and progress in the history of medicinal chemistry than the advances made in the chemotherapy of infectious diseases. The way was paved by Gerhard Domagk's discovery of the therapeutic effects of a sulfonamide that was introduced into medical therapy under the trade name of Prontosil. This compound, synthesized by Josef Klarer and Fritz Mietzsch, was a giant leap towards the therapeutic control of bacterial infections. For this discovery, Gerhard Domagk received the Nobel Prize for Medicine in 1939 (Figs. 5, 6).

Domagk and his colleagues also made major contributions to the chemotherapy of tuberculosis. Conteben (synthesized by Behnisch, Mietzsch, and Schmidt) was launched in 1946. Shortly afterwards, Offe and Siefken discovered isoniazid (Neoteben). These drugs were the first effective agents against the scourge of tuberculosis. The development of these drugs had a significant impact on pharmacotherapy in medicine, and our company is certainly proud of the significance of her scientific contributions.

The progress and the demands of modern pharmaceutical research made the laboratories located within the Bayer plant at Wuppertal obsolete. During the early 1960s, the building of the Pharma Research Center at Wuppertal was started. Today, more than 1300 employees work in pharmaceutical research and development. Of these, 260 are senior scientists. In addition, about 30 scientists are professors and exercise teaching responsibilities at universities.

The expenditure of time and finances for the development of new drugs has increased steadily over the years. Of approx. 10000 new chemical compounds synthesized and tested in our laboratories, only 1 will make it to the market place – and hopefully will also become a commercial success. Today, investment for the research and development of a new drug has reached a volume of DM 250–400 million. At the same time, the time needed to develop a drug has significantly increased to more than 10 years after discovery. It is clear that such high expenditure requires us to focus upon

Fig. 5. Professor Gerhard Domagk (1895–1964) in his laboratory in Wuppertal-Elberfeld in 1935

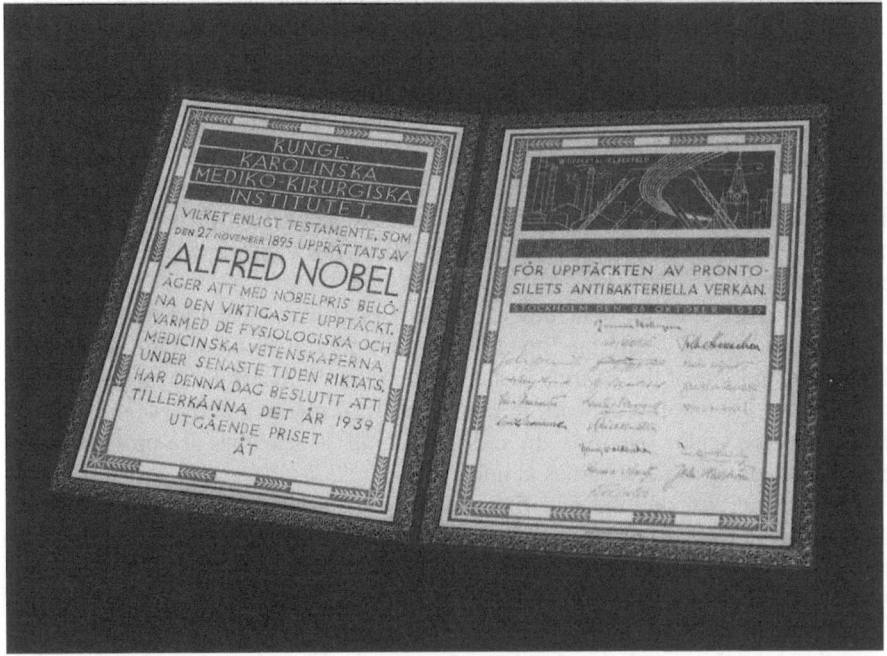

Fig. 6. Certificate confirming the award of the Nobel Prize for Medicine for the drug Prontosil in 1939

Fig. 7. The Prix Galien, the highly regarded French award named after Galen (The physician of Antiquity), was conferred on the therapeutic agent Adalat in 1980

a limited number of areas in pharmaceutical research. We have, therefore, concentrated our efforts on the following indications:
(a) cardiovascular diseases, (b) infectious diseases, (c) metabolic diseases, (d) diseases of the CNS, and (e) chronic inflammatory diseases and arthritis.

The dramatic increase in cardiovascular diseases over the past decades and the discovery and successful development of drugs to treat these diseases have encouraged us to further intensify our research efforts in this field. A breakthrough was achieved with the discovery of a new therapeutic principle – the calcium antagonism of the dihydropyridines – by Bossert and Vater. This principle was introduced into medical therapy in 1975 under the trade name of Adalat (nifedipine). Adalat was awarded the Prix Galien for excellent achievements in pharmacotherapy in Paris in 1980 (Fig. 7).

New dihydropyridines, such as nitrendipine and nisoldipine, are being developed for the treatment of hypertension and coronary heart disease. A further dihydropyridine, nimodipine, exhibits selective activity in cerebrovascular disorders. Clinical trials on a variety of disorders of the CNS (such as stroke, dementia, and Alzheimer's disease) are in progress.

In the field of anti-infective therapy, a wide variety of new drugs with improved activity and tolerability have been developed for the treatment of bacterial infections since the first sulfonamides and penicillins. These new drugs include mezlocillin (Baypen), a broad-spectrum penicillin that is also effective against bacterial pathogens outside the range of ampicillin, and azlocillin (Securopen), which additionally exhibits very good activity against *Pseudomonas*. A recent development is ciprofloxacin, an antibacterial of the quinolone class. This drug can be applied by the oral and parenteral routes, to treat a broad spectrum of bacterial infections.

Another breakthrough was achieved with the discovery of the antimycotic activity of azoles by K. H. Büchel and M. Plempel, which led to the compounds clotrimazole and bifonazole. Both drugs have set new standards in the chemotherapy of mycoses, clotrimazole being the first broad-spectrum antifungal drug.

The development of praziquantel (Biltricide) for the therapy of schistosomiasis in cooperation with E. Merck, Darmstadt, carried on Bayer's tradition in the therapy of tropical diseases. The cure of schistosomiasis – which affects 200 – 300 million people – by a single-dose treatment is regarded as a significant achievement. This therapeutic advance was honored by the renowned Prix Galien in 1987, the second time that a Bayer drug had received this award.

The research activities of our company have been extended into new indications and new technologies. Our new Miles Research Center (West Haven, USA) was opened in April 1988. This research center will house research groups dedicated to the study of autoimmune diseases, molecular pharmacology, and molecular diagnostic methods.

Despite all the impressive achievements and progress in pharmacotherapy, only one-third of known diseases can be attacked causally. Two-thirds of all diseases either cannot be treated or can only be treated palliatively, i.e., by the relief of symptoms. Due to the increase in the size of the elderly population in the Western world, the incidence of age-related diseases (such as cardiovascular, degenerative and chronic inflammatory diseases, and malignant tumors) is expected to rise dramatically. The urgent need for new drugs is best demonstrated by the rapid spread of the human immunodeficiency virus and its sequelae AIDS. We have taken up this challenge by cooperating with Hoechst AG in order to maximize our resources in drug research and development.

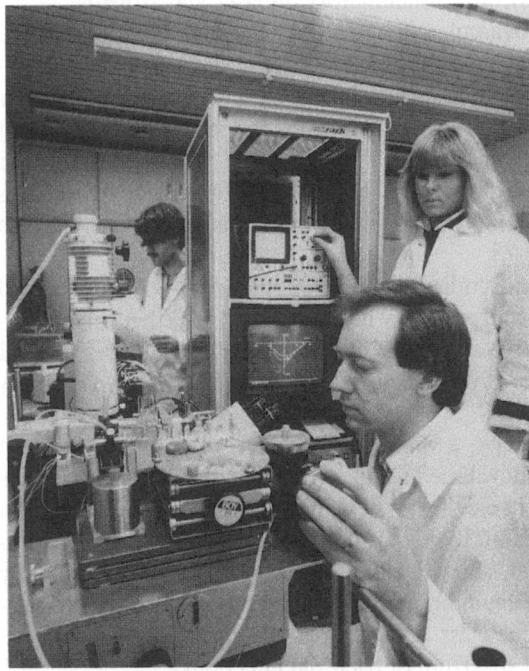

Fig. 8. The search for innovative drugs, utilizing cell biology and computer technology

The complex diseases with a multifactorial pathogenesis represent the important challenges for pharmaceutical research today. The availability of new technologies, especially recombinant DNA techniques and monoclonal antibodies, and the intense utilization of cellular biology and immunology will lead us to the molecular basis of diseases, thus enabling us to develop specific drugs (Fig. 8). Besides these more futuristic approaches, a vast number of regulatory factors have been identified and isolated by the new technologies, e. g. insulin, the interferons, and the clotting factor Factor VIII. These new biological agents are produced in cell culture rather than by chemistry. Within our company, we have developed production methods for the recombinant Factor VIII, and the early-phase clinical trials are scheduled to start this year. This project is regarded as an important milestone in modern pharmaceutical technologies at Bayer.

The new molecular pharmacology allows not only the identification of specific receptors but also the characterization and elucidation of their chemical and three-dimensional structure (Fig. 9). We expect that this new knowledge, in combination with computer technology, will guide us to a more rational design of new and specific drugs.

Detailed structural information now available on important calcium-binding proteins, receptor molecules, and ion channels is increasingly improving our understand-

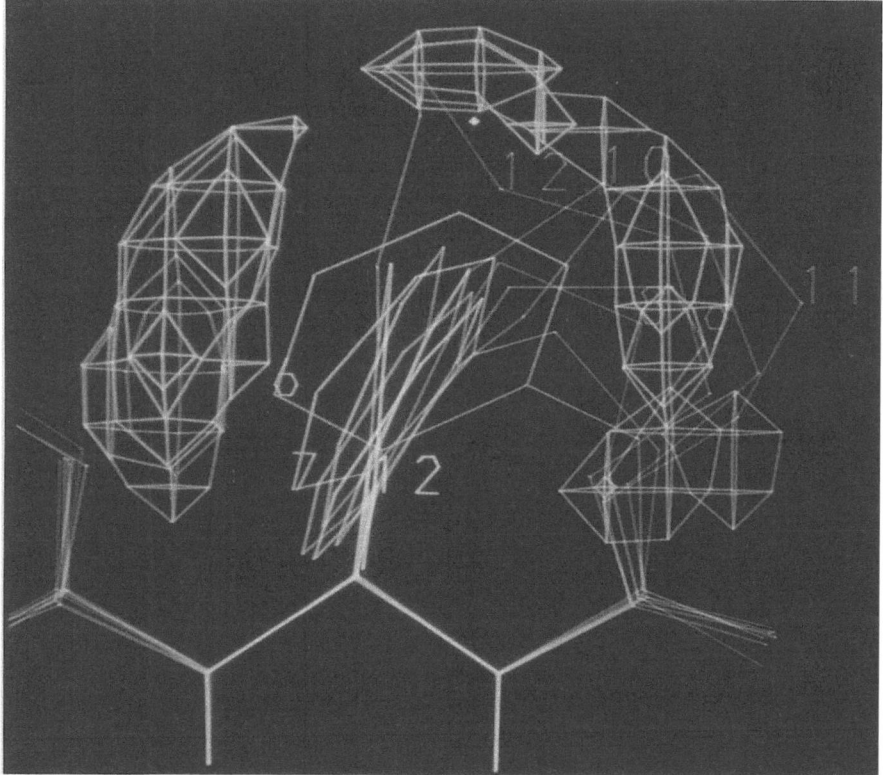

Fig. 9. Drug design: modelling of dihydropyridine molecules with computer graphics. The active dihydropyridine structures are surrounded by the receptor pocket

ing of their physiological function. This Centenary Symposium, *The Calcium Channel: Structure, Function, and Implications,* will express the most recent developments in this rapidly expanding field.

With further scientific progress, advances in diagnosis, biomedical technologies, and pharmacotherapy, tomorrow's patients will benefit from today's research efforts. The fact that we are celebrating our centenary with this scientific meeting indicates that we remain firmly committed to innovation and research in this important field.

The Effect of Lime and Potash on the Frog Heart: An Imaginative Reconstruction of a Paper Presented by Sydney Ringer to the Physiological Society on 9th December 1882

R.A. Chapman

The Laboratory of Cellular Cardiology, School of Veterinary Science, Park Row,
The University of Bristol, BS1 5LS, England

At the time when the Bayer Company was in its infancy, Sydney Ringer (Fig. 1), in middle age and already a distinguished physician, was performing a series of experiments which established the importance of calcium in the regulation of cellular processes. Always forgotten when Ringer's solutions are made, he is best remembered for his work on the isolated heart. He also investigated the contraction of skeletal muscle, the behaviour of peripheral nerve, the clotting of blood and the coagulation of milk. Even with the frog heart, he went far beyond the action of calcium, potassium and sodium, performing experiments on the effects of strontium, barium, rubidium, caesium, ammonium, and drugs such as local anaesthetics and atropine.

At the Bayer Centenary Symposium on "The Calcium channel", it was thought appropriate that Ringer should be remembered in some way. In the *Minutes of The*

Fig. 1. The photograph of Sydney Ringer which accompanied his obituary in the *Proceedings of the Royal Society* (reproduced with permission)

Physiological Society for 1882, it is recorded that Ringer gave a paper at the Scientific Meeting at University College, London, on 9th December. Other papers were given at this meeting by A. D. Waller, F. J. M. Page and Burdon Sanderson, E. A. Schaefer, G. J. Romanes, W. North, F. Gotch and C. Roy. The title of Ringer's paper was "Effects of lime and potash on the frog heart". Although Scientific Meetings of this Society, as opposed to Dinner Meetings, had begun in 1880, it was agreed only in 1883 that abstracts of the communications should be printed in the *Journal of Physiology*. As a consequence, there is no record of what Ringer presented. It was, however, decided to reconstruct this communication and present it before the dinner at the Bayer Centenary Symposium. The reconstruction had, therefore, to be made from the papers published by Ringer at this time or shortly afterwards (Ringer 1882, 1883 a, b; Ringer and Sainsbury 1883) and from other accounts of how Ringer was supposed to have discovered the importance of calcium.

In undertaking this task, I find that History has been somewhat equivocal about this discovery. This would stem from Sir Henry Dale's article "Accident and opportunism in medical research" (1948) which would seem to show some inconsistencies, and which has certainly been misquoted subsequently. Dale reports that Ringer first found that when a solution of salt replaced a blood mixture, the heart beat rapidly failed. Later, an apparently similar solution maintained the mechanical activity for several hours. Dale lays the blame at the door of Ringer's lab boy, Fielder. Fielder explained to Dale, many years later, that he did not want to spend all his time distilling water, so he substituted tap water, expecting Ringer not to notice. In fact, the change in the behaviour of the heart caused Ringer to become very puzzled before he identified the cause. This account would seem to be inconsistent with what Ringer himself reported (Ringer 1882, 1883a). The earlier paper concerns saline solutions made, unbeknown to Ringer, with tap water, and the later paper contains an explanation of the different results he obtained when the saline was made with distilled water, i.e. opposite to Dale's description. Furthermore, in the earlier paper Ringer reports the effects of distilled water upon the heart, but he makes no reference to this effect in the second paper. However, he explains that the distilled water was prepared by Hopkins and Williams. From this I deduce, first, that the distilled water was unlikely to have been made by Fielder at the time of these particular experiments but probably became routine later in the laboratory; secondly, that the experiments reported in the first paper were done with tap water saline which Ringer believed to be made with distilled water; and thirdly, that as soon as Ringer discovered this, he repeated the experiments with distilled water saline and reported that work in the second paper. Fielder's memory might have dimmed over the years, or the story became more elaborate with the telling.

This is the way in which Ringer may have approached the problem, and it is consistent with Ringer's personality as reported in his various obituaries and by the reminiscences of others. In the obituary in the *British Medical Journal* (1910) we find:

It was interesting to observe the different attitude of mind with which he attacked clinical problems on the one hand and those of the laboratory on the other. He never confirmed his diagnosis by avoidance of the post-mortem room; on the contrary he there sought confirmation of its correctness or conviction of its error ... He seemed, in the clinical setting, cautious in drawing conclusions, rarely entering into the pitfall of the clinician –

an explanation. In the laboratory he was fond of advancing therories and testing them with experiment. He was quick to see analogies, fond of expatiating on the different kinds of antagonism and drawing general conclusions. Of those who worked nearby, Osler, a Canadian who spent some time at Burdon Sanderson's laboratory, recalled: *"I always felt Ringer missed his generation and suffered from living in advance of it."*

However, Ringer could become angry, especially if he felt he was being misrepresented: in the Preface to the second edition of his book *The temperature of the body as a means of diagnosis and prognosis in Phthisis (1873)*, he noted: *"My book has been curiously misread or perverted. It would seem indeed that some authors who have cited my views can hardly have read my book."*

Ringer led a rigorous and very regular life: *He would rise early, dispatch a hasty breakfast at eight, and the next few minutes would see him on his way to hospital always on foot ... The hospital visit would generally, in his pharmacological days, conclude with a quick-change appearance in the physiological laboratory – Ringer the physician was transformed into Ringer the pharmacologist – a tracing taken, various suggestions made and off he was again to the morning's consulting. The afternoon would be filled with visits, consultations, the hospital round, a post-mortem and again for as long as possible, a visit to the laboratory. His attraction for the workshop was quite wonderful. Nothing could withstand it, not even the college palings, over which he clambered in the dark, upon one occasion at least, when the gates would not yield (Br. Med. J. 1910).* He therefore managed to combine the three activities of teacher (he wrote several books, his *Handbook of Therapeutics (1869)* running into 13 editions), clinician, some claiming that he would have obtained an eminence similar to Sir William Jenner if he had not devoted so much of his time to a third activity, that of a scientist.

Ringer apparently never travelled outside England, he loved painting and music and was passionately punctual. This all seems to add up to a man who appears to be curiously contemporary, whose company would have been much enjoyed at this symposium: he would have delighted in the experiments, the ideas and the discussion. He would also have had an eye for the promising young scientist, for as Starling (1910) noted: *"The generosity of his nature and the kindliness of his disposition were exemplified in many ways, and in numerous instances the persons whom he assisted never knew the name of their benefactor."*

Since I am the current Honorary Secretary of the Physiological Society, one might expect the following communication to conform to the current and rigorously applied rules. In 1882, there were virtually no rules, and Ringer would have handed around the traces of his experiment and discussed them in a more or less informal way. I am sure that he would have emphasized the ideas of antagonism and interference between substances in their action on the tissues that preoccupied him at this time.

Ringer's Communication

Gentlemen, I wish to offer for your consideration experiments designed to ascertain the influence each constituent of the blood exercises on the contraction of the ventricle. The observations were made with Roy's tonometer, and the frog's ventricle was tied to a cannula as close as possible to the auriculoventricular groove. The first set of records (Fig. 2) shows the experiments made in April and May. Figure 2 part OA shows the response of the ventricle of the frog when exposed to dried bullock's blood dissolved in water and diluted with five parts of 0.75% saline solution. On changing to saline solution, the ventricular beats first become more complete, the trace soon becomes broader, its summit rounder (Fig. 2 part OB), and a slight persistent spasm develops (Fig. 2 part OE). Next, the period of relaxation becomes prolonged, and the whole trace is permanently raised above the base line (Fig. 2 part OF). These are effects that resemble those produced by veratria. The reapplication of the blood mixture obviates the changes, as seen in Fig. 2 part 2, where A is blood mixture, B the change to saline solution, and C and D the return to blood mixture. Next I found that potassium chloride added in small quantities (0.6–1 cc 1% potassium chloride) to saline will completely and speedily obviate the character of the trace occurring in saline solution, as is well shown in Fig. 2 part 5.

It is singular that whilst necessary for the speedy dilatation of the ventricle, potash salts appear less necessary for the proper contraction of the ventricle. I next tested distilled water on the ventricle. Water throws the ventricle into complete and permanent systole (Fig. 2 part 4). This water rigor is removed by adding sodium chloride, but blood mixture is more efficient in restoring the contraction, as seen on either side of the arrow in Fig. 2 part 4. Other salts of soda are effective in removing water rigor, while potassium chloride in physiological doses or greater does not remove water rigor. This observation conflicts with the action of additional potash salts when added to the blood mixture, where the heart is arrested in diastole. At this time, it seemed to me that a saline solution, to which is added one ten-thousandth part of potassium chloride, makes an excellent circulating fluid in experiments with the detached heart. Subsequently, I found that I did not always obtain exactly similar results when these experiments were repeated. This I initially ascribed to the effects of the seasons, as cold affects the responsiveness of the ventricle. But then I discovered that the saline solution which I had used was not made with distilled water (supplied by Hopkins and Williams), but with pipe water supplied by the New River Water Company. I at once tested the action of saline solution made with distilled water and did not get the effects that I have just described. The pipe water contains minute traces of inorganic substances expressed as parts per million as follows: calcium 38.3, magnesium 4.5, sodium 23.3, potassium 7.1, combined carbonic acid 78.2, sulphuric acid 55.8, chlorine 15, silicates 7.1 and free carbonic acid 54.2.

I find that calcium, in the form of lime water, bicarbonate of lime or chloride of calcium, even in minute doses in saline made with distilled water, produces the changes in the beat seen previously when a blood mixture was replaced by pipe water saline, and that these effects are completely removed by a minute trace of a potassium salt. Saline made with distilled water is incapable of sustaining the contractions of the heart, which speedily grow weaker and may cease altogether, such that no contraction can be excited even by a strong break induction shock (Fig. 3 part 1A–D, in the second set of traces of experiments made in October and November). Upon the addition of saline containing a lime salt, the contractions reappear but soon become prolonged (Fig. 3 part 1E). The further addition of potassium chloride hastens dilatation and restores the shape of the beat (Fig. 3 part 1G). On replacing blood

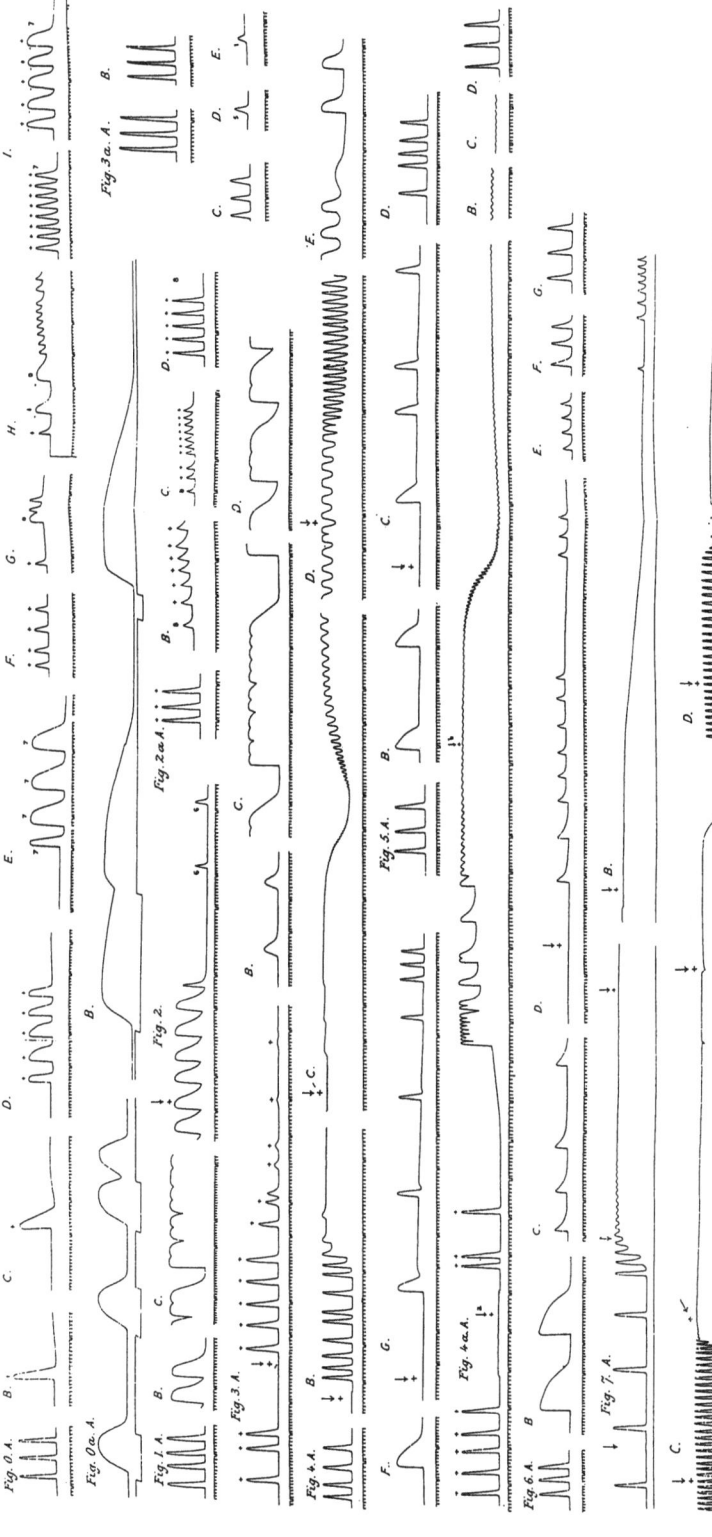

Fig. 2. Complete figure from Ringer (1882). The relevant parts, mentioned in the text above, are described as follows: *Fig. OA.* Trace with blood mixture. *Fig. OB.* Twelve minutes after saline solution was substituted for blood mixture. *Fig. OE.* From another experiment. Thirty-six minutes after saline was substituted for blood mixture. *Fig. 2A.* Trace with blood mixture; *B.* Sixty-three minutes after saline solution substituted for blood mixture; *C.* About six minutes after saline solution replaced by blood mixture; *D.* Forty-two minutes after saline solution was replaced by blood mixture. *Fig. 4.* Showing the effects of water on the ventricle and the antagonizing effect of sodium cloride. *Fig. 5.* Shows the effect of physiological doses on the saline trace: *A.* Trace taken with blood mixture; *B.* Fifteen minutes after blood mixture was replaced by saline solution; *C.* Effect of adding 0.7 cc of 1 per cent potassium chloride to the 100 cc of circulation saline solution. The *arrow* indicates the time the potassium cloride was added; *D.* Effect ten minutes later. (Reproduced with permission)

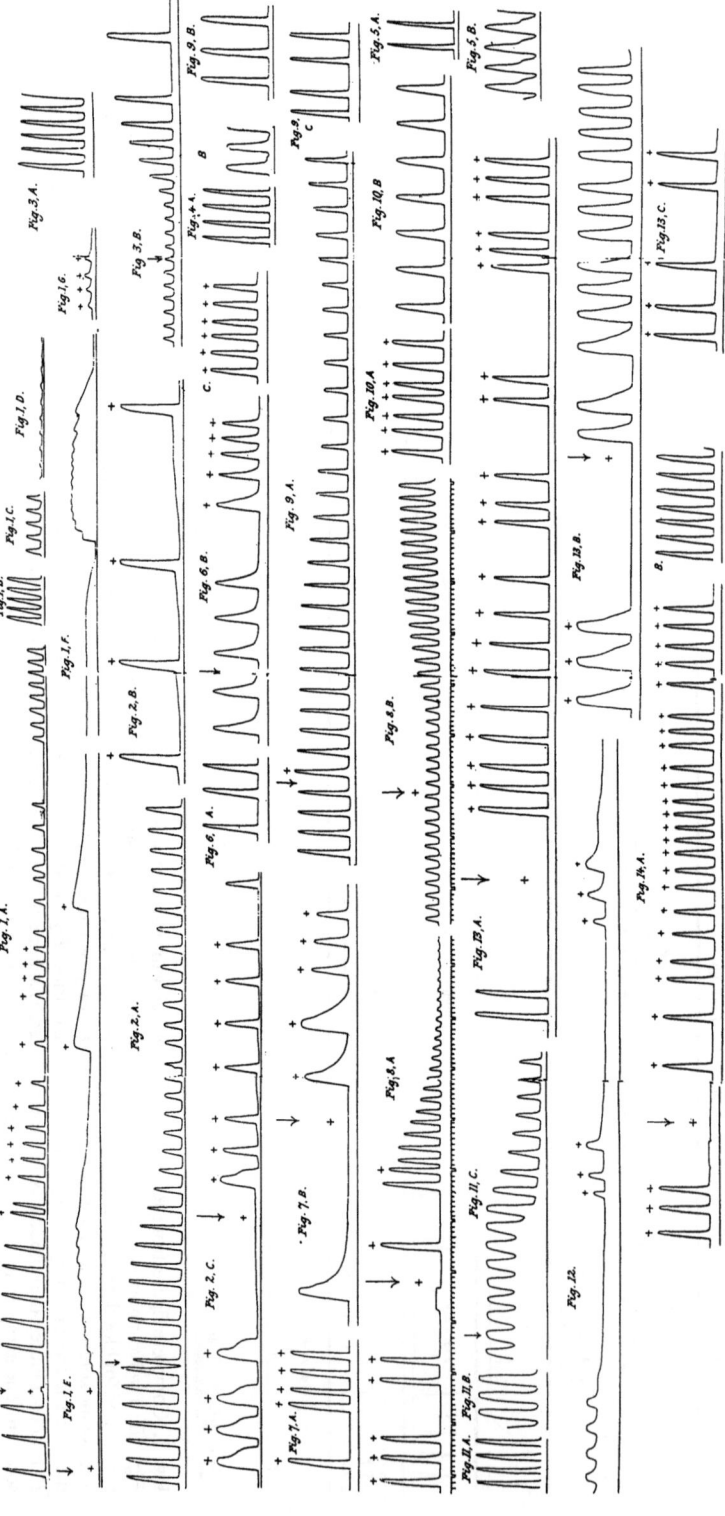

Fig. 3. Complete figure from Ringer (1883a). The relevant parts, mentioned in the text, are as follows: *Fig. 1. A.* Trace with blood mixture and showing the effect of replacing blood mixture by saline solution (made with distilled water). The replacement was made at the point indicated by the *arrow*, x ± indicates induction shock. *Fig. 1. B.* Trace eight minutes after replacing blood by saline solution. *Fig. 1. C.* Fourteen minutes and *Fig. 1. D.* eighteen minutes after the replacement; *Fig. 1. E.* Shows the effect of adding 5 cc of calcium chloride solution to 100 cc of circulating saline solution. *Fig. 1. G.* Shows the effect of 12 minims of potassium chloride solution to 100 cc of the circulating fluid. *Fig. 3. A. Trace with blood mixture*; *Fig. 3. B.* Trace seven minutes after replacing the blood mixture with saline solution. At the point indicated by an *arrow* 0.5 cc of lime water was added to the 100 cc of circulating saline and the trace rapidly increased in height. (Reproduced with permission)

mixture with a saline solution containing potassium chloride and sodium bicarbonate, the contractions quickly grow weak and soon stop; this is seen even with these solutions mixed in various proportions. If calcium chloride or other calcic salts are added to such solutions, contractility is restored, and good spontaneous beats will continue for a long time (Fig. 3 part 3). A mixture of saline with 2.5 parts per 10000 of sodium bicarbonate and calcium chloride in 1 in 19500 with 1 in 10000 of potassium chloride makes a good artificial fluid in which the ventricle will beat perfectly.

I must therefore conclude that the heart's contractility cannot be sustained by saline, by saline with potassium chloride, by saline containing bicarbonate of soda, nor by a mixture of all three. The addition of calcic salt to any of these solutions will sustain contractility. I conclude therefore that a lime salt is necessary for the maintenance of muscular contractility. However, while calcium salts are necessary, it would seem also necessary that potash and sodium salts be present in the appropriate amounts for the ventricle to develop normal beats. If unantagonised by potassium salts, the beats would become so broad and the diastolic dilatation so prolonged, that much fusion of the beats would occur and the ventricle would be thrown into a state of tetanus. This antagonism between potash and lime salts may provide an area of research for others in the future.

It is generally held that chemical similarity indicates similarity in physiological and therapeutic action. However, as I have shown, the physiological action of potash salts differs greatly from that of the salts of sodium. A correspondence between chemical and physiological action might be shown if the salts of barium and strontium are able to sustain muscular contraction in the same way as calcium salts, and if rubidium and caesium show a similar antagonism to that produced by potassium. My recent experiments with Harrington Sainsbury have shown this to be, at least in part, the case. However, interference arises when chemically similar salts are present together. As the matter stands, it is of little importance how these phenomena are named, but we must not forget that, in the chemistry of the tissues, as interference of antagonism obtains even between salts similar in their action when present alone, the mass of the salt present becomes an element requiring consideration.

References

Dale H (1948) Accident and opportunism in medical research. Br Med J 11:451–455
Fye WB (1984) Sydney Ringer, calcium, and cardiac function. Circulation 69:849–853
Ringer S (1869) Handbook of Therapeutics, HK Lewis, London
Ringer S (1873) On the temperature of the body as a means of diagnosis and prognosis in phthisis. Lewis, London
Ringer S (1882) Concerning the influence exerted by each of the constituents of the blood on the contraction of the ventricle. J Physiol 2:380–393
Ringer S (1883a) A further contribution regarding the influence of different constituents of the blood on the contraction of the heart. J Physiol 4:29–42
Ringer S (1883b) An investigation regarding the action of rubidium and caesium salts compared with the action of potassium salts on the ventricle of the frog's heart. J Physiol 4:371–379
Ringer S (1910) Obituary. Br Med J 2:1384–1386
Ringer S (1910) Obituary. Nature 84:540
Ringer S (1910) Obituary. Lancet 2:1386
Ringer S, Sainsbury H (1883) An investigation regarding the action of strontium and barium salts compared with the action of lime on the ventricle of the frog's heart. The Practitioner 11:81–93
Starling EH (1910) Sydney Ringer. Proc R Soc B 84:i–iii

Homology of Calcium-Modulated Proteins: Their Evolutionary and Functional Relationships

R. H. Kretsinger[1], N. D. Moncrief[1, 2, 3], M. Goodman[3], and J. Czelusniak[3]

[1] Department of Biology, University of Virginia, Charlottesville, VA 22901, USA
[2] Center for Molecular Biology, Wayne State University, Detroit, MI 48201, USA
[3] Department of Anatomy and Cell Biology, Wayne State University School of Medicine, Detroit, MI 48201, USA

Introduction

In 1972 Kretsinger published a "Gene Triplication Deduced from the Tertiary Structure" of parvalbumin. He then proposed (1975) that "Calcium Modulated Proteins Contain EF-Hands". Calcium-modulated proteins, as a distinct superfamily of all the proteins that bind calcium, are defined by two characteristics. They are found within the cytosol or on a membrane facing the cytosol. They bind calcium with a dissociation constant about 10^{-6} M under cytosolic conditions, i.e., about 10^{-3} M free Mg^{2+} ion. They are inferred to be involved in transmitting the information inherent in calcium's functioning as a cytosolic messenger.

To date members of this homolog family have been found only in the cytosol; all, or most, calcium-modulated proteins are members of this homolog family. Baba et al. (1984) constructed a phenogram relating the 50 calcium-modulated proteins whose amino acid sequences were available at that time. They concluded that an interaction between gene duplication and natural selection resulted in the evolution of six distinct subfamilies identified at that time: calmodulin (CAM), troponin C (TNC), enzymatic light chain of myosin (ELC), regulatory light chain of myosin (RLC), parvalbumin (PV), and the M_r 9000 intestinal calcium-binding protein (ICBP). Since their publication, 100 additional amino acid sequences of calcium-modulated proteins have become available. Many of these represent new subfamilies. In addition, there are numerous genomic DNA and complementary DNA sequences published as well as new crystal structures for the ICBP (Szebenyi and Moffat, 1986), TNC (Herzberg and James, 1985; Satyshur et al. 1988), and CAM (Babu et al. 1985; Kretsinger et al. 1986). Using this wealth of information we have constructed several phenograms to investigate relationships among the subfamilies and among the proteins within these subfamilies. Interpretations of our preliminary results lend new insights into the evolution, functions, and structures of these fascinating calcium-modulated proteins.

Methods

Data Base and Alignment

Most of the amino acid sequences used in this study have been published; all references will be cited in a more detailed description of this work. We reference recent publications, reviews, or unpublished sequences.

For all of these proteins two to six EF-hand domains were recognized by the authors. Kretsinger (1987) has discussed in detail the characteristics of the EF-hand domain (Fig. 1) and its numerous variations. In summary, the canonical domain consists of 29 residues in a helix, loop, helix conformation:

```
                    1 1 1 1 1 1 1 1 1 1 2 2 2 2 2 2 2 2 2 2
1 2 3 4 5 6 7 8 9 0 1 2 3 4 5 6 7 8 9 0 1 2 3 4 5 6 7 8 9
n     n n       n X Y Z -Y  -X    -Z n       n n         (n)
E                 D       G  I      E
```

The first residue of the first α-helix ("E" of the EF-hand) frequently begins with Glu (E). The residues on the inner (n) aspect of this helix are usually hydrophobic; for instance, Phe is often observed at positions 6, 9, and 22. The second helix also usually begins with Glu (21), which also coordinates the calcium ion at the -Z vertex. This "F" helix, so named for the α-helices of PV – parvalbumin-AB, CD, and EF – also has hydrophobic residues at the inner (n) positions 22, 25, 26, and sometimes at 29. The Ca^{2+} ion, if bound, is coordinated by six residues, whose positions are approximated by the vertices of an octahedron. Five of these – X, Y, Z, -X, and -Z – usually have oxygen-containing side chains – Asp (D), Asn (N), Ser (S), Thr (T), Glu (E), or Gln (Q). The oxygen at position 16 (-Y) comes from the main chain and can be supplied by any amino acid. Asp is usually found at position 10 and Glu at 21. Gly (G) at 15 permits a sharp bend ($\Phi = 90°$, $\Psi = 0°$) in the calcium-binding loop. Ilu (I), Leu (L), or Val (V) at 17 attaches the loop to the hydrophobic core of the molecule. By placing

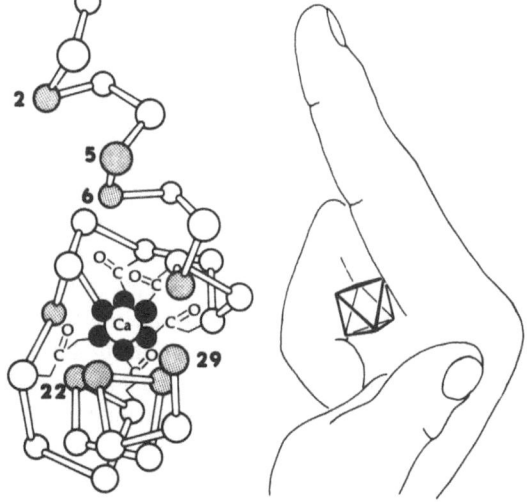

Fig. 1. The calmodulin fold, or EF-hand, consists of an α-helix, symbolized by the forefinger of a right hand, loop about the Ca^{2+} ion, and second α-helix, symbolic thumb. Amino acids 1 through 11 comprise the first helix and 19 through 29 the second. The *stipled α-carbons* – 2, 5, 6, 9, 22, 25, 26, and 29 – usually have hydrophobic side chains. They point inward, as does the side chain of residue 17, interacting with the homologous residues of another EF-hand to form a stable hydrophobic core. The calcium ion, when present, is coordinated by an oxygen atom, or by a water molecule bridged to an oxygen atom, of the side chains of residues 10, 12, 14, 18, or 21; the carbonyl oxygen of residue 16 also coordinates calcium

added weight on these 16 positions one has greater confidence in identifying and aligning domains. A broad range of variation will be discussed for each subfamily, but this basic pattern is observed in all homologs that bind calcium. Some EF-hands do not have the ability to bind calcium; they have greater variations in sequence and in structure. Most EF-hands occur as members of pairs in which the hydrophobic inside residues form a stable core.

The interdomain regions have a much broader range of length and sequence. We assigned optimal alignments for the interdomains by first establishing the relationships among the individual proteins within and between subfamilies based on the sequences within their domains only. Then, holding the ending and starting positions of domains fixed, we adjusted putative insertions and deletions of the interdomain regions to obtain optimal alignments of members of subfamilies. When the relationships among these proteins are subsequently calculated using the complete amino acid sequences, they closely resemble the phenogram calculated using only the domain information. That is, our method of aligning interdomain sequences does not bias the results. Other schemes for alignments of interdomain regions often yield results in conflict with known functional or phylogenetic information.

Computations of Phenograms

The sequence of programs that we used to compute phenograms depends on correctly aligned sequences, as discussed in the preceeding section. Evolutionary relationships among aligned homologous sequences are determined with the aid of five computer algorithms: MMD, UPGMA, FTE, MPAA, MPAL8. A matrix of minimum mutation distances for the aligned sequences is constructed according to Jukes (1963) and Fitch and Margoliash (1967) using the algorithm MMD. From the MMD matrix, an initial tree is constructed by UPGMA (unweighted pairgroup method with arithmetic averaging). This algorithm executes the clustering procedure of Sokal and Michener (1958). The UPGMA tree is built from the smallest and least divergent branches in a series of clustering cycles in which the last cycle roots the tree.

An alternate initial tree can be constructed from the same MMD matrix by the FTE algorithm described by Farris (1972). This algorithm starts like UPGMA by joining the two sequences with the smallest degree of divergence. Then it scans the remaining sequences to join the sequence that adds the least length to the original pair. The procedure is continued with each consecutive sequence being joined to the network (an unrooted tree) on the basis of least length; lengths are apportioned to the links of the growing network by calculations of the type used to construct an additive tree (Cavalli-Sforza and Edwards 1967; Moore et al. 1973a).

The two initial trees produced from the MMD matrix, one by UPGMA and the other by FTE, serve as starting points in the search by branch-swapping computer programs MPAA and MPAL8 for the most parsimonious tree, i.e., the tree of lowest length of NR (nucleotide substitutions that cause amino acid replacements). Each of these programs uses a maximum parsimony algorithm (Moore et al. 1973b; Moore 1976, Goodman et al. 1979) that calculates from amino acid sequences the lowest NR length for each tree examined. MPAA examines all "nearest neighbor single step changes" (NNSSCs) in the network topology, i.e., it examines for each two adjacent interior nodes the three alternative arrangements for the four branches originating from the two nodes. For the N exterior nodes (contemporary sequences) of the unrooted tree, there are 2 (N–3) alternative trees in each round of swaps. The alternative having the lowest NR count is the starting tree for the next round of swaps. The program terminates when two consecutive rounds of NNSCs fail to discover any new trees of lower NR count. At this time, new input trees with more extensive changes in branching arrangement are tried as initial starting points to check the possibility of even shorter trees.

MAPL8 employs a different branch-swapping procedure. Starting with a tree of lowest known NR length, such as the end point of the MPAA iterative search procedure, the investigator divides this tree into eight subtrees. Program MPAL8 then computes the NR score of all possible trees with these eight subtrees (there are 10395 unrooted trees) and lists the trees and their distribution from lowest to highest NR score. In calculating the NR score for each of these 10395 trees, MPAA computes the maximum parsimony solution for the ancestor of each subtree, i.e., in effect deals with these eight ancestral sequences as if they were terminal taxa.

In the search for the lowest NR length tree, MPAA examines many thousands of alternative trees. An unrooted tree at the lowest NR score reached during this phase of the search is converted into a number of rooted trees or dendrograms, each having a different set of eight subtrees but the same network. All possible trees for each eight-subtree dendrogram are then scored by program MPAL8. Network changes that further lowered the NR score are incorporated into new starting dendrograms, and the search continues for the most and next most parsimonious trees. After continued searching fails to find shorter trees than those already found, the search is discontinued.

The lowest NR length tree found as a result of this extensive heuristic search is considered to be a reasonable approximation of the correct geneolgical arrangement or cladogram. The most parsimonious geneological arrangement maximizes the genetic likeness associated with common ancestry while minimizing the incidence of convergent mutations. Since common ancestry rather than convergent evolution is the most probable explanation for any extensive matching of amino acid or nucleotide sequences between species, tree reconstruction algorithms based on maximum parsimony offer a way of using Occam's razor to find the preferred geneological hypothesis from sequence data.

The maximum parsimony procedure, however, does not generate an origin or "root" for the computed cladogram. To do so requires additional information about rates of evolution, and times of speciation. We will treat these in subsequent publications.

Results

Relationships Within Subfamilies

We have indicated ten subfamilies in Fig. 2. Proteins within a subfamily are more closely related to other proteins within that subfamily than they are to any protein outside that subfamily. Even so, the definition of subfamily is somewhat arbitrary for

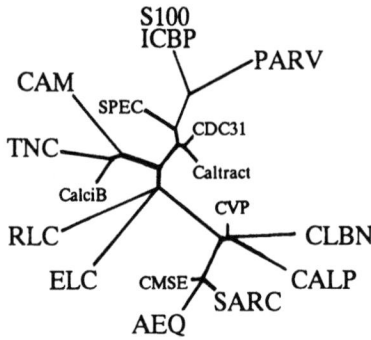

Fig. 2. The phenogram shows the topological relationships among the ten subfamilies of EF-hand homolog proteins: *ELC*, enzymatic light chain of myosin; *RLC* regulatory light chain of myosin; *TNC*, troponin C; *CAM*, calmodulin; S-100*ICBP, several two-domain families including the M_r 9000 intestinal calcium-binding protein; *PARV*, parvalbumin; *CLBN*, calbindin and calretinin; *CALP*, calpain; *SARC*, sarcoplasm calcium-binding protein; *AEQ*, aequorin and luciferin binding protein. Also shown are the relationships to these ten subfamilies of six unique proteins: *Calci B*, B subunit of calcineurin; *SPEC*, the group of ectodermal proteins from *Strongylocentrotus purpuratus*; the CDC31 gene product from *Saccharomyces cerevisiae*; *Caltract*, caltractin from *Chlamydomonas reinhardtii*; *CVP*, calcium vector protein from *Branchiostoma lanceolatum*; *CMSE*, the calcium-binding protein from *Streptomyces erythraeus*.

two reasons. Some subfamilies might be either divided in two or joined to an adjacent subfamily with its member proteins still obeying the preceding definition. Frequently such division depends on functional information. However, the assigment of subfamily does not alter the structure of the phenogram, only its interpretation. Six EF-hand proteins are unique and not yet placed in subfamilies. This may simply reflect a lack of data. As more sequences are determined, and as more functional information becomes available, these unique proteins may be placed in their own subfamilies. We discuss the characteristics of each subfamily before considering the relationships among subfamilies and then the unique proteins.

It is fundamentally important to determine whether a gene duplication and fixation event occurred prior to, or following a speciation event (Patterson 1987). If, for example, the gene duplication event that gave rise to the α and β forms of PV occurred

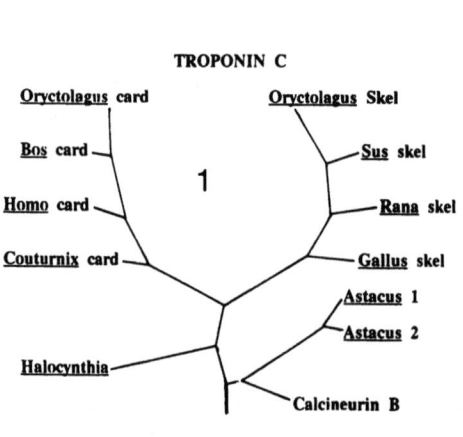

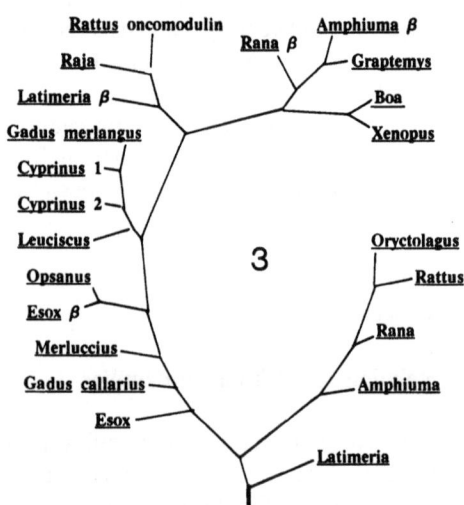

Fig. 3. Legends s. p. 23

prior to the speciation event that gave rise to fish and to tetrapods, then, excluding gene loss, one would expect to see α and β forms in all fish and in all amphibians and reptiles, as well as mammals and birds. The α and the β forms are referred to as paralogs; the β forms from two species are orthologs. If, as in fact has occurred in PV, there are subsequent duplication events to generate isoforms β_1, β_2, etc., then these in turn are paralogs of one another. The phenogram of the α forms, as well as the phenogram of the β forms, reflects the phylogeny of the host species. The splitting of α and of β forms reflect the gene duplication that preceded the fish, tetrapod divergence. In contrast, if one had, for example, compared α forms from fish and from amphibians with a β form from reptiles, one might have been mislead to conclude that reptiles diverged from a common fish, amphibian line.

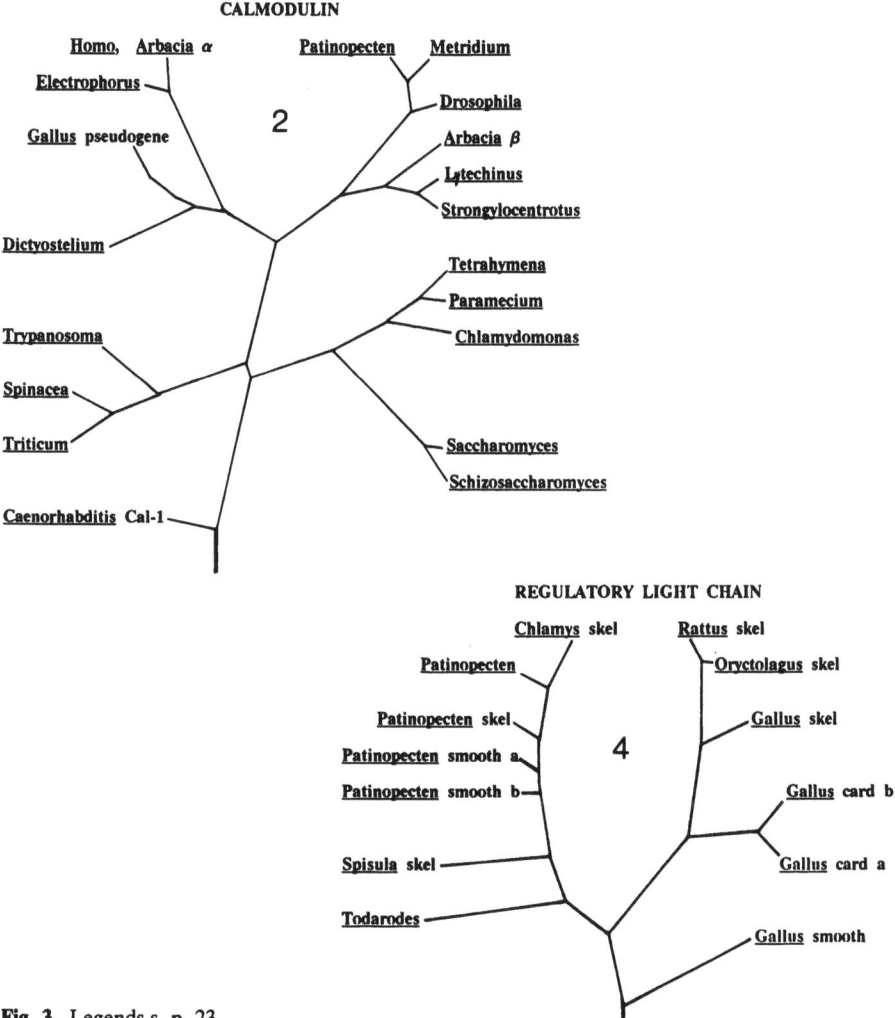

Fig. 3. Legends s. p. 23

CALBINDIN & CALRETININ

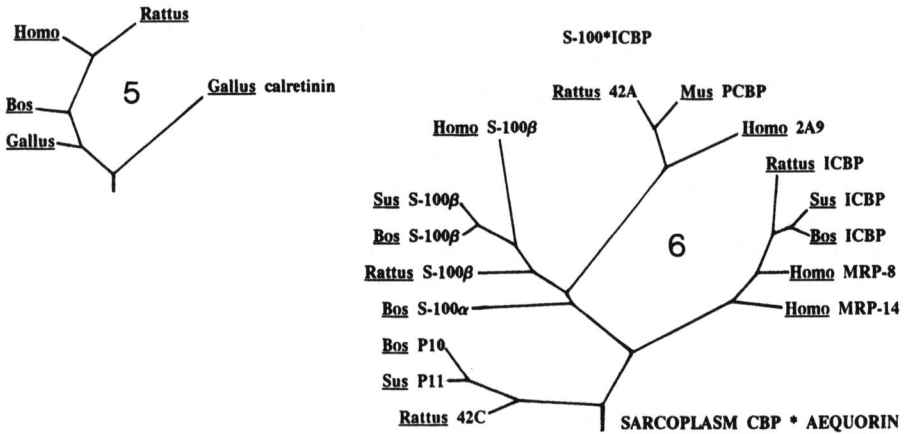

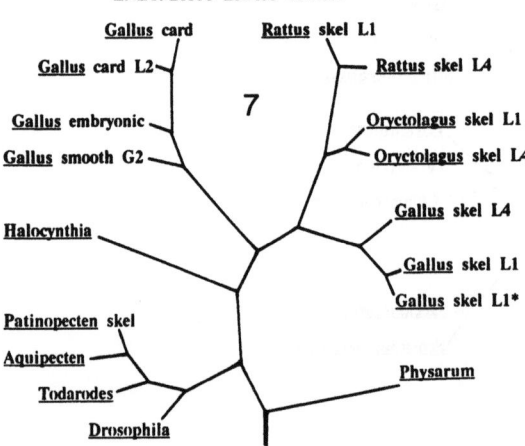

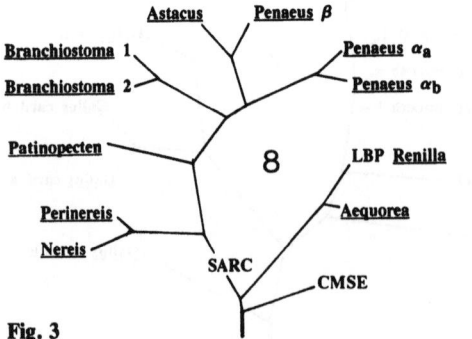

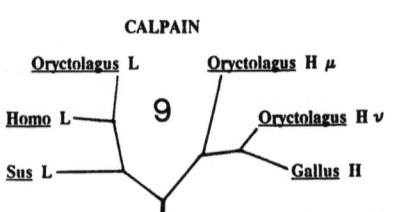

Fig. 3

In general one anticipates seeing the same pattern of evolution, or phylogeny, derived from comparison of molecular structures and sequences, as from a study of morphological, behavioral, and paleontological data. We emphasize, however, that this inference is valid only if one compares orthologs (Fitch 1977). It is often difficult to establish that one has orthologs; how can one ever refute the argument that, following a recent speciation, organism 1 deleted isoform β_1 and organism 2 deleted isoform β_2? For instance, Hardy et al. (1988) sequenced two CAMs in the sea urchin *(Arbacia punctulata)*; the inferred phylogeny of the echinoderms is significantly different when constructed from the α form and from the β form (Fig. 3). Within most of the subfamilies there is good correspondence between the traditional views of organismal phylogeny and the phenogram relating the EF-hand proteins within that subfamily. The exceptions are especially interesting and usually indicate the creation of a new protein, a paralog, with a different function.

The ELCs (or essential, or alkali light chains) of myosin do not bind calcium even though they contain four EF-hand domains (Table 1). Within the vertebrates the skeletal and the smooth/cardiac muscle paralogs arose prior to the divergence of birds and mammals. From the available sequences one infers that there are not distinct skeletal and smooth muscle forms in nonvertebrate animals such as *Drosophila* (Falkenthal et al. 1984). The light chain from *Physarum* (Kobayashi et al. 1988) more closely resembles the ELCs than the RLCs of vertebrates – even though it does bind one calcium ion in domain 1, as do the RLCs.

The RLCs (or DTNB light chains) of myosin also have four EF-hand domains, the first of which binds calcium. The duplication event producing cardiac and skeletal forms in vertebrates appears to have occurred after the vertebrate – mollusc divergence, whereas the RLC from smooth muscle branches prior to the inferred vertebrate – mollusc divergence.

The TNC subfamily has, within the vertebrates, two distinct groups, the skeletal TNCs, with four calcium-binding domains, and the cardiac TNCs, in which domain 1 has an additional amino acid and lacks the ability to bind calcium. Within both groups the phenograms contradict organismal phylogeny. Only single isoforms of skeletal TNC have been identified. If the sequenced forms are paralogs, the other form(s) may have been deleted in some species. The TNC from the body of a protochordate *(Halocynthia roretzi;* Takagi and Konishi 1983) branches prior to the cardiac–skeletal

---

Fig. 3. The phenograms of each of the ten major subfamilies are rooted because each can be related to protein sequences outside their own subfamilies. For most subfamilies all member proteins are identified by the genus names of their organismal sources. *Oryctolagus, Bos, Rattus,* and *Xenopus* calmodulins have amino acid sequences identical to those from humans, as does the α form from *Arbacia*. The sequences from *Metridium* and from *Renilla* are identical. Cal-1 appears not to function as a calmodulin but is most closely related to the calmodulins. Oncomodulin is closely related to the parvalbumins. Several authors explicitly suggested parvalbumins to be β isoforms, as indicated in the drawing. All of the two-domain EF-hand proteins comprise one subfamily, S-100*ICBP. Both genus name and protein designation are given since a standard nomenclature has yet to be established. Within the calpains are found light (L) forms and two heavy forms, Hμ and Hυ. One calretinin is part of the calbindin–calretinin subfamily. The sarcoplasm calcium-binding proteins *(SARC)* subfamily and the aequorin luciferin binding subfamily are closely related. The calcium-binding protein from the prokaryote *(Streptomyces; CMSE)* is closely related to these two subfamilies. Common names for the binomial nomenclature are listed in the appendix to this chapter.

Table 1. Calcium binding and interdomain lengths

CaM	Ca	7	Ca	8	Ca	7	Ca			
TNC sk	Ca	7	Ca	11	Ca	7	Ca			
TNC cd	O	7	Ca	11	Ca	7	Ca			
TNC Hr	O	7	Ca	15	O	7	Ca			
TNC Ap	O	7	Ca	11	O	7	Ca			
RLC	Ca	6*7	O	7*8	O	6*7	O			
ELC	O	7	O	9*10	O	7	O			
ELC Pp	Ca	7	O	9	ca	7	ca			
PARV	–		O	6	Ca	10	Ca			
S-100	CA	11	Ca		–		–			
ICBP	CA	8	Ca		–		–			
42V*p10	O	7*11	Ca		–		–			
42A*2A9	CA	7*11	Ca		–		–			
Calbindin	Ca	13	O	16	Ca	14	Ca	15	Ca	12	O
Calretinin	Ca	18	Ca	15	Ca	14	Ca			
CALP	Ca	1	Ca	6	O	1	O			
AEQ	Ca	12	O	23	Ca	7	Ca			
LBP	Ca	11	O	23	Ca	7	Ca			
SRC PvNd	Ca	19	O	15	Ca	5	Ca			
SARC Bl	Ca	22	Ca	16	Ca	7	O			
SARC P?	Ca	23	Ca	15	ca	4	O			
SRC P?Al	Ca	23	Ca	15	ca	4	O			
SARC Py	Ca	19	O	15	Ca	7	O			
Cal-1	Ca	7	Ca	7	Ca	7	Ca			
CalciB	Ca	3	Ca	8	Ca	12	Ca			
CVP	O	8	O	7	Ca	8	Ca			
SPEC	Ca	6*7	Ca	6*9	Ca	7*8	ca			
CDC31	Ca	7	O	8	O	7	Ca			
Caltractin	Ca	7	Ca	8	Ca	7	Ca			
CM Se	Ca	22	Ca	8	Ca	5	Ca			

The domain structures of the subfamilies and unique proteins of the EF-hand homolog family are indicated as follow: Ca, the domain is demonstrated or strongly inferred to bind calcium; CA, the domains 1 of the S-100*ICBP subfamily coordinate calcium with four peptides carbonyl oxygen atoms; ca, although the domain does not conform to the canonical description, it might bind calcium; O, the domain is demonstrated or strongly inferred not to bind calcium; –, indicates the absence of a domain as in parvalbumin or in S-100*ICBP. The calbindins are unique in having six domains. The numbers indicate the number of amino acids between domains, as defined in the text, 29 residues long for the canonical domain. Subfamilies are abbreviated as in Fig. 2; sk, skeletal; cd, cardiac. Genus and species are abbreviated, for example; Nd, *Nereis diversicolor;* see the Fig. 3.

division of the vertebrates. Its first domain more closely resembles that of cardiac TNC as it has an inserted amino acid and does not bind calcium. The most parsimonious interpretation of these data is that the precursor TNC did not bind calcium in its first domain, and that the ability to bind calcium in the skeletal paralog was acquired subsequent to gene duplication in the vertebrate ancestor. The two similar TNCs from crayfish *(Astacus)* have the full complement of ligands to bind calcium in domains 2 and 4 only; however, Wnuk et al. (1986) reported that only one equivalent of calcium is bound. Even so, both isoforms "restore Ca^{2+} sensitivity to skinned rabbit adductor fast-twitch fibres, devoid of endogenous TNC."

Calcineurin B, discussed as a unique protein, is most closely related to the TNCs and branches early from the *Astacus* TNC stem. This topology has been confirmed in several constructions and is not easily interpreted. All four domains of calcineurin B bind calcium. The important linker joining domains 2 and 3 is 11 amino residues long in vertebrate TNCs, 5 in *Halocynthia* TNC, 11 in *Astacus,* and 8 in calcineurin B (Table 1).

CAM is generally considered to be the prototype calcium-modulated protein. It activates at least 20 different target enzymes or structural proteins and apparently for this reason is highly conserved. It is distributed throughout the eukaryotes, probably found in every cell of every organism. Hence we infer that a common precursor of all contemporary CAMs was present in the organism that was a common ancestor of all of the eukaryotes. Because of its broad distribution and slow evolution CAM is an excellent character for evaluating the relationship among kingdoms and phyla. Note, however, that the two isotypes from *Arbacia* (CAMAPα and CAMAPβ) lie with the α and β subfamilies, respectively – as always, one must distinguish orthologs from paralogs. The pseudogenes of *Gallus* (Putkey et al. 1985) are inferred to have arisen early in the evolution of the α form; one might expect to find them in other animals. CAM from *Dictyostelium* does not group with that from other fungi – *Saccharomyces* and *Schizosaccharomyces* – but instead is most closely related to the pseudogenes from *Gallus.* CAM from *Trypanosoma* groups with the plants instead of the other protocists – *Paramecium, Tetrahymena,* and *Chlamydomonas.* Again one suspects paralogs; other isoforms of calmodulin may be (or have been) present in these organisms.

The CAM-resembling protein (Cal-1) from *Caenorhabditis elegans* is closely related to the CAMs and is inferred to bind four calcium ions (Salvato et al. 1986). However, it is obviously a paralog; there is a distinct CAM within *Caenorhabditis.* Cal-1 has not been identified in other organisms; however, as paleontologists emphasize, "absence of evidence is not evidence of absence." By our calculations, Cal-1 branched off the CAM line prior to the divergence of any other known CAMs. From this result one might infer that the duplication event that produced Cal-1 and CAM occurred early, prior to the divergence of protocist, fungi, plants, and animals. If so, Cal-1 should be accorded equal weight with all of the CAMs in determining the nodal sequence to be compared with nodal sequences representing other subfamilies. Somewhat arbitrarily we include Cal-1 in the CAM subfamily yet consider calcineurin B as a unique protein. As we refine branch lengths, these classifications can be made more objective.

All of the two-domain homologs belong to one distinct subfamily (S-100* ICBP); within it there are four groups. The α and β chains of the dimeric S-100 arose from a gene duplication that preceded the origin of the mammals. The three proteins – 42A from *Rattus*, PCBP from *Mus*, and 2A9 from humans – appear to be closely related and probably deserve the same name. The next group contains not only the well-known ICPB but also several more recently discovered homologs. Further down this limb arise MRP-8 and MRP-14; both are specific for cells of myeloid origin. "In acutely inflamed tissues, macrophages can express MRP-14 but not MRP-8" (Odink et al. 1987). MRP-14 concentrations are about 10 ng ml^{-1} in the plasma of most normal people, 1000 ng ml^{-1} in the homozygous patients suffering from cystic fibrosis, and 200 ng ml^{-1} in symptom-free heterozygous carriers. The fourth group, nearest the node of the subfamily, consists of p10 from cow, p11 from pig, and 42C from rat; they appear not to bind calcium in domain 1 (Table 1). Both p10 and p11 surely perform the same function, providing two of the subunits of the tetrameric endonexin (p10*p36)$_2$. At least ten paralogous forms of these two domain proteins appear in man; one should not be surprised at the discovery of more. To date, functions are known for none of these two domain proteins.

The PVs hold an esteemed position in the history of calcium-modulated proteins. They were the first purified (Henrotte 1952), the first of known amino acid sequence (Pechère et al. 1971) and crystal structure (Kretsinger et al. 1971), and the first to have a recognized gene triplication (Kretsinger 1972). Their function(s) remains unknown, but they are inferred to function as a kinetic buffer of calcium in fast-twitch muscle (Gillis et al. 1982). Their more recent discovery in brain and skin indicates alternate or additional functions. The generation of isoforms within the PVs seems almost as florid as within the S-100*ICPB subfamily. The only isoforms generally recognized are α (on the right in Fig. 3), and β (on the left). Even within the α branch, *Amphiuma* branches prior to *Rana* PV. Numerous such contradictions are seen on the β branch. The duplication that generated α and β forms occurred early, apparently in the common ancestor, to all higher vertebrates but after the divergence of the coelocanth. Subsequently some organisms deleted one or the other; some generated additional isoforms, often, as is the case of *Cyprinus carpio*, quite recently on an evolutionary time scale. Oncomodulin is found primarily, if not solely, in tumor tissue to date isolated from rats and humans (Gillen et al. 1987). It definitely clusters with the β forms. *Rattus norvegicus* has an α form. If the rat has no other β form, oncomodulin might be considered to be the β form; however, this is not yet established. If, as seems more probable, it is a paralog of the β form, one questions whether the β form has been deleted from rats and other mammals, or whether it has simply not succumbed to identification and purification by the biochemist.

Calbindin (CLBN) is the name given to the M_r 30000 homolog from mammalian intestine (Parmentier et al. 1987); it contains six EF-hand domains. Its function is unknown. Calretinin is found in retina; its gene is expressed in neurons (Rogers 1987). It contains four EF-hands. The placement of the CLBN family was established by comparing the first four N-terminal domains and by the total sequence through the first four domains. The comparisons within CLBN were based on both the entire amino acid sequence and six complete domains. The CLBNs and calretinins were formed prior to the avian–mammalian divergence. We cannot preclude the possibility that the relative placement of this subfamily would be different if there has been a complex rearrangement or deletion of domains.

The calpains (CALPs) are muscle proteases. The N-terminal portion contains a cysteine protease homologous to other such proteases. The C-terminal third consists of four EF-hand domains (Suzuki et al. 1987). The CALPs have evolved by splicing together proteins from two distinct homolog families. There are two major forms of CALP; the light (M_r 30000) and the heavy (M_r 80000). Within the heavy form there are μ and υ isotypes. Since the rabbit and chicken μ isotypes cluster together, one infers that the duplication to make μ and υ occurred prior to the emergence of mammals and birds, and that the light–heavy duplication occurred prior to that.

The SARC*AEQ family is an especially intriguing double family, containing the sarcoplasm calcium-binding proteins (SARCs; Jauregui-Adell et al. 1987) of protochordates and invertebrates and the aequorins (AEQs; Inouye et al. 1985) and luciferin binding protein (LBP; Charbonneau et al. 1988) of coelenterates. The relationships among the SARCs are complex. Two SARCs (SCBP P α_a and α_b) from shrimp (*Penaeus* sp.) cluster together near another pair, one from crayfish *(Astacus leptodactylus)* and the β form from shrimp. The two from the amphioxus *(Branchiostoma)* branch next, preceded by the SARC form scallop *(Patinopecten)*. The earliest branch gave rise to the two forms from the sandworms *Perinereis* and *Nereis*. These relationships accord with phylogeny; the function(s) of these proteins remains unknown. They were initially discovered in a search for an analog or homolog of PV in the invertebrates. We made our initial phenograms using only the 29 amino acid regions of unambiguously recognized EF-hand domains as judged to be:

	domain	1	2	3	4
SCBP P α-$_a$		OK	OK	—	—
SCBP P α-$_b$		OK	OK	—	—
SCBP Al		OK	OK	—	—
SCBP P β		OK	OK	—	—
SCBP Bl 1		OK	OK	OK	—
SCBP Bl 2		OK	OK	OK	—
SCBP Py		OK	—	OK	—
SCBP Pv		OK	—	OK	OK
SCPB Nd		OK	—	OK	OK
AEQ		OK	—	OK	OK
LBP		OK	—	OK	OK

Subsequent phenograms were calculated using complete amino acid sequences whose alignments had been guided by our domain-only phenograms. The two phenograms are very similar. The nearest subfamilies are CALP and CLBN in all of these calculations. The 45 amino acid region between domains 1 and 3 appears to function as the luciferin binding region in LBP and as the luciferase of AEQ. If we compare the nodal sequence of AEQ*LBP with that of the SARCs or of the CALPs, we find no statistically significant alignment. The same conclusion is reached if we do not use nodal sequences but instead align one-to-one the two AEQ, or one LBP, domains "2" with each of the nine SARC and six CALPs, i.e., 45 alignments. From this we conclude that the luciferin-binding domain of AEQ diverged from the precursor of domain 2 so far as to be unrecognizable or else it evolved from a different protein spliced into three EF-hands.

Relationships Among Subfamilies

Figure 2 shows the relationships among the ten major subfamilies of calcium-modulated proteins and the three ectodermal proteins from *Strongylocentrotus purpuratus* (SPECs) and five individual proteins not now judged to be members of an identified subfamily. Note that we have not assigned a "root" or "origin" to this phenogram. In contrast, the phenograms of each of the ten subfamilies (Fig. 3) is rooted. Some of the topological relationships may change for the unique proteins when more data is available since some of the branch lengths (not shown) that determine the positions of the unique proteins are very short.

The ELCs and RLCs of myosin branch from the same node. Both associate with the heavy chain of myosin. The first of the four domains of RLC bind calcium; the remaining three appear not to. Note that the four domains of most ELCs do not bind calcium. In contrast, the light chain from *Physarum* (Kobayashi et al. 1988) more closely resembles the ELCs than the RLCs; however, it appears competent to bind calcium in domain 1, as do the RLCs. Calcineurin B splits early from the TNC branch; one infers that the myosin light chain in the common ancestor of the fungi and animals contained four domains, the first of which could bind calcium. It will be interesting to determine whether contemporary fungi have distinct ELCs and RLCs, or whether this duplication occurred later in the animal ancestor.

The TNC and CAM subfamilies are closest neighbors, as previously deduced by Baba et al. (1984). One infers that the common precursor contained four domains competent to bind calcium. It would be especially interesting to know whether the TNCs are found outside the animal kingdom.

The S-100*ICBP subfamily contains all of the known two-domain proteins. We evaluated its placement by aligning its two domains with domains 1 and 2 and with domains 3 and 4 of its other multidomain homologs. Both alignments placed this subfamily at the same position in the overall phenogram; not surprisingly, the internal structure of this subfamily varies little between the two treatments. Whether using domains only or the entire sequences of the ICBPs, we found phenograms of lower NR length when the ICBPs were aligned with domains 1 and 2 of the other proteins. This supports the conclusion that the ICBPs evolved from a four-domain precursor following deletion of domains 3 and 4.

The PVs, as anticipated from their biochemistry, comprise a well-defined subfamily. They consist of three domains. Our tentative interpretation is that an early animal had a four-domain precursor of both PV and S-100*ICBP. Following gene duplication domain 1 was deleted to form the PV precursor, and domains 3 and 4 were deleted to form the S-100*ICBP precursor. It is interesting that the unique proteins, CDC31 from *Saccharomyces* (Baum et al. 1986) and caltractin from *Chlamydomonas* (Huang et al. 1988), are found earlier on this branch.

The CLBN (Fig. 2) contain six domains. The present phenogram is based on comparison of their first four domains with the other homologs. These first four domains give the optimal alignment with calretinin, a four-domain member of the same subfamily. We are now evaluating the origin of the fifth and sixth domains. Although the M_r 28000 CLBNs and the M_r 9000 ICBPs are both found primarily in the epithelium of the intestine, they do seem distantly related and deserving of different names.

The CALP subfamily is also well defined and distinct from its neighboring subfamilies, CALBN and SARC*AEQ. We need more data to propose sequences for putative intermediate forms because the nodal sequences for these subfamilies show a high degree of two- and threefold ambiguity.

The close relationship between SARCs and the LBP and AEQs was unanticipated and is obtained comparing the total amino acid sequence or with various subsets thereof. The close relationship of the calcium-binding protein CMSE from the prokaryote *Streptomyces* is puzzling. We have yet to understand the evolutionary implications.

Unique Proteins

Several proteins are unique. As noted in the description of subfamilies, this uniqueness may reflect an unusual function, a recent origin, a rapid evolution, or simply an inadequate sampling of tissues and phyla. For whatever reason, at the present time six (groups of) proteins seem to be significantly different from the ten identified subfamilies.

Calcineurin is a general protein phosphatase found in mammalian brain. It is a heterodimer with the A subunit (M_r 61000) having the catalytic activity and the B subunit (M_r 20000) imparting calcium sensitivity to the dimer (Aitken et al. 1984). Following this activation calcineurin can be further activated by binding one equivalent of CAM. The N-terminal glycine of calcineurin B is myristilated. Calcineurin B is slightly closer to the TNC subfamily than to CAM; however, the intervening branch is very short. It cannot replace CAM as an activator of cyclic nucleotide phosphodiesterase nor of calcineurin itself.

The SPEC cDNAs were originally derived from clones from polyA RNA from the dorsal ectoderm of *Strongylocentrotus* embryos (Hardin et al. 1985). This group of genes encodes about ten proteins of two classes, SPEC 1 and SPEC 2. One infers from the deduced amino acid sequences that all four EF-hand domains bind calcium. As is seen throughout the EF-hand superfamily, the sequences within domains bear greater resemblance to other EF-hand proteins than do the sequence between domains. SPEC 1 and SPEC 2 resemble one another more closely than they do the other subfamilies, being most closely related to the PVs and to S-100*ICBP; however, the branch distances involved are so short that these relationships might change as more data become available. It is also interesting that the SPECs are closely related to CDC31 *(Saccharomyces)* and to caltractin *(Chlamydomonas)*. Their function(s) remains unknown; however, light chains, CAM, and TNC are also found in these sea urchins. To date SPECs have not been identified in other organisms.

CDC31 is a yeast *(Saccharomyces cerevisiae)* gene whose product is required for spindle pole body duplication (Baum et al. 1986). The first and the fourth domains of the CDC31 protein probably bind calcium. Its nearest neighbor in the phenogram is caltractin, followed by the SPECs.

Caltractin has been, by indirect immunofluorescence, "localized in interphase *Chlamydomonas* cells to the striated fibers which connect the basal body pair and to two striated roots extending from the basal bodies to the nucleus" (Huang et al. 1988). Its near relationship to CDC31 hardly reflects a phylogenetic proximity since *Chlamydomonas* is in the protocist kingdom and *Saccharomyces* in the fungi.

The calcium vector protein (CVP) from the amphioxus *Branchiostoma lanceolatum* contains four domains, as do all of these unique proteins. It interacts with a M_r 36000 protein, which may be a target analogous to those of CAM. It binds two Ca^{2+} ions, apparently in the two C-terminal domains. It contains a disulfide link between Cys-16 (domain 1, number 2) and Cys-78 (domain 2, number 26). With only slight distortion this S-S bond could be accommodated in the crystal structures of PV, TNC, or CAM. Kobayshi et al. (1987) have noted that the inferred stabilization may obviate the need for calcium binding in domains 1 and 2. In the phenogram CVP branches near the CLBNs and the CALPs.

The last of the unique EF-hand proteins, that from the prokaryote *Streptomyces erythraeus* is perhaps the most fascinating. Swan et al. (1987) inferred four calcium-binding sites and further suggested that:

"... The EF-hand motif may have arisen in an ancient protein before the divergence of the eukaryotes and prokaryotes. The similarity of the bacterial protein to CAM rather than to other known EF-hand proteins supports the view that the eukaryotic EF-hand superfamily has diverged from a common calmodulin-like ancestor with four calcium-binding domains. Less likely alternatives are that convergent evolution, or later gene transfer from eukaryote to prokaryote may be involved."

These points are well taken; however, our phenogram confirms the subsequent interpretation of Cox and Bairoch (1988) that the SARCs (and by our analysis AEQ) are the nearest known neighbor of CMSE.

Discussion

Classification is, in and of itself, a major goal of science. The phenograms of Fig. 2 and 3 represent a major advance in the ongoing effort to order the EF-hand homolog family. In a sense they are self-explanatory; however, a brief description of the work remaining to be done should serve as a caution against over-interpretation.

Branch lengths are now being refined and are not included in Fig. 2 and 3. As previously emphasized, some branch lengths are very short, for instance, those defining the placement of CMSE and of calcineurin B. These lengths will inevitably change slightly as more sequences become available. Even such small changes might change some topological relationships and better define whether seemingly unique proteins should be included in established subfamilies.

Several analyses will yield additional insights into the evolution and functions of the EF-hand proteins. We have aligned the domains of all of the proteins, from N-terminus sequentially to C-terminus, excepting PV, whose three domains are numbered 2, 3, 4. Our phenograms based on individual domains should resemble those presented here if these alignments are correct. Further, comparisons between domains 1 through 4, or through 6 for CLBN, should reveal the sequence of duplications that gave rise to the multiple domains.

The rates of evolution vary enormously among subfamilies (Baba et al. 1984). These rates can be placed on an absolute time scale if one is dealing with proven orthologs, and if one knows from paleontology the times of speciation. Lacking such information one can determine relative rates of evolution of any two proteins, or domains thereof, by counting the number of amino acid sequence changes, or encoding nucleotide changes, from the common node to each protein.

As one gets ever more sequence data from diverse organisms, one can construct nodal sequences of greater reliability. These provide a most probable course of evolution, and they allow us to synthesize putative precursor proteins whose chemical characteristics may reveal their early functions. Further, such nodal sequences provide a sensitive reference for comparing distantly related proteins that may be either analogs or homologs of these established EF-hand proteins. All of the proteins included in this study contain readily identified EF-hands. The inference is very strong that they are homologs and have evolved from a common precursor. These analyses will allow us better to understand analogs and why their evolution may have converged toward the EF-hand.

For many of these proteins complementary DNA and genomic DNA sequences are available. From these one can view shorter time scales. For example the amino acid sequences of many vertebrate CAMs are identical even though their cDNA sequences vary 5%–10%. On a longer time scale we can see the patterns of intron positions from the genomic DNAs.

There obviously remains much to do; yet, even at this preliminary stage we summarize several insights. First, the obvious, there are very many, very different EF-hand proteins distributed throughout the fungi, protocists, plants, and animals. The heavy representation of homologs from the animals, and especially from *Homo*, reflects more the anthropomorphic bent of the funding agencies than the distribution in nature. Of the ten subfamilies only the light chains of myosin and CAM have representatives outside the animal kingdom. The S-100*ICBPs, PARVs and calbindin have so far been found only in vertebrates. The so-called unique proteins come from fungi (CDC31, *Saccharomyces*), protocists (caltractin, *Chlamydomonas*), and several animals. One anticipates an equal richness of EF-hand proteins in the plants, fungi, and protocists. The diversity and divergence of the known proteins indicates a broad range of functions and ancient gene duplications and fixations. Hence, we cannot, with the presently available data, easily establish a root of the phenogram relating subfamilies, nor unambiguously establish the sequence of genetic events leading to the presently known distribution of homologs.

The most parsimonious interpretation of domain evolution is that a single ur-domain duplicated to form a pair of domains probably stabilized by a hydrophobic core as seen in contemporary proteins. In the ancestor of eukaryotes a four-domain protein was formed by gene duplication. In its decendent sequences domains 1 and 3 are closely related as are domains 2 and 4. From one such four-domain precursor domain 1 was deleted to form the PV precursor, and from a related precursor domains 3 and 4 were deleted to form the S-100*ICBP subfamily. On another major branch, two domains were added to form the CLBNs. A cysteine protease was spliced to the N-terminus of the CALP precursor. Whether a domain that binds luciferin replaced domain 2 of AEQ, or whether domain 2 diverged to assume this function has yet to be resolved. Whether *Streptomyces* acquired a eukaryotic gene or whether CMSE evolved from a precursor in the common ancestor to eukaryotes and to eubacteria is very important because it lends insight into the question of whether prokaryotes use calcium as a cytosolic messenger.

The EF-hand can bind a Ca^{2+} ion with high affinity and selectivity and subsequently undergo a change in conformation that has functional significance. Such binding is realized only if the EF-hand has the appropriate sequence of oxygen-containing side chains, and if the domain is a member of a pair of domains (Table 1). An examination of the phenogram of subfamilies (Fig. 2) does not reveal a systematic loss or gain of such function. The enabling or disabling mutations and fixations surely occurred numerous times during evolution.

The interdomain residues linking domains 2 and 3 form an α-helix as seen in the crystal structures of TNC and of CAM. The CAM linker bends to allow the two pairs of domains to enfold the target enzyme (Persechini and Kretsinger 1988) and is crucial to the function of CAM. The 2–3 interdomain lengths range from 6 to 16, or to 23 if one accepts domain 2 of AEQ as a homolog. The most parsimonious interpretation is that the putative precursor near the central node had a linker 7–9 residues long. However, as is the case for calcium binding, numerous changes have occurred during evolution.

Conclusions

Given the diverse functions regulated by cytosolic calcium, it is not surprising that there be so many calcium-modulated proteins. We are just beginning to order this diversity and to understand why all (or most) calcium-modulated proteins contain EF-hands. Very many different extracytosolic proteins, of diverse evolutionary origin, bind calcium, some with high affinity and selectivity; they do not contain EF-hands.

Of special interest are the membrane proteins, pumps, and channels. They are imbedded in lipids, in contrast to the known EF-hand proteins. It is difficult to assign a chemical potential to the Ca^{2+} ion in this hydrophobic environment between aqueous solutions $10^{-2.8}$ M $[Ca^{2+}]^{out}$ and $10^{-7.2}$ M $[Ca^{2+}]^{cytosol}$. Whether any or all of them contain EF-hands remains to be seen. In any case, this classification of the calcium-modulated proteins provides a valuable reference point when studying the proteins that control the concentrations of messenger calcium.

References

Aitken A, Klee CB, Cohen P (1984) The structure of the B subunit of calcineurin. Eur J Biochem 139:663–671

Baba ML, Goodman M, Berger-Cohn J, Demaille JG, Matsuda G (1984) The early adaptive evolution of calmodulin. Mol Biol Evol 1:442–455

Babu YS, Sack JS, Greenbough TJ, Bugg CE, Means AR, Cook WJ (1985) Three-dimensional structure of calmodulin. Nature 315:37–40

Baum P, Furlong C, Byers B (1986) Yeast gene required for spindle pole body duplication: homology of its product with Ca^{2+}-binding proteins. Proc Natl Acad Sci 83:5512–5516

Cavalli-Sforza LL, Edwards AWF (1967) Phylogenetic analysis: models and estimation procedures. Evolution 550–570

Charbonneau H, Kumar, Walsh KA, Cormier MJ, MS in preparation

Cox J, Bairoch A (1988) Sequence similarities in calcium binding proteins. Nature 331:491–492

Falkenthal S, Parker VP, Mattox WW, Davidson N (1984) Drosophila melanogaster has only one myosin alkali lt chain gene which encodes a protein with considerable amino acid sequence homology to chicken myosin alkali light chains. Mol Cell Biol 4:956–965

Farris JS (1972) Estimating phylogenetic trees from distance matrices. Amer Naturalist 106:645–668

Fitch WM, Margoliash E (1967) The construction of phylogenetic trees – a generally applicable method utilizing estimates of the mutation distance obtained from cytochrome C sequences. Science 155:279–284

Fitch WM (1977) The phyletic interpretation of macromolecular sequence information: simple methods. In: Hecht MK, Goody PC, Hecht BM (eds) Major patterns in vertebrate evolution. Plenum Press, New York, pp 169–204

Gillen MF, Banville D, Rutledge RG, Narang S, Seligy VL, Whitfield JF, MacManus JP (1987) A complete complementary DNA for the oncodevelopmental calcium-binding protein, oncomodulin. J Biol Chem 262:5308–5312

Gillis JM, Thomason DB, LeFevre J, Kretsinger RH (1982) Parvalbumin and muscle relaxation: a computer simulation study. J Muscle Res Cell Motil 3:377–398

Goodman M, Pechère JF, Haiech J, Demaille JG (1979) Evolutionary diversification of structure and function in the family of intracellular calcium-binding proteins. J Mol Evol 13:331–352

Hardin SH, Carpenter CD, Hardin PE, Bruskin AM, Klein WH (1985) Structure of the spec 1 gene encoding a major calcium-binding protein in the embryonic ectoderm of the sea urchin, strongylocentrotus purpuratus. J Mol Biol 186:243–255

Hardy DO, Bender PK, Kretsinger RH (1988) Two calmodulin genes are expressed in Arbacia punctulata. An ancient gene duplication is indicated. J Mol Biol 199:223–227

Henrotte JG (1952) A crystalline constituent from myogen of carp muscle. Nature 169:968–969

Herzberg O, James MNG (1985) Structure of the calcium regulatory muscle protein troponin-C at 2.8 A resolution. Nature 313:653–659.

Huang B, Mengersen A, Lee VD (1988) Molecular cloning of cDNA for caltractin, a basal body-associated Ca^{2+} binding protein: homology in its protein sequence with calmodulin and the yeast CDC31 gene product. J Cell Biol 107:133–140

Inouye S, Noguchi M, Sakaki Y, Takagi, Miyata T, Iwanaga S, Miyata T, Tsuji F (1985) Cloning and sequence analysis of cDNA for the luminescent protein aequorin. Proc Natl Acad Sci 82:3154–3158

Jauregui-Adell J, Wnuk W, Cos JA (1987) Amino-acid sequence of the sarcoplasmic calcium-binding protein I (SCP I) from crayfish (Astacus leptodactylus). J Muscle Res Cell Motility 8:92–93

Jukes TH (1963) Some recent advances in studies of the transcription of the genetic message. Adv Biol Med Phys 9:1–41

Kobayashi T, Takagi T, Konishi K, Cox JA (1987) The primary structure of a new M_r 18,000 calcium vector protein from amphioxus. J Biol Chem 262:2613–2623

Kobayashi T, Takagi T, Konishi K, Hamada Y, Kawaguchi M, Kohama (1988) Amino acid sequence of the calcium-binding light chain of myosin from the lower eukaryote, Physarum polycephalum. J Biol Chem, in press

Kretsinger RH, Nockolds CE, Coffee CJ, Bradshaw RA (1971) The structure of a calcium binding protein from carp muscle. Cold Spring Harbor Symp Quant Biol 36:217–220

Kretsinger RH (1972) Gene triplication deduced from the tertiary structure of a muscle calcium binding protein. Nature New Biol 240:85–88

Kretsinger RH (1975) Hypothesis: calcium modulated proteins contain EF hands. In: Carafoli E, Clementie F, Drabikowski W, Margreth A (eds) Calcium transport in contraction and secretion. North-Holland Publishing Co., Amsterdam, pp. 469–478

Kretsinger RH, Rudnick SE, Weissman LJ (1986) Crystal structure of calmodulin. J Inorg Biochem 28:289–302

Kretsinger RH (1987) Calcium coordination and the calmodulin folds divergent versus convergent evolution. Cold Spring Harb Symp Quant Biol 52:449–510

Moore GW, Goodman M, Barnabas J (1973a) An iterative approach from the standpoint of the additive hypothesis to the dendrogram problem posed by molecular data sets. J Theor Biol 38:423–457

Moore GW, Barnabas J, Goodman M (1973b) A method for constructing maximum parsimony ancestral amino acid sequences on a given network. J Theor Biol 38:459–485

Moore GW (1976) Proof for the maximum parsimony ("red king") algorithm. In: Goodman M, Tashian RE (eds) Molecular anthropology. Plenum Press, New York, pp 117–137

Odink K, Cerletti N, Bruggen J, Clerc RG, Tarcsay L, Zwadlo G, Gerhards G, Schlegel R, Sorg C (1987) Two calcium-binding proteins in infiltrate macrophages of rheumatoid arthritis. Nature 330:80–82

Patterson C (1987) Introduction. In: Patterson C (ed) Molecules and morphology in evolution: conflict or compromise? Cambridge University Press, Cambridge, pp 1–22

Parmentier M, Lawson DE, Vassart G (1987) Human 27-kDa calbindin complementary DNA sequence: evolutionary and functional implications. Eur J Biochem 170:207–215

Persechini A, Kretsinger RH (1988) Toward a model of the calmodulin-myosin light chain kinase complex: implications for calmodulin function. J Cardiovascular Pharm 12 (in press)

Pechère JF, Capony JP, Ryden L, Demaille J (1971) The amino acid sequence of the major parvalbumin from hake muscle. Biochem Biophys Res Commun 43:1106–1111

Putkey JA, Slaughter GR, Means AR (1985) Bacterial expression and characterization of proteins derived from the chicken calmodulin cDNA and a calmodulin processed gene. J Biol Chem 260:4707–4712

Rogers JH (1987) Calretinin: a gene for a novel calcium-binding protein expressed in neurons. J Cell Biol 105:1343–1353

Salvato M, Sulston J, Albertson D, Brenner S (1986) A novel calmodulin-like gene from the nematode Caenorhabditis elegans. J Mol Biol 190:281–290

Satyshur KA, Rao ST, Pyzalska D, Drendel W, Greaser M, Sundaralingam M (1988) Refined structure of chicken skeletal muscle troponin C in the two calcium state at 2A resolution. J Biol Chem 263:1628–1647

Sokal RR, Michener CD (1958) A statistical method for evaluating systematic relationships. Univ Kansas Sci Bull 38:1409–1438

Suzuki K, Ohno S, Emori Y, Imajoh S, Kawasaki H (1987) Primary structure and evolution of calcium-activated neutral protease (CANP). J Protein Chem 6:7–15

Swan DG, Hale RS, Dhillon D, Leadlay PF (1987) A bacterial calcium-binding protein homologous to calmodulin. Nature 329:84–85

Szenbeyi DME, Moffat K (1986) The refined structure of vitamin D-dependent calcium-binding protein from bovine intestine. Molecular details, ion binding, and implications for the structure of other calcium-binding proteins. J Biol Chem 261:8761–8777

Takagi T, Konishi K (1983) Amino acid sequence of troponin C obtained from ascidian (Halocynthia roretzi) body wall muscle. J Biochem 94:1753–1760

Wnuk W, Schoechlin M, Kobayashi T, Takagi T, Konishi K, Hoar PE, Kerrick WGL (1986) Two isoforms of troponin C from crayfish. Their characterization and a comparision of their primary structure with the tertiary structure of skeletal troponin C. J Muscle Res Cell Motil 7:67–68

Appendix

The common and scientific names used in the text are:

Common name	Scientific name
jellyfish	*Aequorea victoria*
two-toed amphiuma	*Amphiuma means*
scallop	*Aquipecten irradians*
sea urchin	*Arbacia punctulata*
crayfish	*Astacus leptodactylus*
crayfish	*Astacus pontasticus*
boa constrictor	*Boa constrictor*
cow	*Bos taurus*
amphioxus	*Branchiostoma lanceolatum*
roundworm	*Caenorhabditis elegans*
flagellate	*Chlamydomonas reinhardtii*
scallop	*Chlamys nipponensis*
quail	*Coturnix coturnix*
carp	*Cyprinus carpio*
cellular slime mold	*Dictyostelium discoideum*
fruit fly	*Drosophila melanogaster*
electric eel	*Electrophorus electricus*
northern pike	*Esox lucius*
cod	*Gadus callarias*
whiting	*Gadus merlangus*
chicken	*Gallus gallus*
map turtle	*Graptemys geographica*
ascidian	*Halocynthia roretzi*
human	*Homo sapiens*
coelacanth	*Latimeria chalumnae*
chub	*Leuciscus cephalus*
sea urchin	*Lytechinus pictus*
hake	*Merluccius merluccius*
sea anemone	*Metridium senile*
mouse	*Mus musculus*
sandworm	*Nereis diversicolor*
toadfish	*Opsanus tau*
rabbit	*Oryctolagus cuniculus*
ciliate	*Paramecium tetraurelia*
scallop	*Patinopecten* sp.
scallop	*Patinopecten yessoensis*
scallop	*Pecten marinus*
brine shrimp	*Penaeus* sp.
sandworm	*Perinereis vancaurica*
plasmodial slime mold	*Physarum polycephalum*
thornback ray	*Raja clavata*
frog	*Rana esculenta*
rat	*Rattus norvegicus*
sea pansy	*Renilla reniformis*
yeast	*Saccharomyces cerevisiae*
yeast	*Schizosaccharomyces pombe*
spinach	*Spinacia oleracea*
surf clam	*Spisula sachalinensis*
actinobacteria	*Streptomyces erythraeus*
sea urchin	*Strongylocentrotus purpuratus*
pig	*Sus scrofa*
ciliate	*Tetrahymena pyriformis*
squid	*Todarodes pacificus*
wheat	*Triticum aestivum*
flagellate	*Trypanosoma brucei gambiense*
african clawed frog	*Xenopus laevis*

1 The Function of Calcium Channel

Kinetics and β-Adrenergic Modulation of Cardiac Ca^{2+} Channels

W. Trautwein and D. Pelzer

II. Physiologisches Institut der Universität des Saarlandes, 6650 Homburg/Saar, FRG

Introduction

Ca^{2+} channels in excitable tissues play a key role in coupling membrane excitation with cellular effector functions. Ca^{2+} current which flows into the cell is the stimulus for hormone or transmitter secretion and the contraction of smooth and cardiac muscle. In the heart, Ca^{2+} current also contributes to the action potential configuration and impulse initiation and conduction, and is intimately connected with certain patterns of dysrhythmias [10, 11, 14, 15].

Two different Ca^{2+} channels have been reported to coexist in cardiac cells [1]. The T- and L-type channels differ in regard to their voltage ranges of activation and inactivation, but also in their kinetics, single-channel conductance, and pharmacology. The major Ca^{2+} supply for the cell seems to be provided by L-type Ca^{2+} channels. Their kinetics and β-adrenergic control will be the topic of this contribution.

Methods

Single ventricular cells were obtained from hearts of adult guinea pigs by enzymatic dissociation [12, 14 15]. An aliquot of the cell suspension was transferred into a cell culture dish, which formed the inlay of the recording chamber. Within about 5 min, most of the cells had settled on the bottom and could be continuously superfused with saline of the following composition (in mM): NaCl 140, KCl 10.8, CaCl$_2$ 3.6, MgCl$_2$ 1, glucose 10, Hepes 5. The pH was adjusted to 7.4, and the saline prewarmed to 32°–36°C. Under these conditions, 30%–60% of the cells were relaxed, quiescent, rod-shaped, and cross-striated without distinct signs of deterioration [9, 15].

Recording pipettes were pulled from thick-walled, hard, borosilicate glass in a two-step process; coated with insulating varnish; and fire-polished afterwards (see [12, 14, 15] for details of the methods and references). Single Ca^{2+}-channel activity was recorded from membrane patches in the cell-attached configuration. The pipette-filling solution (external membrane patch solution) contained (in mM) BaCl$_2$ 90, NaCl 2, KCl 4, TTX 0.02, Hepes 5 (pH 7.4); its composition eliminated K- and Na-channel currents. When immersed in the bath solution (see above), pipettes had resistances between 3 and 5 MΩ. Tight pipette-membrane seals were obtained by gentle suction after mechanical contact between the pipette tip and the membrane surface of the cell [12]. Typically, seal resistances ranged between 10 and 100 GΩ. The patch membrane potential was changed by clamping the intrapipette potential with reference to the bath while leaving the absolute cell resting membrane potential (RP) free. Although the RP across the membrane patch is not known in the cell-attached configuration, it is assumed to be similar to the -65 ± 8 mV ($n = 15$) RP measured with KCl-filled pipettes in similar cells. Since a zero potential in the recording pipette corresponds to the RP across the membrane patch, de- or hyperpolarizations are expressed as positive or negative voltage displacements from the pipette zero potential, respectively.

Whole-cell Ca^{2+}-channel currents were recorded after establishment of the seal and subsequent rupture of the membrane patch with short pulses of suction (see [12] for details of the method). For comparison with single-channel average currents, pipettes were filled with (in mM) CsCl 112, $MgCl_2$ 8, Na_2ATP 5, Cs_2EGTA 10, Hepes 10 (pH 7.2), and the bath was perfused with an external solution (temperature 34°–36°C) containing 90 mM Ba (same as the external membrane patch solution for single Ca^{2+}-channel recordings, see above). The resistances of the pipettes ranged between 1 and 3 MΩ. For intracellular dialysis, drugs of enzymes were added to the pipette solution (see legends), which contained (in mM) potassium aspartate 80, KCl 50, KH_2PO_4 10, $MgSO_4$ 1, Hepes 5, Na_2ATP 3, EGTA 0.1 (pH 7.3). The external solution in these experiments contained (in mM) NaCl 112, $NaHCO_3$ 24, KCl 5.4, $CaCl_2$ 1.8, $MgCl_2$ 1.0, glucose 10, Hepes 5 (pH 7.4, temperature 35°–37°C), or NaCl 140, KCl 10.8, $CaCl_2$ or $BaCl_2$ 3.6, $MgCl_2$ 1, glucose 10, Hepes 5 (pH 7.4, temperature 34°C). Voltage and corresponding current signals were collected on FM tape. Subsequent automated analysis of the digitized, leakage- and capacitance-corrected records was carried out with procedures similar to those previously described (see [3] for details and references). All presented single-channel data come from membrane patches with only one functional Ca^{2+} channel, as judged by the absence of multiple conductance levels.

Results and Discussion

Whole-Cell and Single-Channel Current

Whole-cell L-type Ca^{2+}-channel current (I_{ca}) in ventricular myocytes is recorded at potentials positive to -50 mV. Upon depolarization under different ionic conditions, I_{ca} reaches a peak within 2–4 ms and thereafter declines at a rate several times slower (Fig. 1). This macroscopic current, at each point in time, describes the average gating behavior of a population of N independent Ca^{2+}-channels, with an elementary single-channel current i and a probability p of being open ($I = N i p$) [11, 14–17]. The rapid ascending phase of whole-cell L-type Ca^{2+}-channel current, carried by either calcium or barium (Fig. 1), is called activation, whereas the slow declining phase reflects a process of inactivation. The whole-cell current I does not give information on N, i, or p (see above). These parameters can, however, be directly studied by recording the activity of a single Ca^{2+}-channel during step depolarization. Figure 2A, a shows 20 consecutive current traces out of 205 sweeps during 300-ms clamp steps applied from the pipette zero potential to RP + 75 mV at 0.5 Hz. Two types of current records are regularly observed: (a) traces with Ca^{2+}-channel activity of various lengths (in the

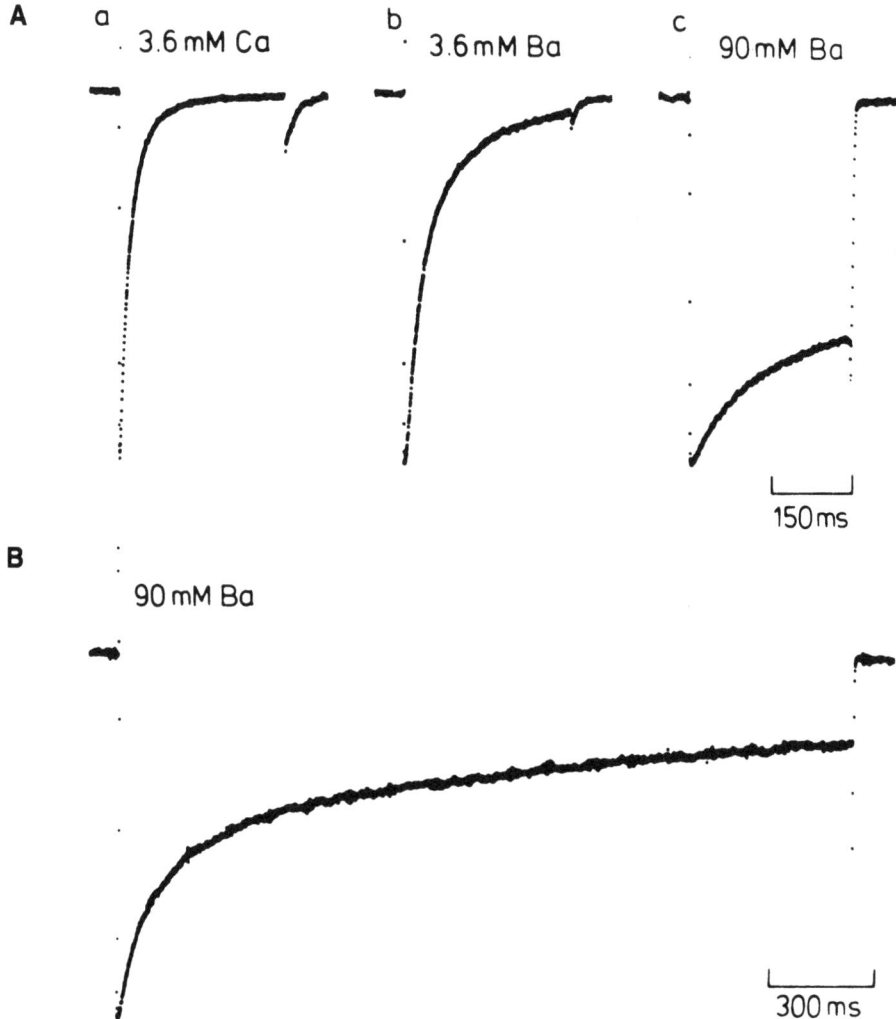

Fig 1A, B. Whole-cell currents from myocytes bathed in solutions containing **A** 3.6 mM Ca (a), 3.6 mM Ba (b), 90 mM Ba (c), and **B** 90 mM Ba. The voltage-clamp pulses were from −50 to +10 mV for 300 ms (**A**) and to +15 mV for 2 s (**B**). The currents have been scaled for comparative reasons. The current densities were 29.6 (3.6 mM Ca), 25 (3.6 mM Ba), and 131 (90 mM Ba) µA/cm² at +10 mV (**A**) and 207 (90 mM Ba) µA/cm² at +15 mV (**B**). (From [8])

form of closely spaced brief pulses of inward current with a unitary amplitude of 1.1 pA) and (b) blank sweeps without any detectable single-channel opening. The records with Ca^{2+}-channel activity show a distinct tendency for openings to occur towards the beginning of the clamp pulse, followed by long periods of silence. The blank sweeps do reflect a condition or conditions where the Ca^{2+} channel is unavailable for opening (see below). In this experiment, only 91 of 205 consecutive depolarization elicited

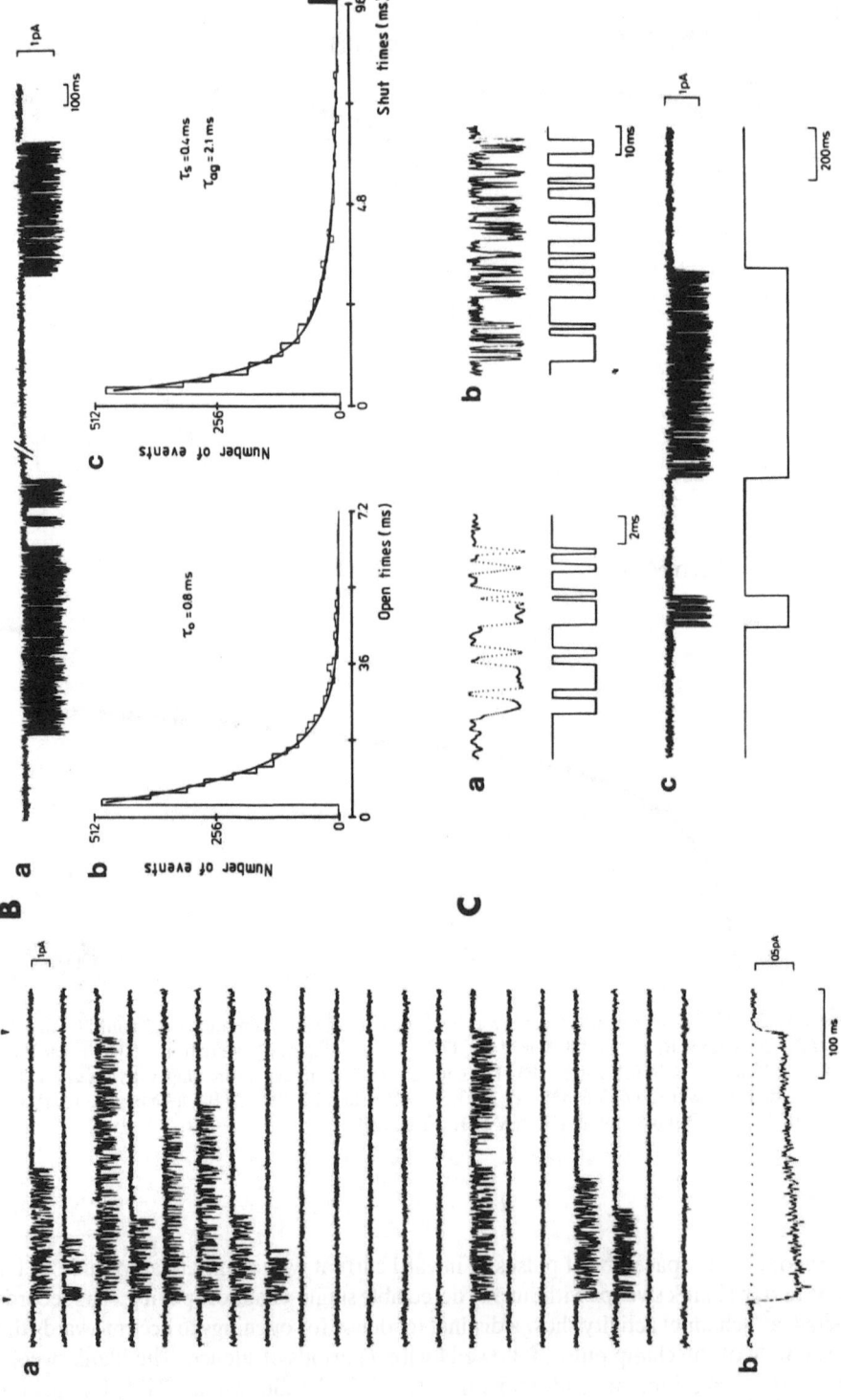

channel activity; the remaining 114 trials gave blank records. Both types of response tend to cluster in groups of consecutive single Ca^{2+}-channel current sweeps.

The average current flowing through the single Ca^{2+} channel during a series of identical voltage steps can be compiled by averaging over the whole ensemble of single-channel current records. The result is the mean current I(t) (Fig. 2A, b) which (a) describes the time course of single-channel activity in the membrane patch, just as a record of whole-cell Ca^{2+}-channel current reflects the average time course of activity of a large number of channels, and (b) displays a rapid rising phase followed by a slow decay, as whole-cell Ca^{2+}-channel currents carried by barium ions do (see Fig. 1). On averaging, the probability of open-state occupancy at each particular time during depolarization, p(t), is determined since all individual channel openings add a unit to I(t) whenever the channel is open. Since (a) the Ca^{2+} channel has only two distinguishable conductance levels, (b) the single-channel current amplitude i does not change with time during the depolarizing pulse, and (c) analysis is restricted to recordings from membrane patches with only one functional Ca^{2+} channel, I(t) is proportional to p(t), where i is a factor of proportionality.

Kinetics of L-Type Ca^{2+}-Channel

The kinetics of Ca^{2+}-channel gating have been studied in records during steady depolarizations of the membrane patch. The reason for not using current traces during depolarization of limited duration has been given elsewhere [3]. In Fig. 2B, a the membrane patch was depolarized to RP + 65 mV. The bursting activity is always interrupted by long intervals of quiescence. In this example, the periods of repetitive discharge range from 48 to 1200 ms and are separated by gaps of between 30 and 2500 ms. This observation suggests the existence of high- and low-frequency fluctuations in the gating behavior of the Ca^{2+} channel.

Open and closed lifetime histograms were compiled from a 25-min current record obtained at RP + 65 mV (Fig. 2B, b, c). The distribution of open times can be described consistently by a single-exponential function (Fig. 2B, b). The best fit yields a time constant, τ_0, of 0.8 ms and includes all observed events (2984). Long openings (lifetimes $> 10 \times \tau_0$) have only been observed in 2 of 71 experiments. Even then, they did not occur frequently enough to produce a detectable deviation from the monoexponential probability density distribution of all opening events. The percentage of long openings in the two cases was 0.49 and 0.86. Substates could not be detected or

---

Fig. 2 A–C. Single-channel currents during consecutive voltage pulses of 75-mV amplitude (RP + 75 mV) applied from the pipette zero potential (equal resting potential of the cell) for 300 ms *(arrows)* at 0.5 Hz *(a)* and the mean current compiled from 205 sweeps *(b)*. Temperature (T) 32°C. Leakage and capacitance currents were subtracted from each current trace. **B** Kinetics of single Ca^{2+}-channel currents during a 25-min steady depolarization to RP + 65 mV. The last two periods of activity *(a)* are separated by a gap of 2.5 s. Sampling rate 2.8 kHz. Open time histogram *(b)*. Closed time histogram *(c)*. The excess of closed times longer than 9.6 ms is given in the *lack black bin*. T 34°C. **C** Single Ca^{2+} channel records at RP + 65 mV *(upper panels)* and corresponding idealizations *(lower panels)* at different time scales. Current records are digitized at 6.6 kHz *(a, b)* or at 1.8 kHz *(c)*. Idealizations represent individual single-channel openings *(a)*, bursts of openings *(b)*, and cluster of bursts *(c)*. Same experiment as in *B*. Cell-attached recording configuration, 90 mM Ba in pipettes. (From [9]).

resolved. Thus, we assume that the L-type cardiac Ca^{2+} channel has only one predominant conducting open state. By contrast, the distribution of the closed times between 0 and 9.6 ms can only be fitted by the sum of two exponential terms with two well-separated time constants. $\tau_s = 0.4$ and $\tau_{ag} = 2.1$ ms (Fig. 2B, c). This two-exponential probability density function, however, still does not allow for 2% of shut intervals longer than 9.6 ms (61 of 2965) represented by the last bin (Fig. 2B, c). Due to the wide scatter in lifetime of the long-lasting gaps (tens of milliseconds to several seconds, mean lifetime = 1291 ms), no conclusion about the underlying probability density function or functions can be drawn. Nevertheless, more than two exponential terms are required for an accurate description of the frequency distribution of all shut lifetimes. The three classes of gaps argue in favour of the existence of at least three distinct closed states of the cardiac Ca^{2+} channel.

The multiexponential shape of the shut lifetime distribution allows the classification of the Ca^{2+}-channel gating transitions on different time scales (Fig. 2C). The long-lasting gaps are prominent only in single-channel current records displayed at low time resolution (Fig. 2C, c). At expanded time scales, the shorter gaps corresponding to τ_{ag} (Fig. 2C. b) and τ_s (Fig. 2C, a) can be resolved. Records like those in Fig. 2C can be divided into groups of openings separated by critical gaps (for definition see [3, 14, 15]). Any deflection of the current trace staying below half the amplitude of the single-channel current for at least 0.15 ms is regarded as a gap separating two openings (Fig. 2C, a). By neglecting the briefest gaps, the observation is confined to the registration of bursts of openings and gaps between bursts. A burst was defined as any series of openings interrupted only by gaps shorter than 2 ms ($= 5 \times \tau_s$), a shut interval for which the probability densities of both exponential terms of the two-exponential distribution function are equal. The schematic trace below the original record (Fig. 2C, b; lower panel) illustrates how the signal is represented by the computer after disregarding all the very short-lived gaps. This procedure has to be extended by describing the bursts of opening themselves as being grouped in clusters of bursts, the gaps between clusters corresponding to the long-lived gaps. For the analysis of clusters, another critical-gap length has to be defined as a limit between the gaps in bursts within clusters and the gaps separating clusters. Since the distribution(s) of the long-lasting gaps is not known, the number of gaps incorrectly classified as shut intervals between clusters was calculated from the probability density function of the slower exponential term for different critical-gap durations. About 21 ms ($= 10 \times \tau_{ag}$) seemed to be a reasonable choice, since the probability of an incorrectly classified shut interval longer than this gap was $\simeq 0.01$. The schematic trace in Fig. 2C,c (lower panel) illustrates how bursts of openings were joined in clusters of bursts by neglecting all gaps shorter than 21 ms. These two critical-gap durations divided the frequency distribution histogram of all closed times into 2157 short closures, 781 intermediate gaps, and 27 long-lasting shut periods. The mean open-state probability within clusters of bursts determined as the time-average of current within clusters divided by the single-channel current amplitude was 0.44.

The findings indicate that the fast gating transitions grouping individual openings into bursts of openings reflect the fluctuations of the Ca^{2+} channel between two short-lived shut states and one open state. The bursting behavior is terminated by the entry of the channel into at least one long-lived shut state, exit from which is slow in comparison to the rapid cycling. As a consequence, bursts of openings are further grouped together in clusters of bursts, the cluster behavior being related to slow gating transitions.

Relation of Single-Channel Activity to Activation and Inactivation of the Mean Current I(t)

Upon a sudden change of membrane potential, the first channel opening occurs with a variable delay (Fig. 3A, a). This time is called first latency or waiting time and gives information about the opening pathway of the Ca^{2+} channel. The probability density distribution of the first latencies (Fig. 3A, b) for a run of pulses to RP+75 mV (cf. examples in Fig. 3A, a) shows a biphasic shape, with a maximum at a time later than zero and very few waiting times longer than 6 ms. The distinct rising phase is a common observation for Ca^{2+} channels [12, 14, 15] and is the expected result for a process in which multiple closed states precede the open state of the channel. Thus, a C_1-C_2-O sequence may best account for the opening pathway of the Ca^{2+} channel, where C_1 and C_2 denote the two short-lived closed states. Integration of the first latency histogram yields the cumulative probability distribution of the waiting times (Fig. 3A, c), i.e., an estimate of the time course of the first entry of the Ca^{2+} channel into the open state. With appropriate scaling, this function superimposes on the ascending phase of the mean current (Fig. 3A, c). It indicates that the ascending phase of I(t) with its sigmoid time course reflects the fast Ca^{2+}-channel gating transitions which cause the rapid bursting behavior in the single-channel current records.

The cluster behavior, resulting from the existence of a third class of long-lived gaps, has frequently been overlooked in earlier studies (see [3] for references). The major reason was that these shut intervals are mostly truncated by the end of the command pulse and thus discarded (cf. Fig. 2A, a). Figure 3B, a illustrates the occurrence of clusters during 900-ms voltage steps to RP+75 mV, applied at 0.2 Hz. The idealized traces, superimposed on the original single-channel current records, are computed by neglecting all shut intervals shorter than a critical gap of 21 ms. The probability of observing more than one cluster per single depolarizing pulse is very small (Fig. 3B, d). On average, only 0.83 clusters occur during a voltage step shorter than or equal to 1 s; i.e., either one cluster of Ca^{2+}-channel activity or a blank sweep is most frequently observed upon depolarization (Fig. 3B, d). Cluster durations were measured from the idealized records and compiled in a frequency histogram (Fig. 3B, c). The distribution of cluster lifetimes is well fitted by a single exponential function with a time constant of 275 ms (Fig. 3B, c), which matches the monoexponential decay of the corresponding mean current I(t) (Fig. 3B, b; solid line). This finding implies that the time constants of the decline of I(t) and the cluster lifetime distribution have to be virtually identical. Thus, the decrease of the probability of channel opening and the resulting decay of I(t) during a step depolarization is due to a transition of the Ca^{2+} channel into the long-lived third type of closed state(s).

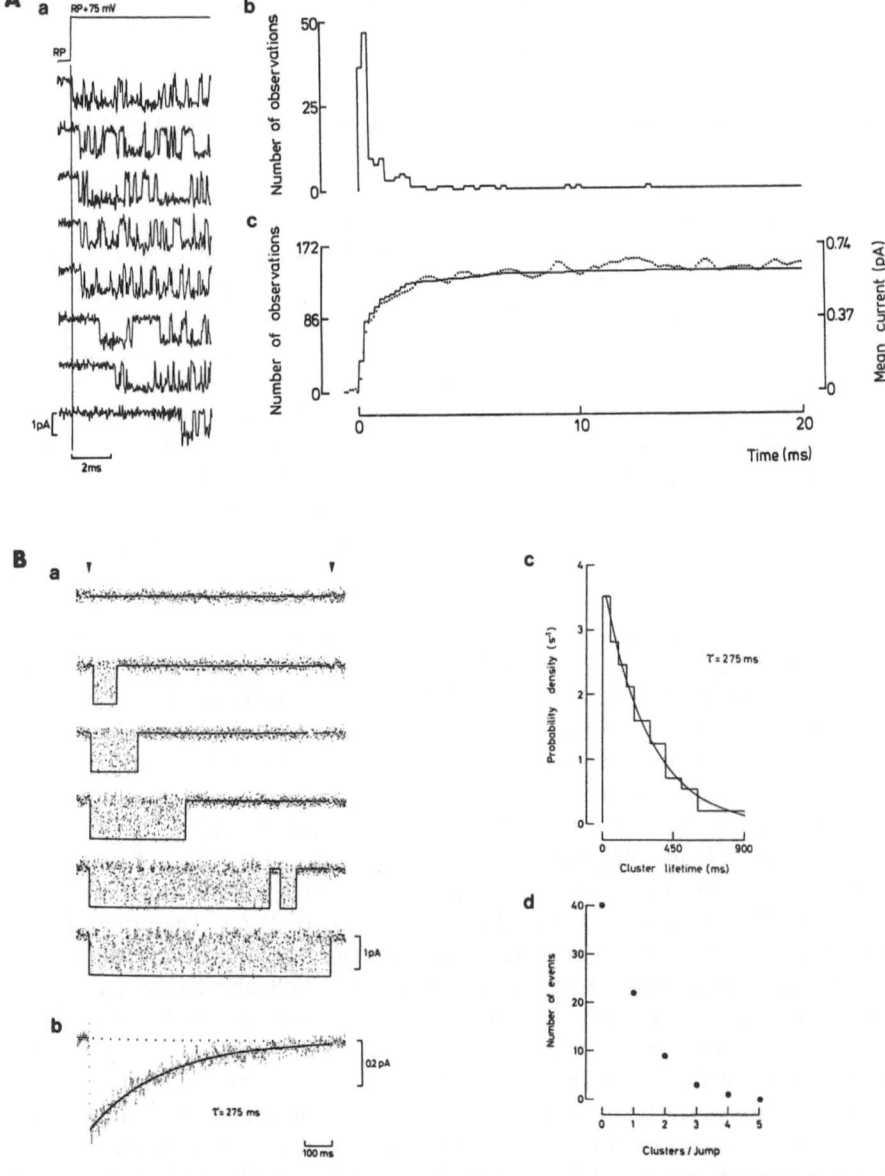

Fig. 3. A Single Ca^{2+}-channel current records (*a*) illustrating the waiting time before the first channel opening. The *vertical line* indicates the onset of the command pulses (300 ms, 0.5 Hz) to RP + 75 mV. The probability density distribution (*b*) of the first latencies obtained from 175 records. The cumulative probability density distribution (*c*) of the fist latencies (*solid line*) superimposed on the ascending phase of the ensemble mean current (*dotted line*). T 32°C. **B** Clusters of bursts (*a; dotted traces*) during 900-ms voltage pulses (*arrows*) to RP + 75 mV at 0.2 Hz, idealized by neglecting shut intervals shorter than 21 ms (*solid traces*). Corresponding mean current (*b; dotted trace*) from 117 single-channel current records. All traces are leakage- and capacitance-subtracted. Cluster lifetime distribution (*c*) with a time constant used for the fit of the declining phase of the mean current (*solid line* in *b*). Number of clusters occurring per voltage step (*d*). T 35°C. Cell-attached recording configuration, 90 m*M* Ba in pipettes. (From [9]).

Steady-State Inactivation and Blanks

The dependence of the fraction of blanks in an ensemble of single-channel current records on the conditioning membrane potential is illustrated in Fig. 4A–D. The patch membrane potential is changed stepwise from different holding potentials to RP+75 mV for 300 ms (Fig. 4A; left panel) at a rate of 0.5 Hz. The open state probability p_o for consecutive single-channel current traces in the entire sample was determined by time-averaging each sweep from the onset of depolarization to the end of the pulse. Examples of single-channel current records and p_o are shown for three samples where the holding potential was RP−30 mV (Fig. 4A), RP+40 mV (Fig. 4B), and RP+60 mV (Fig. 4C). Empty sweeps have zero values of p_o. Depolarization of the conditioning potential results in an increase in the fraction of blank current traces; p_o of the remaining current records with Ca^{2+}-channel activity is essentially unaffected. Also, quantitative analysis of such records did not reveal a detectable effect on the holding potential on the bursting kinetics of the Ca^{2+} channel during the test pulse [15]. The availability of the Ca^{2+} channel in a series of trials was determined as the ratio of the traces containing channel activity over the total number of records in the ensemble. This estimate is equivalent to the probability P that the open state of the Ca^{2+} channel is occupied at least once during a voltage pulse. P decreases at positive holding potentials with a sigmoidal dependence on the potential of the conditioning depolarization (Fig. 4D). The decreasing availability of the Ca^{2+} channel resulted in smaller amplitudes of the corresponding mean currents [14].

In the context of inactivation, it is important to determine whether the Ca^{2+} channel can be activated during the occupancy of any one of the three types of shut states (cf. Fig. 2C). This problem was approached by testing the channel's response to voltage pulses (to RP+75 mV) applied from RP+60 mV at different stages of channel activity (Fig. 4E). Step depolarizations ($n = 53$) always elicit channel openings if they are applied either immediately after an opening (Fig. 4E; upper panel) or after a shut period corresponding by duration to an interburst gap (Fig. 4E; middle panel). By contrast, voltage pulses ($n = 97$) applied after periods of channel inactivity which normally separate clusters of bursts are never able to trigger channel openings (Fig. 4E; lower panel). These findings suggest that the Ca^{2+} channel can only be activated if a step depolarization is applied during the occupancy of the two short-lived shut states. If the long-lived shut state(s) is being occupied, the Ca^{2+} channel is unavailable for opening, and a blank sweep is observed in response to a voltage pulse. On the basis of these data, P would be a global estimate of the steady-state occupancy of the short-lived activatable kinetic states; the fraction of blank single-channel current records in the sample, 1-P, would correspond to the steady-state occupancy of the long-lived nonactivatable shut state(s). An implication of this conclusion is that the modulation of P by the holding membrane potential would represent the voltage dependence of the slow gating transitions of the Ca^{2+} channel. It should be noted, however, that P, even in the most favorable case (Fig. 4D), never reaches unity (see also [3]), suggesting that P is controlled by additional factors distinct from membrane potential (see below).

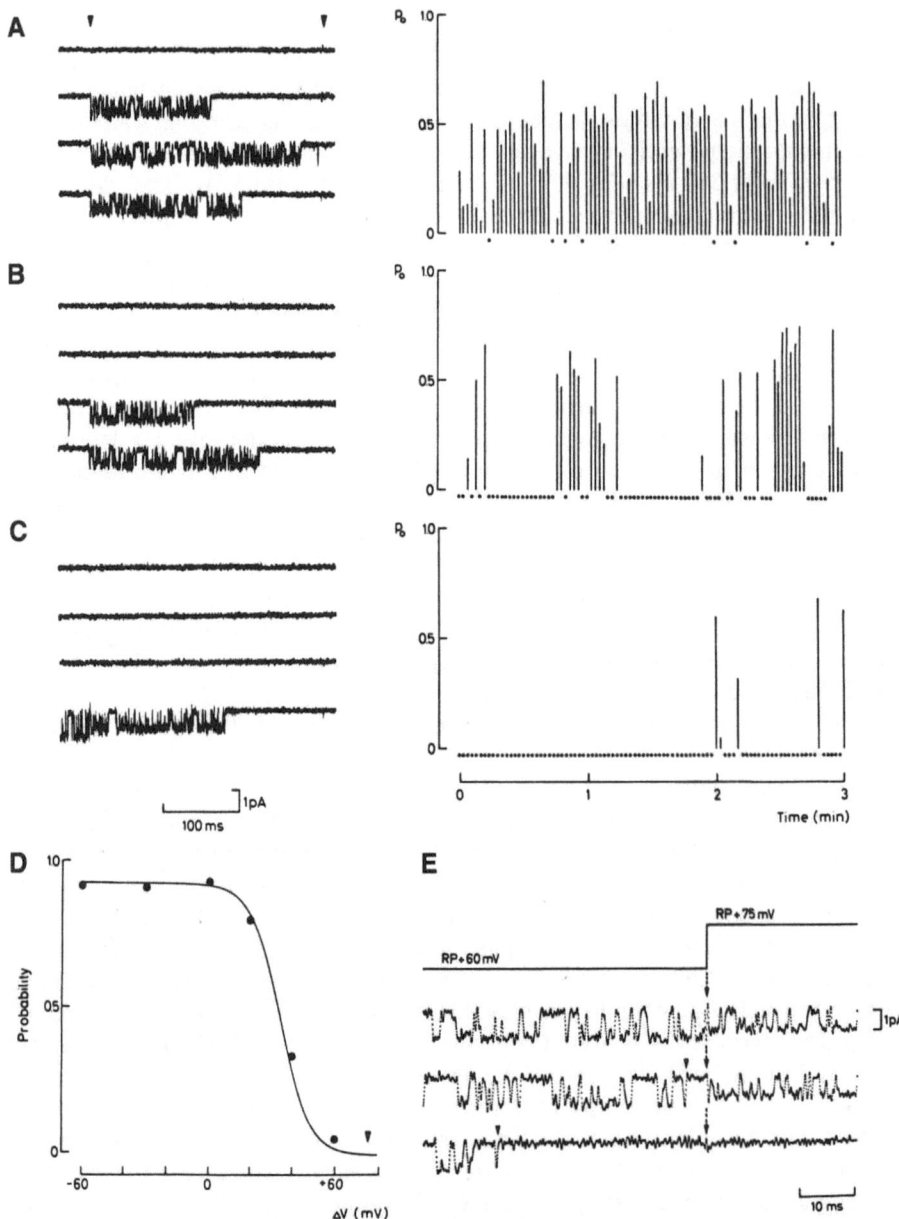

Fig. 4 A–C. Single Ca^{2+}-channel current *(left panels)* and p_o of consecutive single-channel current traces *(right panels)* during 300-ms voltage pulses *(arrows)* to RP + 75 mV from RP – 30 mV **(A)**, RP + 40 mV **(B)** and RP + 60 mV **(C)**. **Dots** in *right panels* indicate empty sweeps. **D** The probabilities that the channel opens at least once per depolarization is calculated as the ratio of the number of current traces with channel activity over the total number of trials *(filled circles)*. It can be described by

$$P = 0.93 / (1 + \exp((V - V_{0.5})/k)),$$

with parameters chosen for the best visual fit: $V_{0.5}$ (potential of half inactivation) = RP + 35 mV; k (slope factor) = 7 mV *(solid line)*. **E** Induction of channel opening by voltage pulses applied during periods of channel inactivity. The *arrows* mark channel closing and the onset of the voltage step. T 32°C. Cell-attached recording configuration, 90 mM Ba in pipettes. (From [9])

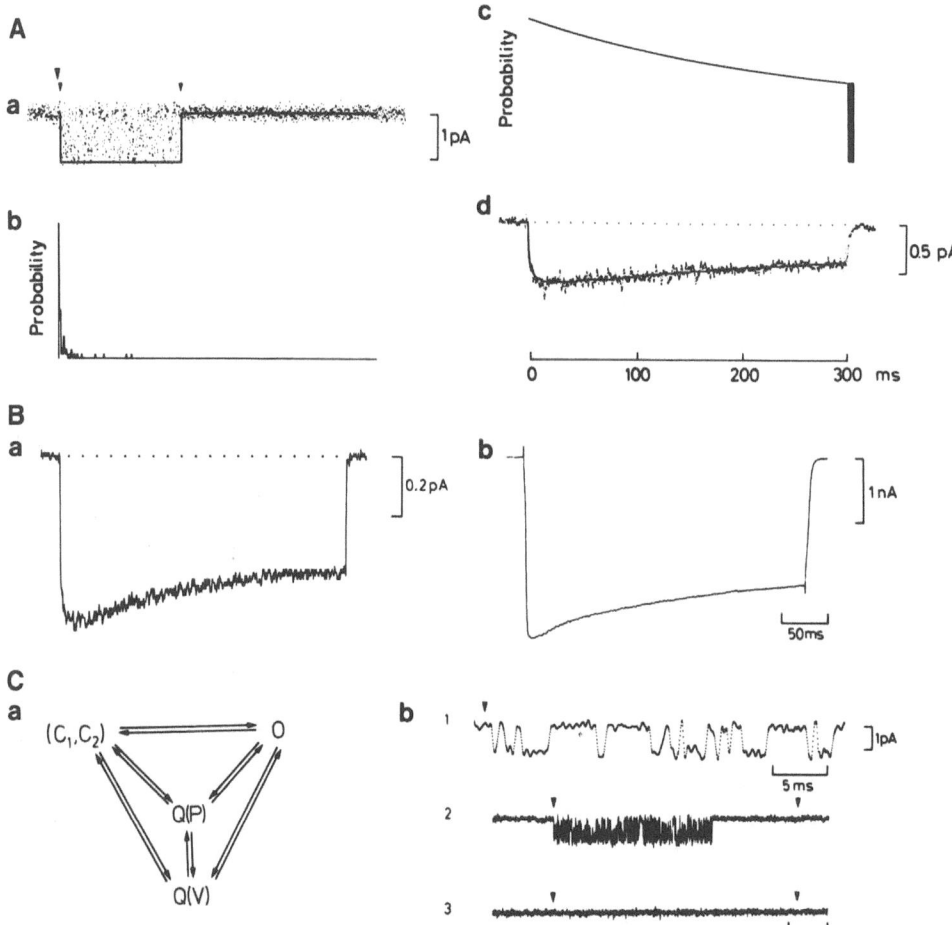

Fig. 5 A–C. First latency (between *first* and *second arrow*) and cluster lifetime (between *second* and *third arrow*) during a 300-ms voltage step to RP + 75 mV *(a)*. First latency distribution *(b)*. Cluster lifetime distribution *(c)*. Corresponding mean current *(d; dotted curve)*. The superimposed *solid curve* shows the theoretical mean current time course computed by the convolution of the first latency and cluster lifetime distributions. Same experiment as in Fig. 2A. **B** Mean current *(a)* from 2064 single Ca^{2+}-channel current records obtained with 90 mM Ba in the pipettes. Whole-cell Ca^{2+}-channel current *(b)* during a 300-ms voltage stop from −50 to +10 mV obtained with 90 mM Ba in the bath. **C** Kinetic model *(a)* of Ca^{2+}-channel gating. First latency and bursts of Ca^{2+}-channel openings *(b1)*, a cluster of bursts *(b2)*, and a blank sweep *(b3)*. (From [9]).

Reconstitution of the Mean and Whole-Cell Current from Single-Channel Data

As already shown (Fig. 3), the ascending phase of the mean current correlates with the first latency distribution (Fig. 5A, a, b,), and the declining phase with the distribution of the cluster lifetimes (Fig. 5A, a, c). It should then be possible to reconstruct the total time course of the mean current by a convolution of both distribution functions.

The convolution integral was obtained by direct numerical integration or by multiplying the Fourier transforms of the first latency and cluster lifetime distributions and taking the inverse transform of the product. In each case, the reconstructed mean current faithfully follows the time course of the original mean current (Fig. 5A, d). This confirms that the ascending and descending phases of I(T) are functions of the first latency and the cluster lifetimes of the Ca^{2+} channel, whereas the absolute amplitude of the reconstructed mean current (records) is mainly determined by P.

In order to compare the activity of single Ca^{2+} channels with the whole-cell current through all Ca^{2+} channels, an ensemble average was compiled from a large number of single-channel current traces recorded from 15 different membrane patches under the same experimental conditions (macroensemble mean current). A total of 2064 single Ca^{2+}-channel current sweeps were pooled from recordings with identical clamp protocols: 300-ms voltage pulses to RP + 75 mV delivered from the pipette zero potential at a rate of 0.5 Hz. Irrespective of a patch-to-patch variability in Ca^{2+}-channel gating behavior [3], the macroensemble mean Ca^{2+}-channel current (Fig. 5B, a) and the whole-cell Ca^{2+}-channel current in 90 mM $[Ba]_o$ (Fig. 5B, b) have a virtually identical time course and overlay each other, when scaled. This finding indicates that the average gating behavior of a single population of individual Ca^{2+} channels in different membrane patches reflects the bulk behavior of all the Ca^{2+} channels in the whole cell.

The fast and slow gating transitions of the cardiac Ca^{2+} channel might be best accounted for by the kinetic scheme shown in Fig. 5C, a. Rapid cycling of the channel among two short-lived shut states (C_1, C_2) and one open state (0), related to the opening pathway, i.e., activation, causes the occurrence of single-channel openings as closely spaced bursts (Fig. 5C, b1) and generates net inward current upon depolarization (cf. Fig. 3A, c). This fast-fluctuating behavior is terminated by the entry into a set of long-lived quiescent state(s) (Q; Fig. 5C, a; Fig. 5C, b2), which results in a decrease of the probability of channel opening and thus is responsible for the decline of net inward current during maintained depolarization, i.e., inactivation (cf. Fig. 3B, b). The time spent by the channel in this state(s) groups bursts of openings into clusters of bursts (Fig. 5C, b2; cf. Fig. 2Cc). The channel is only activatable if a depolarization is applied during the occupancy of any one of the three short-lived kinetic states. If the Q state(s) is already occupied before depolarization, the channel is unavailable for opening, and blank sweeps are observed upon depolarization (Fig. 5C, b3; cf. Fig. 4B). The transitions into and out of the Q state(s) are controlled by the membrane potential (cf. Fig. 4A–D) and a cAMP-dependent phosphorylation reaction (Fig. 6). Thus, the long-lived periods of quiescence should comprise at least two nonactivatable Ca^{2+}-channel states, one of them corresponding to voltage deactivation [Q(V)] and the other to dephosphorylation [Q(P)] (Fig. 4C, a).

β-Adrenergic Modulation and Phosphorylation Hypothesis

β-adrenergic agonists increase the amplitude of L-type I_{c_a} (Fig. 6A, C). Binding of agonist to the β-receptor or elevation of intracellular cAMP largely eliminates nulls (channels in the nonfunctional state), raising the proportion of nonblank sweeps (channels in the functional state). After β-stimulation, the nonblank traces have a

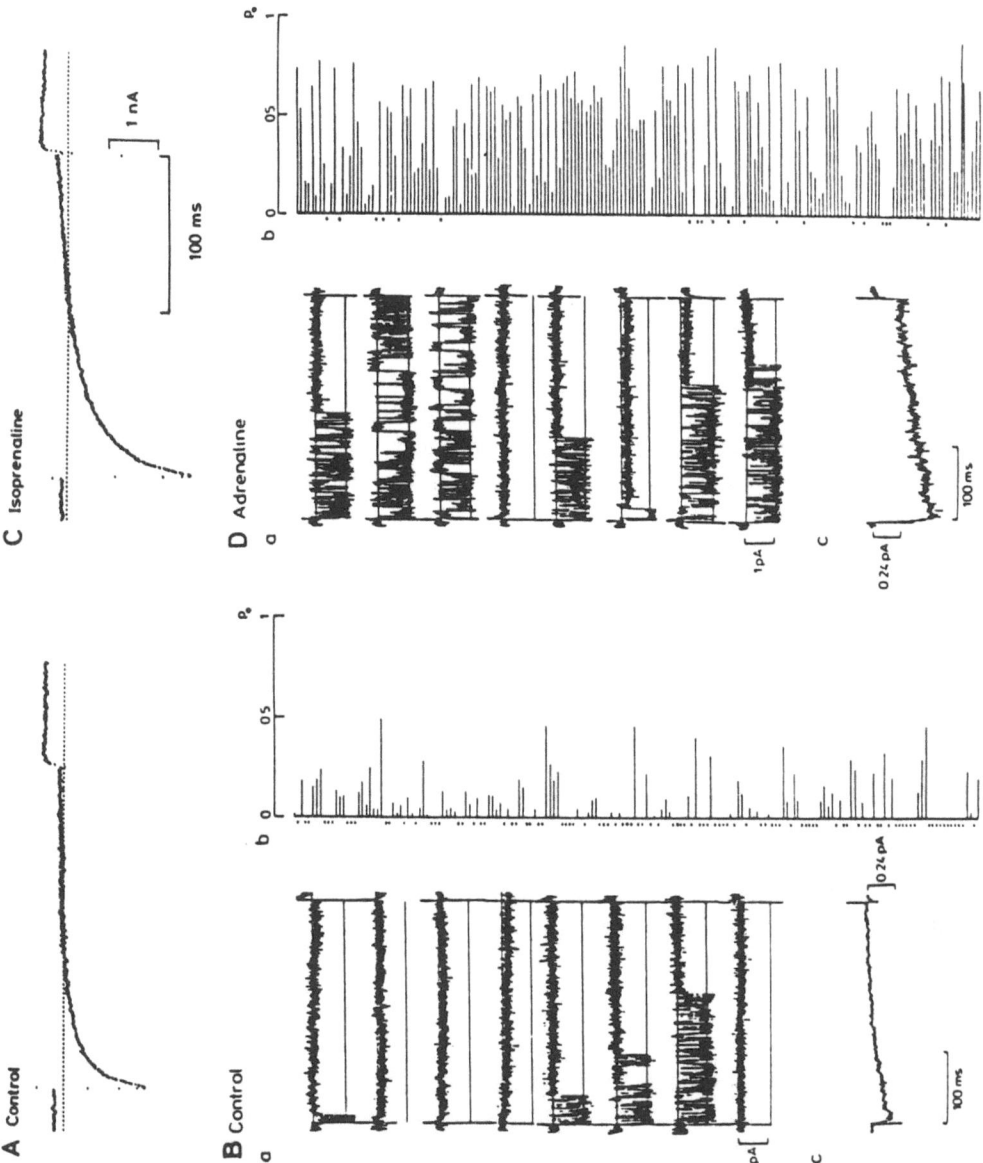

Fig 6 A–D. A, C. Whole-cell Ca^{2+} current before **(A)** and after **(C)** application of 5×10^{-8} M Isp. **B, D** Single Ca^{2+}-channel activity in response to step depolarizations *(a)* before **(B)** and in presence of 1 mM adrenaline *(D)*. *Horizontal lines* indicate the amplitude of the single-channel current determined by amplitude histograms. Opening probability *(b;* integrated activity of each trace) on successive depolarizations. *Dots* mark blanks. Average currents *(c)* of 180 traces. (From [13]).

slightly higher value of p_o because of a moderate prolongation of open times and a distinct abbreviation of gaps between individual channel openings and bursts of openings (Fig. 6B, a, b; 6D, a, b). However, the altered millisecond kinetics can account for only a 1.5- to 2.0-fold increase in whole-cell I_{Ca}. Thus, the changes in the channel availability (p_f) resulting in an abbreviated nonfunctional dormant period of the channel are clearly the dominant factor in the overall increase in activity (up to sixfold) (Fig. 6B, c; D, c). Changes in i or N_r have never been observed (Fig. 6B, D; see [10, 13, 15, 17] for references and more details).

Support for the phosphorylation hypothesis was obtained by dialyzing the cell with low concentrations of cAMP and the catalytic subunit of the cAMP-dependent protein kinase (cAMP-PK). Both agents increased I_{Ca} to the same maximal amplitude (about the factor of 3.5) in a range of concentrations which agree fairly well with those measured in biochemical studies [5, 6, 13]. These results supported the concept of signal transmission along a cascade via stimulation of the adenylate cyclase, leading to an increase in cellular concentration of cAMP and activation of cAMP-PK, which finally phosphorylates the channel protein [4]. Indeed, all interventions which block the cascade suppress the effect of the agonist on I_{Ca}. Thus, when cells were dialyzed with the endogenous inhibitor (PKI) of the cAMP-PK, or with Rp-cAMP(s) for which cAMP-PK is no substrate, isoprenaline (ISP) failed to increase the amplitude of I_{Ca} [6, 13]. Suppression of the effect of β-adrenergic stimulation was also seen when cells were dialyzed with adenylimidodiphosphate (AMP-PNP), a nonhydrolyzable ATP analog [5]. It should be noted that interventions which block the cascade only slightly reduce the amplitude of control I_{Ca} (not β-adrenergic-stimulated). These studies strongly favor the hypothesis that cAMP-dependent protein phosphorylation is only a modulator, but no prerequisite, of Ca^{2+} channel opening.

Specific phosphatases have been found to reduce the amplitude of the β-adrenergic-stimulated cell to control [7]. Pyrophosphatase (PPase) 1 and 2a as well as calcineurin in similar concentrations had about similar effects. Acid and alkaline phosphatases were without effect. These specific phosphatases also prolong the wash-out time of the effect of β-adrenergic agonists.

Conclusion

The kinetic properties and the β-adrenergic modulation of the L-type Ca^{2+} channel in cardiac myocytes were described. A reaction scheme of the channel gating behavior was formulated, whose rate constants depend on voltage, phosphorylation, and probably other $[Ca]_i$-dependent (enzyme) reactions.

References

1. Bean BP (1985) Two kinds of Ca channels in canine atrial cells. J Gen Physiol 86:1–30
2. Brum G, Osterrieder W, Trautwein W (1984) β-adrenergic increase of the calcium conductance of cardiac myocytes studied with the patch clamp. Pflügers Arch 401:111–118
3. Cavalié A, Pelzer D, Trautwein W (1986) Fast and slow gating behaviour of single calcium channels in cardiac cells. Pflügers Arch 406:241–258

4. Flockerzi V, Oeken HJ, Hofmann F, Pelzer D, Cavalié A, Trautwein W (1986) Purified dihydropyridine-binding site from skeletal muscle t-tubules is a functional calcium channel. Nature 323:66–68
5. Kameyama M, Hofmann F, Trautwein W (1985) On the mechanism of β-adrenergic regulation of the Ca channel in guinea-pig heart. Pflügers Arch Eur J Physiol 405:285–293
6. Kameyama M, Hescheler J, Hofmann F, Trautwein W (1986a) Modulation of Ca current during the phosphorylation cycle in the guinea pig heart. Pflügers Arch 407:123–128
7. Kameyama M, Hescheler J, Mieskes G, Trautwein W (1986b) Protein phosphatases regulate the Ca channel in guinea pig heart. Pflügers Arch Eur J Physiol 407:461–463
8. McDonald TF, Cavalié A, Trautwein W, Pelzer D (1986) Voltage-dependent properties of macroscopic and elementary calcium channel currents in guinea pig ventricular myocytes. Pflügers Arch Eur J Physiol 406:437–448
9. Pelzer D, Cavalié A, Trautwein W (1986) Activation and inactivation of single calcium channels in cardiac cells. Experimental Brain Research, Series 14, Springer Verlag, Berlin, Heidelberg
10. Reuter H (1983) Calcium channel modulation by neurotransmitters, enzymes, and drugs. Nature 301:569–574
11. Reuter H (1984) Ion channels in cardiac cells membranes. Annu Rev Physiol 46:473–484
12. Sakmann B, Neher E (eds) (1983) Single channel recording. Plenum, New York
13. Trautwein W, Kameyama M, Hescheler J, Hofmann F (1986) Cardiac calcium channels and their transmitter modulation. Fortschr d Zoolog 33:163–182
14. Trautwein W, Pelzer D (1985a) Gating of single calcium channels in the membrane of enzymatically isolated ventricular myocytes from adult mammalian hearts. In: Zipes DP, Jalife J (eds) Cardiac electrophysiology and arrhythmias. Grune and Stratton, New York, pp 31–42
15. Trautwein W, Pelzer D (1985b) Voltage-dependent gating of single calcium channels in the cardiac cell membrane and its modulation by drugs. In: Marme D (ed) Calcium and cell physiology. Springer, Berlin Heidelberg New York Tokyo, pp 53–93
16. Tsien RW (1983) Calcium channels in excitable cell membranes. Annu Rev Physiol 45:341–358
17. Tsien RW, Bean BP, Hess P, Nowycky M (1983) Calcium channels: mechanisms of beta-adrenergic modulation and ion permeation. Cold Spring Harbor Symposia on Quantitative Biology 48:201–212

Dihydropyridines, G Proteins, and Calcium Channels

A.M. Brown[1], L. Birnbaumer[1,2], and A. Yatani[1]

Departments of [1]Physiology and Molecular Biophysics, and [2]Cell Biology,
Baylor College of Medicine, One Baylor Plaza, Houston, TX 77030, USA

Introduction

Organic calcium-channel blockers, in particular the dihydropyridine (DHP) class, have been extremely useful as therapeutic agents and as probes for structure-function studies. One example that was important to our present work on calcium channels was the observation that guanine nucleotides changed DHP binding in cell-free membrane preparations [1, 2]. This pointed to a possible direct G-protein effect on calcium channels, and we have recently proved this to be the case [3–5]. Over the past several years, we have used DHPs in numerous other studies, and this review summarizes our results.

Results and Discussion

How Many DHP Binding Sites are There?

DHPs act on cardiac calcium currents as if they had more than one binding site. Figure 1 shows that the stimulatory effect of nitrendipine occurs over a concentration range of 5 log units and can be fit as the sum of two one-to-one adsorption isotherms [6]. Additional support for a two-site model may come from the well-known dual stimulatory and inhibitory effects of DHP stereoisomers [7, 8]. However, the dual effects could also be explained using a single-site model by invoking the well-known potential dependence of the effects of DHPs [7–10]. Binding studies using isolated membranes have usually been consistent with single-site models, but a recent binding study using intact cells has pointed to cooperative effects between two DHP-binding sites [11]. Functional studies of calcium current kinetics have been modelled using single sites [11–13] or two sites that are cooperative [14, 15]. Binding to the open state explains the well-known prolongation of open time of single calcium-channel currents produced by DHP agonists, but binding to two sites (as shown in Fig. 2) was necessary to explain the concentration-dependent changes in latency to first opening and burst duration that we observed [15].

Our results also argued against the mode hypothesis of calcium-channel gating, which was used to explain DHP effects on single calcium-channel current kinetics [16, 17]. In the mode model, DHPs stabilize calcium channels in two (mode 0 and mode 2) of three allowable modes that calcium channels can adopt, but they do not change the

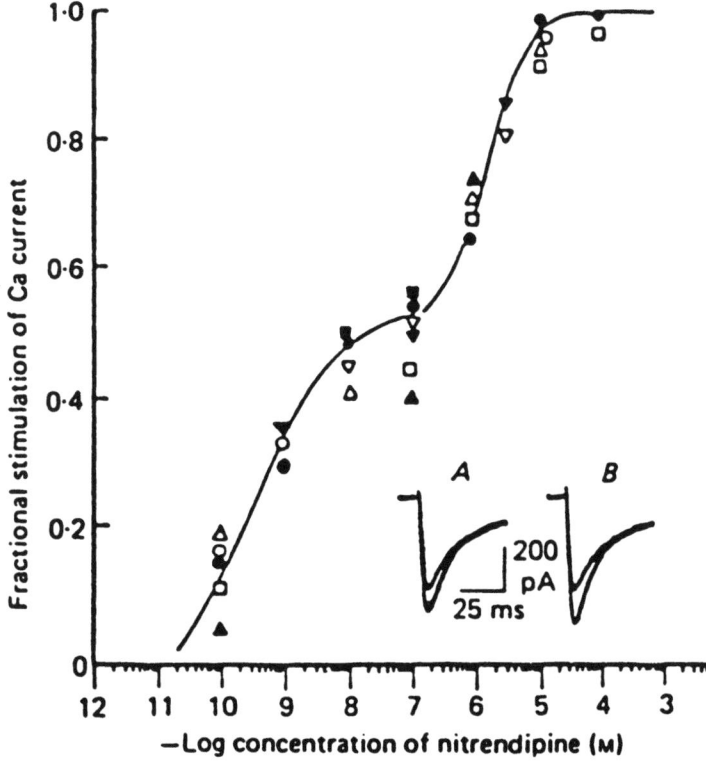

Fig. 1. Dose-response relation of the stimulatory effect of nitrendipine on calcium currents of adult guinea-pig ventricular cells as measured in eight cells. Each symbol represents data from a different cell. *Open symbols* show the results in normal Tyrode solution containing 5×10^{-5} M tetrodotoxin. *Filled symbols* show those results obtained in sodium- and potassium-free solutions. The holding potential was between -90 and -120 mV, and the test pulse, 50 ms in duration, was applied to 0 or $+10$ mV every 30 or 45 s. Changes of calcium current 3 min after drug application were plotted as a fraction of the maximal increase above the calcium current in the control drug-free solution. Thus, the effect was proportional to occupancy, with the maximal effect occurring at maximum occupancy. The maximum response was equal to 1·4 times the control. The calcium current was measured from the zero current level. In sodium- and potassium-free solutions, the holding current level represented the zero current level, as leak current was negligibly small compared to the calcium-current amplitude. The smooth curve is best least-squares fitted to an occupancy model having two independent binding sites. The model is based on one-to-one stoichiometry at each binding site and is given by

$$Y = \frac{\text{Effect}_{max1}}{1 + \text{ED}_{50.1}/X} + \frac{\text{Effect}_{max2}}{1 + \text{ED}_{50.2}/X}$$

where X is the free-drug concentration, Y is the fractional effect, and Effect$_{max}$s are the fractions of the combined maximum effect. The best least-squares fit gave ED$_{50.1}$ = $1·0 \times 10^{-9}$ M, ED$_{50.2}$ = $1·4 \times 10^{-6}$ M, Effect$_{max1}$ = 50% of the total effect. *Inset* shows activation of calcium current by 10^{-6} M (**A**) and 10^{-5} M (**B**) nitrendipine. Control currents and currents 3 min after addition of each concentration of nitrendipine were superimposed. In this experiment, tetrodotoxin-containing normal Tyrode solution was used. The holding potential was -90 mV, and the holding current represented the zero current level. The test potential was $+10$ mV, and 50-ms pulses were applied every 30 s. (42, from the *Journal of Physiology* with permission)

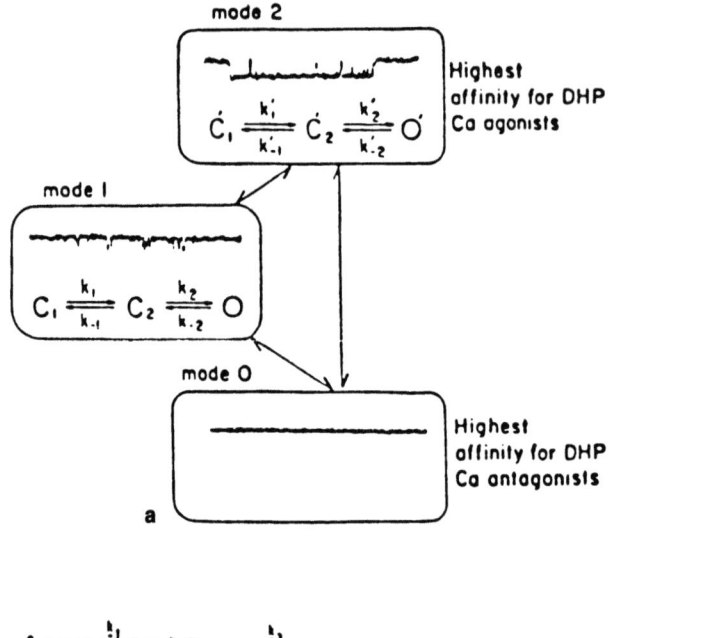

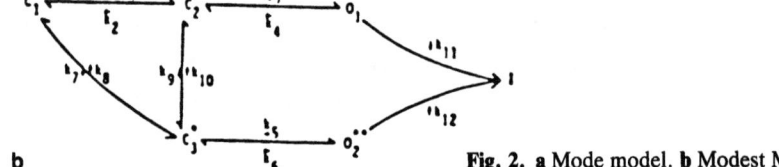

Fig. 2. a Mode model. **b** Modest Markov model

gating within a mode. Switching of modes was undefined but had to be much slower than gating within modes for the modes to be recognizable. In control, prolonged openings occurred less than 1% of the time, a result similar to that obtained by Cavalié et al. [18] but smaller than the frequency reported elsewhere [16, 17]. The prolonged openings became frequent in Bay K8644 at micromolar concentrations [16, 17, 19]. We divided our records into those with long openings (mode 2) and those with regular openings (mode 1). We found that for regular and prolonged openings first latencies (Fig. 3) and burst durations (data not shown) were shortened and lengthened respectively by increases in Bay K8644 concentration. A similar effect of Bay K8644 on first latencies was observed in adrenal chromaffin cells [20]. Hence, DHPs can clearly alter gating of closed states with or without changes in open time, a result that is inconsistent with a simple stabilization effect. Open time frequency distributions were selected from records having two or more openings, at least one of which was four times the normal open time τ. These had 27% of the openings belonging to the normal distribution, whereas the mode model would allow about 1% of the openings to be brief. The results render the mode hypothesis untenable, and we propose as a minimum model the six-state modest Markov model shown in Fig. 2, in which cooperative binding between two sites occurs [15]. The kinetic studies, therefore, would support the idea of two DHP-binding sites per calcium channel.

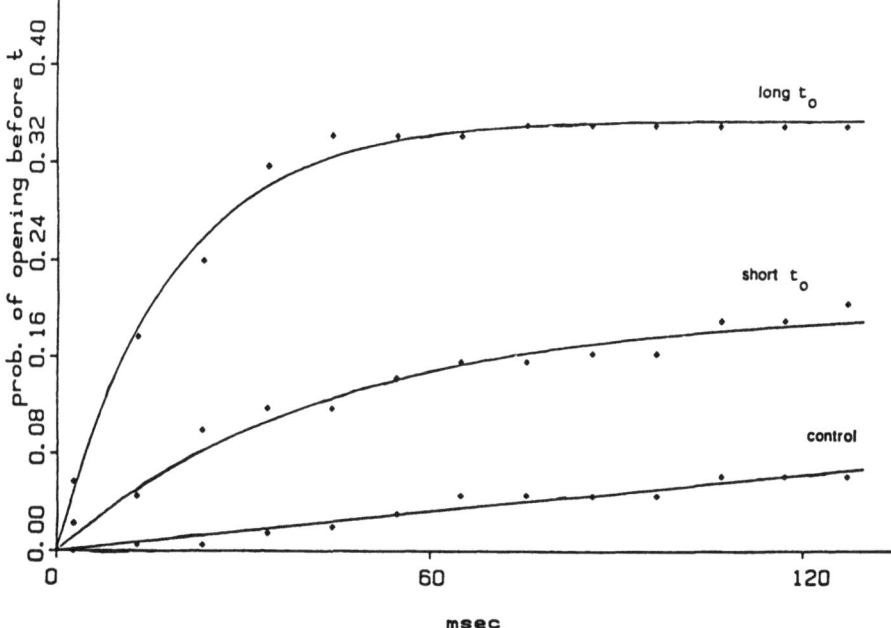

Fig. 3. *Cumulative waiting time distributions for long and short openings (t_o)* in 1 μM (−) Bay K8644 and in control from a cell-attached patch on a neonatal rat ventricular myocyte, with 140 mM potassium aspartate depolarizing solution in the bath and 100 mM $BaCl_2$ solution in the pipette. The first opening of each record in 1 μM DHP is sorted into long- and short-opening sets of data based on its open time: all openings longer than four times the fast open time distribution τ are assigned to the long-opening set of data. Fast and slow open time τs were 0.64 and 13.99 ms respectively. Relative areas from maximum likelihood fits of the open time distributions were 0.44 and 0.56 for fast and slow components respectively, and the total number of openings was 316. The waiting time distributions have been corrected for two channels in the patch and include nulls. Nulls for the sorted sets of data were defined as null records and records with a first opening which did not meet the selection criteria. Control open time distribution data were not sorted by open time. The waiting time distribution for long openings, short openings, and control were fit by nonlinear least-squares with single τs of 18 ms (52 openings), 44 ms (33 openings), and 666 ms (12 openings) respectively. Control data were obtained from 100 records, 1 μM (−) Bay K8644 data from 90 records. Records were analog-filtered by an eight-pole Bessel filter at 3 kHz (-3 dB) and during channel detection by a zero-phase four-pole filter at 1 kHz. In another neonatal ventricular myocyte analyzed in this way, we obtained similar results

The DHP-binding and whole-cell current studies are complicated by the fact that there are many different calcium channels, some of which are DHP insensitive. Single-channel current studies are not limited in this way since the channel type can be identified with far more certainty. The low-threshold or T channel in neurons [21, 22], cardiac muscle [23], and smooth muscle [24], and the "fast" channel in skeletal-musclelike cells [25] are not affected by DHPs, and neither is another subset, the N channels of neurons. Binding of $[3^+]$-(+)-PN200–110 to membrane preparations occurs at subpicomolar K_ds in nervous tissue, cardiac muscle, and smooth muscle; however, the functional effects are very different [26], and it is possible that the differences are due to relative tissue-specific differences in the ratios of DHP-sensitive and -insensitive calcium channels. Furthermore, DHPs bind with low affinity to mitochondria [27] and the nucleotide transporter [28] and, as we discuss next, sodium channels [29, 30].

Pharmacological Similarities Between Sodium and Calcium Channels

It is now known that the cDNA for the DHP receptor is similar to that for the sodium channel [31], expecially for the predicted membrane-spanning portions. This result was anticipated from functional studies. We have found that the sodium-channel blocker diphenylhydantoin [32] also blocks calcium channels, although higher concentrations are required, and nitrendipine blocks sodium channels, again at higher concentrations [29]. We wondered if the effects were nonspecific and related to the lipophilicity of these agents, and so we examined whether the DHP effects on sodium channels were stereospecific. We found that the stereoisomers of Bay K8644 and PN200-110 which have stereospecific effects on calcium channels had exactly the same stereospecific effects on sodium channels, although at higher concentrations [30] (Fig. 4). The result indicates that DHP-binding sites are present in sodium channels, and some DHP binding to these channels (probably low-affinity) must occur. Another calcium antagonist, Bepridil, also blocks sodium channels [33], probably for similar reasons.

Direct G-Protein Gating of Calcium Channels

β-Adrenoreceptor agonists such as isoproterenol increase high-threshold, DHP-sensitive calcium currents via the guanine nucleotide-binding G protein, G_s, the stimulatory regulator of adenylyl cyclase and the cAMP cascade. This is an example of indirect G-protein gating involving soluble cytoplasmic messengers. There are also numerous examples of transmitter-mediated inhibition of calcium currents which are blocked by pertussis toxin (summarized in [34]), and the similarity of these effects to those produced by phorbol ester may implicate protein kinase C. In fact, phorbol ester and diacylglycerol analogs can in some cells inhibit low-threshold, DHP-insensitive calcium currents as well as high-threshold calcium currents [35]. Numerous results (including the effects of GTP, GTPγS, and GDPβS) point to G-protein involvement, and Hescheler et al. [36] have reconstituted opioid inhibition of calcium currents in PTX-treated NG-108 cells with the G protein G_o purified from porcine brain. These results did not exclude a direct G-protein effect on calcium channels and since we had recently demonstrated that G proteins could directly gate potassium channels in heart [37, 38] and neuroendocrine cells [39, 40], we wondered if direct G protein gating of calcium channels also occurred. The experiments were designed so as to exclude cytoplasmic messengers (including phosphorylating enzymes) from playing a role. We used excised membrane patches or vesicles from cardiac sarcolemma or skeletal muscle T tubules that had been incorporated into lipid bilayers, and we found that the G protein G_s or its α-subunit specifically activated high-threshold calcium channels (Fig. 5) [3-5]. The effect occurred at picomolar concentrations and was concentration dependent. The G protein G_k that directly activates muscarinic potassium channels

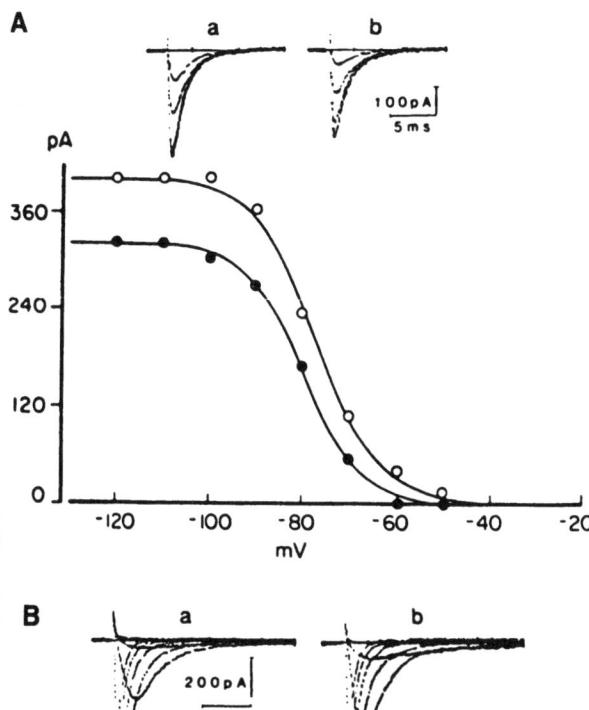
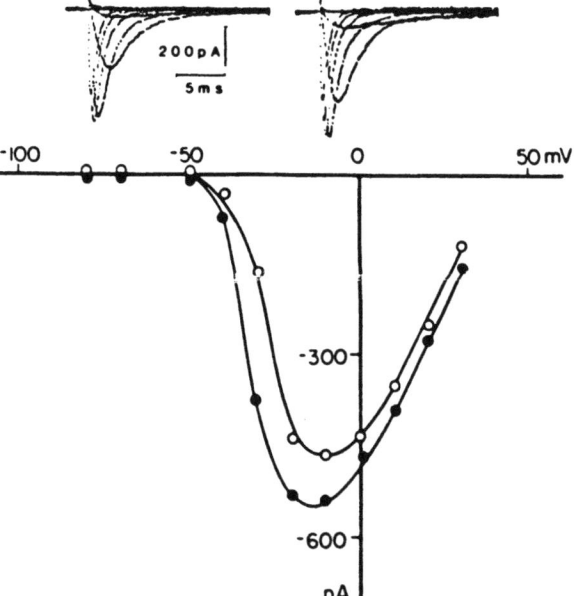

Fig. 4A, B. A Effects of (+) Bay K8644 (10^{-6} M) on availability of sodium currents. For control (o), V_h (potential at which availability is ½ maximal) = -75.0 mV and k (slope factor of availability – potential relationship) = 6.8 mV; for (+) Bay K8644 (•), V_h = -80.3 mV and k = 6.0 mV. *Inset* Control sodium currents elicited from conditioning potentials of -100, -90, -80, and -70 mV, (**a**); sodium current elicited from conditioning pulses of -120, -100, -80, and -70 mV, in presence of drug (*b*). **B** Effects of (−) Bay K8644 (10^{-6} M) on the I-V relationships for whole-cell sodium currents. *Top* Currents produced by 20-ms depolarizing voltage steps of 10-mV increments from -50 to $+10$ mV from a holding potential of -100 mV: control (*a*) and in the presence of the drug (*b*). *Bottom* I–V curves for currents before (o) and after drug (•) (43, from *Am J Physiol*, with permission)

had no effect even at nanomolar concentrations. Reconstitution with G_s mimicked the effects of isoproterenol and GTP, or GTPγS, on the single-channel currents. The G-protein effect was additive to the effects of both protein kinase A and Bay K8644, indicating that the three sites are distinct. Unlike the case for muscarinic potassium channels, where the G protein is obligatory for channel gating, the G- protein effect

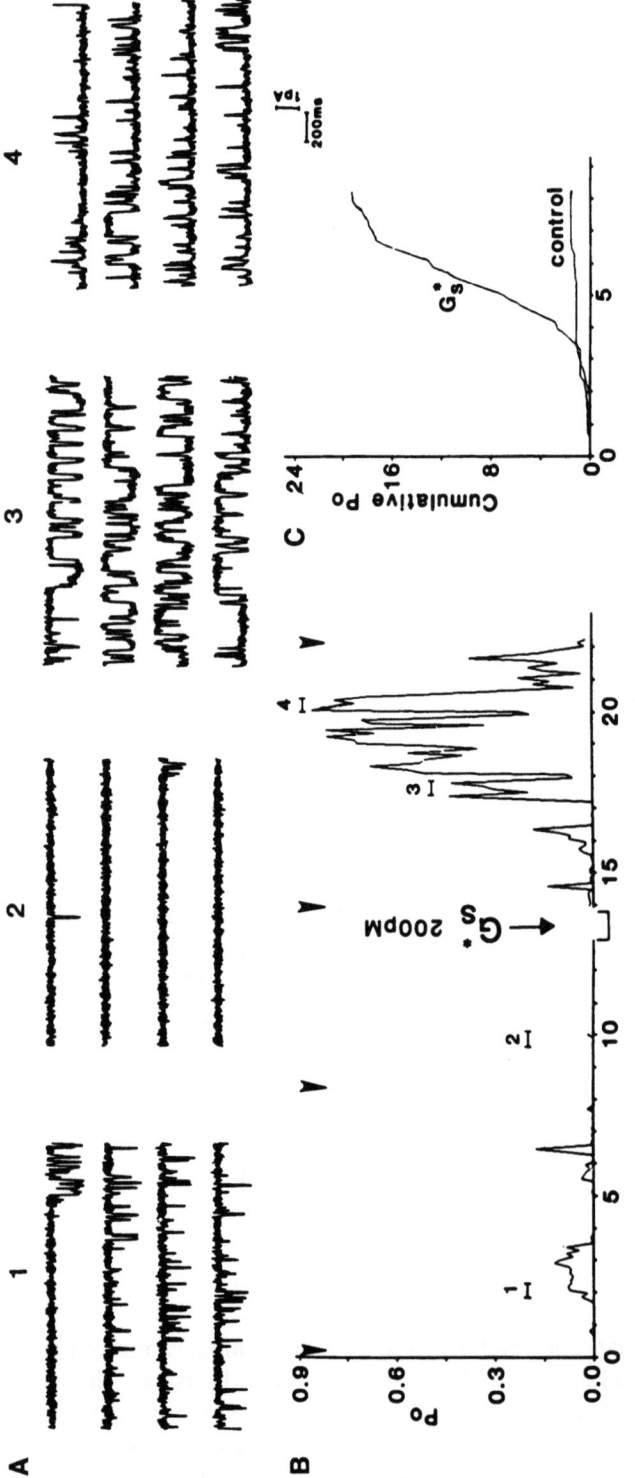

Fig. 5 A–C. Effects of G_s on single cardiac calcium channels incorporated into a planar lipid bilayer. The number of functional calcium channels incorporated into the bilayer was usually less than three and often one. As originally reported, cardiac sarcolemmal calcium channels incorporated into this type of bilayer had properties similar to those of high-threshold or L-type cardiac calcium channels. In the presence of Bay K8644 and with 100 mM barium as the charge carrier, the conductance between -50 and $+20$ mV was ~ 20 pS. The channel opened in bursts, and the probability of opening was voltage dependent. Brief openings that dominate under control conditions could not be detected at the recording bandwidth, and the mean open times were fit to an exponential distribution ($\tau \sim 12 \pm 3$ ms, $n = 9$, at test potentials between 0 and $+20$ mV). The values are consistent with those obtained for Bay K8644-stimulated channels in cell-attached patches. Current traces produced by depolarizing clamp steps to 0 mV from a holding potential of -40 mV are shown in **A** before (*1*, *2*) and after (*3*, *4*) addition of G_s. Pulses were applied every 30 s for 20 s. Leakage and capacitive currents were subtracted. Traces *1*, *4* were taken at the times indicated in **B** and are *2*-s segments from the 20-s pulses. **B** The entire experimental record. Note the decrease in activity with time in control in **B** before addition of G_s^* (100 pM) to the trans chamber. The ordinate is given as P_o because N was 1 in this experiment. **C** Cumulative P_os obtained between the *arrows* in **B**. Cum P_o (G_s)/cum P_o (control) is 11. G_s^*: significant preactivation of G_s with GTPγS. (44, from *Science* with permission)

on calcium channels is modulatory – G_s acts on calcium channels that have been opened by depolarization. The physiological significance of this direct G protein-calcium-channel pathway is unknown. It could be involved in basal regulation, and it might have a role in transmitter-mediated, cAMP-independent increases in calcium currents such as those reported for adrenal glomerulose cells [41]. The results also show that one G protein can have more than one effector, and that G proteins may be integrators as well as transducers in cell signalling.

References

1. Galizzi JP, Fossett M, Lazdunski M (1984) Properties of receptors for Ca^{2+} channel blocker verapamil in transverse tubule membranes of skeletal muscle. Eur J Biochem 144:211–215
2. Triggle DJ, Skattebøl A, Rampe D, Joclyn A, Genjo P (1986) Chemical Pharmacology of Ca^{2+} Channel Ligands. In: Poste G, Crooke ST (eds) New insights into cell and membrane transport processes. Plenum Publishing Corp, New York, pp 215
3. Yatani A, Codina J, Imoto Y, Reeves JP, Birnbaumer L, Brown AM (1987) A G protein directly regulates mammalian cardiac calcium channels. Science 238:1288–1292
4. Imoto Y, Yatani A, Reeves JP, Codina J, Birnbaumer L, Brown AM (1988) The α subunit of G_s directly activates cardiac calcium channels in lipid bilayers. Am J Physiol in Press
5. Yatani A, Imoto Y, Codina J, Hamilton SL, Brown AM, Birnbaumer L (1988) The stimulatory G protein of adenylyl cyclase, G_s, also stimulates dihydropyridine-sensitive Ca^{2+} channels. J Biol Chem 263(20):9887–9895
6. Brown AM, Kunze DL, Yatani A (1986) Dual effects of dihydropyridines on whole cell and unitary calcium currents in single ventricular cells of guinea pig. J Physiol 379:495–514
7. William JS, Grupp IL, Grupp GG, Vaghy PL, Dumont L, Schwartz A, Yatani A, Hamilton SL, Brown AM (1985) Prifole of the oppositely acting enantiomers of the dihydropyridine 202–791 in cardiac preparations: receptor binding, electrophysiological, and pharmacological studies. Biochem Biophys Res Comm 131(1):13–21
8. Sanguinetti MC, Kass RS (1984) Regulation of cardiac calcium channel current in calf and contractile activity by Bay K8644 is voltage-dependent. J Mol cell Cardiol 16:667–670
9. Bean BP (1984) Nitrendipine block of cardiac calcium channels: high affinity binding to the inactivated state. Proc Natl Acad Sci, USA 81:6388–6392
10. Kunze DL, Hamilton SL, Hawkes MJ, Brown AM (1987) Dihydropyridine binding and calcium channel function in clonal rat adrenal medullary tumor cells. Mol Pharm 31:401–409
11. Kokubun S, Prod'hom B, Becker C, Porzig H, Reuter H (1986) Studies on Ca channels in intact cardiac cells: voltage-dependent effects and co-operative interactions of dihydropyridine concentrations. J Pharm Exp Therap 30:511–584
12. Sanguinetti MC, Krafte DS, Kass RS (1986) Voltage-dependent modulation of Ca^{2+} channel current in heart cells by Bay K8644. J Gen Physiol 88:369–392
13. Bechem M, Schramm M (1987) Calcium-agonists. J Mol Cell Cardiol 19(II):63–75
14. Lacerda AE, Brown AM (1986) Atrotoxin increases probability of opening of single calcium channels in cultured neonatal rat ventricular cells. Biophys J 49:174a
15. Lacerda AE, Brown AM (1988) Electrophysiological properties of cardiac calcium channels as revealed by dihydropyridine effects. J Gen Physiol, in revision
16. Hess P, Lansman JB, Tsien RW (1984) Different modes of Ca^{2+} channel gating behavior favoured by dihydropyridine Ca^{2+} agonists and antagonists. Nature 311:538–544
17. Nowycky MC, Fox AP, Tsien RW (1985) Long-opening mode of gating of neuronal calcium channels and its promotion by the dihydropyridine calcium agonist Bay K8644. Proc Natl Acad Sci USA 82:2178–2182
18. Cavalié A, Pelzer D, Trautwein W (1986) Fast and slow gating behavior of single calcium channels in cardiac cells. Relation to activation and inactivation of calcium-channel current. Pflügers Arch 406:241–158
19. Brown AM, Kunze DL, Yatani A (1984) The agonist effect of dihydropyridines on Ca^{2+} channels. Nature 311:570–572

20. Hoshi T, Smith SJ (1987) Large depolarization induces long openings of voltage-dependent calcium channels in adrenal chromaffin cells. J Neuro 7:571–580
21. Carbone E, Lux HD (1984) A low voltage-activated, fully inactivating Ca^{2+} channel in vertebrate sensory neurons. Nature 310:501–502
22. Nowycky MC, Fox AP, Tsien RW (1985) Three types of neuronal calcium channels with different calcium agonist sensitivity. Nature 316:440–443
23. Nilius B, Hess P, Lansman JB, Tsien RW (1985) A novel type of cardiac calcium channel in ventricular cells. Nature 316:443–446
24. Yatani A, Seidel CL, Allen J, Brown AM (1987) Whole-cell and single channel Ca currents of isolated smooth muscle cells from saphenous vein. Circ Res 60(4):523–553
25. Caffrey JM, Brown AM, Schneider MD (1987) Mitogens and oncogenes can block the induction of specific voltage-gated ion channels. Science 236:570–573
26. Rampe D, Luchowski E, Rutledge A, Janis RA, Triggle DJ (1987) Comparative aspects and temperature dependence of [^{3}H] 1,4-dihydropyridine Ca^{2+} channel antagonist and activator binding to neuronal and muscle membranes. Can J Physiol Pharm 65:1452–1460
27. Brush KL, Perez M, Hawkes MJ, Pratt DR, Hamilton SL (1987) Low affinity binding sites for 1,4-dihydropyridines in mitochondria and in guinea pig ventricular membranes. Biochem Pharmacol 36:4153–4161
28. Striessnig J, Zernig G, Glossmann H (1985) Human red-blood-cell Ca^{2+}-antagonist binding sites. Eur J Biochem 150:67–77
29. Yatani A, Brown AM (1985) The calcium channel blocker nitrendipine blocks sodium channels in neonatal rat cardiac myocytes. Circ Res 56(6):868–875
30. Yatani A, Kunze DL, Brown AM (1988) Effects of dihydropyridine calcium channel modulators on cardiac sodium channels. Am J Physiol 254:H140–H147
31. Tanabe T, Takeshima H, Mikami A, Flockerzi V, Takahashi H, Kangawa K, Kojima M, Matsuo H, Hirose T, Numa S (1987) Primary structure of the receptor for calcium channel blockers from skeletal muscle. Nature Lond 328:313–318
32. Yatani A, Hamilton SL, Brown AM (1986) Diphenylhydantoin blocks cardiac calcium channels and binds to the dihydropyridine receptor. Circ Res 59:356–361
33. Yatani A, Brown AM, Schwartz A (1986) Bepridil block of cardiac Ca^{2+} and sodium channels. J Pharmacol Exp Ther 237:9–17
34. Dunlap K, Holz GG, Rane SG (1987) G proteins as regulators of ion channel function. TINS 10:241–244
35. Marchetti C, Brown AM (1988) The protein kinase activator 1-oleyl-2-acetyl-sn-glycerol inhibits two types of calcium currents in GH_3 cells. Am J Physiol 254:C206–C210
36. Hescheler J, Rosenthal W, Trautwein W, Schultz G (1987) The GTP-binding protein G_o regulates neuronal calcium channels. Nature Lond 325:445–447
37. Yatani A, Codina J, Brown AM, Birnbaumer L (1987) Direct activation of mammalian atrial muscarinic potassium channels by GTP regulatory protein, G_k. Science 235:207–211
38. Codina J, Yatani A, Grenet D, Brown AM, Birnbaumer LB (1987) The *alpha* subunit of G_k opens atrial potassium channels. Science 236:442–445
39. Yatani A, Codina J, Sekura RD, Birnbaumer L, Brown AM (1987) Reconstitution of somatostatin and muscarinic receptor mediated stimulation of K^+ channels by isolated G_k protein in clonal rat anterior pituitary cell membranes. Mol Endo 1(4):283–289
40. Codina J, Grenet G, Yatani A, Birnbaumer L, Brown AM (1987) Hormonal regulation of pituitary GH_3 cell K^+ channels by G_k is mediated by its *alpha* subunit. FEBS Lett 216(1):104–106
41. Hescheler J, Rosenthal W, Hensch K-D (1988) Angiotensin-induced stimulation of voltage-dependent Ca^{2+} currents in an adrenal cortical cell line. EMBO J 7:619–624
42. Brown AM, Kunze DL, Yatani A (1986) Dual effects of dihydropyridines on whole cell and unitary calcium currents in single ventricular cells of guinea-pig. J Physiol 379:495–514
43. Yatani A, Kunze DL, Brown AM (1988) Effects of dihydropyridine calcium channel modulators on cardiac sodium channels. Am J Physiol 23(1):H443–H451
44. Yatani A, Codina J, Imoto Y, Reeves JP, Birnbaumer L, Brown AM (1987) A G protein directly regulates mammalian cardiac calcium channels. Science 238:1288–1292

Electrophysiology of Dihydropyridine Calcium Agonists

M. Bechem and M. Schramm

Institute of Pharmacology, Bayer AG, D-5600 Wuppertal 1, FRG

Introduction

In recent years reports have been published on five structurally different 1,4-dihydropyridine (DHP) calcium agonists, the chemical structures of which are shown in Fig. 1 (for review see [14]. Each has an asymmetric carbon atom in position 4 of the dihydropyridine ring, resulting in the existence of two enantiomers for each compound. For 202–791, H160/51, and Bay K8644 the enantiomers have been separated; calcium agonism was found to be associated with only one enantiomer, while calcium antagonism was displayed by the other. Fortunately, for the best investigated calcium agonist – Bay K8644 – the potency of the calcium-antagonistic (+)-R-enantiomer is much lower than that of the calcium-agonistic (−)-S-enantiomer [4]. Therefore, results obtained from investigations with racemic Bay K8644 can be considered as

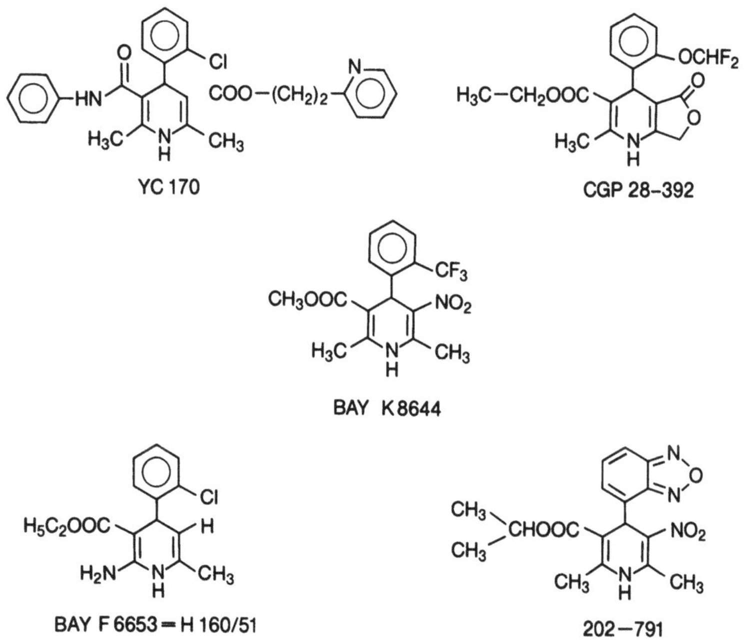

Fig. 1. Chemical structure of calcium agonists for which pharmacological data are available

effects of a "pure" calcium agonist. Estimated from the pharmacological profile, the same seems to be true for CGP 28–392, although this has not been proven.

It is now well established that DHP calcium agonists increase the current through calcium channels due to a direct effect on voltage-dependent gating; for Bay K8644 [7], CGP 28–392 [11], and the calcium-agonistic enantiomer (+)−202 791 [12] a drug-induced prolongation of the mean open time has been demonstrated in single-channel recordings. However, most of these observations were obtained with very high saturating concentrations, much higher than the K_D values found in receptor binding studies. The molecular basis of drug action thus remains poorly understood.

We investigated the effect of moderate drug concentrations on the voltage-dependent kinetics of the whole-cell calcium current to obtain a better insight into the drug-protein interaction.

Methods

Whole-cell voltage clamp measurements were performed on cultured atrial myocytes from hearts of adult guinea pigs by means of single tight-seal patch-clamp pipettes [5]. The isolation and culture conditions of the atrial myocytes have been described in detail elsewhere [2]. The cells were grown in Dulbecco's MEM supplemented with 1% FCS and 0.5% Ultroser (Gibco) for 2–7 days and afterwards used for the voltage-clamp experiments. The media contained either gentamycin (25 µg/ml) or penicillin (100 U/ml) plus streptomycin (100 µg/ml). The cells were cultured on uncoated 35-mm tissue dishes (Nunclon, Delta, Denmark) in an incubator at 37 °C, 90% humidity, and 5% CO_2.

For the experiments, culture medium was replaced by a solution of the following composition (in mM): NaCl 130; KCl 2; $MgCl_2$ 1; $CaCl_2$ 1; $NaHCO_3$ 5; HEPES NaOH (pH 7.4) 20. TTX (10 µg/ml) was added to block sodium currents. The culture dish was placed on the stage of an inverted microscope and kept at room temperature (21–23 °C).

Patch clamp pipettes were made from pyrex glass and filled with the following solution (in mM): K_3-Citrat 60; CsCl 30; $MgCl_2$ 1; cAMP 0.1; ATP 4; EGTA 2; HEPES NaOH (pH 7.2) 20. For voltage control and current amplification a patch clamp amplifier (List LM/EPC7) was used. The voltage protocol was delivered by a PDP11 microcomputer system, which also served for the recording of the filtered (at 3 kHz) currents at a sampling rate of 20 kHz. All recordings were corrected for linear leakage and capacity currents, which were monitored throughout the experiment.

Under the chosen experimental conditions the calcium current can be measured without the interference of other ionic currents. In addition, as inactivation of the calcium current is substantially reduced (normally inactivation is less than 20% for a 100-ms test pulse), voltage-dependent activation and deactivation of the calcium current can be easily studied.

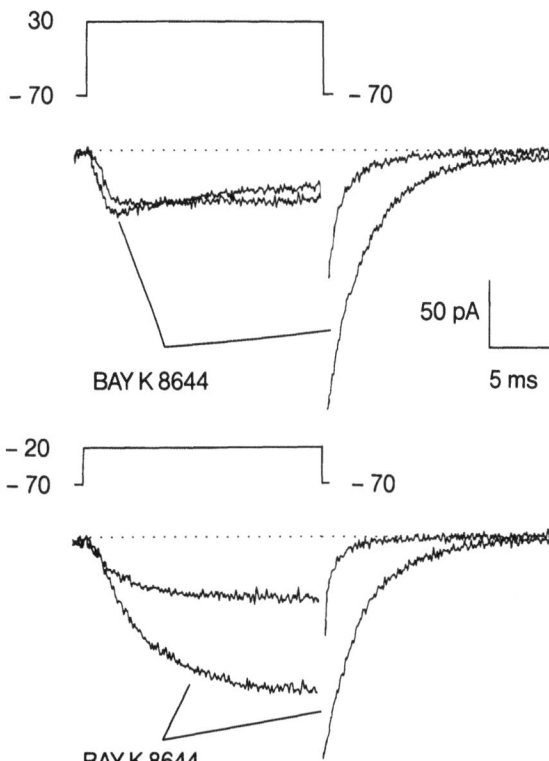

Fig. 2. Calcium current recordings under control conditions and under 3 × 10⁻⁸ M K8644. The voltage-clamp protocol is given above the appropriate current traces

Results and Discussion

Effect of Calcium Agonists on Peak Calcium Current

Figure 2 shows the effect of 3×10^{-8} M Bay K8644 on the calcium current. While the peak current $I_{Ca}(V)$ during the voltage step is strongly increased at a clamp potential of -20 mV, it is unchanged at $+30$ mV. The kinetics of calcium current activation seems to be unaffected, while the deactivation kinetics is strongly slowed down and obviously independent of the clamp potential. A quantitative description of these effects is given below.

The peak calcium current-voltage relations for control and two concentrations of Bay K8644 are shown in Fig. 3. In the presence of 3×10^{-8} M Bay K8644 $I_{Ca}(V)$ is increased, especially at negative membrane potentials. At 3×10^{-6} M it is generally reduced; only at membrane potentials lower than -20 mV it is increased. The reversal potential V_{rev} is about 60 mV and is not affected by the drug (for mean values see [1]). The continuous lines represent best χ^2 fits of

$$I_{Ca}(V) = G(V) \cdot (V - V_{rev})$$

to the measured current values $(G(V) = \text{cell conductance})$.

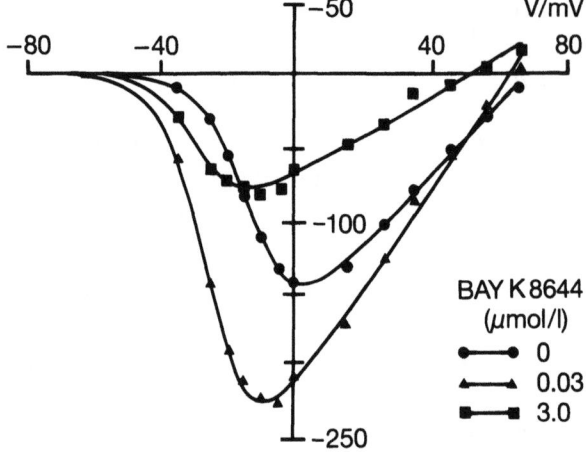

Fig. 3. Influence of Bay K8644 on peak calcium current-voltage relation. The holding potential was −70 mV. The continuous lines were obtained by fitting the experimental data to the formulae given in the text

If n is the number of available, noninactivated calcium channels – which is dependent on resting potential only under our experimental conditions – and g_{ch} is the voltage-independent single calcium channel conductance, then

$$G(V) = n \cdot g_{ch} \cdot P_o(V)$$

with the probability $P_o(V)$ that the single calcium channel is open. Since gating may be considered as thermodynamic fluctuations of a charged particle within an electrical field, $P_o(V)$ was assumed to obey the Boltzmann distribution with a maximum value of 1 ($P_o(V) = 1/(1 + \exp(V_{0.5}-V)/_K)$ with $V_{0.5}$ = potential at half-maximal calcium channel activation; k = steepness). This nonlinear fitting procedure, using V_{rev}, $n \cdot g_{ch}$, $V_{0.5}$, and k as free parameters, resulted in the P_o curves given in Fig. 4: Bay K 8644 dose-dependently shifts the open probability curve to negative membrane potentials (maximal shift ≈ 20 mV) without changing its steepness. The shift of the open probability curve is a monotonic function of the drug concentration reflecting recep-

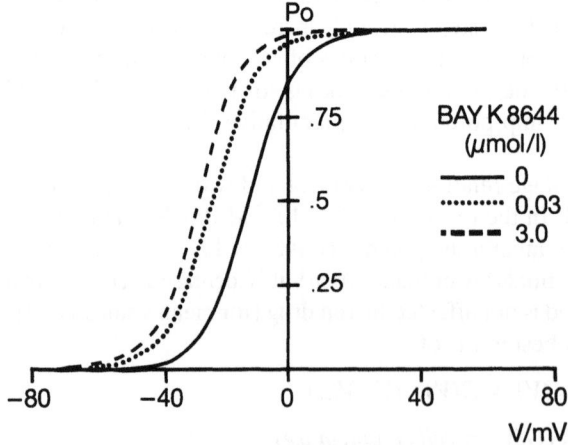

Fig. 4. Shift of the single-channel open probability curve Bay K8644. For explanation see text

tor occupancy much more directly than the calcium current does, which shows a biphasic behavior (cf. Fig. 3).

Effect of Calcium Agonists on Calcium Current Kinetics

As suggested by Fig. 2, the calcium current activation seems to be unaffected by the compound. Omitting the first 1 ms after the voltage step from the analysis (possibly containing the information on a short-lived closed state, which cannot be resolved in our investigations; cf. [3]), calcium current activation can be well fitted by a monoexponential, which is not affected by calcium agonists. In contrast, calcium agonists strongly influence calcium current deactivation kinetics, as can be seen in Fig. 2 for Bay K8644 and in Fig. 5 for CGP 28–392. Under control conditions and in the presence of high drug concentrations, tail currents can be fitted by monoexponentials, while two exponentials are necessary to describe the deactivating current at low drug concentrations, one being identical to the control conditions and one with the slow time constant at high drug concentrations.

This biphasic decay of tail currents under low drug concentrations is interpreted as a result of the fractional receptor occupancy. The K_D values representing half-maximal receptor occupancy have been estimated from these tail current analyses being about 5 nM for Bay K8644 and 100 nM for CGP 28–392. These values are in good agreement with those of receptor binding studies [8, 10].

Assuming a simple two-state model for the voltage-dependent relaxation of channels from the open to the resting state and vice versa, one can calculate the mean single channel open (τ_o) and closed times (τ_c) by means of the formula

$$\tau_o (V) = \tau_R (V)/(1-P_o (V)) \text{ and } \tau_c (V) = (V) = \tau_R (V)/P_o (V)$$

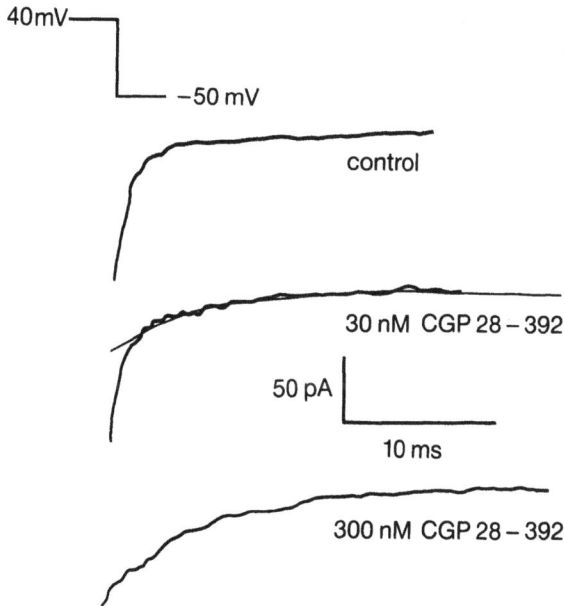

Fig. 5. Effect of CGP 28–392 on calcium current decay induced by a voltage step from 40 mV to −50 mV. Note the biexponential time course at 30 nM CGP 28–392

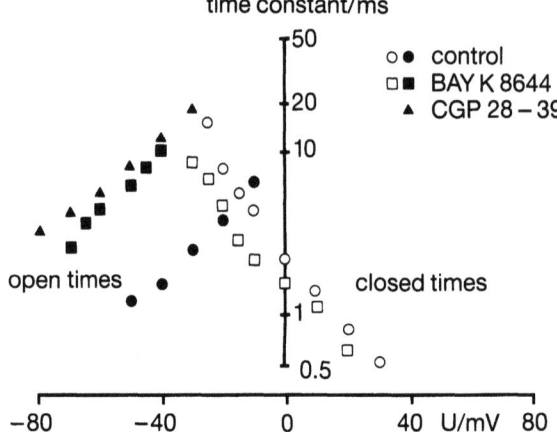

Fig. 6. Influence of 3 μM Bay K8644 and CGP 28–392 on the single-channel open and closed times. For explanation see text

in which τ_R represents the relaxation time constant of calcium current activation and deactivation. These time constants are given in Fig. 6 for control conditions and high concentrations of Bay K8644 and CGP 28–392. Both open and closed time constants show an exponential voltage dependence as expected for the gating process. The steepness of the curves suggests that during gating the equivalent of nearly four charges is moved across the electrical field of membrane.

Within the experimental error, the closed times are unaffected by calcium agonists. In contrast, the open times are strongly increased. For a given membrane potential of −40 mV open times are increased by a factor of 6 from about 1.7 ms under control conditions to about 10 ms. As this factor seems to be identical for all membrane potentials investigated so far, voltage dependence itself is unaffected by the drug. There is only a parallel shift of the open time curve by 40 mV to more negative membrane potentials.

This linear shift can be interpreted thermodynamically as the amount of energy that must be added to the drug-bound channel to enable the closing reaction of the channel protein. For Bay K8644 and CGP 28–392 we calculated this amount to be 2 kcal/mol. The discrepancy between this low energy and the total binding energy of about 11 kcal/mol for Bay K8644 (calculated from the K_D values) can be attributed to the high membrane enrichment of the compound (partition coefficient > 10000; see [6]).

The simple reaction scheme shown in Fig. 7 reflects our idea of the mode of action of calcium agonists. The calcium channel is supposed to be either in the resting state (composed possibly of several closed states) or in the open state, if inactivation is absent. Binding of a drug molecule can occur only if the channel is open. After binding of a drug molecule the channel is in the drug-bound open state from which it can move back to the rested state only after dissociation of the drug molecule, resulting in a

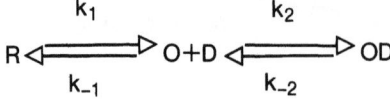

Fig. 7. Reaction model of the action of Bay K8644

markedly reduced closing rate. Therefore, the open times calculated from the tail current analysis can be interpreted as the dissociation time constants of the calcium agonists from the DHP receptor. For a membrane potential of -40 mV, we calculated the following rate constants for Bay K8644:

$k_1 = 50$ s^{-1}
$k_{-1} = 600$ s^{-1}'
$k_2 = 10^{11}$ mol^{-1} s^{-1}
$k_{-2} = 100$ s^{-1}

Although we have obtained no data about the process of inactivation from the drug-bound open state, this is a very interesting point to consider. If inactivation is assumed to take place predominantly from the open channel, an increased open probability would automatically result in an increased inactivation, even if the rate constants of the inactivation process are not influenced directly by the calcium agonists. This might simply explain the calcium agonist induced reduction in calcium channel availability [9] and the calcium-antagonistic effect of the compounds under partly depolarized conditions [13] or high drug concentrations [15] (cf. Fig. 3).

The proposed model attributes the prolonged open time of the channel to the dissociation time of the drug molecule from its receptor. Therefore, we feel that the amount of this prolongation strongly depends on the chemical structure of the DHP used despite the similar effects of the two investigated compounds.

References

1. Bechem M, Schramm M (1987) Calcium agonists. J Mol Cell Cardiol 19 (suppl II):63–75
2. Bechem M, Pieper F, Pott L (1985) Guinea-pig atrial cardioballs. Basic Res Cardiol 80 (suppl):19–22
3. Cavalie A, Pelzer D, Trautwein W (1986). Fast and slow gating behaviour of single calcium channels in cardiac cells. Pflügers Arch 406:241–258
4. Franckowiak G, Bechem M, Schramm M, Thomas G (1985) The optical isomers of the 1,4-dihydropyridine Bay K8644 show opposite effects on the Ca-channel. Eur J Pharmacol 144:223–226
5. Hamill OP, Marty A, Neher E, Sakmann B, Sigworth FJ (1981) Improved patch-clamp techniques for high-resolution current recording from cells and cell-free membrane patches. Pflügers Arch 391:85–100
6. Herbette LG, Chester DW, Rhodes DG (1986) Structural analysis of drug molecules in biological membranes. Biophys J 49:91–94
7. Hess P, Lansmann JB, Tsien RW (1984) Different modes of calcium channel gating behaviour favoured by dihydropyridine Ca agonists and antagonists. Nature 311:538–544
8. Janis RA, Rampe D, Sarmiento JG, Triggle DJ (1984) Specific binding of a Ca channel activator Bay K8644 to membranes from cardiac muscle and brain. Biochem Biophys Res Comm 121:317–323
9. Kass RS (1987) Voltage-dependent modulation of cardiac Ca channel by the optical isomers of Bay K8644: implications for channel gating, Circ Res 61 (suppl II):63–75
10. Laurent S, Kim D, Smith TW, Marsh JD (1985) Inotropic effect, binding properties and calcium flux effects of the calcium channel agonist CGP 28392 in intact cultured embryonic chick ventricular cells. Circ Res 56:676–682
11. Kokubun S, Reuter H (1984) Dihydropyridine derivatives prolong the open state of Ca channels in cultured cardiac cells. Proc Natl Acad Sci USA 81:482–527

12. Kokubun S, Prod'hom B, Becker C, Porzig H, Reuter H (1986) Studies on Ca channels in intact cardiac cells: voltage-dependent effects and cooperative interactions of dihydropyridine enantiomers. Mol Pharmacol 30:571–584
13. Sanguinetti MC, Kass RS (1984) Regulation of cardiac calcium channel current and contractile activity by the dihydropyridine Bay K8644 is voltage-dependent. J Mol Cell Cardiol 16:667–670
14. Schramm M, Towart R (1988) Calcium channels as drug receptors. In: Baker PF (ed). Calcium in drug actions. Springer, Berlin Heidelberg New York London Paris Tokyo pp 89–114 (Handbook of experimental pharmacology, vol 83)
15. Thomas G, Gross R, Schramm M (1984) Calcium channel modulation: ability to inhibit or promote calcium influx resides in the same dihydropyridine molecule. J Cardiovasc Pharmacol 6:1170–1176

Proton-Induced Transformation of Ca^{2+} Channel: Possible Mechanism and Physiological Role

M. Morad and G. Callewaert

University of Pennsylvania, Department of Physiology, Philadelphia, PA 19104, USA

Introduction

Voltage-gated ionic channels are generally identified by the ionic species most permeable through the channel. Ca^{2+} channels, for instance, are identified as such because they are highly selective to Ca^{2+}, even though the channel conducts significant Na^+ when the $[Ca^{2+}]_o$ is below micromolar concentrations (Almers et al. 1984; Hess and Tsien 1984; Hess et al. 1986). Chemically gated channels, on the other hand, tend to be less selective, although they do select between cations and anions. Gating mode of the channel appears to be an inherent property of the channel protein. For instance, the ACh-channel is gated only chemically and not by voltage. On the other hand, Na^+ or Ca^{2+} channels are only voltage gated, with no known chemical gating. In this contribution we will show that neuronal Ca^{2+} channels can have at least two modes of gating. The channel may be voltage or proton gated. The channel can exist in one of two states which are mutually exclusive. In the voltage-gated state the channel is highly selective to divalents, while in the proton-gated state the channel conducts monovalents. Proton sensitivity of the Ca^{2+} channel may play an important role in a number of physiological and pathological processes. We shall present evidence for a possible role of protons, released from the neurosecretory vesicles of chromaffin cells, in regulation of Ca^{2+} channel.

Proton-Gated Na^+ Current in Neurons

In a variety of central and peripheral neurons, rapid extracellular elevation of protons (pH 7.5 to 6.7) activates a large Na^+ current, $I_{Na(H)}$, which inactivates within 1.5–2.0 s (Fig. 1; Krishtal and Pidoplichko 1980; Konnerth et al. 1987; Davies et al. 1988). In outside-out patches of dorsal root ganglion (DRG) neurons, $I_{Na(H)}$ activates within 1–2 ms. In the presence of high $[H^+]_o$, inactivation proceeds slowly, with a time constant of 300 ms. Small elevations in the $[H^+]_o$ which by themselves failed to activate $I_{Na(H)}$ or any channel openings in outside-out patches strongly suppressed $I_{Na(H)}$, suggesting that inactivation can be reached from the open or the closed state of the channel. $I_{Na(H)}$ may be deactivated by step reduction of extracellular proton concentrations. The proton dependence of activation of $I_{Na(H)}$ is steep, with Hill coefficient of 4, suggesting binding of four protons to the receptor site to activate $I_{Na(H)}$.

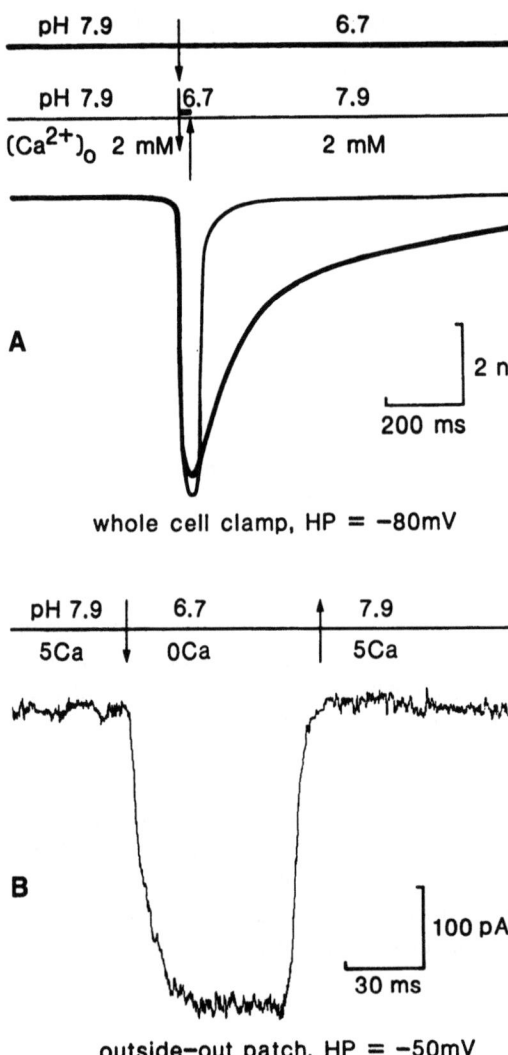

Fig. 1A, B. Activation and deactivation of $I_{Na(H)}$ in chick DRG neurons. **A** The *heavy trace* shows a recording of $I_{Na(H)}$ induced in a whole cell following a step in pH from 7.9 to 6.7, as indicated above the recordings. In the continued presence of pH 6.7, the response inactivated completely within 1–2 s. If, following the pH step from 7.9 to 6.7, the pH was returned after a brief period to 7.9 *(lower protocol)*, then the response was rapidly deactivated, as shown by the *light trace*. **B** Activation and deactivation of $I_{Na(H)}$ in an outside-out patch following the sequence described above the recording. Note that the time course of the events was much faster in the isolated patch than in the whole cell. In this example, the half-time of activation was 2.9 ms, and following a prompt return to pH 7.9, deactivation to a residual level occurred with a half-time of 1.3 ms. Holding potential was −80 and −50 mV, and $[Ca^{2+}]_o$ was 2 and 0.1 mM in **A** and **B**, respectively. (Modified from Davies et al. 1988)

Comparison of activation of $I_{Na(H)}$ in the inside-out, outside-out, and cell-attached patches suggests that the site of activation of $I_{Na(H)}$ is at the external side of the membrane. In fact, high intracellular proton concentrations (pH 6.0) had no significant effect on the magnitude or the time course of $I_{Na(H)}$ activated by step elevation of $[H^+]_o$.

Do Ca^{2+} Channels Transport $I_{Na(H)}$?

We found that $I_{Na(H)}$ was blocked by organic (diltiazem, verapamil) and inorganic (Cd^{2+} and Ni^{2+}) Ca^{2+}-channel blockers. Generally, however, higher concentrations

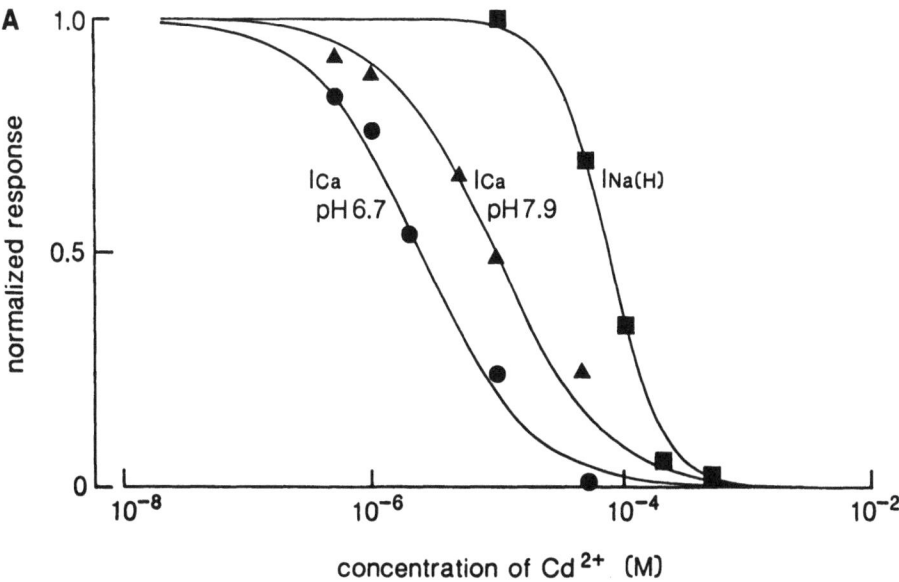
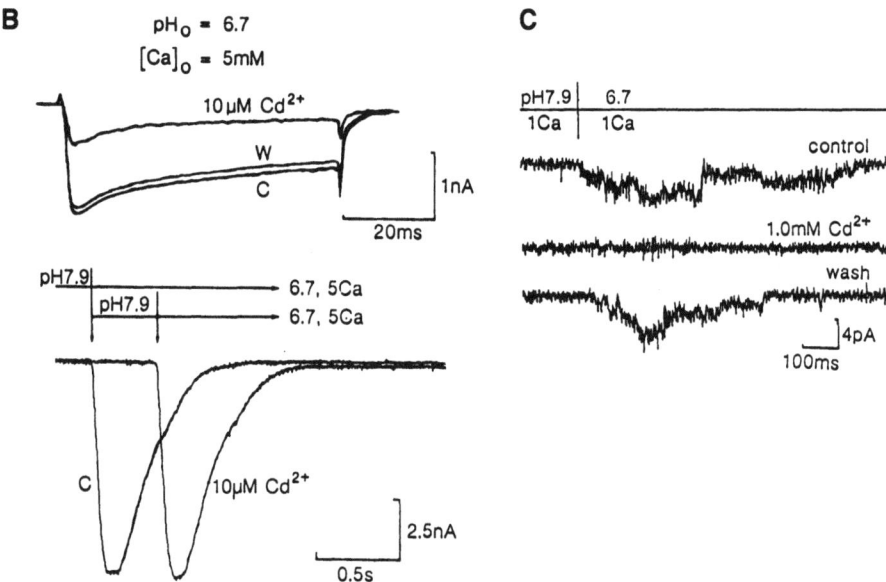

Fig. 2A–C. Block of I_{Ca} and $I_{Na(H)}$ by Cd^{2+} in chick DRG neurons. **A** Dose-response relationship of the block of I_{Ca} (●, pH 6.7; ▲, pH 7.9) and $I_{Na(H)}$ (■) by Cd^{2+} in whole cells. The data were obtained from seven cells where the external Ca^{2+} concentration was 5 mM. The standard deviations are within the points. The curves were fitted assuming a 2:1 stoichiometry of block of $I_{Na(H)}$ and a 1:1 block of I_{Ca}. Half-blocking concentrations were 3×10^{-6} M for I_{Ca} at pH 6.7, 1×10^{-5} M for I_{Ca} at pH 7.9, and 8×10^{-5} M for $I_{Na(H)}$. **B** Comparison of the effect of 10 μM Cd^{2+} on I_{Ca} and $I_{Na(H)}$, both in 5 mM Ca^{2+} and pH 6.7. I_{Ca} was induced following a 60-ms pulse to +10 mV from a holding potential of −80 mV. In the presence of 10 μM Cd^{2+}, I_{Ca} was reduced by 63%, while $I_{Na(H)}$ was unaffected. C, control; W, washout. **C** Block of $I_{Na(H)}$ by 1 mM Cd^{2+} in an outside-out patch at a holding potential of −80 mV, $I_{Na(H)}$ returned after removing Cd^{2+}. (From Davies et al. 1988)

were required to block $I_{Na(H)}$. Figure 2 compares the dose dependence of Cd^{2+}-induced block of I_{Ca} and $I_{Na(H)}$. Note that half-maximal block of voltage-gated Ca^{2+} current occurs at 10^{-5} M, while $I_{Na(H)}$ is blocked by concentrations 100 times higher. The stoichiometry of the block of the two current types by Cd^{2+} is also different; 1:1 block for I_{Ca} and 2:1 block for $I_{Na(H)}$. The specificity of Ca^{2+}-channel blockers in suppressing $I_{Na(H)}$ suggests either that $I_{Na(H)}$ is transported by Ca^{2+} channels or that the proton-activated current flows through a channel which is also blocked by Ca^{2+}-channel blockers.

To differentiate between these two possibilities, we measured I_{Ca} during the time course of $I_{Na(H)}$. The implicit assumption for these experiment was that if $I_{Na(H)}$ and I_{Ca} flow through separate and independently operating channels, then $I_{Na(H)}$ and I_{Ca} should *coexist* simultaneously. Figure 3 illustrates an experiment in which I_{Ca} is activated by depolarizing pulses from −80 to −10 mV prior to, during, and following the activation of $I_{Na(H)}$ by step elevation of $[H^+]_o$ to pH 6.7. In Fig. 3B, the magnitude of I_{Ca} at −10 mV is plotted during the time course of $I_{Na(H)}$. The data show that as $I_{Na(H)}$ is activated, I_{Ca} decreases rapidly and recovers only after significant inactivation of $I_{Na(H)}$ has occurred. I_{Ca} only partially recovered because of the well-known suppressive effects of steady-state elevation of $[H^+]_o$ on I_{Ca} (Iijima et al. 1986; Davies et al. 1988; Prod'hom et al. 1987). In order to verify that I_{Ca} and $I_{Na(H)}$ were mutually exclusive, we reversed the Na^+ gradients by dialyzing the cell with 60 mM Na^+, so that at potentials which activated a large *inward* Ca^{2+} current, the proton-induced Na^+ current was in the *outward* direction. The results consistently showed that as $I_{Na(H)}$ was activated, I_{Ca} decreased and recovered only after $I_{Na(H)}$ was mostly inactivated (Konnerth et al. 1987). Since tetrodotoxin (TTX)-sensitive Na^+ current could be simultaneously activated during the time course of $I_{Na(H)}$, we concluded that only the Ca^{2+} channel was involved in the transport of proton-induced Na^+ current. The complete suppression of I_{Ca} during activation of $I_{Na(H)}$ and similarity of the pharmacological blockers of the two current systems suggests that $I_{Na(H)}$ and I_{Ca} are carried through the same channel, making them temporally and mutually exclusive of each other. This finding provides strong support for the idea that $I_{Na(H)}$ flows through a transformed state of the Ca^{2+} channel. The proton-induced modification of Ca^{2+} channel must involve alteration of both gating and the ionic selectivity of the Ca^{2+} channel.

Although Ca^{2+}-channel blockers uniformly blocked I_{Ca} and $I_{Na(H)}$, we failed to observe a consistent suppressive effect of nifedipine on $I_{Na(H)}$ or I_{Ca} in DRG neurons. This may imply that the DRG high-threshold channels are somewhat modified compared to the heart high-threshold Ca^{2+} channels, where nifedipine has a marked suppressive effect. In this respect, it is of interest that although we could activate $I_{Na(H)}$ in a variety of central and peripheral neurons (Davies et al. 1988), we failed to activate $I_{Na(H)}$ in rat ventricular myocytes.

Is Proton-Induced Transformation Associated with More Than One Ca^{2+}-Channel Type?

The faster run-down of high-threshold I_{Ca} compared to $I_{Na(H)}$ in DRG neurons (Konnerth et al. 1987) may be considered inconsistent with the close correlation between $I_{Na(H)}$ and I_{Ca}. There are, however, two likely explanations for this apparent

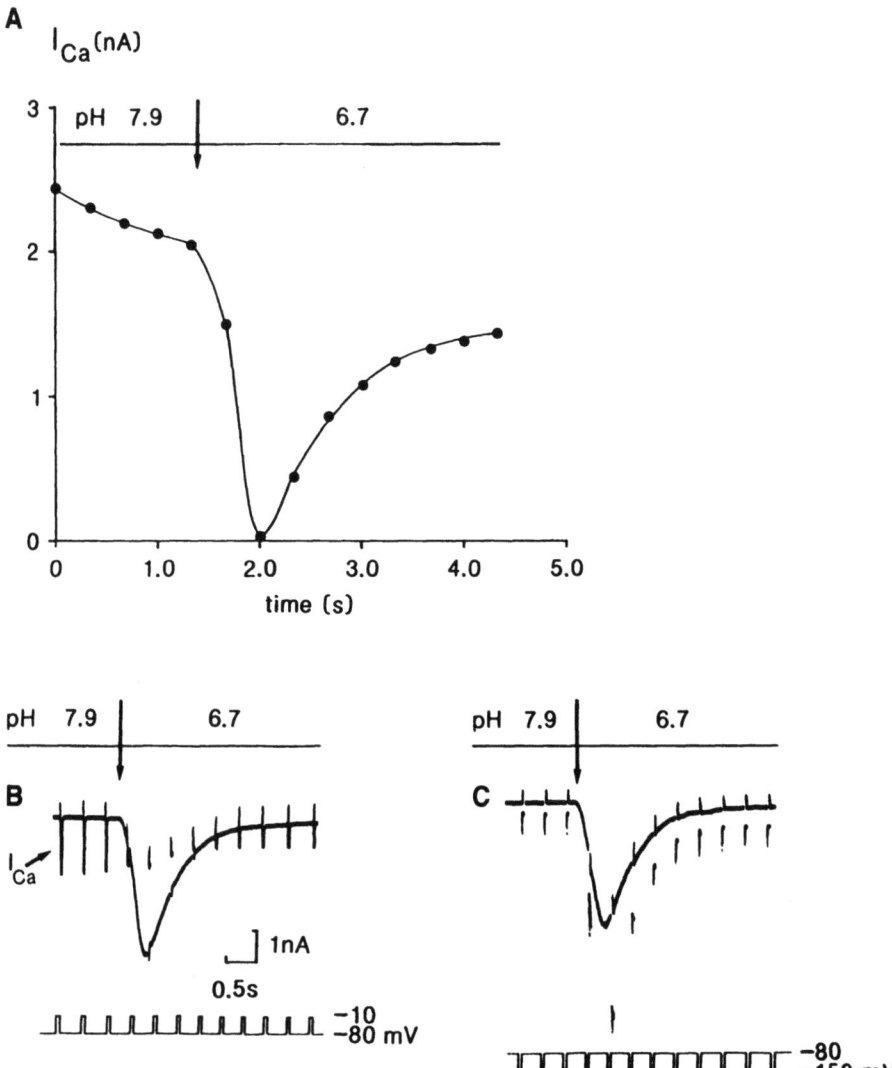

Fig. 3A–C. Transient transformation of Ca^{2+} channel from a voltage-gated to a nongated state. Inward I_{Ca} was activated at a frequency of 3 Hz by short, 30-ms depolarizing pulses before and during the time course of activation of $I_{Na(H)}$ (**B**). Prior to application of the pH step, I_{Ca} was corrected by subtracting the leak current measured during hyperpolarizing pulses (**C**). To determine the amplitude of I_{Ca} during the step in pH, $I_{Na(H)}$ was subtracted from the total current during the pulses. A linear I–V relation of $I_{Na(H)}$ was assumed in line with earlier observations (see Konnerth et al. 1987). I_{Ca} thus determined was plotted prior to and during the application of the pH step (**A**). Note that at the *arrow* (when the pH step was applied), there was a rapid decrease in the amplitude of I_{Ca} as $I_{Na(H)}$ was activated. Slow relaxation of $I_{Na(H)}$ occurred simultaneously with recovery of the voltage-gated I_{Ca}. The decrease in magnitude of I_{Ca} in solutions buffered at pH 6.7 represents the depressive effect of low-pH solutions on the voltage-gated Ca^{2+} channel. $[Ca^{2+}]_o = 5$ mM, [TTX] = 2 µM, room temperature, $pH_i = 7.3$, $[Ca^{2+}]_i = 10^{-9}$ M. (From Konnerth et al. 1987)

Table 1. Properties of $I_{Na(H)}$ vs I_{Ca} in DRG neurons

	$I_{Ca(L)}$	$I_{Ca(T)}$	$I_{Na(H)}$
Run-down	Fast	Slow	Slow
Nifedipine	No effect[a]	No effect	No effect[a]
Amiloride	No effect	Suppression	Suppression at $-V_m$
ω-Conotoxin	Suppression	No effect	No effect
Conductance (pS)[b]	28	25–30	20–25
Inactivation (ms)[b]	> 500	20–25	300

[a] $I_{Na(H)}$ is not present in heart where $I_{Ca(L)}$ is strongly suppressed by nifedipine.
[b] Measured in the Na-conducting state.

discrepancy: (a) the run-down phenomenon may involve the voltage-gating mechanism itself (i.e. proton-gating mechanism is more resistant to run-down, and (b) $I_{Na(H)}$ may involve transformation of low-threshold channels which are more resistant to run-down. Measurements of $I_{Na(H)}$ in isolated patches showed primarily the presence of one class of channels with a single-channel conductance of 24–28 pS (Davis et al. 1988). This conductance can be associated with the Na$^+$ conductance of either the high- or low-threshold Ca^{2+} channels (Carbone and Lux 1987; see also Table 1). In support of the idea that $I_{Na(H)}$ may involve primarily the transformation of low-threshold channels, it has recently been shown that amiloride, which specifically blocks low-threshold I_{Ca} in DRG and neuroblastoma cells, also suppressed $I_{Na(H)}$ (Tang et al. 1988). However, the suppressive effect of amiloride on $I_{Na(H)}$ was strongly voltage dependent, such that the effect was less pronounced at positive potentials. This voltage-dependent effect of amiloride may, in part, be responsible for this drug's lack of significant suppressive effect on the high-threshold Ca^{2+} current. The specificity of amiloride suppression of low-threshold I_{Ca} and $I_{Na(H)}$ is not sufficiently compelling to rule out the involvement of the high-threshold Ca^{2+} channels in proton-induced transformation process. Thus, in the absence of conclusive experimental evidence for association of $I_{Na(H)}$ with a specific Ca^{2+}-channel subtype (see Table 1), we assume that proton-induced transformation involves both the high- and the low-threshold channels.

Possible Mechanism for Transformation of Ca^{2+} Channel

Since Ca^{2+}-binding sites on proteins have generally high affinity for protons, we propose that step elevation of $[H^+]_o$ diplaces Ca^{2+} from the extracellular sites closely associated with the Ca^{2+} channel. Displacement of Ca^{2+} from such a site increases the likelihood of Na$^+$ permeating through the channel, giving rise to $I_{Na(H)}$. Such a model predicts that (a) elevation of Ca^{2+} should suppress $I_{Na(H)}$ and block proton-induced transformation and (b) reduction of $[Ca^{2+}]_o$ should enhance $I_{Na(H)}$ and its rapid removal should activate $I_{Na(H)}$. These predictions were tested, and it was found that elevation of $[Ca^{2+}]_o$ completely blocked $I_{Na(H)}$, with half-maximal block at 5 mM (Davies et al. 1988). The rapid removal of Ca^{2+} activated $I_{Na(H)}$ in solution buffered at pH 7.3, but not at pH 7.9 (Konnerth et al. 1987), suggesting that the availability of protons rather than the removal of Ca^{2+} is critical for activation of $I_{Na(H)}$. Thus, it appears that binding of protons is a necessary step in transformation of Ca^{2+} channel and activation of $I_{Na(H)}$. Consistent with this hypothesis, the rate of activation of $I_{Na(H)}$ was highly proton dependent (Davies et al. 1988).

Activation of $I_{Na(H)}$ appears to require the binding of four protons to the activation site (Konnerth et al. 1987). This observation is consistent with the finding that Cd^{2+} and Ca^{2+} block $I_{Na(H)}$ with 2:1 stoichiometry (Davies et al. 1988). Thus, it appears that occupation of two Ca^{2+}-binding sites is necessary for activation or block of $I_{Na(H)}$. Inactivation of $I_{Na(H)}$ probably reflects in part the binding of Ca^{2+} to these external sites. This is consistent with the findings that (a) the time constant of inactivation of $I_{Na(H)}$ increases with reduction of $[Ca^{2+}]_o$ and (b) inactivation is weakly voltage dependent (100 mV/e-fold change) (Davies et al. 1988).

Possible Physiological Implications

Rapid increases in the $[H^+]_o$ may have a regulatory role in exocytosis and release of neurotransmitter. In chromaffin cells, for instance, the secretory adrenergic vesicles contain high proton concentration (pH 5.3) (Johnson 1987). The Ca^{2+}-dependent

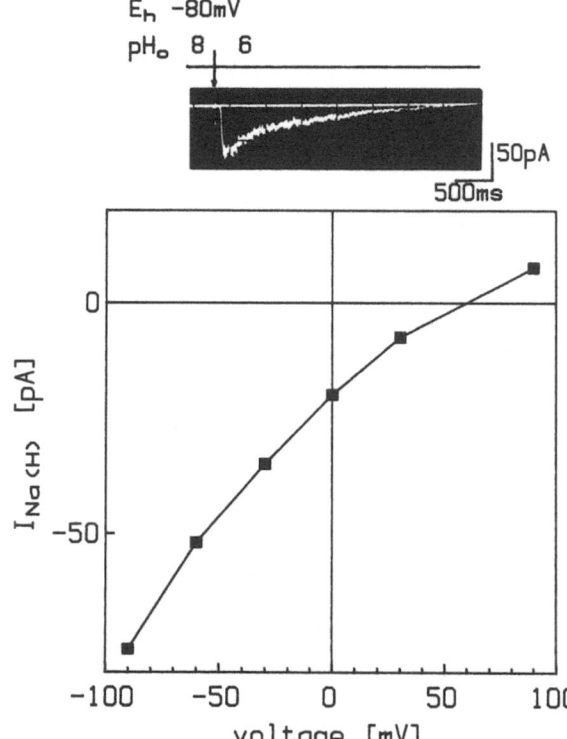

Fig. 4. Voltage dependence of $I_{Na(H)}$ in cultured chromaffin cells. The holding potential was changed for at least 30 s prior to activation of $I_{Na(H)}$. Currents reversed directions at potentials positive to +60 mV. The calculated Na^+ equilibrium potential (E_{Na}) was +64 mV. $[Na+]_i = 10$ mM, $[Na^+]_o = 130$ mM, $[Ca^{2+}]_o = 2$ mM. *Inset* shows activation of $I_{Na(H)}$ at a holding potential (E_h) of −80 mV with a change of pH_o from 8 to 6

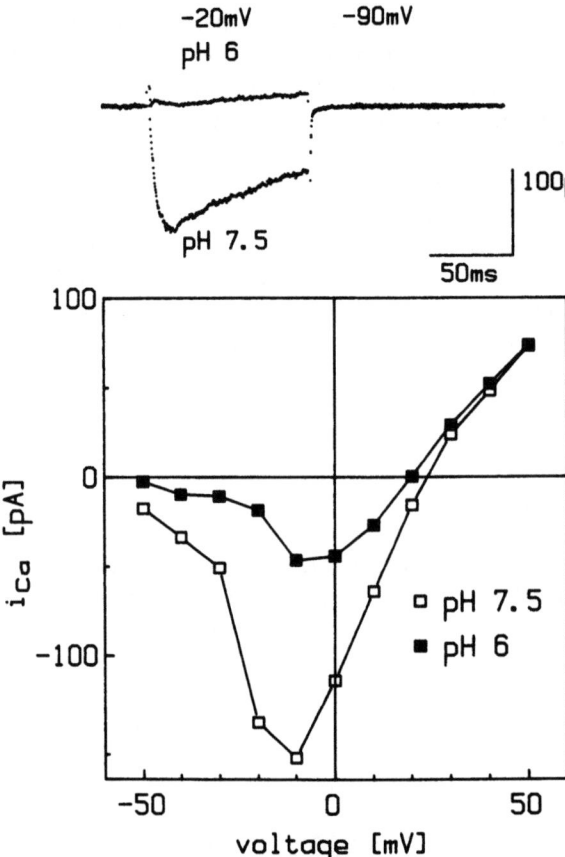

Fig. 5. pH sensitivity of I_{Ca} in cultured chromaffin cells, and current-voltage relation of peak Ca^{2+} current in pH 7.5 and pH 6. Test pulses were applied from a holding potential of –90 mV at a frequency of 0.05 Hz. Leakage current was not subtracted. *Inset* shows time course of I_{Ca} in pH 7.5 and 6 during a step depolarization from –90 to –20 mV. $[Ca^{2+}]_o = 5$ mM, $[Cs^+]_i = 120$ mM, 3 µM TTX

exocytosis and release of transmitter must therefore be accompanied by local (paracellular) increases in $[H^+]_o$. This accompanying release of H^+ may regulate the Ca^{2+} channel locally either by rapid decrease in I_{Ca} or by transformation of the voltage-gated Ca^{2+} channels into proton-activated and Na^+-transporting channels. In either case, the release of protons would reduce Ca^{2+} influx and serve as a negative feedback mechanism for the release of the transmitter. We therefore examined the effect of both rapid and steady-state elevation of $[H^+]_o$ on the Ca^{2+} channels of bovine chromaffin cells. Figure 4 shows that $I_{Na(H)}$ can be activated in whole-cell clamped single chromaffin cells. The voltage dependence of $I_{Na(H)}$ is similar to those obtained in DRG neurons and suggests that the channel under these conditions is highly Na^+ selective. Figure 5 shows that elevation of $[H^+]_o$ to pH 6.0 suppresses about 70% of the I_{Ca} in the physiological range of membrane potentials. Thus, it appears that Ca^{2+} channels of the chromaffin cells are highly sensitive to $[H^+]_o$.

To examine the possible effect of neurosecretion of I_{Ca}, secretion was induced by elevation of K^+ or application of nicotine. I_{Ca} was recorded either from an isolated single chromaffin cell or from a cell within a cluster of cells. Figure 6 shows that Ca^{2+} current ist markedly reduced in a cell within the cluster when the cluster is subjected to

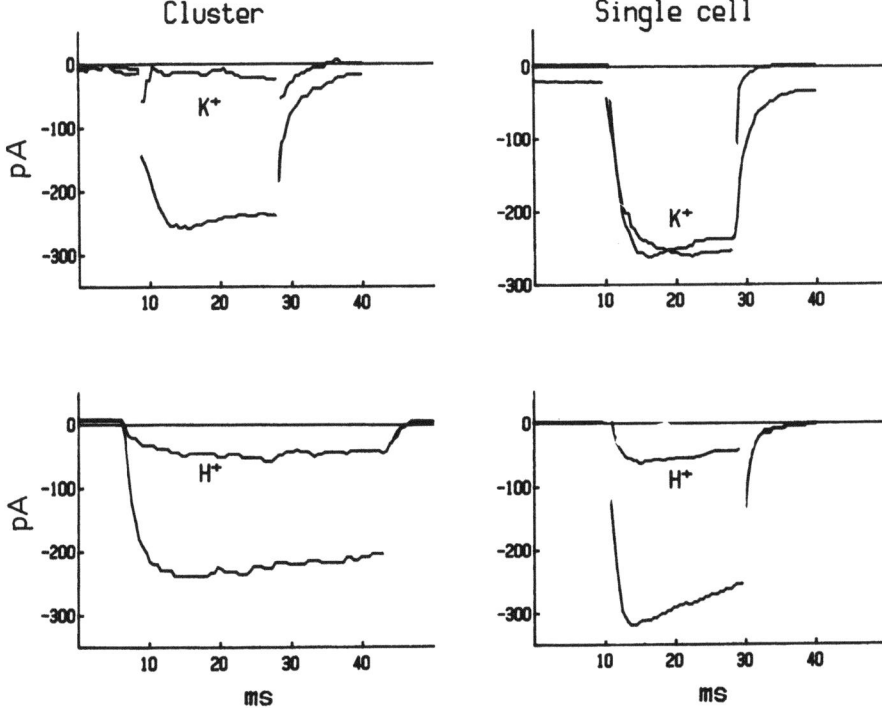

Fig. 6. Effects of step changes in $[H^+]_o$ or $[K^+]_o$ on Ca^{2+} current recorded in cultured chromaffin cells. *Bottom* Lowering the external pH from 7.4 to 6.3 produces a Ca^{2+}-current block in an isolated single cell and in a cell within a cluster of cells. *Top* Increasing the external K^+ from 5.4 to 35.4 mM produces a Ca^{2+}-current block in the cell within the cluster but has no effect on the current recorded in the isolated single cell. Whole-cell recording with 5 mM external Ca^{2+} and Cs^+ as the main internal cation. Holding potential −90 mV, test pulse to 0 mV

rapid elevation of K^+. On the other hand, the elevation of K^+ had little or no effect on I_{Ca} measured in the single isolated cell. Note that whether the cell was isolated or was within a cluster, elevation of $[H^+]_o$ markedly suppressed the Ca^{2+} current. We also examined the effect of the secretogogues, ATP (10 mM), adrenaline (10^{-7} M), and neuropeptides on the I_{Ca} and found no significant suppressive effects on Ca^{2+} current. Thus, the only agent in the secretory vesicles that appeared to inhibit the Ca^{2+} current was H^+. We suggest, therefore, that release of H^+ from the secretory vesicles in the cluster leads to the suppression of Ca^{2+} current in the patched cell, helping to regulate secretion by modulation of Ca^{2+}-channel activity.

References

Almers W, McCleskey EW, Palade PT (1984) A non-selective cation conductance in frog muscle membrane blocked by micromolar external calcium ions. J Physiol 353:565–583

Carbone E, Lux HD (1987) Single low-voltage-activated calcium channels in chick and rat sensory neurons. J Physiol 386:571–601

Davies NW, Lux HD, Morad M (1988) Single channel analysis of proton-induced sodium current in chick dorsal root ganglion neurones: possible site and mechanisms of activation. J Physiol 400:301–331

Hess P, Tsien RW (1984) Mechanism of ion permeation through calcium channels. Nature 309:453–456

Hess P, Lansman JB, Tsien RW (1986) Calcium channel selectivity for divalent and monovalent cations: voltage and concentration dependence of single channel current in vertricular heart cells. J Gen Physiol 88:293–319

Iijima T, Ciani S, Hagiwara S (1986) Effects of external pH on Ca channels: experimental studies and theoretical considerations using a two-site, two-ion model. Proc Natl Acad Sci USA 83:654–658

Johnson RG (1987) Proton pumps and chemiosmotic coupling as a generalized mechanism for neurotransmitter and hormone transport. Ann NY Acad Sci 493:162–177

Krishtal OA, Pidoplichko VI (1980) A receptor for protons in the nerve cell membrane. Neuroscience 5:2325–2327

Konnerth A, Lux HD, Morad M (1987) Proton-induced transformation of calcium channel in chick dorsal root ganglion cells. J Physiol 386:603–633

Prod'hom B, Pietrobon D, Hess P (1987) Direct measurement of proton transfer rates to a group controlling the dihydropyridine-sensitive Ca^{2+} channel. Nature 329:243–246

Tang CM, Presser F, Morad M (1988) Amiloride selectively blocks the low threshold (T) calcium channel. Science 240:213–215

Membrane Potential and Dihydropyridine Block of Calcium Channels in the Heart: Influence of Drug Ionization on Blocking Activity

R. S. Kass and J. P. Arena

University of Rochester School of Medicine and Dentistry, Rochester, NY 14642, USA

Introduction

The purpose of this contribution is to review the experiments that have contributed to our understanding of the mechanisms of action of nisoldipine and other dihydropyridine (DHP) calcium-channel antagonists in the heart. Electrophysiological experiments have shown that membrane potential modulates drug activity in a manner that is consistent with differential binding of this drug to voltage-determined states of the calcium channel (Bean 1984; Sanguinetti and Kass 1984). The predicted binding affinities from these electrical experiments agree well with data obtained from radioligand studies of partially depolarized cells (Kokubun et al. 1986; Janis and Triggle 1983). One theoretical framework in which these data have been interpreted is the modulated receptor hypothesis (Hille 1977a, b; Hondeghem and Katzung 1977). Although this model was formulated to explain the interactions of local anesthetic molecules with receptors associated with sodium channels, this theory has been remarkably useful in understanding the mechanism of action of DHP compounds with regard to calcium-channel modulation. This is not, however, the only theory that has been proposed to explain these interactions (see Hess et al. 1984).

Most experiments designed to investigate the interactions of DHP compounds with calcium channel have been carried out with compounds that are neutral at physiological pH. These drugs (such as nifedipine, nisoldipine, and nitrendipine) all have pK_as less than 4.0 and are predominantly neutral at physiological pH. They can reach receptors that lie in or near the calcium-channel pore via hydrophilic or hydrophobic pathways. In classic studies of local anesthetic block of sodium channels, Hille and coworkers (see Hille 1977a, b; Schwartz et al. 1977) systematically varied the ionization state of blocking molecules to determine whether drug charge would influence access to drug receptor in a predictable manner.

In this contribution experiments in which the ionization of the DHP molecule is varied are discussed in the context of the modulated receptor hypothesis. The results of these experiments, also consistent with this theory, may be useful in determining the location of the DHP receptor within the calcium-channel pore.

Methods

Most of the work described was carried out in this laboratory using two types of voltage clamp procedures: the two-microelectrode voltage clamp of calf cardiac Purkinje fibers (Kass et al. 1979) and the whole-cell arrangement of the patch clamp (Hamill et al. 1981) for studies in guinea pig ventricular myocytes. In both cases, membrane currents were measured in response to controlled voltage pulses. Solutions were designed to minimize or eliminate overlapping ion-channel currents. In the Purkinje fiber experiments, tetrabutylammonium (TBA) was injected to block potassium-channel currents, and tetrodotoxin (TTX, 10–50 µM) was present to block sodium-channel currents. In the single-cell experiments, potassium was replaced by cesium in the pipette solutions. During an experiment, the cell under investigation was dialyzed by this solution, and potassium-channel currents were found to be blocked (Kass and Krafte 1987).

Single ventricular cells were isolated as decribed in Kass and Krafte (1987), and shortened bundles of Purkinje fiber cells were obtained from calf hearts as described in Sanguinetti and Kass (1984). Additional details of experimental protocols may be found in these two papers.

Three drugs were used in this study. Nisoldipine (a gift from Bayer AG) was dissolved in polyethylene glycol 400 (10 mM) stock solutions and diluted down to experimental concentrations for a given experiment. The pK_a of nisoldipine has been estimated to be less than 3. Nicardipine (a gift from Syntex, Inc.) has a pK_a of 7.0. Amlodipine (a gift from Pfizer, Inc.) has a pK_a of 8.6.

Results

Voltage-Dependent Block by Nisoldipine: A Neutral DHP

Figure 1 shows the results of the simplest test for voltage-dependent modulation of nisoldipine block of I_{Ca}. It shows currents measured in response to pulses applied very infrequently (once every 2 min), but from different holding potentials. Records were obtained in control solutions and then in the presence of 200 nM nisoldipine.

When pulses were applied from the negative holding potential, the drug was very ineffective at blocking I_{Ca} (Fig. 1a). In contrast, if currents were measured from depolarized holding potentials, almost all available current was blocked (Fig. 1b). The results of this type of experiment had several important implications. First, it was clear that one drug concentration could have dramatically different effects on I_{Ca} due only to the cell membrane potential. Thus, in cells with negative resting potentials (less than −70 mV), this nisoldipine concentration would barely inhibit I_{Ca}, and an experiment done under these conditions would suggest that higher concentrations of drug would be needed to cause significant decrease of calcium influx. Second, in cells with moderately depolarized resting potentials, this drug concentration would be assayed as more than sufficient to cause a reduction in I_{Ca}.

Figure 2 shows the relative contributions of holding and pulse potential in the development of voltage-dependent I_{Ca} block by nisoldipine. In this experiment, the development of I_{Ca} block by nisoldipine was determined using three different voltage

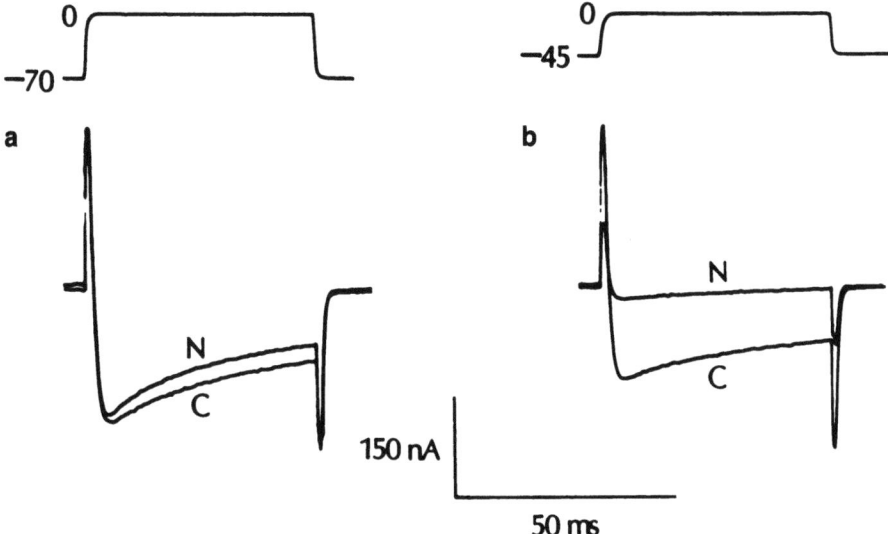

Fig. 1a,b. Holding potential influences nisoldipine block of I_{Ca}. Current is shown in response to pulses to 0 mV applied from −70 (**a**) and −45 (**b**) mV holding potentials. Pulses were applied after 2-min pulse-free interval at specified holding potential. Currents are shown in the absence *(C)* and presence *(N)* of 200 nM nisoldipine. (From Sanguinetti and Kass 1984)

protocols that were designed to discriminate between pulse potential- and holding potential-dependent block. In one method, the holding potential was changed from −70 mV to −45 mV, and block was measured as a function of time at the more positive voltage. As seen in Fig. 2, block of current develops slowly using this procedure. After 30 s at −45 mV, about 60% of the available channels are blocked. When we added very brief (20-ms) pulses applied once every 2 s to 0 mV (a voltage that ensures the opening of calcium channels) to the change in holding potential, there was little difference in either the amount or the time course of developed block. If longer pulses were applied at the same frequency when the holding potential was changed, a small amount of additional block was found to develop.

Results of this experiment were interpreted to mean that voltage enhances nisoldipine block of I_{Ca} at potentials where calcium channels show very low probabilities of opening (−45 mV holding potential). Thus, these results argued against the necessity of channels first opening before block could develop.

Block by Amlodipine and Nicardipine: Two Charged DHP Derivatives

Similar experiments were carried out with nicardipine, a DHP compound (pK_a 7.0) which is 70% neutral at physiological pH. Figure 3 gives an example of these experimental results, illustrating currents measured with voltage protocols similar to those applied in the experiment of Fig. 2; however, in this case the block caused by a change in holding potential alone was about 40% less than the block caused by simultaneous changes in holding potential and application of test pulses that opened channels. This, and other similar experiments, suggested that the charged form of a DHP compound could block channels, but that, perhaps, channel openings were necessary before this block could occur.

More recently, experiments have been carried out to further test this possibility. This work was undertaken with amlodipine (pK_a 8.6). This compound was of interest because it was possible to change the relative fraction of charged vs. neutral drug molecules over a testable pH range. Thus, some experiments were carried out at pH_o

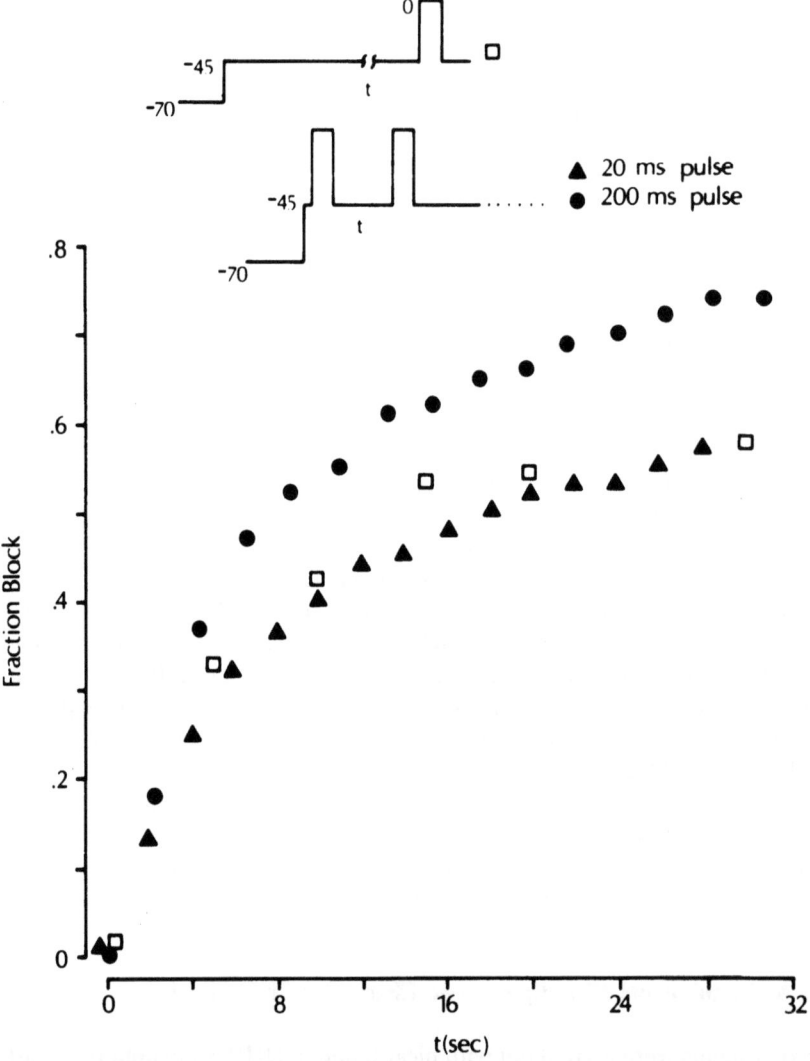

Fig. 2. Relative influence of holding and pulse potential on I_{Ca} block by nisoldipine (200 nM). *Inset* shows voltage protocols. Plot is fraction of current blocked vs time at −45 mV for each protocol. For train protocols, block was measured during application of 20-ms *(triangles)* or 200-ms *(circles)* pulses to 0 mV applied once every 2 s. For single-pulse protocol, block was measured during application of a single pulse to 0 mV *(squares)* after the time indicated along the abscissa at −45 mV. (From Sanguinetti and Kass 1984)

7.4 (94% of drug molecules are neutral) and some at pH_o 10.0 (96% of drug molecules are neutral).

Figures 4, 5 compare the effects of amlodipine on membrane currents measured under steady-state conditions from –80 mV and –40 mV holding potentials. Figure 4 shows currents measured in pH_o 7.4, and Fig. 5 illustrates currents measured in pH_o 10.0. In both cases, currents are inhibited much more effetively by given drug concentrations at the more positive holding potential than they are at –80 mV. Thus,

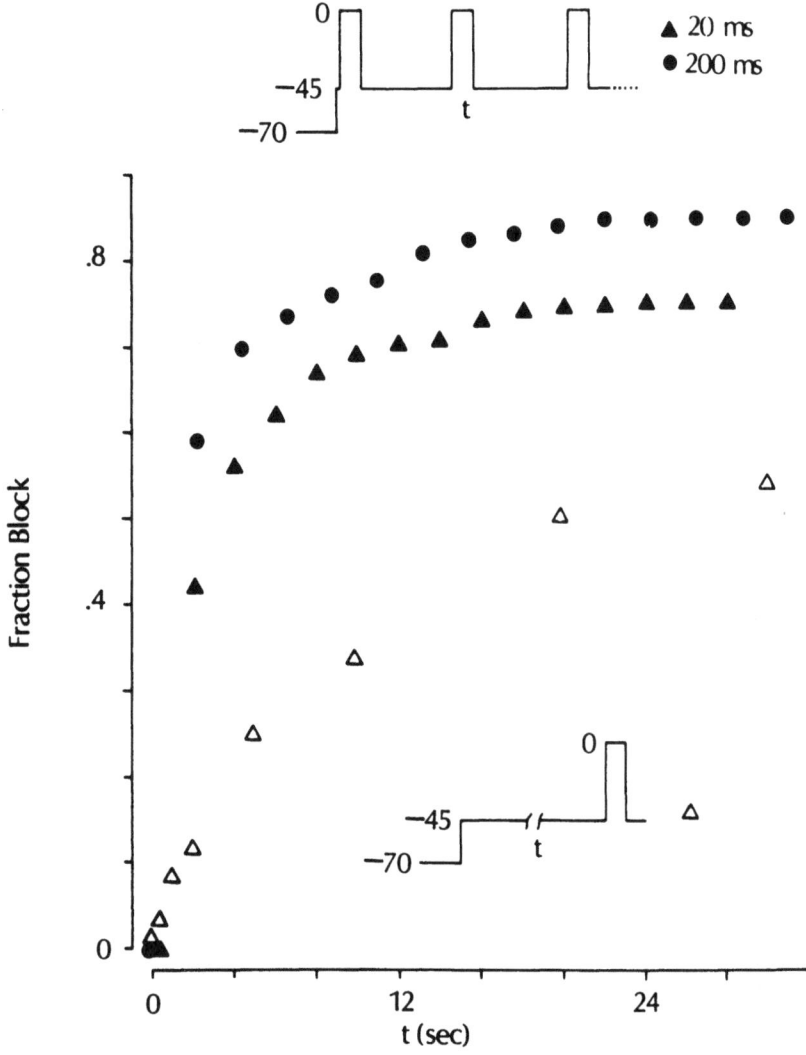

Fig. 3. Relative influence of holding and pulse potential on nicardipine block of I_{Ca}. Same protocol as in Fig. 2 *(insets)*. Fraction of current blocked by nicardipine (200 nM) was measured for 20-ms *(filled triangles)* and 200-ms *(circles)* train protocols and for a single-pulse protocol *(open triangles)*. Holding potential was –75 mV, and train conditioning potential was –45 mV. (From Sanguinetti and Kass 1984)

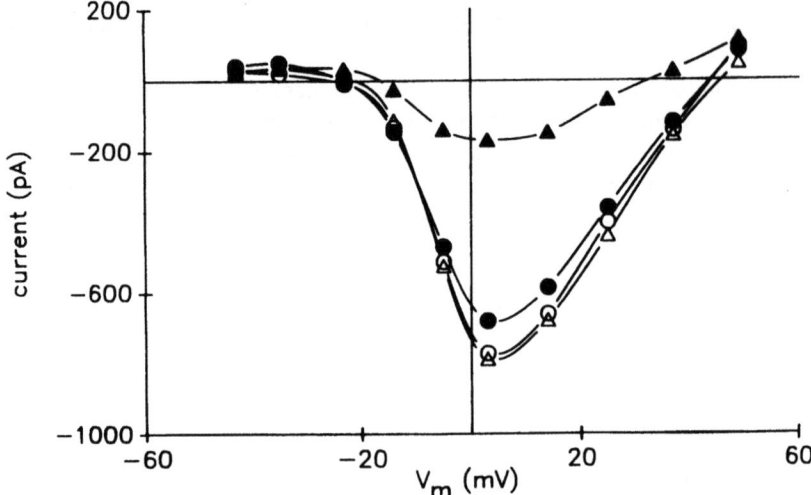

Fig. 4. Influence of holding potential on amlodipine block (pH$_o$ 7.4). Membrane potential was held at −80 mV and then at −40 mV, and a series of voltage pulses to different membrane potential (Vm) was applied to measure I$_{Ca}$. The membrane was kept at each holding potential for at least 3 min between runs. The plot shows peak currents measured from −80 mV *(circles)* and −40 mV *(triangles)* in the absence *(open symbols)* and presence *(filled symbols)* of amlodipine (1 μM). Cell 1152. (From Kass et al. 1988)

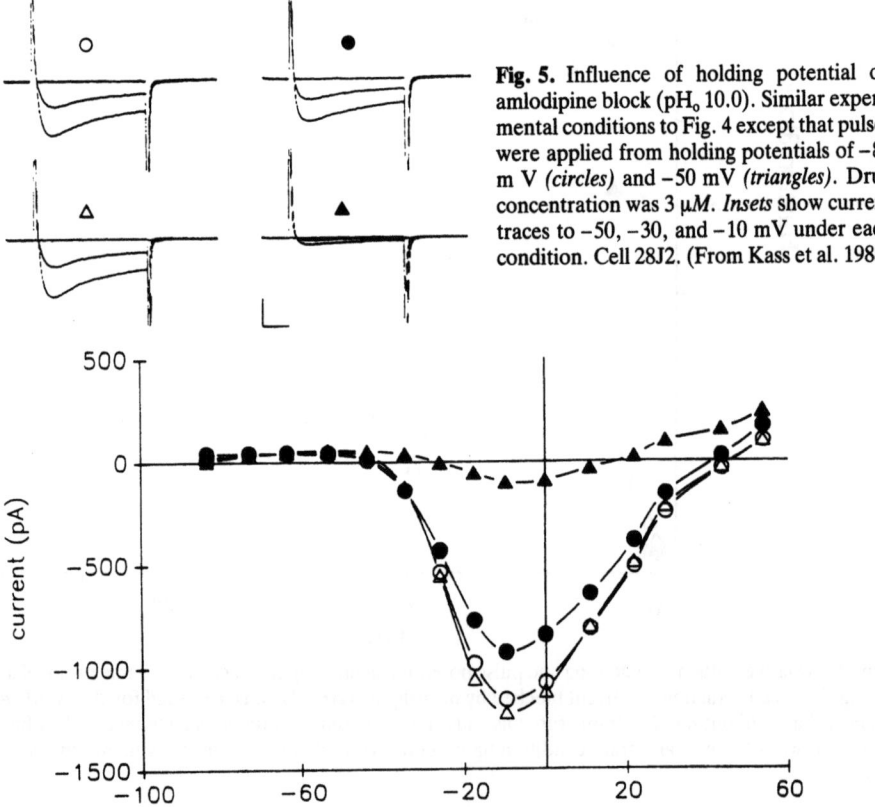

Fig. 5. Influence of holding potential on amlodipine block (pH$_o$ 10.0). Similar experimental conditions to Fig. 4 except that pulses were applied from holding potentials of −80 mV *(circles)* and −50 mV *(triangles)*. Drug concentration was 3 μM. *Insets* show current traces to −50, −30, and −10 mV under each condition. Cell 28J2. (From Kass et al. 1988)

in the steady state, this drug acts very similarly to nisoldipine, although higher drug concentrations are needed to cause comparable inhibition of current.

Figures 6, 7 show that the kinetics of drug action differ greatly in pH_o 10.0 and pH_o 7.4, indicating that drug charge does, in fact, affect the interaction with the DHP receptor. In each experiment, the cell membrane was held at −80 mV while amlodipine was introduced into the extracellular solution. As demonstrated in Figs. 4, 5, there was very little drug-induced reduction of I_{Ca} in either pH_o 10.0 or 7.4 under these conditions. Then, in the presence of the drug, the membrane potential was changed to −50 mV, and trains of test pulses were imposed.

The test pulses were applied to 0 mV, a voltage that causes calcium channels to open. Two sets of pulse trains were applied: the first train consisted of 20-ms pulses,

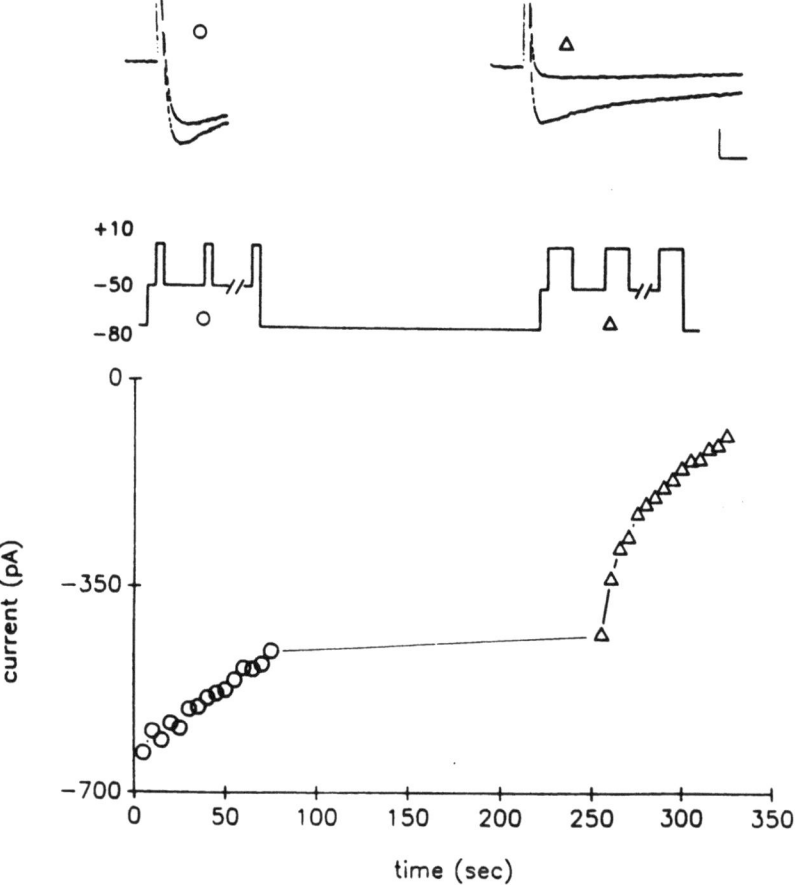

Fig. 6. Kinetics of amlodipine block in pH_o 7.4. Amlodipine (3 μM) was washed in while the cell was held at −80 mV without pulsing. The holding potential was then changed to −50 mV, and 20-ms pulses to +10 mV were applied every 5 s. Peak inward current in response to these pulses is plotted *(circles)* against time of the pulse train. The membrane was returned to −80 mV for 3 min without pulsing and then returned to −50 mV with 200-ms pulses *(triangles)* to 0 mV being applied every 5 s. Peak currents in response to the second train of pulses are plotted against time (including the period at rest). *Insets* show superpositions of the first and last pulses of each pulse train. The first 66 ms of data are shown for the 200-ms pulse. Cell 9301. (From Kass et al. 1988)

and the second set comprised 200-ms pulses. In each case, the pulses were applied once every 5 s. The cell was held at −80 mV without stimulation for 2–3 min between the sets of pulse trains in order to allow time for calcium channels to recover from drug block. Under these experimental conditions, clear differences emerged between block by the charged and that by the neutral form of the drug molecules.

Figure 6 illustrates the effects of ionized amlodipine under these conditions. During the application of a train of brief pulses from a −50 mV conditioning voltage, block developed slowly. Peak current was reduced 25% in 75 s. This block was not relieved at −80 mV despite a 3-min pulse-free period at this voltage. Development of block continued with the application of 200-ms pulses from the −50 mV conditioning voltage. In this case, approx. 80% of the remaining current was blocked in the same time period. Thus, prolongation of pulse duration enhances block, but equally as important, I_{Ca} does not recover from amlodipine block in pH_o 7.4 over the time frame tested here.

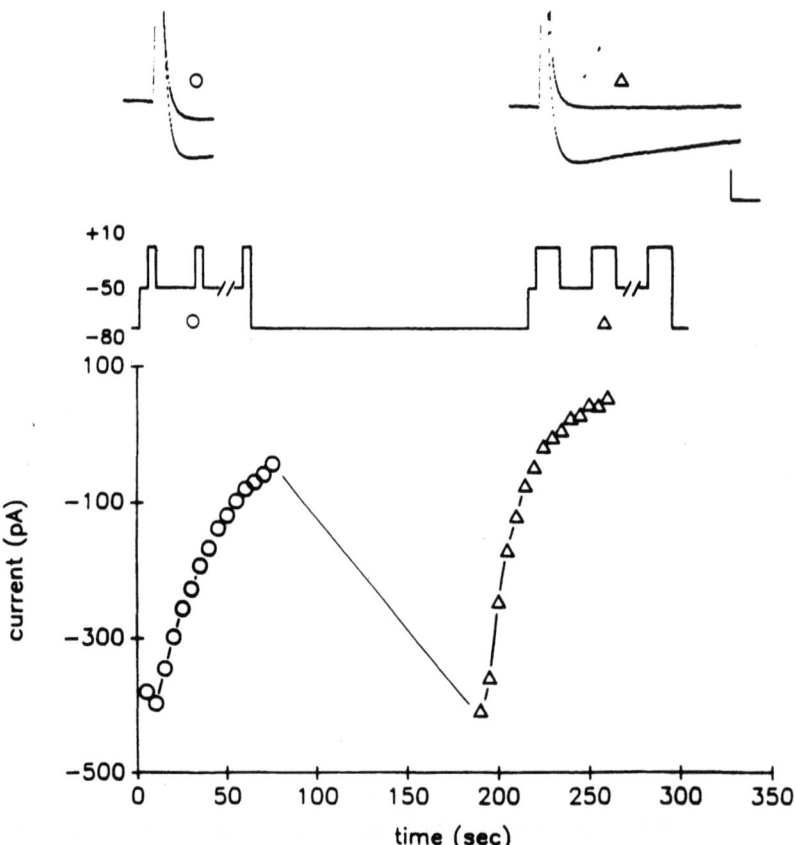

Fig. 7. Kinetics of amlodipine block in pH_o 10.0. Similar experimental protocol to that in Fig. 6. Currents in response to 20-ms pulses applied from conditioning potential of −50 mV are plotted *(circles)*, as well as currents in response to 200-ms pulses *(triangles)*. The period between pulse trains was 2 min, and the holding potential for this time was −80 mV. Note complete recovery from block between trains of pulses. *Insets* show superpositions of the first and last pulses of each pulse train. The first 66 ms of data are shown for the 200-ms pulse. Cell 28J82. (From Kass et al. 1988)

A similar experiment carried out in a solution buffered to pH 10.0 is shown in Fig. 7. Here, block develops rapidly during application of brief pulses applied from a –50 mV conditioning voltage, and increasing pulse duration only causes minor increases in blocking activity during a subsequent pulse train. However, there is a dramatic contrast between recovery from block in this solution and that shown in Fig. 6. Block by the neutral form of the drug is completely relieved during the 2-min interval between pulse trains in this experiment.

In summary, depolarization enhances amlodipine-induced block of I_{Ca} in solutions of physiological and alkaline pH_o. In pH_o 7.4, depolarized holding potentials promote channel block, but block is enhanced by depolarized pulses that open and inactivate channels. Recovery from block is very difficult to reverse. In pH_o 10.0, development of block is less sensitive to voltage pulses but is much more reversible. Recovery from block is rapid and complete after cell membrane potential is returned to values near –80 mV.

Discussion

Previous models of DHP binding predict a low affinity for the receptor when channels are in the rested state and a high affinity for inactivated and open channels (Sanguinetti and Kass 1984; Cohen and McCarthy 1987). Amlodipine shares the property of low affinity during the rested state because no tonic block of I_{Ca} at negative (–80 mV) membrane potentials occurred. This was the case for both the charged and neutral forms of the drug. It also appears that amlodipine has a high affinity for open and/or inactivated channels because depolarization in both the neutral and charged forms enhances block. However, differences in recovery at negative holding potentials and onset of block during trains of pulses from depolarized holding potentials indicate differences in the accesses of charged and neutral drug to the DHP receptor.

Block by the charged drug is consistent with a drug that accesses the DHP receptor through open channels because block is enhanced by prolongation of pulse duration during application of pulse trains that open and then inactivate channels. This pattern is consistent with binding to open channels and/or the open channels allowing access of drug to bind to inactivated channels. These experiments do not distinguish between these two possibilities.

Block by neutral drug molecules is less sensitive to pulse duration during repetitive pulse application. Instead, block develops rapidly at depolarized potentials whether brief or long pulses are applied. This is consistent with block via the membrane pathway because block occurs in protocols with minimal channel openings.

The modulated receptor hypothesis put forward by Hille (1977a) and Hondeghem and Katzung (1977) was proposed to explain the interaction of local anesthetics with sodium channels. According to this model, charged and neutral drug molecules can bind to receptors associated with sodium channels. Drug molecules can interact with the receptor, which is in the sodium-channel pore, via the membrane phase (hydrophobic pathway) or via the inner mouth of the channel (hydrophilic pathway). Block is enhanced by channel openings and transitions into the inactivated state for neutral drug. Charged drugs can access receptors only via open channels.

In addition, Hille (1977b) and Schwarz et al. (1977) found that I_{Na} recovery from block by ionized drug molecules was markedly slower than recovery from neutral drug block. This was accounted for by assuming that bound drug cations are trapped in the channel until they lose their charge or the channel opens.

This scheme predicts the effects reported in the present contribution remarkably well. Accordingly, channel block by ionized amlodipine is enhanced by channel openings, and recovery from block by the charged form of the drug is much slower than recovery from block by neutral amlodipine.

The results under conditions in which a fraction of drug molecules is neutral and a fraction charged could then be explained by mixed activities of the neutral and ionized drug molecules. For example, at pH_o 7.4, amlodipine is 94% charged and 6% neutral. Under these conditions, the slow holding potential-dependent blocking activity is most likely promoted by neutral drug molecules. The enhanced rate of block caused by the simultaneous application of pulses is probably due to ionized drug molecules. Similarly, the failure to relieve block by charged drug at negative potentials could be due to the very low probability of channel openings at these voltages (see Reuter 1984).

The formalism of the modulated receptor hypothesis was built on known structural properties of the sodium channel. These properties, along with the experimental observations of local anesthetic block of the channel, placed restrictions on the locations of the receptors for these drugs. Block occurred via an intracellular pathway and was very difficult to reverse for charged drug molecules, very much like our results for amlodipine. Thus, the local anesthetic receptor was postulated to lie within the channel pore but between the "selectivity filter" and the inner mouth of the channel (Hille 1977b; Schwarz et al. 1977). Such a model can account for the development of, and recovery from, block by the charged form of the drug.

It remains to be determined whether the DHP receptor is located in a region of the L-type calcium channel that resembles its counterpart for local anesthetics in sodium channels. One test will be to determine whether the receptor location allows rapid titration by extracellular protons and whether the drug-receptor complex can be titrated by extracellular pH. The results of such future studies will be useful in further characterizing the structure of this important ion channel.

References

Bean BP (1984) Nitrendipine block of cardiac calcium channels: high-affinity binding to the inactivated state. Proc Natl Acad Sci 81:6388–6392

Cohen CJ, McCarthy RT (1987) Nimodipine block of calcium channels in rat anterior pituitary cells. J Physiol (London) 387:195–225

Hamill OP, Marty A, Neher E, Sakmann B, Sigworth FJ (1981) Improved patch-clamp techniques for high-resolution current recording from cells and cell-free membrane patches. Pflugers Arch 391:85–100

Hess P, Lansman JB, Tsien RW (1984) Different modes of gating behavior favored by dihydropyridine agonists and antagonists. Nature 311:538–544

Hille B (1977a) The pH-dependent rate of action of local anesthetics on the Node of Ranvier. J Gen Physiol 69:475–496

Hille B (1977b) Local anesthetics: hydrophilic and hydrophobic pathways for the drug-receptor reaction. J Gen Physiol 69:497–515

Hondeghem LM, Katzung BG (1977) Time- and voltage-dependent interactions of antiarrhythmic drugs with cardiac sodium channels. Biochim Biophys Acta 472:373–398

Janis RA, Triggle DJ (1983) New developments in Ca channel antagonists. J Med Chem 26:775–785

Kass RS, Krafte DS (1987) Negative surface charge density near heart calcium channels: Relevance to single channel studies and block by dihydropyridines. J Gen Physiol 89:837–839

Kass RS, Siegelbaum SA, Tsien RW (1979) Three-micro-electrode voltage clamp experiments in calf cardiac Purkinje fibres: is slow inward current adequately measured? J Physiol (Lond) 290:201–225

Kass RS, Arena JP, DiManno D (1988) Block of heart calcium channels by amlodipine: Influence of drug charge on blocking activity. J Cardiovasc Pharmacol, in press

Kokubun S, Prod'hom B, Becker C, Porzig H, Reuter H (1986) Studies on Ca channels in intact cardiac cells: Voltage-dependent effects and cooperative interactions of dihydropyridine enantiomers. Mol Pharmacol 30:571–584

Reuter H (1984) Electrophysiology of Calcium Channels in the Heart. In: Opie LH (ed) Perspectives in Cardiovascular Research. Raven Press, New York, 43–51

Sanguinetti MC, Kass RS (1984) Voltage-dependent block of calcium channel current in the calf cardiac Purkinje fiber by dihydropyridine calcium channel antagonists. Circulation Res 55:336–348

Schwarz W, Palade PT, Hille B (1977) Local anesthetics. Effect of pH on use-dependent block of sodium channels in frog muscle. Biophys J 20:343–368

Dihydropyridine-Sensitive and -Insensitive Ca^{2+} Channels in Normal and Transformed Fibroblasts*

C. Chen[1], M. J. Corbley[2,3], T. M. Roberts[2,4], and P. Hess[1]

[1] Department of Cellular and Molecular Physiology and Program in Neuroscience, Harvard Medical School, Boston, MA 02115, USA
[2] Dana-Farber Cancer Institute, and Departments of
[3] Biological Chemistry and Molecular Pharmacology, and
[4] Pathology, Harvard Medical School, Boston, MA 02115, USA

Introduction

The importance of voltage-sensitive Ca^{2+} channels in the surface membrane of excitable cells is undisputed and well studied. Ca^{2+} ions entering the cytoplasm through Ca^{2+} channels provide the message which leads to neurotransmitter release, contraction in cardiac and smooth muscle, and secretion in many endocrine and exocrine cells. More recently, it has become clear that intracellular Ca^{2+} ions are also involved in the control of a variety of cellular functions which are not restricted to electrically excitable cells, such as cell adherence, cytoskeletal organization, motility, phagocytosis, and even control of cell growth.

The most commonly used cells for the study of cell growth are 3T3 fibroblast cell lines. These cells offer high plating efficiency, rapid growth, sensitivity to contact inhibition (Todaro et al. 1964), and a high frequency of transformation when infected or transfected with specific oncogenes; thus, they provide a convenient preparation to study the regulation of both normal and abnormal growth (Todaro et al. 1965).

The mechanism of regulation of intracellular Ca^{2+} (Ca_i) in fibroblasts is poorly understood, and very little is known about Ca^{2+}-specific transport systems across the plasma membrane. Here, we demonstrate that 3T3 fibroblasts contain two types of voltage-sensitive Ca^{2+} channels. The two Ca^{2+}-channel types were identified by their unitary properties and pharmacological sensitivities. While the two Ca^{2+}-channel types were consistently present in all control 3T3 cells, one type of Ca^{2+} channel was selectively suppressed in 3T3 cells transformed by activated c-H-*ras*, EJ-*ras*, *src*, *v*-fms, or polyoma middle T oncogenes. The unexpected finding of voltage-sensitive Ca^{2+} channels in these nonexcitable cells and the control of their functional expression by transforming oncogenes raises important questions about their role in the control of calcium-sensitive processes in nonexcitable cells.

Methods

3T3 fibroblasts were cultured in DMEM and 10% fetal calf serum. Patch clamp recordings were performed with standard techniques (Hamill et al. 1981). For the

---

* This work was supported by grants from the United States Public Health Service and the American Cancer Society.

whole-cell recordings, invidual cells were obtained by brief (5–10 min) trypsinization (0.05% trypsin + 0.5 mM EDTA). The cell suspension was allowed to settle in the recording chamber, and currents were measured from individual cells not connected to their neighbors. Appropriate ionic substitutions were used to identify currents through Ca^{2+} channels and to suppress any other ionic conductances. The internal solution contained 135 mM CsCl, 10 mM EGTA, 10 mM Hepes, 5 mM $MgCl_2$, 4 mM ATP (pH 7.5), the external solution 20 mM $BaCl_2$, 135 mM TEACl, 10 mM Hepes (pH 7.5). Depolarizing pulses were delivered every 3–4 s. Linear leak and capacity currents were substracted digitally. In some experiments, chloride ions were replaced by aspartate both internally and externally.

Single-channel recordings were obtained directly from cells in culture dishes, as well as from cells dispersed as described for the whole-cell recordings. We noted no differences in unitary channel properties between the two conditions. The cell membrane potential was zeroed (Hess et al. 1986) by the bath solution, which contained 145 mM potassium aspartate, 5 mM EGTA, 10 mM Hepes (pH 7.5). The pipette solution contained 110 mM $BaCl_2$ or 110 mM $CaCl_2$, 10 mM Hepes (pH 7.5). Linear leak and capacity currents were subtracted digitally. The traces were sampled at 5 kHz and filtered at 1 kHz (−3 dB, eight-pole Bessel filter). All recordings were obtained at room temperature (22°C).

To obtain clonal cell lines containing and expressing a particular oncogene, cell lines were grown up from neomycin/G418-resistant colonies selected after infection with a recombinant retrovirus (Cepko et al. 1984; Korman et al. 1987) carrying the neomycin/G418 resistance gene and the following inserts:
1. No second gene (control virus-infected cells)
2. Activated c-H-*ras*-1 cDNA encoding leucine in place of glutamine at codon 61 (M. J. C. Orbley, unpublished work; Der et al. 1986; Yuasa et al. 1983; Fujita et al. 1985)
3. v-*fms*, the transforming gene of the McDonough strain of feline sarcoma virus (D. K. Morrison and T. M. Roberts, unpublished work)
4. Polyoma virus middle tumor antigen (middle T; Cherington et al. 1986)
5. v-*src*, the transforming gene of the Rous sarcoma virus
6. Polyoma virus large tumor antigen (large T), simian virus 40 large tumor antigen (SV40 large T)

NIH 3T3 cells were also grown up from a transformed focus after transfection with EJ-*ras* DNA (Tabin et al. 1982; Taparowsky et al. 1982).

The dexamethasone-inducible cell line was obtained by inserting the polyoma middle tumor gene into an expression vector and placing it under the control of the dexamethasone-inducible promoter from the mouse mammary tumor virus long terminal repeat (MMTV-LTR; see e. g., Huang et al. 1981; Raptis et al. 1985). A clonal line was then derived from NIH 3T3 cells cotransfected with this plasmid and a neomycin resistance marker. The pure enantiomer of the dihydropyridine Ca^{2+}-channel agonist (+) 202–791 was kindly provided by Dr. Hof, Sandoz AG, Switzerland.

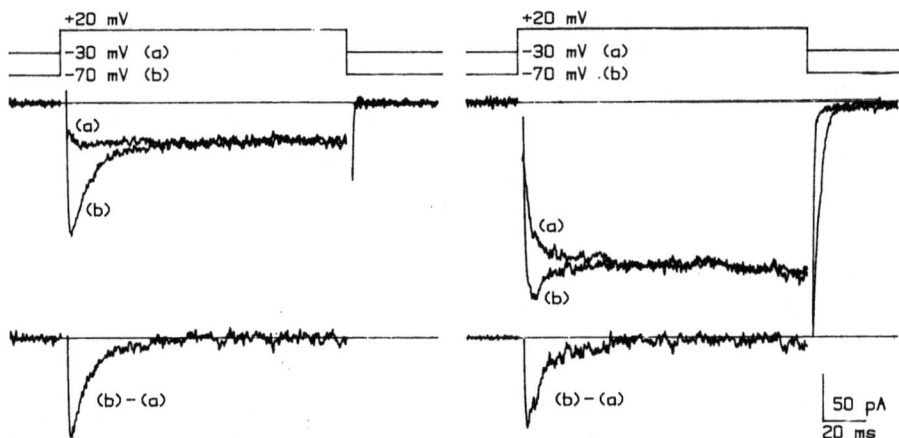

Fig. 1. Two types of Ca^{2+} channels in Swiss 3T3 fibroblasts. Whole-cell currents carried by Ba^{2+} (20 mM) in response to voltage clamp steps to +20 mV are shown before *(left)* and after *(right)* addition of the DHP Ca^{2+}-channel agonist (+) 202–791 (1 µM). Voltage protocol is shown above the current traces. Two currents elicited from holding potentials of −30 *(a)* and −70 mV *(b)* are superimposed. The extra component of current activated from the more negative holding potential is shown as the difference between the two current traces *(bottom)*. Linear leak and capacity currents were subtracted. Cell capacitance is 141 pF. (From Chen et al. 1988, with permission)

Results

Figure 1 shows the two types of Ca^{2+}-channel currents that can be recorded from a 3T3 fibroblast. A test depolarization to +20 mV from a holding potential of −30 mV elicits an inward current which activates within 5–10 ms and remains well maintained throughout the depolarization. An additional transient component of inward current is activated at the same test potential when the fibroblast is held at −70 mV prior to the depolarization. This current activates rapidly to a peak and then decays to zero within about 50 ms (Fig. 1, bottom trace, left panel). The two inward currents differ not only in their kinetics and dependence on the holding potential but also in their sensitivity to drugs. Figure 1 shows that the dihydropyridine (DHP) Ca^{2+}-channel agonist (+) 202–791 (Hof et al. 1985; Kokubun et al. 1986) selectively increases the maintained current component but fails to affect the transient current.

Figures 2, 3 show the corresponding unitary events. Single-channel currents underlying the maintained whole-cell Ca^{2+}-channel component are shown in Fig. 2: at +20 mV, brief openings of ~ 1.2-pA amplitude occur in bursts throughout the depolarizing pulse, giving rise to a well-maintained average current (Fig. 2a, bottom trace). As expected from the whole-cell recordings, (+) 202–791 greatly increases the averaged current (Fig. 2b, bottom trace). The single-channel currents in the presence of (+) 202–791 demonstrate the mechanism of drug action: the appearance of long-lasting channel openings, which occur rarely in the absence of drug (Fig. 2a, second trace), is greatly enhanced by the DHP Ca^{2+}-channel agonist (Hess et al. 1984; Kokubun and Reuter 1984), thus leading to an overall increase in the channel-opening probability.

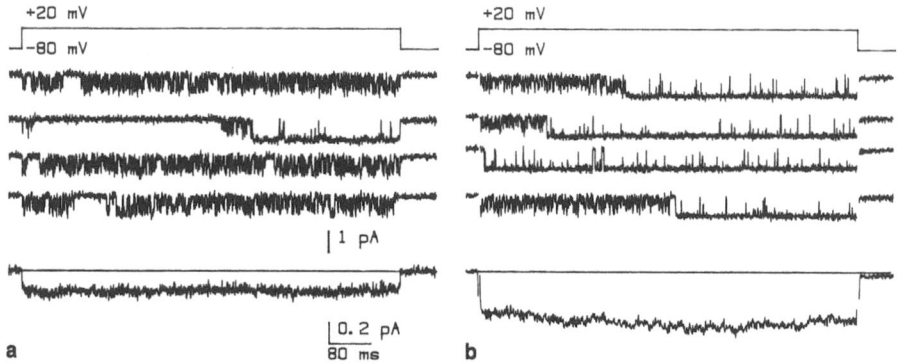

Fig. 2a, b. A single L-type Ca^{2+} channel in a 3T3 fibroblast. Selected individual current traces from a cell-attached recording are shown before (**a**) and after (**b**) addition of 1 μM (+) 202–791 to the bath. The patch potential is given above the current traces. Below each set of current records, the mean current (obtained as the average of 50–200 sweeps in each condition) is shown. Current traces with no detectable openings are not shown. The pipette solution contained 110 mM BaCl$_2$. Linear leak and capacity currents were subtracted. The unitary conductance was 25 pS (From Chen et al. 1988, with permission)

The unitary properties of the second Ca^{2+}-channel type are shown in Fig. 3, which was obtained from another patch. Here, depolarizations to a more negative test potential (−20 mV) elicit inward currents of ~0.4-pA amplitude, which tend to be grouped towards the beginning of the depolarization. The time course of the averaged current corresponds nicely to the transient whole-cell component, and as with the whole-cell current, the elementary activity is greatly reduced by holding potentials positive to −60 mV and is insensitive to DHP drugs (not shown).

The macroscopic and microscopic kinetics, elementary conductances, and DHP sensitivities of the two Ca^{2+}-channel types in 3T3 cells are very similar to the L- (long-lasting, large conductance) and T-type (transient) Ca^{2+} channels described in much greater detail for excitable cells such as sensory neurones (Carbone and Lux 1984); Nowycky et al. 1985; Fedulova et al. 1985; Dupont et al. 1986), cardiac muscle (Nilius et al. 1985; Bean 1985; Mitra and Morad 1986; Bonvallet 1987), smooth muscle (Bean et al. 1986; Benham et al. 1987), and anterior pituitary cell lines (Armstrong and Matteson 1985; Cohen and McCarthy 1987).

In addition to L- and T-type Ca^{2+} channels, we also found several classes of K channels, Cl channels, and nonselective cation channels in 3T3 fibroblasts. One of these showed measurable permeability to divalent ions but little selectivity for di- over monovalent ions (nonselective cation channel; C. Chen and P. Hess, unpublished work).

If intracellular Ca^{2+} plays an important role in the control of cell growth (Moolenaar et al. 1984; Hesketh et al. 1985; Bravo et al. 1985; Mix et al. 1984; Lopez-Rivas and Rozengurt 1983; Morgan and Curran 1986), one would predict an altered regulation of Ca^{2+} in oncogenically transformed cells (Weinberg 1985; Bishop 1987). We therefore compared Ca^{2+} channels in normal 3T3 fibroblasts with those of transformed fibroblasts. While the current densities for the L-type Ca^{2+} channels

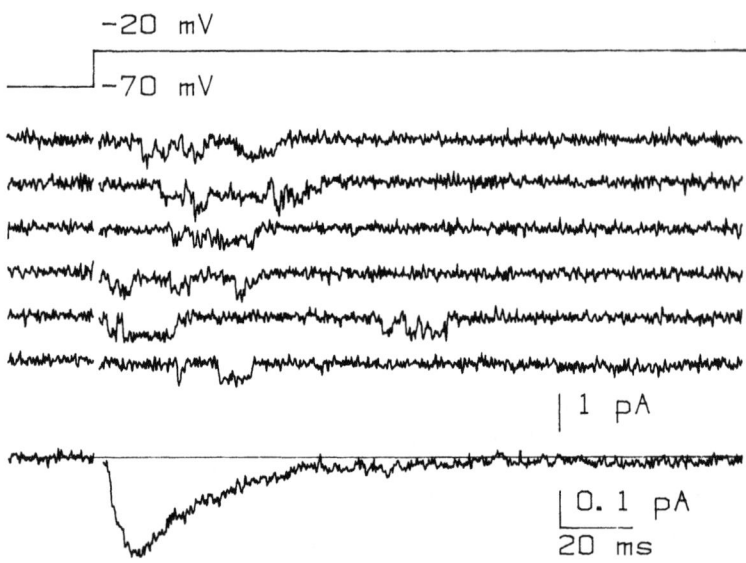

Fig. 3. Cell-attached patch containing several T-type Ca^{2+} channels. Experimental conditions were the same as in Fig. 2, except that the pipette solution was 110 mM $CaCl_2$, with the conductance being 8 pS. Unlike the L-type channel (Hess et al. 1986), the T-type channel has similar conductance and kinetics with Ca^{2+} and Ba^{2+} as the charge carrier (Nilius et al. 1985; Carbone and Lux 1987) (From Chen et al., with permission)

were not significantly different between normal and transformed cells, cells transformed by oncogenes whose protein products are localized in the cell membrane (activated c-H-*ras*, EJ-*ras*, v-*fms*, *src*, polyoma middle T oncogenes) specifically lacked the T-type currents found in control 3T3 cells. In contrast, the two cell lines which carried oncogenes whose protein products are largely confined to the nucleus (polyoma large T, SV40 large T) still contained T-type Ca^{2+} channels, but at a significantly lower density than control, nontransformed 3T3 cells. Examples of current traces and a summary of the current densities in control and transformed 3T3 cells are shown in Fig. 4.

In an effort to correlate the time course of the functional suppression of T-type Ca^{2+} currents with the establishment of the transformed phenotype, we used a cell line which contained the polyoma middle T oncogene under the control of a dexamethasone-inducible promoter (Raptis et al. 1985; Kaplan et al. 1986). In this cell line, which has a nontransformed phenotype in the absence of glucocorticoids, maximal levels of middle T antigen and morphological transformation can be detected within 24–48 h of addition of 1 µM dexamethasone to the culture medium (Raptis et al. 1985). Figure 5 shows that T-channel density was not significantly decreased 1 week following the induction of transformation. However, after a month, only 50% of the cells still expressed T-type channels, and the mean density of T-type current had decreased to less than 15% of that found in control cells.

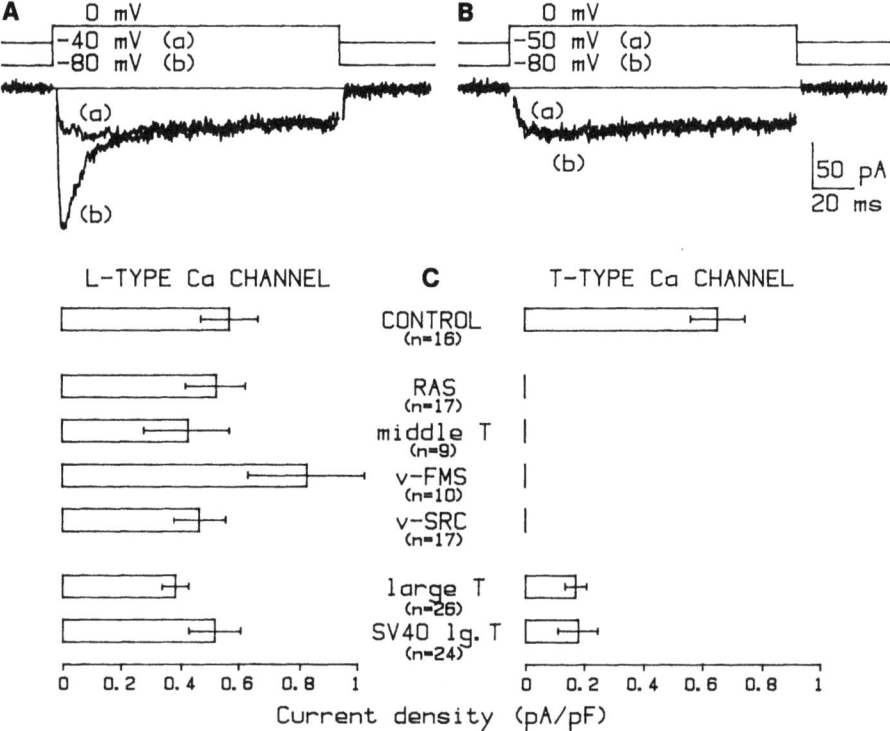

Fig. 4A–C. Current densities through Ca^{2+} channels in normal and transformed 3T3 fibroblasts. **A, B** Superposition of whole-cell Ba currents elicited from two holding potentials *(a, b)*. Voltage protocol is shown above the current traces. Control virus-infected NIH 3T3 cell (**A**) has both components of Ca^{2+}-channel current. Cell capacitance is 95 pF. The two current traces recorded from different holding potentials in a NIH 3T3 cell transformed by the activated c-H-*ras* oncogene (**B**) superimpose perfectly, indicating a complete lack of T-type current. Capacitance is 76 pF. **C** Mean values ± SEM of the current densities of the two Ca^{2+}-channel types in control and transformed cells, measured in whole-cell recordings like those illustrated in **A, B**. The number of cells studied *(n)* is indicated for each group. The mean densities of the L-type current did not differ significantly between control and transformed cells. Solutions and patch clamp methods were the same as in Fig. 1, except that Cl ions were replaced by aspartate (Cs aspartate, TEA aspartate) and acetate (Ba acetate). Cell capacitance was measured from the capacity current elicited by a small depolarization. Currents were measured after leak subtraction at the peak of the current-voltage relation (−20 mV and +20 mV for T and L currents respectively), with 20 mM Ba as the charge carrier. Part of this figure is reproduced with permission from Chen et al. (1988)

Discussion

The presence of voltage-dependent Ca^{2+} channels in nonexcitable cells (Chen et al. 1987, 1988) was unexpected, even though voltage-dependent Na channels had previously been described in certain fibroblast lines (Pouyssegur et al. 1980). Ca^{2+} channels in fibroblasts have now been found by two other groups (Peres et al. 1988; Lovisolo et al. 1988). The density of Ca^{2+} channels in 3T3 fibroblasts is considerably lower than

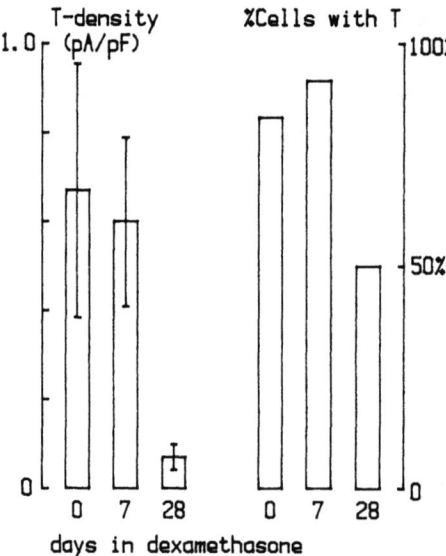

Fig. 5. Densities of Ba currents carried through T-type Ca^{2+} channels in 3T3 fibroblasts containing a dexamethasone-inducible polyoma middle T oncogene (Raptis et al. 1985). Recording conditions and measurements of current densities are the same as in Fig. 4. The mean values ± SEM of the T-current densities *(left)* are plotted in the absence of dexamethasone (time 0), and 7 and 28 days after induction of transformation by addition of 5 µM dexamethasone to the culture medium. The bars on the *right* indicate the percentage of cells which have a measurable T-type current at the time indicated

that in heart or neurons, but roughly comparable to that in certain vascular smooth muscle cells (Bean et al. 1986; Benham et al. 1987) With 20 mM Ba as the charge carrier, the maximal current (I) carried by either channel type in fibroblasts is about equal and averages 0.5–0.6 pA/pF cell capacitance (see Fig. 4). With values of the elementary current (i) and the opening probability (p) obtained from our single-channel recordings, and a specific capacitance of 1 µF/cm^2, we can estimate the density of functional channels ($n = I/ip$) as approx. 1 per 10–20 µm^2 cell membrane for each type of Ca^{2+} channel. We detected no significant differences between channel densities in fibroblasts obtained from exponentially growing or confluent, growth-arrested cultures.

The role of voltage-dependent Ca^{2+} channels in fibroblasts is not obvious. We have evidence that under physiological ionic conditions, the membrane potential can be as negative as −70 mV and can oscillate between −70 and ∼−10 mV (Chen et al., unpublished work; Henkart and Nelson 1979; Okada et al. 1982), thus covering the physiological activation range for both channel types. Activation of Ca^{2+} channels in fibroblasts is likely to contribute to such Ca^{2+}-sensitive processes as control of secretion, shape change, motility, and phagocytosis.

We cannot determine at what level the functional expression of T-type Ca^{2+}-channel current is blocked in transformed cells. A direct effect of an oncogene product on the channel seems unlikely because the products of the oncogenes tested differ in their direct cellular actions: the ras product (p21^{ras}) is a guanine nucleotide binding protein (Shih et al. 1980), whereas the other oncogene products directly (src, fms) or indirectly (middle T, through binding to pp60^{src}) lead to increased tyrosine kinase activity (Rettenmier et al. 1987; Hunter and Sefton 1980; Collett et al. 1980; Levinson et al. 1980; Courtneidge and Smith 1983; Bolen et al. 1984). A direct action on the T-type channel by a hypothetical messenger molecule formed later in the cascade of transforming biochemical events presumed common to all oncogene products is also unlikely, since even after a week of induction of the middle T oncogene (Fig. 5), the T-type current is not measurably reduced.

The fact that cells transformed by oncogenes encoding proteins with primarily nuclear location also have a reduced T-channel density further points toward a nonspecific correlation between channel expression and transformation. The incomplete suppression of T-channel function in cells transformed by polyoma large T or SV40 large T may somehow be related to the generally lower potency for morphological transformation of this class of oncogenes (Land et al. 1983).

The results of the inducible middle T oncogene allows additional conclusions:
1. Loss of functional expression of T-type Ca^{2+} channels is a consequence, and not the cause, of tranformation.
2. The unaltered density of T-type channels 1 week (about 10 cell cycles) after induction means that the transformed cells are still actively producing the channel, therefore compensating for the effect of "dilution" associated with each cell division.
3. The significant overall reduction of T channel density, together with a complete loss of the channel in 50% of the cells 1 month after induction of transformation, may instead point to a slight selective growth advantage for cells with few or no T-type Ca^{2+} channels, which would then permit them to outgrow the rest of the population over long periods. A direct effect of dexamethasone on T-channel expression seems unlikely because such effects would be expected to happen rapidly (within hours) and be manifest after a week, at a time when the T-channel density is still unchanged.

The observed suppression of Ca^{2+}-channel activity in transformed fibroblasts runs contrary to the generally accepted positive correlation between Ca^{2+} and cell growth and may point to a negative feedback mechanism for the control of cell growth by Ca^{2+}. Several explanations for the unexpected selective elimination of T-type Ca^{2+} current in transformed fibroblasts are worth considering:
1. Ca^{2+} influx through T-type channels may be directly or indirectly coupled to one of the normal regulatory mechanisms preventing uncontrolled growth, such as contact inhibition or anchorage dependence.
2. T-type channels may depend for their correct insertion and functional expression on elements of the cytoskeleton or other membrane components which are altered during morphological transformation.

Our results differ in an important respect from the only other published study (Caffrey et al. 1987) of a correlation between oncogene tranformation and channel expression. Caffrey et al. (1987) found that in a muscle cell line, Ca^{2+} channels were only expressed when the cells were allowed to differentiate into muscle cells. Prevention of this differentiation by transforming oncogenes also prevents channel expression. In contrast, our results show that L-type Ca^{2+} channels can be expressed even in fully transformed nonexcitable cells, and that different types of Ca^{2+} channels can be affected differentially by oncogenic transformation.

Whatever the reason for the selective elimination of the T-type Ca^{2+} current by cell transformation, our results point out new differences between the two Ca^{2+}-channel types and further strengthen the view that each channel type is a separate structural and functional entity.

Acknowledgements. We would like to thank Drs. Deborah K. Morrison and Van Cherington for help in preparing the transformed cell lines, and Drs. Howard Green and Olaniyi Kehinde for advice and for letting us use their culture facilities.

References

Armstrong CM, Matteson DR (1985) Two distinct populations of calcium channels in a clonal line of pituitary cells. Science 227:65–67

Bean BP (1985) Two kinds of calcium channels in canine atrial cells. J Gen Physiol 86:1–30

Bean BP, Sturek M, Puga A, Hermsmeyer K (1986) Calcium channels in muscle cells isolated from rat mesenteric arteries: modulation by dihydropyridine drugs. Circ Res 59:229–235

Benham CD, Hess P, Tsien RW (1987) Two types of calcium channels in single smooth muscle cells from rabbit ear artery studied with whole-cell and single-channel recordings. Circ Res 61:I-10–I-16

Bishop JM (1987) The molecular genetics of cancer. Science 235:305–311

Bolen JB, Thiele CJ, Israel MA, Yonemoto W, Lipsich LA, Brugge JS (1984) Enhancement of cellular src gene product associated tyrosyl kinase activity following polyoma virus infection and transformation. Cell 38:767–777

Bonvallet R (1987) A low threshold calcium current recorded at physiological Ca concentration in single frog atrial cells. Pfluegers Arch 408:540–542

Bravo R, Burckhardt J, Curran T, Muller R (1985) Stimulation and inhibition of growth by EGF in different A431 cell clones is accompanied by the rapid induction of c-fos and c-myc proto-oncogenes. EMBO J 4:1193–1197

Caffrey JM, Brown AM, Schneider MD (1987) Mitogens and oncogenes can block the induction of specific voltage-gated ion channels. Science 236:570–573

Carbone E, Lux HD (1984) A low voltage activated, fully inactivating Ca channel in vertebrate sensory neurones. Nature 310:501–511

Carbone E, Lux HD (1987) Kinetics and selectivity of a low-voltage-activated calcium current in chick and rat sensory neurones. J Physiol 386:547–570

Cepko CL, Roberts BE, Mulligan RC (1984) Construction and applications of a highly transmissible murine retrovirus shuttle vector. Cell 37:1053–1062

Chen C, Hess P (1987) Calcium channels in mouse 3T3 and human fibroblasts. Biophys J 51:226a

Chen C, Corbley MJ, Roberts TM, Hess P (1988) Voltage-sensitive calcium channels in normal and transformed 3T3 fibroblasts. Science 239:1024–1026

Cherington V, Morgan B, Spiegelman B, Roberts T (1986) Recombinant retroviruses that transduce individual polyoma tumor antigens: Effects on growth and differentiation. Proc Natl Acad Sci USA 83:4307–4311

Cohen CJ, McCarthy RT (1987) Nimodipine block of calcium channels in rat anterior pituitary cells. J Physiol 387:195–225

Collett MS, Purchio AF, Erikson RL (1980) Avian sarcoma virus-transforming protein, pp60v-src, shows protein kinase activity specific for tyrosine. Nature 285:167–169

Courtneidge SA, Smith AE (1983) Polyoma virus transforming protein associates with the product of c-src cellular gene. Nature 303:435–438

Der CJ, Finkel T, Cooper GM (1986) Biological and biochemical properties of Human rasH genes mutated at codon 61. Cell 44:167–176

Dupont JL, Bossu JL, Feltz A (1986) Effect of internal calcium currents in rat sensory neurones. Pflugers Archiv 406:433–435

Fedulova SA, Kostyuk PG, Veselovsky NS (1975) Two types of calcium channels in the somatic membrane of new-born rat dorsal root ganglion neurones. J Physiol 359:431–446

Fujita J, Srivastava SK, Kraus MH, Rhim JS, Tronick SR, Aaronson SA (1985) Frequency of molecular alterations affecting ras proto-oncogenes in human urinary tract tumors. Proc Natl Acad Sci USA 82:3849–3853

Hamill OP, Marty A, Neher E, Sakmann B, Sigworth FJ (1981) Improved patch-clamp techniques for high-resolution current recording from cells and cell-free membrane patches. Pflugers Arch 391:85–100

Henkart M, Nelson P (1979) Evidence for an intracellular calcium store releasable by surface stimuli in fibroblasts (L cells). J Gen Physiol 73:655–673

Hesketh T, Moore JP, Morris JD, Taylor MV, Rogers J (1985) A common sequence of calcium and pH signals in the mitogenic stimulation of eukaryotic cells. Nature 313:481–484

Hess P, Lansman JB, Tsien RW (1984) Different modes of Ca channel gating behaviour favoured by dihydropyridine Ca agonists and antagonists. Nature 311:538–544

Hess P, Lansman JB, Tsien RW (1986) Calcium channel selectivity for divalent and monovalent cations. J Gen Physiol 88:293–319

Hof RP, Ruegg UT, Hof A, Vogel A (1985) Stereoselectivity at the calcium channel: opposite action of the enantiomers of a 1,4-dihydropyridine. J Cardiovasc Pharmacol 7:689–693

Huang AL, Ostrowski MC, Berard D, Hager GL (1981) Glucocorticoid regulation of the Ha-MuSV p21 gene conferred by sequences from mouse mammary tumor virus. Cell 27:245–255

Hunter T, Sefton BM (1980) Transforming gene product of Rous sarcoma virus phosphorylates tyrosine. Proc Natl Acad Sci USA 77:1311–1315

Kaplan DR, Whitman M, Shaffhausen B, Raptis L, Garcea RL, Pallas D, Roberts TM, Cantley L (1986) Phosphatidylinositol metabolism and polyoma-mediated transformation. Proc Natl Acad Sci USA 83:3624–3628

Kokubun S, Reuter H (1984) Dihydropyridine derivatives prolong the open state of Ca channels in cultured cardiac cells. Proc Natl Acad Sci USA 81:4824–4827

Kokubun S, Prod'hom B, Becker C, Prozig H, Reuter H (1986) Studies on Ca channels in intact cardiac cells: voltage-dependent effects and cooperative interactions of dihydropyridine enantiomers. Mol Pharmacol 30:571–584

Korman AJ, Frantz JD, Strominger JL, Mulligan RC (1987) Expression of human class II major histocompatibility complex antigens using retrovirus vectors. Proc Natl Acad Sci USA 84:2150–2154

Land H, Parada LF, Weinberg RA (1983) Cellular Oncogenes and Multistep Carcinogenesis. Science 222:771–778

Levinson AD, Oppermann H, Varmus HE, Bishop JM (1980) The purified product of the transforming gene of Avian Sarcoma virus phosphorylates tyrosine. J Biol Chem 11973:11980–

Lopez-Rivas A, Rozengurt E (1983) Serum rapidly mobilizes calcium from an intracellular pool in quiescent fibroblastic cells. Biochem Biophys Res Comm 114:240–247

Lovisolo D, Alloatti G, Bonelli G, Tessitore L, Baccino FM (1988) Potassium and calcium currents and action potentials in mouse Balb/c 3T3 fibroblasts. Pfluegers Arch in press

Mitra R, Morad M (1986) Two types of calcium channels in guinea-pig ventricular myocytes. Proc Natl Acad Sci USA 83:5340–5344

Mix LL, Dinerstein RJ, Villereal ML (1984) Mitogens and melittin stimulate an increase in intracellular free calcium concentration in human fibroblasts. Biochem Biophys Res Commun 119:69–75

Moolenaar WH, Tertoolen LG, de Laat SW (1984) Growth factors immediately raise cytoplasmic free Ca^{2+} in human fibroblasts. J Biol Chem 259:8066–8069

Morgan JI, Curran T (1986) Role of ion flux in the control of c-fos expression. Nature 322:552–555

Nilius B, Hess P, Lansman JB, Tsien RW (1985) A novel type of cadiac calcium channel in ventricular cells. Nature 316:443–446

Nowycky MC, Fox AP, Tsien RW (1985) Three types of neuronal calcium channel with different calcium agonist sensitivity. Nature 316:440–443

Okada Y, Tsuchiya W, Yada T (1982) Calcium channel and calcium pump involved in oscillatory hyperpolarizing responses of L-strain mouse fibroblasts. J Physiol 327:449–461

Peres A, Zippel R, Sturani E, Mostacciuolo G (1988) A voltage dependent calcium current in mouse Swiss 3T3 fibroblasts. Pfluegers Arch in press

Pouyssegur J, Jacques Y, Lazdunski M (1980) Identification of a tetrodotoxin sensitive Na^+-channel in a variety of fibroblast lines. Nature 286:162–164

Raptis L, Lamfrom H, Benjamin T (1985) Regulation of cellular phenotype and expression of polyomavirus Middle T antigen in rat fibroblasts. Mol Cell Biol 5:2476–2485

Rettenmier CW, Jackowski S, Rock CO, Roussel MF, Sherr CJ (1987) Transformation by the v-fms oncogene product: An analog of the CSF-1 receptor. J Cell Biochem 33:109–115

Shih TY, Papageorge AG, Stokes PE, Weeks MO, Scolnick EM (1980) Guanine nucleotide-binding and autophosphorylating activities associated with the p21src protein of Harvey murine sarcoma virus. Nature 287:686–691

Tabin CJ, Bradley SM, Bargmann CI, Weinberg RA, Papageorge AG, Scolnick EM, Dhar R, Lowy DR, Chang EH (1982) Mechanism of activation of a human oncogene. Nature 300:143–149

Taparowsky E, Suard Y, Fasano O, Shimizu K, Goldfarb M, Wigler M (1982) Activation of the T24 bladder carcinoma transforming gene is linked to a single amino acid change. Nature 300:762–765

Todaro GJ, Green H, Goldberg BD (1964) Transformation of properties of an established cell line by SV40 and Polyoma virus. Proc Natl Acad Sci USA 51:66–73

Todaro GJ, Lazar GK, Green H (1965) The initiation of cell division in a contact-inhibited mammalian cell line. J Cell Comp Physiol 66:325–334

Weinberg RA (1985) The action of oncogenes in the cytoplasm and nucleus. Science 230:770–776

Yuasa Y, Srivastava SK, Dunn CY, Rhim JS, Reddy EP, Aaronson SA (1983) Acquisition of transforming properties by alternative point mutations within c-bas/has human proto-oncogene. Nature 303:775–779

Protein Phosphorylation and the Inactivation of Dihydropyridine-Sensitive Calcium Channels in Mammalian Pituitary Tumor Cells

D. Kalman[1] and D. L. Armstrong[2]

[1] Department of Biology, University of California, 405 Hilgard Avenue, Los Angeles, CA 90024, USA
[2] Laboratory of Cellular and Molecular Pharmacology, National Institute of Environmental Health Sciences, P.O. Box 12233, Research Triangle Park, NC 27709, USA

Introduction

We describe here the results of our recent experiments on the inactivation of the dihydropyridine-sensitive calcium channel in mammalian pituitary tumor cells from the GH_3 clonal line (Kalman et al. 1987a, b, 1988).

Structure-Function of Dihydropyridine-Sensitive Calcium Channel. The primary structure of the dihydropyridine-sensitive calcium channel protein that has been purified recently from vertebrate skeletal muscle and reconstituted in artificial membranes shows remarkable similarity to the voltage-activated sodium channel (Tanabe et al. 1987). Nevertheless, in cells that contain both these channels, they appear to serve very different functions. Sodium channels underlie electrical signalling, and the voltage-dependent inactivation of those channels permits high-frequency, unidirectional propagation of the action potential. In contrast, calcium channels appear to be involved primarily in chemical signalling by rapidly raising the intracellular concentration of calcium (Schlegel et al. 1987). From that point of view, their function might be better served if inactivation were regulated by calcium rather than depolarization.

Inactivation of Voltage-Activated Calcium Channels. In a variety of excitable cells from invertebrates (for review see Eckert and Chad 1984) and vertebrate cardiac muscle (Mentrard et al. 1984; Bechem and Pott 1985), the macroscopic calcium current appears to inactivate in a calcium-dependent manner. Thus, replacing calcium with barium as the charge carrier, injecting exogenous calcium buffers like EGTA into the cell, or depolarizing sufficiently close to the calcium equilibrium potential to reduce calcium influx all reduce inactivation. In contrast, none of the voltage-activated calcium channels observed with patch-clamp techniques on vertebrate neurons appear to inactivate in a calcium-dependent manner (for review see Miller 1987). Some channels inactivate in a clearly voltage-dependent manner, but the dihydropyridine-sensitive channels that are characterized by a larger conductance in barium, a higher activation threshold, and faster deactivation after repolarization showed little or no inactivation on a millisecond time scale. In the patch-clamp experiments, however, either barium was used as the charge carrier to resolve unitary currents or EGTA was included in the solution bathing the cytoplasmic side of the membrane, so any calcium-dependent processes would have been largely inhibited.

Dihydropyridine-sensitive calcium channels do exhibit a rapid loss of activity, or "run-down," that is accelerated by calcium in cell-free patches exposed to standard physiological saline solutions (Fenwick et al. 1982; Cavalie et al. 1983; Nilius et al. 1985). We have recently demonstrated that run-down can be prevented by cAMP-dependent protein phosphorylation in cell-free patches from GH_3 cells when proteolysis is inhibited by leupeptin (Armstrong and Eckert 1987; Chad et al. 1987). Because the calcium-inactivating current in molluscan neurons shows a similar phosphorylation dependence (Doroshenko et al. 1982, 1984; Eckert et al. 1985; Chad and Eckert 1986), we have reinvestigated the mechanism of inactivation of the dihydropyridine-sensitive calcium channels in GH_3 cells under conditions that optimize detection of calcium-dependent processes.

Results

Inactivation of Dihydropyridine-Sensitive Channels. Figure 1 illustrates the time course of the voltage-activated calcium current recorded with a patch pipette in the whole-cell configuration (Hamill et al. 1981). When GH_3 cells are dialyzed with a minimal saline solution containing 5 mM EGTA to buffer the concentration of calcium below 10 nM on the cytoplasmic side of the membrane, the pharmacologically isolated current elicited by depolarization from -40 to 0 mV (Fig. 1A) rapidly reaches a peak and remains constant throughout the step (Fig. 1B). In contrast, when EGTA is omitted from the intracellular solution the current inactivates with a half-time of approx. 10 ms to a steady-state level 40%–75% smaller than the peak (Fig. 1C). This relaxation cannot be attributed to activation of a voltage-dependent outward current. In the absence of EGTA, the peak current amplitude runs down quickly until no further current can be elicited by depolarization from -40 mV (Fig. 1C). The current elicited by depolarization from -40 mV is also blocked by 1 μM nimodipine when it is applied before run-down is complete (Fig. 1B). On average, less than 8% of the current elicited by depolarization from -40 mV is impervious to run-down or block by dihydropyridines. Thus, dihydropyridine-sensitive calcium channels in GH_3 cells, like the predominant voltage-activated calcium channels in a wide variety of excitable cells (Eckert and Chad 1984), inactivate rapidly when channel activation leads to calcium ion accumulation inside the cell.

Further evidence for a calcium-dependent mechanism is illustrated in Fig. 2. When barium is substituted for calcium as the charge carrier in the external solution (Fig. 2A), the current does not inactivate, even though EGTA is omitted from the internal solution. However, when the cell membrane is held at -80 mV between steps, depolarization evokes an additional transient component of inward current. Unlike the sustained current in barium, the transient component does not run down, but it inactivates almost completely at -40 mV. This transient component of the macroscopic calcium current has been interpreted as evidence for a second class of voltage-activated calcium channels that inactivate by a voltage-dependent mechanism (Armstrong and Matteson 1985) rather than a voltage-dependent component to inactiva-

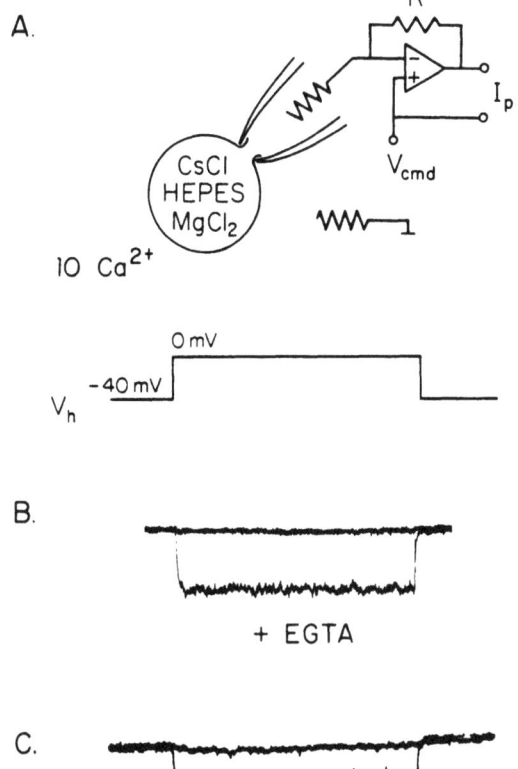

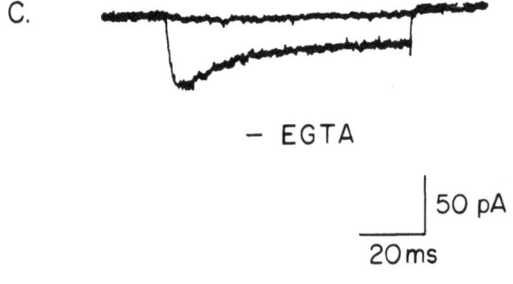

Fig. 1A–C. Calcium-current inactivation in GH_3 cells recorded with patch pipettes in the whole-cell configuration. **A** Recording configuration an voltage protocol. The cell was bathed in a solution containing 10 mM Ca^{2+}, 20 mM TEA, and 2 µM TTX to isolate pharmacologically the calcium current elicited by depolarization from a holding potential of −40 mV. **B** Calcium current recorded under such conditions when calcium in the internal solution was buffered at 10 nM with 5 mM EGTA. Addition of 1 µM nimodipine completely blocks the inward current and reveals no outward current activated by voltage. **C** Calcium current recorded without adding EGTA to the internal solution. In the absence of EGTA, the dihydropyridine-sensitive current inactivates with a time constant of approx. 10 ms to a steady-state level 50% smaller than the peak. In addition, the peak amplitude gradually decreases, or runs down, over 10 min (not shown) until no further inward current can be elicited by depolarization

tion of the dihydropyridine-sensitive channels. We have confirmed this hypothesis in GH_3 cells with single-channel recordings of the two classes (Armstrong and Eckert 1987; Armstrong and Kalman 1988).

In some cells, EGTA does not eliminate inactivation of the calcium current, which has been interpreted as evidence against a calcium-dependent mechanism (Matteson and Armstrong 1984; Bean 1985). We have also observed that EGTA does not completely suppress inactivation in all GH_3 cells; however, replacing external calcium with barium does eliminate inactivation of the dihydropyridine-sensitive channels (Fig. 2B). Thus, inactivation is specific to calcium. Furthermore, inactivation is blocked in all cells (e.g., Fig. 2C) when intracellular calcium is buffered with BAPTA, a more effective buffer of rapid calcium transients (Tsien 1980; Neher and Marty 1985). Therefore, we ascribe the inactivation observed in the presence of

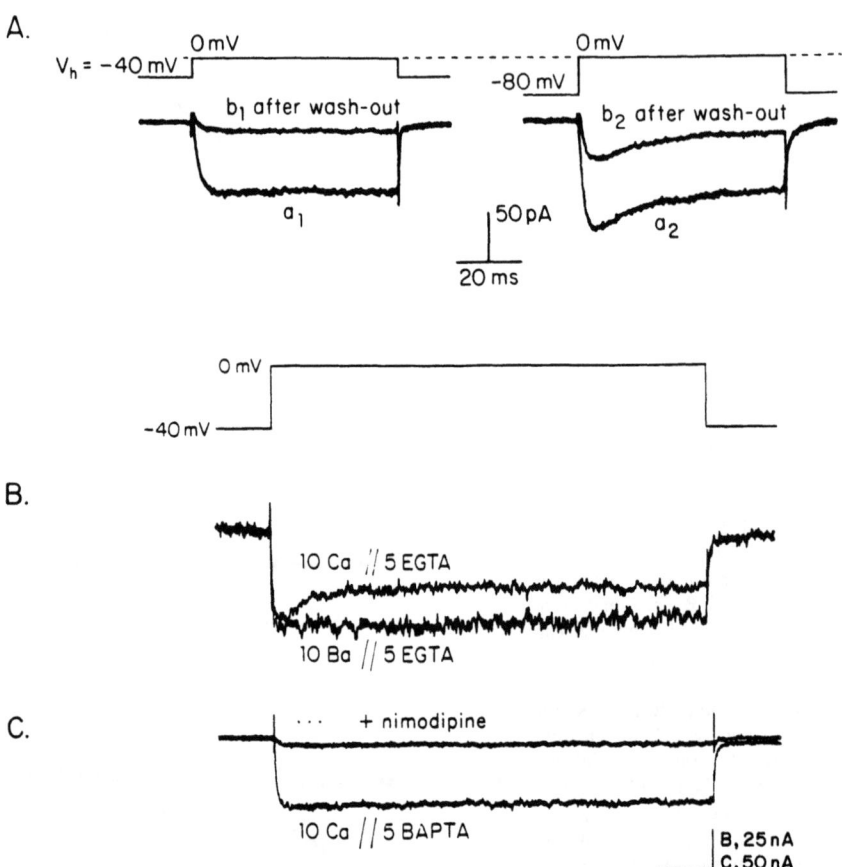

Fig. 2A–C. Inactivation of the dihydropyridine-sensitive current is a calcium-dependent process. **A** Although no EGTA has been added to the internal solution, the barium current elicited by depolarization from −40 mV (trace a^1) does not inactivate. Nevertheless, it runs down completely after several minutes (trace b^1). An additional transient component of inward barium current is elicited by depolarization from a more negative holding potential, −80 mV (trace a^2). This component does not run down (trace b^2) but inactivates completely at −40 mV (trace b^1). Thus, the sustained component in barium can be studied in isolation by holding at −40 mV. **B** In some cells, the calcium current elicited by depolarization from −40 mV inactivates despite the addition of EGTA to the internal solution; however, subsequently replacing calcium with barium as the charge carrier in the external solution does eliminate inactivation, so the process appears to be calcium specific. **C** When 5 mM BAPTA (Tsien 1980) is used to buffer calcium to 10 nM, inactivation of the dihydropyridine-sensitive calcium current is never observed

EGTA to its ineffectiveness as a calcium buffer on a millisecond time scale rather than to any intrinsic voltage dependence of the inactivation of dihydropyridine-sensitive channels.

Voltage Dependence of Inactivation. The analysis of calcium-channel inactivation has always been complicated by the voltage dependence of channel activation and,

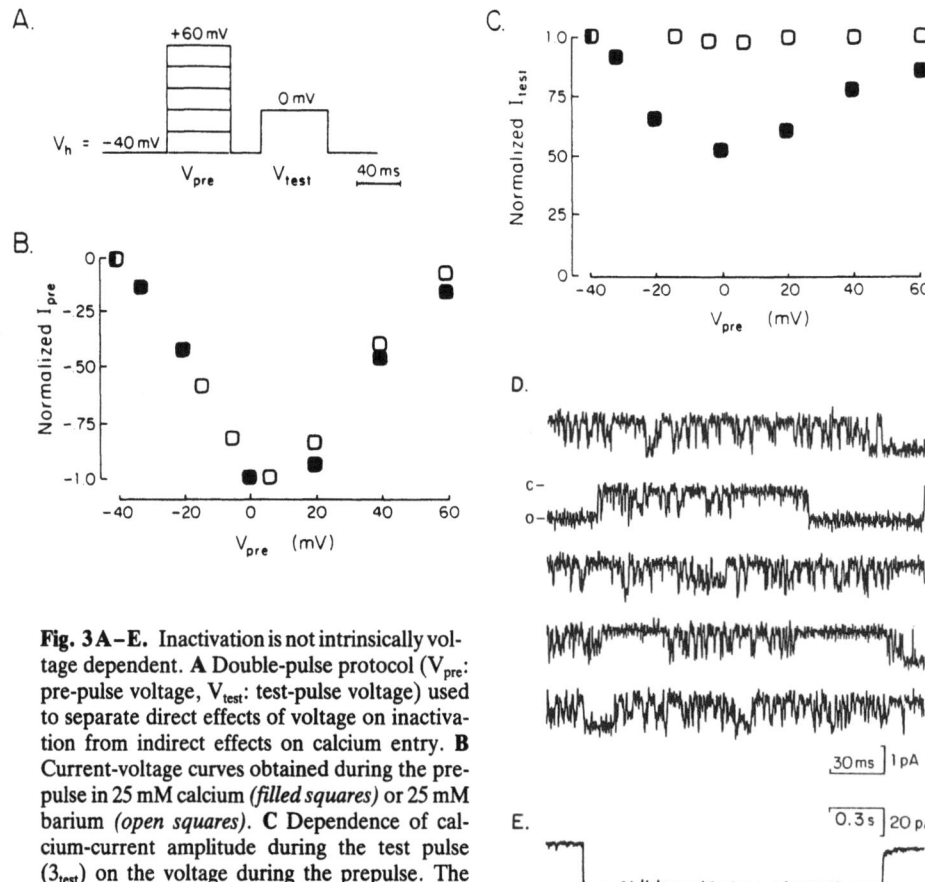

Fig. 3A–E. Inactivation is not intrinsically voltage dependent. **A** Double-pulse protocol (V_{pre}: pre-pulse voltage, V_{test}: test-pulse voltage) used to separate direct effects of voltage on inactivation from indirect effects on calcium entry. **B** Current-voltage curves obtained during the prepulse in 25 mM calcium *(filled squares)* or 25 mM barium *(open squares)*. **C** Dependence of calcium-current amplitude during the test pulse (3_{test}) on the voltage during the prepulse. The peak current during the test pulse is decreased in direct proportion to the amount of calcium entry during the prepulse but is unaffected by barium entry during the prepulse. **D** Five sequential records at 3-s intervals of unitary barium currents across a cell-free patch of membrane clamped to 0 mV (c closed, o open). Note that the channel does not inactivate under these conditions when run-down is prevented by adding reagents to support cAMP-dependent protein phosphorylation on the cytoplasmic side of the membrane (cf. Armstrong and Eckert 1987). **E** Whole-cell barium current elicited by depolarization from −40 to 0 mV for 1.5 s

therefore, calcium entry. The double-puls protocol illustrated in Fig. 3A was designed to separate any direct effects of voltage on inactivation from its indirect effect on calcium entry (Eckert and Tillotson 1981). As expected for a purely calcium-dependent process, the amplitude of the test pulse was reduced (Fig. 3C) in direct proportion to the amount of calcium entering the cell during the prepulse (Fig. 3B). Thus, inactivation only appears voltage-dependent to the extent that calcium entry depends on voltage. Additional evidence against an intrinsically voltage-dependent inactivation process was obtained by examining barium currents on a longer time scale (Fig. 3D, E). No inactivation of barium currents was observed over several seconds at 0 mV in both whole-cell recordings (Fig. 3E) and single-channel recordings from cell-

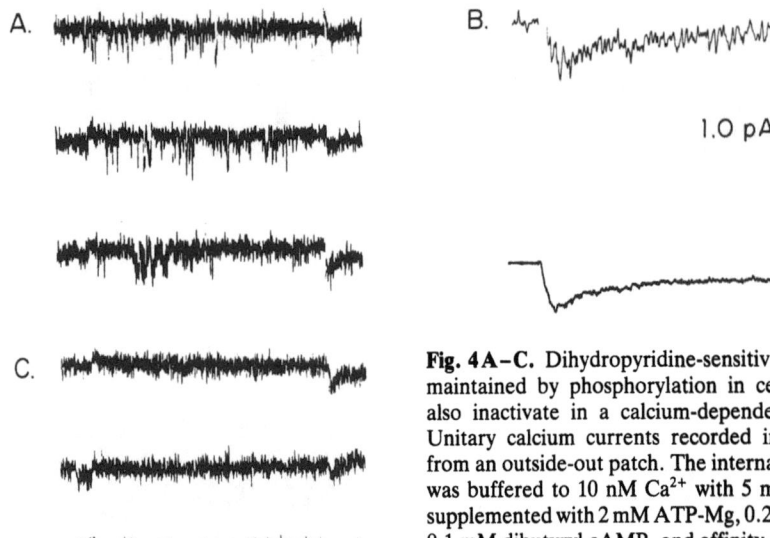

Fig. 4A–C. Dihydropyridine-sensitive channels maintained by phosphorylation in cell-free patches also inactivate in a calcium-dependent manner. **A** Unitary calcium currents recorded in 90 mM Ca^{2+} from an outside-out patch. The internal CsCl solution was buffered to 10 nM Ca^{2+} with 5 mM EGTA and supplemented with 2 mM ATP-Mg, 0.2 mM leupeptin, 0.1 mM dibutyryl cAMP, and affinity-purified catalytic subunit of the cAMP-dependent protein kinase (2 µg/ml). Unitary calcium currents are easily resolved. **B** An ensemble average of such activity *(upper trace)* from another outside-out patch buffered to 5 µM Ca^{2+} with 5 mM dibromo-BAPTA has the same time course as the whole-cell current in 90 mM Ca^{2+} recorded before patch excision *(lower trace)*. **C** Despite the presence of EGTA and reagents to support cAMP-dependent phosphorylation in the solution on the cytoplasmic side of the membrane of the patch in **A**, unitary calcium-current activity eventually ceases completely

free patches (Fig. 3D; see below). We did observe a slow inactivation of dihydropyridine-sensitive barium current over tens of seconds, like the effects in heart muscle cells reported by Cavalie et al. (1986); however, this effect is much too slow to contribute to the inactivation which occurs in calcium on a millisecond time scale.

Inactivation of Channels Maintained by Phosphorylation. Dihydropyridine-sensitive calcium-channel activity can be maintained by cAMP-dependent phosphorylation in cell-free patches of membrane when proteolysis is inhibited by EGTA and/or leupeptin (Armstrong and Eckert 1987; Chad et al. 1987). Under those conditions, unitary calcium currents can also be resolved (Fig. 4A). When the calcium at the cytoplasmic side of the membrane is buffered with 5 mM dibromo-BAPTA (pCA = 5.3), the ensemble average of activity (Fig. 4B, upper trace) closely resembles the inactivation of the whole-cell current recorded in 90 mM Ca^{2+} from the same cell before the outside-out patch was formed (Fig. 4B), lower trace). However, even when calcium is buffered below 10 nM with 5 mM EGTA or BAPTA, activity never lasts indefinitely in calcium (Fig. 4C).

Unlike run-down, this persistent loss of activity in calcium occurs despite the presence of exogenous reagents to support cAMP-dependent phosphorylation (Fig. 5B). Like inactivation, however, the loss is removed by substituting barium for calcium as the charge carrier (Fig. 5C). Thus, the loss of activity in calcium may reflect a persistent inactivation of the phosphorylation-dependent channels. Further evidence for that hypothesis has been obtained in studies of dihydropyridine action.

1. Barium

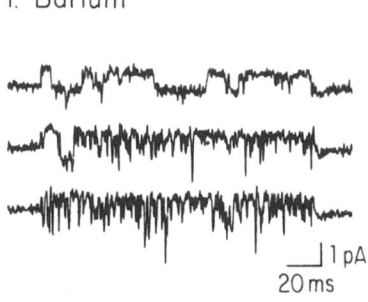

2. Calcium

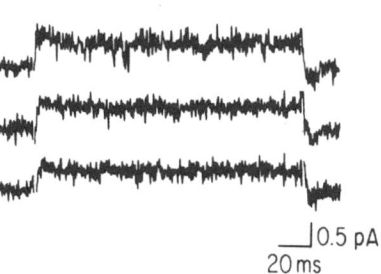

3. Barium

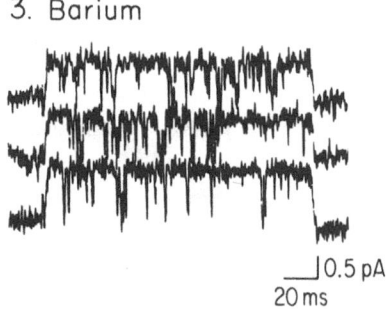

Fig. 5. Like inactivation of the whole-cell current, the persistent loss of phosphorylation-dependent activity in cell-free patches is removed by replacing calcium with barium as the charge carrier. Same conditions as in Fig. 4A

Removal of Inactivation by Bay K8644. We have observed that the dihydropyridine agonist Bay K8644 slows run-down significantly in cell-free patches (Armstrong and Eckert 1987; Chad et al. 1987), so we investigated the effects of Bay K8644 on inactivation (Fig. 6). Inactivation of both the whole-cell current (Fig. 6A) and the unitary currents in cell-free patches (Fig. 6B, bottom trace of ensemble acitivity) were blocked in the presence of 2 µM Bay K8644. In addition, Bay K8644 also prevented the persistent loss of activity in 90 mM Ca^{2+}.

Removal of Inactivation by cAMP-Dependent Phosphorylation. Figure 7 illustrates the effect on inactivation of stimulating adenylate cyclase in GH_3 cells with vasoactive intestinal peptide (VIP): the dihydropyridine-sensitive current elicited by depolarization from -40 mV inactivates more slowly. Initially, VIP also increases the amplitude of the peak current (not shown), but in dialyzed cells without EGTA present, rundown quickly resumes. This allows the rate of inactivation to be compared directly on currents of equal amplitude. Thus, the same initial influx of calcium is much less effective at promoting inactivation after stimulating the cell with VIP. Because both forskolin and exogenous dibutyryl cAMP produce similar effects at all voltages, we attribute the effects of VIP to raising the intracellular concentration of cAMP (Kalman et al. 1988).

Discussion

We have demonstrated that the inactivation of the dihydropyridine-sensitive calcium current in GH_3 cells, which occurs on a millisecond time scale in the absence of

A.

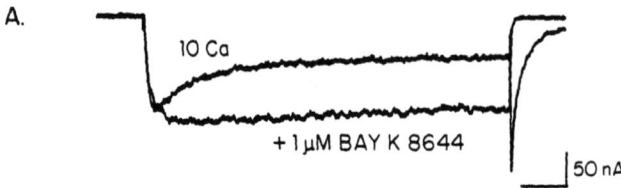

B.

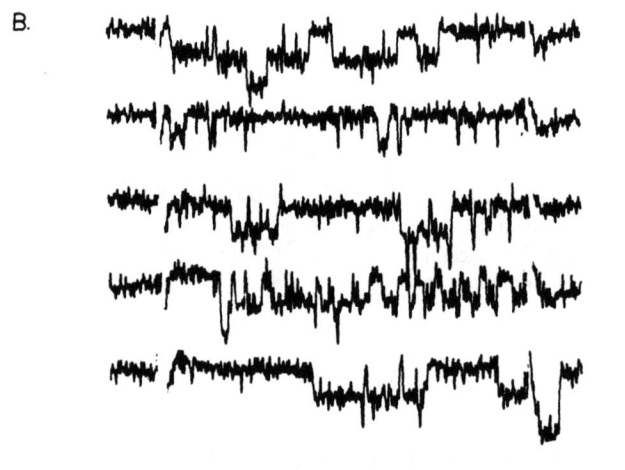

C.

Fig. 6A–C. Bay K8644 slows inactivation of calcium currents recorded in dialyzed cells (**A**) and cell free patches (**C**). Records were obtained as before in the presence or absence of 2 μM Bay K8644. **B** Representative traces of unitary calcium currents obtained 15 min after patch formation during steps from −40 to mV in 2μM Bay K8644. Note that Bay K8644 also prevents the persistent loss of activity in calcium

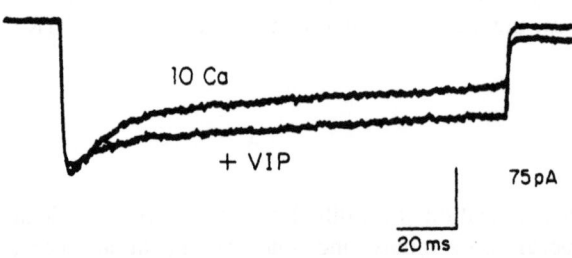

Fig. 7. Calcium-dependent inactivation slowed in the presence of 100 nM vasoactive intestinal peptide *(VIP)*. Note that the same peak calcium current produces much less inactivation in the presence of VIP

exogenous calcium buffers, is a purely calcium-dependent process. When contamination by other voltage- and calcium-activated channels is excluded, we find no evidence for any voltage-dependent processes on this time scale other than the voltage dependence of calcium entry through active channels. We suggest that calcium-dependent inactivation is a general property of dihydropyridine-sensitive channels in mammalian cells.

Enzymatic Mechanism for Calcium-Dependent Inactivation. In many invertebrate cell types, the channels that inactivate in a calcium-dependent manner are not selectively modulated by dihydropyridines; nevertheless, they closely resemble the dihydropyridine-sensitive channels in mammalian cells in many other respects, notably the run-down of activity that is accelerated by calcium in standard physiological saline solutions (Byerly and Hagiwara 1988). Eckert and Chad (1984) have proposed that calcium stimulates both inactivation and run-down by activating an endogenous phosphatase to dephosphorylate the calcium-channel protein or a closely associated regulatory molecule in the membrane. An alternative hypothesis proposed by Doroshenko et al. (1982), that intracellular accumulation of calcium reduces the phosphorylation-dependent calcium current by stimulating a cAMP phosphodiesterase, seems less likely after the demonstration that calcium-dependent inactivation can be reconstituted in molluscan neurons perfused internally with a constant excess (5 mM) of dibutyryl cAMP (Chad and Eckert 1986).

Several of our results on GH_3 cells are consistent with the Eckert and Chad hypothesis. Like the run-down in dialyzed molluscan neurons (Doroshenko et al. 1982, 1984; Chad and Eckert 1986), run-down in cell-free patches from GH_3 cells can be prevented by reagents that support cAMP-dependent protein phosphorylation (Armstrong and Eckert 1987). Similarly, stimulating cAMP production with VIP or forskolin in dialyzed cells increased the peak current but decreased the rate of inactivation considerably. This result clearly differs from the effect observed when the peak current is increased by other means, such as voltage or changes in the extracellular calcium concentration. In those cases, the larger calcium current inactivates more rapidly, presumably because a larger calcium influx results in a larger calcium transient on the cytoplasmic side of the membrane (Chad et al. 1984). Finally, the channels maintained by cAMP-dependent phosphorylation in cell-free patches reversibly lose activity when calcium carries the current instead of barium. Thus, the loss of activity in calcium may reflect a persistent inactivation of the phoshorylation-dependent channels; however, as noted above, neither EGTA nor BAPTA prevented the persistent loss of activity in calcium. Therefore, if the loss results solely from calcium-dependent inactivation, one must postulate that the buffers are saturated continually by the calcium flux through active channels or the leak when 90 mM calcium is used as the charge carrier.

GH_3 cells do contain a calcium-dependent protein phosphatase bound to the membrane (Wolff et al. 1987) that could dephosphorylate the dihydropyridine-sensitive channels. Calcineurin, the major calmodulin-binding protein in mammalian brain (Klee et al. 1979), has been demonstrated to be a calcium-activated phosphatase, IIb (Stewart et al. 1982). Calcineurin dephosphorylates the mammalian dihydropyridine-binding protein in vitro (Hosey et al. 1986), and it accelerates inactivation of the phosphorylation-dependent calcium current in molluscan neurons in a calcium-dependent manner (Chad and Eckert 1986). Its basal, calcium-independent activity may also be responsible for the loss of activity we observe in cell-free patches in the absence of ATP (Armstrong and Eckert 1987) even though we buffer the calcium ion concentration below 10 nM on the cytoplasmic side of the patch. The dephosphorylation of one channel every few minutes by a calcium-dependent phosphatase in the absence of calcium is not inconsistent with the dephosphorylation of hundreds of channels every few milliseconds when macroscopic calcium currents raise intracellular calcium levels in the absence of exogenous buffers. Nevertheless, in the absence of a specific inhibitor of the enzyme in situ, we cannot exclude the alternative that calcium binds directly to the channel at a site that is modified allosterically by phosphorylation. Other calcium-independent phosphatases have been reported to decrease the phosphorylation-dependent current in mammalian heart cells (Kameyama et al. 1986; Hescheler et al. 1987).

Implications for Mechanism of Dihydropyridine Action. We have observed that Bay K8644 does not stimulate the activity of dihydropyridine-sensitive channels in cell-free patches from GH_3 cells when it is applied in the absence of reagents to support protein phosphorylation (Armstrong and Eckert 1987; Chad et al. 1987). On the other hand, calcium channels modulated by Bay K8644 run-down more slowly in those solutions. Therefore, we have postulated that Bay K8644 binds preferentially to the phosphorylated state of the channel and protects it from dephosphorylation (Armstrong et al. 1987; Armstrong 1988). If inactivation of the dihydropyridine-sensitive channels were caused by dephosphorylation of the channel, one would predict that Bay K8644 should also slow the rate of calcium-dependent inactivation, and that is what we observe in both dialyzed cells and cell-free patches (Fig. 6). This result leads directly to the speculation that dihydropyridine antagonists inhibit channel activity by binding with higher affinity to the dephosphorylated channel and stabilizing the inactivated state (Armstrong 1988).

Implications for Calcium Regulation. In summary, the dependence of dihydropyridine-sensitive channel activity on protein phosphorylation and dephosphorylation reactions provides a simple framework for understanding calcium-channel modulation by endogenous neurotransmitters, fluctuating levels of intracellular calcium, and clinically important drugs like the dihydropyridines. In addition, the calcium dependence of inactivation seems well suited to the apparent role of these voltage-activated channels in regulating the intracellular concentration of calcium.

Summary

The inactivation of dihydropyridine-sensitive calcium channels has been studied in rat pituitary tumor cells (GH_3) voltage-clamped with patch pipettes at a holding potential of -40 mV to minimize contamination by other voltage-activated calcium channels. When exogenous calcium buffers are omitted from the solution bathing the cytoplasmic side of the membrane, the calcium current recorded at 0 mV inactivates with a half-time of ~ 10 ms to a steady-state level 40%–75% smaller than the peak. In double-pulse experiments, inactivation showed the same voltage dependence as calcium entry.

Substituting barium for calcium as the charge carrier in the external solution, reducing the rise in intracellular calcium with exogenous buffers, or adding Bay K8644 all blocked inactivation. No evidence was obtained for an intrinsically voltage-dependent component of inactivation on a millisecond time scale. The dihydropyridine-sensitive channels maintained by cAMP-dependent protein phosphorylation in cell-free patches also show calcium-dependent inactivation. Conversely, raising cAMP levels in dialyzed cells slows calcium-dependent inactivation of calcium current. Thus, our results support the hypothesis that inactivation results from dephosphorylation of the dihydropyridine-sensitive channels by an endogenous, calcium-activated phosphatase.

References

Armstrong CM, Matteson DR (1985) Two distinct populations of calcium channels in a clonal line of pituitary cell. Science 227:65–676

Armstrong D (1988) Calcium channel regulation by protein phosphorylation in a mammalian tumor cell line. Biomed Res, in press

Armstrong D, Eckert R (1987) Voltage activated calcium channels that must be phosphorylated to respond to membrane depolarization. Proc Natl Acad Sci USA 84:2818–2522

Armstrong D, Erxleben C, Kalman D (1987) Calcium channels modulated by Bay K8644 appear less susceptible to dephosphorylation. Biophys J 51:233a

Armstrong D, Kalman D (1987) The role of protein phosphorylation in the response of dihydropyridine-sensitive calcium channels to membrane depolarization in mammalian pituitary tumor cells. In: Grinnell A, Armstrong D, Jackson M (eds) Calcium and ion channel modulation. Plenum New York pp 215–227

Bean PB (1985) Two kinds of calcium channels in canine artrial cells. J Gen Physiol 86:1–30

Bechem M, Pott L (1985) Removal of Ca current inactivation in dialysed guinea pig atrial cardioballs by calcium chelators. Pfluegers Archiv 404:10–20

Byerly L, Hagiwara S (1987) Calcium channel diversity. In: Grinell A, Armstrong D, Jackson M (eds) Calcium and ion channel modulation. Plenum, New York pp 3–17

Cavalie A, Ochi R, Pelzer D, Trautwein W (1983) Elementary currents through Ca^{2+} channels in guinea pig myocytes. Pfluegers Archiv 398:284–297

Cavalie A, Pelzer D, Trautwein W (1986) Fast and slow gating behavior of single calcium channels in cardiac cells. Pfluegers Archiv 406:241–258

Chad JE, Eckert R (1986) An enzymatic mechanism for calcium current inactivation in dialyzed *Helix* neurons. J Physiol (Lond) 378:31–51

Chad JE, Eckert R, Ewald D (1984) Kinetics of Ca-dependent inactivation of calcium current in neurones of *Aplysia californica*. J Physiol (Lond) 347:279–300

Chad JE, Kalman D, Armstrong D (1987) The role of cAMP dependent phosphorylation in the maintenance and modulation of voltage activated calcium channels. In: Eaton DC, Mandel LJ (eds) Cell calcium and the control of membrane transport. The Rockefeller University Press, New York pp 167–186 (Society of General Physiologists Series, vol 42)

Doroshenko PA, Kostyuk PG, Martynyuk AI (1982) Intracellular metabolism of adenosine 3' −5'-cyclic monophosphate and calcium inward current in perfused neurones of *Helix pomatia*. Neuroscience 7:2125–2134

Doroshenko PA, Kostyuk PG, Martynyuk AI, Kursky MD, Vorobetz ZD (1984) Intracellular protein kinase and calcium inward currents in perfused neurones of the snail *Helix pomatia*. Neuroscience 11:263–267

Eckert R, Chad JE (1984) Inactivation of calcium channels. Progr Biophys Molec Biol 44:215–267

Eckert R, Chad JE, Kalman D (1986) Enzymatic regulation of the calcium current in dialyzed and intact molluscan neurones of *Aplysia californica*. J Physiol (Paris) 81:318–324

Eckert R, Tillotson D (1981) Calcium-mediated inactivation of the calcium conductance in caesium-loaded giant neurones of *Aplysia californica*. J Physiol (Lond) 314:265–280

Farber LH, Wilson FJ, Wolff DJ (1987) Calmodulin-dependent phosphatases of PC12, GH$_3$ and C$_6$ cells: physical, kinetic and immunochemical properties. J Neurochem 49:404–414

Fenwick EM, Marty A, Neher E (1982) Sodium and calcium channels in bovine chromaffin cells. J Physiol (Lond) 331:599–635

Hamill OP, Marty A, Neher E, Sakmann B, Sigworth FJ (1981) Improved patch clamp techniques for high-resolution current recording from cells and cell-free membrane patches. Pfluegers Archiv 398:284–297

Hescheler J, Kameyama M, Trautwein W, Mieskes G, Soling HD (1987) Regulation of the cardiac calcium channel by protein phosphatases. Eur J Biochem 165:261–266

Hosey MM, Borsetto M, Lazdunski M (1986) Phosphorylation and dephosphorylation of dihydropyridine-sensitive voltage-dependent calcium channel in skeletal muscle membranes by cAMP- and Ca-dependent processes. Proc Natl Acad Sci USA 83:3733–3737

Kalman D, Erxleben C, Armstrong D (1987a) Inactivation of the dihydropyridine-sensitive calcium current in GH$_3$ cells is a calcium-dependent process. Biophys J 51:432a

Kalman D, O'Lague PH, Armstrong D (1987b) Increasing the intracellular concentration of cAMP reduces Ca-dependent inactivation of Ca channels. Soc Neurosci Abs 13:104

Kalman D, O'Lague P, Erxleben C, Armstrong D (1988) Calcium-dependent inactivation of the dihydropyridine-sensitive calcium channels in GH$_3$ cells. J Gen Physiol, accepted for publication

Kameyama M, Hescheler J, Mieskes G, Trautwein W (1986) The protein-specific phosphatase I antagonizes the β-adrenergic increase of the cardiac Ca current. Pfluegers Archiv 407:461–463

Klee CB, Crouch TH, Krinks MH (1979) Calcineurin: a calcium- and calmodulin-binding protein of the nervous system. Proc Natl Acad Sci USA 76:6270–6273

Matteson DR, Armstrong CM (1984) Na and Ca channels in a transformed line of anterior pituitary cells. J Gen Physiol 83:371–394

Mentrard D, Vassort G, Fischmeister R (1984) Calcium-mediated inactivation of the calcium conductance in cesium-loaded frog heart cells. J Gen Physiol 83:105–131

Miller RJ (1987) Calcium channels in neurons. In: Venter JC, Triggle D (eds) Structure and Physiology of the slow inward calcium channel. Liss, New York pp 161–246

Neher E, Marty A (1985) BAPTA, unlike EGTA, efficiently suppresses Ca transients in chromaffin cells. Biophys J 47:278a

Nilius B, Hess P, Lansman JB, Tsien RW (1985) A novel type of cardiac calcium channel in ventricular cells. Nature 316:443–446

Schlegel W, Winiger BP, Mollard P, Vacher P, Wuarin F, Zahnd GR, Wollheim CB, Dufy B (1987) Oscillations of cytosolic Ca^{2+} in pituitary cells due to action potentials. Nature 329:719–721

Stewart AA, Ingbretsen TS, Manalan A, Klee CB, Cohen P (1982) Discovery of a Ca- and calmodulin-dependent protein phosphatase: probable identity with calcineurin (CaM-BP$_{80}$). FEBS Lett 137:80–84

Tanabe T, Takeshima H, Mikami A, Flockerzi V, Takahashi H, Kangawa K, Kojima M, Matsuo H, Hirose T, Numa S (1987) Primary structure of the receptor for calcium channel blockers from skeletal muscle. Nature 328:313–318

Tsien RY (1980) New calcium indicators and buffers with high selectivity against magnesium and protons: design, synthesis, and properties of prototype structures. Biochem 19:2396–2404

Sodium Currents Through Neuronal Calcium Channels: Kinetics and Sensitivity to Calcium Antagonists*

E. Carbone[1] and H. D. Lux[2]

[1] Dipartimento di Anatomia e Fisiologia Umana, Corso Raffaello 30, I-10125 Torino, Italy
[2] Max-Planck-Institut für Psychiatrie, Abteilung Neurophysiologie, Am Klopferspitz 18A, D-8033 Planegg, FRG

Introduction

In contrast to their physiological role, calcium channels of various tissues become highly permeable to Na^+ ions when the free Ca^{2+} concentration is lowered below micromolar levels (Kostyuk et al., 1983; Almers and McCleskey, 1984; Fukushima and Hagiwara, 1985; Hess et al., 1986; Carbone and Lux, 1987c). This feature might derive either from the presence of a regulatory site of high calcium specificity (Kostyuk et al., 1983; Kostyuk and Mironov, 1986; Lux and Carbone, 1987) whose occupation by Ca^{2+} inhibits Na^+ permeation or, alternatively, from the existence of two binding sites within the channel pore whose simultaneous occupation by Ca^{2+} promote Ca^{2+} flow by ion-ion interactions (Almers and McCleskey, 1984; Hess and Tsien, 1984). Recent observations based on single-channel measurements (Lux and Carbone, 1987; Carbone and Lux, 1987a) and rapid exchange of extracellular Ca^{2+} and H^+ concentration (Hablitz et al., 1986; Konnerth et al., 1987; Davies et al., 1988) have provided further support for the former view, i.e., calcium channels might possess two distinct permeability states for Ca^{2+} and Na^+ ions.

To obtain a more detailed picture of the gating properties and pharmacology of calcium channels in their Ca^{2+}- or Na^+-permeable mode, we will review here some recent findings on the blocking action of calcium antagonists on calcium and sodium currents flowing through two types of neuronal calcium channels. In particular, we will focus on the recent observation that the neuronal calcium-channel blocker ω-conotoxin (ω-CgTX) acts differently on calcium and sodium currents (Carbone and Lux, 1988), as if the toxin were able to distinguish different permeability states of calcium channels.

Types of Calcium Channel

Activation-Inactivation Gating, Permeability and Channel Density

Presently, there is overwhelming evidence for the existence of at least two classes of calcium channels in a number of neurons (Llinas and Yarom, 1981; Carbone and Lux,

---

* This work was partially supported by NATO (grant no. 0576/87) and the Consiglio Nazionale delle Ricerche (grant no. 87.00068.04).

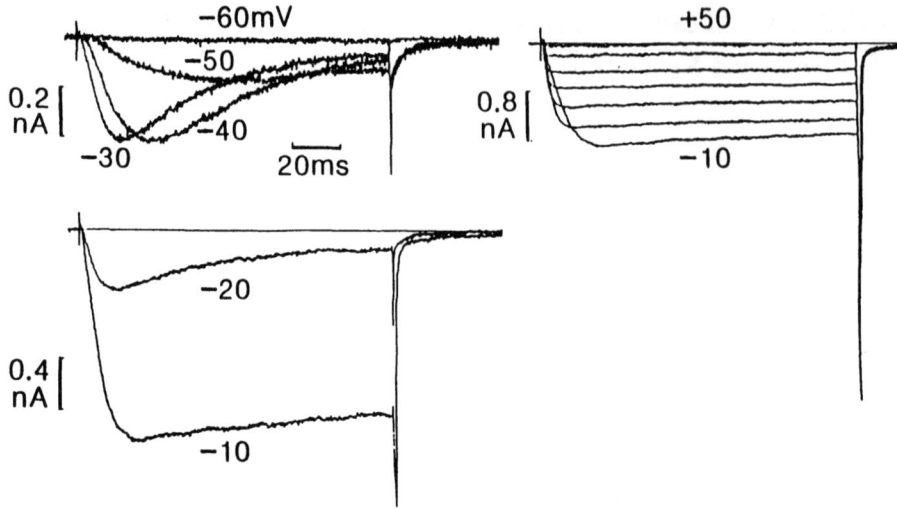

Fig. 1. Time course of calcium currents recorded from a freshly dissociated rat DRG neuron at the potentials indicated. Details on cell preparation, patch-clamp protocol, and pipette fabrication are given elsewhere (Hamill et al., 1981; Carbone and Lux, 1984a; 1986a; 1987b). Holding potential −80 mV. The bath contained (in mM): 120 Choline Cl, 5 CaCl$_2$, 2 glucose, 10 Na-Hepes (pH 7.3), and 3 μM TTX. The pipette filling solution was (in mM): 130 CsCl, 20 TEACl, 0.25 CaCl$_2$, 5 EGTA, 10 glucose, 10 Cs-Hepes (pH 7.3). (Modified from Carbone & Lux 1986b)

1984a, b; Fedulova et al., 1985; Bossu et al., 1985; Nowycky et al., 1985; Armstrong and Matteson, 1985). One class, low-voltage-activated (LVA, T, or low-threshold calcium channels, give origin to calcium currents which turn on between −60 and −30 mV, reach a peak around 10–50 ms, and quickly inactivate in a voltage-dependent manner (Fig. 1, top left). These channels are metabolically stable and remain functional for long periods of time in excised membrane patches (Carbone and Lux, 1984a, 1987c; Nilius et al., 1985; Armstrong and Eckert, 1987). They are present in central and peripheral neurons at a density of two to five channels/μm^2 and possess a single-channel conductance (γ) of 4–8 pS in 95 mM Ba^{2+} (Nowycky et al., 1985; Carbone and Lux, 1987c). The value of γ remains nearly unchanged when Ba^{2+} is replaced by Ca^{2+} or Sr^{2+}.

The second type of calcium channels, high-voltage-activated (HVA, L, or high-threshold), turn on between −20 and −10 mV, inactivate slowly, and their rate of inactivation appears to depend on calcium entry into the cell (Fig. 1, top right; see also Eckert and Chad, 1984). There is now strong evidence that the functional states of these channels are linked to metabolic inputs such as ATP, GTP, and cyclic nucleotides (Fedulova et al., 1985; Chad and Eckert, 1986). L channels are present in nearly all types of neurons at a density of 10–30 channels/μm^2 and possess a single-channel conductance of 15–20 pS in 95 mM Ba^{2+} which nearly halves when Ba^{2+} is replaced by Ca^{2+}.

In contrast to other cell types, peripheral and central neurons apparently possess a third type of calcium channels (N) which activate at high voltages (−10 to 0 mV) and differ from the L channels by their faster voltage-dependent inactivation and smaller single-channel conductance (10–15 pS in 95 mM Ba^{2+}; Nowycky et al., 1985). There is still some doubt, however, regarding the existence and properties of this channel. This is mainly due to (a) the absence of pharmacological tools acting selectively on N channels (see below), (b) the lack of clarity concerning the nature of N channel inactivation, whether voltage- or calcium-dependent, and (c) technical difficulties in measuring the activities of single calcium channels under near-physiological conditions for a sufficiently long period of time. Surprisingly, there is no strong support for the idea that the activity of this channel is somehow related to the macroscopic inactivation of HVA calcium currents.

Sensitivity to Drugs

In the last few years the pharmacology of neuronal calcium channels has shown considerable progress. There are presently several compounds in routine laboratory use which allow efficient separation of low-threshold (T) and high-threshold (L and N) calcium channels. No agent selective for the N channel has so far been found. Cd^{2+} ions (Fox et al., 1987) and the neurotoxic peptide ω-conotoxin (ω-CgTX, Feldman et al., 1987; McCleskey et al., 1987) specifically block HVA calcium channels with little effects on the LVA type. The opposite is true for Ni$^+$ (Fox et al., 1987; Carbone et al., 1987), phenytoin (Yaari et al., 1987), long-hydrocarbon-chain alcohols (Llinas and Yarom, 1986), and the potassium-sparing diuretic amiloride (Tang et al., 1988), which selectively block LVA channels in central and peripheral neurons. Noteworthy are the effects of menthol (a cyclic alcohol derived from peppermint oil) on calcium currents (Swandulla et al., 1987): it is shown to speed up the inactivation kinetics of HVA channels and to depress, in a dose-dependent manner, the amplitude of LVA calcium currents in vertebrate sensory neurons.

High-threshold calcium currents can be selectively depressed by verapamil (Boll and Lux, 1985; Fedulova et al., 1985), but are little affected by micromolar concentrations of dihydropyridine derivatives at normal resting potentials. Nifedipine inhibits high-threshold L-type calcium channels only after prolonged depolarizations positive to −40 mV (Rane et al., 1987). Bay K8644, reported to selectively affect HVA channels in cardiac cells, has proved ineffective when applied to chick sensory neurons (Swandulla and Armstrong, 1988).

Sodium Currents Through Calcium Channels

Kinetic Properties

Recent observations in mollusc neurons (Kostyuk et al., 1983), lymphocytes (Fukushima and Hagiwara, 1985), skeletal muscles (Almers and McCleskey, 1984) cardiac cells (Lansmann et al., 1986), and dorsal root ganglion (DRG) neurons (Carbone and Lux, 1987a, c) have shown that low- and high-threshold calcium currents become largely permeable to Na$^+$ when the external free Ca^{2+} concentration is reduced to submicromolar levels. Figure 2 shows the typical effects on LVA and

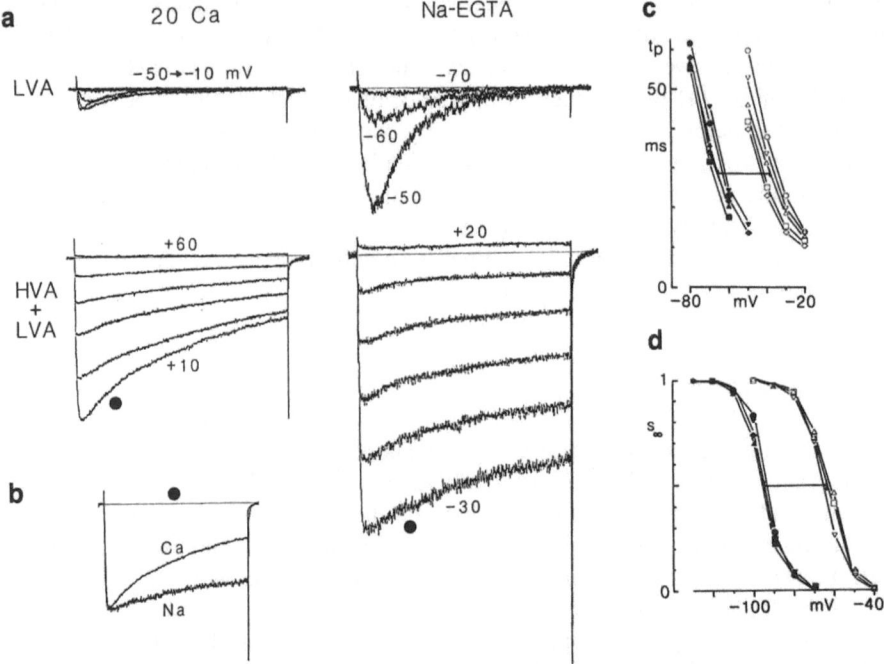

Fig. 2a–d. Time course of calcium and sodium currents through low- and high-threshold calcium channels in cultured chick DRG neurons. **a** LVA *(top)* and HVA *(bottom)* currents recorded before *(left)* and during *(right)* application of 140 mM Na$^+$ and 10^{-7} M Ca^{2+} (Na-EGTA, pCa 7). Step depolarizations were as indicated. Holding potential −90 mV. The internal pipette filling solution contained (in mM): 130 CsCl, 20 TEACl, 2 MgCl$_2$, 10 EGTA, 10 glucose, 10 Na-Hepes (pH 7.3). The bath composition was (in mM): 120 NaCl, 20 CaCl, 2 MgCl$_2$, 10 Hepes (pH 7.5) or 140 NaCl, 4 CaCl$_2$, 5 EGTA, 10 Na-Hepes (pH 7.5). **b** Comparison of HVA-current inactivation in normal-Ca^{2+} and low-Ca^{2+} *(Na-EGTA)* solutions. The two overlapped traces were taken from **a** *filled circles* and normalized to the peak amplitude. **c, d** Negative-voltage shift of the time to peak (t_p) and steady-state slow inactivation (s_∞) of LVA currents caused by the application of 10^{-7} M Ca^{2+} *(filled symbols)*. *Empty symbols* represent values obtained in 20 mM Ca^{2+}. Different symbols represent different cells. The time to peak was determined as described elsewhere (Carbone & Lux, 1987b). In **d** are reported the normalized peak currents plotted against the holding potential. Depolarizations were to −60 mV (20 mM Ca^{2+}) and −30 mV (10^{-7}) M Ca^{2+}). (Modified from Carbone & Lux, 1988)

HVA currents in chick sensory neurons induced by lowering external Ca^{2+} concentration (see also Carbone and Lux, 1988). Lowering external Ca^{2+} from 20 to 0.1 μM causes:

1. a two-to fourfold increase in LVA and HVA peak current amplitude (Fig. 2a),
2. a −35 mV voltage shift of the voltage-dependent kinetic parameters (Fig. 2c, d), and
3. a reduction of the current reversal potential.

As suggested by observations based on single-channel recordings (Lansmann et al., 1986; Levi and DeFelice, 1986; Carbone and Lux, 1987c), the increased current amplitude at low Ca^{2+} is mainly a consequence of the increased permeability of both types of calcium channels to Na⁺ ions. At low external Ca^{2+}, the unitary current amplitude of single LVA channels in excised patches of DRG neurons is about 5 times larger than at 20 mM Ca^{2+} (Carbone and Lux, 1987c). Similar findings are reported for the high-threshold L channel of beating (Levi and DeFelice, 1986) and dihydropyridine-treated (Lansmann et al.,1986) heart cells.

As shown in Fig. 2a, the time course of HVA-channel inactivation at low external Ca^{2+} is somewhat lengthened compared to that at 20 mM Ca^{2+} (Lux et al., 1987). This is consistent with the view that HVA-channel inactivation is slow and partly related to the rate of calcium entry into the cell. This is evident from Fig. 2b, where the HVA calcium current recorded at +10 mV is compared after normalization of the peak amplitude to the corresponding HVA sodium current recorded at -30 mV. The time constant of HVA-channel inactivation increases about twofold (from 98 to 188 ms) when external Ca^{2+} is replaced by Na⁺. From the present data, however, the possibility cannot be ruled out that lengthening of HVA-channel inactivation at low Ca^{2+} reflects structural changes of the channel. Recent observations using "caged" Ca^{2+} in DRG neurons suggest that Na⁺ ions affect the Ca^{2+}-induced inactivation of HVA channels by either interfering with the access of Ca^{2+} to an internal regulatory site or inducing changes in the channel structure (Morad et al., 1988).

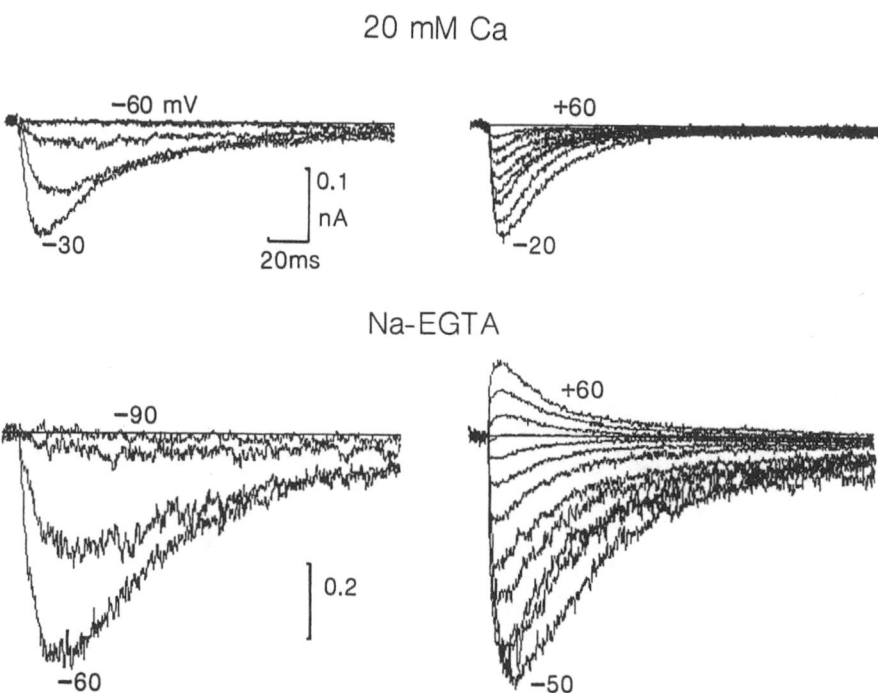

Fig. 3. Calcium and sodium currents through LVA channels in a chick DRG neuron. The calcium currents *(top)* were recorded in 20 mM Ca^{2+} at the membrane potential indicated. Depolarizations were separated by steps of 10 mV. The sodium currents *(bottom)* were recorded in Na-EGTA (pCa 8), starting from -90 mV. The cell was not treated with any drug and apparently possessed very few HVA channels. Holding potential -100 mV

In contrast to the situation in HVA channels, the speed of current inactivation in LVA channels at low external Ca^{2+} is comparable to that at normal Ca^{2+}, reinforcing the view that LVA-channel inactivation is strictly voltage-dependent (Carbone and Lux, 1984a, 1987b). Figure 3 shows more convincingly that LVA-channel gatings in chick DRG do not undergo drastic changes when lowering external Ca^{2+}. Activation and inactivation kinetics of LVA currents, recorded from a cell possessing a large proportion of LVA channels, appear little affected by lowering $[Ca^{2+}]$ from 20 mM (top) to 10^{-8} M (bottom, Na-EGTA, pCa 8). Sodium current inactivation remains fast and complete over a wide range of potentials (-50 to $+60$ mV) and independent of both the direction (bottom right of Fig. 3) and type (Fig. 4) of current carrying ion.

Sensitivity to Calcium Antagonists

LVA and HVA calcium channels apparently preserve the same pharmacological sensitivity to drugs independently of their Ca^{2+} or Na^+ permeability. This is shown in Fig. 5, which depicts the effects of verapamil and amiloride, two calcium antagonists selective for HVA and LVA channels respectively (Boll and Lux, 1985; Tang et al., 1988). As for calcium currents, application of 20 µM verapamil in chick DRG produces a large depression of HVA sodium currents (trace V in Fig. 5a, bottom) with little effect on the LVA current component (Fig. 5a, top). The same is true for amiloride (500 µM), which is shown to inhibit selectively the low-threshold calcium and sodium currents in neuroblastoma cells (Tang et al., 1988) and chick DRG

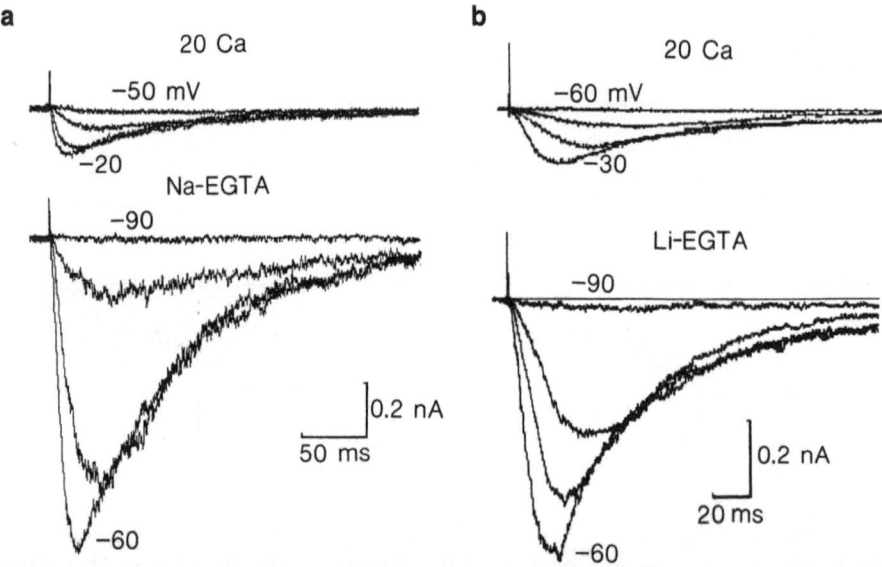

Fig. 4a, b. Time course of sodium (**a**) and lithium (**b**) currents through LVA channels in low-Ca^{2+} solutions recorded from two different chick DRG neurons. Membrane depolarizations were as indicated. The bath was 20 mM Ca^{2+} at control *(top traces)* and was changed to Na-EGTA or Li-EGTA (pCa 8). Holding potential -100 mV

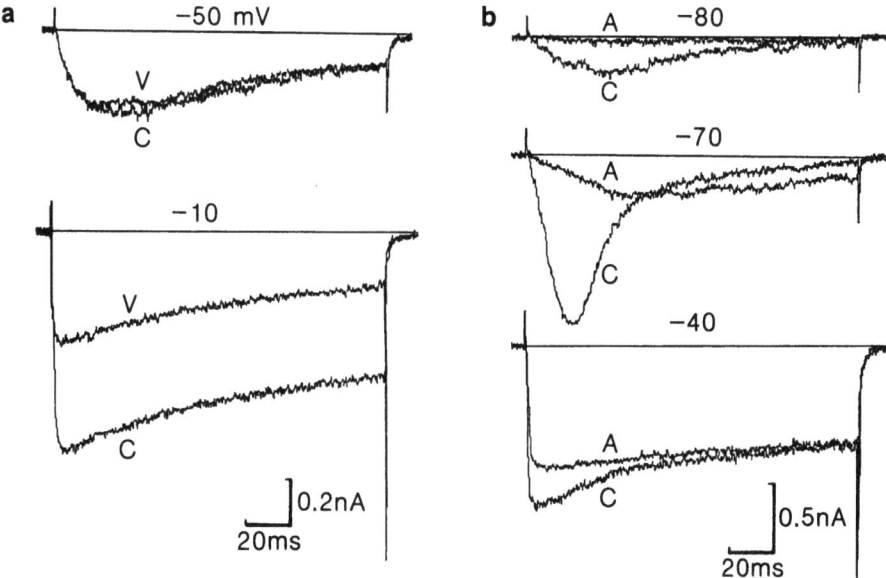

Fig. 5a, b. Effects of calcium antagonists on LVA and HVA sodium currents. **a** Sodium currents recorded before *(C)* and during *(V)* application of 20 µM verapamil to the bath at the membrane potential indicated. Note the marked depression of HVA sodium currents at −10 mV. **b** Sodium currents recorded at three different membrane potentials before *(C)* and during *(A)* application of 500 µM amiloride to the bath. Note the marked depression of LVA sodium currents at −80 and −70 mV and the disappearance of the transient LVA component at −40 mV. In both experiments the bath was Na-EGTA (pCa 8). Holding potential −90 mV. (From Carbone and Morad, unpublished)

(Carbone and Morad, unpublished). Figure 5b shows the action of the drug on LVA sodium currents of chick sensory neurons. Amiloride blocks almost fully the transient LVA sodium current at various membrane potentials (−80 to −40 mV) with little or no effect on the slowly inactivating sodium current activated at −40 mV.

Blockade of Calcium Channels by ω-Conotoxin is Mediated by the Permeant Ion

As pointed out elsewhere (Cruz and Olivera, 1986; Feldman et al., 1987; McCleskey et al., 1987), the blocking action of ω-CgTX on Ca currents deviates considerably from that of verapamil and other calcium antagonists. ω-CgTX blocks selectively and with high affinity the HVA calcium currents in a number of neurons. The action of the toxin, however, is slow. With 5µM of toxin the onset of the block occurs within several minutes and is nearly irreversible. Prolonged washing produces almost no recovery of persistently blocked calciums channels. Due to these features, ω-CgTX represents an ideal tool for isolation and biochemical purification of the high-threshold L-type calcium channel in neurons (Barhanin et al., 1988).

ω-CgTX shows another interesting property: its blocking action can be modulated by the type of permeant cation (Carbone and Lux, 1988). This is shown in Fig. 6.

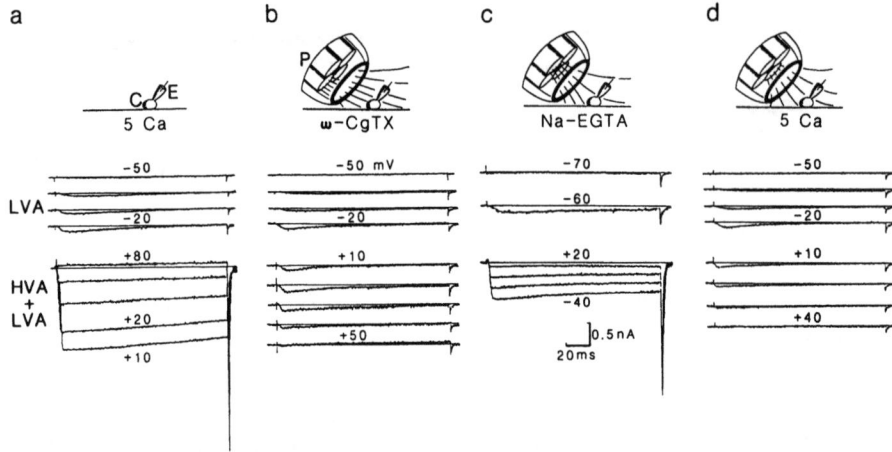

Fig. 6a–d. Relief of the ω-CgTX blockade in low-Ca^{2+} solutions. *Upper part:* Schematic representation of the experimental conditions. C, cell; E, patch-electrode; P, four-way perfusion pipette. *Lower part:* LVA and HVA currents recorded from a chick DRG cell in 5 mM Ca^{2+} **(a)**, after application of 5 µM ω-CgTX **(b)**, during washing with Na-EGTA (PCa 7) **(c)**, and on returning to 5 mM Ca^{2+} **(d)**. Step depolarizations as indicated. Holding potential −90 mV. (Modified from Carbone & Lux, 1988)

Application of 5 µM ω-CgTX to the bath inhibits HVA calcium currents in a dose-dependent manner with little change to the size and time course of LVA calcium currents (Fig. 6b). HVA-channel blockade by ω-CgTX is persistent and hardly recovers after washing for several hours. However, blockade of HVA channels is promptly relieved when calcium is replaced by a toxin-free low-Ca^{2+} solution (Fig. 6c). Despite the persistent block of the toxin, HVA sodium currents can be resumed at potentials positive to −40 mV. The size of these currents was found to vary from cell to cell and to be linked to the slow loss of calcium channels induced by cell dialysis (Fenwick et al., 1982; Cavalié et al., 1983).

The appearance of HVA sodium currents in ω-CgTX-treated cells is unlikely to be a consequence of unbinding of the toxin from its receptor site. On returning to normal-Ca^{2+} toxin-free solution (Fig. 6d), HVA calcium currents are shown to be blocked despite continuous washing. If toxin unbinding occurred during application of toxin-free solutions, at least a small fraction of HVA calcium currents should have been detected on returning to normal calcium conditions. This, however, was never observed.

Several line of evidence suggest that the appearance of HVA sodium currents is likely the consequence of a reversible modification of the channel-toxin complex and not the result of the turning on of some ω-CgTX-insensitive current or due to a direct interaction of EGTA molecules with calcium channels. The former possibility is ruled out by the finding (Fig. 7a) that ω-CgTX-insensitive sodium currents possess the same kinetics and pharmacology as HVA sodium currents recorded in non-toxin-treated cells (Fig. 2). The latter is contradicted by the following arguments: If the appearance

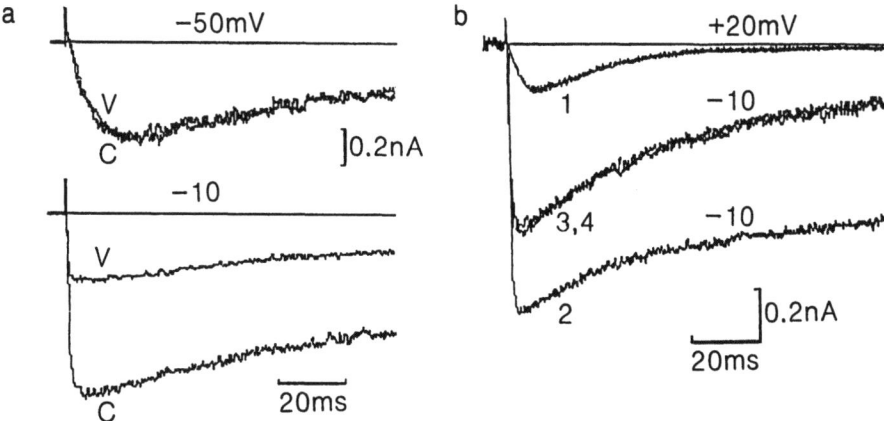

Fig. 7a, b. a Action of verapamil on ω-CgTX-resistant sodium currents in a chick DRG neuron. Sodium currents were recorded before *(C)* and during *(V)* application of 20 μM verapamil at the potential indicated in Na-EGTA (pCa 7). The cell was pretreated with 5 μM ω-CgTX in 5 mM Ca^{2+} and then washed with Na-EGTA (pCa 7) *(C)*. Notice the marked depression of HVA sodium currents at −10 mV *(V)* and the absence of any effect at −50 mV. Holding potential −90 mV. **b** Two sites of action of ω-CgTX on HVA channels. Trace *1* was recorded from a DRG cell treated with 10 μM ω-CgTX in 5 mM Ca^{2+}. HVA calcium currents were fully blocked under these conditions. Trace *2* represents the corresponding ω-CgTX-resistant sodium current, recorded after changing the bath to Na-EGTA (pCa 9) toxin-free solution. Traces *3* and *4* are sodium currents recorded 5 and 10 s after addition of 2 μM ω-CgTX at −10 mV. Holding potential −90 mV. (Modified from Carbone & Lux, 1988)

of toxin-resistant sodium currents were linked to the formation of a EGTA-ω-CgTX complex, the first order reaction

$$\text{EGTA} + \omega\text{-CgTX} \underset{k_{-a}}{\overset{k_a}{\rightleftarrows}} \text{EGTA} - \omega\text{-CgTX} \qquad (1)$$

predicts that sodium currents appear at a rate ([EGTA] $k_a + k_{-a}$) higher than that responsible for calcium-current block (k_{-a}). This, however, is at variance with our observation that ω-CgTX-resistant sodium currents develop at a much lower rate than that at which calcium currents are blocked.

Two Sites of Action of ω-Conotoxin

ω-Conotoxin-resistant sodium currents which appear during application of toxin-free Na-EGTA solutions can be quickly blocked by further additions of the toxin. Compared to the persistent action on HVA calcium currents, blocking of HVA Na currents occurs reversibly and faster ($K_d \simeq 0.5$ μM). The onset of sodium-current blockade develops exponentially, with time constants that decrease with increasing toxin concentration. The partial block of sodium currents reaches steady-state conditions in 30 s with 1 μM of toxin and nearly 10 s with 10 times more toxin. The offset of ω-CgTX inhibition also proceeds exponentially, but independently of the degree of sodium current blockade (τ_{off} about 34 s). The results depicted in Fig. 7b indicate that binding of ω-CgTX in low-Ca^{2+} solutions is likely to be independent of the site controlling the persistent block of the toxin, suggesting the existence of a second independent binding site for the toxin. HVA calcium currents are first blocked by 10 μM ω-CgTX (trace 1) to obtain complete and persistent saturation of the ω-CgTX receptors available in normal Ca^{2+}. Then the cell is washed with an Na-EGTA solution to evoke ω-CgTX-resistant sodium currents (trace 2) which are quickly depressed by addition of 2 μM ω-CgTX to the bath (traces 3, 4).

Interestingly, the ability of ω-CgTX to block sodium currents through HVA calcium channels allows the pharmacological dissection of LVA sodium currents over a wide range of membrane potentials (−80 to +70 mV). As shown in Fig. 8a, a strong depression of steady-state HVA sodium currents by ω-CgTX (5 μM) reveals LVA sodium currents of large amplitudes ($I_{max} \simeq 0.7$ nA) with strictly voltage-dependent kinetic and permeability properties (panels b, c).

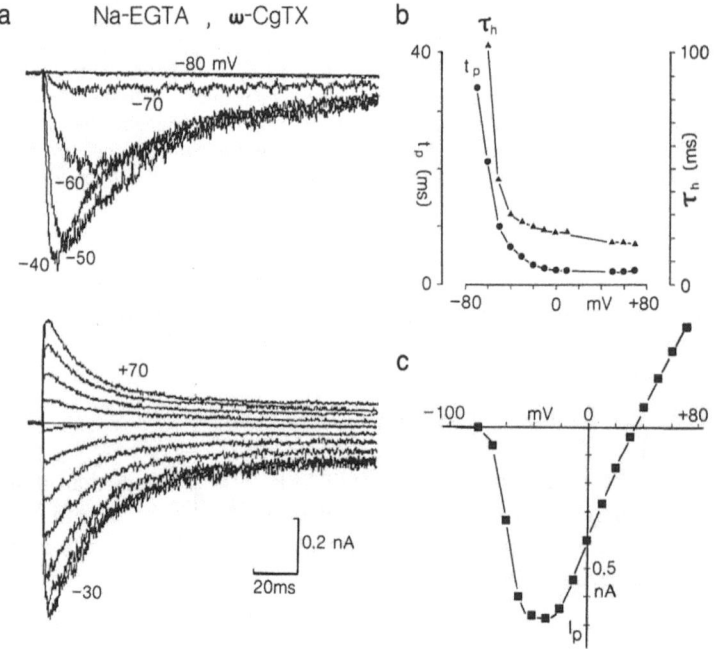

Fig. 8a–c. Time course and kinetic parameters of LVA sodium currents in sensory neurons poisoned with ω-CgTX. **a** Family of LVA sodium currents recorded at the potential indicated in an Na-EGTA (pCa 9) solution containing 5 μM ω-CgTX. *Top:* −80 to −40 mV. *Bottom:* −30 to +70 mV, steps of 10 mV. Holding potential −100 mV. **b** Time to peak *(t_p, circles)* and inactivation time constant *(τ_h, triangles)* vs voltage. τ_h was estimated by curve-fitting the sodium current decay with a single exponential function (see Carbone & Lux, 1987b). **c** Relationship between current and voltage at the peak of the LVA current. Note the steepness in the potential region of negative conductance (−70 to −40 mV) and the linear relationship between −20 and +70 mV

Discussion and Conclusions

The present data reinforce the view that calcium channels might possess two distinct modes of ion permeation (Ca and Na mode) which can be distinguished pharmacologically by ω-CgTX. In contrast to this, there seems to be little support to the idea that the macroscopic activation-inactivation kinetics of calcium channels is changed drastically by changing the type of permeating ion. The time course of activation and inactivation of LVA channels is hardly affected by replacing normal Ca^{2+} with Na-EGTA solutions. Some uncertainty remains about the lengthening of HVA calcium channel inactivation in conditions of low Ca^{2+}, which might be related either to conformational changes of the inactivation process to the entry of Ca^{2+} ions into the cell (Eckert and Chad 1984). Apart from this, our observations are consistent with the view that activation and inactivation gating of calcium channels are somehow located at the inner side of the channel and that structural changes induced by lowering external Ca^{2+} might involve only the more external part of the pore and its selectivity filter (Lauger, 1985; Kostyuk and Mironov, 1986; Lux and Carbone, 1987).

The idea that calcium channels possess two modes of ion permeation fits a number of experimental observations. First, it explains why micromolar increases in external Ca^{2+} concentration reduce the size of macroscopic sodium currents by increasing the probability that calcium channels shut (Ca mode) rather than by reducing the average time the channel remains open for Na^+ ions (Na mode) (Lux and Carbone, 1987). This is consistent with previously held views that the Ca^{2+} specificity of HVA channels derives from the presence of an external Ca^{2+}-regulatory site which is occupied by Ca^{2+} when the channel is nonconductive for Na^+ (Kostyuk et al., 1983; Kostyuk and Mironov, 1986). Second, the concept of two gating modes is in line with recent observations that sudden elevations of H^+ ion concentration (Konnerth et al., 1987; Davies et al., 1988) or stepwise decreases of extracellular Ca^{2+} concentration (Hablitz et al., 1986) produce a transient inward sodium current flowing through calcium channels which are transformed from a voltage-gated Ca^{2+}-permeable state to a proton-gated Na^+-permeable state. Third, other calcium antagonists have been found to become calcium agonists depending on the current-carrying ion. Thus, in heart cells (Nilius et al., 1985) and chick sensory neurons (Boll and Lux, 1985), Bay K8644 enhances the activity of the high-threshold (L) calcium channel when Ba^{2+} is the main current-carrying cation, but depresses it if Ba^{2+} is replaced by Ca^{2+}.

Acknowledgements. We are grateful to Dr. M. Morad for suggesting the use of amiloride as a Ca-channel blocker. We thank also Mrs. H. Tyrlas and R. Meixner for valuable technical assistance.

References

Almers W, McCleskey EW (1984) Non-selective conductance in calcium channels of frog muscle: calcium selectivity in a single-file pore. J Physiology (Lond) 353:585–608

Armstrong D, Eckert R (1987) Voltage-activated calcium channels that must be phosphorylated to respond to membrane depolarization. Proc Natl Acad Sci (USA) 84:2518–2522

Armstrong CM, Matteson DR (1985) Two distinct populations of calcium channels in a clonal line of pituitary cells. Science 227:65–67
Barhanin J, Shimd A, Lazdunski M (1988) Properties of structure and interaction of the receptor for ω-Conotoxin, a polypetide active on Ca^{2+} channels. Biochem Biophys Res Commun 150:1051–1062
Boll W, Lux HD (1985) Action of organic antagonists on neuronal calcium currents. Neuroscience Lett 56:335–339
Bossu JL, Feltz A, Thomann JM (1986) Depolarization elicits two distinct calcium currents in vertebrate sensory neurons. Pflugers archiv Europ J Physiol 403:360–368
Carbone E, Lux HD (1984a) A low voltage-activated fully inactivating calcium channel in vertebrate sensory neurons. Nature (Lond) 310:501–503
Carbone E, Lux HD (1984b) A low voltage-activated calcium conductance in embryonic chick sensory neurons. Biophys J 46:413–418
Carbone E, Lux HD (1986a) Sodium channels in cultured chick dorsal root ganglion neurons. Eur Biophys J 13:259–271
Carbone E, Lux HD (1986b) Low- and high-voltage activated Ca channels in vertebrate cultured neurons: properties and function. In: Heinemann U, Klee M, Neher E, Singer W (eds) Exp Brain Res (Suppl)14:1–8
Carbone E, Lux HD (1987a) External Ca^{2+} ions block unitary Na^+ currents through Ca channels of cultured chick sensory neurones by favoring prolonged closures. J Physiology (Lond) 382:125P
Carbone E, Lux HD (1987b) Kinetics and selectivity of a low voltage-activated Ca current in chick and rat sensory neurones. J Physiology (Lond) 386:547–570
Carbone E, Lux HD (1987c) Single low voltage-activated calcium channels in chick and rat sensory neurones. J Physiology (Lond) 386:571–601
Carbone E, Lux HD (1988) ω-Conotoxin blockade distinguishes Ca- from Na-permeable state in neuronal Ca channels. Pflugers Arch Eur J Physiol (in press)
Carbone E, Morad M, Lux HD (1987) External Ni^{2+} selectively blocks the low-threshold Ca^{2+} currents of chick sensory neurons. Pflugers Archiv Europ J Physiol 408:R60
Cavalié A, Ochi R, Pelzer D, Trautwein W (1983) Elementary currents through Ca channels in guinea pig myocytes. Pflugers Arch Europ J Physiol 398:284–297
Chad JE, Eckert R (1986) An enzymatic mechanism for calcium current inactivation in dialysed *Helix* neurones. J Physiology (Lond) 378:31–51
Cruz LJ, Olivera BM (1986) Calcium channel antagonists. ω-Conotoxin defines a new high-affinity site. J Biol Chem 261:6230–6233
Davies NW, Lux HD, Morad M (1988) Site and mechanism of activation of proton-induced sodium current in chick dorsal root ganglion neurons. J Physiology 399:195–225
Eckert R, Chad JE (1984) Inactivation of Ca channels, Prog Biophys Molec Biol 44:215–267
Fedulova SA, Kostyuk PG, Veselovsky NS (1985) Two types of calcium channels in the somatic membrane of new-born rat dorsal root ganglion neurons. J Physiology (Lond) 359:431–446
Feldman DH, Olivera BM, Yoshikami D (1987) Omega *Conus geographus* toxin: a peptide that blocks calcium channels. FEBS Lett 214:295–300
Fenwick EM, Marty A, Neher E (1982) Sodium and calcium channels in bovine chromaffine cells. J Physiology (Lond) 331:599–635
Fox AP, Nowycky MC, Tsien R (1987) Kinetic and pharmacological properties distinguishing three types of calcium currents in chick sensory neurones. J Physiology 394:173–200
Fukushima Y, Hagiwara S (1985) Currents carried by monovalent cations through calcium channels in mouse neoplastic B lymphocytes. J Physiology (Lond) 358:255–284
Hablitz JJ, Heinemann U, Lux HD (1986) Step reductions in extracellular Ca activate a transient inward current in chick dorsal root ganglion cells. Biophys J 50:753–757
Hamill OP, Marty A, Neher E, Sakmann B, Sigworth FJ (1981) Improved patch-clamp techniques for high-resolution current recording from cells and cell-free membrane patches. Pflugers Arch Europ J Physiol 319:85–100
Hess P, Lansmann JF, Tsien RW (1986) Calcium channel selectivity for divalent and monovalent cations: voltage and concentration dependence of single channel current in guinea pig ventricular heart cells. J Gen Physiol 88:293–319
Hess P, Tsien RW (1984) Mechanism of ion permeation through calcium channels. Nature (Lond) 309:453–456

Konnerth A, Lux HD, Morad M (1986) Proton-induced transformation of calcium channel in chick dorsal root ganglion cells. J Physiology (Lond) 386:603–633

Kostyuk PG, Mironov SL, Shuba YM (1983) Two ion-selecting filters in the calcium channel in chick dorsal root ganglion cells. J Membrane Biol 76:83–93

Kostyuk PG, Mironov SL (1986) Some predictions concerning the calcium channel model with different conformational states. Gen Physiol Biophys 6:649–659

Lansmann JB, Hess P, Tsien RW (1986) Blockade of calcium current through single calcium channels by Cd^{2+}, Mg^{2+} and Ca^{2+}: voltage and concentration dependence of calcium entry into the pore. J Gen Physiol 88:321–347

Lauger P (1985) Ionic channels with conformational substrates. Biophys J 47:581–590

Levi R, DeFelice LJ (1986) Sodium-conducting channels in cardiac membranes in low calcium. Biophys J 50:11–19

Llinas R, Yarom Y (1981) Properties and distribution of ionic conductances generating electroresponsiveness of mammalian inferior olivary neurons in vitro. J Physiology (Lond) 315:569–584

Llinas R, Yarom Y (1986) Specific blockage of the low-threshold calcium channel by high molecular weight alcohols. Neurosciences (Abstract) 12:174

Lux HD, Carbone E (1987) External Ca ions block Na conducting Ca channel by promoting open to closed transitions. In: Ovchinnikov YA, Hucho F (eds) Receptors and ion channels, Walter de Gruyter & Co, Berlin, pp 149–155

Lux HD, Carbone E, Davies N (1987) Monovalent cation currents through low- and high-threshold Ca channels in chick sensory neurons. Pflugers Arch 408 (suppl):R60

McCleskey EW, Fox AP, Feldman DH, Cruz LJ, Olivera BM, Tsien RW, Yoshikami D (1987) ω-Conotoxin: direct and persistent blockade of specific types of calcium channels in neurons but not in muscle. Proc Natl Acad Sci USA 84:4327–4331

Morad M, Kaplan JM, Davies NE, Lux HD (1988) Kinetics of Ca^{2+}-induced Ca^{2+} channel inactivation and block, using "caged" Ca^{2+} in dorsal root ganglion neurons. Biophys J 53:21a

Nilius B, Hess P, Lansman JB, Tsien RW (1985) A novel type of cardiac calcium channel in ventricular cells. Nature 316:443–446

Nowycky MC, Fox AP, Tsien RW (1985) Three types of neuronal calcium channels with different calcium agonist sensitivity. Nature (Lond) 316:440–443

Rane SG, Holz GG, Dunlap K (1987) Dihydropyridine inhibition of neuronal calcium current and substance P release. Pflugers Archiv Eur J Physiol 409:361–366

Swandulla D, Armstrong CM (1988) Fast deactivating calcium channels in chick sensory neurons. J gen Physiology (in press)

Swandulla D, Carbone E, Schafer K, Lux HD (1987) Effect of menthol on two types of Ca currents in cultured sensory neurons of vertebrates. Pflugers Archiv Europ J Physiol 409:52–59

Tang C, Presser F, Morad M (1988) Amiloride selectively blocks the low threshold (T) calcium channel. Science 240:213–215

Yaari Y, Hamon B, Lux HD (1987) Development of two types of calcium channels in cultured mammalian hippocampal neurons. Science 235:680–682

Block of Sodium Currents Through a Neuronal Calcium Channel by External Calcium and Magnesium Ions

H. D. Lux[1], E. Carbone[2], and H. Zucker[1]

[1] Abteilung Neurophysiologie, Max-Planck-Institut für Psychiatrie, D-8033 Planegg, FRG
[2] Dipartimento di Anatomia e Fisiologia Umana, I-10125 Torino, Italy

Introduction

Calcium transport through calcium channels is vital for a variety of cellular functions. For its duty, the channels must select calcium ions against usual majority ions such as those of sodium. Indeed, calcium channels become permeable to sodium and other alkaline ions if the external level of free Ca is below micromolar values [1, 16, 17, 20, 21, 28]. Current views of the mechanisms [1, 19, 22] by which calcium ions interfere with the flux of alkaline ions through calcium channels employ specifically strong binding of calcium ions to channel sites, but differ in kinetic aspects of ion permeation. For studying the block by divalent cations of sodium currents through calcium channels, it is useful to consider that divalent cations, whether or not permeant, may enter the channel in a voltage-dependent manner. If entry of blocking divalent cations is favored by a "regulatory" site at the external mouth of the channel [22], the channel block is expected to be weakly or not voltage dependent. This has been suggested to apply in a limited range of membrane potentials in snail neurons [22]. However, a strongly voltage-dependent block of sodium currents was shown on calcium channels of mouse neoplastic B-lymphocytes [16]. Indeed, maximum block rather than lack of voltage dependency was observed in a comparable region of membrane potential.

The voltage dependence of the block by divalent ions can provide information about the location of the block inside the electric field experienced by the blocking ion. There is also the question of the pathway by which the block is exerted. This could be direction specific or dependent on externally and internally located energy barriers of the channel. The results can be compared with predictions from theories on the selectivity of the calcium channel for which the exclusion of monovalent ion passage is an obvious prerequisite.

Kinetic details of the block of sodium currents by divalent cations can be deduced from the behavior of single channels. It has been reported [6, 25] that micromolar elevation of external Ca causes prolonged closures and shortens open times for Na^+ currents through neuronal calcium channels with little effect on single-channel conductance. Open times for Na^+ decrease with increased $[Ca^{2+}]_o$ in a manner suggesting that single entering calcium ions may block the channel for sodium permeation. The probability that channels reopen for Na is markedly reduced already at micromolar levels of $[Ca^{2+}]_o$, indicating an efficiently increased occupation of closed states. The block of sodium currents and thus the calcium specifity of this channel could be explained by calcium-promoted changes in state occupancies of the channel.

Preparation and Condition

Currents through calcium channels of chick dorsal root ganglionic cells were recorded under conditions in which ordinary sodium and potassium conductances were blocked by drugs (tetrodotoxin, 3 µM; tetraethylammonium, 20 mM) and absence of potassium ions. In this and other preparations, high-threshold calcium currents are known to run down with time [3–5, 7, 10, 13, 14]. Their block was supported in most experiments by applying 50 µM verapamil [3, 13] or 5 µM ω-conotoxin [9, 26], which preferentially affects this calcium current. The neuronal fast-inactivating calcium channel that is activated at lower voltages (LVA channel) [2, 3–5, 7, 8, 12, 13, 27] persists under these conditions and can be studied in isolation (Fig. 1). Reduction of $[Ca^{2+}]_o$ in the presence of usual $[NA^+]_o$ or $[Li]_o$ generated large inward currents. These currents were activated at more negative membrane potentials, showing a voltage shift of about -35 mV (see also Fig. 1a), likely due to reduced external charge screening in low $[Ca^{2+}]_o$. Most of this shift occurred in the range of Ca^{2+} concentrations from 2×10^{-2} to 1×10^{-4} M. The slope of the activation-voltage relationship at 5×10^{-5} M $[Ca^{2+}]_o$ was not observed to differ significantly (within 4 mV) from that recorded in low $[Ca^{2+}]_o$ ($< 1 \times 10^{-9}$ M). For already effectively blocking calcium (and magnesium) concentrations up to 10^{-4} M as considered here, membrane potential shifts appeared small compared with the investigated range of membrane potentials (see Figs. 1, 2).

Replacement of Na^+ (or Li^+) by choline removed the inward currents in low $[Ca^{2+}]_o$, as did the millimolar addition of metallic divalent calcium channel blockers Ni, Co, and Cd. Sample averages of the currents showed activation and inactivation time courses similar to those of LVA calcium currents. The currents largely disappeared due to steady inactivation with holding potentials positive to -70 mV. These results indicate that the currents were carried by the alkaline ions flowing through LVA calcium channels.

Block of Sodium Currents by External Calcium and Magnesium

A sigmoidal, monotonically decreasing blocking efficacy with increased positivity of the internal membrane potential characterizes the action of externally applied Mg^{2+} (Fig. 2a). At negative membrane potentials (around -100 mV), the sodium current block by Mg is slightly stronger than that by Ca, as seen in the deactivation or "tail" currents of Fig. 1. At more positive potentials, the block is relieved, contrary to that by external calcium ions which exert a maximal block at around -20 mV. Increasingly positive membrane potentials also relieve the calcium block of sodium currents (see Fig. 2a). Unblocking is more prominent at potentials positive to zero. The dependence on voltage is strong. This is manifested by the necessity to assume that nearly half of the field across the channel is utilized (H.D. Lux and H. Zucker, unpublished observations), when calcium or magnesium ions are thought to bind inside the channel and thus block single-file sodium passage. The difference between the action of $[Mg^{2+}]_o$ and that of $[Ca^{2+}]_o$ (see also [16]) can be attributed by the fact that calcium ions but not magnesium ions pass the channel to carry significant currents.

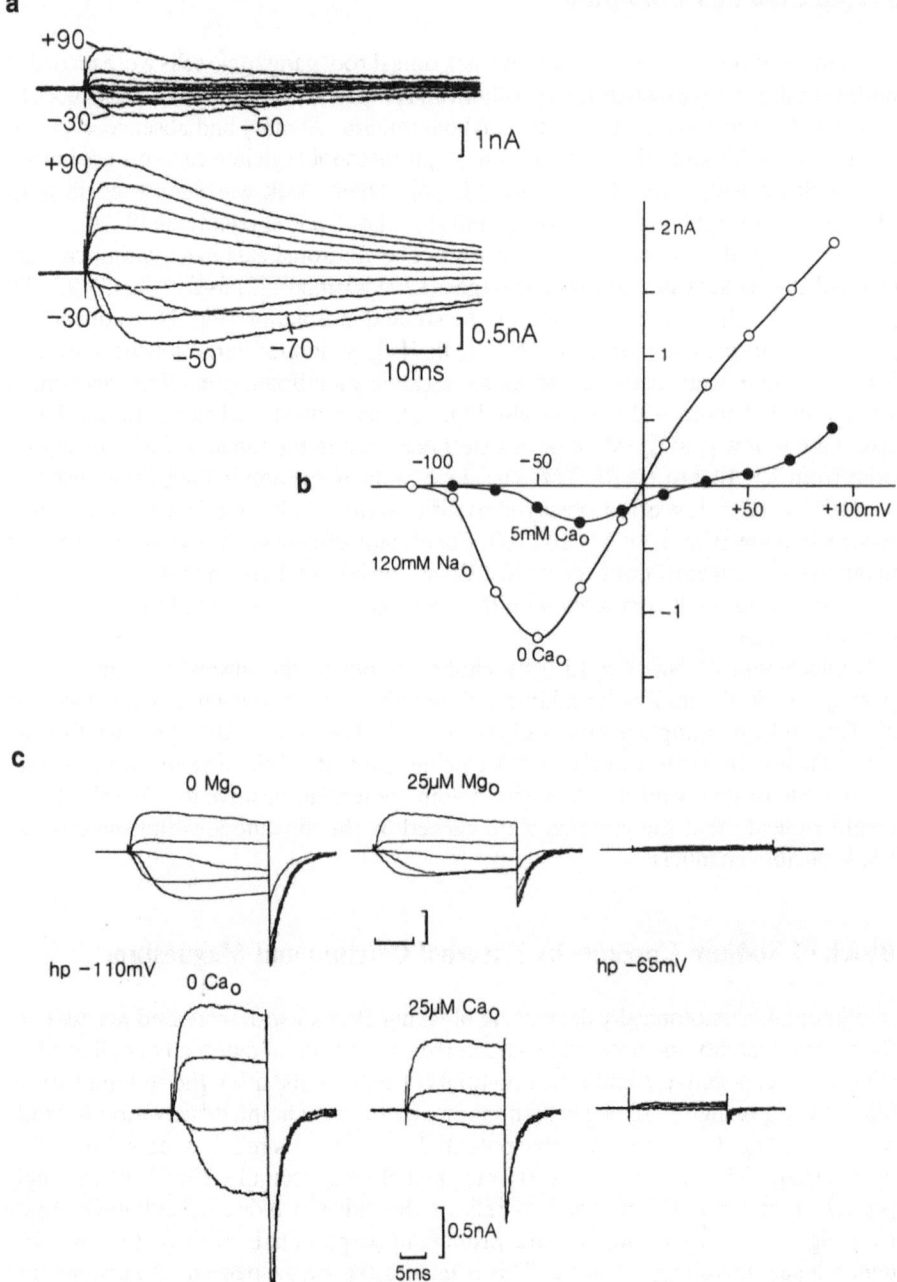

Fig. 1a–c. Calcium and sodium currents through the LVA channel. **a** *upper panel*, currents with 5 mM [Ca^{2+}]$_o$ successively recorded from −70 to +90 mV with voltage increments of 20 mV. Same membrane potentials indicated when superimposed currents overlap. *Lower panel*, sodium currents of same cell in the presence of 10 mM external ethyleneglycoltetraacetic acid (EGTA) with nominally zero [Ca^{2+}]$_o$. External Na$^+$ as well as internal Na$^+$ (perfusing patch pipette) was 120 mM in both

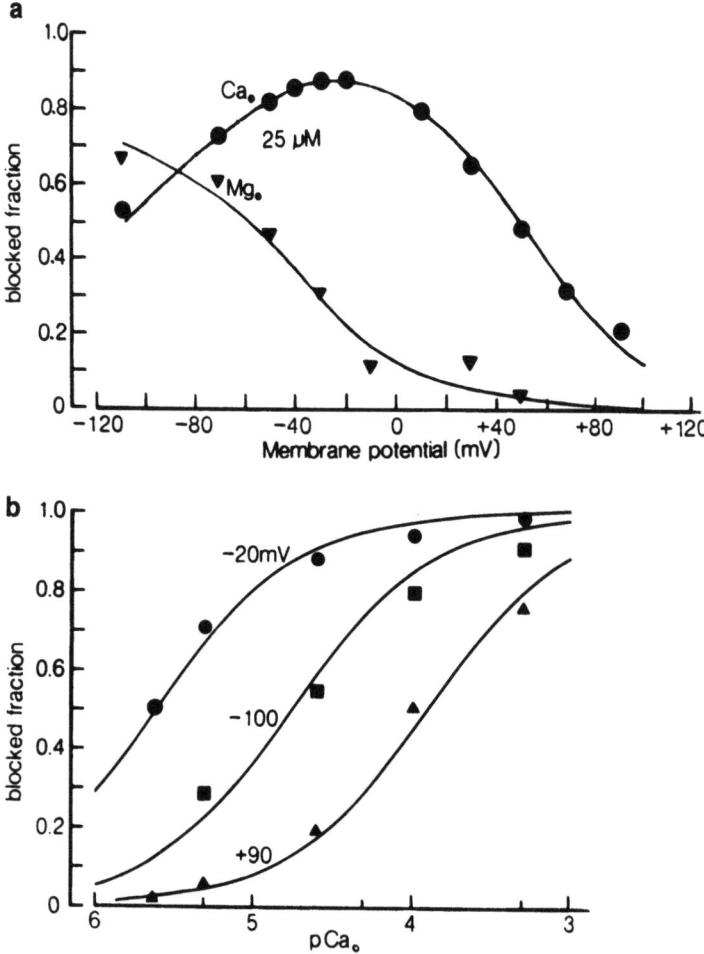

Fig. 2a, b. Voltage dependence of the block of sodium currents through LVA calcium channels by $[Ca^{2+}]_o$ and $[Mg^{2+}]_o$. **a** Block by 25 μM $[Ca^{2+}]_o$ (●) and $[Mg^{2+}]_o$ (▼). The block is normalized by determining the fraction of unblocked currents relative to those recorded in the absence of Ca and Mg (see Fig. 1c). **b** Effects of different $[Ca^{2+}]_o$ on LVA sodium currents at three membrane potentials as indicated. *Solid lines* represent titration curves with 1:1 stoichiometry with apparent K_D values of 2.4 μM (−20 mV), 25 μM (−100 mV), and 130 μM (+90 mV). Data points are mean values received from three to eight cells

---

situations. **b** Current-voltage relationship in the presence (●) and the absence (○) of external $[Ca^{2+}]$. Holding potential was −100 mV. Note the reduced sodium outward currents and the shifted reversal potential with 5 mM Ca_o. **c** Sodium currents through LVA channel in the presence of 25 μM free $[Mg^{2+}]_o$ *(upper panel)* and 25 μM $[Ca^{2+}]_o$ *(lower panel)* during successively applied voltage pulses; for Mg: to −70, −50, −30, +10, +50 mV; for Ca: to −50, −10, +50, +90 mV membrane potential. Controls in the absence of external divalent ions, different cells. Note the predominant effect of Mg_o on the current trace at −70 mV and on tail currents at −110 mV $[Ca]_o$ primarily affects sodium currents during intermediate depolarizing steps. Currents (at 0 Ca_o) are inactivated *(right)* at lowered holding potential (hp −65 mV)

With inside negative potentials magnesium ions merely become attracted to the channel interior, and the probability that they leave the channel toward the external side is reduced because of the applied voltage field. Even if barriers at internal channel sites are lowered by the field, they are high enough to prevent inward magnesium passage, as is suggested by the lack of significant magnesium inward currents at strong negative potentials. By contrast, calcium currents, once activated, increase with increased inside negativity, and this increases the fraction of ions which leave the channel at its interior site. With more positive membrane potentials the block by magnesium ions is reduced because the voltage field favors their external exit. The block by calcium ions, however, resumes a maximum in an intermediate region of voltage (-30 to 0 mV) before its significant reduction becomes obvious with increasingly inside positive potentials. It appears reasonable to assume that the maximum of the block is due to the longest average sojourn of calcium ions inside the channel. The assumption of a potential dependent balance between rates of entry (from the external side) and exits (at both sides of the channel) was used to describe the potential dependence of the block (Fig. 2; H. D. Lux and H. Zucker, unpublished observations).

At any given membrane potential, the block of sodium currents through this calcium channel by $[Ca^{2+}]_o$ (Fig. 2b) appeared to be approximated by a 1:1 stoichiometry, in line with the assumption of binding of a single divalent ion to a specific blocking site inside the channel. A voltage-independent component of the block as predicted by the assumption of an externally located "regulatory site" [22] could not be verified. If present, a tendency of the blocking curve to attain a constant level, different from zero, should be observed with sufficiently large negative or positive membrane potentials. Calcium ions are, in general, more powerful than magnesium ions in blocking sodium (or lithium) currents through the LVA calcium channel, but a quantitative comparison can only be made for given membrane potentials. At potentials with maximum inward currents (-55 to -40 mV), the blocking efficacy of calcium versus magnesium is about 5. An important finding was the insensitivity of the sodium currents to internally applied calcium and magnesium concentrations [8]. To exert a sodium current block in this way, about 100-fold stronger calcium concentrations than those effective at the outside were found necessary [8]. Millimolar $[Mg^{2+}]_i$ was similarly ineffective.

The potential dependent apparent K_D for Ca_o was maximally 2.3 μM at -40 mV [8]. Even including voltage-dependent changes, it is evident that these K_D values are 2–3 orders of magnitude smaller than those derived for calcium currents in this preparation [8], as is true for others (see [1, 16, 18–22]). In fact, the high affinity of the channel for calcium ions as deduced from their blocking capability would not allow a significant transport of calcium ions. Attempts to remove this discrepancy necessitate assumptions such as ion – ion repulsion inside the channel [1, 19] or specific ion – channel interactions (see [11, 23, 24]).

Observations on Single-Channel Activities

At Ca^{2+} below 0.1 μM, the single-channel Na^+ currents (Fig. 3) lasted on average 0.26 $\pm$ 0.06 ms, a far shorter time than that of unitary Ca^{2+} currents [5, 8]. Openings

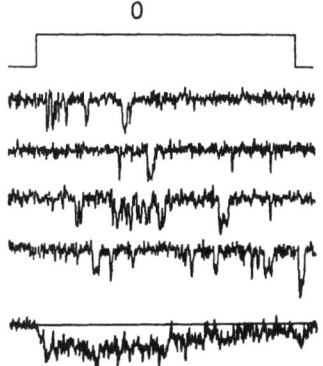
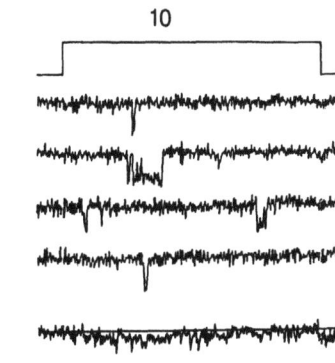
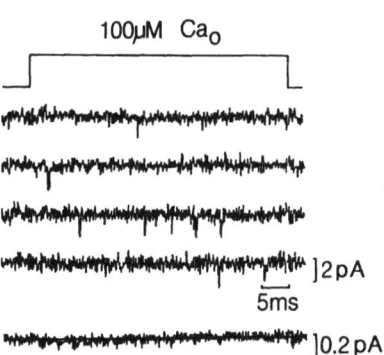

Fig. 3. Block by external Ca^{2+} of Na^+ currents through a single LVA calcium channel. Example of single-channel Na^+ currents in an outside-out patch of chick dorsal root ganglionic cells recorded during successive depolarizations to -40 mV in the presence of 10 mM $EGTA_o$ *(left)*, 1×10^{-5} *(middle)*, and 1×10^{-4} M $[Ca^{2+}]_o$ *(right)*. Bottom traces are sample averages of the currents recorded during 25 (0 Ca_o) to 73 (100 μM Ca_o) trials. Note the increase in silent periods already at 10 μM $[Ca^{2+}]_o$ with little effect on open durations. Rather short-lasting openings are observed with 100 μM $[Ca^{2+}]_o$, and traces with activity were selected for display, since "blanks" became particularly frequent with stronger $[Ca^{2+}]_o$. Holding potential was -110 mV in all cases. Cut-off frequency 4 kHz

occurred frequently in bursts, with intermediate closures of similar mean duration. The average amplitude of the single-channel events during depolarizations to -40 mV was $1.85 \pm$ pÅ, which was about five times larger than that recorded with 20 mM $[Ca^{2+}]$ at corresponding potentials. Generally, these depolarizations produced sufficiently large currents, as seen in sample averages or whole-cell recordings (see Fig. 1).

Increasing $[Ca^{2+}]_o$ from nanomolar to 10 μM values strongly reduced the current amplitude of averaged samples. However, the amplitude of single-channel currents remained unchanged (see Fig. 3a). The main effect on unitary currents was a decrease of mean open times with their values halved at 50–60 μM $[Ca^{2+}]_o$ (Fig. 3b). This concentration is about 20 times larger than that required (2.3 μM) to produce half-block of the averaged currents of samples and of whole cells (Fig. 4a) at this membrane potential. However, closed times increased, and numbers of openings per activating pulse decreased in proportion to the reduction of averaged currents (see Fig. 4b). With $[Ca^{2+}]_o \geq 10^{-4} M$, closures frequently overlasted the pulse duration, and "blanks" were common. Thus, the primary reason for the reduction of averaged single-channel currents as well as of whole-cell currents with increased $[Ca^{2+}]_o$ appears to be an increase of the periods during which channels are closed for sodium permeation.

Barium ions were found to act as calcium ions at similar concentrations, but magnesium was less effective at these membrane potentials, in accordance with the results presented in Fig. 2. Membrane patches subjected to high internal free Ca^{2+} (100–500 μM) behaved like patches dialyzed with low free $[Ca^{2+}]$ ($< 10^{-9}$ M) showing the same changes in the distributions of closed and open states (see also [8]). Internal calcium ions are thus rather ineffective in blocking sodium currents through single LVA calcium channels. This applies also to millimolar $[Mg^{2+}]_i$.

It could be of interest to consider the number of charges transported during an average opening for Na. The half-reduced average open time at a $[Ca^{2+}]_o$ of 50–60 μM coincidentally compares with the average time interval by which about one calcium ion would be found among transported sodium ions, if the probability of the entry of calcium and sodium ions corresponds to their relative bulk concentrations. Of course, the concentration ratio Ca/Na at the channel entry may differ from that in the bulk solution, and entering of sodium and calcium ions may experience specific barrier conditions. Nevertheless, the correspondence is in line with the view that the block of sodium passage may be mediated by single calcium ions. The continuous line in Figure 4a represents the calculated mean open time of the channel under such assumption. The asymptotic value of the mean open time (0.26 ms at $[Ca^{2+}]_o < 10^{-9}$ M) and the Ca^{2+} concentration at the half open time would result in an estimate of 7×10^7 $M^{-1}s^{-1}$ for the rate of Ca^{2+} entry. A minimum for the rate by which Ca^{2+} leaves the binding site of about 3 ms^{-1} can be deduced from the fast time constant of the closed states in low and micromolar $[Ca^{2+}]_o$. Although this results in a somewhat higher apparent dissociation constant for the Ca^{2+}-channel site complex than estimated from whole-cell currents, it is a value too small to be compared with observed single-channel calcium transport, in the order of 10^3 ms^{-1} [8, 15].

Despite the possibility of early reopenings, the fraction and the mean time of long-lasting closures increased markedly with micromolar increase in $[Ca^{2+}]_o$, as shown by the distribution of closed times in Figure 4. From the dependence of the closures on $[Ca^{2+}]_o$, the rate of calcium entry into long-lasting Ca^{2+}-mediated closed states (for Na^+) was estimated to be of the order of 10^8 $M^{-1}s^{-1}$. For this, it was assumed that the backward rates to the open state are Ca^{2+} independent and comparable to the inverse of the fast time constant of closing. Such sensitivity to external Ca is, in fact, predicted from earlier arguments (see [1]), using probable target sizes of the calcium channel entry. These results suggest that a particularly high Ca^{2+} affinity of the conductive state is not a prerequisite for the calcium selectivity of the LVA channel. Selectivity can also be provided by a high Ca^{2+} affinity of nonconductive states.

Conclusions

Present models [1, 19, 22] are well capable to account for the selectivity changes of calcium channels but predict a decrease of single-channel conductance for Na^+, with increasing $[Ca^{2+}]_o$ rather than increased closed times of the channel. In an elegant recent version, the block of sodium conductance is assumed to persist for considerable time due to high-affinity binding of one calcium ion to either one of two similar sites of calcium channels. Double occupancy by calcium ions, however, mobilizes calcium binding due to electrostatic repulsion [1, 19]. The mechanism as proposed would unblock the channel for sodium transport after the leaving of calcium ions. Thus, an increased number of unblocking events, expressed as short closures, should be

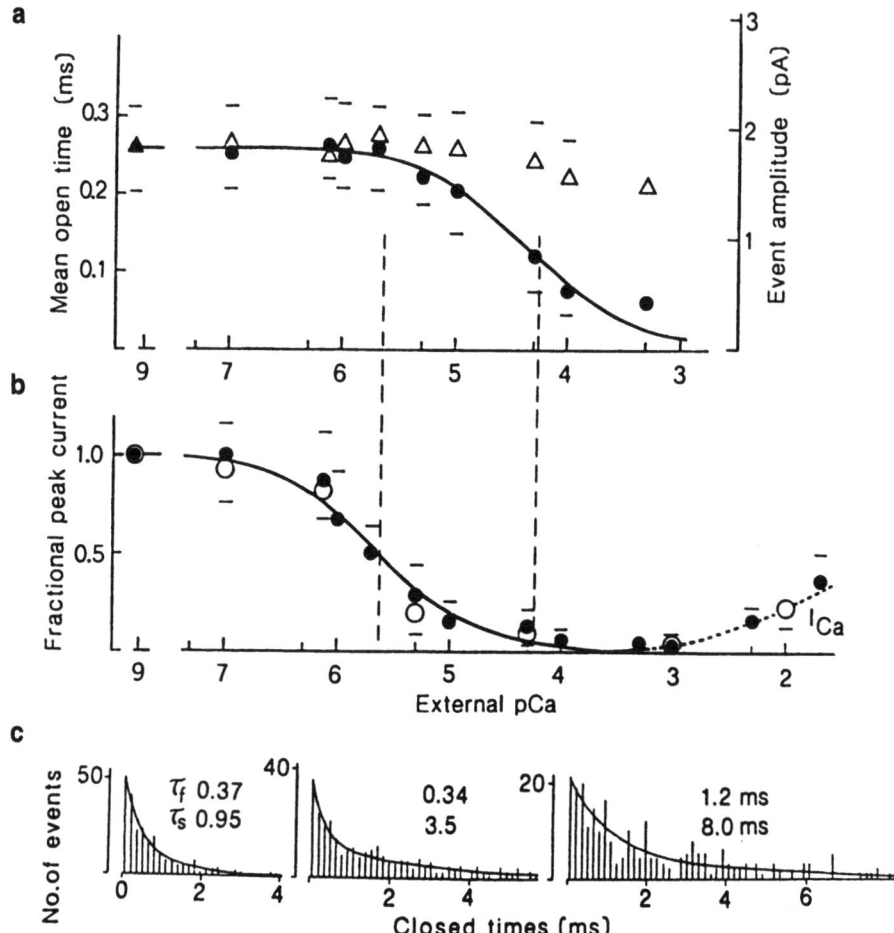

Fig. 4a–c. Parameters of Na$^+$ currents through LVA calcium channels as a function of $-\log[Ca^{2+}]_o$. **a** Amplitudes ($\triangle$) and mean open times ($\bullet$) with SD *(bars)* of four to seven determinations each. Data at 0.5 mM are means from two experiments with sufficient collections of openings. All patches were sequentially exposed to three significantly different levels of $[Ca^{2+}]_o$. Half-value of mean open time is at $5.5 \times 10^{-5} M [Ca^{2+}]_o$ *(dashed line)* with assumed 1:1 stoichiometry *(solid line)*. **b** Averages of current samples ($\bullet$) of experiments in **a** means of currents from whole DRG cells ($\circ$). Currents are given as fraction of values received at nominally zero $[Ca^{2+}]_o$ with data sampled over a period of 40 ms including peaks. The continuous fitting line determined as in **a** but with half-block at $[Ca^{2+}]_o = 2.3$ µM. **c** Closed times received from single-channel events of a patch at three levels of $[Ca^{2+}]_o$. Double exponentials with minimum χ^2 were fitted to the distributions with time constants indicated

expected with increased $[Ca^{2+}]_o$ and enhanced probability of double occupancy (see [22, 23]). Rather, the contrary was observed on the LVA calcium channel. It was also found difficult to reconcile the strong release of the current block at positive membrane potentials (see Fig. 2) with a decreased probability of calcium entry and thus of double occupation as theoretically demanded.

Since internal access of calcium ions is negligible in our case, the general double binding site, three-barrier model [1, 19] can be greatly simplified and tested for its applicability to the behavior of mean open times and to the occupancy of the sodium-permeable state that is depicted by the averaged currents. It turns out that this model invariably predicts a higher sensitivity to increased $[Ca^{2+}]_o$ of open times than of the general occupancy of the sodium permeable state. Since the observations on LVA calcium channels contradict in essential aspects the theoretical results from the two-binding site model, we propose that ion – channel instead of ion – ion interactions should be considered. In fact, open and closed times for sodium and calcium passage are known to differ greatly [8], which supports the view that the channel fluctuates in ion-specific modes.

In line with the concept of ion-specific modes of channel permeation (on fluctuating barrier assumptions, see [11, 23, 24]), the channel would, however, conduct small calcium currents for significant times when it is closed for Na. Transitions from sodium to calcium conductive states of the channel may thus underlie the block of monovalent ion passage and the calcium selectivity of the LVA calcium channel. The idea that calcium channels possess different modes of ion permeation also receives support from observations of proton-induced changes in calcium channel permeability (for review see M. Morad, this volume) and from recent pharmacological results (E. Carbone and H. D. Lux, this volume).

Acknowledgement. We thank Mrs. R. Meixner for perfect technical assistance.

References

1. Almers W, McCleskey EW (1984) Non-selective conductance in calcium channels of frog muscle: calcium selectivity in a single-file pore. J Physiol (Lond) 353:585–608
2. Armstrong CM, Matteson DR (1985) Two distinct populations of calcium channels in a clonal line of pituitary cell. Science 227:65–67
3. Boll W, Lux HD (1985) Action of organic antagonists on neuronal calcium currents. Neurosci Lett 56:335–339
4. Bossu JL, Feltz A, Thomann JM (1985) Depolarization elicits two distinct calcium currents in vertebrate sensory neurons. Pflügers Arch 403:360–368
5. Carbone E, Lux HD (1984) A low voltage-activated, fully inactivating Ca channel in vertebrate sensory neurones. Nature (Lond) 310:501–502
6. Carbone E, Lux HD (1986) External Ca^{2+} ions block unitary Na^+ currents through Ca^{2+} channels of cultured chick sensory neurones by favouring prolonged closures. J Physiol (Lond) 382:124P
7. Carbone E, Lux HD (1987) Kinetics and selectivity of a low voltage-activated calcium current in chick and rat sensory neurones. J Physiol (Lond) 386:547–570
8. Carbone E, Lux HD (1987) Single low voltage-activated calcium channels in chick and rat sensory neurones. J Physiol (Lond) 386:571–601
9. Carbone E, Lux HD (1988) ω-Conotoxin blockade distinguishes Ca from Na permeable states in neuronal calcium channels. Pflügers Arch (in press)
10. Cavalié A, Ochi R, Pelzer D, Trautwein W (1983) Elementary currents through Ca^{2+} channels in guinea pig myocytes. Pflügers Arch 398:284–297
11. Ciani S (1984) Coupling between fluxes in one-particle pores with fluctuating energy profiles. Biophys J 46:249–252
12. DeRiemer SA, Sakmann B (1986) Two calcium currents in normal rat anterior pituitary identified by a plaque assay. Exp Brain Res (Suppl) 14:139–154

13. Fedulova SA, Kostyuk PG, Veselovsky NS (1985) Two types of calcium channels in the somatic membrane of new-born rat dorsal root ganglion neurones. J Physiol (Lond) 359:431–446
14. Fenwick EM, Marty A, Neher E (1982) Sodium and calcium channels in bovine chromaffin cells. J Physiol (Lond) 331:599–635
15. Fox AP, Nowycky MC, Tsien RW (1987) Single-channel recordings of three types of calcium channels in chick sensory neurones. J Physiol (Lond) 394:173–200
16. Fukushima Y, Hagiwara S (1985) Currents carried by monovalent cations through calcium channels in mouse neoplastic B lymphocytes. J Physiol (London) 358:255–284
17. Garnier D, Rougier O, Gargouil YM, Coraboeuf E (1969) Analyse électrophysiologique du plateau des réponses myocardiques, mise en évidence d'un courant lent entrant en absence d'ions bivalents. Pflügers Arch 313:321–342
18. Hess P, Lansman JB, Tsien RW (1986) Calcium channel selectivity for divalent and monovalent cations. Voltage and concentration dependence of single channel current in ventricular heart cells. J Gen Physiol 88:293–319
19. Hess P, Tsien RW (1984) Mechanism of ion permeation through calcium channels. Nature (Lond) 309:453–456
20. Kostyuk PG, Krishtal OA (1977) Effects of calcium and calcium-chelating agents on the inward and outward current in the membrane of mollusc neurones. J Physiol (Lond) 270:569–580
21. Kostyuk PG, Mironov SL, Doroshenko PA (1982) Energy profile of the calcium channel in the membrane of mollusc neurons. J Membrane Biol 70:181–189
22. Kostyuk PG, Mironov SL, Shuba YM (1983) Two ion-selecting filters in the calcium of the somatic membrane of mollusc neurons. J Membrane Biol 76:83–93
23. Läuger P (1985) Ionic channels with conformational substates. Biophys J 47:581–590
24. Läuger P, Stephan W, Frehland E (1980) Fluctuations of barrier structure in ionic channels. Biochem Biophys Acta 602:167–180
25. Lux HD, Carbone E (1987) External Ca ions block Na conducting Ca channel by promoting open to closed transitions. In: Ovchinnikov YA, Hucho E (eds) Receptors and ion channels. de Gruyter, Berlin pp 149–155
26. McCleskey EW, Fox AP, Feldman DH, Cruz LJ, Olivera BM, Tsien RW, Yoshikami D (1987) ω-Conotoxin: direct and persistent blockade of specific types of calcium channels in neurons but not in muscle. Proc Natl Acad Sci USA 84:4327–4331
27. Nowycky MC, Fox AP, Tsien RW (1985) Three types of neuronal calcium channels with different agonist sensitivity. Nature (Lond) 316:440–443
28. Reuter H, Scholz H (1977) A study of the ion selectivity and kinetic properties of calcium-dependent slow inward current in mammalian cardiac muscle. J Physiol (Lond) 264:17–47

The Voltage Sensor of Skeletal Muscle Excitation-Contraction Coupling: A Comparison with Ca^{2+} Channels

G. Pizarro, G. Brum, M. Fill, R. Fitts, M. Rodriguez, I. Uribe, and E. Rios

Department of Physiology, Rush University, 1750 W. Harrison St., Chicago, IL 60612, USA

Introduction

A recent development in the study of Ca^{2+} channels is the realization [2, 44] that skeletal muscle Ca^{2+} channels have a number of properties in common with the voltage sensor of excitation-contraction (EC) coupling, the molecule or structure of skeletal muscle membrane that has the role of sensing the action potential to control the opening of the Ca^{2+} release channels of the sarcoplasmic reticulum (SR).

Specifically, it was proposed that the high-affinity dihydropyridine (DP) receptor (DPR) isolated from T-tubule membrane fractions of skeletal muscle is the voltage sensor of EC coupling [44]. If verified, this hypothesis would have a troubling implication, i.e., that the best studied DPRs – those of skeletal muscle – might not be functional Ca^{2+} channels. On the other hand, the implications for EC coupling are exciting: the voltage sensor of EC coupling has until now been a largely hypothetical entity, the existence of which is affirmed on the basis of theoretical necessity (muscle must have a voltage sensor) and the measurement of small intramembrane charge movements [46], tentatively ascribed to conformational changes of the voltage sensor. The proposal that the voltage sensors are DPRs brings a promise of biochemical and molecular approaches to the study of charge movements and the voltage sensors.

Since this hypothesis was formulated, more evidence in its favor has appeared, which is, however, only indirect: the DPRs have been shown to reside preferentially in the junctional membrane of the T tubules [48], and preliminary results obtained by Knudson et al. [27] demonstrate the existence of an interaction between DPRs (T-tubular) and ryanodine receptor proteins (Ca^{2+} release channels of the sarcoplasmic reticulum membrane [25, 32]). Additionally, Franzini-Armstrong and coworkers have generated freeze-fracture images of junctional T-membrane particles [12] which are not inconsistent with freeze-dried images of the purified DPR [35]. Finally, K. Campbell's biochemical and immunological studies of the T-SR junction have failed to demonstrate proteins large enough to account for the T-membrane particles, other than the large subunits of the DPR (personal communication).

This and other evidence of a close similarity between Ca^{2+} channels and voltage sensors has led us and others to at least consider a strict version of this DPR hypothesis, according to which the voltage sensors of EC coupling are, at the same time, Ca^{2+} channels.

The work to be described is a more extensive comparison between phenomena related to the voltage sensors of EC coupling and membrane Ca^{2+} currents (I_{Ca}). Two classes of experiments were performed: (a) exposure of muscle fibers to known antagonists and agonists of the slow Ca^{2+} channel to evaluate their effect on EC coupling and (b) direct comparisons between biophysical properties of I_{Ca} and EC coupling phenomena (voltage dependence of activation and time dependence of recovery after inactivation).

Our conclusion is that the EC coupling phenomena and I_{Ca} have very similar drug sensitivity, voltage sensitivity, and time dependence of recovery. There are, however, subtle, interesting differences, which seem to indicate that the voltage sensors of EC coupling are not the pathways of I_{Ca}.

Methods

The experiments were performed on fast-twitch fibers of the musculus semitendinosus of *Rana pipiens*. The techniques of fiber dissection and mounting have been extensively described. Briefly, a 0.5-mm segment of fiber is under voltage clamp in a double vaseline gap device (described in [4, 5]). In this preparation, we performed, in parallel, the measurement of four variables (which change as a consequence of imposed membrane voltage pulses):

1. Total *membrane current*. Ionic substitutions were performed in the extra- and intracellular media to prevent most ionic currents. K^+ and Na^+ channels were blocked by drugs, and there was no Cl^- present. In most cases, the only time-dependent ionic current allowed was a Ca^{2+} current (or a current carried by another ion through Ca^{2+} channels). We refer to the measured Ca^{2+} current as I_{Ca}, as we have not made any effort to separate slow and fast currents.
2. *Intramembrane charge movements*. As most time-dependent ionic currents were eliminated, the remaining time-dependent currents are capacitive. By subtracting the linear component of the capacitive current (which is obtained at very positive potentials [4] or at potentials negative to -100 mV [22]), a nonlinear capacitive current (charge movement) is obtained. (A curious jargon has evolved in which "charge movement" means nonlinear capacitive current as well as the fundamental phenomenon underlying this current, that is, the electrically driven conformational change of membrane-resident proteins.)
3. *Ca^{2+} transients* is the accepted name for the time resolved measurement of the transient increase in myoplasmic Ca^{2+} concentration ($[Ca^{2+}]_i$). This measurement is based on the measurement of changes in absorbance of a Ca^{2+}-sensitive dye introduced intracellularly [30, 31] by diffusion from the cut ends of the fiber segment.
4. *Ca^{2+} release flux*. The changes in $[Ca^{2+}]_i$ associated with depolarizing pulses are due to opening of Ca^{2+}-release channels in the SR. A time-resolved measurement of Ca^{2+} flux through these channels (proportional to their Ca^{2+} current) was derived from the Ca^{2+} transients by a method developed by Schneider and coworkers [37, 38].

Solutions

Most experiments were performed in an internal solution, which is essentially isotonic cesium glutamate, to which 0.8 mM Antipyrylazo III is added (details in [5]); the nominal [Ca^{2+}] is 50 nM. The external medium is an isotonic tetraethylammonium (TEA) metanesulfonate with tetrodotoxin (TTX), 3,4-diaminopyridine, and 2 mM [Ca^{2+}]. Low concentrations of drugs used in these experiments were added externally. All experiments were performed at temperatures between 7° and 11°C.

Results

A first group of results concerns the effects of antagonist dihydropyridines [nifedipine, (−)202–791] and their interaction with low [Ca^{2+}]$_o$. Then the effects of the antagonist phenylalkylamine D600 and those of an agonist DP were explored. Finally, a comparison was made of the voltage dependence and time dependence of recovery of I_{Ca} and EC coupling phenomena.

Effects of Ca^{2+} Antagonists

Nifedipine and Holding Potential

Effects of nifedipine on EC coupling have been described by Lamb [33], Ríos and Brum [44], and Dulhunty and Gage [10]. Figure 1 shows that the effects depend on the

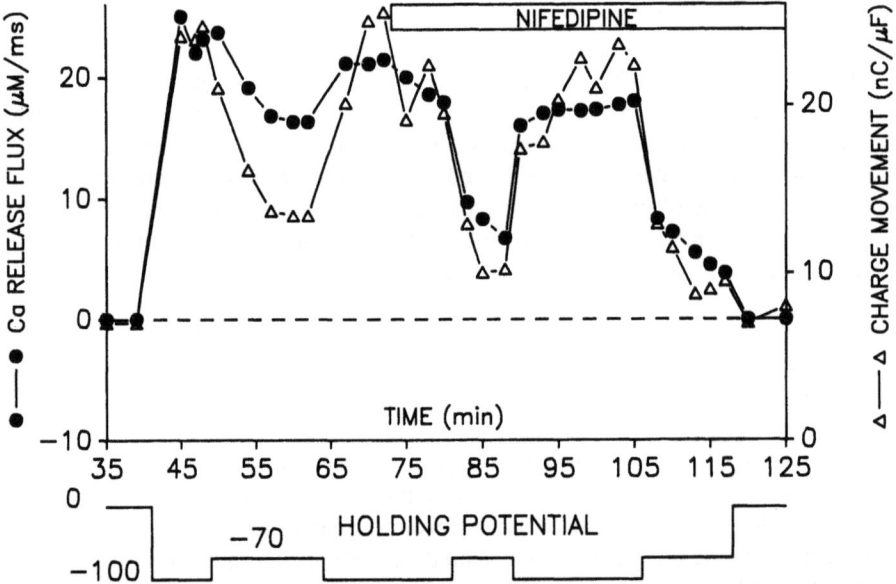

Fig. 1. Peak Ca^{2+}-release flux *(circles)* and total charge moved *(triangles)* during individual pulses from −70 mV (prepulse level) to 0 mV. *Abscissa* indicates time after beginning of experiment. *Top box* represents 500 nM nifedipine admitted in the external solution. *Lower trace* shows changes in holding potential. Fiber diameter (D) 55 μm, length (L) 540 μm, linear capacitance (C_L) 7.2 nF, temperature (T) 11°C

holding potential (HP). A fiber was subjected to short depolarizing pulses that cause charge movement and Ca^{2+} release flux. The peak of Ca^{2+} release flux and the total charge moved during the pulse are plotted for the successive pulses as a function of time in the experiment. The holding potential was initially 0 mV (the fiber was inactivated). Consequently, there was no release, and only 7 nC/μF intramembrane charge moved during the test pulse (which always went from a prepulse level of −70 mV to 0 mV). When the HP was changed to −100 mV, the fiber "reprimed": Ca^{2+} release appeared, and charge movement increased substantially. A partial depolarization of the HP to −70 mV caused partial inactivation; this effect reversed when HP was brought back to −100 mV. Nifedipine (500 nM, extracellular) caused only a slight decrease of both charge movement and Ca^{2+} release; however, when the fiber was depolarized to −70 mV, both variables decreased markedly. Repolarization to −100 mV brought release and charge back, close to reference levels. In summary, the effect of nifedipine is large in the partially inactivated fiber and small in the polarized situation. The effects on charge movement and Ca^{2+} release are comparable; however, the effect of depolarization per se on charge movement appears to be greater than the effect on Ca^{2+} release [44, 34].

Nifedipine and Low Ca^{2+}

Figure 2 demonstrates antagonism between nifedipine and extracellular Ca^{2+}: the fiber was initially hyperpolarized (HP = −110 mV) in a reference solution ($[Ca^{2+}]_o$ = 2 mM), and a pulse to −10 mV elicited the Ca^{2+} transient record A1. At this hyper-

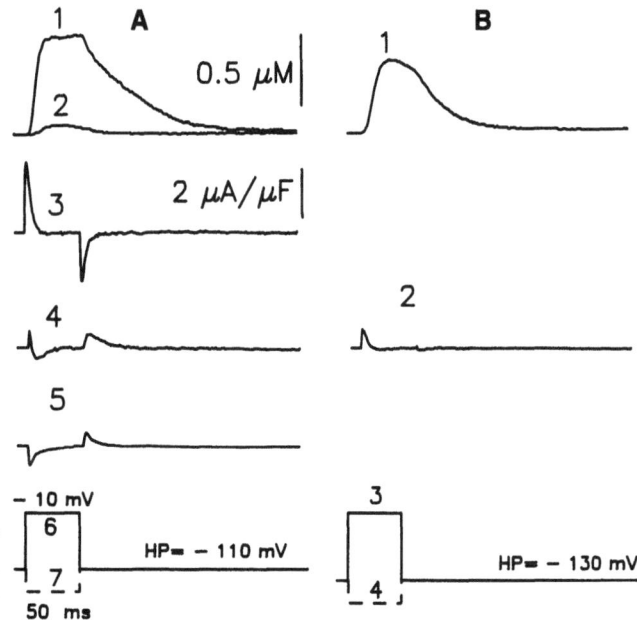

Fig. 2A, B. Effects of nifedipine and low Ca^{2+}. A Ca^{2+} transient (1) obtained in reference solution with nifedipine (1μM) a pulse (6) from −110 mV (holding potential) to −10 mV. Ca^{2+} transient (2) in a solution with nifedipine (1μM) no Ca^{2+}, 5 mM EGTA, and Mg added to a free concentration of 2.5 mM. Corresponding charge movement records (3, 4) were obtained by a P/2.5 procedure [with test (6) and control (7) pulses]. Direct difference (control current in nifedipine minus control current in reference) is also shown (5). **B** Effect of changing HP to −130 mV in Ca^{2+} transients (1) and charge movement (2). Note that new test and control pulses start from −130 mV. D 60 μm, C_L (as measured with pulse 4) 8.3 nF, L 540 μm, T 7°C. See text for further explanations

polarized holding potential, neither nifedipine (as shown in Fig. 1) nor low $[Ca^{2+}]_o$ (as shown by Brum et al. [5]) appreciably inhibit the Ca^{2+} release function. As shown in record A2, 3 min after exposure to a solution with nifedipine (1 μM) and 2.5 mM magnesium substituted for Ca^{2+} ("nifedipine + 2.5 Mg"), the Ca^{2+} transient almost disappeared. In this and four other experiments, Ca^{2+} release flux was reduced to less than 10%; the effect demonstrates potentiation between the effects of nifedipine and lack of extracellular Ca^{2+} (or, equivalently, antagonism between nifedipine and Ca^{2+}). This is, again, an effect secondary to the interference with charge movement. Record A3 in Fig. 2 shows charge movement in the reference situation. It was obtained by a conventional (P/2.5) procedure, in which the current during a -40 mV "control" pulse was scaled up and subtracted from the current during the $+100$ mV "test" pulse. When the same procedure was attempted in the nifedipine +2.5 Mg situation, the paradoxical record A4 was obtained, in which the charge movement current goes negative at the ON. The simple explanation of this paradox is that there is proportionally more charge movement during the negative-going control pulse than during the test pulse in nifedipine +2.5 Mg; this results in a negative current difference.

The charge movement during the test depolarization was reduced by the effect of nifedipine +2.5 Mg. Additionally, the charge movement during the negative-going control *increased,* as demonstrated by record A5, which is the direct difference: (control current in nifedipine +2.5 Mg) − (control current in reference solution). In quantitative terms, 5.8 nC/μF of charge appeared in the -40 mV control in nifedipine +2.5 Mg. This is analogous to the effects of low $[Ca^{2+}]_o$ on charge 2 (see Table 1), demonstrated by Brum et al. [6].

Figure 2B demonstrates that the joint effect of nifedipine and lack of Ca^{2+} is reduced by further hyperpolarization. The fiber of Fig. 2, almost completely inhibited in nifedipine +2.5 Mg, was hyperpolarized to a HP of -130 mV. The Ca^{2+} transients recovered rapidly and substantially. This was accompanied by the reappearance of charge movement (B2, obtained by a similar procedure, the pulses now starting from -130 mV). However, the charge movement was still small (record B2), and frequently the charge displaced during the ON transition seemed greater than at the OFF.

The experiments of Figs. 1, 2 show that the effects of nifedipine, $[Ca^{2+}]_o$, and HP are interrelated; a model described in the subsequent discussion accounts for these interactions. Finally, the experiments indicate that all these effects are primarily on the T-tubular molecule that generates charge movement (the hypothetical voltage sensor).

(−)202−791

We conducted experiments with the (−) enantiomer of the dihydropyridine 202−791 [28] and the phenylalkylamine D600 [41]. At 100 nM, (−)202−791 had effects similar to nifedipine. We could not obtain reliable estimates of a half-effect concentration, as we had especially severe problems of contamination of our Lucite chambers with this drug. Contamination is revealed by an inhibitory effect that lingers over several experiments and can only be eliminated by cleaning of the chamber with rather destructive procedures (a combination of prolonged soaking in ethanol or mineral oil and UV irradiation). After several attempts to use this drug, we were left with the impression that it had profound inhibitory effects, perhaps HP-dependent, and that very low concentrations were required.

D600

Eisenberg et al. [11] demonstrated a radical use-dependent inhibition of EC coupling by D600, which they called "paralysis," and showed that an intervening K contracture was needed for manifestation of the effect. In the present studies, we applied D600 at either 30 µM or 1.5 µM. Figure 3A shows the pulse protocol: instead of intervening K contractures, the 100-ms test pulses were separated by long (500-ms) "conditioning" depolarizations. Ca^{2+} transients, Ca^{2+} release flux, and charge movement were measured during the successive test pulses. Charge movement and Ca^{2+}-release

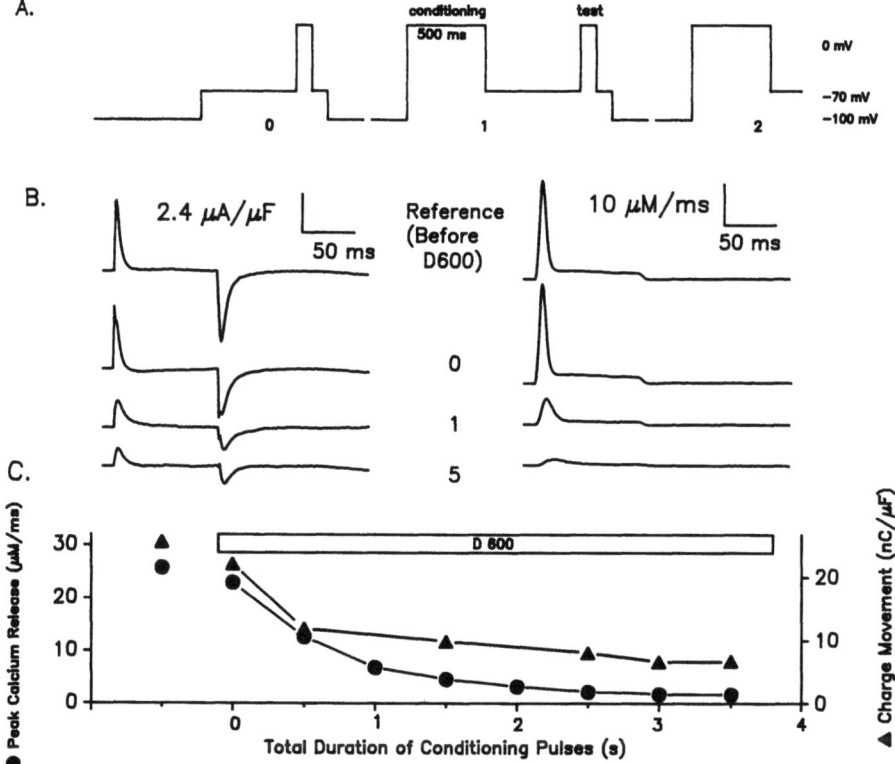

Fig. 3A–C. Onset of the effect of D600. **A** Pulse protocol. **B** Selected traces of charge movement *(left)* and Ca^{2+}-release flux *(right)* elicited by test pulses from −70 mV (prepulse level in **A**) to 0 mV. **C** Peak release *(circles)* and charge moved *(triangles)* during successive test pulses. The fiber is initially in reference solution; *horizontal bar* shows application of 30 µM D600, and *abscissa* represents cumulated duration of intercalated depolarizations (0.5 s each). D600 per se has little effect *(points of abscissa 0)*; successive conditioning depolarizations bring release to near 0 and charge to 6 nC/µF. D 65 µm, C_L 8.8 nF, T 7°C. See text for further explanations

records (represented in Fig. 3B) were reduced only slightly by the presence of 30 µM D600; however, the intercalated long depolarizations had the consequence that successive test pulses gave less Ca^{2+} release and charge movement. After five repeats (Fig. 3B), the waveforms were much smaller. In Fig. 3C, peak value of the release waveform and total charge moved are plotted as a function of cumulative duration of the conditioning depolarizations. Both variables decayed roughly in parallel, reinforcing the idea [23] that decay in release is a consequence of the reduction in charge. As with nifedipine before, the charge movement (measured with positive controls [4]) did not decay to 0, but to about 6 nC/µF.

In Figure 4, a different pattern of pulses was used to explore charge movements in a voltage range negative to the holding potential (pulse protocol and records labeled *charge 2*), as well as in the depolarizing voltage range (labeled *charge 1*; for timing details see Table 1). In this and five other fibers (Table 1), the obvious decay in charge 1 that accompanies D600 paralysis is paralleled by an increase in charge 2 (Fig. 4B).

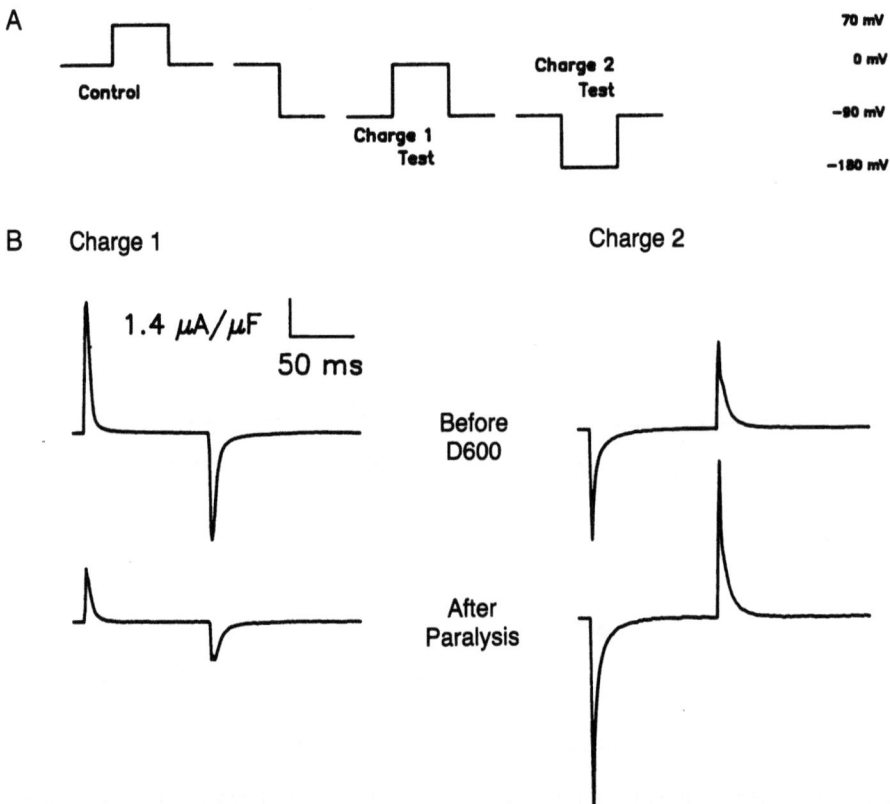

Fig. 4A, B. Effects of D600 on charge 1 and charge 2. A fiber is subjected to symmetric 90-mV pulses from a HP of −90 mV, and the charge moved (charge 1 with the positive-going pulse and charge 2 with the negative pulse) is measured using as control of linear capacitance the current recorded at HP 0 mV with a pulse to 70 mV. The procedure is repeated after paralysis (caused by a 20-s depolarization to 0 mV in 30 µM D600). D 79 µm, L 520 µm, C_L 7.1 nF, T 7°C

Table 1. Effects of D600 on EC coupling

Fiber no.	Charge 1 (nC/μF)		Charge 2 (nC/μF)		Peak release (μM/ms)	
	Before drug	After paralysis	Before drug	After paralysis	Before drug	After paralysis
264	12.68	1.74	7.53	13.49	4.29	0.08
266	8.67	4.03	15.84	22.53	10.81	0.30
267	11.64	6.07	24.37	26.19	13.37	0.05
268	19.35	8.03	19.47	20.61	3.1	0.05
273	7.82	1.85	12.78	22.31	25.71	0.17
325	25.5	12.7	9.3	14.1	25.7	1.8
Average difference[a] (mean ± SD)	−8.54 ± 3.53		5.06 ± 3.17			

Charge 1 and Ca^{2+} release were measured during pulses from −70 mV to 0 mV, charge 2 during pulses from −70 mV to −160 mV. All fibers were in reference solution except no. 325, which was in a solution with 10 mM [Ca^{2+}]. Data after paralysis were obtained in 30 μM D600 (except no. 264 in 100 μM), after a 1- to 5-min depolarization to 0 mV. Holding potential, −90 mV, was imposed at least 5 min before the test pulses.
[a] (charge after paralysis) − (charge before drug).

The overall increase in charge 2, in the voltage range explored by these pulses, was 5.06 nC/μF (SD 3.17), whereas the increase in charge 1 was greater in all fibers (average 8.54 nC/μF, SD 3.53). In summary, D600 blocks Ca^{2+} release in a strongly use-dependent manner; the effect is probably due to an interaction with the voltage sensor.

Melzer and Pohl have studied the effects of D600 with similar techniques and found similar results [39]; however, in experiments in which the recovery of release and charge 1 was followed, upon hyperpolarization in the presence of D600, they did not find a change in charge 2 (personal communication). We cannot explain the discrepancy, otherwise than by the fact that our slightly different procedure (to measure charge 2 before and after paralysis by a depolarizing pulse) permitted comparisons that were closer in time.

D600 and I_{Ca}

The effect of D600 on membrane Ca^{2+} current had the same remarkable use dependence that has been described for the effects on EC coupling. In the experiment of Fig. 5, large test pulses were applied that activated substantial "slow" Ca^{2+} current (as demonstrated by the first inset record). Of course, these test pulses also generated a large flux of Ca^{2+} release (records not shown). The graph plots peak of Ca^{2+} release flux and amplitude of the membrane Ca^{2+} current, as measured by the OFF tail. The reference values of both variables are high (first points in the graph); the mere admission of D600 in the bath did not have an important effect on either variable. A 60-s depolarization was then applied; the subsequent test pulses caused negligible Ca^{2+} release and I_{Ca} (second inset). The use-dependent abolition of Ca^{2+} release is

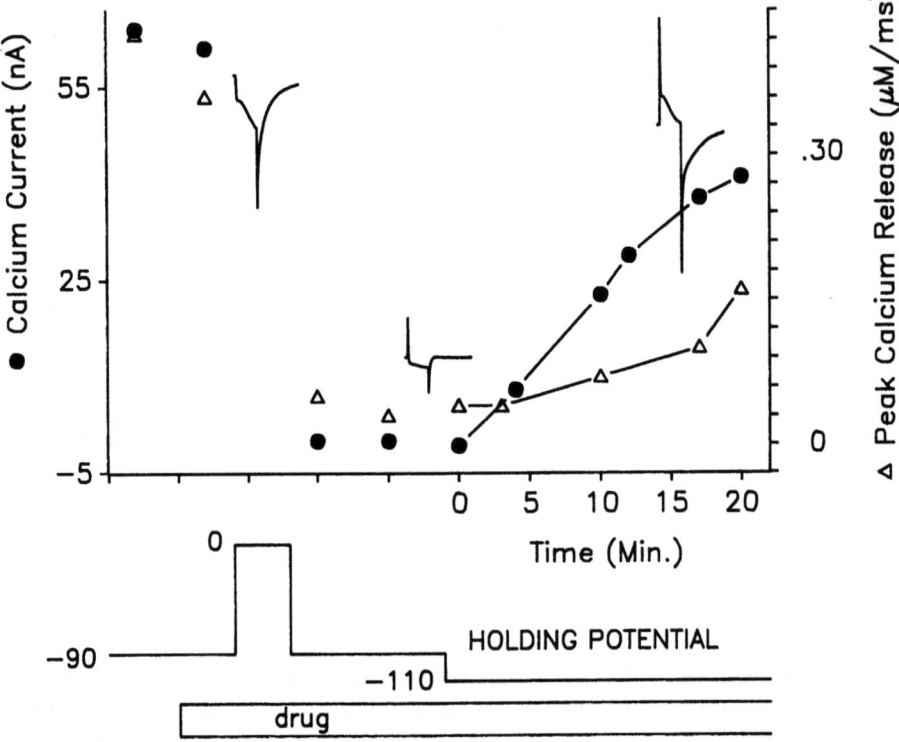

Fig. 5. Paralysis of I_{Ca}: asymmetric current at the end of a 200-ms pulse to 0 mV *(triangles)* and peak Ca^{2+}-release flux *(circles)* in successive pulses. Bottom box marks application of 1.5 µM D600, which per se has little effect on current *(inset)* and release. Bottom trace represents HP during the experiment. After a 60-s depolarization, both current and release remain reduced. Both recover upon hyperpolarisation to −110 mV. D 71 µm, L 520 µm, C_L 9.7 nF, T 9°C

therefore accompanied by a similar block of the membrane Ca^{2+} current, at a drug concentration that does not have an effect [40] without the intervening long depolarization.

The ensuing manipulation in the experiment illustrates further the striking parallelism between both effects: when the holding potential was changed by 20 mM in the hyperpolarizing direction, both I_{Ca} and Ca^{2+} release flux recovered slowly in successive test pulses. All the properties of the effect of D600 on EC coupling are thus repeated in the effect on I_{Ca}.

Effects of a Ca^{2+} Agonist

The Ca^{2+} agonist dihydropyridine (+)202–791 was tested as illustrated in Fig. 6. The effects on EC coupling were inhibitory in a way that depended strongly on HP. Figure 6 shows effects of 100 nM drug at HP of −90 mV: the Ca^{2+} transients were minimally affected by the drug and the derived Ca^{2+} release flux was reduced by about 10%. I_{Ca}

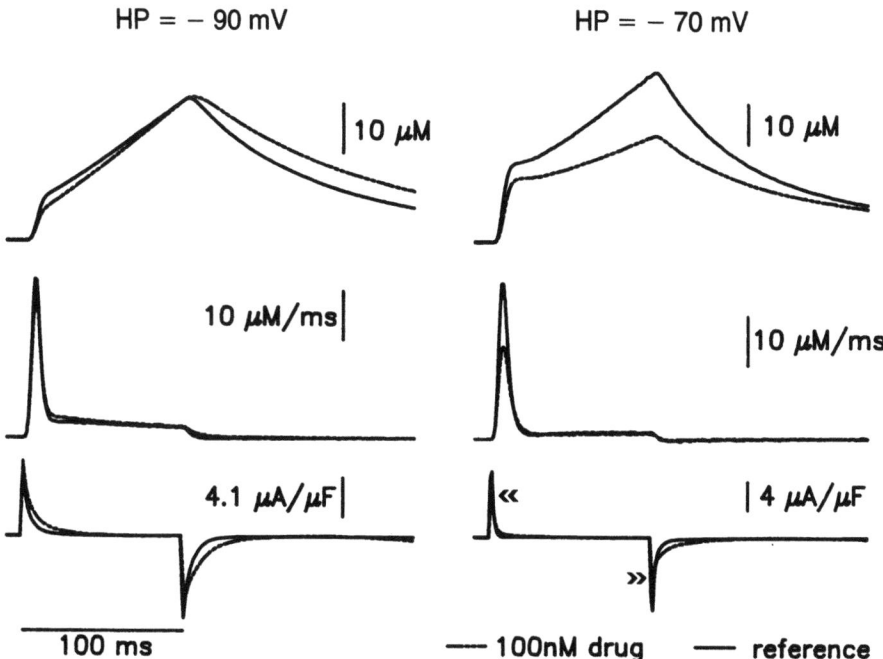

Fig. 6. Effects of (+)202–791. *Top* Ca^{2+} transients. *Middle* Ca^{2+}-release flux. *Bottom* Charge movement currents. Pulses to 0 mV in reference solution *(continuous lines)* or in 100 nM (+)202–791 *(dash-dot lines; arrows* mark peaks). Release flux is reduced at HP −70 mV *(right)* but not changed at HP −90 mV *(left)*. D 86 μm, L 540 μm, C_L 9.6 nF, T 8°C

was increased by about 150% (records not shown). The records of charge movement appear to have changed little, though the potentiated I_{Ca} made charge movement measurements unreliable.

The situation was different when the fiber was partially depolarized (HP −70 mV, Fig. 6): Ca^{2+} transients and Ca^{2+} release flux were reduced by 50% or more (56% average after 10 min in the drug, SD 16%, n = 4 fibers). Charge movement appears to be smaller in the presence of the drug, although the problems in the measurement of charge movement introduced by the potentiated I_{Ca} were present in the partially depolarized fibers as well.

To summarize the results presented so far: all Ca^{2+} agonists and antagonists tested have an effect on EC coupling at concentrations comparable to those that affect Ca^{2+} channels. However, the Ca^{2+} agonist used did not cause potentiation of Ca^{2+} release. Other differences between the gating of Ca^{2+} release channels and that of T-membrane Ca^{2+} channels are demontrated by the following series of experiments.

Voltage Dependence

Figure 7 plots the voltage dependence of peak Ca^{2+} release flux (averages ± SEM) in seven fibers. The experiments were conducted in a 100 mM Ca^{2+} external solution in

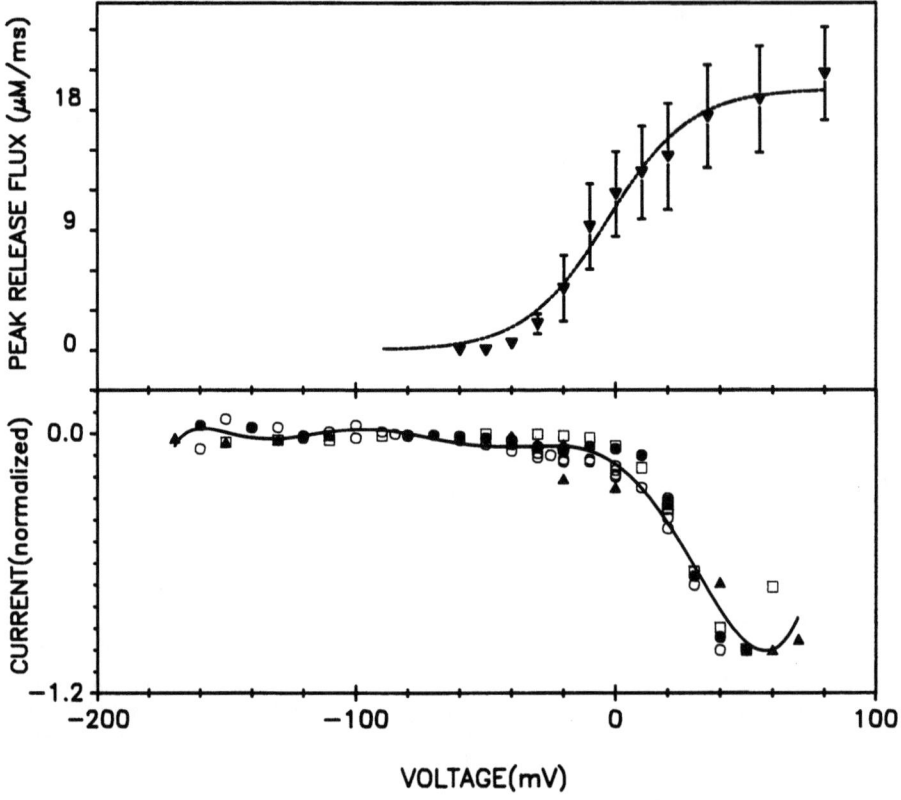

Fig. 7. Voltage-dependence of I_{Ca} and Ca^{2+} release flux. *Top* Peak Ca^{2+} release flux (averages ± SEM) in seven fibers in an external solution containing 100 mM Ca methanesulfonate (plus pH buffer and TTX as in [5]). *Dash-dot line* shows two-state canonical [46] distribution function, fitted to the averages. Best-fit parameters: maximum release 19.6 µM/ms, center voltage −3 mV, slope factor 16 mV. *Bottom* Asymmetric current, at the end of 200-ms pulses, recorded in five of the seven fibers above. The *continuous line* is a 7th-order polynomial fitted to all points. The asymmetric current was generated scaling and subtracting from the total current a control obtained with a 10-mV pulse from HP −90 mV

order to make I_{Ca} large, thus permitting a more sensitive measurement of the activation of the membrane Ca^{2+} channels. The lower graph in Fig. 7 is a plot of membrane current, presumably carried by Ca^{2+}, at the end of 200-ms pulses. Even though the two phenomena are not strictly comparable (as the driving force for I_{Ca} changes with pulse voltage), it is clear that their voltage dependence is not the same; specifically, there is Ca^{2+} release at or above −30 mV, whereas the major component of I_{Ca} starts to activate at about 0 mV. (There is a hint of an inward current, variable from fiber to fiber, at lower potentials.)

The high $[Ca^{2+}]_o$ used causes a shift of all voltage-dependent processes to higher voltages by about 30 mV. In a solution with 2 mM Ca^{2+}, the threshold of Ca^{2+} release is near − 60 mV, and that of Ca^{2+} current is −30 mV.

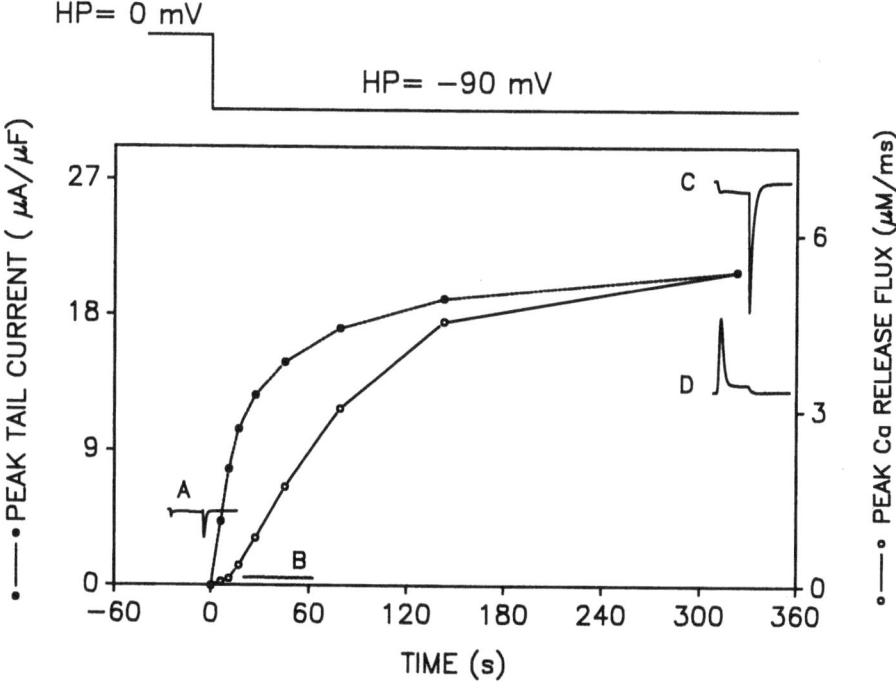

Fig. 8. Recovery after inactivation. *Top trace* indicates holding potential. Tail current *(filled circles)* and peak release *(open circles)* for successive pulses to 0 mV are shown. *Abscissa* shows time after repolarization. *Insets* show current *(A)*, and Ca^{2+} release *(B)* at 4 s and corresponding records *(C, D)* after 320 s. The external solution contains 100 mM Li methanesulfonate, 1 mM EGTA, and no Ca^{2+} added. D 76 µm, L 540 µm, C_L 7.5 nF, T 11°C

Recovery After Inactivation

Figure 8 shows yet another dissociation between gating of Ca^{2+} release and of I_{Ca}. We have demonstrated [3, 6, 42] that Ca^{2+} release is possible in solutions containing no Ca^{2+} provided that high concentrations of other metal ions are present. In a solution with no Ca^{2+} and high concentrations of monovalent cations, the ionic currents through Ca^{2+} chanels become huge [1]; this fact permits the experiment illustrated, carried out in an external solution containing 100 mM Li^+ as the sole metal ion. The holding potential was initially 0 mV, which caused complete inactivation of both Ca^{2+} release and membrane Ca^{2+} current. At time 0, the holding potential was set to -90 mV, and the repriming of both Ca^{2+} release flux and ion current through the (T-membrane) Ca^{2+} channels was followed in time. As shown in Fig. 8, the membrane current was very large after recovery (inset C); it is a current carried by Li^+ through the Ca^{2+} channel. The graph plots peak Ca^{2+} release flux and peak tail current (I_{Li}) and shows that I_{Li} recovers more rapidly than the release flux. This dissociation, seen in three experiments, again shows that release through the SR Ca^{2+} channel and gating of T-tubular Ca^{2+} channels are not simultaneous under all circumstances.

Discussion

Comparison with Previous Work

The present experiments demonstrate substantial similarities between the pharmacology of EC coupling and that of Ca^{2+} channels. Table 2 puts the present results in perspective: it compares available data on drug effects on membrane Ca^{2+} currents and on various manifestations of EC coupling. (I_{Ca} has been the subject of many pharmacological studies, and the entries in Table 2 are only pointers to that literature.)

Table 2 shows that, in general, all Ca^{2+}-active drugs have inhibitory effects on EC coupling. This includes the agonist and antagonist DPs, the phenylalkylamines, and the benzothiazepines. Additionally, several studies have reported potentiating

Table 2. Effects of Ca^{2+}-active drugs on I_{Ca} and EC coupling

	Effects			
	DP antagonists	Diltiazem	D600	DP agonists
I_{Ca}	Inhibition [1] 1 µM	Inhibition [19] 50 nM [18] 50 µM	Inhibition [40] 15 µM [ª] 1 µM [41] 70 nM	Potentiation [16, 34] 50 µM Inhibition [45]ᵉ
K-contractures	Inhibition [18] 50 nM [13] 16 nM [43] 30 nM [10]ᵇ Potentiation [10]ᵇ	Inhibition [18] 1 µM	Inhibition [2, 11] 30 µM	Inhibition [10] 50 µM
Tension by twitch, tetanus, or voltage clamp	Inhibition [10]ᵇ [24] 10 µM Potentiation [10, 17]ᵇ No effect [36] 100 µM	Inhibition [17] 25 µM Potentiation [19] 1 µM	Inhibition [13,14,17] 10 µM [29]ᶜ Potentiation [2, 9] 10 µM	Inhibition [10] 50 µM [24] 100 nMᵈ Potentiation [24] 100 nMᵈ
Calcium transients	Inhibition [44,ª] 30 nM	NA	Inhibition [ª] 1 µM [9, 39] 30 µM	Inhibition [ª] 100 nM
Charge movement (charge 1)	Inhibition [33] 10 µM [44] 30 nM	Potentiation [47] 250 µM	Inhibition [ª,7,23,39] 30 µM	Inhibition [ª] 100 nM

Half-effect or typical concentrations are given after reference citations.
NA, not available
ª Results from this paper
ᵇ Potentiation or inhibition depending on external free [Ca].
ᶜ Verapamil instead of D600 was used.
ᵈ Inhibition of a fast phase, potentiation of a slow phase of tension under voltage clamp.
ᵉ Potentiation or inhibition depending on voltage.

effects. Most of these potentiating effects may be rationalized as a left shift of the activation curves of EC coupling (that is, the curves relating EC coupling events to test voltage).

The fact that all Ca^{2+} channel antagonists have effects on EC coupling is evidence of a close similarity between the voltage sensors of EC coupling and Ca^{2+} channel proteins. Our laboratory has recently provided additional evidence of this similarity along very different lines. First, we showed that EC coupling could proceed in the absence of $[Ca^{2+}]_o$, provided that *metal ions* were present [3]. We later showed that the results could be explained by the existence of an essential cation-binding site (the "priming site") at the voltage sensor, which has to be occupied in order to permit EC coupling [4, 6]. Interpreting the efficacy of the ions to support EC coupling as a measure of their affinity for the priming site, we estimated relative binding affinities for series IA and IIA metal ions. We found that the sequence was similar to the permeability sequence found by Hess et al. [20] for the L-type cardiac Ca^{2+} channel. Since this sequence is believed to reflect the affinity of the ions for intrapore permeation sites, the similarity between sequences reveals chemical similarities between the priming site of the skeletal muscle voltage sensor and the intrapore Ca^{2+} channel site(s).

Can this similarity be taken to imply that the voltage-sensing molecules are actual, functioning Ca^{2+} channels? Other results in this contribution and in the literature show that the processes of Ca^{2+} release from the SR and Ca^{2+} current through the T-membrane channel have different properties, including the following:

1. I_{Ca} is potentiated by (+)202–791, in parallel with an inhibition of Ca^{2+} release flux.
2. I_{Ca} and Ca^{2+} release flux have different voltage dependence of activation.
3. They recover at different rates after inactivation.

These differences virtually rule out the possibility that the voltage sensors of EC coupling are the channels underlying I_{Ca}. The existence of a two-way dissociation – EC coupling without I_{Ca} gating (Fig. 7, at voltages between –30 and 0 mV) and I_{Ca} in the absence of EC coupling (Fig. 8 at times early in the recovery process) – indicates that voltage sensing for EC coupling and I_{Ca} gating are carried out by different proteins, however close their similarity may be.

Interdependent Effects of Drugs, Ca^{2+}, and Holding Potential

These experiments stress the interdependence among three agents: the drugs, extracellular free Ca^{2+}, and the holding potential. In the following, we elaborate on an already existing framework of theory that explains this interdependence. This theory includes the assumption of specific interactions between Ca^{2+} and the voltage sensor, different from the surface charge-modifying role that Ca^{2+} ions have traditionally been given in classical biophysics [21]. We feel that this theory, which explains a large body of experimental results in EC coupling, should be given consideration to explain some effects of Ca^{2+}, other ions, and Ca^{2+} antagonists on Ca^{2+} channels (like those recently tackled by Kass and Krafte [26]).

Based on the properties of charge 1 and charge 2, Brum and Ríos [4] proposed that the voltage sensor of EC coupling exists in at least four states, interconnected as follows:

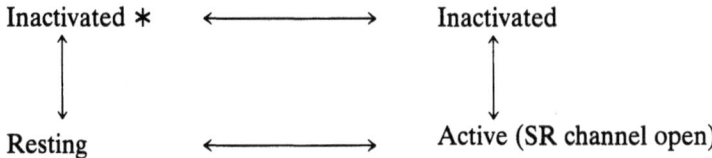

They proposed that the horizontal transitions are fast and generate observable charge movement; the Resting⟷Active transition generates charge 1, moving with a central potential of $V_1 \sim -30$ mV. The Inactivated ⟷ Inactivated * transitions generate charge 2, centered at $V_2 \sim -120$ mV. In this view, the vertical transitions are slow and not directly driven by the membrane electric field.

To explain the known fact that $[Ca^{2+}]_o$ antagonizes inactivation, Brum et al. [6] proposed the following addition to the model:

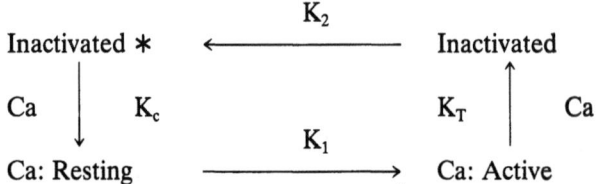

In this view, Ca^{2+} binds preferentially to the two lower ("primed") states, thus stabilizing them and antagonizing inactivation. In the absence of Ca^{2+}, the system inactivates (unless other ions are present and take its place, binding with lower affinity to the priming site). The Ks are the equilibrium ratios, defined in the direction of the arrows; for example,

$$K_T = I/Ca:A = K/Ca \tag{1}$$

where I and Ca:A are probabilities of the Inactivated and Ca:Active states, Ca is $[Ca^{2+}]_o$, and K is the dissociation constant of the Ca^{2+}-binding reaction:

$$K = I.Ca/Ca:A \tag{2}$$

The assumption that the horizontal transitions are voltage sensitive amounts to equating

$$K_1 = \exp[(V-V_1)/k] \tag{3}$$
$$K_2 = \exp[-(V-V_2)/k] \tag{4}$$

where V_1 and V_2 are the midpoint voltages of the horizontal transitions, assumed to be distributed following a Boltzmann function, and $k = RT/z'F$, with z' representing the apparent valence of the sensor.

The condition of microscopic reversibility of the transitions around the loop of four states requires the following relationship between the equilibrium ratios:

$$1/K_c = K_T \exp[(V_2-V_1)/k] \tag{5}$$

Finally, drug action is introduced as follows:

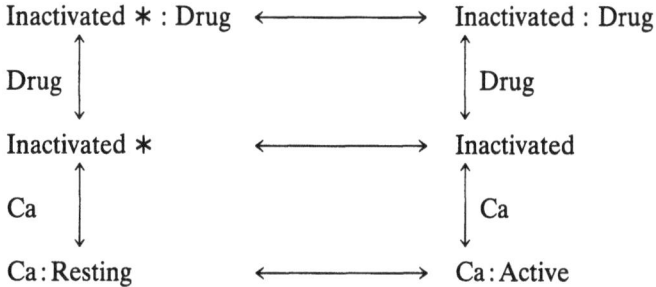

This scheme assumes that the drug only binds with high affinity to the inactivated states. Consequently, the drugs "cause" inactivation (the "induced fit" phenomenon [8]). The drugs are antagonized by all interventions that push the receptors away from inactivation (like hyperpolarization or high $[Ca^{2+}]_o$). The scheme is formalized below to explain other observations, including the apparent reduction in DP-binding sites when a fiber is polarized [47].

Brum et al. [6] derived the mathematical expression of the "inactivation curve" of the model. In the following, we generalize their expression to include the effect of Ca^{2+} and Ca^{2+} antagonists.

The equilibrium distribution of the inactivation variable, that is, the sum of the probabilities of the noninactivated states Ca:Resting and Ca:Active, is given by Eq. 9 of [6]

$$Ca:R + Ca:A = \langle 1 + K_T(1 + \exp[-(V-V_2)/k]/(1 + \exp[-(V-V_1)/k])\rangle^{-1} \quad (6)$$

In the presence of drug, the equations will keep the form of Eq. 6, but the equilibrium constant between inactivated and activated sensors will be:

$$K_T' = (I + D:I)/Ca:A \quad (7)$$

where D:I is the probability of the Drug: Inactivated state. In turn,

$$D:I = D \cdot I/K_D \quad (8)$$

where K_D is the dissociation constant of drug binding. The effect of the drugs on the inactivation process is thus embodied in the following equation, derived from Eqs. 7, 8:

$$K_T' = (I/Ca:A)(1 + D/K_D) =$$
$$= K_T \cdot (1 + D/K_D) \quad (9)$$

Finally, the effect of $[Ca^{2+}]_o$ is introduced by substitution of K from 2 in (9)

$$K_T' = K(1 + D/K_D)/Ca \quad (10)$$

This equation justifies the observation that the drugs enhance, and Ca^{2+} antagonizes the tendency to inactivation, given by the equilibrium ratio K_T'.

Brum et al. [6] also showed that the inactivation expression of Eq. 6 is a sigmoidal (Boltzmann) function with negative slope, and a central voltage V given by

$$V/k = \ln \langle [\exp(V_1/k) + K_T \exp(V_2/k)] / (1 + K_T) \rangle \tag{11}$$

that is, V is a (logarithmic) mean of V_1 and V_2 with weights 1 and K_T. In the presence of drug or low Ca^{2+}, K_T' (which takes the place of K_T) increases according to Eq. 9. Therefore, the effect is to shift the midvoltage of inactivation towards V_2, which is −120 mV. This is in excellent agreement with the present results, as −120 mV is precisely the voltage that was necessary to surpass, in the negative direction, in order to achieve repriming in the presence of nifedipine and low Ca^{2+} (Fig. 2).

Ca^{2+}-Drug Antagonism

This model also permits one to express quantitatively the antagonism between Ca^{2+} and drug, in terms of an apparent dissociation constant for drug binding. We start defining the apparent dissociation constant (K_D') as follows:

$$K_D'/D = (I + I* + Ca:A + Ca:R) / (D:I + D:I*) \tag{12}$$

All probabilities in Eq. 12 may be expressed in terms of I by the use of Eqs. 1–5. After some rearrangements, an expression is found for K_D':

$$K_D' = K_D \langle 1 + (Ca/K) [1 + \exp \langle (V_1-V)/k \rangle] / [1 + \exp \langle (V_2-V)/k \rangle] \rangle.$$

This equation summarizes the antagonistic effects of Ca^{2+} and voltage on drug binding: Ca^{2+} directly increases K_D'; at large and positive V, K_D' tends to $K_D(1 + Ca/K)$. At large negative voltages, it tends to the limit $K_D \cdot (Ca/K) \cdot \exp[(V_1-V_2)/k]$; this is typically much greater than the positive voltage limit, as $\exp[(V_1-V_2)/k]$ has a value between 50 and 150.

The model does not explain the various observations of potentiation, which in general amount to a left shift of the activation curve. The model would account for this effect if the sensor in the Active state had a finite drug-binding affinity greater than in the Resting state. We have not pursued this generalization, as the potentiating effects seem minor by comparison.

Conclusion

The voltage sensor of EC coupling is sensitive to all types of Ca^{2+}-active organic molecules tested, which in general cause a voltage-dependent inhibition, antagonized by external Ca^{2+}. Some differences in drug effects, voltage sensitivity, and kinetics suggest that the voltage sensors are not at the same time Ca^{2+} channels contributing to I_{Ca}. The similarities indicate that the proteins underlying both functions are chemically related, to the extent that the voltage sensors may include a porelike region. Studies of the voltage sensor have led to the understanding of the role of extracellular Ca^{2+} as a specific stabilizer of non-inactivated states; it is possible that Ca^{2+} plays a similar role in Ca^{2+} channels.

Acknowledgement. This work was supported by a grant from the Muscular Dystrophy Association to E. R. and by grants from the National Institutes of Health, USA, to E. R., R. F. and Dr. Philip M. Best. We are grateful to Dr. Best for his support.

References

1. Almers W, McCleskey EW, Palade PT (1984) A non-selective cation conductance in frog muscle membrane blocked by micromolar external calcium ions. J Physiol 353:565–583
2. Berwe D, Gottschalk G, Luttgau HC (1987) The effects of the Ca-antagonist gallopamil (D600) upon excitation-contraction coupling in toe muscles of the frog. J Physiol 385:693–708
3. Brum G, Fitts R, Pizarro G, Ríos E (1987) A Ca–Mg–Na site must be occupied for intramembrane charge movement and Ca release in frog skeletal muscle. Biophys J 51:552a
4. Brum G, Ríos E (1987) Intramembrane charge movement in skeletal muscle fibres. Properties of charge 2. J Physiol 387:489–517
5. Brum G, Ríos E, Stéfani E (1988a) Effects of extracellular calcium on the calcium movements of excitation-contraction coupling. J Physiol 398:441–473
6. Brum G, Fitts R, Pizarro G, Ríos E (1988b) Voltage sensors of the frog skeletal muscle membrane require calcium to function in excitation-contraction coupling. J Physiol 398:475–505
7. Caputo C, Bolaños P (1988) Effect of D600 and La^{+++} on charge movement in depolarized muscle fibers. Biophys J 53:604a
8. Colquhoun D, Hawkes AG (1977) Relaxation and fluctuations of membrane currents that flow through drug-operated channels. Proc R Soc Lond B 199:231–262
9. Dorrscheidt-Kafer M (1977) The action of D600 on frog skeletal muscle: facilitation of excitation-contraction coupling. Pflugers Arch 369:259–269
10. Dulhunty AF, Gage PW (1988) Effects of extracellular calcium concentration and dihydropyridines on contraction in mammalian skeletal muscle. J Physiol in press
11. Eisenberg RS, McCarthy RT, Milton RL (1983) Paralysis of frog skeletal muscle fibres by the calcium antagonist D600. J Physiol 341:495–505
12. Ferguson DG, Schwartz HW, Franzini-Armstrong C (1984) Subunit structure of junctional feet in triads of skeletal muscle: a freeze-drying, rotary shadowing study. J Cell Biol 99:1735–1742
13. Fill DM (1987) Excitation-contraction coupling: The effect of calcium channel antagonistic drugs on reprimed and inactivated skinned skeletal muscle fibres. Thesis, University of Illinois at Urbana
14. Fill DM, Fitts R, Pizarro G, Rodríguez M, Ríos E (1988) Effects of Ca agonists and antagonists on EC coupling in skeletal muscle fibers. Biophys J 53:603a
15. Fill DM, Best PM (1988) Paralysis of contracture in skinned frog skeletal muscle fibers by calcium antagonists. J Gen Physiol (in press)
16. Franckowiak G, Bechem M, Schramm M, Thomas G (1985) The optical isomers of the 1,4-dihydropyridine Bay K8644 show opposite effects on calcium channels. Eur J Pharmacol 114:223–226
17. Gallant EM, Goettl VM (1985) Effects of calcium antagonists on mechanical responses of mammalian skeletal muscles. Eur J Pharmacol 117:259–265
18. Gamboa-Aldeco R, Huerta M, Stéfani E (1988) Effect of Ca channel blockers on K contractures in twitch fibres of the frog *(Rana pipiens)*. J Physiol 397:389–399
19. González-Serratos H, Valle-Aguilera R, Lathrop DA, García MdelC (1982) Slow inward calcium currents have no obvious role in muscle excitation-contraction coupling. Nature 298:292–294
20. Hess P, Lansman JB, Tsien RW (1986) Calcium channel selectivity for divalent and monovalent cations. Voltage and concentration dependence of single channel current in ventricular heart cells. J Gen Physiol 88:293–320
21. Hille B (1984) Ionic channels in excitable membranes. Sinauer, Sunderland, Massachusetts, pp 316–320
22. Horowicz P, Schneider MF (1981a) Membrane charge movement in contracting and non-contracting skeletal muscle fibres. J Physiol 314:565–593
23. Hui CS, Milton RL, Eisenberg RS (1984) Charge movement in skeletal muscle fibres paralyzed by the calcium entry blocker D600. Proc Nat Acad Sci USA 81:2582–2585

24. Ildefonse M, Jacquemond V, Rougier O, Renaud JF, Fosset M, Lazdunski M (1985) Excitation contraction coupling in skeletal muscle: evidence for a role slow Ca channels using Ca channel activators and inhibitors in the dihydropyridine series. Biochem Biophys Res Comm 129:904–909
25. Inui M, Saito A, Fleischer S (1987) Purification of the ryanodine receptor and identity with feet structures of junctional terminal cisternae of sarcoplasmic reticulum from fast skeletal muscle. J Biol Chem 262:1740–1747
26. Kass RS, Krafte D (1987) Negative surface charge density near heart Ca channels. Relevance to block by dihydropyridines. J Gen Physiol 89:629–644
27. Knudson CM, Imagawa T, Kahl SD, Gaver MG, Leung AT, Sharp AH, Jay SD, Campbell KP (1988) Evidence for physical association between junctional sarcoplasmic reticulum ryanodine receptor and junctional transverse tubular dihydropyridine receptor. Biophys J 53:605a
28. Kongsamut S, Kamp TJ, Miller RJ, Sanguinetti MC (1985) Calcium channel agonist and antagonist effects of the stereoisomers of the dihydropyridine 202-791. Biochem Biophys Res Comm 130:141–148
29. Kotsias BA, Muchnik S, Paz CAO (1986) Co^{2+}, low Ca^{2+} and verapamil reduce mechanical activtity in rat skeletal muscles. Am J Physiol 250:C40–56
30. Kovács L, Ríos E, Schneider MF (1979) Calcium transients and intramembrane charge movement in skeletal muscle fibres. Nature 279:391–396
31. Kovács L, Ríos E, Schneider MF (1983) Measurement and modification of free calcium transients in frog skeletal muscle fibres by a metallochromic indicator dye. J Physiol 343:161–196
32. Lai FA, Erickson HP, Rousseau E, Liu QY, Meissner G (1988) Purification and reconstitution of the calcium release channel from skeletal muscle. Nature 331:315–320
33. Lamb G (1986) Components of charge movement in rabbit skeletal muscle. The effects of tetracaine and nifedipine. J Physiol 376:85–100
34. Lamb G, Walsh T (1987) Calcium currents, charge movement and dihydropyridine binding in fast- and slow-twitch muscle of rat and rabbit. J Physiol 393:595–617
35. Leung AT, Imagawa T, Block B, Franzini-Armstrong C, Campbell KP (1988) Biochemical and ultrastructural characterization of the 1,4-dihydropyridine receptor from rabbit skeletal muscle. Evidence for a 52,000-Da subunit. J Biol Chem 263:994–1001
36. McCleskey EW (1985) Calcium channels and intracellular calcium release are pharmacologically different in frog skeletal muscle. J Physiol 361:231–249
37. Melzer W, Ríos E, Schneider MF (1984) Time course of calcium release and removal in skeletal muscle fibres. Biophys J 45:637–641
38. Melzer W, Ríos E, Schneider MF (1987) A general procedure for determining calcium release in skeletal muscle fibers. Biophys J 51:849–863
39. Melzer W, Pohl B (1987) Effects of D600 on the voltage sensor for Ca release in skeletal muscle fibres of the frog. J Physiol 390:151P
40. Palade PT, Almers W (1985) Slow calcium and potassium currents in frog skeletal muscle: their relationship and pharmacologic properties. Pflugers Arch 405:91–101
41. Pelzer D, Trautwein W, McDonald TF (1982) Calcium channel block and recovery from block in mammalian ventricular muscle treated with organic channel inhibitors. Pflugers Arch 394:97–105
42. Pizarro G, Fitts R, Ríos E (1988) Selectivity of a cation-binding membrane site essential for EC coupling in skeletal muscle. Biophys J 53:645a
43. Rakowski RF, Olszewska E, Paxson C (1987) High affinity effect of nifedipine on K-contracture in skeletal muscle suggests a role for calcium channels in skeletal muscle. Biophys J 51:550P
44. Ríos E, Brum G (1987) Involvement of dihydropyridine receptors in excitation-contraction coupling in skeletal muscle. Nature 325:717–720
45. Sanguinetti MC, Krafte DS, Kass RS (1986) Voltage dependent modulation of Ca channel current in heart cells by Bay K8644. J Gen Physiol 88:369–392
46. Schneider MF, Chandler WK (1973) Voltage dependent charge movement in skeletal muscle: a possible step in excitation-contraction coupling. Nature 242:244–246
47. Schwartz LM, McCleskey EW, Almers W (1985) Dihydropyridine receptors in muscle are voltage-dependent but most are not functional calcium channels. Nature 314:747–751
48. Sharp AH, Imagawa T, Leung AT, Campbell KP (1987) Identification and characterization of the dihydropyridine-binding subunit of the skeletal muscle dihydropyridine receptor. J Biol Chem 262:12309–12315
49. Walsh KB, Bryant SH, Schwartz A (1987) Suppression of charge movement by calcium antagonists is not related to calcium channel block. Pflugers Arch 409:217–219

2 The Structure of the Calcium Channel

Biochemistry and Molecular Pharmacology of Ca^{2+} Channels and Ca^{2+}-Channel Blockers

J. Barhanin, M. Fosset, M. Hosey, C. Mourre, D. Pauron, J. Qar, G. Romey, A. Schmid, S. Vandaele, and M. Lazdunski*

Centre de Biochimie du Centre National de la Recherche Scientifique, Parc Valrose, 06034 Nice Cedex, France.

Most of what we now know of the structure of the Ca^{2+} channel comes from work that has been carried out with skeletal muscle membranes. This is because it was shown a number of years ago that the richest source of receptors for Ca^{2+}-channel blockers and particularly for 1,4-dihydropyridines (DHP) ist the skeletal muscle T-tubule membrane system (Fosset et al. 1983). There are two main types of Ca^{2+} channels in skeletal muscle: one of them is the low-threshold, or T-type, Ca^{2+} channel, the other one the high-threshold, or L-type, Ca^{2+} channel (Cognard et al. 1986a, 1986b). Patch-clamp analysis has actually identified two different subtypes of L-type Ca^{2+} channels in mammalian skeletal muscle (Cognard et al. 1986a). Both are blocked by 1,4-dihydropyridines, but they have different voltage sensitivities of the activation process and different inactivation rate kinetics.

In muscular dysgenesis, a mouse muscle disease corresponding to an absence of contraction, receptor sites for Ca^{2+}-channel blockers are nearly absent in muscle membranes, and functional expression of the Ca^{2+} channel is lacking (Pinçon-Raymond et al. 1985; Beam et al. 1986; Romey et al. 1986). In vitro innervation of diseased myotubes by spinal cord cells of normal animals has been shown to restore both Ca^{2+}-channel activity and contraction (Rieger et al. 1987). The lack of Ca^{2+}-channel and contractile activity in the mutant is due to the absence of intact triadic structures (Pinçon-Raymond et al. 1985; Rieger et al. 1987). The diseased muscle only contains pseudo-diads (Pinçon-Raymond et al. 1985). Reinnervation restores triadic organization, together with Ca^{2+}-channel activity and contraction. Therefore, an intact triadic organization is essential for the expression of both Ca^{2+}-channel activity and contraction in skeletal muscle. Innervation seems to be essential to reach a correct stage of triad differentiation.

The Ca^{2+}-channel blocker receptor of skeletal muscle has been photoaffinity labelled and purified. These two approaches have shown that the two main components in its structure have a (M_r) of about 170 kDa (Hosey and Lazdunski 1988). One of these components (M_r 165 kDa) has been photoaffinity labelled by diltiazem, bepridil, PN200–110, azidopine and azidopamil (Galizzi et al. 1986a; Hosey et al. 1987; Sharp et al. 1987; Sieber et al. 1987; Striessnig et al. 1987; Takahashi et al. 1987; Vaghy et al. 1987); its mobility is unaltered upon reduction. The other component, which also has a M_r of about 170 kDa, is cleaved by reduction into two components of 140 kDa and

---

* This work was supported by the Centre National de la Recherche Scientifique, the Fondation pour la Recherche Médicale, and the Association des Myopathes de France.

about 30 kDa respectively (Borsotto et al. 1985; Schmid et al. 1986a; Barhanin et al. 1987). The first component is often known as the α_1-component, while the other one is known as the α_2-component. The α_2-component is heavily glycosylated (Barhanin et al. 1987), while the α_1-component is only very partially glycosylated (Hosey et al. 1987). The α_1-component, which bears all binding sites for Ca^{2+}-channel blockers, is also the target of cAMP-dependent phosphorylation (Fig. 1; Hosey et al. 1986, 1987), which is known to activate the skeletal muscle Ca^{2+} channel (Schmid et al. 1985) in the same way as the cardiac Ca^{2+} channel (Reuter 1983). The α_1-component is the subunit that has recently been cloned and sequenced, and which has an extensive homology with the Na^+-channel protein (Tanabe et al. 1987). Antibodies against the different polypeptide elements of the α_2-subunit reveal that these are also present in heart, brain and smooth muscle (Schmid et al. 1986a, 1986b; Cooper et al. 1987). The 140 kDA or 30-kDa subunits are not detected in tissues that lack Ca^{2+}-blocker receptors. Moreover, a monoclonal antibody against the α_2-subunit immunoprecipitates both DHP and phenylalkylamine binding activity (Vandaele et al. 1987).

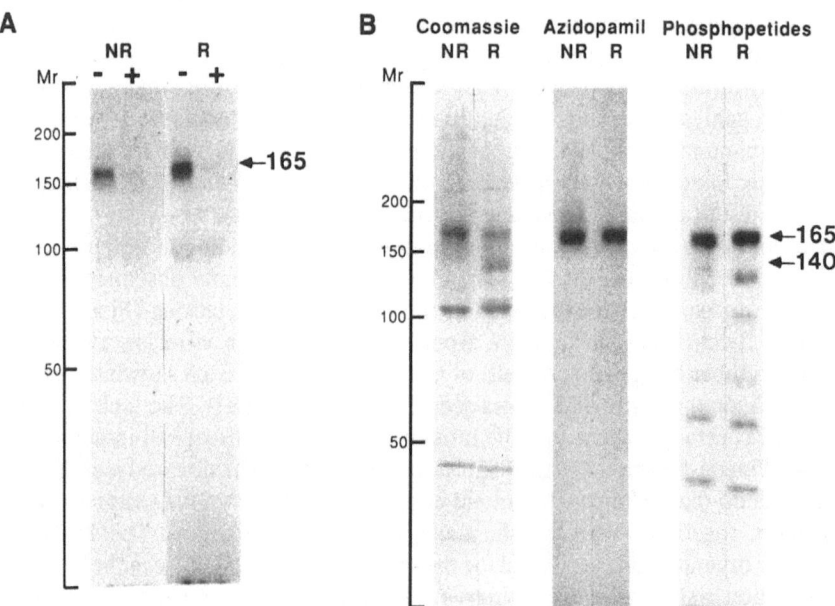

Fig. 1A, B. Photoaffinity labelling and phosphorylation of peptides present in membranes and partially purified preparations of DHP- and phenylalkylamine-sensitive Ca^{2+} channels from rabbit skeletal muscle. **A** T- tubule membranes were photoaffinity labelled as described in the text in the absence (−) or presence (+) of 1 μM (±)verapamil, and the labelled peptides were analysed by gel electrophoresis and fluorography. The fluorogram shown was exposed for 14 days. **B** Partially purified preparations obtained after WGA-Affigel chromatography were analysed by gel electrophoresis under reducing (*R*) or non-reducing (*NR*) conditions. Shown are the Coomassie bluestained gels depicting the peptide content, fluorograms showing the [^{3}H]azidopamil-labelled peptides, and autoradiograms depicting the peptides phosphorylated by cAMP-dependent protein kinase. The fluorogram was obtained after a 5-day exposure using Kodak XAR film, while the autoradiogram was obtained after a 6-h exposure using Kodak X-Omat S film. Note that the small reduction-induced increase in the apparent molecular weight of the 165-kDA peptide was not evident in the Coomassie-stained gel because of the overlap in mobility of the 170- and 165-kDA peptides under non-reducing conditions

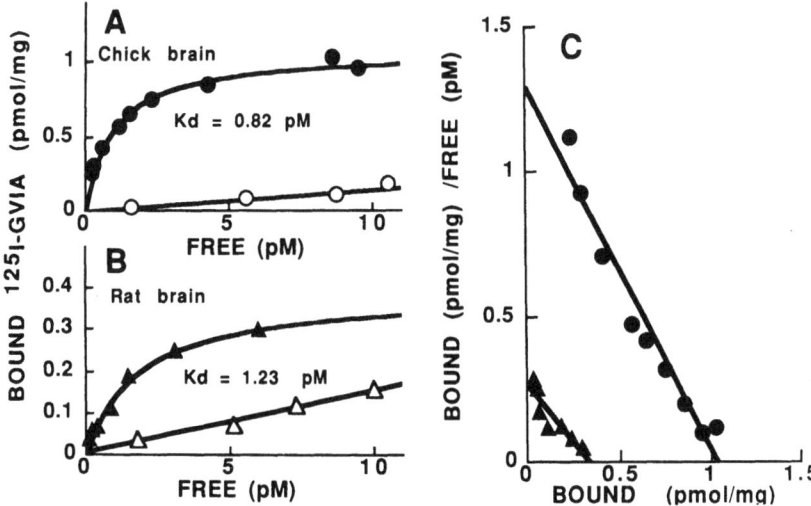

Fig. 2 A–C. Equilibrium binding of ^{125}I-GVIA to chick and rat brain synaptosomes. Chick (**A**) or rat (**B**) brain synaptosomes (1 and 3 µg protein per millilitre, respectively) were incubated with various concentrations of ^{125}I-GVIA for 4 h at 25 °C (final volume 4.5 ml). Each value is the mean of duplicate filtrations of 2 ml. Non-specific binding (○, △) was measured in the presence of 10^{-9} M unlabelled toxin. Specific binding (●, ▲) is the difference between total and non-specific binding. **C** Scatchard plot of the data (● chick, ▲ rat). The specific activity of the ^{125}I-GVIA used was 2000 cpm/fmol.

1,4-Dihydropyridines and phenylalkylamines have been important tools providing a molecular access to the structure of skeletal muscle Ca^{2+} channels. Receptors for these different categories of drugs are also present in the central nervous system (CNS) (Cortes et al. 1984; Mourre et al. 1987), in which they have been localized by quantitative autoradiography at different stages of brain development (Mourre et al. 1987); however, the function of the potential Ca^{2+} channels associated to these receptors is not yet clear (Miller 1987; Hirning et al. 1988). Conversely, a polypeptide toxin, ω-conotoxin, has recently been shown to be a very active blocker of both L- and N-type Ca^{2+} channels (Olivera et al. 1984; Reynolds et al. 1986; McCleskey et al. 1987) in the CNS. This toxin forms very high affinity complexes with its receptor (Barhanin et al. 1988) (K_d in the picomolar range), and the stoichiometry of binding is particularly large in chick brain membranes (Fig. 2). The toxin dissociates slowly from its receptor ($k_{-1} = 9.4 \times 10^5$ s^{-1}, i.e. $t_{1/2} \simeq 2$ h), and the binding is antagonized by 10 mM Ca^{2+}. Cross-linking experiments (Barhanin et al. 1988) have shown that a large peptide of 210–220 kDa was labelled using the azidonitrobenzoyloxy derivative of ω-conotoxin (ANB-NOS), while disuccinimidyl suberate (DSS) specifically cross-linked the toxin to a 170-kDa component comprising a 140-kDa peptide disulfide linked to a 30-kDa peptide (Figs. 3, 4). The 210- to 220-kDa peptide might be the equivalent in the brain of the α_1-subunit identified for the skeletal muscle Ca^{2+} channel, while the 170-kDa peptide (140 + 30 kDa) might be the equivalent of the α_2-component of the receptor for Ca^{2+}-channel blockers that is present in T tubules.

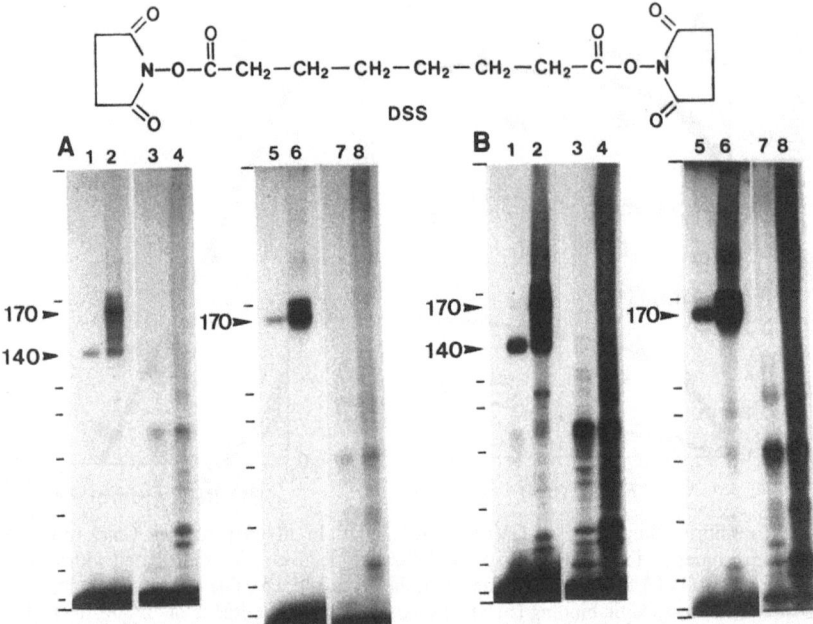

Fig. 3 A, B. Covalent cross-linking of ^{125}I-GVIA to chick brain receptor components with DSS (*top*). The autoradiograms depict the peptides labelled by ^{125}I-GVIA using 30 µM DSS (lanes *1, 3* and *5, 7*) or 300 µM DSS (lanes *2, 4* and *6, 8*). Lanes *3, 4* and *7, 8* were obtained for experiments performed in the presence of 10^{-9} M unlabelled GVIA. Gels were run under reducing (lanes *1–4*) and non-reducing conditions (lanes *5–8*). **A** Autoradiograms exposed for 26 h at -70 °C on Kodak X-Omat films with a Cronex Hi-plus intensifying screen (Dupont). **B** Autoradiograms from the same gels as in **A** exposed for 140 h. Positions of molecular weight markers (BioRad) are indicated by *horizontal bars*: myosin (200000), β-galactosidase (116500), phosphorylase B (96500), bovine serum albumin (66200), ovalbumin (45000), carbonic anhydrase (31000) and soybean trypsin inhibitor (21500)

Table 1. Effects of Ca^{2+}-channel antagonists on $(-)$-[^{3}H]D888 binding to insect nervous system and mammalian skeletal muscle

	K_I(nM)	
	Insect nervous system	Mammalian skeletal muscle
$(-)$D888	0.052	2
$(+)$D888	0.190	
$(-)$Verapamil	0.005	20[a]
$(+)$Verapamil	2.3	
$(-)$D600	0.052	40[a]
$(+)$D600	4.7	
Fluspirilene	12– 14	0.4
$(-)$Bepridil	17– 26	20[a]
$(+)$Bepridil	17	
d-cis-Diltiazem	190–270	60
l-cis-Diltiazem	190	900

[a] K_I values (inhibition constant) obtained from racemic compound instead of pure enantiomers.

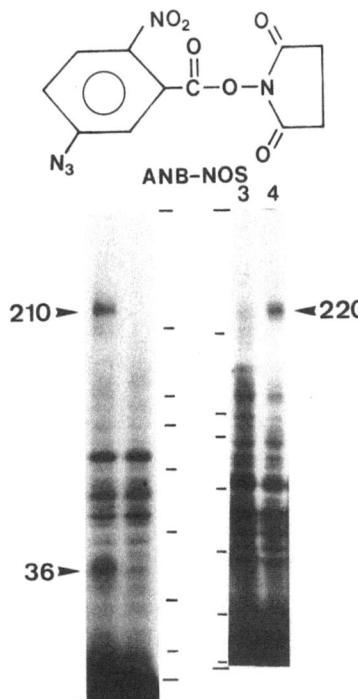

Fig. 4. Analysis of the peptides photoaffinity labelled using ANB-^{125}I-GVIA on rat and chick brain membranes. Preparations labelled with ANB-^{125}I-GVIA in the absence (lanes *1, 4*) or presence (lanes *2, 3*) of 10^{-9} M unlabelled GVIA were analysed under reducing conditions. Lanes *1, 2* show rat brain membranes, lanes *3, 4* chick brain membranes. Molecular weight markers *(horizontal bars)* are the same as in Fig. 3

A particularly interesting case is found in the nervous system of *Drosophila* (Pauron et al. 1987). Binding studies using (−)-[^{3}H]D888 and (±)-[^{3}H]verapamil have identified a single class of very high affinity binding sites for these ligands ($K_d = 0.1$–0.4 nM, $B_{max} = 1600$–1800 fmol/mg protein) (Table 1). The most potent molecule in the phenylalkylamine series was (−)verapamil, with a K_d value as exceptionally low as 4.7 pM. Molecules in the benzothiazepine and diphenylbutylpiperidine series of Ca^{2+}-channel blockers, as well as bepridil, inhibited (−)-[^{3}H]D888 binding, suggesting a close correlation, as in the mammalian system, between these receptor sites and those recognized by phenylalkylamines. It is of particular interest that unlike what has been observed in the mammalian system, 1,4-DHPs are without effect on phenylalkylamine binding. This observation strongly suggests that this is a case of a protein which has the receptors for Ca^{2+}-channel blockers, including very high affinity receptors for phenylalkylamines, but lacks receptors for DHPs. The protein has been affinity labelled with a tritiated (arylazido)phenylalkylamine ($K_d = 0.24$ nM), and a protein of M_r 135 ± 5 kDa was found to be specifically labelled (Fig. 5). This protein would correspond to the α_1-subunit of the neuronal Ca^{2+} channel in *Drosophila*.

Another interesting case is found in plant membranes (Graziana et al. 1988). Ca^{2+}-channel inhibitors of the phenylalkylamine and diphenylbutylpiperidine series, as well as other blockers such as bepridil, inhibit $^{45}Ca^{2+}$ influx into carrot protoplasts. The corresponding plasma membrane has a high density (120 pmol/mg protein) of sites for the phenylalkylamine (−)-[^{3}H]D888 ($K_d = 85$ nM). For ten different Ca^{2+}-channel blockers, there was a good correlation between efficacy of blockade of $^{45}Ca^{2+}$

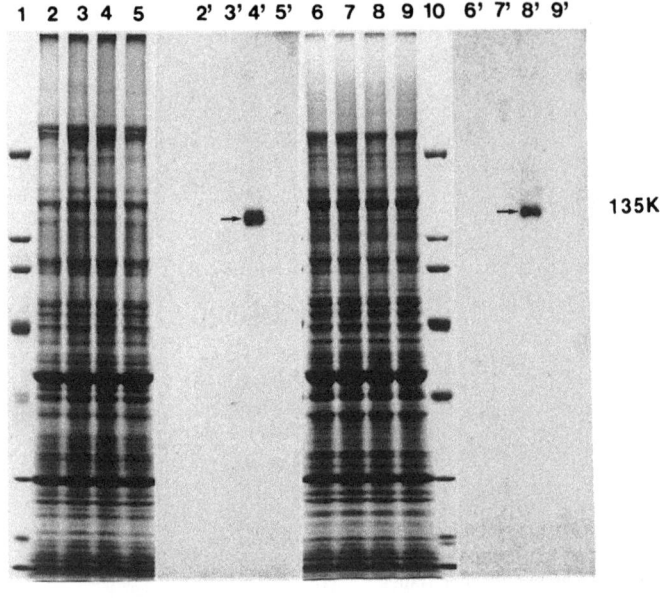

Fig. 5. Photoaffinity labelling of the phenylalkylamine receptor with [^{3}H]LU49888. Labelling experiments were performed as described in Pauron et al. (1987). Lanes *1–5* and *6–10* show Coomassie blue staining of the gel, lanes *2'–5'* and *6'–9'* autoradiography of the gel. Lanes 2, 4, 6, 8 and 2', 4', 6', 8' show incubation in the absence of unlabelled (±) verapamil, lanes 3, 5, 7, 9 and 3', 5', 7', 9' incubation in the presence of unlabelled (±)verapamil. Lanes 2, 3, 2', 3' and 6, 7, 6', 7' are from incubations which where not irradiated. Lanes *1, 10* are molecular weight markers from BioRad (from *top* to *bottom*): myosin (200000), β-galactosidase (116500), phosphorylase B (96500), bovine serum albumin (66200), ovalbumin (45000), carbonic anhydrase (31000), soybean trypsin inhibition (24500) and lysozyne (14400). *β-Me*, β-mercaptoethanol (disulfide reducing conditions); *IAA*, iodoacetamide (non-reducing conditions). Despite the relatively large non-specific binding of [^{3}H]LU49888 after UV illumination, no non-specifically labelled bands appeared in the autoradiography because all the non-specific incorporation occurred in phospholipids and migrated ahead ot the dye front of the gel

influx into protoplasts and efficacy of binding of the [^{3}H]ligand to membranes. Binding sites for DHPs are absent, and no blockade of Ca^{2+} influx was observed with several molecules in this series, such as (+)-PN200–110, nifedipine or nitrendipine.

DHPs have been very important compounds opening the way for a potent pharmacology of L-type Ca^{2+} channels. New molecules are emerging now which may have other properties. Neuroleptics of the diphenylbutylpiperidine series include molecules such as fluspirilene, penfluridol, pimozide and clopimozide. This class of drugs has the ability to relieve schizophrenic symptoms, but it is also known for its anti-anxiety properties in psychovegetative disorders. Fluspirilene at very low doses influences mental state in the direction of anxiolysis, relaxation and increased self-confidence. Fluspirilene blocks L-type Ca^{2+}-channel activity (Galizzi et al. 1986b; Qar et al. 1987). [^{3}H]fluspirilene binds to skeletal muscle T-tubule membranes with a high affinity (K_d = 0.1 nM), and the fluspirilene receptor appears to be distinct from other well-identified receptors (Galizzi et al. 1986b). Diphenylbutylpiperidines bind to heart, smooth muscle and brain membranes with K_d values in the range of 10–100 nM. Binding activity and blockade of Ca^{2+} channels are parallel in smooth muscle cells (Qar et al. 1987). It may be that the anti-anxiety properties of fluspirilene which are observed at very low doses of the drug are due to blockade, with a very high affinity, of Ca^{2+} channels of specific populations of neurones involved in the control of anxiety. The neuroleptic properties of this drug are of course due to its action on D_2 receptors.

Benzolactams are another series of new compounds which also bind with high affinity with receptors associated to Ca^{2+}-channel function since they block Ca^{2+}-channel activity in a nanomolar range of concentration (Qar et al. 1988). Although this series of molecules is competitive in binding with DHPs and although it has an action which, like that of DHPs, is voltage dependent, it binds to a receptor site which is new and allosterically related to receptors for all other Ca^{2+}-channel blockers (Qar et al. 1988).

The voltage-sensitive Na^+ channel has multiple receptors for natural toxins and other compounds of pharmacological importance. At least six different receptor sites have been identified for natural toxins (Lazdunski et al. 1987). The situation may turn out to be the same for the L-type Ca^{2+} channel. If there are numerous receptors for drugs chemically synthesized by pharmaceutical companies, it may be that there are endogenous ligands corresponding to each category of these receptors.

Acknowledgements. We thank C. Roulinat-Bettelheim for expert secretarial assistance.

References

Barhanin J, Coppola T, Schmid A, Borsotto M, Lazdunski M (1987) The calcium channel antagonists receptor from rabbit- skeletal muscle. Reconstitution after purification and subunit characterization. Eur J Biochem 164:525–531

Barhanin J, Schmid A, Lazdunski M (1988) Properties of structure and interaction of the receptor for ω-conotoxin, a polypeptide active on Ca^{2+} channels. Biochem Biophys Res Commun 150:1051–1062

Beam KG, Knudson CM, Powell JA (1986) A lethal mutation in mice eliminates the slow calcium current in skeletal muscle cells. Nature 320:168–170

Borsotto M, Barhanin J, Fosset M, Lazdunski M (1985) The 1,4-Dihydropyridine receptor associated with the skeletal muscle voltage-dependent Ca^{2+} channel. Purification and subunit composition. J Biol Chem 260:14255–14263

Cognard C, Lazdunski M, Romey G (1986a) Different types of Ca^{2+} channels in mammalian skeletal muscle cells in culture. Proc Natl Acad Sci USA 83:517–521

Cognard C, Romey G, Galizzi JP, Fosset M, Lazdunski M (1986b) Dihydropyridine-sensitive Ca^{2+} channels in mammalian skeletal muscle cells in culture: electrophysiological properties and interactions with Ca^{2+} channel activation (Bay K8644) and inhibitor (PN 200–110). Proc Natl Acad Sci USA 83:1518–1522

Cooper CL, Vandaele S, Barhanin J, Fosset M, Lazdunski M, Hosey MM (1987) Purification and characterization of the dihydropyridine-sensitive voltage-dependent calcium channel from cardiac tissue. J Biol Chem 262:509–512

Cortes R, Supavilai P, Karobath M, Palacios JM (1984) Calcium antagonist binding sites in the rat brain: quantitative autoradiographic mapping using the 1,4-dihydropyriridines [^{3}H]PN200–110 and [^{3}H]PY 108–068. J Neural Trans 60:169–197

Fosset M, Jaimovich E, Delpont E, Lazdunski M (1983) [^{3}H]Nitrendipine receptors in skeletal muscle. J Biol Chem 258:6083–6092

Galizzi JP, Borsotto M, Barhanin J, Fosset M, Lazdunski M (1986a) Characterization and photoaffinity labeling of receptor sites for the Ca^{2+} channel inhibitors d-cis-diltiazem, (±)bepridil, desmethoxyverapamil, and (+)PN200–110 in skeletal muscle transverse tubule membranes. J Biol Chem 261:1393–1397

Galizzi JP, Fosset M, Romey G, Laduron P, Lazdunski M (1986b) Neuroleptics of the diphenylbutylpiperidine series are potent calcium channel inhibitors. Proc Natl Acad Sci USA 83:7513–7517

Graziana A, Fosset M, Ranjeva R, Hetherington A, Lazdunski M (1988) Ca^{2+} channel inhibitors that bind to plant cell membranes block Ca^{2+} entry into protoplasts. Biochemistry 27:764–768

Hirning LD, Fox, AP, McCleskey EW, Olivera BM, Thayer SA, Miller RJ, Tsien RW (1988) Dominant role of N-type Ca^{2+} channels in evoked release of norepinephrine from sympathetic neurons. Science 239:57–61

Hosey MM, Lazdunski M (1988) Calcium channels: molecular pharmacology, structure and regulation. J Membrane Biol in press.

Hosey MM, Borsotto M, Lazdunski M (1986) Phosphorylation and dephosphorylation of dihydropyridine-sensitive voltage-dependent Ca^{2+} channel in skeletal muscle membranes by cAMP- and Ca^{2+}-dependent processes. Proc Natl Acad Sci USA 83:3733–3737

Hosey MM, Barhanin J, Schmid A, Vandaele S, Ptasienski J, O'Callahan C, Cooper C, Lazdunski M (1987) Photoaffinity labelling and phosphorylation of a 165 kilodalton peptide associated with dihydropyridine and phenylalkylamine-sensitive calcium channels. Biochem Biophys Res Commun 147:1137–1145

Lazdunski M, Frelin C, Barhanin J, Lombet A, Meiri H, Pauron D, Romey G, Schmid A, Schweitz H, Vigne P, Vijverberg HPM (1987) Polypeptide toxins as tools to study voltage-sensitive Na^+ channels. In: Tetrodotoxin, saxitoxin and the molecular biology of the sodium channel. N Y Acad Sci 479:204–220

Mc Cleskey EW, Fox AP, Feldman DH, Cruz LJ, Olivera BM, Tsien RW, Yoshikami D (1987) ω-Conotoxin: Direct and persistent blockade of specific types of calcium channels in neurons but not muscle. Proc Natl Acad Sci USA 84:4327–4331

Miller RJ (1987) Multiple calcium channels and neuronal function. Science 235:46–52

Mourre C, Cervera P, Lazdunski M (1987) Autoradiographic analysis in rat brain of the postnatal ontogeny of voltage-dependent Na^+ channels, Ca^{2+}-dependent K^+ channels and slow Ca^{2+} channels identified as receptors for tetrodotoxin, apamin and (−)-desmethoxyverapamil. Brain Res 417:21–32

Olivera BM, McIntosh JM, Cruz LJ, Luque FA, Gray WR (1984) Purification and sequence of a presynaptic peptide toxin from *Conus geographus* venom. Biochemistry 23:5087–5090

Pauron D, Qar J, Barhanin J, Fournier D, Cuany A, Pralavorio M, Bergé JB, Lazdunski M (1987) Identification and affinity labeling of very high affinity binding sites for the phenylalkylamine series of Ca^{2+} channel blockers in the *Drosophila* nervous system. Biochemistry 26:6311–6315

Pinçon-Raymond M, Rieger F, Fosset M, Lazdunski M (1985) Abnormal transverse tubule system and abnormal amount of receptors for Ca^{2+} channel inhibitors of the dihydropyridine family in skeletal muscle from mice with embryonic muscular dysgenesis. Dev Biol 112:458–446

Qar J, Galizzi JP, Fosset M, Lazdunski M (1987) Receptors for diphenylbutylpiperidine neuroleptics in brain, cardiac, and smooth muscle membranes. Relationship with receptors for 1,4-dihydropyridines and phenylalkylamines with Ca^{2+} channel blockade. Eur J Pharmacol 141:261–268

Qar J, Barhanin J, Romey G, Henning R, Lerch U. Oekonomopulos R, Urbach H, Lazdunski M (1988) A novel high affinity class of Ca^{2+} channel blockers. Mol Pharmacol 33:363–369

Quirion R (1985) Characterization of binding sites for two classes of calcium channel antagonists in human forebrain. Eur J Pharmacol 117:139–142

Reuter H (1983) Calcium channel modulation by neurotransmitters, enzymes and drugs. Nature 301:569–574

Reynolds IJ, Wagner JA, Snyder SH, Thayer SA, Olivera BM, Miller RJ (1986) Brain voltage-sensitive calcium channel subtypes differentiated by ω-conotoxin fraction GVIA. Proc Natl Acad Sci USA 83:8804–8807

Rieger F, Bornaud R, Shimahara T, Garcia L, Pinçon-Raymond M, Romey G, Lazdunski M (1987) Restoration of dysgenic muscle contraction and calcium channel function by co-culture with normal spinal cord neurons. Nature 330:563–566

Romey G, Rieger F, Renaud JF, Pinçon-Raymond M, Lazdunski M (1986) The electrophysiological expression of Ca^{2+} channels and of apamin-sensitive Ca^{2+}-activated K^+ channels is abolished in skeletal muscle cells from mice with muscular dysgenesis. Biochem Biophys Res Commun 136:935–940

Schmid A, Renaud JF, Lazdunski M (1985) Short-term and long-term effects of β-adrenergic effectors and cyclic AMP on nitrendipine-sensitive voltage-dependent Ca^{2+} channels of skeletal muscle. J Biol Chem 260:13041–13046

Schmid A, Barhanin J, Coppola T, Borsotto M, Lazdunski M (1986a) Immunochemical analysis of subunit structures of 1,4-dihydropyridine receptors associated with voltage-dependent Ca^{2+} channels in skeletal, cardiac and smooth muscles. Biochemistry 25:3492–3495

Schmid A, Barhanin J, Mourre C, Coppola T, Borsotto M, Lazdunski M (1986b) Antibodies reveal the cytolocalization and subunit structure of the 1,4-dihydropyridine component of the neuronal Ca^{2+} channel. Biochem Biophys Res Commun 139:996–1002

Sharp AH, Imagawa T, Leung AT, Campbell KP (1987) Identification and characterization of the dihydropyridine binding subunit of the skeletal muscle dihydropyridine receptor. J Biol Chem 262:12309–12315

Sieber M, Nastainczyk W, Zubor V, Wernet W, Hofmann F (1987) The 165-KDa peptide of the purified skeletal muscle dihydropyridine receptor contains the known regulatory sites of the calcium channel. Eur J Biochem 167:117–122

Striessnig J, Knaus HG, Grabner M, Moosburger K, Seitz W, Lietz H, Glossmann H (1987) Photoaffinity labelling of the phenylalkylamine receptor of the skeletal muscle transverse-tubule calcium channel. FEBS Lett 212:247–253

Takahashi M, Seagar MJ, Jones JF, Reber BFX, Catterall WA (1987) Subunit structure of dihydropyridine-sensitive calcium channels from skeletal muscle. Proc Natl Acad Sci USA 84:5478–5482

Tanabe T, Takeshima H, Mikami A, Flockerzi V, Takahashi H, Kangawa K, Kojima M, Matsuo H, Hirose T, Numa S (1987) Primary structure of the receptor for calcium channel blockers from skeletal muscle. Nature 328:313–318

Vaghy PL, Striessnig J, Miwa K, Knaus HG, Itagaki K, McKenna E, Glossmann H, Schwartz A (1987) Identification of a novel 1,4-dyhydropyridine- and phenylalkylamine-binding polypeptide in calcium channel preparations. J Biol Chem 262:14337–14342

Vandaele S, Fosset M, Galizzi JP, Lazdunski M (1987) Monoclonal antibodies that coimmunoprecipitate the 1,4-dihydropyridine and phenylalkylamine receptors and reveal the Ca^{2+} channel structure. Biochemistry 26:5–9

The Structure of the Ca^{2+} Channel: Photoaffinity Labeling and Tissue Distribution*

H. Glossmann[1], J. Striessnig[1], L. Hymel[2], G. Zernig[1], H. G. Knaus[1], and H. Schindler*[2]

[1] Institut für Biochemische Pharmakologie, Universität Innsbruck, A-6020 Innsbruck, Austria
[2] Institut für Biophysik, Universität Linz, A-4020 Linz, Austria

Introduction

Photoaffinity labeling is an important, extremely valuable tool for the characterization of pharmacological receptors. Radioactive photoaffinity probes allow, even in crude preparations with very low receptor densities, identification of the polypeptide(s) which bind the ligand. In general, the photoaffinity protection profile (obtained by adding unlabeled compounds interacting competitively or allosterically with the receptor) should reflect the reversible binding interaction profile. In Ca^{2+}-channel research, three arylazide photoaffinity probes are useful, namely (−)- or (±)-[^{3}H]azidopine (1,4-dihydropyridines), [N-methyl-^{3}H]LU49888 (a phenylalkylamine), and an azido derivative of [^{125}I]ω-conotoxin GVIA (CgTx). The first two ligands are well characterized by reversible binding experiments and interact in a stereoselective manner and with high affinity with L-type Ca^{2+} channels [1–5]. The ^{125}I-iodinated CgTx photoaffinity probe presumably incorporates into the N-type channel components, exclusively found in neuronal tissues [6]. L-type Ca^{2+} channels have distinct drug-receptor domains for different classes of drugs (e.g., the 1,4-dihydropyridines, phenylalkylamines, and benzothiazepines) linked to each other (and to high- and low-affinity divalent cation binding sites) by reciprocal allosteric coupling mechanisms [7]. Recently, the benzothiazinones [8] and diphenylbutyl-piperidines [9, 10] have been suggested to act at yet another site. Thus, identification of L-type Ca^{2+}-channel drug-receptor-carrying polypeptides by photoaffinity labeling is aided by the great variety of other drugs which stimulate or inhibit reversible binding of the photoaffinity labels – often in a stereoselective manner. L-type Ca^{2+} channels are most abundant in skeletal muscle transverse (T)-tubule membranes [11, 12], and the 1,4-dihydropyridine-receptor-carrying polypeptide (termed α_1) was first identified in this preparation [13, 14]. Not unexpectedly, α_1 also carries the binding domain for the phenylalkylamines, as shown by [N-methyl-^{3}H]LU49888 photolabeling [1, 2, 15]. High-affinity stereoselective phenylalkylamine receptors exist on a 135-kDa [N-methyl-^{3}H]LU49888-photolabeled polypeptide in membrane extracts from *Drosophila* heads. This polypeptide does not bind 1,4-dihydropyridines or benzothiazepines [16]. If shown to be a component of a functional phenylalkylamine-

---

* Research of H. G. was supported by the Deutsche Forschungsgemeinschaft, Dr. Legerlotz Foundation, and Jubiläumsfonds der Nationalbank. H. G. and H. S. are funded by a Schwerpunktprogramm of the Austrian Fonds zur Förderung der Wissenschaftlichen Forschung.

sensitive Ca^{2+} channel, the *Drosophila* system is the first direct evidence that a drug-receptor domain (which is colocalized with and allosterically coupled to 1,4-dihydropyridine receptors on α_1) occurs independently from the other drug receptors.

The photoaffinity labels for L-type Ca^{2+} channels are also useful for identifying other sites which bind Ca^{2+} antagonists. So far, three distinct mammalian membrane proteins, all involved in transport and/or linked to nucleoside or nucleotide metabolism, have been identified which reversibly bind 1,4-dihydropyridines and phenylakylamines: the plasma membrane glycoprotein gp 170 (an ATPase), coded by the multidrug-resistance *(mdr)* gene [17]; the plasma membrane nucleoside transporter [18]; and polypeptides on the inner mitochondrial membrane [19]. The gp 170 and the mitochondrial polypeptides have been photolabeled with $(-)$-$[^3H]$azidopine. Plasma membranes from *Chlamydomonas reinhardtii* bind verapamil and other Ca^{2+} antagonists with dissociation constants in the submicromolar range [20–22], suggesting that high-affinity sites for these compounds existed very early in evolution. It is unknown, at present, if these sites are also on "Ca^{2+} channels." In any event, the wide distribution and early evolutionary appearance of Ca^{2+}-antagonist binding (and receptor!) sites is surprising and poses an interesting question: Are these other sites related to the receptors on L-type Ca^{2+} channels, and if so, why were they retained in this highly regulated (G proteins, phosphorylation) structure? Photoaffinity probes may be useful in this context to identify directly specific regions within the primary amino acid sequences of the different Ca^{2+}-antagonist-binding proteins, including, of course, the L-type Ca^{2+} channel.

A major concern in the isolation and reconstitution of receptors is proteolytic cleavage. Neither a perfect reversible binding profile nor function (after reconstitution) guarantees structural integrity of a purified preparation. The isolated Ca^{2+} channel from skeletal muscle is a good example. Photoaffinity probes were an indispensable tool for clarifying the subunit composition of the channel and proving structural preservation of α_1 throughout the purification procedure [23]. In the following, we will give examples of the usefulness of the Ca^{2+}-channel photoaffinity probes; discuss the behavior of our purified, reconstituted L-type Ca^{2+} channel; and present results on a novel 1,4-dihydropyridine binding site, localized on mitochondrial membranes.

Results and Discussion

Arylazide Photoaffinity Ligands for Ca^{2+}-Antagonist- and Toxin-Sensitive Ca^{2+} Channels

For photoaffinity labeling of Ca^{2+}-antagonist binding sites in different tissues, both arylazides and nonarylazides have been used. Most studies with nonarylazides were unsuccessful in labeling a high-affinity, Ca^{2+}-channel-associated polypeptide. Instead, several non-Ca^{2+}-channel-related, low-affinity, high-capacity binding sites (M_rs 30000–45000) were characterized (for review see [24]). We have introduced two enantiomerically pure, tritiated, high-affinity arylazide photoaffinity ligands for biochemical characterization of L-type Ca^{2+}-channel drug receptors in excitable tissues: the 1,4-dihydropyridines (DHPs) $(\pm)$- and $(-)$-$[^3H]$azidopine, and the phe-

Table 1. Properties of $(-)$-[^{3}H]azidopine and [N-methyl-^{3}H]LU49888

	$(-)$-[^{3}H]azidopine	[N-methyl-^{3}H]LU49888
Specific radioactivity (Ci/mmol)	40–50	80–85
Stability in aqueous solutions	Good	Good
Loss of binding activity due to UV irradiation[a] (%)		
1 min	4	N.D.
1.5 min	8	
2 min	11	
3 min	18	
Dissociation half-life of complex at 2°–4 °C (h)		
Skeletal muscle	> 5[b]	> 1[c]
Hippocampus	No dissociation after 4 h	> 1[c]
Dissociation constants (nM)		
Skeletal muscle		
Particulate	0.35	2.0
Solubilized	0.70	> 40
Purified	3.1	62
Particulate		
cardiac muscle	0.03	N.D.
Particulate		
hippocampus	0.09	1.4
Specific photoincorporation[d]		
Particulate	0.6–1.5	0.2–1.7
Purified	2–11	0.8–6.0
Binding to low-affinity, non-Ca^{2+}-channel-linked sites		
	High (in cardiac tissue)	Low

[a] A cheap blacklight lamp (Philips TL40W/08) was used for irradiation
[b] Determined with racemic [^{3}H]azidopine
[c] Dissociation induced by dilution.
[d] Percentage of specifically incorporated ligand compared to specific (reversible) binding prior to photolysis
For references see [1, 2, 14, 23, 40, 61]

nylalkylamine [N-methyl-^{3}H]LU49888. Their properties are summarized in Table 1. $(-)$-[^{3}H]azidopine binds with subnanomolar dissociation constants to neuronal, skeletal muscle, and cardiac 1,4-DHP receptors. At ambient temperature, binding is reversible in the absence of ultraviolet (UV) light. However, decreasing the incubation temperature to 2°C considerably slows the dissociation rate of the complex, resulting in apparently irreversible binding, e. g., to brain membranes (see Table 1). Therefore, photoactivation of the $(-)$-[^{3}H]azidopine-receptor complex is usually carried out on ice. As neither the binding activity of the ligand $(-)$-[^{3}H]azidopine (Table 1) nor that of the receptor protein (not shown) is severely affected by our mild photolysis conditions, UV light-induced dissociation of the complex is prevented. The covalent incorporation of $(-)$-[^{3}H]azidopine can be expected to occur near its binding site even if photoactivated, reactive species of longer half-life are generated.

The affinity of [N-methyl-^{3}H]LU49888 to its receptor site in different tissues is lower than for $(-)$-[^{3}H]azidopine. However, the dissociation of the [N-methyl-

^{3}H]Lu49888-receptor complexes at 2°C is also favorable for photoaffinity labeling (Table 1). The reversible binding properties of [N-methyl-^{3}H]LU49888 are nearly indistinguishable from those of $(-)$-[^{3}H]desmethoxyverapamil. [N-methyl-^{3}H]LU49888 is therefore a highly suitable probe for the reversible characterization of phenylalkylamine receptors. Phenylalkylamine receptor interaction is inhibited by divalent cations, whereas 1,4-DHP receptors show an absolute requirement for, e.g. Ca^{2+} or Mg^{2+}, to bind 1,4-DHPs with high affinity [25–27]. Figure 1 shows the inhibition of [N-methyl-^{3}H]LU49888 binding to membrane-bound skeletal muscle T-tubule Ca^{2+} channels by different mono-, di-, and trivalent cations. The rank order of potency ($Cd^{2+} > Zn^{2+} > La^{3+} > Mn^{2+} > Ca^{2+} > Sr^{2+} > Ba^{2+} > Mg^{2+} >> Na^+ = K^+$)

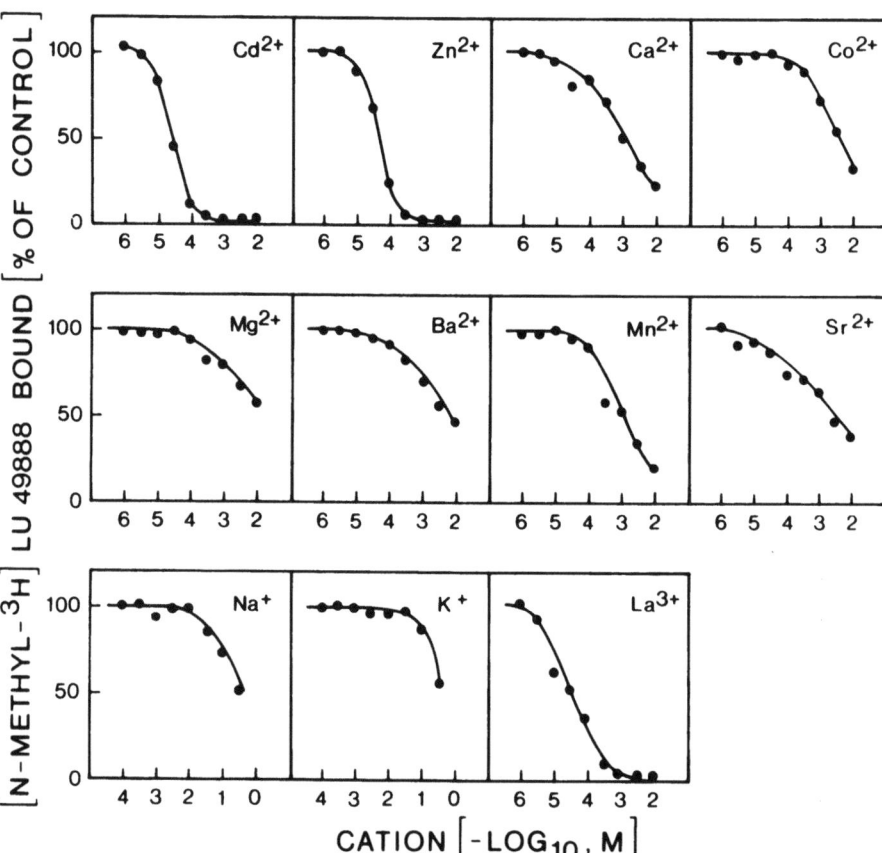

Fig. 1. Inhibition of [N-methyl-^{3}H]LU49888 binding to skeletal muscle Ca^{2+}-antagonist receptors by cations. [N-methyl-^{3}H]LU49888 (1.1–2.5 nM) was incubated at 25°C with 3–4 μg/ml partially purified skeletal muscle T-tubule membrane protein in the absence or presence of increasing concentrations of the cations shown. Data from two experiments were pooled and computer-fitted to the general dose-response equation. The following binding parameters were obtained: Cd^{2+} IC_{50} = 15.9 ± 1.1 μM, n_H = 1.89 ± 0.19; Zn^{2+} IC_{50} = 50.2 ± 3.2 μM, n_H = 1.31 ± 0.09; Ca^{2+} IC_{50} = 820 ± 270 μM, n_H = 0.66 ± 0.09; Co^{2+} IC_{50} = 4.1 ± 1.0 mM, n_H = 0.71 ± 0.11; Mg^{2+} IC_{50} > 10 mM; Ba^{2+} IC_{50} > 5 mM; Mn^{2+} IC_{50} = 1.2 ± 0.4 mM, n_H = 0.61 ± 0.12; Sr^{2+} IC_{50} = 743 ± 540 μM, n_H = 0.35 ± 0.08; Na^+, K^+ IC_{50} > 100 mM; La^{3+} IC_{50} = 29.7 ± 11.3 μM, n_H = 0.73 ± 0.18
n_H, apparent Hill slope

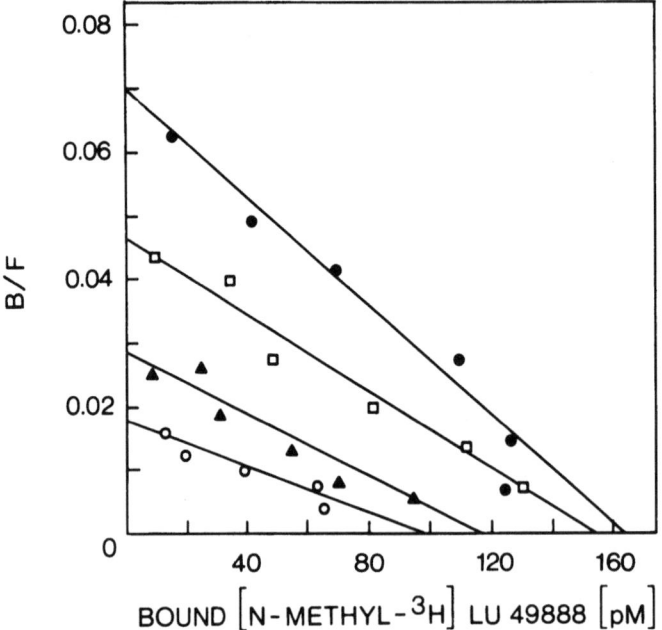

Fig. 2. Mechanism of inhibition of [N-methyl-^{3}H]LU49888 binding to guinea pig skeletal muscle T-tubule Ca^{2+} channels by Cd^{2+}. Saturation analysis was performed by incubating (60 min at 25°C) increasing concentrations of [N-methyl-^{3}H]LU49888 with 0.01 mg/ml membrane protein in the absence or presence of different Cd^{2+} concentrations. Scatchard transformation of the data is shown. K_d and B_{max} values were calculated by linear regression analysis: control (●) K_d = 2.3 nM, B_{max} = 161 pM; 10 μM Cd^{2+} (□) K_d = 3.2 nM, B_{max} = 151 pM; 20 μM Cd_2^+ (▲) K_d = 4.1 nM, B_{max} = 115 pM; 30 μM Cd^{2+} (○) K_d = 5.5 nM, B_{max} = 96.7 pM
B/F = bound/free ligand concentration

is very similar to the results previously obtained with (−)-[^{3}H]desmethoxyverapamil in brain and skeletal muscle membranes [28, 29]. Saturation studies (Fig. 2) reveal a mixed-type (K_d increase, B_{max} decrease) inhibition for Cd^{2+}. The interaction of both [N-methyl-^{3}H]LU49888 and the cations with the channel is completely reversible. Removal of the divalent cation, e. g., Cd^{2+} (Fig. 3a), from its binding site by chelation with EDTA relieves its allosteric inhibitory effect and immediately restores [N-methyl-^{3}H]LU49888 binding. Conversely, occupation of the inhibitory divalent cation binding site causes dissociation of the performed [N-methyl-^{3}H] LU49888-receptor complex (Fig. 3b).

Recently, a marine snail toxin, ω-conotoxin GVIA (CgTx), has been introduced into Ca^{2+}-channel research [30–34]. This 27-amino-acid polypeptide toxin blocks only neuronal L-, N- (irreversibly), and T-type Ca^{2+} channels (reversibly; see [31]). It is therefore not expected to be an L-type channel-selective tool, as are the organic Ca^{2+} antagonists. The toxin can be easily iodinated with ^{125}I to high specific activity with chloramine-T. The mono-[^{125}I]-iodinated toxin binds with very high affinity to receptors in guinea pig brain (apparent half-saturation at 1.5–2.5 pM; the conditions

The Structure of the Ca^{2+} Channel: Photoaffinity Labeling and Tissue Distribution   173

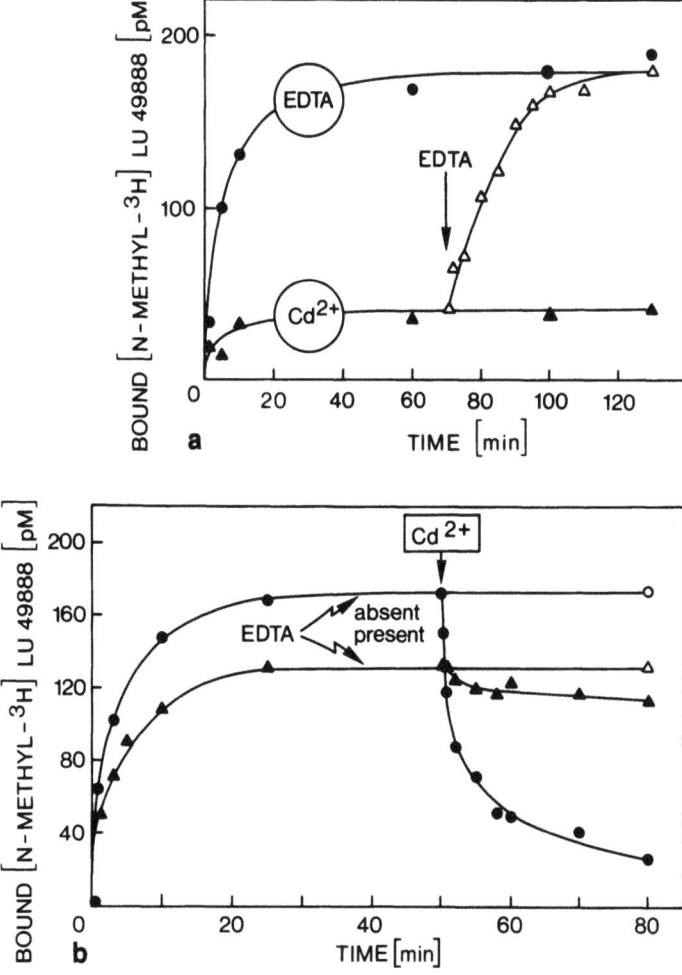

Fig. 3a, b. Reversible interaction of Cd^{2+} with the skeletal muscle Ca^{2+}-antagonist receptor. **a** Association kinetics: [N-methyl-^{3}H]LU49888 was incubated with partially purified guinea pig skeletal muscle T-tubule membrane protein (0.02 mg/ml) in the presence of 3 mM EDTA or a submaximal inhibitory concentration of CdCl$_2$ (0.1 mM). Specific binding was measured after the times indicated by rapid filtration of the incubation mixture over glass-fiber filters. After equilibrium was reached, EDTA (3 mM final concentration) was added to the incubation mixture containing Cd^{2+}. The recovery of [N-methyl-^{3}H]LU49888 binding was followed. **b** Dissociation kinetics: [N-methyl-^{3}H]LU49888-Ca^{2+}-channel complexes were formed as described in **a** (3.2 nM [N-methyl-^{3}H]LU49888, 0.01 mg/ml membrane protein) in the absence and presence of 3 mM EDTA. Dissociation of the complex at equilibrium was initiated by addition of Cd^{2+} (0.1 mM final concentration). For the samples with no EDTA, dissociation rate constant k$_{-1}$ was 0.127 min^{-1}(t$_{1/2}$ = 3.9 min)

of the binding assays are given in Fig. 4 and in [33]. The maximal density of [^{125}I]CgTx-labeled sites exeeds the density of (+)-[^{3}H]PN200–110-labeled 1,4-DHP receptors (< 350 fmol/mg protein) at least five fold, suggesting that neuronal L-type channels are only a minority of all labeled sites. Since the binding of [^{125}I]CgTx is quasi-irreversible (see Fig. 4), as is the inhibition of neuronal L- and N-type channels in

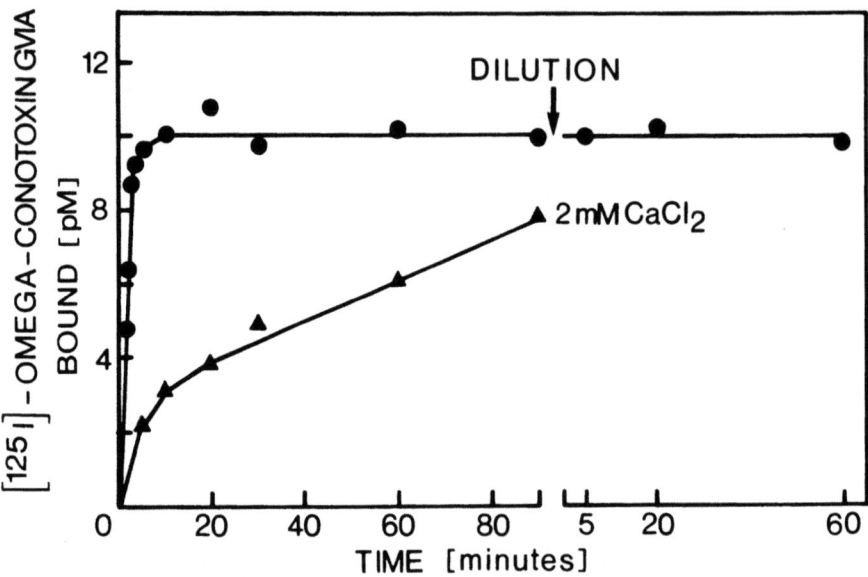

Fig. 4. Association kinetics of [^{125}I]ω-conotoxin GVIA. [^{125}I]CgTx (44 rM) was added to 0.024 mg/ml guinea pig cerebral cortex membrane protein preincubated (30 min at 25°C) with (total binding) or without (nonspecific binding) 30 nM unlabeled toxin. Bound toxin was separated from free toxin after the times indicated by rapid filtration over GF/C Whatman filters, using the polyethylene glycol buffer method as described elsewhere [23]. The experiment was repeated under exactly the same conditions with 2 mM CaCl$_2$ added to the membranes during preincubation. Note that the inhibitory action of the reversible ligand Ca^{2+} is overcome by the quasi-irreversible binding of the toxin

electrophysiological experiments, we conclude that most of the binding of the toxin is to N-type Ca^{2+} channels.

As with the binding of phenylalkylamines (Fig. 1) or benzothiazepines to L-type Ca^{2+} channels, formation of the [^{125}I]CgTx-receptor complex is prevented by relatively high concentrations of monovalent cations (IC$_{50}$ values for NaCl and KCl are > 100 mM) and low concentrations of divalent cations (IC$_{50}$ values for Cd^{2+}, Ni^{2+}, Co^{2+}, Ca^{2+}, Ba^{2+} and Mg^{2+} were 236 ± 25, 474 ± 50, 999 ± 94, 1620 ± 80, 2030 ± 190, and 3800 ± 240 μM; means ± asymptotic SD, $n = 3$). As shown in Fig. 4, the presence of 2 mM CaCl$_2$ significantly slows the formation to the [^{125}I]CgTx-receptor complex in guinea pig cerebral cortex membranes. The experiment reveals the reversible mode of interaction of the cation with the receptor which is overcome by the (quasi-)irreversible binding of the toxin. This cannot be explained by the formation of covalent bonds between the toxin and its binding domain. The complex can be easily dissociated, e.g., upon SDS-PAGE, even if carried out under disulfide-bond-protecting conditions. Therefore, cross-linking techniques have to be employed to attach the toxin covalently to channel components.

We synthesized an azido[^{125}I]CgTx derivative very similar to the probe used by Abe et al. [6] by introducing an azidobenzoate moiety into the ^{125}I-iodinated toxin molecule (Fig. 5). Azido[^{125}I]CgTx still binds to the receptor with high affinity and can be covalently incorporated after mild UV irradiation.

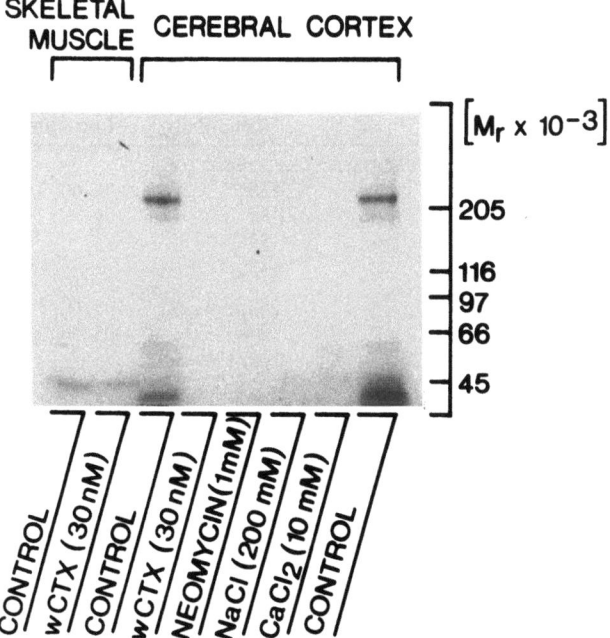

Fig. 5. Photoaffinity labeling of guinea pig cerebral cortex membranes with azido-[^{125}I]ω-conotoxin GVIA. Azido[^{125}I]CgTx was synthesized by adding an excess of N-hydroxysuccinimidyl-azidobenzoate in dimethylsulfoxide to [^{125}I]CgTx(1 μCi) dissolved in 0.15 M boric acid/NaOH, pH 8.5. The mixture was incubated for 60 min at 25°C. The reaction was quenched by dilution in assay buffer (50 mM Tris-Cl, pH 7.4). The label (38000 dpm) was incubated (60 min, 25°C) with guinea pig cerebral cortex (0.24 mg) or partially purified guinea pig skeletal muscle T-tubule (0.3 mg) membrane protein in a final assay volume of 0.25 ml. Unlabeled toxin *(wCTX,* 30 nM final concentration), neomycin (1 mM), NaCl (0.2 M), or CaCl$_2$ (10 mM) were absent *(control)* or present in the incubation mixture. Free toxin was subsequently removed by centrifugation of the labeled membranes through a cushion of 5% sucrose (1 ml). The pellets were resuspended in 0.2 ml assay buffer, irradiated for 30 min with UV light in plastic petri dishes, and spearated on SDS-PAGE (8% gel). An autoradiogram of the dried gel is shown. The mobility of molecular weight standards is indicated. Note that the photoaffinity labeling protection profile is similar to the binding inhibition profile (see [33])

Table 2 summarizes the results of photoaffinity-labeling experiments carried out with (−)-[^{3}H]azidopine, [N-methyl-^{3}H]LU49888, and azido[^{125}I]CgTx. The specificity of labeling was demonstrated by comparing the pharmacological profile of reversible binding with the protection profile in photoaffinity labeling (see references given in Table 2). In skeletal muscle, both tritiated enantiomers of [^{3}H]azidopine were employed to follow the enrichment and structural integrity of the 155-kDa polypeptide through several purification steps [23]. Although poly- and monoclonal antibodies have been raised against polypeptides copurifying with the skeletal muscle 1,4-DHP binding activity [35–39], in other excitable tissues a channel-associated Ca^{2+}-antagonist receptor (similar to α$_1$ in skeletal muscle) has so far only been characterized by means of photoaffinity labeling (Table 2).

Table 2. Photoaffinity labeling with $(-)$-[^{3}H]azidopine, [N-methyl-^{3}H]LU49888, and azido [^{125}I]CgTx

Tissue	Ligand	Photolabeled polypeptides (kDa)		SDS-PAGE	References
		High-affinity	Low-affinity		
Skeletal muscle T-tubule membranes					
Nonpurified	LU	155	92, 60, 33	NR, R	[1, 2]
	AZ	158	99	R	[14]
Purified	LU	155		NR, R	[1, 2]
	AZ	155		NR, R	[2]
Hippocampus	LU	265, 190–195	95, 60, 35	NR, R	[40]
membranes	AZ	190–195	95, 56, 33	NR	[40]
	CTX	265, 190–195	45	NR, R	[63]
Cardiac muscle membranes					
	LU	N.D.			
	AZ	165	39, 35	NR	[61]

Conditions for photoaffinity labeling and SDS-PAGE are given in the references cited
NR, nonreducing conditions; R, reducing conditons; LU, [N-methyl-^{3}H]LU49888; AZ, $(-)$-[^{3}H]azidopine; CTX, azido[^{125}I]CgTx

Table 3 describes a special feature of $(-)$-[^{3}H]azidopine photoaffinity labeling in brain and skeletal muscle under our experimental conditions. The formed covalent bonds (at least a significant fraction thereof) are sensitive to nucleophilic reagents. The covalently attached ligand can be released, e.g., by dithiothreitol, β-mercaptoethanol, and hydroxylamine. These (nucleophile-sensitive) bonds are only observed with the Ca^{2+}-channel-associated 1,4-DHP receptors, but not with other binding sites, including an antiazidopine immunoglobulin serum [40]. Such bonds were also never observed after specific photoincorporation of [N-methyl-^{3}H]LU49888. This unique property of $(-)$-[^{3}H]azidopine might be of some help in distinguishing between L-type Ca^{2+}-channel-associated and unrelated photolabeled structures in future experiments.

In addition to a 190- to 195-kDa polypeptide, [N-methyl-^{3}H]LU49888 labels another high-M_r-polypeptide (M_r 265000) with high affinity and a Ca^{2+}-channel-typical pharmacological profile in guinea pig hippocampus membranes. Interestingly, azido-[^{125}I]CgTx specifically labels two high-M_r polypeptides in guinea pig cerebral cortex, which cannot be distinguished by size (in SDS-PAGE) from the [N-methyl-^{3}H]LU49888-labeled bands. Apparently, Ca^{2+} channels in neurons and elsewhere are a family of very large polypeptides. We entertain the hypothesis [40] that the 260-kDa [N-methyl-^{3}H]LU49888-labeled polypeptide may represent T-type channels, which are known to be insensitive to 1,4-DHPs.

Biochemical Characterization of Purified Ca^{2+} Channel from Guinea Pig Skeletal Muscle T-Tubule Membranes

Purification of the 155-kDa Ca^{2+}-channel polypeptide from skeletal muscle T-tubule membranes can be easily followed with reversible 1,4-DHP and phenylalkylamine

Table 3. Sensitivity of $(-)$-$[^3H]$azidopine photolabeling against nucleophilic compounds

Sample			Incorporation (fmol)	% of control
Experiment 1				
Control		T	64.1	100
DTT		NS	16.9	19.1
	0.003 mM	T	64.2	100
	0.01 mM	T	46.7	72.9
	0.03 mM	T	46.1	71.8
	0.1 mM	T	38.2	59.5
	0.3 mM	T	25.2	39.3
	1.0 mM	T	15.6	17.5
	3.0 mM	T	12.6	14.2
	10.0 mM	T	11.5	12.9
	10.0 mM	NS	6.5	10.1
Experiment 2				
Control		T	95.8	100
DTT		NS	26.3	27.4
	10.0 mM	T	24.4	25.5
	10.0 mM	NS	12.6	13.2
β-ME				
	0.25 M	T	20.6	21.5
	0.25 M	NS	10.8	11.2
HA				
	0.1 M	T	74.4	83.7
	0.1 M	NS	23.5	24.5

Purified skeletal muscle Ca^{2+} channels were photoaffinity labeled with (C)-$[^3H]$azidopine in the absence (T) or presence (NS) of 1 µM unlabeled PN200–100, and 0.1-ml portions were preincubated for 15 min at 57°C with a final concentration of 10 mM N-ethylmaleimide (control) or increasing concentrations of dithiothreitol (DTT), β-mercaptoethanol (β-ME), or hydroxylamine (HA) at neutral pH. The sample was then precipitated with chloroform/methanol in the presence of carrier protein, and the incorporated radioactivity determined.

binding or by quantitative photoaffinity labeling. The specific activity of the 155-kDa $(-)$-$[^3H]$azidopine-photolabeled band, as well as of reversible 1,4-DHP binding, increases about 100- to 150-fold after lection affinity chromatography and sucrose gradient centrifugation of the digitonin-solubilized membranes (1–1.5 nmol $[^3H]$PN200–110 binding sites per milligram of protein). Only one single polypeptide (α_1) is photolabeled in the purified preparation, whereas other polypeptides are additionally labeled in the membranes [14, 23]. However, analysis of the proteins by reducing SDS-PAGE followed by Coomassie (Fig. 6, lane 2) or silver staining (see e.g. [1, 2, 5, 36]) reveals the presence of at least four polypeptides that always copurify together under nondenaturing conditions, indicating their tight noncovalent association [5]. These polypeptides are termed α_1-(155 kDa), α_2-(135kDa), β-(55kDa), and γ-subunit (33 kDa) according to their apparent molecular mass in reducing SDS-PAGE (5%–15% gradient gels).

The α_2-polypeptide is covalently linked to several smaller 20- to 30-kDa polypeptides (termed δ-subunits; see [5]) by disulfide bonds, and it migrates with a lower

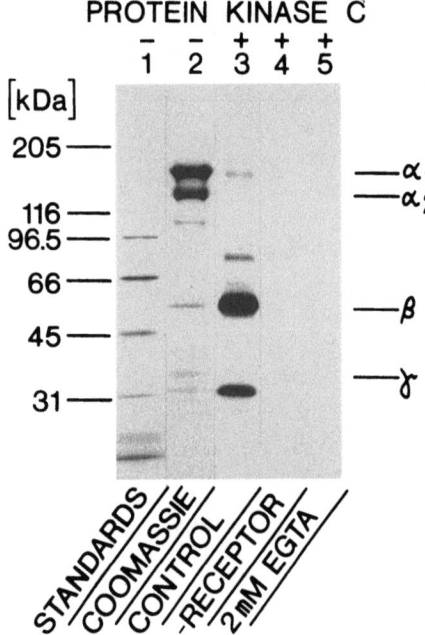

Fig. 6. Phosphorylation of the purified skeletal muscle Ca^{2+}-antagonist receptor with protein kinase C. Purified Ca^{2+}-antagonist receptor protein (0.2 μg; 1–1.5 nmol (+)-[^{3}H]PN200–110 binding sites per milligram of protein were phosphorylated with protein kinase C (1.25 nM) in the presence of l-phosphatidyl-l-serine (20 μg/ml), 1,2-dioleyl-rac-glycerol (5 μg/ml), 5 mM $MgCl_2$, 0.5 mM $CaCl_2$, 1 mM dithiothreitol, and 20 μM ^{32}P-γ-ATP (specific activity 21.2 dpm/fmol). Incubation was carried out for 30 min at 30 °C in a total volume of 25 μl. The reaction was terminated by addition of an equal volume of electrophoresis sample buffer (containing 10 mM dithiothreitol). The sample was electrophoresed on a 5%–15% polyacrylamide gradient gel. Molecular weight standard proteins (Biorad) and 10μg purified receptor protein were run on the same gel and stained with Coomassie blue. The stained proteins (lanes 1, 2) are shown in comparison to the autoradiogram of the dried gel exposed to Kodak XAR film for 24 h (lanes 3–5)

mobility than the α-subunit under nonreducing conditions of SDS-PAGE. Association of the δ-subunits with the $α_2$ core polypeptide can be demonstrated by two-dimensional SDS-PAGE (Fig. 7). Separation in the first dimension is carried out under nonreducing conditions (proteins were denatured in the presence of 10 mM N-ethylmaleimide). Separation in the second dimension is then carried out after equilibration of the excised gel slice with 10 mM dithiothreitol at room temperature (reducing conditions). A silver stain after the second-dimension run is shown in Fig. 7. The $α_2$-polypeptide does not migrate on the diagonal like the other stained polypeptides, reflecting its large shift in apparent molecular mass after reduction. The δ-polypeptides released from $α_2$ by reduction cannot be easily stained with silver, and we have visualized them by immunostaining. To this end, the purified Ca^{2+} channel was denatured by boiling for 3 min in the presence of 10 mM dithiothreitol and 5% SDS, and separated on a size exclusion column (see Fig. 8). Polypeptides eluting in fractions 50–55 (comprising γ- and δ-subunits) were employed for raising polyclonal antibodies in rabbits. After transfer to nitrocellulose, these antisera clearly visualize the δ-, γ-, and $α_2$-, but not the β- and $α_1$-subunits (Fig. 7). The staining of the $α_2$-polypeptide may indicate the presence of residual antigenic sites on $α_2$ which were not dissociated under the (mild) conditions used for reduction in the second-dimension gel electrophoresis run. We suggest that they were dissociated when the Ca^{2+} channel was denatured and reduced for size exclusion chromatography under much harsher conditions.

As the $α_2$- and γ-subunits are heavily glycosylated in rabbit preparations [3, 5] and very weak glycosylation ($α_1$) is observed in the guinea pig ($α_2 >> γ > α_1$; see Fig. 9), we determined the "true" molecular weights using Ferguson plot analysis (Fig. 10).

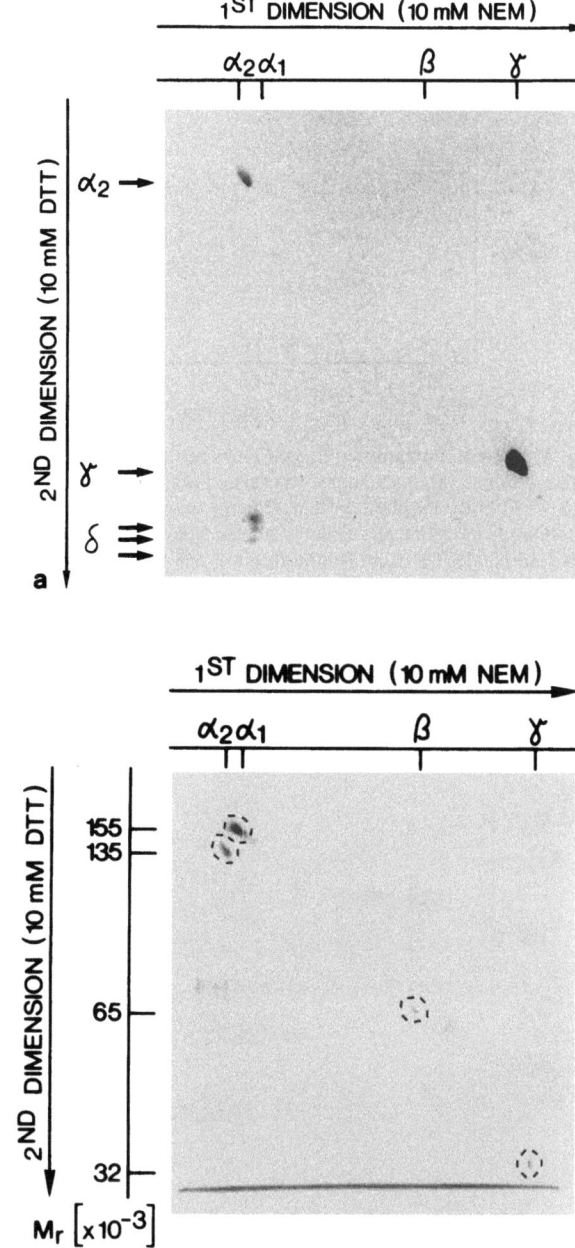

Fig. 7a, b. Detection of α-associated (disulfide bridge-linked) polypeptides by two-dimensional SDS-PAGE. Purified skeletal muscle Ca^{2+}-antagonist receptor protein (2 μg) was solubilized in sample buffer containing 10 mM N-ehtylmaleimide *(NEM)* and separated on an 8% polyacrylamide slab gel. One lane was stained with silver to identify the individual bands (not shown). Two lanes from the same gel were cut out, equilibrated with 10 mM dithiothreitol *(DTT)* for 30 min at 25°C, and mounted on 8% polyacrylamide gels. After electrophoresis, the gels were either stained with silver (**a**) or blotted onto nitrocellulose and immunostained (**b**) employing a polyclonal rabbit antiserum raised against the denatured 30-kDa region (γ and δ-subunits of purified Ca^{2+} channels). No staining was observed using preimmune serum (not shown)

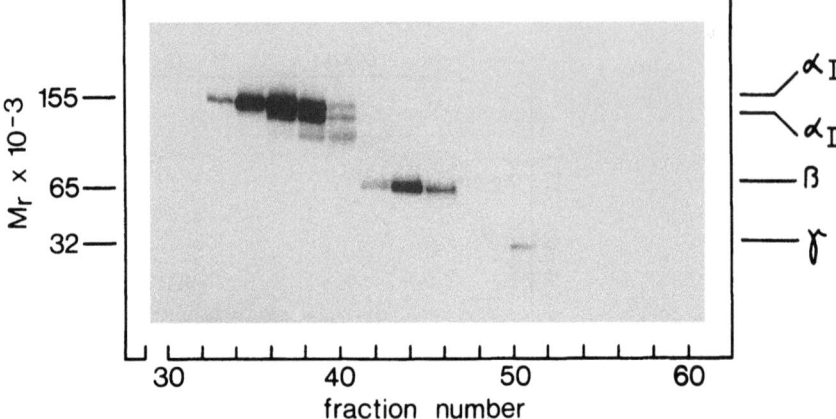

Fig. 8. Subunit separation of the purified skeletal muscle Ca^{2+} channel under denaturing conditions. Size exclusion chromatography was carried out as described using a Superose 6 prep grade column (1,6 × 50 cm). Purified protein (25 µg) was dialyzed against a solution of 10 mM $NaHCO_3$, lyophilized, and iodinated with ^{125}I-labeled Bolton and Hunter reagent as described [23]. The sample was resuspended in 10 mM dithiothreitol, 5% SDS, 0.1 M Tris-Cl, pH 7.4, and boiled for 3 min. Thereafter, it was loaded onto the column, equilibrated in 10 mM sodium phosphate buffer (pH 7.0), 0.15 M NaCl, 0.1% SDS, 0.1 mM dithiothreitol, and eluted at room temperature at a flow rate of 0.3 ml/min. Aliquots of the eluted fractions were separated on 5%–15% polyacrylamide gels. An autoradiogram of the dried gel is shown. Fractions 50–55 were taken for raising polyclonal antisera in rabbits (see text for details)

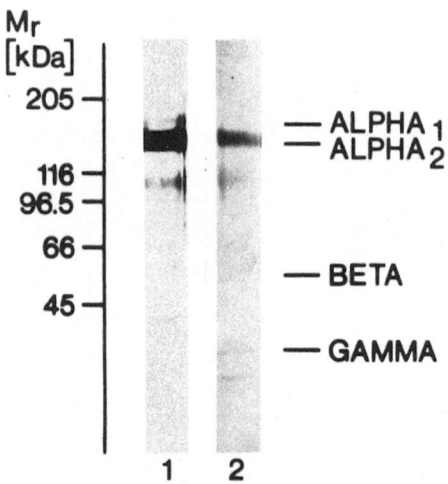

Fig. 9. Glycoprotein staining of the purifed Ca^{2+}-antagonist receptor from guinea pig skeletal muscle. Purified skeletal muscle Ca^{2+}-antagonist receptor protein (2 µg) was separated under reducing conditions on 8% gels and blotted onto nitrocellulose. Glycoproteins were stained with either concanavalin A (lane 1) or wheat germ agglutinin (lane 2). Staining with concanavalin A was carried out as described [23]. For detection of wheat germ lectin binding, the nitrocellulose strip was blocked with 2% (w/v) polyvinylpyrrolidone in TBS (20 mM Tris-Cl, pH 7.4, 150 mM NaCl) and incubated with wheat germ lectin (0.01 mg/ml) in blocking solution (BS). The nitrocelluose was then washed three times for 10 min with BS followed by incubation (90 min) with rabbit anti-wheat germ lectin antiserum (Sigma) diluted in BS. The strip was washed as before, and the bound antibody was visualized using peroxidase-linked anti-rabbit immunoglobulin. Staining of proteins was specific, as it was completely absent when incubation with lectins was carried out in the presence of specific sugars (5% (w/v)α-methyl-mannoside and N-acetyl-glucosamine, respectively). Note that α_1 is slightly stained by concanvalin A. A so far unidentified band (M_r 110000) is stained by both lectin probes

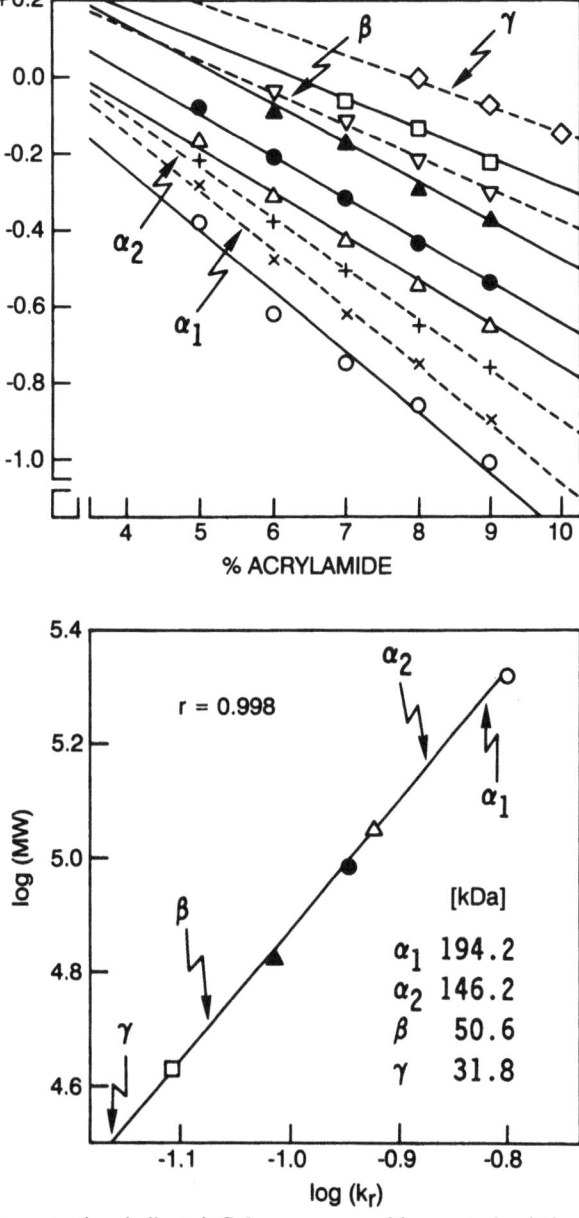

Fig. 10. Molecular weight determination of the purified Ca^{2+}-antagonist receptor polypeptides by Ferguson plot analysis. Purified skeletal muscle Ca^{2+}-antagonist receptor and molecular weight standard proteins (Biorad) were seperated on SDS-PAGE under reducing conditions (10 mM dithiothreitol in the sample buffer) at the polyacrylamide concentrations indicated. Gels were prepared from a stock solution of 30% (w/v) of polyacrylamide, 0.8% (w/v) N, N'-methylenebisacrylamide. The log of the R_f values of each standard protein was calculated and plotted against the acrylamide concentration *(left)*. Data points are means from two to three determinations. The slopes of the regression lines (retardation coefficient, k_R) were calculated and replotted as a function of the molecular weight (M_r) for the standard proteins *(right)*. The M_r of the Ca^{2+}-antagonist receptor polypeptides (α_1, α_2, β, γ) was obtained by fitting their slopes to the calibration line. Molecular weight standards: myosin (○), β-galactosidase (△), phosphorylase b (●), bovine serum albumin (▲), and ovalbumin (□). The free mobility (defined as the R_f at 0% acrylamide) is the extrapolated y-axis intercept and was similar for standard (0.45 ± 0.05, SD n = 5) and receptor polypeptides (0.47 ± 0.04, SD n = 4)

This type of analysis often corrects for the fact that glycoproteins behave anomalously in SDS-PAGE due to decreased binding of detergent compared to nonglycoproteins. Unexpectedly, the free mobility of all polypeptides investigated was very similar to that of the standard proteins used. The $M_r(194200)$, however, determined for the α_1-polypeptide was somewhat larger than estimated from 5%–15% or 8% gradient gels but is in good agreement with the values calculated from its deduced amino acid sequence (212 k, [41]).

As is exemplified below, the functional activity of the isolated skeletal muscle Ca^{2+} channels is strongly dependent on cAMP-dependent phosphorylation. This has previously been shown to be a major regulatory mechanism for cardiac L-type Ca^{2+} channels (for review see [42]). In in vitro phosphorylation studies, the α_1-subunit is the major substrate for the cAMP-dependent protein kinase [42–44]. In contrast, the β-subunit and (less so) the α_1-subunit are substrates for in vitro phosphorylation by the protein kinase C (Fig. 6; see also [45]). The physiological and functional importance of this phosphorylation remains to be established.

In summary, the α_1-subunit carries regulatory sites (Ca^{2+}-antagonist receptors, cAMP-dependent phosphorylation sites) expected for L-type Ca^{2+} channels. The α_1-subunit, is a transmembrane protein with high structural similarity to the voltage-dependent Na^+ channel [41]. It is not known whether α_1 alone suffices to form functional channels. The physiological role, if any, of the other, tightly associated, copurifying "subunits" is the object of ongoing research.

Characterization of Solubilized Brain Ca^{2+} Channels

We addressed the question whether [^{125}I]CgTx and Ca^{2+}-channel antagonist receptors are localized on the same or different polypeptides in brain. Therefore, guinea pig cerebral cortex membranes were prelabeled with (+)-[^{3}H]PN200–110 and solubilized with digitonin. The solubilized material was loaded onto a sucrose density gradient and spun for 1.5 h at 210000 × g (Fig. 11). The peak of specific (+)-[^{3}H]PN200–110 binding activity migrated with a sedimentation coefficient of 20 S, in good agreement with findings reported elsewhere [46]. The 1,4-DHP receptor prelabeled gradient material was investigated for [^{125}I]CgTx binding activity by postlabeling. Interestingly, the [^{125}I]CgTx binding activity partially overlaps with the 1,4-DHP binding peak, but in three different experiments the [^{125}I]CgTx binding peak was always significantly larger than 20 S. Our data clearly show that a major fraction of the [^{125}I]CgTx binding activity is behaving differently by this type of analysis than the L-type Ca^{2+} channel. As seen from Fig. 11, sucrose gradient centrifugation constitutes a valuable step in the purification of the [^{125}I]CgTx-labeled Ca^{2+} channels as seven fold enrichment is achieved. The polypeptides specifically photolabeled in guinea pig brain membranes by the azido[^{125}I]CgTx probe were also photolabeled in the partially purified gradient material (not shown). As emphasized for the L-type channel, the structural integrity of the N-type channel polypeptides must be followed with irreversible probes (preferably by photolabels) upon purification, and the azido[^{125}I]CgTx is well suited for this purpose.

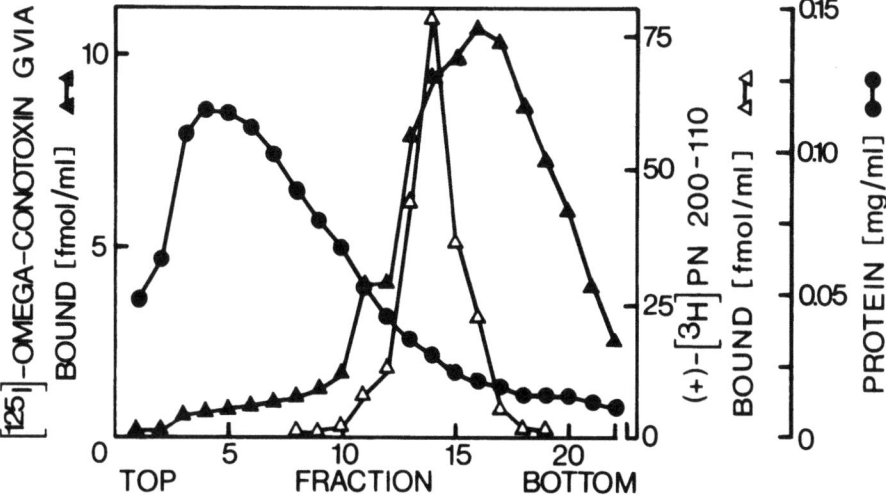

Fig. 11. Separation of solubilized brain 1,4-DHP and ^{125}I-ω-conotoxin GVIA binding activity by sucrose density gradient centrifugation. Guinea pig cerebral cortex membranes (6.9 mg protein) were prelabeled with (+)-[^{3}H]PN200–110 at 37°C in the absence (total binding) and presence (nonspecific binding) of 300 nM unlabeled (+)-PN200–110. After 60 min, the membranes were cooled to 2°C and washed twice with ice-cold Tris buffer (20 mM Tris-Cl, pH 7.4, 0.1 mM phenylmethylsulfonyl-fluoride). The pellets were solubilized for 30 min in ice-cold Tris buffer containing 1% (w/v) digitonin and 0.2 M NaCl. Insoluble material was removed by centrifugation of the incubation mixture for 60 min at 45 000 × g. The supernatants containing solubilized total or nonspecific (+)-[^{3}H]PN200–110 binding activity were layered onto linear sucrose gradients [5%–20% (w/w) sucrose, 20 mM Tris-Cl, 0.1% (w/w) digitonin] and spun in a VTI50 rotor at 210 000 × g for 90 min, 22 fractions being collected from the bottom of the gradients. From each fraction 0.2-ml aliquots were precipitated using the polyethylene glycol precipitation technique to determine precipitable (+)-[^{3}H]PN200–110 binding activity; 0.1-ml aliquots of the total binding gradient were also assayed for total and nonspecific ^{125}I-ω-Conotoxin GVIA binding activity. Nonspecific binding was determined in the presence of 30 nM unlabeled toxin

Reconstitution and Modulation of Ca^{2+}-Channel Activity

There have been several reports on the reconstitution of Ca^{2+}-channel activity using purified skeletal muscle 1,4-DHP receptor preparations [47–50], but the relevance of the Ca^{2+}-antagonist receptors as functioning Ca^{2+} channels in the skeletal muscle T-tubule has not found universal acceptance [51–53]. Crucial to a complete understanding of the physiological role of the receptor as a Ca^{2+} channel is its regulation in the planar bilayer by modulatory agents whose effects on intact systems are well known. We describe here effects of divalent cations, phenylalkylamines, 1,4-DHPs, and cAMP-dependent protein kinase on purified Ca^{2+}-antagonist receptor-induced Ca^{2+}-channel activity. The channel preparations used contain (as proven by photoaffinity labeling) intact $α_1$-subunits, indistinguishable in size from those labeled in freshly isolated skeletal muscle T-tubule membranes.

Methodology – An Unbiased Approach. Sophisticated study of purified channel molecules involves inserting them in a defined way into planar lipid bilayers. In order to obtain meaningful qualitative or quantitative analysis of channel properties, a method is required which enables investigation of a known number of channel molecules. To date, only one method is available which can accomplish this task: the vesicle-derived, septum-supported bilayer (VSB) method [54]. Using this method, planar membranes are obtained which have the same protein-to-lipid ratio (within a factor of 2) as the vesicle suspension used to form them, and the channel proteins are incorporated in an initially random lateral distribution.

VSB membranes have the highest specific capacitance ($0.75\ \mu F/cm^2$), specific resistance ($10^8\ \Omega/cm^2$), and break-down voltage ($> 360\ mV$ with $10\ mV/s$ ramp) of any known planar lipid bilayer system. The baseline is completely stable for long periods ($> 30\ min$) and free of artifacts up to holding potentials of at least $\pm 250\ mV$ (several minutes). The resolution product, which is the product of time resolution (filter setting) and amplitude resolution (2-σ width of baseline noise), is $1 \times 10^{-15}\ A \times s$ for 100 μm apertures, permitting registration of channel events as low as 1 pS and less, depending on the channel open time. This is in contrast to black lipid films, which have a specific capacitance of about $0.4\ \mu F/cm^2$ and cannot withstand holding potentials of more than 100 mV for more than a few seconds. Further, the often-used "dipping" method with patch pipettes [47, 55] is fraught with channel-like shifts in the baseline, making it difficult to sort out channel activity from artifacts [56].

The relevance of methods for which the observed channel activity is unrelated to the amount of channel protein added [48, 57] must be limited to the observation of already expected effects. With such a (biased) methodology, no statement can be made as to the relative functionality of the biochemical preparation. However, since the very purpose of reconstitution studies with purified channel proteins is to assign functional properties to specific components, the wisdom of this approach must be questioned. Moreover, we believe that since individual channel molecules often behave erratically (as every „channelologist" knows), modulatory effects to be demonstrated at the single-channel level must be supported by statistics in order to be meaningful. We have chosen a simple basis for channel statistics: measurements performed with 30–100 channel polypeptides (i.e., 1,4-DHP binding sites) per bilayer. Although individuals gating events can no longer be resolved in such activity, and more detailed analysis of molecular channel characteristics is necessarily deferred to future studies, we find this approach superior to showing individual events selected for their beauty but lacking statistics.

Activation of Ca^{2+}-Selective Conductance. Ca^{2+}-channel activity was observed inconsistently in most but not all receptor preparations measured if the material was incorporated into the bilayer without further treatment after purification. Its single-channel conductance was complex, revealing a smallest conductance of about 0.9 pS in 100 mM $BaCl_2$ and a variety of often time-correlated states up to at least 50 pS [50]. The activity was Ca^{2+}- or Ba^{2+}-selective ($P_{Ba}/P_{Na} = 30$).

We found that phosphorylation of the α_1-polypeptide of the purified preparation via cAMP-dependent protein kinase led to a dramatic activation of the membrane Ba^{2+} conductance (G_m), even for preparations which showed no activity without phosphorylation (Fig. 12A, B; [50]). The activation was rapid (1 min), highly reproduc-

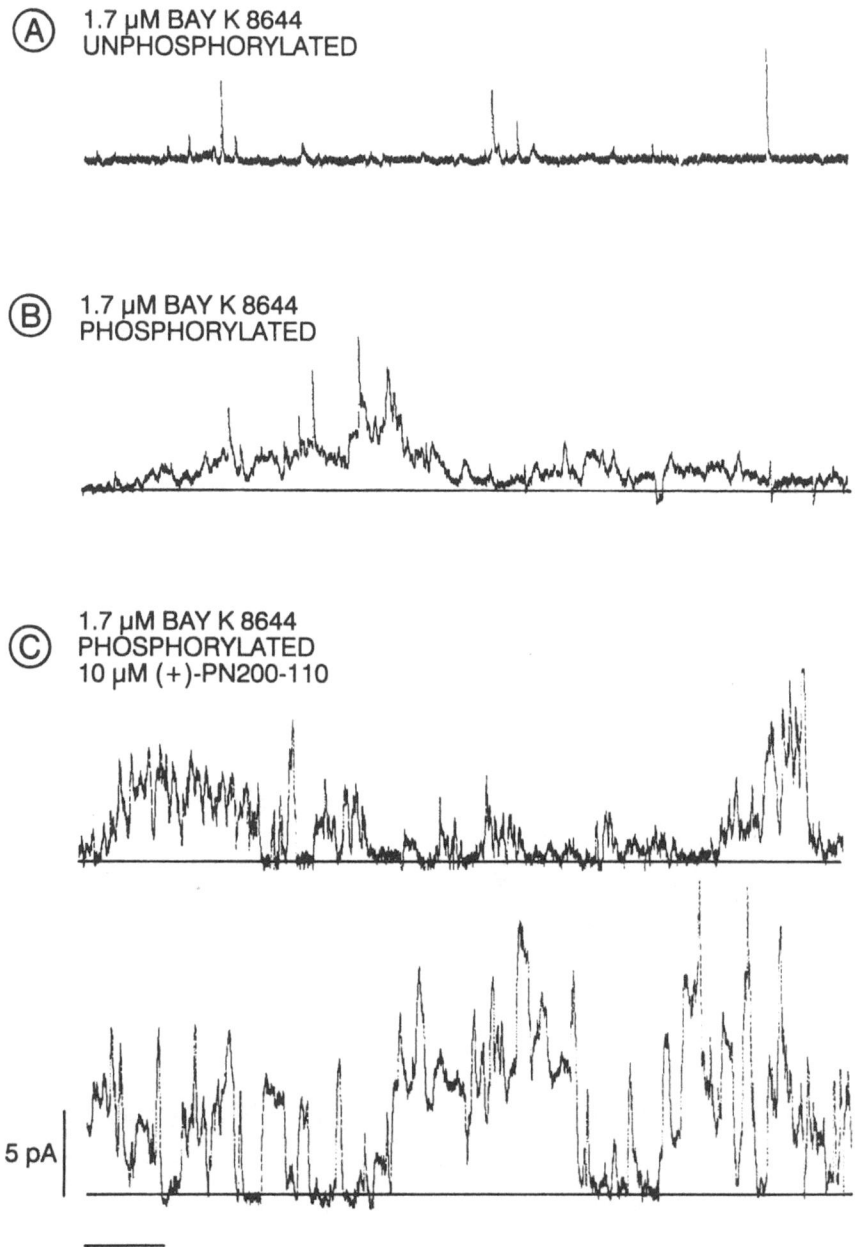

Fig. 12A–C. Activation of CA^{2+}-antagonist receptor-induced Ba^{2+} current by cAMP-dependent protein kinase and (+)-PN200–110. Ba^{2+} current induced by 100 1,4-DHP receptor sites incorporated (cis side) into a 100-μm planar bilayer of purified soybean phospholipid/cholesterol (6/1; w/w) in 110 mM NaCl, 10 mM Hepes/Tris (pH 7.4) (both sides), and 110 mM $BaCl_2$ (cis side >) at a holding potential of (cis side) + 50 mV **(A, B)** and + 70 mV **(C). A** Unphosphorylated channels (as isolated) show essentially no activity. **B** Same as **A**, 5 min after addition of 1 mM ATP-γ-S, 1 mM $MgCl_2$, and 250 units of the catalytic subunit of cAMP-dependent protein kinase per milliliter. **C** Same membrane, 8 min *(upper trace)* and 10 min *(lower trace)* after addition of (cis) 10 μM (+)-PN200–110

ible, and dependent on protein kinase, ATP, and divalent cation (Mg^{2+}). (±)-Bay K8644, added to the same bilayers, was without noticeable effect both prior to and after phosphorylation. However, the 1,4-DHP antagonist (+)-PN200–110 (10 μM) induced interesting qualitative and quantitative changes in the activity of channels which had already been activated by phosphorylation (Figure 12C; [50]). At 8–10 min after addition of (+)-PN200–110, the "single-channel" gating properties became significantly more "all-or-none" cooperative, reducing fast fluctuation between substates. Amplitude histogram analysis [50] indicated many of the conductance states and transitions to be multiples of 4 of the smallest conductance, 0.9 pS, up to at least 60 pS.

Inhibition of Ca^{2+}-Channel Activity. Ba^{2+} currents were blocked by phenylalkylamines and by Cd^{2+}, but not by 1,4-DHP antagonists, under the conditions used. (D890 5 μM; Fig. 13A) was able to completely block the Ba^{2+} current within 1.5 min from the same side of the membrane to which the phosphorylation mix had been previously added (defining that side functionally as the intracellular side). Specific binding sites for divalents are demonstrated by the ability to block the Na^+ current

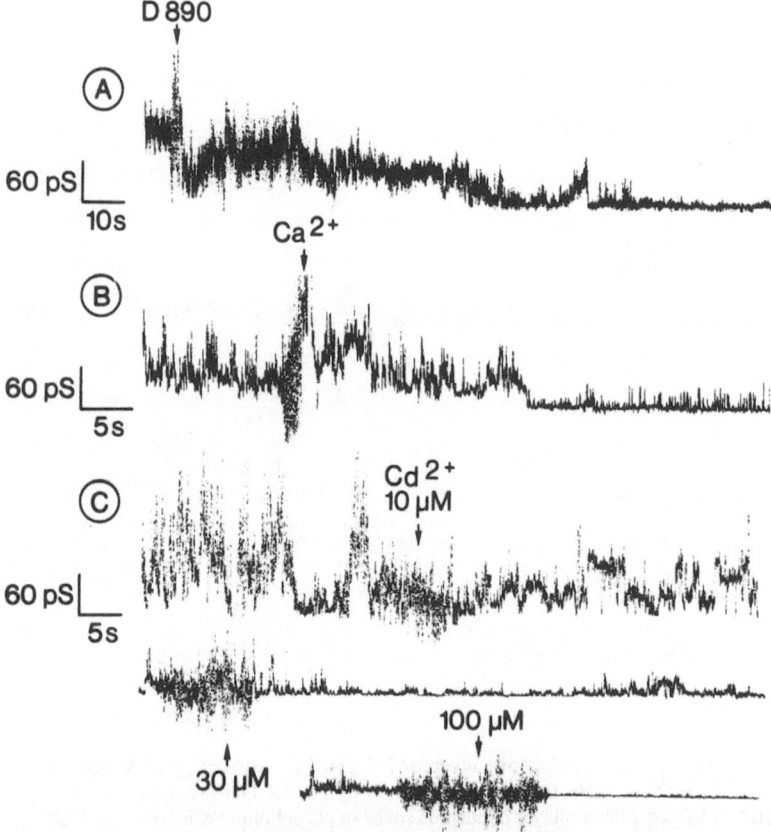

Fig. 13A–C. Inhibition of Ca^{2+}-antagonist receptors by D890 and divalent cations. Conditions are the same as in Fig. 12B. **A** Addition of 5 μM D890 (cis side) rapidly induces blockage of the Ba^{2+} current. **B** Ca^{2+}-channel activity measured without $BaCl_2$ in the cis solution. Addition of 100 μM $CaCl_2$ (both sides) blocks the Na^+ current. **C** Block of Ca^{2+}-channel activity by Cd^{2+}

through the channel by 100 µM $CaCl_2$ (Fig. 13B) and blockage of the Ba^{2+} current by 100 µM Cd^{2+} (Fig. 13C).

Significance of Modulation. In summary, we have shown that the purified and reconstituted Ca^{2+}-antagonist receptor from skeletal muscle possesses many of the physiological regulatory mechanisms expected for an L-type Ca^{2+} channel in skeletal muscle. Channel activity was found to be Ca^{2+}-selective, but conductive to monovalent cations in the absence of divalent cations. It was blocked by Cd^{2+} [58] and in the expected concentration range by phenylalkylamines from the cytoplasmic side. The channel was dramatically activated by phosphorylation (using the cAMP-dependent protein kinase) in such a way as to indicate that this regulatory step is obligatory for channel opening, with further control in the phosphorylated state by membrane potential [50]. Finally, the 1,4-DHP antagonist (+)-PN200–110 was unexpectedly found to stabilize the open and closed states of the channel, making the channel gating more highly cooperative and revealing its structure as an oligochannel.

Photoaffinity Labeling of Ca^{2+}-Antagonist Binding Sites Unrelated to L-Type Ca^{2+} Channels

In guinea pig heart, [^{3}H]nitrendipine interacts with the 1,4-DHP receptor domain of the L-type Ca^{2+}-channel as well as another class of binding sites displaying a density roughly 1000-fold higher (1.8 nmol/mg protein; [19]) than the Ca^{2+}-channel-associated stereoselective [^{3}H]nitrendipine binding sites (0.3–1.6 pmol/mg protein; [59–61]). The majority of these high-capacity [^{3}H]nitrendipine binding sites are located on the mitochondrial inner membrane and are biochemically and pharmacologically distinct from the Ca^{2+}-channel-associated DHP domain [59]. These sites are also found in guinea pig liver and kidney mitochondria. 1,4-DHPs and phenylalkylamines interact nonstereoselectively with the mitochondrial binding sites. Their dissociation constants are in the submicromolar and micromolar range, whilst the benzothiazepine d-*cis*-diltiazem is without any effect up to 100 µM. Most interestingly, [^{3}H]nitrendipine binding is inhibited by purine and less so by pyrimidine nucleotides. The rank order of binding inhibition potency is: ATP (IC_{50} 11.8 µM) = ATP-γ-S > AppNHp > ADP >> GTP = ITP = CTP > UTP > GT_4P > GTP-γ-S > GppNHp > IDP > CDP > GDP (Fig. 14). ATP decreases both the association and the dissociation rate of the [^{3}H]nitrendipine binding site complex, suggesting negative heterotopic allosteric interaction. This is confirmed by equilibrium binding studies, where 0.3 mM ATP decreases the B_{max} from 1.62 ± 0.08 nmol/mg protein to 0.73 ± 0.24 nmol/mg protein whilst only moderately decreasing the K_d from 561 ± 54 nM to 352 ± 43 nM. The ATP inhibition of mitochondria [^{3}H]nitrendipine binding is not due to phosphorylation or ATPase activity. It is completely reversible by wash-out. The complex interaction of nucleotides with the high-capacity low-affinity 1,4-DHP binding site suggests its involvement in the homeostasis of energy-rich phosphates, although the ADP/ATP carrier of the mitochondrial inner membrane can be excluded on the basis of functional assays [62].

As (+)- and (−)-azidopine competitively inhibit reversible [^{3}H]nitrendipine binding with K_d values of 1342 ± 412 nM and 834 ± 154 nM displaying Hill slopes of 0.88

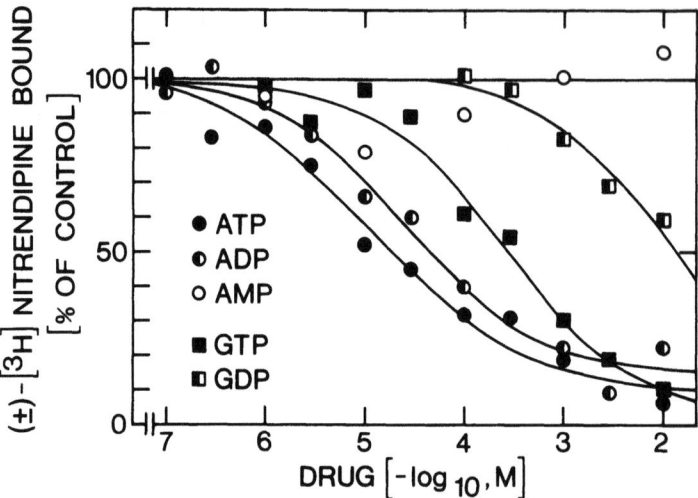

Fig. 14. The high-capacity, low-affinity [³H]nitrendipine sites on the mitochondrial inner membrane: Inhibition by purine nucleotides and analogs. Cardiac mitochondrial membrane protein (7.3–28.4 µg) was incubated with 1.5–4.3 nM (±)-[³H]nitrendipine in the presence or absence of varying concentrations of purine nucleotides for 60 min at 37°C. Inhibitors shown are ATP, ADP, AMP, GTP, and GDP. Values shown are means of two to eight experiments each performed in duplicate

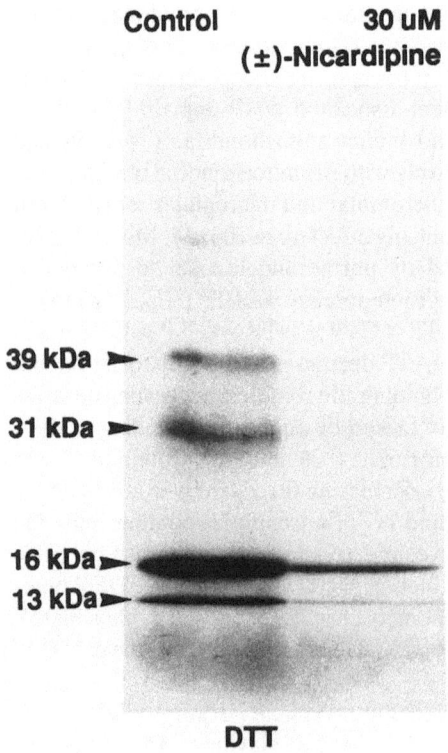

Fig. 15. (±)-[³H]azidopine photoaffinity labeling of guinea pig liver mitochondria. (±)-[³H]azidopine (8.88 nM) was incubated with 2.4 mg mitochondrial membrane protein in an assay volume of 10 ml for 60 min at 30°C. Total binding prior to UV irradiation was 375100 dpm/ml *(control)*, and binding in the presence of 30 µM (±)-nicardipine amounted to 274500 dpm/ml *(nonspecific binding)*. The samples were transferred to 10-cm-diameter petri dishes on ice and subjected to UV irradiation by a Philips TL40W/08 blacklight lamp at a distance of 10 cm for 5 min. After removal of unbound ligand by centrifugation (39000 × g for 10 min at 2°C) and extraction of lipids with chloroform, 86.3 µg protein containing 113700 dpm (total incorporation) and 94.2 µg protein containing 48200 dpm (nonspecific incorporation) were denatured under reducing conditions for 2 min in a boiling stop solution containing 10% SDS, 0.313 M Tris-HCl (pH 6.8), 10% glycerol, 0.1 mM PMSF (phenylmethylsulfonylfluoride), and 10 mM DTT, and separated on 10%–15% polyacrylamide gradient slab gel. The gel was stained with Coomassie blue, impregnated with Amplify (Amersham), and exposed to Hyperfilm (Amersham) for 4 weeks

and 1.11, respectively, the use of $(\pm)$-$[^3H]$azidopine as a photoaffinity probe for the mitochondrial high-capacity low-affinity 1,4-DHP binding site seemed promising. A fluorogram of the photoincorporation pattern of $(\pm)$-azidopine into mitochondrial membranes is shown in Fig. 15. Four distinct polypeptides are labeled: a 39-kDa protein, a 31-kDa protein, and two polypeptides in the 16- to 13-kDa range. $(\pm)$-Nicardipine (30 µM; also used as a definition of nonspecific binding in the reversible $[^3H]$nitrendipine binding assay) completely protects only the 39-kDa and the 31-kDa protein, the two peptides in the 16- to 13-kDa range being only partially protected. Photoaffinity labeling of both the 39-kDa and the 31-kDa protein occurs regardless of whether the proteins are separated under alkylating (10 mM N-ethylmaleimide) or reducing (10 mM dithiothreitol) conditions. In addition, the mitochondrial membrane-photoincorporated azidopine, in contrast to the α_1-subunit(s) of L-type Ca^{2+} channel, is completely stable to nucleophilic reagents. The two photolabeled mitochondrial proteins might very well represent the target structures characterized by reversible $[^3]$nitrendipine binding. Their functional significance is still unknown.

Acknowledgements. The autors would like to thank the chemists and pharmacologists of Bayer AG for generous support and help in 1,4-DHP research, especially Drs. Hoffmeister, Kazda, Schramm, Meyer, Wehinger, Kinast, Schwenner, and Traber. The chemists and pharmacologists of Knoll AG provided us with the tools to identify phenylalkylamine receptors. Sandoz AG gave us the optical enantiomers of PN200–110, and Goedecke AG aided us in ω-conotoxin GVIA label development and supplied highly purified protein kinase C.

References

1. Striessnig J, Knaus HG, Grabner M, Moosburger K, Seitz W, Lietz H, Glossmann H (1987) Photoaffinity labeling of the phenylalkylamine receptor of the skeletal muscle transverse-tubule calcium channel. FEBS Lett 212:247–253
2. Vaghy PL, Striessnig J, Miwa K, Knaus HG, Itagaki K, McKenna E, Glossmann H, Schwartz A (1987) Identification of a novel 1,4-dihydropyridine- and phenylalkylamine-binding polypeptide in calcium channel preparations. J Biol Chem 262:14337–14342
3. Leung AT, Imagawa T, Campbell KP (1987) Structural characterization of the 1,4-dihydropyridine receptor of the voltage-dependent calcium channel from rabbit skeletal muscle. Evidence for two distinct high molecular weight subunits. J Biol Chem 262:7943–7946
4. Flockerzi V, Oeken HJ, Hofmann F (1986) Purification of a functional receptor for calcium channel blockers from rabbit skeletal muscle microsomes. Eur J Biochem 161:217–224
5. Takahashi M, Seagar MJ, Jones JF, Reber BFX, Catterall WA (1987) Subunit structure of dihydropyridine-sensitive calcium channels from skeletal muscle. Proc Natl Acad Sci USA 84:5478–5482
6. Abe T, Saisu H (1987) Identification of the receptor for omega-conotoxin in brain. J Biol Chem 262:9877–9882
7. Glossmann H, Ferry DR, Rombusch M (1984) Molecular pharmacology of the calcium channel: Evidence for subtypes, multiple drug-receptor sites, channel subunits, and the development of a radioiodinated 1,4 dihydropyridine calcium channel label. [^{125}I]iodipine. J Cardiovasc Pharmacol 6:608–621
8. Striessnig J, Meusburger E, Grabner M, Knaus HG, Glossmann H, Kaiser J, Schölkens B, Becker R, Linz W, Henning R (1988) Evidence for a distinct calcium antagonist receptor for the novel benzothiazinone compound HOE 166. Naunyn Schmiedeberg's Arch Pharmacol 337:331–340

9. Galizzi JP, Fosset M, Romey G, Laduron P, Lazdunski M (1986a) Neuroleptics of the diphenylbutylpiperidine series are potent calcium channel inhibitors. Proc Natl Acad Sci USA 83:7513–7517
10. King VF, Garcia ML, Kaczorowski GJ (1988) Interaction of fluspirilene with cardiac L-type calcium channels. Biophys J 53:557a
11. Fosset M, Jaimovich E, Delpont E, Lazdunski M (1983) [^{3}H]Nitrendipine receptors in skeletal muscle. J Biol Chem 258:6068–6092
12. Glossmann H, Ferry DR, Boschek CB (1983) Purification of the putative calcium channel from skeletal muscle with the aid of [^{3}H]nimodipine binding. Nauny Schmiedeberg's Arch Pharmacol 323:1–11
13. Ferry DR, Rombusch M, Goll A, Glossmann H (1984) Photoaffinity labeling of Ca^{2+} channels with [^{3}H]azidopine. FEBS Lett 169:112–118
14. Ferry DR, Kämpf K, Goll A, Glossmann H (1985) Subunit composition of skeletal muscle transverse tubule calcium channel evaluated with the 1,4-dihydropyridine photoaffinity probe, [^{3}H]azidopine. EMBO J 4:1933–1940
15. Sieber M, Nastainczyk W, Zubor V, Wernet W, Hofmann F (1987) The 165-kDa peptide of the purified skeletal muscle dihydropyridine receptor contains the known regulatory sites of the calcium channel. Eur J Biochem 167:17–122
16. Pauron D, Qar J, Barhanin J, Fournier D, Cuany A, Pralavorio M, Berge JB, Lazdunski M (1987) Identification and affinity labeling of very high affinity binding sites for the phenylalkylamine series of calcium channel blockers in the Drosophila nervous system. Biochemistry 26:6311–6315
17. Pastan I, Gottesmann M (1987) Multiple-drug resistance in human cancer. N Engl J Med 316:1388–1393
18. Striessnig J, Zernig G, Glossmann H (1985) Human red-blood-cell calcium antagonist binding sites. Eur J Biochem 150:67–77
19. Zernig G, Glossmann H (1988) A novel 1,4-dihydropyridine-binding site on mitochondrial membranes from guinea-pig heart, liver and kidney. Biochem J 252: in press
20. Dolle R, Nultsch W (1988) Specific binding of the calcium channel blocker [^{3}H]verapamil to membrane fractions of Chlamydomonas reinhardtii. Arch Microbiol 149:451–458
21. Dolle R, Nultsch W (1988) Characterization of d-[^{3}H]cis-diltiazem binding to membrane fractions and specific binding of calcium channel blockers to isolated flagellar membrane of Chlamydomonas reinhardtii. Biologists in press
22. Dolle R (1988) Isolation of plasma membrane and binding of the calcium antagonist nimodipine in chlamydomonas reinhardtii. Physiologia Plant in press
23. Striessnig J, Moosburger K, Goll A, Ferry DR, Glossmann H (1986) Stereoselective photoaffinity labeling of the purified 1,4-dihydropyridine receptor of the voltage-dependent calcium channel. Eur J Biochem 161:603–609
24. Glossmann H, Ferry DR, Striessnig J, Goll A, Moosburger K (1987) Resolving the structure of the calcium channel by photoaffinity labeling. Trends Pharmacol Sci 8:95–100
25. Glossman H, Ferry DR, Lübbecke F, Mewes R, Hofmann F (1981) Calcium channels: direct identification with radioligand binding studies. Trends Pharmacol Sci 3:431–437
26. Triggle DJ, Swamy VC (1983) Calcium antagonists. Circ Res 52:17–28
27. Janis RA, Silver PJ, Triggle DJ (1987) Drug action and cellular calcium regulation. Advances Drug Res 16:309–591
28. Reynolds IJ, Snowman AD, Snyder SH (1986) (−)-[^{3}H]Desmethoxyverapamil labels multiple calcium channel modulator receptors in brain and skeletal muscle: Differentiation by temperature and dihydropyridines. J Pharmacol Exp Ther 237:731–738
29. Glossmann H, Ferry DR, Goll A, Striessnig J, Zernig G (1985) Calcium channels: Basic properties as revealed by radioligand binding studies. J Cardiovasc Pharmacol 7:S20–S30
30. Cruz LJ, Olivera BM (1986) Calcium channel antagonist. J Biol Chem 261:6230–6233
31. McCleskey EW, Fox A, Feldman DH, Cruz LJ, Olivera BM, Tsien RW, Yoshikami D (1987) Omega-conotoxin: Direct and persistent blockade of specific types of calcium channels in neurons but not muscle. Proc Natl Acad Sci USA 84:4327–4331
32. Abe T, Koyano K, Saisu H, Nishiuchi Y, Sasakibara S (1986) Binding of omega-conotoxin to receptor sites associated with the voltage-sensitive calcium channel. Neurosci Lett 71:203–208

33. Knaus HG, Striessnig J, Koza A, Glossmann H (1987) Neurotoxic aminoglycoside antibiotics are potent inhibitors of [^{125}I]-omega-conotoxin binding to guinea-pig cerebral cortex membranes. Naunyn Schmiedeberg's Arch Pharmacol 336:583–586
34. Wagner JA, Snowman AD, Snyder SH (1987) Aminoglycoside effects on voltage-sensitive calcium channels and neurotoxicity. N Engl J Med 317:1669
35. Morton ME, Froehner SC (1987) Monoclonal antibody identifies a 200 kDa subunit of the dihydropyridine-sensitive calcium channel. J Biol Chem 262:11904–11907
36. Leung AT, Imagawa T, Block B, Franzini-Armstrong C, Campbell KP (1988) Biochemical and ultrastructural characterization of the 1,4-dihydropyridine receptor from rabbit skeletal muscle. J Biol Chem 263:994–1001
37. Takahashi M, Catterall WA (1987) Identification of an alpha-subunit of dihydropyridine-sensitive brain calcium channels. Science 236:88–91
38. Schmid A, Barhanin J, Coppola T, Borsotto M, Lazdunski M (1986) Immunochemical analysis of subunit structures of 1,4 dihydropyridine receptors associated with voltage-dependent calcium channels in skeletal, cardiac, and smooth muscles. Biochemistry 25:3492–3495
39. Vandaele S, Fosset M, Galizzi JP, Lazdunski M (1987) Monoclonal antibodies that coimmunoprecipitate the 1,4 dihydropyridine and phenylakylamine receptors and reveal the calcium channel structure. Biochemistry 26:5–9
40. Striessnig J, Knaus HG, Glossmann H (1988) Photoaffinity labeling of the calcium-channel-associated 1,4-dihydropyridine and phenylakylamine receptor in guinea-pig hippocampus. Biochem J 252 in press
41. Tanabe T, Takeshima H, Mikami A, Flockerzi V, Takahashi H, Kangawa K, Kojima M, Matsuo H, Hirose T, Numa S (1987) Primary structure of the receptor for calcium channel blockers from skeletal muscle. Nature 328:313–318
42. Hofmann F, Nastainczyk W, Röhrkasten A, Schneider T, Sieber M (1987) Regulation of the L-type calcium channel. Trends Pharmacol Sci 8:393–398
43. Curtis BM, Catterall WA (1985) Phosphorylation of the calcium antagonist receptor of the voltage-sensitive calcium channel by cAMP-dependent protein kinase. Proc Natl Acad Sci USA 82:2528–2532
44. Glossmann H, Striessnig J, Hymel L, Schindler H (1987) Purified L-type calcium channels: only one single polypeptide (alpha$_1$ subunit) carries the drug receptor domains and is regulated by protein kinases. Biomed Biochem Acta 46:S351–356
45. Nastainczyk W, Röhrkasten A, Sieber M, Rudolph C, Schächtele C, Marme D, Hofmann F (1987) Phosphorylation of the purified receptor for calcium channel blockers by cAMP kinase and protein kinase C. Eur J Biochem 169:137–142
46. Curtis BM, Catterall WA (1983) Solubilization of the calcium antagonist receptor from rat brain. J Biol Chem 258:7280–7283
47. Flockerzi V, Oeken HJ, Hofmann F, Pelzer D, Cavalie A, Trautwein W (1986a) The purified dihydropyridine-binding site from skeletal muscle T-tubulus is a functional calcium channel. Nature (London) 323:66–68
48. Smith JS, McKenna EJ, Ma J, Vilven J, Vaghy PL, Schwartz A, Coronado R (1987) Calcium channel activity in a purified dihydropyridine receptor preparation. Biochemistry 26:7182–7188
49. Talvenheimo JA, Worley III JF, Nelson MT (1987) Heterogeneity of calcium channels from a purified dihydropyridine receptor preparation. Biophys J 52:891–899
50. Hymel L, Striessnig J, Glossmann H, Schindler H (1988) Purified skeletal muscle 1,4 dihydropyridine receptor forms phosphorylation-dependent oligomeric calcium channels in planar bilayers. Proc Natl Acad Sci USA 85: in press
51. Schwartz LM, McCleskey EW, Almers W (1985) Dihydropyridine receptors in muscle are voltage-dependent but most are not functional calcium channels. Nature 314:747–751
52. Rios E, Brum G (1987) Involvement of dihydropyridine receptors in excitation-contraction coupling in skeletal muscle. Nature 325:717–720
53. Agnew WS (1987) Dual roles for DHP receptors in excitation-contraction coupling. Nature 328:297
54. Schindler H (1988) Planar lipid-protein membranes: Strategies for formation and of detecting dependencies of ion transport functions on membrane conditions. Methods Enzymol 171 in press
55. Coronado R, Latorre R (1983) Phospholipid bilayers made from monolayers on patch-clamp pipettes. Biophys J 43:231–236

56. Ehrlich BE, Schen CR, Garcia ML, Kaczorowski GJ (1986) Incorporation of calcium channels from cardiac sarcolemmal membrane vesicles into planar lipid bilayers. Proc Natl Acad Sci USA 83:193–197
57. Imagawa T, Smith JS, Coronado R, Campbell KP (1987) Purified ryanodine receptor from skeletal muscle sarcoplasmic reticulum is the calcium-permeable pore of the calcium release channel. J Biol Chem 262: 16636–16643
58. Lansman JB, Hess P, Tsien RW (1986) Blockade of current through single calcium channels by Cd^{2+}, Mg^{2+} and Ca^{2+}. J Gen Physiol 88:321–347
59. Bellemann P, Ferry D, Lübbecke F, Glossmann H (1981) [^{3}H]Nitrendipine, a potent calcium antagonist, binds with high affinity to cardiac membranes. Drug Res 31 (II): 2064–2067
60. Garcia MK, King VF, Siegl PKS, Reuben JP, Kaczorowski GJ (1986) Binding of Ca^{2+}-entry blockers to cardiac sarcolemmal membrane vesicles. J Biol Chem 261:8146–8157
61. Ferry DR, Goll A, Glossmann H (1987) Photoaffinity labeling of the cardiac calcium channel. Biochem J 243:125–135
62. Zernig G, Moshammer T, Graziadei I, Glossmann H (1988) Regulation of the novel mitochondrial dihydropyridine binding site by nucleotides. Naunyn-Schmiederberg's Arch Pharmacol (abstr): in press
63. Glossmann H, Striessnig J (1988) Calcium channels. Vitam Horm in press

Site-Specific Phosphorylation of the Skeletal Muscle Receptor for Calcium-Channel Blockers by cAMP-Dependent Protein Kinase*

A. Röhrkasten[1], H. E. Meyer[2], T. Schneider[1], W. Nastainczyk[1], M. Sieber[1], H. Jahn[1], S. Regulla[1], P. Ruth[1], V. Flockerzi[1], and F. Hofmann[1]

[1] Institut für Physiologische Chemie, Medizinische Fakultät, Universität des Saarlandes, D-6650 Homburg/Saar, FRG
[2] Intitut für Physiologische Chemie, Abt. Biochemie Supramolekularer Systeme, Ruhr Universität Bochum, D-4630 Bochum, FRG

Perspectives and Overview

Voltage-operated calcium channels are important factors in excitation-contraction coupling of cardiac and smooth muscle. The cardiac L-type calcium channel was the first channel known to be modified by hormones [11, 21, 31, 32]. The activation of the cardiac β-adrenergic receptor stimulates cAMP levels and increases three- to four-fold voltage-dependent opening (P_o) of the channel. A similar increase in P_o has been observed in hippocampal neurons [10]. Catecholamines also stimulate the calcium influx into myotubes [25], suggesting that stimulation of L-type calcium current by catecholamines is a widespread phenomenon. The stimulation of the calcium current is probably caused by the activation of cAMP-dependent protein kinase. The perfusion of a cardiac myocyte with active cAMP-dependent protein kinase elicits the same response as stimulation of the β-receptor [15], suggesting that phosphorylation of the calcium channel itself or a closely associated protein modifies the open probability of the channel. The P_o of cardiac and skeletal muscle calcium channels may be affected also by the α-subunit of the GTP-binding protein G_s [33, 34].

These results raised the possibility that the function of voltage-dependent calcium channels is affected directly by reversible covalent modifications of the channel protein. The verification of this hypothesis has been difficult since the cardiac calcium channel has not been purified completely. However, the receptor for organic calcium-channel blockers (CaCB) has been purified in large quantities from rabbit skeletal muscle [7, 27]. This allowed some of the possibilities raised by the above hypothesis to be verified. An important question was whether or not the purified receptor is phosphorylated by cAMP-dependent protein kinase. This contribution will mainly focus on the peptides phosphorylated by cAMP-dependent protein kinase. Other aspects of the CaCB receptor have been summarized in several recent articles [11, 12].

---

* This work was supported by a grant from Deutsche Forschungsgemeinschaft and Fonds der Chemischen Industrie.

Structure of CaCB Receptor from Striated Muscle

It is well established that the L-type calcium channel binds the three types of organic calcium channel blockers: the 1,4-dihydropyridines, the phenylalkylamines, and the benzothiazepines (see [14] for further references). The highest density of the high-affinity binding sites for these drugs was found in the transverse (T) tubulus of skeletal muscle [9]. These membranes contain between 10 and 80 pmol binding sites per milligram T-tubulus protein. This high density has facilitated the biochemical characterization and purification of the binding sites for calcium channel blockers. The purified CaCB receptor from rabbit skeletal muscle contains three proteins with apparent molecular weights of 165, 55, and 32 kd [27]. These proteins copurify with a constant ratio, suggesting that they are subunits of the same functional macromolecule. The receptor preparation is contaminated to a variable degree with a fourth protein, which contains an internal disulfide bond and yields two peptides of 130 and 28 kd in the presence of reducing agents. This protein is heavily glycosylated and probably a peripheral membrane protein [17, 28]. Its relationship to the other three subunits is unclear. Recently, the 165-kd receptor protein from rabbit skeletal muscle has been cloned [30]. Its primary structure shows that it belongs to the family of voltage-regulated ion channels.

The cardiac muscle L-type calcium channel and the cardiac CaCB receptor differ considerably from their skeletal muscle counterparts. Photoaffinity labelling of the cardiac CaCB receptor partially purified from bovine heart shows that the receptor is a 195-kd protein [26]. A copurification of other proteins with the 195-kd protein has not been observed. This suggests that the cardiac and skeletal muscle CaCB receptors are different entities. This interpretation is in agreement with their different physiological properties and functions, and the different size of their mRNA in Northern blots.

Phosphorylation of CaCB Receptor from Skeletal Muscle

As discussed above, strong evidence has been accumulated indicating that L-type calcium channels are regulated by phosphorylation. Reconstitution experiments of the purified CaCB receptor showed that these proteins are able to form functional calcium channels which share some properties with the calcium channels of living cells [8, 29]. The P_o of the reconstituted channel increases after phosphorylation of the receptor proteins by cAMP-dependent protein kinase [4]. These functional data are supported by in vitro phosphorylation experiments [5, 7, 19]. Cyclic AMP-dependent protein kinase phosphorylates the 165-kd subunit rapidly and with a slower time course also the 55-kd subunit (Table 1). Up to 2 mol phosphate per mol 165-kd protein is incorporated during a 1- to 2-h incubation [19, 24]. About half of the phosphate is incorporated during the first 5 min, suggesting that one or two peptides are phosphorylated very rapidly. The identity of the rapidly phosphorylated peptide was unclear. Seven potential cAMP phosphorylation sites (seryl residues 687, 1502, 1575, 1757, 1772, and 1854, and threonyl residue 1552) have been predicted from the primary structure of the CaCB receptor [30]. The sequence of the in vitro modified phosphopeptides has been determined [23].

Table 1. Time course of the phosphorylation of the purified CaCB receptor by cAMP-dependent protein kinase

Incubation time (min)	^{32}P as % of maximal incorporated phosphate (2 h)	
	165-kd protein	55-kd protein
5	52	14
15	70	44
60	100	92
120	100	100

The purified CaCB receptor (134 µg/ml) was phosphorylated by cAMP-dependent protein kinase (0.9 µM). Thereafter, the subunits were separated by SDS gel electrophoresis. The gel was sliced, and the amount of ^{32}P associated with the 165-kd and 55-kd proteins was measured.

Phosphopeptides of the 165-kd Subunit

The isolated [27] and phosphorylated 165-kd subunit was digested with three different proteinases to identify the seryl and threonyl residues phosphorylated by cAMP-dependent protein kinase: L-1-tosylamide-2phenylethyl-chloro-methylketone (TPCK)-treated trypsin, endoproteinase Lys-C, and endoproteinase Glu-C [23]. The digests were fractionated by reverse-phase HPLC in each case (Fig. 1). Only two major phosphopeptides were separated by HPLC regardless of which proteinase was used. The separation of the tryptic phosphopeptides after various incubation times

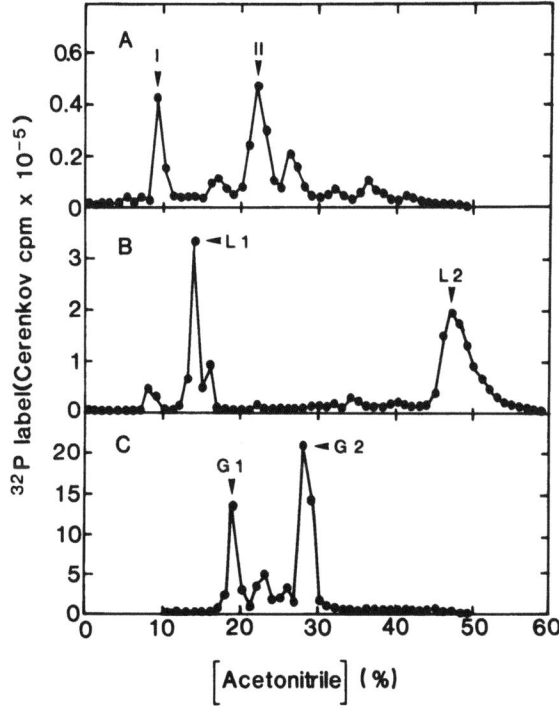

Fig. 1A–C. HPLC separation of ^{32}P-phosphopeptides generated by proteolytic digestion of the 165-kd protein. The purified CaCB receptor was phosphorylated by cAMP-dependent protein kinase for 2 h. The CaCB-receptor subunits were separated by gel exclusion chromatography as described [27]. The isolated and phosphorylated 165-kd protein was digested with TPCK-treated trypsin (**A**), endoproteinase Lys-C (**B**), or endoproteinase Glu-C (**C**). The digests were fractionated on a 25 cm×4.6 mm LiChrosorb RP-18 HPLC column, using a linear gradient of water/acetonitrile in 0.1% (v/v) trifluoracetic acid. No radioactive phosphate was eluted above 60% acetonitrile. The overall recovery of applied counts was 90%

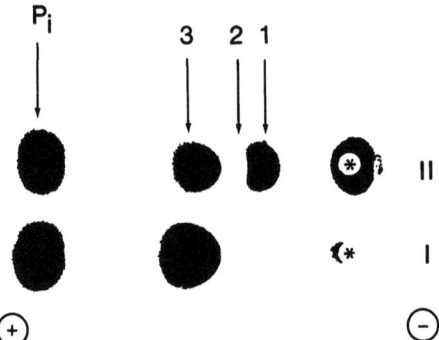

Fig. 2. Electrophoretic separation and autoradiography of partial acid hydrolysates of tryptic peptides I and II. The peptides I and II (see Fig. 1) were subjected to partial acid hydrolysis in 6 N HCl at 110°C for 2 h. Phosphoamino acids were then separated by thin-layer cellulose electrophoresis in 7% formic acid at pH 1.7. Marker phosphoamino acids were located by ninhydrin. The radioactive amino acids were detected by autoradiography. P_i, phosphate; *1*, phosphothreonine; *2*, phosphotyrosine; *3*, phosphoserine. The *asterisks* represent the origins

showed that peptide I was phosphorylated very rapidly (within 15 min), whereas phosphorylation of peptide II was completed only after a 1-h incubation. Phosphoamino acid analysis of phosphopeptides I and II yielded phosphoserine, and phosphotyrosine and some phosphothreonine, respectively (Fig. 2). Initial attempts to purify and sequence the tryptic phosphopeptides I and II failed since too many tryptic trypsin peptides were present in the rechromatographed peptides. In contrast, the peptides L1, G1, and G2 obtained from digestion with endoproteinase Lys-C and Glu-C were successfully purified and sequenced. Phosphoserine residues were detected as the phenylthiohydantoin (PTH) derivative of dithiothreitol (DTT)-dehydroalanine, as described by Meyer et al. [18]. The sequences of L1, G1, and G2 are summarized in Table 2.

Table 2. The phosphopeptides of the 165-kd subunit of the CaCB receptor modified by cAMP-dependent protein kinase

Peptide	Digestion enzyme	Fragment	Position in sequence	Length	Phosphorylated residues
I	TPCK-treated trypsin	MSR	Met-686 to Arg-688	3 aa	Ser-687[a]
L1	Endoproteinase Lys-C	MSRGLPDK	Met-686 to Lys-693	8 aa	Ser-687[b]
G1	Endoproteinase Glu-C	RKRRKMSRGLPDKTEEE	Arg-681 to Glu-697	17 aa	Ser-687[b]
II	TPCK-treated trypsin	TNSLPPVMANQR	Thr-1615 to Arg-1626	12 aa	Ser-1617[a]
L2	Endoproteinase Lys-C	DTVQIQAGLRTIEEEAAPE...	Asp-1551 to ?	≤ 166 aa	[a]
G2	Endoproteinase Glu-C	RTNSLPPVMANQRPLQFAE	Arg-1614 to Glu-1632	19 aa	Ser-1617[b]

The amino acid (aa) sequences of major phosphopeptides shown in Fig. 2 and the residue numbers in the primary structure of the 165-kd subunit as published by Tanabe et al. [30] are listed.
[a] Deduced
[b] Determined

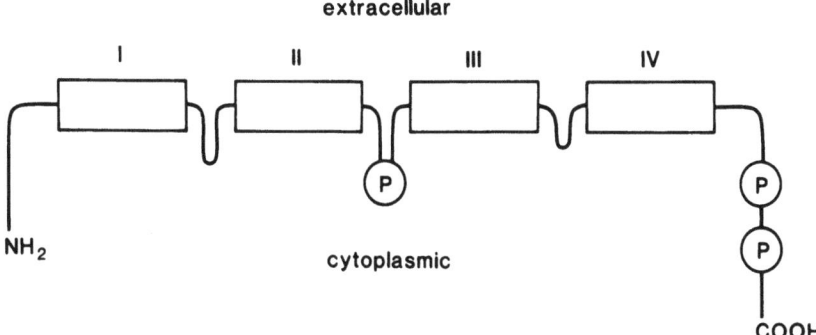

Fig. 3. A model of the transmembrane topology of the 165-kd CaCB-receptor subunit and of the localization of the phosphorylation sites. The structure of the 165-kd CaCB-receptor protein is shown as given by Tanabe et al. [30]. The four membrane-spanning regions are shown as *boxes*. The in vitro-determined phosphorylation sites *(P)* of cAMP-dependent protein kinase are indicated. The rapidly phosphorylated site, Ser-687, is located between membrane-spanning regions, II and III. The other two sites are in the carboxyterminal part

Phosphorylation Sites

The sequences shown in Table 2 allowed the localization of the phosphorylated amino acid on the known primary structure of the 165-kd CaCB-receptor subunit [30]. The rapidly phosphorylated serine present in I, L1, and G1 is Ser-687, and the slowly phosphorylated aminio acids are Ser-1617 and probably Thr-1615. Evidence for the phosphorylation of Thr-1615 was obtained by the phosphoamino acid analysis of the corresponding tryptic peptide II (see also Fig. 2). The sequence around Ser-687 and -1617 is in agreement with the known primary-sequence requirements for the cAMP-dependent protein kinase [6]. The rapid phosphorylation of Ser-687 suggests that its phosphorylation might be physiologically important. More than 67% of the totally incorporated phosphate was incorporated within 5 min. Ser-687 is located on the cytoplasmic domain of the 165-kd subunit between the membrane-spanning regions II and III, whereas the more slowly phosphorylated Ser-1617 is present in the carboxy-terminal part of the CaCB receptor (Fig. 3).

Concluding Remarks

Of the two phosphorylation sites detected in this study only Ser-687 fulfills some of the requirements of a potentially important in vivo phosphorylation site: (a) Ser-687 was phosphorylated rapidly; (b) the phosphorylation site is exposed to the water on an intermembrane loop; and (c) according to the model predicted by Tanabe et al. [30] this loop is present on the cytoplasmic side of the membrane. It is therefore quite possible that phosphorylation at Ser-687 occurs in vivo and increases the probability of the calcium channel opening upon depolarization of the membrane. The localization of this phosphorylation site appears ideally suited for regulatory effects since phosphorylation of Ser-687 could affect the transmembrane regions II and III, which are thought to be part of the membrane-spanning pore.

It is not clear at present whether the 165-kd subunit functions as a regular calcium channel in vivo. Detailed analysis of the excitation-contraction coupling in vertebrate skeletal muscle has not revealed strong evidence that activation of an L-type calcium channel is a prerequisite for muscle contraction [2, 3, 16, 22]. Good evidence has been produced to suggest that part of the charge-shifting system of T tubulus can be blocked by calcium channel blockers [2, 16, 22]. It is therefore possible that in vivo the skeletal muscle CaCB receptor triggers directly in a voltage-dependent fashion the opening of the calcium release channel. It is not known whether calcium influx through the CaCB receptor is needed for a sustained contraction [1, 13, 20] or not [3]. L-type calcium channels have been detected in skeletal muscle [1, 3, 13, 20]. It is therefore conceivable that the CaCB receptor may function in two ways in skeletal muscle, namely, as voltage sensor and charge carrier, and as calcium channel. The particular function may depend on the presence of other regulatory proteins and covalent modification of the receptor. Protein phosphorylation may be an important link between these two functions. Further work is needed to resolve these questions. The apparent discrepancies between in vivo and in vitro results will be solved in the near future, when cDNA clones of calcium channels from other tissues become available.

Acknowledgements. We thank Mrs. Poesch for typing the manuscript, Mrs. Siepmann for the graphical work, Mrs. Ernert for excellent technical support, and Mr. Korte for running the sequencer.

References

1. Avila-Sakar AJ, Cota G, Ramboa-Aldeco R, Garcia J, Huerta M, Muniz J, Stefani E (1986) Skeletal muscle Ca^{2+} channels. J Muscle Res Cell Motil 7:291–298
2. Berwe D, Gottschalk G, Lüttgau CH (1987) Effects of the calcium antagonist gallopamil (D600) upon excitation-contraction coupling in toe muscle fibres of the frog. J Physiol 385:693–707
3. Brum G, Rios E, Stefani E (1988) Effect of extracellular Ca^{2+} on Ca^{2+}-movements of E–C coupling in frog skeletal muscle fibres. J Physiol 398:441–473
4. Cavalié A, Flockerzi V, Hofmann F, Pelzer D, Trautwein W (1987) Two types of calcium channels from rabbit fast skeletal muscle transverse tubules in lipid bilayers: differences in conductance, gating behaviour and chemical modulation. J Physiol (London) 390:82 p
5. Curtis BM, Catterall WA (1985) Phosphorylation of the calcium antagonist receptor of the voltage-sensitive calcium channel by cAMP-dependent protein kinase. Proc Natl Acad Sci USA 82:2528–2532
6. Edelmann AM, Blumenthal DK, Krebs EG (1987) Protein serine/threonine kinases. Ann Rev Biochem 56:567–613
7. Flockerzi V, Oeken H-J, Hofmann F (1986) Purification of a functional receptor for calcium channel blockers from rabbit skeletal muscle microsomes. Eur J Biochem 161:217–224
8. Flockerzi V, Oeken H-J, Hofmann F, Pelzer D, Cavalié A, Trautwein W (1986) The purified dihydropyridine binding site from skeletal muscle T-tubules is a functional calcium channel. Nature 323:66–68
9. Galizzi D-P, Borsotto M, Barhanin J, Fosset M, Lazdunski M (1986) Characterization and photoaffinity labeling of receptor sites for the Ca^{2+} channel inhibitors d-cis-diltiazem, (±)-bepridil, desmethoxyverapamil, and (+)-PN 200–110 in skeletal muscle transverse tubule membranes. J Biol Chem 261: 1393–1397
10. Gray R, Johnston D (1987) Noradenaline and β-adrenoceptor agonists increase activity of voltage-dependent calcium channels in hippocampal neurons. Nature 327:620–622
11. Hofmann F, Nasainczyk W, Röhrkasten A, Schneider T, Sieber M (1987) Regulation of the L-type calcium channel. TIPS 8:393–398

12. Hofmann F, Schneider T, Röhrkasten A, Nastainczyk W, Sieber M, Ruth P, Flockerzi V (1988) Calcium channels: structure and function of the skeletal muscle calcium antagonist receptor. Arzneim-Forsch/Drug Res, in press
13. Ildefonse M, Jacquemond V, Rougier O, Renaud JF, Fosset M, Lazdunski M (1985) Excitation contraction coupling in skeletal muscle: Evidence for a role of slow Ca^{2+} channels using Ca^{2+} channel activators and inhibitors in the dihydropyridine series. Biochem Biophys Res Common 129:904–909
14. Janis RA, Silver PJ, Triggle DJ (1987) Drug action and cellular calcium regulation. Adv Drug Res 16:309–591
15. Kameyama M, Hescheler J, Mieskes G, Trautwein W (1986) The protein-specific phosphatase 1 antagonizes the β-adrenergic increase of the cardiac Ca current. Pflügers Arch 407:461–463
16. Lamb GD, Walsh T (1987) Calcium currents, charge movement and dihydropyridine binding in fast- and slow-twitch muscle of rat and rabbit. J Physiol 393:595–617
17. Leung AT, Imagawa T, Block B, Franzini-Armstrong C, Campbell KP (1988) Biochemical and ultrastructural characterization of the 1,4-dihydropyridine receptor from rabbit skeletal muscle. J Biol Chem 263:994–1001
18. Meyer HE, Hoffmann-Posorske E, Korte H, Heilmeyer, jr LMG (1986) Sequence analysis of phosphoserine-containing peptides. Modification for picomolar sensitivity. FEBS Letters 204:61–66
19. Nastainczyk W, Röhrkasten A, Sieber M, Rudolph C, Schächtele C, Marme D, Hofmann F (1987) Phosphorylation of the purified receptor for calcium channel blockers by cAMP kinase and protein kinase C. Eur J Biochem 169:137–142
20. Palade PT, Almers W (1985) Slow calcium and potassium currents in frog skeletal muscle: their relationship and pharmacologic properties. Pflügers Arch 405:91–101
21. Reuter H (1984) Ion channels in cardiac cell membranes. Ann Rev Physiol 46:473–484
22. Rios E, Brum G (1987) Involvement of dihydropyridine receptors in excitation-contraction coupling in skeletal muscle. Nature 235:717–720
23. Röhrkasten A, Meyer HE, Nastainczyk W, Sieber M, Hofmann F (1988) cAMP-dependent protein kinase rapidly phosphorylates serine 687 of the skeletal muscle receptor for calcium channel blockers, (submitted)
24. Röhrkasten A, Nastainczyk W, Sieber M, Jahn H, Regulla St, Hofmann F (1988) Phosphorylation of the purified CaCB-receptor. J Cardiovasc Pharmacol, in press
25. Schmid A, Renaud J-F, Lazdunski M (1985) Short Term and Long Term Effects of β-adrenergic effectors and cyclic AMP on nitrendipine-sensitive voltage-dependent Ca^{2+} channels of skeletal muscle. J Biol Chem 24:13041–13046
26. Schneider T, Hofmann F (1988) The bovine cardiac receptor for calcium channel blockers is a 195 kDa protein. Eur J Biochem, in press
27. Sieber M, Nastainczyk W, Zubor V, Wernet W, Hofmann F (1987) The 165-kDa peptide of the purified skeletal muscle dihydropyridine receptor contains the known regulatory sites of the calcium channel. Eur J Biochem 167:117–122
28. Takahashi M, Seagar MJ, Jones JF, Reber BFX, Catterall WA (1987) Subunit structure of dihydropyridine-sensitive calcium channel from skeletal muscle. Proc Natl Acad Sci USA 84:5478–5482
29. Talvenheimo JA, Worley III JF, Nelson MT (1987) Heterogeneity of calcium channels from a purified dihydropyridine receptor preparation. Biophys J 52:891–899
30. Tanabe T, Takeshima H, Mikami A, Flockerzi V, Takahashi H, Kangawa K, Kojima M, Matsuo H, Hirose T, Numa S (1987) Primary structure of the receptor for calcium channel blockers from skeletal muscle. Nature (London) 328:313–318
31. Trautwein W, Kameyama M, Hescheler J, Hofmann F (1986) Cardiac calcium channels and their transmitter modulation. Progress in Zoology 33:163–182
32. Tsien RW, Bean BC, Hess A, Lansmann JB, Nilius B, Nowycky MC (1986) Mechanisms of calcium channel modulation by β-adrenergic agents and dihydropyridine calcium agonists. J Mol Cell Cardiol 18:691–710
33. Yatani A, Codina J, Imoto Y, Reeves JP, Birnbaumer L, Brown AM (1987) A G Protein Directly Regulates Mammalian Cardiac Calcium Channels. Science Vol 238:1288–1292
34. Yatani A, Imoto Y, Codina J, Birnbaumer L, Brown AM (1988) The G protein, Gs, directly activates skeletal muscle Ca^{2+} channels. Biophysical J 53:20a

Molecular Properties of Dihydropyridine-Sensitive Calcium Channels from Skeletal Muscle

M. J. Seagar, M. Takahashi, and W. A. Catterall

Department of Pharmacology, University of Washington, Seattle, WA 98195, USA

Introduction

Voltage-sensitive calcium channels provide an essential link between transient changes in membrane potential and a variety of cellular responses. In neurons, calcium influx couples membrane depolarization to transmitter release at the synaptic terminal [1], whereas in the cell body it may activate calcium-dependent potassium channels and thus modulate repetitive firing patterns [2]. In cardiac and smooth muscle tissue, calcium channels mediate excitation-contraction coupling, although in skeletal muscle their physiological role is still a matter for debate [3, 4].

Skeletal muscle fibers have substantial voltage-activated calcium currents that have been measured under voltage clamp [5]. These currents originate almost entirely in the transverse (T) tubular system and are blocked by dihydropyridine calcium antagonists [6]. They have been presumed to play a role in excitation-contraction coupling, although direct evidence for such a role has not been obtained. T-tubule membranes can be extensively purified from skeletal muscle by a combination of differential and density gradient centrifugation [7]. Antibodies prepared against the most highly purified fractions of T-tubule membranes stain only T-tubules and not sarcolemma or sarcoplasmic reticulum of intact muscle, indicating a high degree of purity of this preparation [7]. Analysis of [^{3}H]-nitrendipine binding to membrane fractions from skeletal muscle reveals a specific localization of the calcium-antagonist receptor in the T-tubule fraction [8, 9]. These membranes contain a ten fold greater concentration of calcium antagonist receptor than any other membrane preparation described to date. Since T-tubule membranes are the most enriched source of calcium-antagonist receptors and they display a substantial voltage-activated calcium current that is blocked by dihydropyridines, they provide a favorable experimental preparation for examination of the molecular properties of the calcium antagonist receptor and its relationship to voltage-sensitive calcium channels. It is anticipated that the T-tubule calcium channel will resemble those of other tissues which are blocked by dihydropyridines and, therefore, that information on the molecular properties of this calcium channel will give insight into others.

Characterization of Detergent-Solubilized Calcium-Antagonist Receptor

Since it is likely to be an intrinsic membrane protein, the first essential step in purification and biochemical characterization of the calcium antagonist receptor is solubilization from an appropriate membrane source and characterization of the solubilized protein. After a survey of several detergents, we concluded that digitonin is the most effective detergent for solubilization of a specific [^{3}H]nitrendipine-receptor complex from brain and skeletal muscle T-tubule membranes [10, 11]. Up to 40% of the receptor-ligand complex is solubilized. The dissociation of bound [^{3}H]nitrendipine from the complex is accelerated by verapamil and slowed by diltiazem through allosteric interactions between the nitrendipine-binding site and the binding sites for those ligands, as previously observed in intact membranes. These results show that the three different binding sites for calcium antagonist drugs remain associated as a complex after detergent solubilization of the calcium-antagonist receptor. The results also provide further support for the conclusion that a specific receptor complex has been solubilized under conditions which allow retention of the functional allosteric regulation of dihydropyridine binding.

Sedimentation of the solubilized [^{3}H]nitrendipine-receptor-digitonin complex through sucrose gradients gives a single peak of specifically bound nitrendipine with a sedimentation coefficient of 19–20 S [10]. Comparison of the sedimentation behavior of the solubilized complex of brain, heart, and skeletal and smooth muscle indicates that they have identical size. These results provide support for the view that the calcium antagonist receptor in different tissues is quite similar.

Many plasma membrane proteins are glycosylated during their synthesis and transport to the cell surface. The solubilized calcium antagonist receptor from brain or skeletal muscle is specifically adsorbed to and eluted from affinity columns with immobilized wheat germ agglutinin or other lectins [10–13]. Evidently, one or more of the subunits of the calcium antagonist receptor are glycoproteins.

Purification and Subunit Composition

We have purified the calcium antagonist receptor, solubilized from T-tubule membranes by digitonin, 330-fold by affinity chromatography on wheat germ agglutinin Sepharose, ion-exchange chromatography on DEAE-Sephadex, and velocity sedimentation through sucrose gradients [11]. Analysis of the purified protein by NaDodSO$_4$/polyacrylamide gel electrophoresis (PAGE) under alkylating conditions and silver staining (Fig. 1, lane 1) revealed three classes of polypeptide, which we have designated α (167 kd), β (54 kd), and γ (30 kd). When disulfide bonds were cleaved with dithiothreitol, the α-band split into two clearly resolved protein populations with molecular weights of 175 and 143 kd (Fig. 1, lane 2). In the initial studies from this laboratory, the anomalous behavior of the α-polypeptide was ascribed to partial cleavage and/or re-formation of intrachain disulfide bonds resulting in a variable fraction of the protein with smaller apparent size [11]. The more recent use of a battery of specific labeling methods has now shown that the 175- and 143-kd polypeptides are two distinct calcium channel subunits, $α_1$ and $α_2$ which have similar size but clearly different properties [14].

A polyclonal antibody (PAC-10) obtained from the ascites fluids of a SJL/J mouse immunized with purified calcium channel selectively labeled the 167-kd polypeptide before reduction of disulfide bonds, but only the 175-kd α_1-polypeptide after reduction (Fig. 1, lane 3). No immunolabeling was observed with preimmune serum or with PAC-10 which had been preadsorbed with purified calcium channel (not shown). These observations indicate that the 175- and 143-kd components are distinct polypeptides.

Subunit Glycosylation

Solubilized [^{3}H]dihydropyridine receptors specifically bind to various immobilized lectins, and affinity chromatography on wheat germ agglutinin (WGA) sepharose is the most efficient purification step [11–13]. These results imply that at least one subunit is glycosylated. The oligosaccharide chains of the purified calcium channel were detected by separating subunits by NaDodSO$_4$/PAGE and probing the resolved polypeptides with [^{125}I]ConA or [^{125}I]WGA. Under alkylating conditions, [^{125}I]ConA labeled the α-polypeptide, but after disulfide reduction only the α_2-subunit was labeled (Fig. 1, lane 4). [^{125}I]WGA bound to both the α- and γ-protein bands in gels run under alkylating conditions, and to the α_2- and γ-subunits after dithiothreitol treatment. Disulfide reduction led to the appearance of two new [^{125}I]WGA-labeled components at 24–27 kd that are clearly distinct from the γ-subunit (Fig. 1, lane 5). These polypeptides were also detected, but much less distinctly, by silver staining (Fig. 1, lane 2). They appear to be disulfide-linked to the α_2-subunit under nonreducing conditions; previous immunochemical evidence [15] suggests that the smaller polypeptide may be proteolytically derived from the larger, so we refer to them collectively as the δ-subunit. No labeling of α_1 or β was detected with either lectin. [^{125}I]ConA and [^{125}I]WGA binding to calcium channel subunits was blocked in the presence of 100 mM α-methylmannoside or N-acetylglucosamine, respectively (not shown).

To determine the extent of glycosylation and the core polypeptide size of the calcium channel subunits, purified channel preparations were labeled with ^{125}I, incubated with glycosidases to remove oligosaccharide chains, and analyzed by NaDodSO$_4$/PAGE and autoradiography. Sequential deglycosylation with neuraminadase and endoglycosidase F caused a reduction in the apparent sizes of the α_2- and γ-subunits, reaching core polypeptide sizes of 105 kd and 20 kd, respectively. Poor iodination of the δ-subunit prevented estimation of its carbohydrate content by this method. No shift in the mobility of the α_1- and β-subunits was noted, confirming the absence of N-linked carbohydrate in these two subunits.

Covalent Labeling of Calcium Channel Subunits

[^{3}H]PN200–110 and [^{3}H]azidopine have been shown to covalently label a 145- to 170-kd polypeptide in T-tubule membranes [16–18] and purified calcium channels [19] that presumably corresponds to one of the two α-subunits. In our preparations, [^{3}H]azidopine was incorporated by UV photolysis into a polypeptide, which migrated as a band of 167 kd before reduction of disulfide bonds and 175 kd after dithiothreitol

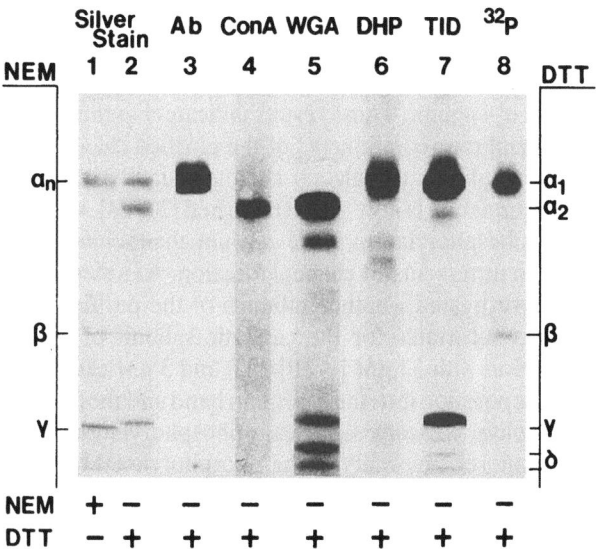

Fig. 1. Polypeptide composition of the dihydropyridine-sensitive calcium channel. Lanes 1, 2 show silver staining. Purified calcium channels were analyzed by NaDodSO$_4$/PAGE and silver staining with or without reduction of disulfide bonds as indicated beneath each lane. Lane 3 shows immunoblotting. Polypeptides separated by NaDodSO$_4$/PAGE, with or without reduction of disulfide bonds as indicated below each lane, were electrophoretically transferred to nitrocellulose strips and immunolabeled by incubation with PAC-10 followed by incubation with [^{125}I]Protein A, washing, and autoradiography. Lanes 4, 5 show glycosylation. Calcium channel subunits were transferred to a nitrocellulose sheet and labeled with [^{125}I]ConA (4) or were labeled directly in the gel with [^{125}I]WGA (5). Lane 6 shows photoaffinity labeling. T-tubule membranes (0.4 mg/ml) in 25 mM Hepes and 1 mM CaCl$_2$ adjusted to pH 7.5 with Tris base were incubated with 6 nM [^{3}H]-azidopine and then irradiated for 15 min at 4°C with a 30-w UV source (λ_{max} 356 nm). The membranes were solubilized in 1% digitonin, 10 mM Hepes, 185 mM NaCl 0.5 mM CaCl$_2$, 0.1 mM phenylsulfonyl fluoride, and 1 mM pepstatin A adjusted to pH 7.5 with Tris base; calcium channels were partially purified by chromatography on WGA-Sepharose and analyzed by NaDodSO$_4$/PAGE and fluorography. Lane 7 shows hydrophobic labeling, [^{125}I]TID (15 Ci/mmol) was prepared, and purified calcium channel was labeled with [^{125}I]TID (100 mCi/ml) in a buffer containing 0.1% digitonin. Lane 8 shows phosphorylation. Purified calcium channel was incubated with 0.3 μM cAMP-dependent kinase catalytic subunit and 0.12 μM carrier free [γ-^{32}P]ATP for 15 min at 37°C as previously described [14] (lanes 6, 7). NEM means N-ethylmaleinimide, DTT means dithiothreitol

treatment (Fig. 1, lane 6). The electrophoretic behavior of this polypeptide identifies it as the α_1-subunit. No labeling was observed in the presence of 2 μM PN200–110 (not shown).

Ion channel-forming polypeptides should contain transmembrane segments which may be detected using the hydrophobic probe [^{125}I]-3-(trifluoromethyl)-3-(m-iodophenyl)diazirine (TID). This photoreactive compound partitions into free detergent micelles and detergent associated with the major hydrophobic domains of integral membrane proteins, and it is specifically incorporated into these regions by photolysis. The α_1- and γ-subunits were prominently labeled by TID, with a much lower level of incorporation into α_2 and δ (Fig. 1, lane 7). The β-subunit was not detectably labeled. Quantitation of [^{125}I]TID in excised protein bands showed that the α_1- and γ-subunits incorporated ten fold more TID per unit mass than the α_2- or δ-subunits, even though, as shown below, nearly all α_1- and γ-subunits are associated with an α_2-subunit. These results indicate that the α_1- and γ-subunits are the principal transmembrane components of the purified calcium channel complex.

The regulation of calcium channels via a pathway involving cAMP and protein phosphorylation is now well established [20–24], although it is not known whether the site of phosphorylation is the calcium channel itself or another intracellular protein which in turn regulates channel function. As a step toward resolving this question, we have investigated whether subunits of the purified calcium-antagonist receptor can serve as substrates for the catalytic subunit of cAMP-dependent protein kinase. Incubation with 33 μM [γ-^{32}P]ATP and 3 μM catalytic subunit led to stoichiometric ^{32}P incorporation into the α-protein band and the β-subunit at rates that are consistent with a physiologically significant phosphorylation reaction [25]. Comparison of the electrophoretic mobility of the phosphorylated bands before and after reduction of disulfide bonds showed that the α_1-subunit is a good substrate for this enzyme, while the α_2- and δ-polypeptides are not labeled (Fig. 1, lane 8). The β-subunit was more weakly labeled at the low ATP concentration used in this experiment (see legend). These results identify the α_1- and β-subunits of the calcium channel as potential sites of regulation of the voltage-sensitive calcium channel by cAMP-dependent phosphorylation.

Analysis of Noncovalent Subunit Interactions

By the use of several labeling techniques, we have established that α_1 has the properties expected of the calcium channel, including a binding site for dihydropyridine calcium antagonists, at least one cAMP-dependent phosphorylation site, and extensive hydrophobic domains. It is important to determine whether other polypeptides present in the purified preparation are persistent impurities or specifically associated components of the oligomeric calcium channel complex.

The data presented in Fig. 1, lane 3 demonstrate that PAC-10 antibodies recognize only the α_1-subunit of NaDodSO$_4$-denatured calcium channel. However, this polyclonal serum produced by immunization with native calcium channel may contain antibodies which bind only to native conformations of the α_2-, δ-, β-, or γ-subunits. To eliminate any antibodies with this specificity, a nitrocellulose strip containing α_1-subunit was used as an affinity matrix to purify anti-α_1 antibodies. Purified anti-α_1 antibodies specifically precipitated [^{3}H]PN200–110-labeled calcium channel, while proteins from PAC-10 that were nonspecifically adsorbed to a bare nitrocellulose strip did not.

Immunoprecipitation of ^{125}I-labeled calcium channel (Fig. 2) showed that only α_1 was precipitated after NaDodSO$_4$ denaturation (lane 7). In contrast, α_1, α_2, β, and γ were precipitated as a complex in 0.5% digitonin (lane 1) or 0.1% 3-[(3-cholamidopropyl) dimethylammonio]-1-propanesulfonate (CHAPS, lane 3), detergent conditions which are known to stabilize dihydropyridine binding, allosteric coupling of the three calcium-antagonist receptor sites, and ion conductance activity. A higher concentration of CHAPS (1%) caused dissociation of α_2 from the complex (data not

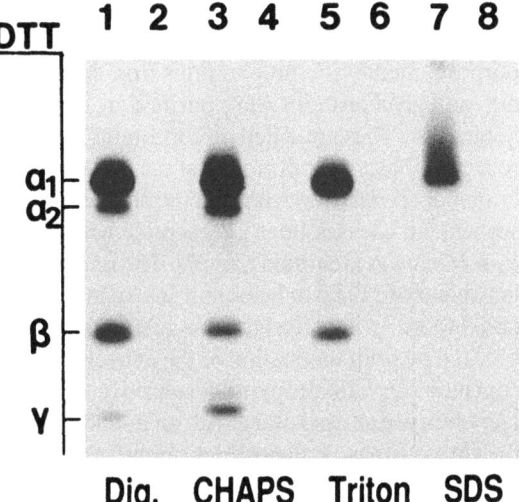

Fig. 2. Immunoprecipitation of calcium channel subunits by anti-α_1 antibodies. ^{125}I-labeled calcium channel was immunoprecipitated as described previously [14, 26] using affinity-purified anti-α_1 antibodies (lanes 1, 3, 5, and 7) or a control preparation (lanes 2, 4, 6, and 8) in an immunoassay buffer containing the detergents indicated: 0.5% digitonin, 0.1% CHAPS, and 1% Triton X-100, or 1% SDS for 2 min at 100°C. This was followed by immunoassay in 0.5% digitonin after detergent exchange by gel filtration on a 2-ml Sephadex G-50 column

shown). In addition, experiments in 1% Triton X-100 (lane 5) showed complete dissociation of the α_2-subunit. The β-subunit and a small fraction of the γ-subunit (not easily seen in Fig. 2) were coimmunoprecipitated with α_1 in Triton X-100. The results of these immunoprecipitation experiments have been confirmed by selective elution of α_2-subunits from immune complexes. Purified calcium channels were immunoprecipitated in 0.5% digitonin as in Fig. 2, lane 1. Resuspension in a buffer containing 1% Triton X-100 caused complete release of the $\alpha_2\delta$ dimer and partial release of γ without loss of β from the precipitate (data not shown). Complementary data supporting these observations have also been obtained using lentil lectin agarose, which has the same specificity as ConA (see Fig. 1) and is a selective probe for the α_2-subunit.

Immunological Detection of Calcium Channel Components in Other Tissues

In order to determine the immunological cross-reactivity of antisera raised against purified rabbit skeletal calcium channel with calcium channels from heart and brain, radioimmune assays were performed. In each tissue, calcium channels were labeled with [^{3}H]PN200–110, solubilized with digitonin, and purified by chromatography on wheat germ agglutinin-sepharose. PAC-10 antiserum which we have used to characterize the α_1-subunit of the calcium channel (see Fig. 1 lane 3, Fig. 2) showed poor cross-reactivity with other tissues. However, another antiserum, PAC-2, recognized [^{3}H]PN200–110 – labeled calcium channel in skeletal muscle, heart, and brain, and the concentration dependence of immunoprecipitation was then determined. Each channel was precipitated to a similar extent by maximum concentrations of antiserum, and the ratio of antiserum for half-maximal immunoprecipitation was 1.0:1.8:7.9 for skeletal muscle, heart, and brain. These results indicate that antigenic determinants present on the skeletal muscle calcium channel are also present on calcium channels in heart and brain.

Since these antibodies recognize calcium channels in heart and brain, it is possible to use them to identify and characterize the related proteins in these tissues. For this purpose, membrane preparations from heart and brain were solubilized with digitonin, and glycoproteins were purified by chromatography on wheat germ agglutinin-Sepharose. These purified glycoproteins were labeled with ^{125}I, and the components recognized by antibodies against the skeletal muscle calcium channel were isolated by immunoprecipitation with specific antiserum and a protein A-Sepharose immunoadsorbent. In each tissue, a polypeptide with the characteristics of the calcium channel α_2-subunit was identified [26–27]. The immunoprecipitated component had a molecular weight of 170 kd in heart and 169 kd in brain in nonreducing conditions and 141 kd and 140 kd, respectively, after reduction. In each case, immunoprecipitation was blocked by prior incubation of the antiserum with purified calcium channel. Thus, we conclude that dihydropyridine-sensitive calcium channels in skeletal muscle, heart, and brain are all associated with an α_2-subunit that is homologous, but not identical in the three tissues. Polypeptides homologous to other subunits of the skeletal muscle calcium channel have not yet been detected using these techniques.

Functional Properties of Purified Calcium-Antagonist Receptor in Phospholipid Vesicles

We purify the calcium-antagonist receptor as a performed complex with the calcium antagonists [^{3}H]nitrendipine or [^{3}H]PN200–110. After solubilization, it retains allosteric interactions among the separate binding sites for verapamil, diltiazem, and dihydropyridines. However, other aspects of calcium channel function cannot be assessed in detergent solution. It is important, therefore, to return the purified calcium-antagonist receptor to a membrane environment, and to determine whether the purified protein is capable of mediating voltage-dependent calcium flux.

In order to maximize the probability of purification of the calcium channel in an active state, the calcium-antagonist receptors in T-tubule membranes were incubated with sufficient [^{3}H]PN200–110 to label approx. 1% of the binding sites. This label was used to identify the calcium-antagonist receptor during purification as in previous studies [11]. The T-tubule membranes were then incubated in an excess of the specific calcium-channel activator Bay K8644 so that the remaining 99% of dihydropyridine sites would be occupied by this agent. The calcium-antagonist receptors were then solubilized in digitonin and purified in the continued presence of Bay K8644 using previously described procedures. Phosphatidylcholine for reconstitution was solubilized in the zwitterionic detergent CHAPS because it is poorly soluble in digitonin. Purified calcium antagonist receptor dispersed in digitonin was then mixed with phosphatidylcholine dispersed in CHAPS, and single-walled phospholipid vesicles were formed by removal of the detergents by molecular sieve chromatography. Analysis of the resulting vesicle preparations by sucrose density gradient sedimentation shows that the purified calcium-antagonist receptors are quantitatively incorporated into phosphatidylcholine vesicles, and 15–25% of vesicles contain at least one calcium-antagonist receptor [28]. These preparations therefore provide a suitable system for analysis of the ion transport properties of this purified protein.

Initial rates of influx of $^{45}Ca^{2+}$ or $^{133}Ba^{2+}$ into reconstituted phosphatidylcholine vesicles were measured under counter transport conditions, in which the intravesicular compartment contains a high concentration of unlabeled Ca^{2+} or Ba^{2+}. These conditions greatly increase the amount of $^{45}Ca^{2+}$ or $^{133}Ba^{2+}$ uptake required to achieve isotopic equilibrium and greatly slow the approach to isotopic equilibrium, as described previously for reconstituted sodium channels. They therefore maximize any ion flux mediated by reconstituted channels. Under these conditions, calcium influx into phosphatidylcholine vesicles containing reconstituted calcium-antagonist receptors was two- to threefold greater than influx into protein-free phosphatidylcholine vesicles. This increase is completely blocked by verapamil at a concentration of 100 μM. If the calcium channel activator Bay K8644 is removed from the vesicle preparation by molecular sieve chromatography, $^{45}Ca^{2+}$ influx is markedly reduced. These results show that at least a fraction of the purified calcium-antagonist receptors can function as calcium channels when incorporated into phosphatidylcholine vesicles [28].

The inhibition of the reconstituted calcium channels by different concentrations of organic calcium channel blockers from 10^{-9} M to 10^{-4} M was examined. Half-maximal inhibition was observed with approx. 1.5 μM verapamil, 1.0 μM D600, or 0.2 μM PN200–110. These concentrations are similar to those that give half-maximal inhibition of voltage-activated calcium currents in intact skeletal muscle fibers, consistent with the conclusion that the calcium influx in reconstituted vesicles is mediated by functional purified calcium channels.

To determine whether the calcium influx stimulated by Bay K8644 and blocked by PN200–110 and verapamil required the presence of the subunits of the calcium-antagonist receptor and not other detectable proteins, purified preparations were sedimented through sucrose gradients, and each fraction was examined for bound [^{3}H]PN200–110, polypeptide composition, and ability to mediate $^{133}Ba^{2+}$ influx when incorporated into phosphatidylcholine vesicles. A close quantitative correlation was observed between the presence of the α-, β-, and γ-subunits of the calcium-antagonist receptor and the ability to mediate barium influx [28]. Thus, these results are also consistent with the conclusion that the purified calcium-antagonist receptor is capable of mediating ion flux with the pharmacologic characteristics expected of voltage-sensitive calcium channels.

An Oligomeric Model for the Dihydropyridine-Sensitive Calcium Channel

On the basis of present knowledge of the structure of the dihydropyridine-sensitive calcium channel, and in analogy with current models of the structure of voltage-sensitive sodium channels [29], we propose a model (Fig. 3) based on a central ion channel-forming element interacting with three other noncovalently associated subunits. The α_1-subunit, which contains the calcium-antagonist-binding sites, cAMP-dependent phosphorylation sites, and the largest hydrophobic domains, appears to be the central ion channel-forming component of the complex. Its apparent molecular weight of 175 kd from NaDodSO$_4$/PAGE is likely to be a reasonable approximation of the true polypeptide molecular weight since no N-glycosylation was detected. The

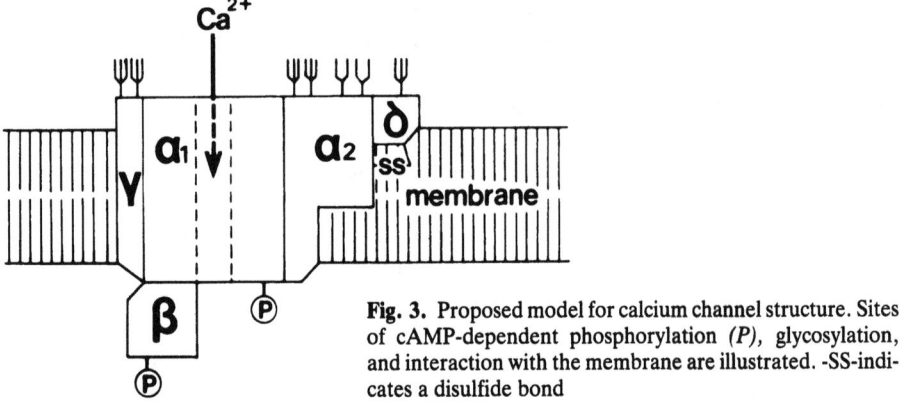

Fig. 3. Proposed model for calcium channel structure. Sites of cAMP-dependent phosphorylation (P), glycosylation, and interaction with the membrane are illustrated. -SS- indicates a disulfide bond

recent cloning and sequencing of this calcium channel subunit [30] has shown that it contains four homologous transmembrane domains analogous to those of the rat brain sodium-channel α-subunit, whose mRNA alone encodes a functional ion channel [31, 32]. Like α_1, the sodium-channel α-subunit also contains cAMP-dependent phosphorylation sites [29] and extensive hydrophobic domains that are efficiently labeled by TID [33].

The β-subunit is also a substrate for cAMP-dependent kinase [25], but hydrophobic labeling indicates that it does not interact with the membrane phase and it is not a glycoprotein. It is probably therefore tightly associated with an intracellular domain of α_1.

The γ-subunit of 30 kd interacts independently with α_1, contains at least one transmembrane segment, and consists of approx. 30% carbohydrate. All these properties are similar to those of the β_1-subunit of the rat brain and skeletal muscle sodium channels [29]. A polypeptide of similar size appears to be associated with the apamin-sensitive calcium-activated potassium channel [33], and it is interesting to speculate that this subunit may be a conserved constituent of voltage- or calcium-dependent ion channels.

The $\alpha_2\delta$ dimer appears to interact with α_1, although the conditions necessary to achieve dissociation result in a loss of dihydropyridine-binding activity. The 105-kd core polypeptide of α_2 contains a heavily glycosylated extracellular domain but displays weak hydrophobic labeling, indicating a limited intramembrane domain. For this reason, it seems unlikely that the ion channel is formed jointly by α_1 and α_2 at their zone of interaction.

The proposed model assumes a complex containing one of each subunit type. Our present results and previous data showing quantitative binding of solubilized calcium channels to ConA [12] suggest that each complex contains at least one α_1- and one α_2-subunit but do not specify the stoichiometry of any subunits. α_1 and α_2 appear to be present in approximately equal amounts on silver-stained gels, and the α_1- and β-subunits incorporate approx. 1 mol ^{32}P per mole of complex. A complete hydrodynamic analysis of the skeletal muscle calcium channel has not been reported. However, a size of 370 kd determined for the rat ventricular muscle dihydropyridine receptor [35] is within reasonable range of the predicted size of the complex represented in Fig. 3 (416 kd). Thus, an assumption of 1 mol of each subunit in the complex is plausible but requires direct experimental verification.

Conclusion

The molecular properties of dihydropyridine-sensitive calcium channels are now being elucidated by following a general strategy that was used previously in studies of voltage-sensitive sodium channels in this laboratory: identification by specific ligand binding and covalent labeling, solubilization and isolation by conventional protein purification methods, and reconstitution of channel function in vitro. The work reviewed here illustrates the substantial progress achieved with this approach to date. Future directions include further definition of the functional properties of the purified calcium-channel complex in reconstituted phospholipid vesicles; analysis of the mechanism of regulation of the channel by protein phosphorylation; determination of the primary structure of the peripheral channel subunits; extension of these molecular studies to calcium channels in heart, brain, and smooth muscle; and comparison of the structural features of calcium channels with those of voltage-sensitive sodium channels in order to define common structural themes underlying the function of voltage-sensitive ion channels in general.

References

1. Katz B, Miledi R (1969) J Physiol (London) 203:459-487
2. Adams DJ, Smith SJ, Thompson SJ (1980) Ann Rev Neurosci 3:141-167
3. Hagiwara S, Byerly L (1981) Ann Rev Neurosci 4:69-125
4. Reuter H (1979) Ann Rev Physiol 41:413-424
5. Sanchez JA, Stefani E (1978) J Physiol (London) 283:197-209
6. Almers W, Funk R, Palade PT (1981) J Physiol (London) 312:177-207
7. Rosemblatt M, Hidalgo C, Vergara C, Ikemoto N (1981) J Biol Chem 256:8140-8148
8. Fosset M, Jaimovich E, Delpont E, Lazdunski M (1983) J Biol Chem 258:6086-6092
9. Glossmann H, Ferry DR, Boschek CB (1983) Naunyn-Schmiedebergs Arch Pharmacol 323:1-11
10. Curtis BM, Catterall WA (1983) J Biol Chem 258:7280-7283
11. Curtis BM, Catterall WA (1984) Biochemistry 23:2113-2118
12. Glossmann H, Ferry DR (1983) Naunyn-Schmiedebergs Arch Pharmacol 323:279-291
13. Borsotto M, Barhanin J, Norman RI, Lazdunski M (1984) Biochem Biophys Res Commun 122:1357-1366
14. Takahashi M, Seagar MJ, Jones JF, Reber BFX, Catterall WA (1987) Proc Natl Acad Sci USA 84:5478-5482
15. Schmid A, Barhanin J, Coppola T, Borsotto M, Lazdunski M (1986) Biochemistry 25:3492-3495
16. Ferry DR, Goll A, Glossmann H (1984) FEBS Lett 169:112-167
17. Ferry DR, Kampf K, Goll A, Glossmann H (1985) EMBO J 4:1933-1940
18. Galizzi JP, Borsotto M, Barhanin J, Fosset M, Lazdunski M (1986) J Biol Chem 261:1393-1397
19. Striessing J, Moosburger K, Goll A, Ferry DR, Glossmann H (1986) Eur J Biochem 161:603-609
20. Reuter H (1974) J Physiol (London) 242:429-451
21. Reuter H (1983) Nature 30:569-574
22. Tsien RW, Giles W, Greengard P (1972) Nature New Biol 240:181-183
23. Brum G, Flockerzi V, Osterrieder W, Trautwein W (1983) Pflugers Archiv 398:147-154
24. Schmid A, Renaud J, Lazdunski M (1985) J Biol Chem 260:13041-13046

25. Curtis BM, Catterall WA (1985) Proc Natl Acad Sci USA 82:2528-2532
26. Takahashi M, Catterall WA (1987) Science 236:88-92
27. Takahashi M, Catterall WA (1987) Biochemistry 26:5518-5526
28. Curtis BM, Catterall WA (1986) Biochemistry 25:3077-3083
29. Catterall WA (1986) Ann Rev Biochem 55:953-985
30. Tanabe T, Takeshima H, Mikami A, Flockerzi V, Takahashi H, Kangawa K, Kojima M, Matsuo H, Hirose T, Numa S (1987) Nature 328:313-318
31. Noda M, Ikeda T, Suzuki H, Takeshima H, Takahashi T, Kuno M, Numa S (1986) Nature 322:826-828
32. Goldin AL, Snutch T, Lubbert H, Dowsett A, Marshall J, Auld V, Downey W, Fritz LC, Lester HA, Dunn R, Catterall WA, Davidson N (1986) Proc Natl Acad Sci USA 83:7503-7509
33. Reber BFX, Catterall WA (1987) J Biol Chem, in press
34. Seagar MJ, Labbe-Julle C, Granier C, Goll A, Glossmann H, Van Reitschoten J, Couraud F (1986) Biochemistry 25:4051-4057
35. Horne WA, Weiland GA, Oswald RE (1986) J Biol Chem 261:3588-3594

Molecular Characterization of the 1,4-Dihydropyridine Receptor in Skeletal Muscle

P.L. Vaghy, E. McKenna, and A. Schwartz

Department of Pharmacology and Cell Biophysics, University of Cincinnati College of Medicine, Cincinnati, Ohio 45267, USA

Introduction

Calcium antagonists have been successfully used for the treatment of several cardiovascular disorders, such as coronary heart disease, supraventricular arrhythmias, and hypertension. The most selectively acting drugs belong to one of three chemical groups: 1,4-dihydropyridines (DHP), phenylalkylamines, and benzothiazepines. These drugs inhibit L-type Ca^{2+} channels [1–3] by binding at distinct but allosterically interacting sites [4–6]. The drug receptor is believed to comprise part of the L-type Ca^{2+} channel. Pharmacological, electrophysiological, and radioligand binding studies have provided evidence for the existence of pharmacologically relevant receptors in various tissues [4–6]. However, until very recently the identity of the DHP-binding polypeptide has been an enigma.

Biochemical Characteristics of the Receptor

Purified skeletal muscle DHP receptor preparations consist of two large molecular weight (160–220 kDa) subunits termed, α_1 and α_2, and several smaller (< 55 kDa) ones [7–18]. This concept of two large subunits represents a revision of the initial ideas about the subunit composition of the DHP receptor [19–27]. Under nonreducing conditions the α_1 and α_2 subunits frequently overlap on sodium dodecyl sulfate polyacrylamide gels. This, and proteolytic degradation of α_1, may account for the description of only one α subunit in the early preparations. The exact subunit composition and nomenclature of the subunits remains controversial [9, 28], but there is agreement on the identity of the drug receptor. The α_1 subunit is the drug receptor; the functional roles of the other subunits (α_2, β, γ and δ) are not known.

Covalent labeling of the drug-binding polypeptide and examination of the purified DHP receptor preparations under both nonreducing and disulfide-reducing conditions proved to be crucial in the recognition of two distinct α subunits (α_1 and α_2). Only the α_1 subunit was photoaffinity labeled with DHP and phenylalkylamine derivatives [7, 9–11, 14, 16–18]. The electrophoretic mobility of the labeled polypeptide, in contrast to α_2, did not change on sodium dodecyl sulfate polyacrylamide gels upon treatment with disulfide-reducing agents. Antibodies selectively recognizing α_1 [8, 9, 13, 29], α_2 [20, 22–24, 30, 31], β [28], γ [31], or δ [20, 30] subunits immunoprecipitated

Table 1. Biochemical characteristics of the putative calcium channel subunits

	α_1	α_2	β	γ	δ
Silver staining	Weak	Strong	Weak	Very weak	Very weak
Apparent M_r (kDa)					
10 mM NEM[a]	165	170 ($\alpha_2 + \delta$)	50	33	170 ($\alpha_2 + \delta$)
10 mM DTT[b] or βME[c]	165	140	50	33	30
Photoaffinity labeling					
[³H]Azidopine	Yes	No	No	No	No
[³H]LU 49888	Yes	No	No	No	No
Phosphorylation by protein kinases A, B, C	Yes	No	Yes	No	No
Glycosylation	Weak	Strong	Weak	Strong	Strong
Hydrophobic domains	Extensive	Moderate	nd	nd	nd

nd, not detected
[a] N-ethylmaleimide
[b] Dithiothreitol
[c] beta-mercaptoethanol

drug-receptor complexes, suggesting that these polypeptides are distinct but closely associated under nondenaturing conditions.

Table 1 summarizes some biochemical characteristics of the five putative subunits. The size estimations were made from silver-stained sodium dodecyl sulfate polyacrylamide gels. The α_1 subunit as well as the β subunit are phosphorylated in vitro by protein kinases A, B, and C [8, 9, 11, 16, 28, 32–34] and are weakly glycosylated [8, 9, 16, 32]. The significance of the protein phosphorylation remains to be established. The α_2, γ, and δ subunits are extensively glycosylated [8, 9, 16] but are not phosphorylated. Under nonreducing conditions, the α_2 and δ subunits are linked by disulfide bond(s) [20]. This could explain the characteristic "shift" in electrophoretic migration of the α_2 subunit upon disulfide reduction.

Primary Structure of the α_1 Subunit

The amino acid sequence of the skeletal muscle α_1 subunit (or DHP receptor) has been deduced from the complete nucleotide sequence of cDNA clones encoding the α_1 subunit ([11]; Ellis et al., unpublished data). The cDNA clones correspond to 1873 amino acids, a large portion of which represents the primary structure of the α_1 subunit. This critical structural information about the drug receptor allows for the construction of models and design of novel experiments. These models can be revised and retested until an understanding of structure-function relationships is achieved. It is probable that serveral isoforms of the α_1 subunit exist, since L-type calcium channel properties and drug-binding affinities are quite variable among tissues. A membrane spanning model (Fig. 1) based upon the relative hydrophobic and hydrophilic charac-

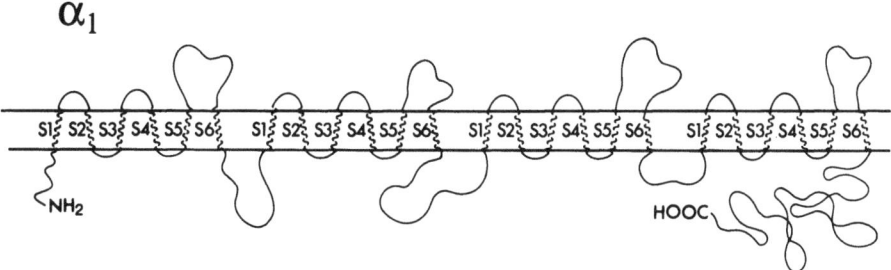

Fig. 1. Topographical representation of the α_1 subunit of the skeletal muscle L-type calcium channel. The model was constructed from the primary structure based on hydrophilicity/hydrophobicity of the 1873 amino acid components

teristics of the amino acids, shows a hypothetical configuration of the α_1 subunit within the membrane. Like the sodium channel [35], there are four motifs, each containing six membrane-spanning regions, S1–S6. A region of particular importance is S4, because it consists of charged amino acids – lysine and arginine – that could conceivably represent the voltage-sensor region that is characteristic of all voltage-dependent channels. This region is highly conserved in isoforms of voltage-dependent channels such as sodium channels [35] and potassium channels [36].

Reconstitution Studies

The presence of functional Ca^{2+} channels in purified skeletal muscle receptor preparations have been demonstrated by reconstitution experiments. When purified skeletal muscle DHP receptors were inserted into liposomes, specific $^{45}Ca^{2+}$ and $^{133}Ba^{2+}$ transport occured [37]. The ion flux was dependent upon the presence of micromolar concentrations of (+)Bay K8644, a Ca^{2+}-channel activator. This cation flux was inhibited by organic Ca^{2+} channel inhibitors such as (+)PN200–110, D600, verapamil, and inorganic inhibitors ($La^{3+} > Cd^{2+} > Ni^{2+} > Mg^{2+}$). The transport rates were slow, however, and only a small percentage of the protein was capable of mediating cation transport.

A number of groups [15, 33, 38, 39] reconstituted functional Ca^{2+} channels into planar lipid bilayers that retained Bay K8644 sensitivity and cation selectivity. These bilayer experiments clearly show that purified receptor preparations from skeletal muscle T-tubules exhibit at least two conductances when incorporated into planar lipid bilayers. Whether these conductance levels represent two distinct channels or two substate conductances of the same channel remains to be determined. The subunit(s) necessary for Ca^{2+} channel activity cannot be determined by reconstitution of a purified multisubunit complex into liposomes or bilayers. For determination of the minimal structural requirement for Ca^{2+} channel function, reconstitution of the individual subunit(s) or combinations of subunits is required.

Summary

The view on the identity of the skeletal muscle 1,4-dihydropyridine receptor has changed during the past 2 years. Previously, this receptor was attributed to a single 170 kDa polypeptide (now called the α_2 subunit) which characteristically reduced its apparent molecular weight to 140 kDa upon disulfide reduction when studied on sodium dodecyl sulfate polyacrylamide gels. Photoaffinity labeling data from a number of laboratories, however, show that 1,4-dihydropyridine (nifedipine-like drugs) bind specifically to a previously unrecognized, approximately 165 kDa polypeptide (α_1 subunit), which does not decrease its apparent molecular weight upon disulfide reduction. The α_1 subunit is phosphorylated in vitro by protein kinases. A, B, and C and its primary structure has been determined using molecular genetic techniques. Voltage-dependent Ca^{2+} channel activity has been reconstituted from purified 1,4-dihydropyridine receptor preparations, but the minimal structural requirement for Ca^{2+} channel function remains to be determined.

References

1. Nowycky MC, Fox AP, Tsien RW (1985) Three types of neuronal calcium channel with different calcium agonist sensitivity. Nature 316:440–446
2. McCleskey EW, Fox AP, Feldman D, Tsien RW (1986) Different types of calcium channels. J Exp Biol 124:177–190
3. Miller RJ (1987) Multiple calcium channels and neuronal function. Science 235:46–52
4. Vaghy PL, Williams JS, Schwartz A (1987) Receptor pharmacology of calcium entry blocking agents. Am J Cardiol 59:9A–17A
5. Triggle DJ, Janis RA (1987) Calcium channel ligands. Ann Rev Pharmacol Toxicol 27:347–369
6. Janis RA, Silver PJ, Triggle DJ (1987) Drug action and cellular calcium regulation. Adv Drug Res 16:309–591
7. Striessnig J, Knaus H-G, Grabner M, Moosburger K, Seitz W, Lietz H, Glossmann H (1987) Photoaffinity labeling of the phenylkalkylamine receptor of the skeletal muscle transverse tubule calcium channel. FEBS Lett 212:247–253
8. Leung AT, Imagawa T, Campbell KP (1987) Structural characterization of the 1,4-dihydropyridine receptor of the voltage-dependent calcium channel from rabbit skeletal muscle. J Biol Chem 262:7943–7946
9. Takahashi M, Seagar MJ, Jones JF, Reber BFX, Catterall WA (1987) Subunit structure of dihydropyridine-sensitive calcium channels from skeletal muscle. Proc Natl Acad Sci USA 84:5478–5482
10. Sieber M, Nastainczyk W, Zubor V, Wernet W, Hofmann F (1987) The 165-kDa peptide of the purified skeletal muscle dihydropyridine receptor contains the known regulatory sites of the calcium channel. Eur J Biochem 167:117–122
11. Tanabe T, Takeshima H, Mikami A, Flockerzi V, Takahashi H, Kangawa K, Kojima M, Matsuo H, Hirose T, Numa S (1987) Primary structure of the receptor for calcium channel blockers from skeletal muscle. Nature 328:313–318
12. Imagawa T, Leung AT, Campbell KP (1987) Phosphorylation of the 1,4-dihydropyridine receptor of the voltage-dependent Ca^{2+} channel by an intrinsic protein kinase in isolated triads from rabbit skeletal muscle. J Biol Chem 262:8333–8339
13. Morton ME, Froehner SC (1987) Monoclonal antibody identifies a 200-kDa subunit of the dihydropyridine-sensitive calcium channel. J Biol Chem 262:11904–11907
14. Vaghy PL, Striessnig J, Miwa K, Knaus H-G, Itagaki K, McKenna E, Glossman H, Schwartz A (1987b) Identification of a novel 1,4-dihydropyridine- and phenylalkylamine-binding polypeptide in calcium channel preparations. J Biol Chem 262:14337–14342

15. Talvenheimo JA, Worley III JF, Nelson MT (1987) Heterogeneity of calcium channels from a purified dihydropyridine receptor preparation. Biophys J 52:891–899
16. Hosey MM, Barhain J, Schmid A, Vandaele S, Ptasienski J, O'Callahan C, Cooper C, and Lazdunski M (1987) Photoaffinity labelling and phosphorylation of a 165 kilodalton peptide associated with dihydropyridine and phenylalkylamine-sensitive calcium channels. Biochem Biophys Res Commun 147:1137–1145
17. Sharp AH, Imagawa T, Leung AT, Campbell KP (1987) Identification and characterization of the dihydropyridine-binding subunit of the skeletal muscle dihydropyridine receptor. J Biol Chem 262:12309–12315
18. Kanngiesser U, Nalik P, Pongs O (1988) Purification and affinity labeling of dihydropyridine receptor from rabbit skeletal muscle membranes. Proc Natl Acad Sci USA 85:2969–2973
19. Curtis BM, Catterall WA (1984) Purification of the calcium antagonist receptor of the voltage-sensitive calcium channel from skeletal muscle transverse tubules. Biochemistry 23:2113–2118
20. Lazdunski M, Schmid A, Romey G, Renaud JF, Galizzi J-P, Fosset M, Borsotto M, Barhanin J (1987) Dihydropyridine-sensitive Ca^{2+} channels: molecular properties of interaction with Ca^{2+} channel blockers, purification, subunit structure, and differentiation. J Cardiovasc Pharm 9:S10–S15
21. Striessnig J, Moosburger K, Goll A, Ferry DR, Glossmann H (1986) Stereoselective photoaffinity labeling of the purified 1,4-dihydropyridine receptor of the voltage-dependent calcium channel. Eur J Biochem 161:603–609
22. Norman RI, Burgess AJ, Allen E and Harrison TM (1987) Monoclonal antibodies against the 1,4-dihydropyridine receptor associated with voltage-sensitive Ca^{2+} channels detect similar polypeptides from a variety of tissues and species. FEBS Lett 212:127–132
23. Takahashi M, Catterall WA (1987a) Identification of an α subunit of dihydropyridine-sensitive brain calcium channels. Science 236:88–91
24. Takahashi M, Catterall WA (1987b) Dihydropyridine-sensitive calcium channels in cardiac and skeletal membranes: Studies with antibodies against the α subunit. Biochemistry 26:5518–5526
25. Nakayama N, Kirley TL, Vaghy PL, McKenna E, Schwartz A (1987) Purification of a putative Ca^{2+} calcium protein from rabbit skeletal muscle: deterination of the amino-terminal sequence. J Biol Chem 262:6572–6576
26. Soldatov NM (1988) Purification and characterization of dihydropyridine receptor from rabbit skeletal muscle. Eur J Biochem 173:327–338
27. Hosey MM, Borsotto M, Lazdunski M (1986) Phosphorylation and dephosphorylation of the major component of the voltage-dependent Ca^{2+} channel in skeletal muscle membranes by cyclic AMP and Ca^{2+}-dependent processes. Proc Natl Acad Sci USA 83:3733–3737
28. Leung AT, Imagawa T, Block B, Franzini-Armstrong C, Campbell KP (1988) Biochemical and ultrastructural characterization of the 1,4-dihydropyridine receptor from rabbit skeletal muscle. Evidence for a 52000 Da subunit. J Biol Chem 263:994–1001
29. Fitzpatrick LA, Chin H, Nirenberg M, Aurbach GD (1988) Antibodies to an α subunit of skeletal muscle calcium channels regulate parathyroid cell secretion. Proc Natl Acad Sci USA 85:2115–2119
30. Vandaele S, Fosset M, Galizzi J-P, Lazdunski M (1987) Monoclonal antibodies that coimmunoprecipitate the 1,4-dihydropyridine and phenylalkylamine receptors and reveal the Ca^{2+} channel structure. Biochemistry 26:5–9
31. Sharp AH, Gaver M, Kahl SD, Campbell KP (1988) Structural characterization of the 32 kDa subunit of the skeletal muscle 1,4-dihydropyridine receptor. Biophys J 53:231a
32. Vaghy PL, Itagaki K, Miwa K, McKenna E, Schwartz A (1987) Mechanism of action of calcium modulator drugs: identification of a unique, labile, drug-binding polypeptide in a purified calcium channel preparation. Ann NY Acad Sci USA 522:176–186
33. Flockerzi V, Oeken H-J, Hofmann F, Pelzer D, Cavalié A, Trautwein W (1986) Purified dihydropyridine-binding site from skeletal muscle t-tubules is a functional calcium channel. Nature 323:66–68
34. Nastainczyk W, Röhrkasten A, Sieber M, Rudolph C, Schachtele C, Marme D, Hofmann F (1987) Phosphorylation of the purified receptor for calcium channel blockers by cAMP kinase and protein kinase C. Eur J Biochem 169:137–142
35. Noda M, Ikeda T, Kayano T, Suzuki H, Takeshima H, Kurasaka M, Takahashi H, Numa S (1986) Existence of distinct sodium channel messenger RNA in rat brain. Nature 320:188–192

36. Tempel BL, Jan YN, Jan LY (1988) Cloning of a probable potassium channel gene from mouse brain. Nature 332:837–839
37. Curtis BM, Catterall WA (1986) Reconstitution of the voltage-sensitive calcium channel purified from skeletal muscle transverse tubules. Biochemistry 25:3077–3083
38. Smith JS, McKenna EJ, Ma J, Vilven J, Vaghy PL, Schwartz A, Coronado R (1987) Calcium channel activity in a purified dihydropyridine receptor preparation of skeletal muscle. Biochemistry 26:7182–7188
39. Glossmann H, Striessnig J, Hymel L, Schindler H (1988) Purification and reconstitution of calcium channel drug-receptor sites. Ann NY Acad Sci 522:150–161

Reconstitution of Solubilized and Purified Dihydropyridine Receptor from Skeletal Muscle Microsomes as Two Single Calcium Channel Conductances with Different Functional Properties*

D. Pelzer[1], A. Cavalié[1], V. Flockerzi[2], F. Hofmann[2], and W. Trautwein[1]

[1] II. Physiologisches Institut and [2] Physiologische Chemie, Medizinische Fakultät, Universität des Saarlandes, D-6650 Homburg/Saar, FRG

Introduction

Calcium entry through voltage-sensitive calcium channels (VSCC) is a vital link between membrane depolarization and cellular function (Hille 1984). In many excitable cells, the entry of Ca^{2+} ions can be inhibited by phenylalkylamines such as verapamil, D600, or D888; the benzodiazepine diltiazem; dihydropyridine (DHP) compounds such as nifedipine, nitrendipine, nimodipine, or PN200–110; and so on (Fleckenstein et al. 1985; Lichtlen 1985; Miller 1987a; Venter and Triggle 1987; Bechem and Schramm 1988; Pelzer et al. 1988). These drugs are often grouped together as Ca^{2+} antagonists (Fleckenstein 1977) or Ca^{2+}-channel blockers (Katz and Reuter 1979), although they represent quite different classes of organic compounds. As therapeutic agents in clinical medicine, they have proven effective in the treatment of cardiovascular disorders (Braunwald 1985). As ligands that bind to different but interacting sites on VSCC with high affinity and specificity, radiolabeled compounds out of each major class have taken the place of natural toxins as molecular probes to count, purify, and characterize VSCC (Catterall 1986; Fosset and Lazdunski 1987; Glossmann et al. 1987).

Further analysis of the receptor protein for the Ca^{2+}-channel ligands then led to insight into the molecular properties and subunit structure of VSCC (Flockerzi et al. 1986a, b; Hosey et al. 1986; Lazdunski et al. 1986; Schmid et al. 1986a, b; Barhanin et al. 1987; Cooper et al. 1987; Ferry et al. 1987; Glossmann et al. 1987; Hosey et al. 1987; Lazdunski et al. 1987; Sieber et al. 1987; Vandaele et al. 1987; Glossmann and Striessnig 1988; Hosey et al. 1988; Lazdunski et al. 1988). Finally, the primary structure of the major 170-kd peptide subunit or purified skeletal muscle DHP receptor (DHPR) was deduced from its cDNA sequence (Tanabe et al. 1987). Electrophysiological studies simultaneously provided information on ion transport, gating behavior, and chemical modulation as fingerprints of multiple kinds of VSCC: the classical L-type VSCC and the novel T-, N-, and B-types (Miller 1987a, b; Tsien et al. 1987; Pelzer et al. 1988). Lipid bilayer recordings from reconstituted DHPR-containing membrane preparations from brain, heart, and skeletal muscle (Nelson et al. 1984; Affolter and Coronado 1985; Ashley et al. 1986; Ehrlich et al. 1986; Rosenberg et al. 1986; Coyne et al. 1987; Ma and Coronado 1987; Talvenheimo et al.

---

* This work was supported by the Deutsche Forschungsgemeinschaft, SFB 246.

1987), and solubilized (sDHPR) and purified DHPR (pDHPR) from skeletal muscle T tubules (Flockerzi et al. 1986b; Cavalié et al. 1987; McKenna et al. 1987; Talvenheimo et al. 1987; Trautwein et al. 1987) also revealed a heterogeneity of VSCC, even from the highly purified DHPR preparations. Here, we will focus on the results from 20 of 252 experiments where the reconstitution of sDHPR and pDHPR from skeletal muscle T tubules into lipid bilayers at the tips of glass patch pipettes was unequivocally associated with the reconstitution of $CoCl_2$-sensitive VSCC. We show that sDHPR and pDHPR preparations exhibit two divalent cation-selective conductance levels of about 9 and 20 pS, either singly or together in the same bilayer. The 20-pS single-channel events have many properties in common with L-type VSCC. By contrast, the 9-pS elementary events seem to represent activity from a distinct type of $CoCl_2$-sensitive VSCC with different functional properties, which are to some degree similar to those of T-type VSCC (Miller 1987a, b; Tsien et al. 1987; Pelzer et al. 1988).

Materials and Methods

Materials and Solutions

Bovine heart phosphatidylethanolamine (PE) and bovine brain phosphatidylserine (PS) were obtained from Avanti Polar Lipids (Birmingham, Ala, USA), and cholesterol was purchased from Sigma Chemical (St. Louis, Mo, USA). Lipid solutions were prepared from a mixture of 70% PE, 15% PS, and 15% cholesterol dissolved in n-hexane (1 mg/ml). ($\pm$)-Methoxyverapamil [($\pm$)-D600], ($\pm$)-Bay K8644, (+)-PN200-110, and adenosine-5-(γ-thio)-triphosphate (ATP-γ-S) were from Knoll AG (Ludwigshafen, FRG), Bayer AG (Leverkusen, FRG), and Serva (Heidelberg, FRG), respectively. The catalytic (C) subunit of the cAMP-dependent protein kinase (cA kinase) was prepared and purified from bovine heart (Beavo et al. 1974; Hofmann 1980). The bath and pipette-filling solutions usually contained 90 mM $BaCl_2$, 5 mM Hepes, and 0.02 mM tetrodotoxin (TTX). When indicated (e.g., Fig. 2), the $BaCl_2$ concentration in the pipette was reduced to 20 mM. The pH was adjusted to 7.4 by $Ba(OH)_2$, and the experiments were carried out at room temperature. All other materials used were of the highest quality and purity available to us.

Purification of DHP Receptor

The DHP receptor was solubilized from microsomal membranes of white rabbit back and hindleg skeletal muscle and purified to an apparent density of 1.5–2.0 nmol binding sites per milligram of protein as described elsewhere (Flockerzi et al. 1986a). Solubilized and purified DHP receptor preparations were reconstituted into lipid vesicles, and the vesicles were stored in small aliquots at −80°C until they were used for elementary current recording after their incorporation into lipid bilayer membranes (Flockerzi et al. 1986b).

Lipid Bilayers

Solvent-free lipid bilayers were formed from monolayers on the tips of thick-walled, unpolished, uncoated, hard borosilicate glass patch pipettes (5–10 MΩ) by the contact method, as described elsewhere (Hanke et al. 1984). For lipid monolayers, 5–15 µl lipid solution was spread at the surface of the salt solution contained in the 1.5-ml bath. Glass-bilayer seal resistances were in the order of 10–50 GΩ and the probability of successful bilayer formation was ≥ 0.75. The reliability of our bilayer assembly was periodically tested with gramicidin that regularly formed pores with elementary conductances of 3–7 pS (100 mM NaCl) in our bilayers; this is in agreement with previous reports in other bilayer set-ups (Finkelstein and Andersen 1981).

Prior to incorporation of DHPR-containing liposomes into the lipid bilayers, their electrical stability was always assured by applying constant potential gradients between −100 and +100 mV across the artificial membranes. Only those bilayers with stable artifact-free current baselines for 15–30 min were considered reliable for further experimentation. With this test at the beginning of each experiment, about 20%–25% of tightly sealed bilayers had to be discarded because of the spontaneous occurrence of single-channel-like current pulses of variable amplitudes. By contrast to protein-related single-channel activity, no sharp peaks could be detected in the current amplitude histograms of such bilayer-related elementary events. Additionally, sequences of single-channel-like bilayer activity were accompanied by irreversible stepwise decrements of the seal resistance. These observations suggest that spontaneous single-channel-like elementary current pulses in pure bilayers may arise from "microruptures" of the bilayer itself and/or of the glass-bilayer seal. Support for this idea comes from experiments which show that monomolecular films of stearic acid properly compressed on the water surface are not completely homogenous, but can include densely scattered small holes (Uyeda et al. 1987).

Incorporation of VSCC into Lipid Bilayers

Reconstituted DHPR samples were incorporated into performed artifact-free lipid bilayers by adding 10–40 µl of the liposome suspension (corresponding to a total of 0.01–0.1 pmol DHP-binding sites) to the bath underneath the bilayer membrane. The responsiveness of incorporated channels to chemical agents was tested by applying the chemicals at desired concentrations to the bath side of the lipid bilayer.

Sidedness of VSCC in Lipid Bilayers

According to Tanabe et al. (1987), all potential cAMP-dependent phosphorylation sites are located on the cytoplasmic side of the DHPR molecule. When DHPR-containing liposomes were incorporated, the functional consequences of VSCC phosphorylation were observed when C subunit of the cA kinase was applied to the bath side of the bilayer in the presence of MgATP (see also Flockerzi et al. 1986b; Trautwein et al. 1987). This implies that DHPR molecules functioning as VSCC are likely to be positioned in our bilayer assembly with their cytoplasmic sides facing the bath, resulting in an inside-out configuration of single-channel recording.

Electrical Recording and Data Analysis

Elementary currents were recorded at constant potentials using a conventional patch-clamp amplifier (L/M-EPC 7, List Medical Electronic, Darmstadt, FRG). The potential across the bilayer membrane was controlled by clamping the pipette potential (V_p) with respect to the bath. According to the recording configuration, a positive membrane potential corresponds to a negative voltage signal on the pipette voltage display and vice versa. In the subsequent discussion of results, pipette polarities are given exclusively. Positive current signals displayed as upward deflections from the zero-current baseline reflect cationic current flow from the pipette into bath. Anionic current flow in the opposite direction was excluded by measurements of the current reversal potential following a decrease of the $BaCl_2$ concentration in the pipette to 20 mM. Single-channel currents were recorded continuously on an FM tape recorder at 7.5 or 15 ips and then played back through a four-pole Bessel filter (−3 dB) at one–fourth of the digitizing rate (1.5–3.6 kHz) for analysis on a Nicolet MED-80 or PDP 11–73 computer, as described elsewhere (Cavalié et al. 1986). The open state probability (P_o) is defined as the time-average current divided by the single-channel current amplitude, and it was calculated from 3- to 35-min records continuously chopped into 2-s segments.

Results

The DHP receptor solubilized and purified from rabbit skeletal muscle microsomes appears to be a multisubunit protein, with five major peptide bands of apparent molecular mass of 165, 130, 55, 32, and 28 kd under reducing conditions (Flockerzi et al. 1986a, b; Sieber et al. 1987). Under nonreducing conditions, the 130-kd and 28-kd peptides migrate as a single peptide of 165 kd, the latter being unrelated to that 165-kd peptide which does not change its molecular mass upon reduction (Sieber et al. 1987). In the attempt to study the functional and regulatory properties of this protein containing Ca^{2+}-antagonist receptors and cAMP-dependent phosphorylation sites in an in vitro recording system, sDHPR and pDHPR were reconstituted into lipid vesicles, and the vesicles were incorporated into lipid bilayer membranes formed at the tips of patch pipettes. In most experiments, either no reconstitution was obtained within 60 min of observation (136/252 ≃ 54%) or reconstitution of doubtful single-channel activity was achieved (96/252 ≃ 38.1%), with individual elementary conductances between 3 and 80 pS. This activity was insensitive to block by (±)−D600, (+)−PN200−110, and even $CoCl_2$ in concentrations up to 40 mM. Only in a low percentage of trials (20/252 ≃ 7.9%) was the reconstitution of sDHPR and pDHPR unequivocally associated with the reconstitution of VSCC activity, as judged from the responsiveness of elementary events to organic VSCC blockers [(±)−D600], VSCC activators [(±)−Bay K8644], and/or cAMP-dependent phosphorylation, or from sensitivity to the inorganic VSCC blocker $CoCl_2$.

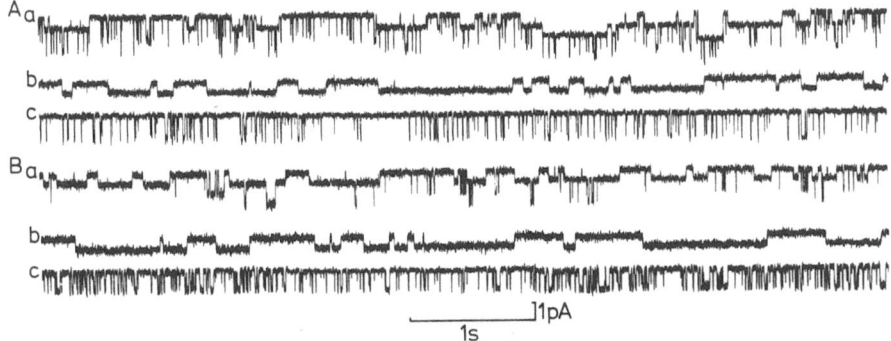

Fig. 1A, B. Single-channel recordings after incorporation of the solubilized (**A**) or purified (**B**) DHP receptor into lipid bilayers. The current records were obtained in six different bilayers. In all cases, activity started within the first 5–7 min after addition of DHPR to the bath side underneath the bilayer membrane and lasted for > 2 h. The pipette potential (V_P) was −75 mV (**Aa**), −70 mV (**Ab, c**), −65 mV (**Ba, b**), and −60 mV (**Bc**). At negative V_P, downward deflections represent channel openings and reflect negative elementary cationic current flowing from the bath into the pipette. The pipette and bath contained 90 mM BaCl$_2$ solutions. The records were digitized at 1.5 kHz

Figure 1 shows typical elementary current records obtained after successful incorporation of sDHPR (Fig. 1A) and pDHPR (Fig. 1B) as VSCC. In both DHPR preparations, two single-channel current levels were regularly identified, either singly (Fig. 1Ab, c, Bb, c) or together (Fig. 1Aa, Ba) in the same bilayer. At a V_p of −70 mV, the larger elementary current level had a unitary current amplitude of 1.45 pA, which roughly corresponds to a conductance of 20 pS (Fig. 1Ac, Bc). By contrast, the unitary amplitude of the smaller single-channel current level was 0.65 pA at the same V_p, which gives a conductance of about 9 pS (Fig. 1Ab, Bb). In addition, the 9-pS events exhibit distinctly longer open times than the 20-pS openings. In the example experiments (Fig. 1Ab, c, Bb, c), the 20- and 9-pS conductance levels were observed singly in different bilayers for more than 2 h without interconversion of one into the other. This observation, which is typical for all stable experiments with only one conductance level to start with, favors the conclusion that the two conductance levels represent activity from two distinct channel molecules rather than the conclusion that they arise from the same molecule. However, because we do not know what factors might control the frequency of appearance of different conductance states, we are aware that we cannot completely rule out the latter alternative (see the discussion below).

Current-Voltage Relations

Figure 2 shows the 20-pS (Fig. 2Aa) and 9-pS (Fig. 2Ba) conductance levels of elementary current separately at constant pipette potentials in symmetrical 90 mM BaCl$_2$ solutions. The elementary current-V_p relationships were linear over a wide potential range, and slope conductances were 19.6 (Fig. 2Ab; sDHPR), 20.1 (Fig.

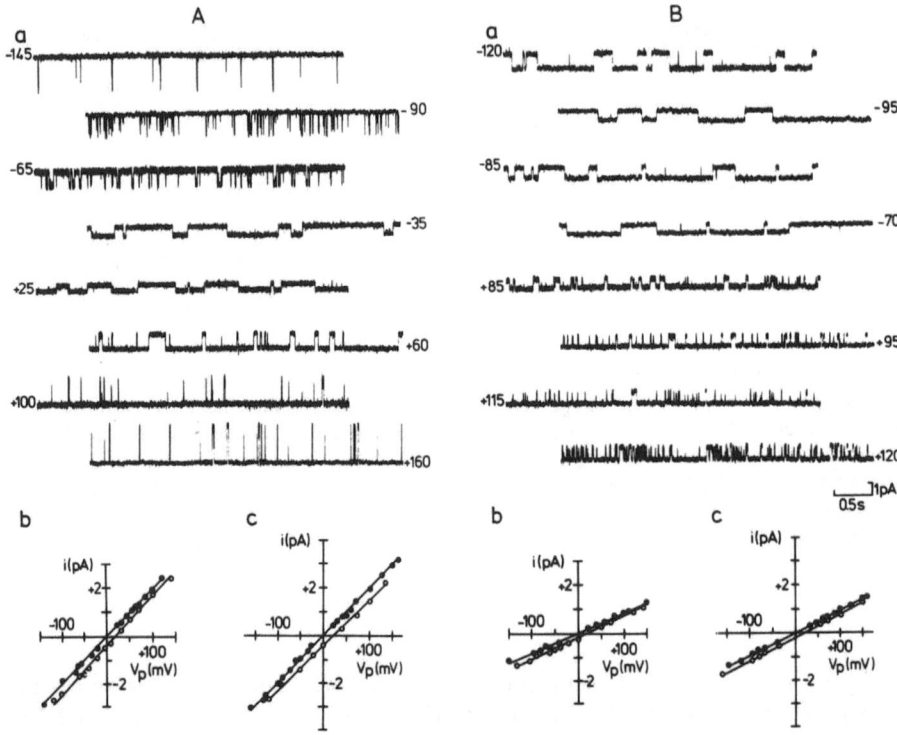

Fig. 2A, B. Current-voltage relations and selectivity of two types of elementary events. The original traces of single-channel current (**Aa, Ba**) were recorded at various pipette potentials (V_P), indicated beside each record, in symmetrical 90 mM BaCl$_2$ solutions. In this and all following figures, downward deflections at negative V_P (positive membrane potentials) represent channel openings, whereas at positive V_P (negative membrane potentials) channel openings are represented by upward deflections of single-channel current. The records shifted to the *left* in **Aa** and **Ba** were obtained from sDHPR, whereas the traces displaced to the *right* were recorded from pDHPR. The sampling frequency was 1.5 kHz. The *lower panels* show elementary current-voltage relations from mean values ($4 \leq n \leq 8$, SD not shown) of single-channel current amplitudes, in symmetrical 90 mM BaCl$_2$ solutions (•) and with 20 mM BaCl$_2$ in the pipette and 90 mM BaCl$_2$ in bath (o), for sDHPR (**Ab, Bb**) and pDHPR (**Ac, Bc**). The *straight lines* are least-squares fits to the data with the following parameters: **Ab** $\gamma = 19.6$ pS, $V_r = +0.9$ mV (•); $\gamma = 20.9$ pS, $V_r = +18.7$ mV (o); **Ac** $\gamma = 20.1$ pS, $V_r = +0.9$ mV (•); $\gamma = 18.9$ pS, $V_r = +22.7$ mV (o); **Bb** $\gamma = 8.5$ pS, $V_r = +0.3$ mV (•); $\gamma = 9.3$ pS, $V_r = +18.8$ mV (o); **Bc** $\gamma = 9.5$ pS, $V_r = +2.1$ mV (•); $\gamma = 10.0$ pS, $V_r = +19.5$ mV (o)

2Ac; pDHPR), 8.5 (Fig. 2Bb; sDHPR), and 9.5 pS (Fig. 2Bc; pDHPR), respectively. The zero-current potential (V_r) was around 0 mV in each case. With 20 mM BaCl$_2$ in the pipette and 90 mM BaCl$_2$ in the bath (Fig. 2Ab, c, Bb, c), the I–V relations remained linear, and the slope conductances were almost identical (20.9, 18.9, 9.3, and 10 pS, respectively) to those obtained in symmetrical 90 mM BaCl$_2$ solutions. However, V_r ranged between $+18.7$ and $+22.7$ mV, as expected if Ba^{2+} ions are the charge carrier of both elementary conductance levels.

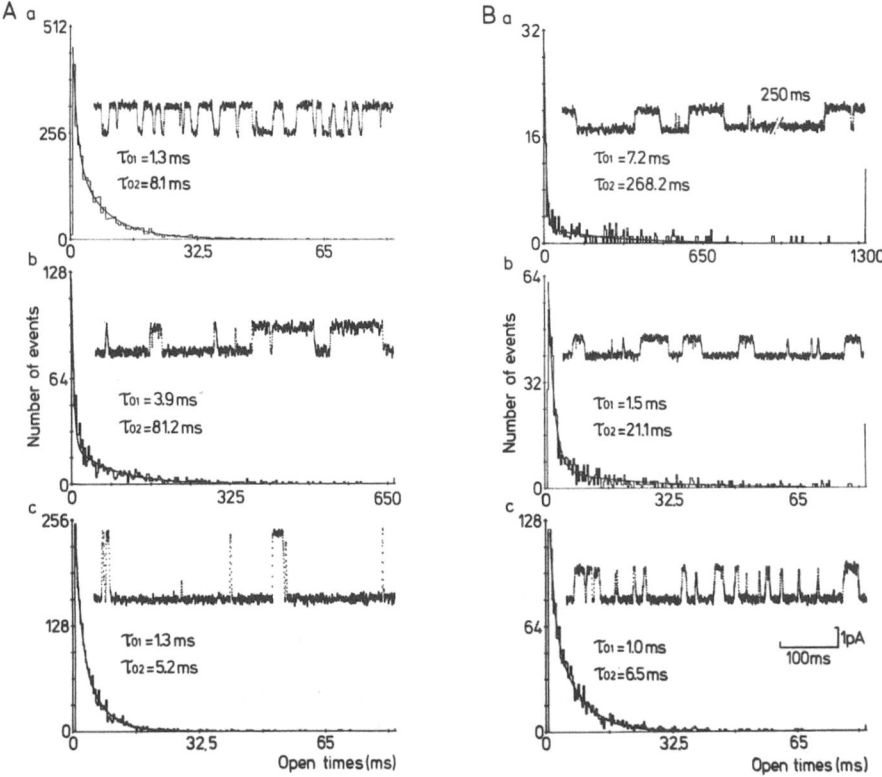

Fig. 3A, B. Open times from single-channel currents with conductances of 20 pS (**A**) and 9 pS (**B**). V_P was −80 mV (**Aa, Ba**), +62 mV (**Ab**), +165 mV (**Ac**), +85 mV (**Bb**), and +150 mV (**Bc**). Open time distributions were compiled from recordings showing only one current level and digitized at 2.0–3.1 kHz. A biexponential function with time constants τ_{01} and τ_{02} was fitted to each open time distribution by a nonlinear, unweighted, least-squares method. The *insets* show representative example records digitized at 1.5 kHz. The pipette and bath contained 90 mM BaCl$_2$ solutions

Gating Kinetics

The voltage dependence of the kinetic behavior of the 20- (Fig. 3A) and 9-pS (Fig. 3B) elementary events was analyzed singly in separate bilayers. In symmetrical 90 mM BaCl$_2$ solutions (Figs. 3, 4), the open and closed time distributions of both conductance levels could be fitted by two exponential functions over a wide potential range. At V_p −80 mV, the time constants τ_{01} and τ_{02} were 1.3 and 8.1 ms for the 20-pS single-channel conductance (Fig. 3Aa), and 7.2 and 268.2 ms for 9-pS events (Fig. 3Ba). Closed time constants (not shown) were 2.5 and 79 ms, and 1.7 and 32 ms, respectively, at the same potential. Typically, at $V_p \simeq$ −80 mV, τ_{02} of the 9-pS events was 30–50 times longer than τ_{02} of the 20-pS events; τ_{01} was at the most ten times longer for the 9-pS single-channel activity. A potential change to V_p +62 mV

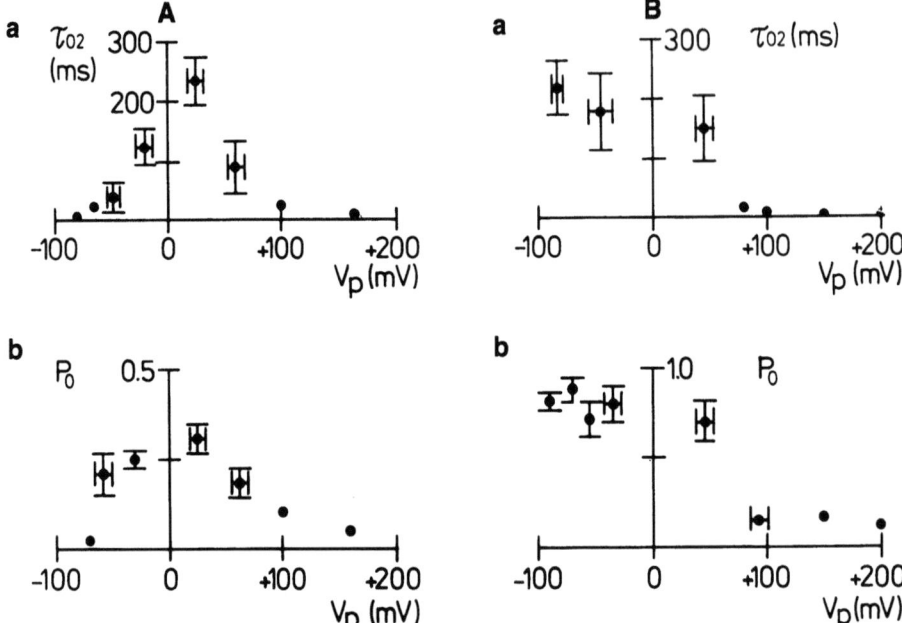

Fig. 4A, B. Voltage dependence of open times (τ_{02}) and open-state probability (P_o) from single-channel events with conductances of 20 pS (**A**) and 9 pS (**B**). Each point represents the mean value of τ_{02}s and P_os determined in three to ten experiments, where only one conductance level was recorded at least at three potentials. The pipette and bath contained 90 mM BaCl$_2$ solutions. The sampling frequency was 2.0–3.6 kHz

prolonged both open times of the 20-pS conductance (Fig. 3Ab), but a further increase to V_p +165 mV shortened them again (Fig. 3Ac). By contrast, similar changes in V_p produced a continuous decrease in both open times of the 9-pS elementary events (Fig. 3Bb, c). In conclusion, the τ_{02}-V_p (Fig. 4Aa) and P_o-V_p (Fig. 4Ab) relations were bell-shaped for the 20-pS elementary conductance events, with a maximum around 0 mV, and sigmoidal for the 9-pS single-channel activity, with the longest τ_{02}s and highest P_os at negative pipette potentials in symmetrical 90 mM BaCl$_2$ solutions (Fig. 4Ba, b). The voltage dependence of τ_{01} for both conductance levels (not shown) was similar to that of τ_{02} (see Fig. 3).

Chemical Responsiveness

Figures 5, 6 show the sensitivity of both conductance levels of single-channel activity occurring singly in separate bilayers to (±)-D600 (Fig. 5Aa, d, Ba, d), (±)-Bay K8644 (Fig. 5Ab, d, Bb, d), cAMP-dependent phosphorylation (Fig. 5Ac, d, Bc, d), and CoCl$_2$ (Fig. 6). Bath application of 25 μM (±)-D600 shortened 20-pS open-channel lifetimes and prolonged closed times between channel openings and/or groups of openings (Fig. 5Aa). With (±)-Bay K8644 (10 μM) (Fig. 5Ab) or C subunit of cA kinase (1 μM plus 1 mM ATP-γ-S and 0.3 mM MgCl$_2$) (Fig. 5Ac)

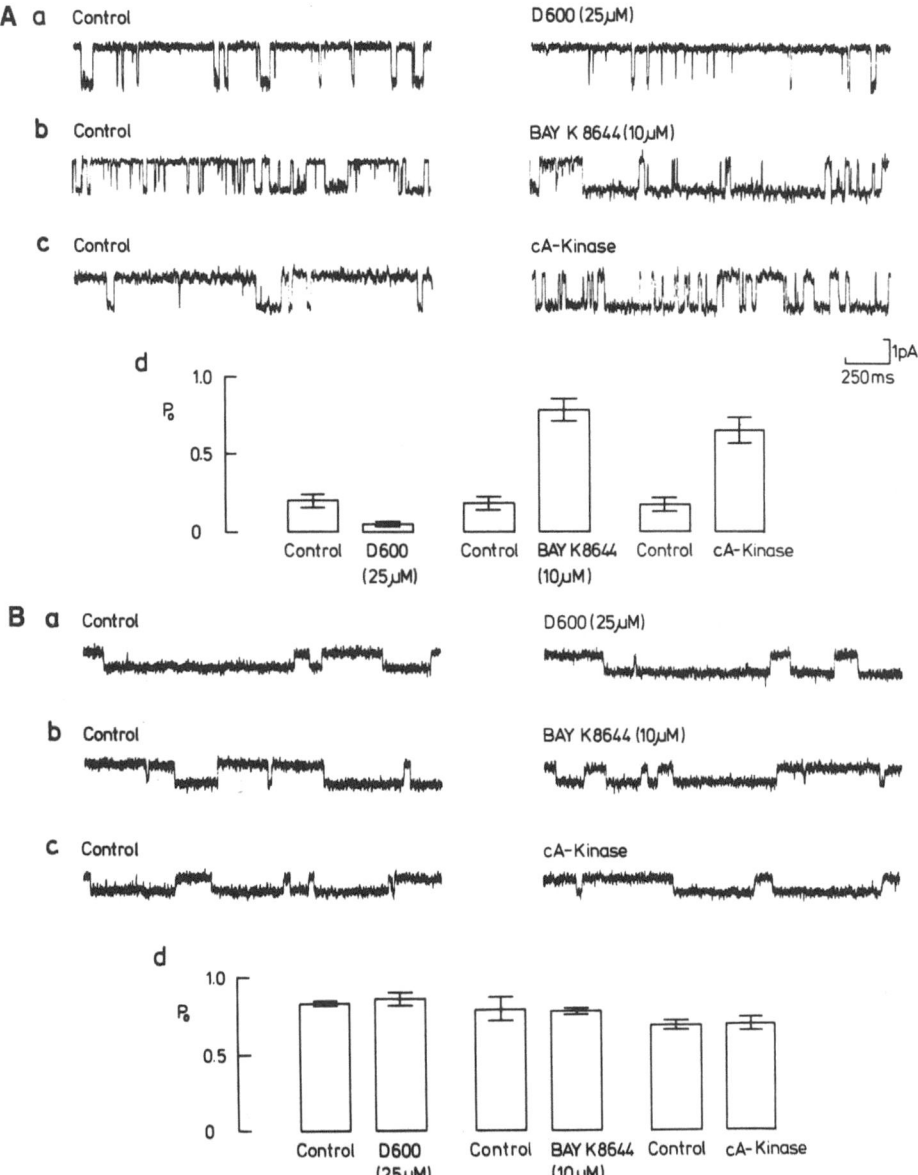

Fig. 5A, B. Responsiveness of the two conductance levels from DHP receptor to (±)-D600, (±)-Bay K8644, and cA kinase in lipid bilayers. Single-channel currents with slope conductances of 20 pS (**A**) and 9 pS (**B**) were recorded singly for 5–10 min. (±)-D600 (25 µM), (±)-Bay K8644 (10 µM), or C subunit (1 µM) of the cA kinase plus ATP-γ-S (1 mM) and $MgCl_2$ (0.3 mM) were then applied to the bath side of the bilayer, and the recording was continued for 10–20 min. The original records were obtained at V_P of −65 mV (**Aa**), −52 mV (**Ab**), −60 mV (**Ac**), −70 mV (**Ba**), −60 mV (**Bb**), and −55 mV (**Bc**). The sampling frequency was 1.5 kHz. The responsiveness of the DHP receptor to (±)-D600, (±)-Bay K8644 and cA kinase (**Ad, Bd**) is expressed in terms of open-state probability (P_o). Each value is the mean P_o from three to five measurements in 5- to 15-min records. V_P was −57.5 ± 7.5 mV (**Ad**), −65 ± 5 mV (**Bd**, control and (±)-D600), −55 ± 5 mV (**Bd**, control and (±)-Bay K8644), and −57.5 ± 2.5 mV (**Bd**, control and cA kinase)

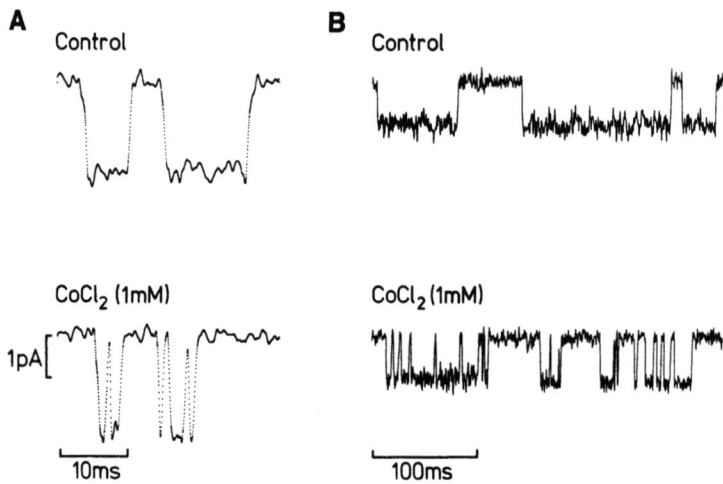

Fig. 6A, B. Responsiveness of the two conductance levels from DHP receptor to CoCl$_2$ in lipid bilayers. Single-channel currents with slope conductances of 20 pS (**A**) and 9 pS (**B**) were recorded singly for 5–10 min. CoCl$_2$ was then applied to the bath side of the bilayer, and the recording was continued for 10–20 min. The example records were obtained at V$_P$ of –100 mV

present in the bath, 20-pS open-channel lifetimes became longer, and gaps between channel openings and/or groups of openings got shorter. As a consequence of these drug-induced changes in gating behavior, P$_o$ of the 20-pS conductance level at V$_p$ between –52 mV and –70 mV was reduced 2.5–6 times by 25 µM (±)–D600 (apparent IC$_{50}$ ≃ 10–14 µM) and enhanced 2.5–6 times by 10 µM (±)–Bay K8644 and cAMP-dependent phosphorylation (1 µM C subunit, 1 mM ATP-γ-S, 0.3 mM MgCl$_2$) (Fig. 5Ad). By contrast, gating behavior and P$_o$ of the 9-pS conductance level were largely unresponsive to these chemicals at similar V$_p$ and concentrations (Fig. 5Ba–d). Both conductance levels of channel activity were sensitive to block by 1 mM CoCl$_2$ (Fig. 6).

Discussion

In this study, we show that sDHPR and pDHPR preparations reconstituted into lipid bilayers formed on the tips of patch pipettes exhibit two divalent cation-selective conductance levels of 9 and 20 pS, similar in single-channel conductance to VSCC reported in a variety of intact preparations (Nilius et al. 1985; Nowycky et al. 1985; Worley et al. 1986; see Tsien et al. 1987 for review). The larger conductance level is similar to the VSCC identified in intact rat T-tubule membranes (Affolter and Coronado 1985) and described in sDHPR and pDHPR preparations (Flockerzi et al. 1986b; Trautwein et al. 1987), and it has many properties in common with L-type VSCC (Tsien et al. 1987; Pelzer et al. 1988). It is sensitive to augmentation by the DHP agonist (±)–Bay K8644 and cAMP-dependent phosphorylation, and to block by the phenylalkylamine (±)–D600 and the inorganic blocker CoCl$_2$. Its open-state probability and open times are increased upon depolarization, as expected for a voltage-dependent activation process. Upon depolarization beyond the reversal potential, however, open-state probability and open times decline again. A reasonable way to explain the bell-shaped dependence of open times and open-state probability on membrane potential is to assume voltage-dependent ion-pore interactions that produce closing of the channel at strong negative and positive membrane potentials. By contrast, the smaller conductance level may be similar to the 10.6-pS T-tubule VSCC described by Rosenberg et al. (1986), and it may best be compared to T-type VSCC (Tsien et al. 1987; Pelzer et al. 1988). It is largely resistant to augmentation by ($\pm$)-Bay K8644 and cAMP-dependent phosphorylation or block by ($\pm$)-D600, but it is sensitive to block by $CoCl_2$. Its open times and open-state probability show a sole dependence on membrane potential, where depolarization increases both parameters sigmoidally from close to zero up to a saturating level. Neither of the elementary conductance levels exhibits significant inactivation over a wide potential range, which may suggest that skeletal muscle VSCC inactivation is either poorly or not at all voltage dependent. This possibility seems in agreement with bilayer recordings on reconstituted intact T-tubule membranes (Affolter and Coronado 1985; Rosenberg et al. 1986) and voltage-clamp recordings on intact fibers (Almers et al. 1981, 1984, 1985). It supports the idea that the decline of Ca^{2+} current in intact skeletal muscle fibers may be due to Ca^{2+} depletion from the T-tubule system (Almers et al. 1981) and/or to inactivation induced by Ca^{2+} release from the sarcoplasmic reticulum (Almers et al. 1985).

We consistently observe two conductance levels of 9 and 20 pS, either singly or together in the same bilayer from solubilized DHPR samples and even highly purified DHPR preparations. This raises the possibilities that skeletal muscle may contain two distinct types of VSCC with closely related molecular structures, or alternatively, that the DHP-sensitive VSCC in skeletal muscle may exhibit two primary conductance levels. The observations that (a) both conductance levels are observed in intact T-tubule membranes (Talvenheimo et al. 1987); (b) the two conductance levels clearly exhibit distinct kinetic, pharmacologic, and regulatory properties (see also Cavalié et al. 1987; Trautwein et al. 1987); and (c) both conductance levels from sDHPR and pDHPR preparations can be observed singly in different bilayers for more than 2 h without interconversion (see also Talvenheimo et al. 1987) strongly argue against the conclusion that the two conductance levels arise from the same molecule. They rather reflect activity from two distinct molecules. If so, the two types of molecules functioning as VSCC are likely to contain sufficient structural homology to explain the presence of both channel types in pDHPR samples. Nevertheless, we cannot completely rule out the possibility that a covalent modification of one and the same channel molecule determines its conductance level and its chemical responsiveness. In this case, the presence of modified channels in our preparation would determine which conductance levels and pharmacologic responses we observe, and these conductance levels would not necessarily be interconvertible under our experimental conditions. Findings on gramicidin and nicotinic ACh-receptor channels may serve as precedents for such a possibility (Siemen et al. 1986; Busath et al. 1987).

Multiple conductance levels of VSCC have been described in a variety of cell types other than skeletal muscle (Nilius et al. 1985; Nowycky et al. 1985; Worley et al. 1986; see Tsien et al. 1987 for review). Moreover, using the voltage clamp or the whole-cell configuration of the patch clamp, two components of macroscopic Ca^{2+} current have been described in intact skeletal muscle fibers (Cota and Stefani 1986) and in cultured skeletal muscle cells (Beam et al. 1986; Cognard et al. 1986); a slow DHP-sensitive component, and a fast, transient DHP-insensitive component. The two conductance levels that we (see also Cavalié et al. 1987) and others (Talvenheimo et al. 1987) have observed from intact T-tubule membranes, sDHPR samples, and pDHPR preparations may well underlie the two components of voltage-dependent Ca^{2+} current in skeletal muscle.

Our results (Flockerzi et al. 1986b; Cavalié et al. 1987; Trautwein et al. 1987) and those of Talvenheimo et al. (1987) firmly establish the presence of functional VSCC in pDHPR preparations. However, the picture has been complicated by the finding that a large fraction of DHP receptors in skeletal muscle do not appear to be VSCC (Schwartz et al. 1985; Rios and Brum 1987). DHP receptors in this tissue appear to have split personalities, capable of acting both as voltage sensors and as VSCC in excitation-contraction coupling (Agnew 1987). Support for this recent appreciation comes from the similarities between the pharmacology of voltage sensors and VSCC in skeletal muscle (Berwe et al. 1987; Rios and Brum 1987). Thus, it is possible that some of the polypeptides that have been purified and reconstituted are in fact the voltage sensor, and that VSCC have copurified with these molecules. The low success rate of reconstitution of skeletal muscle sDHPR and pDHPR as functional VSCC may also indicate that only a few percent of DHPR in skeletal muscle T-tubule membranes function as VSCC. Our success rate is infact in good accord with the estimate of skeletal muscle DHPR acting as VSCC (Schwartz et al. 1985). The low number of successful experiments definitely cannot be explained by technical limitations of our bilayer reconstitution assembly since, for example, Ca^{2+}-antagonist-receptor-containing preparations from *Drosophila* brain (Pauron et al. 1987) consistently yield success rates of VSCC reconstitution of $\geq 90\%$ (Pelzer and Lazdunski, unpublished observation). The low success rate of skeletal muscle DHPR reconstitution as VSCC is also unlikely to be explained by dephosphorylation of a per se functional molecule, since (a) unphosphorylated DHPR preparations and prephosphorylated DHPR samples give similar success rates and (b) channel activity could never be induced by cAMP-dependent phosphorylation after a sufficient waiting time (≥ 30 min) following the bath addition of DHPR samples to allow for incorporation in still mute bilayers. An additional explanation for the low success rate which has not yet been explored would be that most reconstituted sDHPR and pDHPR from skeletal muscle are in fact silent VSCC but lack direct activation by regulatory G proteins (Yatani et al. 1987) and/or another essential (structural?) component (Vilven et al. 1988).

Acknowledgements. The authors wish to thank Mr. H. Ehrler for skillfull electronics support and Ms. S. Bastuck for competent technical assistance. We also thank Drs. S. Pelzer and A. O. Grant for discussions, comments on the manuscripts and assistance with some of the experiments.

References

Affolter H, Coronado R (1985) Biophys J 48:341–347
Almers W, Fink R, Palade PT (1981) J Physiol (Lond) 312:177–207
Almers W, McCleskey EW, Palade PT (1984) J Physiol (Lond) 353:565–583
Almers W, McCleskey EW, Palade PT (1985). In Rubin RP, Weiss GB, Patney JW Jr (eds) Calcium in biological systems. Plenum, New York, pp 312–330
Agnew WS (1987) Nature (Lond) 328:297
Ashley RH, Montgomery RAP, Williams AJ (1986) J Physiol (Lond) 381:115 P
Barhanin J, Coppola T, Schmid A, Borsotto M, Lazdunski M (1987) Eur J Biochem 164:525–531
Beavo JA, Bechtel PJ, Krebs EG (1974) Methods Enzymol 38:299–308
Beam KG, Knudson CM, Powell JA (1986) Nature (Lond) 320:168–170
Bechem M, Schramm M (1988). In Piper HM, Isenberg G (eds) Isolated adult cardiomyocytes, Vol II. CRC Press Inc, Bota Raton, in press
Berwe D, Gottschalk G, Lüttgau HC (1987) J Physiol (Lond) 385:693–707
Braunwald E (1985). In Lichtlen PR (ed) Recent aspects in calcium antagonism. Schattauer, Stuttgart, pp 203–210
Busath DD, Andersen OS, Koeppe II RE (1987) Biophys J 51:79–88
Cavalié A, Pelzer D, Trautwein W (1986) Pflügers Arch 406:241–258
Cavalié A, Flockerzi V, Hofmann F, Pelzer D, Trautwein W (1987) J Physiol (Lond) 390:82 P
Catterall WA (1986). In Lüttgau HC (ed) Membrane control of cellular activity. Fischer, Stuttgart, pp 3–27
Cognard C, Lazdunski M, Romey G (1986) Proc Natl Acad Sci USA 83:517–521
Cooper CL, Vandaele S, Barhanin J, Fosset M, Lazdunski M, Hosey MM (1987) J Biol Chem 262:509–512
Cota G, Stefani E (1986) J Physiol (Lond) 370:151–163
Coyne MD, Dagan D, Levitan IB (1987) J Membrane Biol 97:205–213
Ehrlich BE, Schen CR, Garcia ML, Kaczorowski GJ (1986) Proc Natl Acad Sci USA 83:193–197
Ferry DR, Goll A, Glossmann H (1987) Biochem J 243:127–135
Finkelstein A, Andersen OS (1981) J Membrane Biol 59:155–171
Fleckenstein A (1977) A Rev Pharmac Tox 17:149–166
Fleckenstein A, Van Breemen C, Groß R, Hoffmeister F (1985) Cardiovascular effects of dihydropyridine type calcium antagonists and agonists. Springer, Berlin Heidelberg New York
Flockerzi V, Oeken HJ, Hofmann F (1986a) Eur J Biochem 161:217–224
Flockerzi V, Oeken HJ, Hofmann F, Pelzer D, Cavalié A, Trautwein W (1986b) Nature (Lond) 323:66–68
Fosset M, Lazdunski M (1987) In Venter JC, Triggle D (eds) Structure and physiology of the slow inward calcium channel. Liss, New York, pp 141–159
Glossmann H, Ferry DR, Striessnig J, Goll A, Moosburger K (1987) TIPS 8:95–100
Glossmann H, Striessnig J (1988) Vitamins and hormones, in press
Hanke W, Methfessel C, Wilmsen U, Boheim G (1984) Bioelectrochem Bioenerget 12:329–339
Hille B (1984) Ionic channels of excitable membranes. Sinauer, Sunderland, Mass.
Hofmann F (1980) J Biol Chem 255:1559–1564
Hosey MM, Borsotto M, Lazdunski M (1986) Proc Natl Acad Sci USA 83:3733–3737
Hosey MM, Barhanin J, Schmid A, Vandaele S, Ptasienski J, O'Callahan C, Cooper C, Lazdunski M (1987) Biochem Biophys Res Commun 147:1137–1145
Hosey MM, Barhanin J, Schmid A, Vandaele S, Ptasienski J, O'Callahan C, Cooper C, Lazdunski M (1988) In Costa E, Biggio G, Toffano G (eds) Voltage-sensitive ion channels: modulation by neurotransmitters and drugs. Liviana, in press
Katz AM, Reuter H (1979) Am J Cardiol 44:188–190
Lazdunski M, Schmid A, Romey G, Renaud JF, Galizzi JP, Fosset M, Borsotto M, Barhanin J (1986) J Cardiovasc Pharmacol 9:S10–S16
Lazdunski M, Schmid A, Romey G, Renaud JF, Galizzi JP, Fosset M, Coppola T, Borsotto M, Barhanin J (1987) In Tucek S (ed) Synaptic transmitters and receptors. Wiley, Chichester, pp 28–39
Lazdunski M, Barhanin J. Borsotto M, Cognard C, Cooper CL, Coppola T, Fosset M, Galizzi JP, Hosey MM, Mourre C, Renaud JF, Romey G, Schmid A, Vandaele S (1988) Ann N Y Acad Sci, in press

Lichtlen PR (1985) Recent aspects in calcium antagonism. Schattauer, Stuttgart
Ma J, Coronado R (1987) Biophys J 51,5a
KcKenna EJ, Smith JS, Ma J, Vilven J, Vaghy P, Schwartz A, Coronado R (1987) Biophys J 51,2
Miller RJ (1987a) Science (Wash DC) 235:46–52
Miller RJ (1987b) In Venter JC, Triggle D (eds) Structure and physiology of the slow inward calcium channel. Liss, New York, pp 161–246
Nelson MT, French RJ, Krueger BK (1984) Nature (Lond) 308:77–80
Nilius BP, Hess P, Lansman JB, Tsien RW (1985) Nature (Lond) 316:443–446
Nowycky MC, Fox AP, Tsien RW (1985) Nature (Lond) 316:440–443
Pauron D, Qar J, Barhanin J, Fournier D, Cuany A, Pralavorio M, Berge JB, Lazdunski M (1987) Biochemistry 26:6311–6315
Pelzer D, Cavalié A, McDonald TF, Trautwein W (1988). In Piper HM, Isenberg G (eds) Isolated adult cardiomyocytes, Vol II. CRC Press, Boca Raton, in press
Rios E, Brum G (1987) Nature (Lond) 325:717–720
Rosenberg RL, Hess P, Reeves JP, Smilowitz H, Tsien RW (1986) Science (Wash DC) 231:1564–1566
Schmid A, Barhanin J, Coppola T, Borsotto M, Lazdunski M (1986a) Biochemistry 25:3492–3495
Schmid A, Barhanin J, Mourre C, Coppola T, Borsotto M, Lazdunski M (1986b) Biochem Biophys Res Commun 139:996–1002
Schwartz LM, McCleskey EW, Almers W (1985) Nature (Lond) 314:747–751
Sieber M, Nastainczyk W, Zubor V, Wernet W, Hofmann F (1987) Eur J Biochem 167:117–122
Siemen D, Hellmann S, Maelicke A (1986) Springer, Berlin Heidelberg New York
Talvenheimo JA, Worley III JF, Nelson MT (1987) Biophys J 52:891–899
Tanabe T, Takeshima H, Mikami A, Flockerzi V, Takahashi H, Kangawa K, Kojima M, Matsuo H, Hirose T, Numa S (1987) Nature (Lond) 328:313–318
Trautwein W, Cavalié A, Flockerzi V, Hofmann F, Pelzer D (1987) Circ Res (suppl I), I17–I23
Tsien RW, Hess P, McCleskey EW, Rosenberg RL (1987) Ann Rev Biophys Chem 16:265–290
Uyeda N, Takenaka T, Aoyama K, Matsumoto M, Fujiyoshi Y (1987) Nature (Lond) 327:319–321
Vandaele S, Fosset M, Galizzi JP, Lazdunski M (1987) Biochemistry 26:5–9
Venter JC, Triggle D (1987) Structure and physiology of the slow inward calcium channel. Liss, New York
Vilven J, Leung AT, Imagawa T, Sharp AH, Campbell KP, Coronado R (1988) Biophys J 53:556a
Worley III JF, Deitmer JW, Nelson MT (1986) Proc Natl Acad Sci USA 83:5746–5750
Yatani A, Codina J, Imoto Y, Reeves JP, Birnbaumer L, Brown AM (1987) Science (Wash DC) 238:1288–1292

1,4-Dihydropyridines as Modulators of Voltage-Dependent Calcium-Channel Activity*

D. W. Chester[1,4] and L. G. Herbette[1-4]

Departments of [1]Medicine, [2]Radiology, [3]Biochemistry, and [4]Biomolecular Structure Analysis Center
University of Connecticut Health Center, Farmington, CT 06032, USA

Introduction

Receptor Stereospecificity

The interaction of 1,4-dihydropyridine (DHP) ligands with their receptors in a variety of tissues has been shown to be highly stereospecific. DHP analogs which differ in substitutions to either the dihydropyridine or aryl ring have markedly different activities and inotropic effects. Moreover, the observation that drug enantiomeric pairs possess opposing inotropic activities suggests a complex interconnection between the ligand structure (orientation/conformation) and the configuration of the receptor-binding domain. In general, these ligands have both high partition coefficients and binding affinities. The pathway of approach for the drug-receptor interaction may dictate the allowable drug conformations and orientations, which would then have an impact on the "success" of drug collisions with a stereoselective binding domain. In particular, ligands which partition into the membrane would be affected by the local microenvironment, producing an energy-minimized eqilibrium conformation and orientation. It is anticipated that the process of bilayer partition is rapid on the receptor-binding time scale, and as such, the conformational/orientational equilibrium would be established long before the actual binding event (see Fig. 3). Knowledge of the contributing factors for bilayer location, orientation, and conformation in concert with structure – function relationships and molecular modelling approaches might facilitate drug design efforts. Drug design, in turn, can have an impact on the clinical efficacy of these ligands as they are used to treat and control cardiovascular abnormalities.

The molecular mechanisms involved in DHP antagonist/agonist interaction with voltage-dependent calcium channels are relatively unknown. Further, there is uncertainty concerning the interplay between drug stereoselectivity (and opposing inotropic activity of DHP enantiomers), membrane potential, receptor activation state, and the membrane itself as modulators of channel function. We have been interested in ascertaining the time-averaged location of these drugs in the bilayer where drug-membrane (both lipid and receptor) interactions are maximized. As stated previ-

---

* This work was supported by research grants HL-33026 from the National Institutes of Health, by a grant from the Whitaker Foundation, and by a gift from the Patterson Trust Foundation and RJR Nabisco, Inc.

ously, we have reasoned that if the membrane bilayer location is an important parameter in facilitating receptor binding, the physical chemical characteristics of that locale could induce drug conformations which differ substantially from that observed in crystal structure determinations where drug-drug interactions are maximized. In fact, molecular modelling calculations for the antiarrhythmic drug, amiodarone, suggest that there are several solvent-dependent minimum energy conformations (including air/water interfacial interactions) (Chatelain et al. 1986). In addition, we have shown that the energy-minimized structure for amiodarone in an environment (low dielectric medium) similar to its membrane bilayer location differs significantly from the crystal structure (see Fig. 5 and Trumbore et al. 1988). More recently, Holtje and Marrer (1987), using a quantum-mechanical calculation approach, have suggested that the minimum energy DHP conformations and the resulting molecular electrostatic potentials may play a role in defining antagonist vs agonist action on calcium-channel gating, even in the case of enantiomeric pairs of DHP.

Drug/Receptor Characteristics

DHP calcium-channel antagonists inhibit voltage-dependent calcium-channel "opening" in cardiac sarcolemma and a variety of other membranes (Janis and Triggle 1984; Hess et al. 1984; Janis and Scriabine 1983). The specific binding of these drugs to the membrane receptor has been shown by Belleman and coworkers (1983) to be highly stereoselective, saturable, and reversible, with K_d values ranging between 0.1 and 5 nM (Janis and Triggle 1984), with intrinsic "on rates" of 10^7 (Ms)$^{-1}$ (Rhodes et al. 1985; Herbette et al. 1986). Moreover, Lee and Tsien (1983) have demonstrated that receptor state is an important parameter in DHP binding. It appears that these ligands may bind more preferentially and with greater affinity to the inactivated state of the receptor, which is modulated by membrane potential (Bean 1984; Lee and Tsien 1983) and, possibly, receptor phosphorylation (Flockerzi et al. 1986). Flockerzi et al. (1986) have shown that phosphorylation of the α_1-subunit (M_r 142000) of the T-tubule DHP receptor effects an increased probability of calcium-channel opening.

High binding affinity and receptor stereoselectivity affect both drug action and activity. Langs and Triggle (1985) have demonstrated that subtle changes in the pyridine ring puckering or torsion angles between the antiparallel pyridine and aryl rings have substantial effects on drug activity as calcium-channel modulators. Crystallographic studies on a series of nifedipine analogs reveal a limited correlation between increasing distortion from pyridine ring planarity and pharmacological activity (Langs and Triggle 1985; Triggle et al. 1980). In addition, the orientation of the carboxylate carbonyl substituent relative to the pyridine ring double bonds appears to be related to antagonist activity (Langs and Triggle 1985). A cis orientation places the ester-linked substituent forward, toward the aryl ring, with potential to stabilize its antiparallel orientation. Further, since no functionally active trans, trans DHP isomers have been identified, the cis orientation must convey some of the ligand activity (e. g., H bond potential).

Drug action is also markedly affected by substitution to both the pyridine and aryl rings. The *agonist* Bay K8644, containing a 3-nitro substituent to the pyridine ring, differs from a molecularly similar *antagonist* analog with a 3-carboxylate substitution

(Langs and Triggle 1985). Drug stereochemistry also impacts on activity since the enantiomeric pair of the DHP, Sandoz 202,791, exhibits both positive [S(+) isomer] and negative [R(−) isomer] inotropic activity (Hof et al. 1985; Kongsamut et al. 1985). Moreover, the S(+) *agonist* activity is converted to *antagonist* by membrane depolarization (Reuter et al. 1985). Thus, drug conformation and orientation are important criteria for consideration in the receptor-ligand interaction, as are the effects of membrane potential on the receptor site and functional state.

A reasonable amount of progress has been made in the characterization, purification, and reconstitution of the voltage-dependent calcium channel from a variety of membrane sources. Initial work by Venter et al. (1983) and Curtis and Catterall (1984, 1986) yielded a DHP receptor of M_r 210000–260000, with an α- (M_r 135000), β- (M_r 50000), and γ- (M_r 33000) subunit structure. While there is disagreement in the literature regarding molecular weights and subunit composition, it appears that the receptor is comprised of a $M_r \sim 175000$ glycoprotein (α_2- or δ-subunit) and nonglycosylated α_1, β-, and γ-subunits of $M_r \sim 170000$, 50000, and 33000, respectively (Sharp et al. 1987; Takahashi and Caterall 1987). The M_r 175000 glycoprotein, as identified by wheat germ peroxidase labelling, has been shown by a number of laboratories to migrate at $M_r \sim 150000$ under reducing conditions (use of dithiothreitol or mercaptoethanol), signifying that this subunit is coupled to a M_r 29000–32000 peptide via disulfide linkage(s).

It is interesting to note, however, that there are differences between DHP receptors isolated from different sources and with different detergent regimes. Horne et al. (1986) have isolated the DHP receptor from rat ventricular muscle with a variety of detergent regimes and determined an oligomeric complex molecular weight (M_r 370000) which differs from that cited above. Rosenberg and coworkers (1986) have demonstrated through combined patch clamp and planar lipid bilayer electrophysiological studies that there are differences in the electrical characteristics of T-tubule and cardiac sarcolemmal voltage-dependent channels. Differences were identified in the gating properties between these two systems, namely, conductance, channel activity and channel activation, and open times. While there appear to be some functional (Rosenberg et al. 1986) and some antigenic (Takahashi and Catterall 1987) differences between skeletal and cardiac DHP receptors, Takahashi and Catterall (1987) have shown that the two receptors share similar sedimentation coefficients, with S values of 21 and 21.6 for skeletal and cardiac channels, respectively. Recently, the complete amino acid sequence for the skeletal muscle DHP receptor was deduced from cDNA sequence analysis (Tanabe et al. 1987). This molecular biology approach offers a number of advantages for probing the receptor and its interaction with antagonists/agonists, not the least of which is the capability to synthesize in vitro the receptor for reconstitution studies. Moreover, it may become possible to examine bilayer spanning domains, possible sites for DHP-receptor binding, and the region that may be associated with the voltage sensor. This approach, in concert with structure, kinetic, and electrophysiological approaches, should allow us to gain a more thorough understanding of the receptor pharmacology.

Several groups have prepared both mono- and polyvalent antibodies against these various components for the purpose of immunoprecipitation. Antibodies (anti-140 and anti-32) prepared by Schmid et al. (1986) and Cooper and coworkers (1987) have demonstrated a clear relationship between the M_r 32000 peptide and the M_r 175000

DHP α_2-receptor glycoprotein. Moreover, they have used these antibodies to explore the structural similarities between receptor sources: skeletal muscle T tubules, smooth muscle, and cardiac sarcolemma. It is interesting that antibodies generated against epitopes on the M_r 170000 peptide (Leung et al. 1987) are capable of activating the calcium channel in planar lipid bilayer studies in the absence of added Bay K8644. However, antibody binding does not appear to interfere with ligand binding, suggesting that the antibody does not have access to the Bay K8644 binding site (Vilven et al. 1988). In yet another very interesting study, monoclonal antibodies derived against the skeletal muscle DHP receptor (α_1) have been shown to specifically inhibit the slow inward DHP-sensitive calcium current in BC3Hl myocytes in cell culture (Morton et al. 1988). Since these workers have demonstrated that their antibody does bind to a polypeptide in BD3Hl cells that is homologous to the DHP receptor in skeletal muscle, they postulate that this peptide is a functionally important component of the voltage-gated calcium channel rather than a voltage sensor.

One very exciting development is the successful development of a high-affinity idiotypic anti-DHP antibody by Campbell et al. (1987), which could be used to probe the receptor-binding domain. This antibody is very specific for DHP, with K_d values in the femtomolar range (60 fM) and no observed binding for other calcium-channel ligands such as phenylalkylamines and benzothiazepines. The production of an antibody which binds DHP with high affinity is similar in many respects to drug design modalities except that, in this case, the "receptor" is being "designed" to bind the ligand. The K_d range for this idiotypic DHP antibody could represent a lower limit for the "receptor"-ligand interaction. It is also very interesting to note that despite the high binding affinity, "soluble receptor" binding occurs, with a time frame required for saturation of 15–30 min.

Receptor activity may be modulated by phosphorylation since Curtis and Catterall (1986) observed that the α_2-subunit is phosphorylated in intact T-tubule membranes, and both α- and β-subunits may be phosphorylated in vitro by the catalytic subunit of the cAMP-dependent protein kinase. This observation has been disputed by Imagawa et al. (1987) since they demonstrated, under their conditions, that the α_2-glycoprotein is not phosphorylated by the kinase. Moreover, they have identified an endogenous kinase, not regulated by either Ca^{2+}/calmodulin or cAMP, that phosphorylates the M_r 170000 α_1- and M_r 52000 β-subunit proteins. Flockerzi et al. (1986) have suggested that cAMP-dependent protein kinase phosphorylation of the α_1-subunit (M_r 142000 in their system) is the functionally active site since the rate and extent of phosphorylation of this subunit is substantially greater than that observed for the β-subunit. In planar lipid bilayer reconstitution studies, they have shown that (a) the isolated receptor is functional as a classical L-type calcium channel, with single-channel conductances of 20 pS; (b) these channels could be modulated by channel agonists (Bay K8644) and antagonists (gallopamil); and (c) phosphorylation of the channel prolongs the channel open time and decreases the time interval between channel openings. These observations are consistent with those of Curtis and Caterall (1986) suggesting that the DHP receptor is the calcium channel itself since reconstituted receptor vesicles are active in facilitating DHP-antagonist-inhibitable calcium transport.

Bilayer Pathway Model for Receptor-Ligand Interaction

Rhodes and coworkers (1985) have proposed a two-step model, the *membrane pathway* model, for DHP-receptor binding, in which these drugs partition into and become oriented within the membrane bilayer before diffusing laterally on approach to the receptor-binding site (Fig. 1). This pathway is in contrast to an "aqueous pathway" involving random diffusion through the bulk water phase outside the bilayer and subsequent binding to an exposed receptor site. Calculations of theoretical diffusion-limited rates for anisotropic ligand-receptor binding indicate that the membrane pathway is 3 orders of magnitude faster than that of the aqueous pathway. The radioligand-measured DHP-binding rates of 10^7 $(Ms)^{-1}$ do not allow clear distinction between these two pathways since these values are within the range of the theoretical forward rate constants calculated for reasonably anisotropic ligands by an aqueous approach. From the preceding discussion, it is clear that neither the ligand nor its binding is isotropic in nature, and receptor binding is dependent upon several parameters in addition to diffusion. The calculated diffusion-limited rates do not take into account these other binding constraints such as receptor state and, as a result, only reflect the theoretical maximum diffusion-limited rates for binding. It also appears from the preceding discussion that the receptor-binding process is not diffusion limited. Therefore, the composite rates of the individual steps in the receptor-binding process must be considered if one is to gain a thorough understanding of the pathway of ligand approach to the receptor. In addition, the apparent anisotrophy of the DHP ligand could be used to explore the drug-bilayer physical chemical interactions required to facilitate the receptor-binding process.

High membrane partition coefficients and an equilibrium position at the hydrocarbon core/water interface observed for these drugs (Herbette et al. 1986) support the

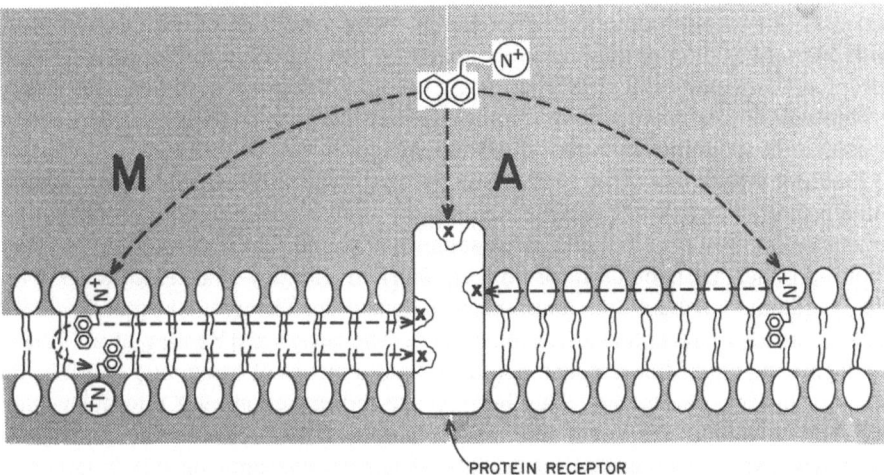

Fig. 1. Drug-receptor-binding models illustrating a single-step aqueous pathway to receptor binding *(A)* and a two-step bilayer pathway *(M)*. The components of the two-step model are ligand partition into the membrane followed by lateral diffusion and subsequent interaction with a binding domain within the membrane bilayer. An additional kinetic component could be associated with diffusion (flip-flop) across the bilayer leaflet

notion that bulk lipid partitioning is an integral first step in the overall receptor-binding mechanism. Moreover, the DHP, nimodipine, has been shown by neutron diffraction studies to diffuse ("flip flop") between the monolayers of the membrane bilayer (Herbette 1986). This observation is consistent with recent studies conducted by Affolter and Coronado (1985, 1986), demonstrating that both DHPs (Bay K8644) and phenylalkylamines (D600) readily cross the bilayer to interact with asymmetrically reconstituted calcium channels in planar lipid bilayer systems.

Drug Interactions with Receptors and Membrane Bilayers: Techniques for Study

Interaction and subsequent effects of drugs on membrane structure and function has been a subject of interest and study for some time. Several biophysical techniques have been used to approach this problem (for comprehensive review, see Goldstein 1984). The various techniques that have been applied involve (^{31}P, ^{2}H, ^{1}H, ^{13}C) nuclear magnetic resonance (NMR), electron spin resonance (ESR), differential scanning calorimetry, fluorescence spectroscopy, and X-ray and neutron diffraction. ^{2}H NMR studies were used to determine the position of the local anesthetic, benzyl alcohol, in the mid-acyl chain region of the membrane bilayer. Another local anesthetic, dibucaine, was shown in ^{31}P NMR studies to interact with the membrane surface (Cullis and De Kruijff 1979). Studies of phenothiazine (antipsychotic ligand, e.g., chlorpromazine)/membrane interactions using differential scanning calorimetry, in concert with ^{13}C and ^{31}P NMR studies, have shown that the phenothiazine dialkylaminoalkyl chains are oriented near the phospholipid headgroups, with the heterocyclic ring structure penetrating just beyond the glycerol backbone into the hydrocarbon region of the membrane (Frenzel et al. 1978). This bilayer position is similar to that observed by diffraction techniques for the β-adrenergic-receptor antagonist, propranolol (Herbette et al. 1983), which has similar structural characteristics. The location determined for the phenothiazine class of drug is consistent with a recent finding pertaining to its noncompetitive binding to the acetylcholine receptor (Giraudat et al. 1987). In these studies, it has been shown that chlorpromazine, as a noncompetitive channel blocker, was photoaffinity-labelled to the hydrophobic spanning domain of the β-subunit.

Traditional radioligand filtration assays have been used to measure receptor-ligand binding interactions. While these techniques are sufficient to study equilibrium conditions, the time-resolved capabilities are somewhat limited. The fluorescence spectroscopic approach has been used for both equilibrium (Bentley et al. 1985; Kolb et al. 1983) and time-resolved (for comprehensive review, see Shaafi and Fernandez 1983) measurements of receptor-ligand interaction. The review by Bentley et al. (1985) highlights the use of fluorescence polarization as a means of examining steady-state receptor-binding parameters. The fluorescence redistribution after photobleaching (FRAP) technique has been successfully employed to ascertain the diffusional dynamics of various molecules [including a calcium-channel antagonist (Chester et al. 1987)] in a variety of membrane systems (Koppel 1979; Shaffi and Fernandez 1983). Moreover, this technique offers the potential of assessing, simultaneously, diffusion rates associated with fast- and slow-diffusing components of the membrane bilayer (Koppel and Sheetz 1983).

We have employed both diffraction and fluorescence spectroscopic techniques in approaching the question of ligand location and kinetics associated with drug-membrane bilayer and receptor interactions. The time-averaged drug location within the bilayer was determined by X-ray and neutron diffraction approaches. In neutron diffraction studies, the use of isomorphous substitution techniques (identical drug structures except for the analogs being deuterated at different specific locations within the molecule) has the potential to allow both orientational and possibly conformational assignments for these ligands in the membrane bilayer and at receptor sites if the receptor protein density in the membrane can be made sufficiently high. These time-averaged structures can then be compared with both calculated quantum-mechanical minimum energy and crystal structures. The fluorescence studies examined the kinetics (time-resolved properties) of the microscopic events associated with the receptor-binding pathway. It should be possible, with tandem use of these techniques, to distinguish between a one-step aqueous route to receptor binding and a membrane bilayer pathway which involves bilayer partition, lateral diffusion, and receptor-ligand interaction for a variety of membrane protein receptors. Moreover, the fluorescence techniques offer the capacity to examine the relative effects of such things as membrane bilayer physical state on the receptor-binding pathway. In a preliminary study, Valdivia and Coronado (1988) have suggested that bilayer lipid composition affects the binding affinities of DHPs to reconstituted receptors in planar lipid bilayers. It is clear that bilayer physical state does affect integral membrane protein function (Chester et al. 1986), and this effect could be translated to the interaction of the ligands with the bilayer and the allowable equilibrium conformations/orientations. In cases where there are profound changes in bilayer physical state (e. g., changes in bilayer fluid dynamics, phospholipid class composition), the interaction between DHP and receptor in these membranes may possibly be significantly more constrained.

The research tools described above could become part of an integrated system of data collection and analysis to allow new breakthroughs in drug design. The processes would involve four major areas: single crystal diffraction studies, membrane diffraction studies in the presence of deuterated ligand, molecular modelling of the obtained data, and kinetic measurements of the rates of ligand interaction with the receptor under defined conditions. Based on the (dis)similarities between different analogs with regard to bilayer vs crystal conformation, membrane location/orientation and binding kinetics, etc. will allow calculations for potential site reactivity and binding affinity–the business ends of the molecular interaction. The molecular modelling routines could then be used to predict, based on the various physical chemical characteristics involved in the interaction, structures which could have enhanced efficacy for receptor binding. In this way, one could (a) effectively determine the structural and functional parameters necessary for drug-receptor interactions, (b) "map" out the essential molecular features necessary for this binding without a detailed structural analysis of the receptor site, and (c) bring together the pharmacokinetic and -dynamic parameters for a drug at the molecular level, all in the relevant environment in which the drugs act, namely, the membrane.

Experimental Methods of Approach

Sample Preparation

Multilayer membrane samples for diffraction and fluorescence (FRAP) studies were prepared by the spin dry method, as previously described in detail (Chester et al. 1987). Briefly, multilamellar vesicle preparations used in these studies were added to lucite sedimentation cells containing either a nonfluorescent Aclar (Dacron, Allied Chemical Co., NJ) or aluminum foil substrate, used respectively for FRAP or diffraction measurements. The vesicles were sedimented onto the substrate, after which the normal bucket caps were replaced with spin dry caps, and the pelleted vesicles spin dried under centrifuge vacuum at $65\,000 \times g$ for 5 h. On completion of the spin dry process, the samples were mounted and rehydrated over saturated salt solutions, which define specific relative humidities.

Diffraction

Small-Angle X-Ray and Neutron Diffraction

X-ray diffraction experiments with membrane lipid bilayers were carried out using a GX-18 rotating anode generator (Marconi Avionics, Borehamstead, England), equipped with a camera utilizing a single Franks' mirror, as previously described in detail (Herbette et al. 1985). The diffraction patterns were collected on either film (Kodak NS-5T) or a Braun position-sensitive proportional counting gas-flow detector (Innovative Technology Inc., South Hamilton, MA).

Samples were equilibrated to different relative humidities in sealed containers at 8°C. Diffraction peaks from samples at various relative humidities were integrated, and these appropriately phased structure factors (phasing performed by a swell analysis) were used to generate one-dimensional electron density profile structures.

Neutron diffraction studies were carried out at the Brookhaven National Laboratory High Flux Beam Reactor (Upton, NY). For these studies, specifically deuterated drug analogs were used as isomorphous replacements for protonated drug to assign ligand location as a function of the deuterium location. The structure factors were phased either by a swell analysis, as in the X-ray case, or by plots of the structure factor against D_2O/H_2O ratios to facilitate sample contrast variation.

Small-Molecule Crystallographic Studies

Selected crystals of different drugs were mounted on a glass fiber. All diffraction was collected using a TEXRAY/Rigaku system equipped with a RU-200 X-ray generator, AFC-5 diffractometer, and TEXRAY control software. The entire system was obtained from Molecular Structure Corp. (College Station, TX). Intensity data was collected by measuring counts at peak positions, and background corrected. Sample degradation was monitored by measuring the intensity of three standard reflections every 150 points. Crystal structures were solved by direct methods using the TEXSAN package of programs (Molecular Structures Corp., College Station, TX), including MITHRIL.

Fluorescence Studies

The fluorescence studies have involved both fluorescence redistribution after photobleaching (FRAP) and spectroscopy techniques as a means of addressing the time-resolved questions of ligand partition, lateral diffusion, and receptor binding. The FRAP studies used multilamellar stacks similar to that in the diffraction experiments. This allows us to work at very low probe-to-lipid concentrations while obtaining a very good signal-to-noise ratio. FRAP measurements of diffusion coefficients for $DiIC_{16}$ (phospholipid) and nisoldipine-lissamine rhodamine (Ns–R) were performed with an excitation wavelength of 5145 Å on an Ortholux II fluorescence microscope (E. Leitz, Inc., Rockleigh NJ) equipped with a vertical illuminator, water-cooled argon laser light source (Lexel) and a galavanometric scanning mirror. The basic design and geometry of the optical system has been previously described (Koppel 1979). The diffusion coefficients were determined as a function of relative bilayer hydration, as detailed by Chester et al. (1987). The initially uniform sample fluorescence was locally depleted (or bleached) by short (40-ms) exposure to a laser beam focused by a 10X objective to a small, circularly symmetric spot. The fluorescence redistribution after photobleaching was followed with a series of 12-point scans with an attenuated monitoring beam. The percent recovery and recovery rate were determined with a 3-parameter nonlinear least-squares analysis of the time decay of the fluorescence depletion monitored coincident with the position of the bleaching pulse.

The fluorescence spectroscopy studies to assess the rates of membrane partition and, ultimately, receptor binding, involve the use of unilamellar vesicles of either native membrane, pure lipid isolated from native membranes, or reconstituted membrane systems. Receptor reconstitution would allow us to address specifically the question of the microscopic binding rate constant for this process. Interestingly, the rate measurements for binding to a soluble receptor (idiotypic antibody) performed by Campbell et al. (1987) indicate that while there is high-affinity binding, approx. 15 min are required for the specific binding event. In this situation, the soluble ligand must diffuse through an aqueous pathway to bind with the receptor in solution. If the membrane functions to "set up" the ligand for binding, this rate would be expected to increase substantially. The fluorescence studies which to date have addressed ligand partition rates have involved the use of an SLM Aminco 8000 C.

Molecular Modelling Calculations of Drug Structures in the Membrane Bilayer

The molecular mechanics program MMP2, which assumes a homogeneous environment for the molecule, was used to estimate the minimum energy conformation of a ligand in the bilayer. Where necessary, due to the lack of needed parameters (e.g., iodine atoms) in the MMP2 parameter set, parameters were based on published values. Initial starting coordinates for these analyses were obtained from crystallographic X-ray diffraction data. In order to determine the minimum energy conformation, energy minima were computed in a dielectric range from k = 2 to k = 80. At k = 2, the dihedral driver option of MMP2 was then used to generate several energy-minimized sets of conformations related to the initial crystallographic conformation. These conformations were generated by torsion in 60° steps about bonds of interest. The total energies of the minimized structures derived from the crystal structures and the dihedral driver option were compared to determine whether a given structure was a local minimum. The conformation with the lowest total energy was deemed to be the global minimum for the study. The structure of lowest energy was then compared with the crystal structure using some common origin for the two structures, and the relative differences were noted. Some graphics and geometric calculations used CHEM-X (developed and distributed by Chemical Design Ltd., Oxford, UK).

Discussion of Experimental Results

As stated above, we have attempted to use the techniques of diffraction and fluorescence to assess both location and kinetics for a number of lipophilic ligands in bilayers. The basic tenet behind these studies is summarized in Fig. 1, which represents a model for the pathways by which these ligands may gain access to their specific receptors. There is a reasonable amount of evidence from our and other laboratories (Kokubun and Reuter 1984; Affolter and Coronado 1985; Vilven et al. 1988) to suggest that ligand partitioning into membrane bilayers is an integral first step in the receptor-binding process. In our studies, we have therefore considered the membrane bilayer pathway, as our basic working assumption.

Figure 2 illustrates the locations identified for a β-adrenergic antagonist (propranolol) and an uncharged DHP calcium-channel antagonist (nimodipine). We have been able to show that both DHP-receptor agonists (Bay K8644) and antagonists (nimodipine and Bay P8857) share a similar position within the bilayer. This is interesting in light of the suggestion that agonists and antagonists compete for the same binding domain on the receptor (Janis et al. 1984; Hamilton et al. 1987). The charged DHP-receptor antagonist, amlodipine, appears to be in the same general location, but it is positioned slightly higher in the bilayer toward the phosphate headgroups. This location is reasonable since the quaternary amine moiety of this ligand would be expected to interact with the zwitterionic charge plane of the headgroup region of the membrane bilayer in a fashion similar to propranolol. Drug partitioning to the discrete locale of the bilayer hydrocarbon core/water interface would suggest that the physical chemical characteristics (charge character, abrupt change in dielectric about the interface, structured water phase, hydrogen bond potential, etc.) of this interfacial region provide a minimum free energy environment for drug interaction. Moreover, these ligands have been shown to have membrane partition coefficients ranging between 1000 and 125000 consistent with the observed membrane locations. In our studies with amiodarone, we have determined that this drug is located below the bilayer interface. This location is consistent with its very high membrane partition coefficient of $\sim 10^6$. The residence half-time for this drug in the membrane bilayer is very long and is comparable to the observed long clinical half-life (Herbette et al. 1988). This would suggest that amiodarone is relatively inaccessible to the bulk aqueous phase. Other ligands which appear to have similar mid-acyl chain

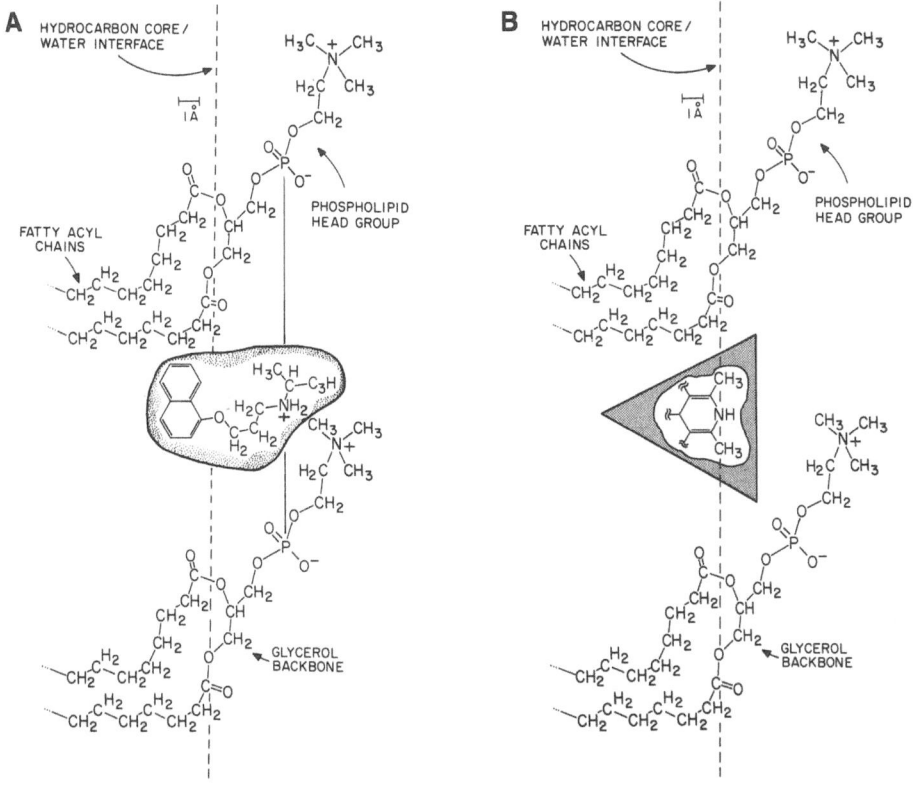

Fig. 2A, B. The relative locations of propranolol (**A**) and nimodipine (**B**) in a membrane bilayer, as determined, by X-ray and neutron diffraction

locations include the local anesthetic, benzyl alcohol (Cullis and De Kruijff 1979), and, in preliminary studies in our laboratory, the phenylalkylamines.

From the structure studies, then, it appears that drugs which interact with the membrane have well-defined time-averaged locations, which may or may not be related to the position of the receptor-binding domain along the bilayer normal. As such, the relative correlation between these two "sites" would have a dramatic effect on the effective ligand concentration at the binding site (see Fig. 3). This concentration gradient could yield membrane drug concentrations on the order of 10^3- to 10^4-fold higher than anticipated, especially when the ligand is distributed in a rather finite region about the hydrocarbon core/water interface of the membrane bilayer. As such, this would have a substantial impact on calculations of pharmacodynamic properties, traditionally derived from total solution ligand concentrations.

Figure 3 presents several model considerations which may have impact on drug pharmacokinetics as well as pharmacodynamics. Our initial hypothesis, highlighted in Fig. 1, is that there are two basic pathways that can be considered with respect to drug-receptor interaction: an aqueous pathway and a membrane bilayer pathway. In the

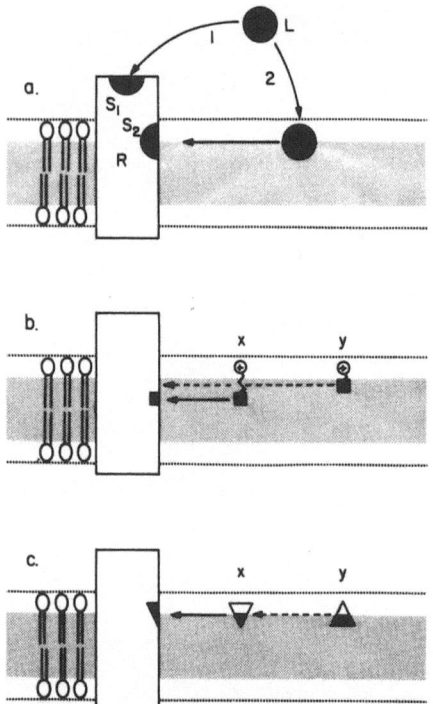
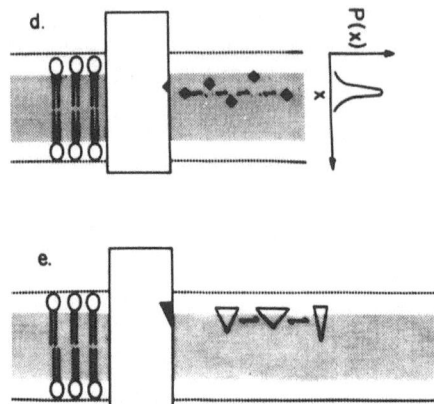

Fig. 3a–e. Molecular parameters that may define the basis for the design of membrane-active drugs. **a** Drug *(L)* may reach a receptor (R) binding site (S_1 or S_2) on the surface of the membrane via an aqueous route *(1)* or by a membrane bilayer pathway *(2)*. **b** The anisotropic physical chemical nature of the membrane along the bilayer normal may constrain the ligands (x or y) depth of penetration to a well-defined region where the local drug concentration is relatively high at the level of the receptor site. **c** Bilayer constraints on the drug (x or y) orientation may effect binding ability. **d** Diffusion about an equilibrium position in the membrane as a function of the ligand concentration gradient centered about the interface. **e** Ligand conformational effects as aa function of bilayer interaction

case of the acetylcholine receptor, the interaction appears to occur via an aqueous route, while the interaction of the noncompetitive inhibitor chlorpromazine occurs via the bilayer route (Giraudat et al. 1987).

Our structure data and those of others clearly support the concept that the membrane bilayer pathway may be operable for several amphipathic ligands and protein receptors in addition to the DHP receptor. With regard to the overall rates of receptor binding, however, there are a number of conditions that must be taken into consideration. McCloskey and Poo (1986) have shown that receptor site density is an important consideration when evaluating the advantage of one pathway over the other. These authors point out that at high receptor site density, binding rates via aqueous or membrane routes would essentially be inseparable. Conversely, at low receptor site density, there would be a substantial rate advantage gained by drug partitioning into the membrane bilayer. This is clearly the case with respect to DHP receptor interaction in cardiac sarcolemma, where the density is approx. one receptor site per square micrometer (Colvin et al. 1985). Moreover, these membranes have a substantially increased lipid-to-protein ratio (approx. 2–3 μmol lipid per milligram protein) compared with that observed with other membrane sources (0.6–0.8 μmol lipid per milligram protein) (Colvin et al. 1985).

Our structure data have also revealed that different drugs do not always occupy the same general membrane locale. While it is clear that the hydrocarbon core/water interface does provide a highly anisotropic environment which accommodates certain drug interactions, it is also clear that other drugs such as amiodarone are not found at this position in the membrane. Phenylalkylamines, as well, appear to interact at a position below the hydrocarbon core/water interface. Thus, for both amiodarone and phenylalkylamines, the bulk of the molecule is below the interfacial region.

Figure 3b illustrates the potential effects that bilayer structure may have on the depth of ligand penetration. Relative correlation between the receptor-binding domain location along the bilayer normal and the time-averaged drug location in the membrane bilayer would have direct impact on the ligand concentration at the level of the binding site. In some recent studies, we have noted that changes in bilayer physical state can affect the observed drug location in the membrane. In one case, the Lβ phase (gel or "frozen" state for the lipid fatty acyl chains) appears to exclude ligand from the membrane, while in another case the ligand is pushed further into the bilayer.

Figure 3c, e illustrates, the potential effects of bilayer location on drug orientation and conformation (see also Fig. 4). A myriad of structure-activity studies have been performed, demonstrating that, at least in the case of DHP, receptor binding is highly stereospecific. This stereospecificity mandates that, upon "presentation" to the receptor, the ligand would be in some reasonably well-defined conformation and orientation. Assuming that the ligand is tumbling free in solution and contains all motional degrees of freedom, an increased collision frequency would be required to obtain a "successful" interaction. This is supported by the observation that extremely high affinity DHP binding to idiotypic antibodies in solution requires approx. 15 min to reach 50% "receptor" saturation (Campbell et al. 1987). As discussed previously, DHP inotropic activity is dependent on ligand structure. The structure-activity relationships observed for DHP suggest that the orientation of specific portions of the molecule, such as the antiparallel configuration of the phenyl and pyridine rings, is important for ligand inotropic effect. The antiparallel configuration of these two rings in the DHP molecule appears to be stabilized by the carboxylate ester substituents and the corresponding orientation of the carbonyl relative to the pyridine ring double bonds (Langs and Triggle 1985).

The physical chemical characteristics of the drug interaction within the bilayer could induce energy-minimized drug conformations which differ substantially from that observed in single-crystal studies used to interpret structure-function correlates. Phospholipid packing constraints, charge characteristics, hydrogen bond potential, counterion interactions, etc. would be anticipated to constrain "allowable" conformations and orientations of the drug in the membrane bilayer. Figure 4a illustrates a situation where the ligand orientation in the bilayer is displaced away from the bilayer normal to accommodate fatty acyl and headgroup packing requirements, as well as quaternary amine alignment with the zwitterionic charge plane. Figure 4b suggests that the torsion angles about the various ring substituents may be altered as a function of interaction at the bilayer interface. Specifically, it may be possible that the carbonyl groups are oriented more toward the water phase, effecting a change in the ester substituent juxtaposition with the nitrophenyl ring and resultant changes in motional constraints on phenyl ring movement. We have reasoned that the equilibrium conformations and orientations would actually facilitate an increase in frequency of success-

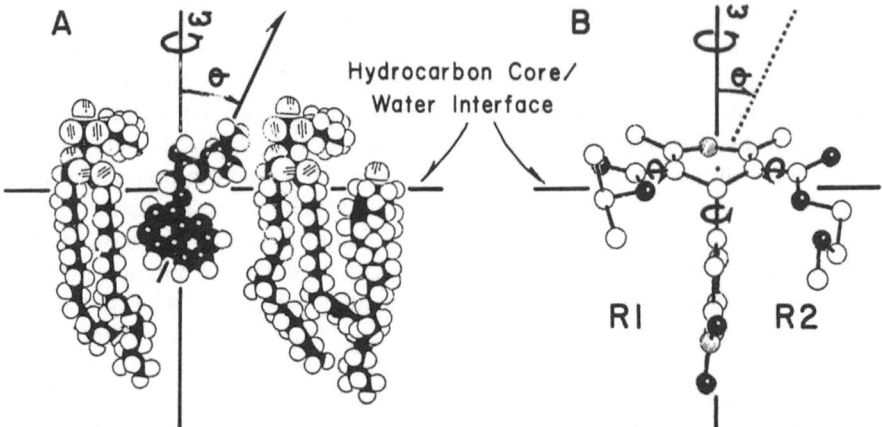

Fig. 4. A A potential drug orientation in the membrane bilayer such that all bilayer packing constraints and charge interactions could be satisfied. In this case, the drug (propranolol) could be tilted with a molecular axis θ° away from the bilayer normal. In addition, the drug could rotate in the bilayer about an axis parallel with the bilayer normal. **B** A DHP located at or near the hydrocarbon core/water interface. In this case, the molecule could be oriented as described in **A**. Additionally, hydrophobic/hydrophilic interactions at the interface could affect the relative orientation of the carboxylate carbonyl groups, and hence, the ester substituents, allowing for alterations in motional constraints on the antiparallel nitrophenyl ring. The combined effects could result in ligand structures which differ from that observed in single-crystal structure studies

ful collisions with the receptor, due in part to the reduced degrees of freedom allowed by drug partition to a discrete location in the bilayer. This discrete location and its relative impact on drug-receptor interaction are highlighted in Figs. 2, 3d. Thus, the orientation and conformation of a drug in a membrane bilayer could be highly coupled to the location that the drug assumes in this membrane. This bilayer structure for the drug (in its most relevant environment) is what the receptor "sees."

It is clear that the physical chemical characteristics of the time-averaged bilayer location of a drug would have some potential ramifications on the orientation and conformation of the ligand in this environment. This conformation could be substantially different than that observed in crystal structure studies, where drug-drug interactions are maximized. While we are attempting to determine directly the conformation and orientation of ligands in membranes using neutron diffraction techniques, it was of interest to examine quantum-mechanically the effects of drug environment on the ligand's conformation. For these studies, minimum energy conformations for amiodarone in a synthetic membrane bilayer system were calculated for several different values of dielectric constant, using the MMP2 program, which assumes a homogeneous environment for the molecule. Crystal structure data were used for the initial starting coordinates. Since electronic interactions made only minor contributions to the total energy of the structure, increasing dielectric constant (k) from $k = 2$ to $k = 80$ had a negligible effect on the calculated conformation. As discussed above, amiodarone assumes a time-averaged location deep within the bilayer. Therefore, $k = 2$ was used to calculate the minimum energy structure of this

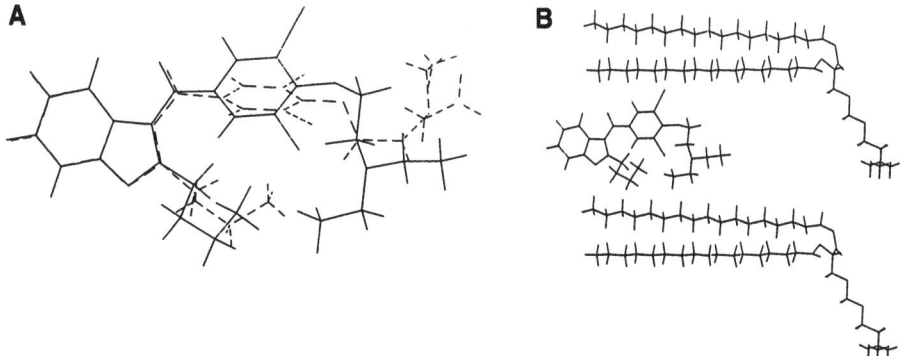

Fig. 5. A The minimum energy structure of amiodarone *(dashed bonds)* calculated in a low dielectric constant medium with the program MMP2 superimposed on the crystal structure *(solid bonds)* of the drug. **B** Amiodarone incorporated into a membrane bilayer, showing the position of the iodine atoms relative to the acyl chains and the size of the molecule relative to the hydrocarbon core region

drug in the bilayer. Our starting assumption is that the bilayer hydrocarbon core can be represented as an isotropic solution. While we realize the limitations to this assumption, changes in the drug conformation under these conditions would be anticipated to be even more constrained when acyl chain interactions are taken into account.

Energy minimization of amiodarone under these conditions demonstrated that significant differences could be anticipated in the membrane-associated drug structure, as compared with the crystallographic drug structure. The lowest energy structure showed an overall RMS shift of 1.36 Å from the initial crystal structure. The largest change was a reorientation of the benzofuran ring with respect to the central 3,5-diiodophenyl ring. In the crystal structure, the torsion angle between these two rings was 81.84°. In the energy-minimized bilayer structure, the benzofuran ring was more nearly parallel to the diiodophenyl ring, with torsion angle of 40.85° as shown in Fig. 5 (Trumbore et al. 1988). Shifts in the conformation of other regions of the molecule were not as pronounced, but were significant. From these observations, we suggest that for membrane-active drugs, the structure of the drug in the bilayer is not necessarily the same as its crystal structure. Thus, drug design modalities for membrane-active drugs that rely on some structure-function activity information should be based on a structure data base that is derived from the relevant environment in which these drugs reside.

Ligand partition to the bilayer hydrocarbon core/water interface would, on a time-averaged basis, produce a steep concentration gradient of drug about this interface, as illustrated in Fig. 6. With regard to our proposed bilayer pathway model, either the membrane is functioning as an effective "sponge" or drug partitioning into the bilayer is an integral first step in the receptor-binding process. Since it appears from a number of studies (Affolter and Coronado 1985, 1986; Kokubun and Reuter 1984) with both DHP and phenylalkylamine calcium-channel blockers that receptor-binding site access must be gained via ligand partitioning into the bilayer, these data strongly support the latter contention for the nonspecific time-averaged drug location; that is,

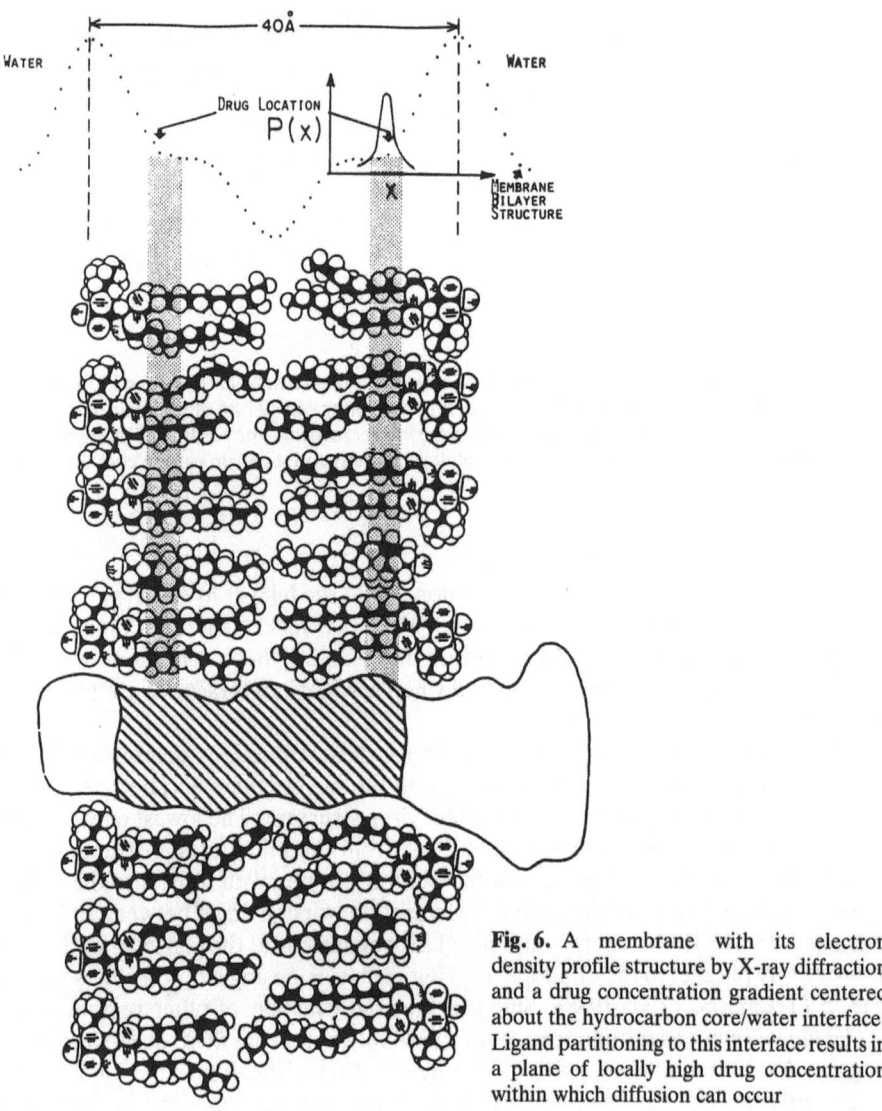

Fig. 6. A membrane with its electron density profile structure by X-ray diffraction and a drug concentration gradient centered about the hydrocarbon core/water interface. Ligand partitioning to this interface results in a plane of locally high drug concentration within which diffusion can occur

partition is the first step in the receptor-binding mechanism. As such, drug partitioning to the membrane bilayer hydrocarbon core/water interface would produce a plane of locally high drug concentration, within which lateral diffusion would occur.

We have used the fluorescence redistribution after photobleaching (FRAP) technique with a fluorescent DHP, nisoldipine-lissamine rhodamine (Ns-R, Fig. 7a), to assess the diffusional dynamics of these ligands in pure cardiac sarcolemmal lipid bilayers. The drug diffusion data were compared with phospholipid diffusion using the probe $DiIC_{16}$ (Fig. 7a). As the bilayers were maximally hydrated, the diffusion coefficients calculated for both the Ns-R and phospholipid were essentially identical, with a value of 3.8×10^{-8} cm^2/s (Fig. 7b). Further studies with this and other probes

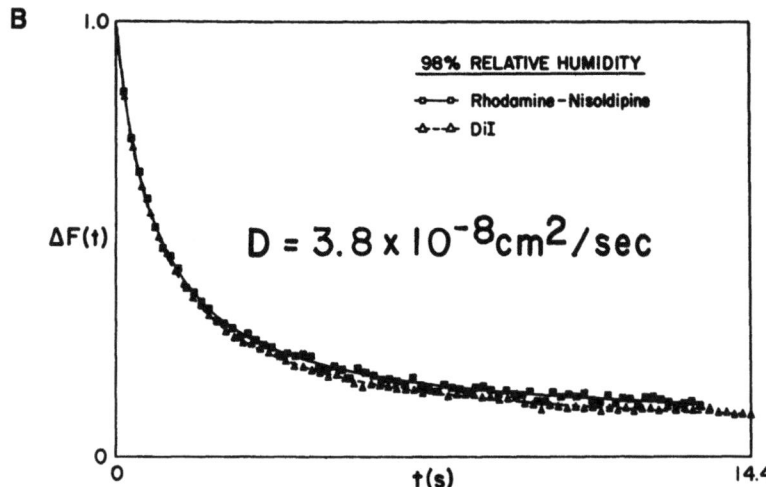

Fig. 7A, B. A Structures for nisoldipine-lissamine rhodamine (Ns–R) and DiIC$_{16}$ probes used in FRAP experiments for determination of drug diffusion coefficients. **B** Three-parameter least-squares fits of data obtained for Ns–R and DiIC$_{16}$ at near-maximum bilayer hydration (98% relative humidity), demonstrating similarity in diffusion coefficients *(D)* for these two probes. Δ F (t) represents change in fluorescence at time t

(Mason and Chester 1988) have demonstrated that this value is a reasonable lower limit for drug diffusion in these membranes. We are currently in the process of determining the partition rates for these compounds into the membrane bilayer. Preliminary data indicate that drug partition is rapid, with a half-time to completion of 1–3 s (as yet unpublished observations). As such, we can begin to assess time constants for ligand partitioning into the bilayer and diffusion within the bulk lipid matrix en route to specific receptor binding.

While we have shown data and speculated on what we interpret is the pathway for drug interaction with membrane receptors, it is interesting to speculate on the mechanism by which certain drugs exhibit differential effects on calcium-channel activity as a function of membrane potential. There have been several cases in which opposing inotropic activity of drug enantiomers has been observed (Hof et al. 1985; Kongsamut et al. 1985; Reuter et al. 1985). Moreover, in some cases (e.g., Sandoz 202,791 compound), changes in membrane potential can change agonist activity into antagonist activity (Reuter et al. 1985). Changes in membrane potential would not be expected to have their greatest effect on the structure or physical chemical environment of the membrane and hence the drug's environment. Thus, the drug orientation and conformation would be expected to remain constant, as constrained by its membrane environment. However, the greatest effect would be expected to have impact on the receptor protein. This effect may occur as a function of a voltage sensor which interacts with the DHP receptor (or is within the receptor structure itself), thereby changing the binding interaction with agonists or antagonists.

Recently, a quantum-mechanical modelling paper showed that, as a function of minimum energy drug conformation, one could differentiate between agonists and antagonist in the electrostatic potential over the 5' carboxylate of the DHP (Höltje and Marrer 1987). They reasoned that the different electrostatic charge potentials for agonists and antagonist could affect the potential of the protein receptor. To examine this hypothesis, they chose to model the interaction of tryptophan with the DHP since this amino acid residue would tend to induce positive electrostatic potentials for a ligand in close proximity. Their conclusion is that the electrostatic charge potential for the tryptophan residue was influenced, in direct proportion, by the electrostatic charge potential of the DHP. One may postulate then that DHP interaction within the receptor-binding domain may involve hydrophobic stacking interactions between the ligand and aromatic amino acid residues in the binding pocket. One may envision a myriad of hydrogen bond potentials between the ligand and charged amino acid residues in the binding pocket as well. Rebek (1987), in a very interesting publication, examined model building for molecular recognition. In his studies, a variety of parameters were assessed, including binding forces, site selectivity, substrate character, microenvironment, etc., which would be expected to affect binding of substrate into a molecular recognition cleft. Some of these considerations may be useful in evaluating the interaction of DHP with their receptors in the membrane and building new hypotheses regarding mechanisms for ligand inotropic activity as well as drug design.

Conclusions

An understanding of the molecular mechanisms of drug binding to membrane proteins requires both that the structure of the drug in different environments be determined and that the kinetics of interaction with these different environments be known. In addition, a static and dynamic picture of how drugs interact with cell membranes can be used in the design of agents which exhibit an optimized pharmacokinetic and -dynamic profile. With the advent of new biophysical approaches, these molecular parameters can now be determined. Thus, a rational, or should we say a molecular, approach to both drug design and drug delivery may eventually be secured.

Acknowledgements. We would like to thank Dr. D.G. Rhodes for his invaluable contribution during discussions pertaining to this manuscript. Also we would like to thank Drs. J. Moring and S.D. Wang for providing the crystallographic coordinants used in the molecular mechanics calculations and Dr. A. Katz for continual discussions regarding drug-membrane interactions which allowed us to relate these results to clinical questions. The Chem X package used in generating Fig. 5 is developed and distributed by Chemical Design LTD Oxford, UK. We would also like to thank our graduate students M. Trumbore and R.P. Mason for their assistance in these studies. They were supported by a Health Center Research Advisory Committee Graduate student fellowship. Dr. L. Herbette acknowledges his current affiliation as an Established Investigator of the American Heart Association. The work was carried out in the Biomolecular Structure Analysis Center at the University of Connecticut Health Center. We would like to thank the staff of the Structure Center for their dedication in keeping the facilities in optimal running condition. We would also like to thank Ms. Tammy Wojtusik for preparing the final form of this manuscript for submission. The Structure Center acknowledges support from the State of Connecticut Department of Higher Education's High Technology Project and Grant Program.

References

Affolter H, Coronado R (1985) Agonists of Bay K8644 and CGP 28392 open calcium channels from skeletal muscle transverse tubules. Biophys J 48:341–357

Affolter H, Coronado R (1986) The sidedness of reconstituted calcium channels from muscle transverse tubules as determined by D-600 and D-890 blockade. Biophys J 49:197a

Bean BP (1984) Nitrendipine block of cardiac calcium channels: high-affinity binding to the inactivated state. Proc Natl Acad Sci USA 81:6388–6392

Belleman P, Schade A, Towart R (1983) Dihydropyridine receptor in rat brain labeled with [^{3}H] nimodipine. Proc Natl Acad Sci USA 80:2356–2360

Bently KL, Thompson LK, Klebe RJ, Horowitz PM (1985) Fluorescence polarization: a general method for measuring ligand binding and membrane microviscosity. Biotechniques 3:356–366

Campbell KP, Sharp AH, Kahl SD (1987) Anti-dihydropyridine antibodies exhibit [^{3}H] nitrendipine binding properties similar to the membrane receptor for the 1,4-dihydropyridine Ca^{2+} channel antagonists. J Cardiovasc Pharm 9:S113–S121

Chatelain P, Ferreira J, Laurel R, Ruysschaert JM (1986) Amiodarone induced modifications of the phospholipid physical state. A fluorescence polarization study. Biochem Pharm 35:3007–3013

Chester DW, Tourtellotte ME, Melchior DL, Romano AH (1986) The influence of saturated fatty acid modulation of bilayer physical state on cellular and membrane structure and function. Biochim Biophys Acta 860:383–398

Chester DW, Herbette LG, Mason RP, Joslyn AF, Triggle DJ, Koppel DE (1987) Diffusion of dihydropyridine calcium channel antagonists in cardiac sarcolemmal lipid multibilayers. Biophys J 52:1021–1030

Colvin RA, Ashavaid TF, Herbette LG (1985) Structure function studies of canine cardiac sarcolemmal membranes. I. Estimation of receptor site densities. Biochim Biophys Acta 812:601–608

Cooper CL, Vandaele S, Barhanin J, Fosset M, Lazdunski M, Hosey MM (1987) Purification and characterization of the dihydropyridine-sensitive voltage-dependent calcium channel from cardiac tissue. J Biol Chem 262:509–512

Cullis PR, De Kruijff B (1979) Lipid polymorphism and the functional roles of lipids in biological membranes. Biochim Biophys Acta 559:399–420

Curtis BM, Catterall WA (1984) Purification of the calcium antagonist receptor of the voltage-sensitive calcium channel from skeletal muscle transverse tubules. Biochemistry 23:2113–2118

Curtis BM, Catterall WA (1986) Reconstitution of the voltage-sensitive calcium channel purified from skeletal muscle transverse tubules. Biochemistry 25:3077–3083

Flockerzi V, Oeken H-J, Hofmann F, Pelzer D, Cavalie A, Trautwein W (1986) Purified dihydropyridine-binding site from skeletal muscle T-tubules is a functional calcium channel. Nature 323:66–68

Frenzel J, Arnold K, Nuhn P (1978) Calorimetric, ^{13}C NMR and ^{31}P NMR studies on the interaction of some phenothiazine derivatives with dipalmitoyl phosphatidylcholine model membranes. Biochim Biophys Acta 507:185–197

Giraudat J, Dennis M, Heidmann T, Haumont PY, Lederer F, Changeux JP (1987) Structure of the high-affinity binding site for noncompetitive blockers of the acetylcholine receptor. [^{3}H]Chlorpromazine labels homologous residues in the beta and delta chains. Biochemistry 26:2410–2418

Goldstein DB (1984) The effects of drugs on membrane fluidity. Ann Rev Pharm Toxicol 24:43–64

Hamilton SL, Yatini A, Brush K, Schwartz A, Brown AM (1987) A comparison between the binding and electrophysiological effects of dihydropyridines on cardiac membranes. Mol Pharm 31:221–231

Herbette LG, Katz AM, Sturtevant JM (1983) Comparison of the interaction of propranolol and timolol with model and biological membrane systems. Mol Pharm 24:259–269

Herbette LG, MacAlister T, Ashavaid TF, Colvin RA (1985) Structure-function studies of canine cardiac sarcolemmal membranes. II. Structural organization of the sarcolemmal membrane as determined by electron microscopy and lamellar X-ray diffraction. Biochim Biophys Acta 812:609

Herbette LG, Chester DW, Rhodes DG (1986) Structural analysis of drug molecules in biological membranes. Biophys J 49:91–93

Herbette LG, Trumbore M, Chester DW, Katz AM (1988) Possible molecular basis for the pharmacokinetics and pharmacodynamics of three membrane-active drugs: propranolol, nimodipine and amiodarone. J Mol Cell Card, in press

Hess P, Lansman JB, Fox AP, Nowycky MC, Nilius B, McCleskey EW, Tsien RW (1985) Calcium channels: mechanisms of modulation and ion permeation. J Gen Physiol 86:5a

Hof RP, Reugg UT, Hof A, Vogel A (1985) Stereoselectivity at the calcium channel: opposite action of the enantiomers of a 1,4-dihydropyridine. J Cardiovasc Pharm 7:689–693

Höltje H-D, Marrer S (1987) A molecular graphics study on structure-action relationships of calcium-antagonistic and agonistic 1,4-Dihydropyridines. J Computer-Aided Mol Design 1:23–30

Horne WA, Weiland GA, Oswald RE (1986) Solubilization and hydrodynamic characterization of the dihydropyridine receptor from rat ventricular muscle. J Biol Chem 261:3588–3594

Imagawa P, Leung AT, Campbell KP (1987) Phosphorylation of the 1,4-dihydropyridine receptor of the voltage dependent calcium channel by an intrinsic protein kinase in isolated triads from rabbit skeletal muscle. J Biol Chem 262:8333–8339

Janis RA, Rampe D, Sarmiento JG, Triggle DJ (1984) Specific binding of a calcium channel activator, [^{3}H] Bay K8644, to membranes from cardiac muscle and brain. Biochem Biophys Res Comm 121:317–323

Janis RA, Scriabine A (1983) Sites of action of Ca^{2+} channel inhibitors. Biochem Pharmacol 32:3499–3507

Janis RA, Triggle DJ (1984) 1,4-Dihydropyridine Ca^{2+} channel antagonists and activators: a comparison of binding characteristics with pharmacology. Drug Dev Res 4:5425–5437

Kokubun S, Reuter H (1984) Dihydropyridine derivatives prolong the open state of Ca^{++} channels in cultured cardiac cells. Proc Natl Acad Sci USA 81:4824–4827

Kolb VM, Koman A, Terenius L (1983) Fluorescent probes for opioid receptors. Life Sci 33:423–426

Kongsamut S, Kamp TJ, Miller RJ, Sanquinetti MC (1985) Calcium channel agonist and antagonist effects of the stereoisomers of the dihydropyridine 202-791. Biochem Biophys Res Comm 130:141–148

Koppel DE (1979) Fluorescence redistribution after photobleaching. A new multipoint analysis of membrane translational dynamics. Biophys J 28:281–292

Koppel DE, Sheetz MP (1983) A localized pattern photobleaching method for the concurrent analysis of rapid and slow diffusion processes. Biophys J 43:175–181

Langs DA, Triggle DJ (1985) Conformational features of calcium channel agonist and antagonist analogs of nifedipine. Mol Pharm 27:544–548

Lee KS, Tsien RW (1983) Mechanism of calcium channel blockade by verapamil, D-600, diltiazem and nitrendipine in single dialysed heart cells. Nature 302:790–794

Leung AT, Imagawa T, Campbell KP (1987) Structural characterization of the 1,4-dihydropyridine receptor of the voltage-dependent Ca^{2+} channel from rabbit skeletal muscle. Evidence for two distinct high molecular weight subunits. J Biol Chem 262:7943–7946

Mason RP, Chester DW (1988) Diffusional dynamics of an active rhodamine-labelled 1,4-dihydropyridine in sarcolemmal lipid multibilayers. Effects of the rhodamine substituent and multibilayer dehydration on ligand diffusion. Submitted Manuscript

McClosky M, Poo M-M (1986) Rates of membrane associated reactions: reduction of dimensionality revisited. J Cell Biol 102:88–96

Morton ME, Caffrey JM, Brown AM, Froehner SC (1988) Monoclonal antibody to the $alpha_1$-subunit of the dihydropyridine-binding complex inhibits calcium currents in BC3H1 myocytes. J Biol Chem 263:613–616

Rebek Jr, J (1987) Model studies in molecular recognition. Science 235:1478–1484

Reuter H, Porzig H, Kokobun S, Prod'hom B (1985) Voltage dependence of dihydropyridine ligand binding and action in intact cardiac cells. J Gen Physiol 86:5a

Rhodes DG, Sarmiento JG, Herbette LG (1985) Kinetics of binding of membrane-active drugs to receptor sites. Diffusion limited rates for a membrane bilayer approach of 1,4-dihydropyridine calcium channel antagonists to their active site. Mol Pharm 27:612–623

Rosenberg RL, Hess P, Reeves JP, Smilowitz H, Tsien RW (1986) Calcium channels in planar lipid bilayers: insights into mechanisms of ion permeation and gating. Science 231:1564–1566

Shaafi RI, Fernandez SM (1983) Fast methods in physical biochemistry and cell biology. Elsevier, Amsterdam

Schmid A, Barhanin I, Coppola T, Borsotto M, Lazdunski M (1986) Immunochemical analysis of subunit structures of 1,4-dihydropyridine receptors associated with voltage-dependent Ca^{2+} channels in skeletal, cardiac, and smooth muscles. Biochemistry 25:3492–3495

Sharp AH, Imagawa T, Leung AT, Campbell KP (1987) Identification and characterization of the dihydropyridine-binding subunit of the skeletal muscle dihydropyridine receptor. J Biol Chem 262:12309–12315

Takahashi M, Catterall WA (1987) Dihydropyridine-sensitive calcium channels in cardiac and skeletal muscle membranes: studies with antibodies against the alpha subunits. Biochemistry 26:5518–5526

Tanabe T, Takeshima H, Mikami A, Flockerzi V, Takahashi H, Kangawa K, Kojima M. Matsuo H, Hirose T, Numa S (1987) Primary structure of the receptor for calcium channel blockers from skeletal muscle. Nature 328:313–318

Triggle AM, Shefter E, Triggle DJ (1980) Crystal structures of calcium channel antagonists: 2,6-dimethyl-3,5-dicarbomethoxy-4- [2-nitro-, 3-cyano, 4-(dimethylamino)-, and 2,3,4,5,6-pentafluorophenyl]-1,4-dihydropyridine. J Med Chem 23:442–445

Trumbore MW, Chester DW, Rhodes D, Herbette LG (1988) The structure and location of amiodarone in a membrane bilayer as determined by molecular mechanics and quantitative X-Ray diffraction. Biophys J, in press

Venter JC, Fraser CM, Schaber JS, Yung CY, Bolger G, Triggle DJ (1983) Molecular properties of the slow inward calcium channel. I. Molecular weight determination by radiation inactivation and covalent affinity labeling. J Biol Chem 258:9344–9348

Valdivia H, Coronado R (1988) Pharmacological profile of skeletal muscle calcium channels in planar lipid bilayers. Biophys J 53:555a

Vilven J, Leung AT, Imagawa T, Sharp AH, Campbell KP, Coronado R (1988) Interaction of calcium channels of skeletal muscle with monoclonal antibodies specific for its dihydropyridine receptor. Biophys J 53:556a

Dihydropyridine Pharmacology of the Reconstituted Calcium Channel of Skeletal Muscle*

H. Valdivia and R. Coronado

Department of Physiology and Molecular Biophysis, Baylor College of Medicine, Houston, TX 77030, USA

Introduction

Gating of dihydropyridine-sensitive calcium channels in cell-free systems such as the planar bilayer is dependent on the constant presence of agonist DHPs, typically micromolar levels of Bay K8644 (Affolter and Coronado 1985, 1986; Coronado and Affolter 1986a, b; Rosenberg et al. 1986; Ehrlich et al. 1986; Coronado and Smith 1987; Coronado 1987; Ma and Coronado 1988; Vilven et al. 1988; Valdivia and Coronado 1988). The agonist not only increases the lifetime of the open channel but, most important, reduces the run-down or time-dependent decrease in activity once a recording is initiated. In one extreme case, the calcium channel of skeletal muscle transverse (T) tubules, Bay K8644 completely eliminated the need for voltage pulse protocols. Evidently, the drug stabilizes the channel in the open conformation for long periods and over a wide range of holding potentials (Affolter and Coronado 1985). In this channel, we have found it possible to record up to 30 min of steady-state activity at 0 mV, with the limitation in most instances being the durability of bilayers.

We have just begun to appreciate the advantages of recording calcium channel under steady state. One purely technical advantage is the shear amount of data that can be collected when open probability is high and invariant with time, as when Bay K8644 is present. From a pharmacological perspective, which this contribution adopts, steady-state activation permitted us to construct dose-response curves of single channels for DHPs and to compare them with radioligand-binding affinities of the same under strictly *identical* conditions, i.e., same solutions, membrane source, and membrane potential. The latter is significant since it has been recognized that binding affinities of DHPs are voltage dependent (Sanguinetti and Kass 1984; Bean 1984). To our knowledge, this type of comparison has not been possible in the past simply because binding experiments are done under stationary conditions but calcium channel measurements are not. Of necessity, calcium currents of cells are measured using pulse protocols, during which channels open transiently and inactivate on a time scale of milliseconds (Tsien 1983). In contrast, a ligand-binding measurement requires equilibration of the free and ligand-bound forms of the receptor. For DHPs,

---

* This work was supported by National Institute of Health grants GM-36852 and HL-37044, Grants-in-Aid from the American Heart Association and the Muscular Dystrophy Association of America, and by an Established Investigatorship from the American Heart Association to R.C.

association and dissociation rates at nanomolar concentrations are slow, on the time scale of seconds (Rhodes et al. 1985).

Ligand-binding parameters and electrophysiological consequences of DHPs have been compared in numerous studies of cardiac cells, but there are large numerical discrepancies. In general, calcium channels are less sensitive to DHPs than expected from ligand-binding experiments (Lee and Tsien 1984; Janis and Triggle 1984; Janis et al. 1984a, b, 1985; Kokubun et al. 1987; Hamilton et al. 1987). Numbers of DHP receptors and numbers of calcium channels per cell differ by orders of magnitude both in skeletal and cardiac muscle (Green et al. 1985; Schwartz et al. 1985). Other complications arise from the fact that some DHPs are racemic mixtures where each pure enantiomer has opposite electrophysiological effects (Williams et al. 1985; Kokubun et al. 1987). Consequently, the pharmacological relation between calcium channels and DHP receptors still remains vague despite many efforts. Our results on purified skeletal muscle membranes indicate a strong quantitative correlation between the pharmacology of single calcium channels and that of DHP receptors for a set of ten DHPs, some racemic and some pure enantiomers. Except for nitrendipine, we find a one-to-one correspondence between binding affinities and dose-response curves of channels at 0 mV. Evidently, during membrane purification and planar bilayer reconstitution, the channel and the drug receptor complex remain tightly coupled. This suggests that the DHP-sensitive calcium channel and DHP receptor may share common components or that they may represent different domains of the same structure.

Materials and Methods

Planar Bilayer Experiments

Lipid bilayers were cast from an equimolar mixture of PE and PS dissolved in decane at a concentration of 20 mg lipid/ml. Lipid solution was spread across a 300-μm-diameter polystyrene aperture separating two aqueous chambers designated *cis* and *trans*. The volume of each chamber was 3.0 ml. T-tubule vesicles (10–50 μg) were added to the cis solution under stirring. Cis and trans solutions were always the same; cis 100 mM BaCl$_2$, 50 mM NaCl, 10 mM Hepes-Tris (pH 7.0), and 1 μM racemic Bay K 8644 (unless otherwise specified); trans 50 mM NaCl, 10 mM Hepes-Tris (pH 7.0). All experiments were performed at room temperature. Cis solution was connected via an Ag/AgCl electrode and an agar/KCl bridge to the head-stage input of a List L/M EPC 7 amplifier (List Electronic DA-Eberstadt, FRG). Trans solution was held at ground potential using the same electrode arrangement. Records were filtered at 0.1 KHz corner frequency on an eight-pole Besel (Frequency Devices, Springfield, MSS), digitized at 1 point/ms on a 12-bit A/D converter (Kiethley Instruments, Cleveland, Ohio), and fed into an IBM PC/AT computer (IBM Instruments, Danbury, Conn). Drugs were added to either side from stock solutions dissolved in 100% high-quality methanol. Final methanol concentration in the chamber was always <1%. Control experiments showed that methanol, at the concentrations used, had no effect on channel activity.

Receptor-binding Assays

In standard equilibrium binding experiments, samples of 15–40 μg protein/ml were incubated at room temperature (26°C) in 1 ml of a solution containing 50 mM NaCl, 10 mM Tris-HCl (pH 7.4), and the required concentration of (+)-(methyl-^{3}H)PN200–110 (0.05–7 nM). Specific binding was defined as the amount of radiolabel that could be displaced competitively by 1 μM cold PN200–110 or nitrendipine. In all binding experiments, (^{3}H)PN200–110 was the last compound added. Incubation time was 40–60 min. For determination of total amount of (^{3}H)PN200–110 present, a small aliquot (20 μl) was removed from samples before filtration. Binding was terminated by rapid filtration on Whatman GF/B or GF/F glass fiber filters. Filters were washed twice with 5 ml of an ice-cold solution containing 20 mM Tris-HCl (pH 7.2) and 200 mM choline chloride. Nonspecific binding to filters was negligible and independent of the presence of unlabeled ligand in the incubation medium. Radioactivity determinations were made with 6 ml of a Beckman HP/b scintillant on a Beckman LS 3801 scintillation counter. All experiments were performed under dim light to avoid photolysis of DHPs.

Binding assays in the presence of divalent cations were performed essentially as described above. Free calcium concentration in the range 1–1000 μM was calculated using a computer program (see Donaldson and Merrick 1975) and verified with a calcium electrode. Ionic strength of solutions containing a high concentration of divalents was kept approximately constant by replacing calcium chloride by choline chloride. Controls showed no effect of choline chloride on (^{3}H)PN200–110 binding. Inhibition of (^{3}H)PN200–110 binding by different drugs was carried out under equilibrium conditions using 0.2 nM (^{3}H)PN200–110 and 20 μg/ml T-tubule protein. Receptor occupancy by the radiolabel was 20%–30%. Incubation solutions varied; solution A: 50 mM NaCl, 10 mM Tris-HCl (pH 7.2); solution B: 100 mM BaCl$_2$, 10 mM Tris-HCl (pH 7.2); solution C and solution D: same as A and B plus 1 μM racemic Bay K8644, respectively. Details are given in figure legends. Protein concentration was determined by the Lowry method, using bovine serum albumin as standard.

Preparation of Vesicles from Rabbit Skeletal Muscle

T-tubule vesicles were prepared from rabbit back and leg white muscle. Light muscle microsomes sedimenting at 10%/20% sucrose interface were used in all experiments. Portions of back and leg muscle are partially homogenized in buffer A [0.3 M sucrose, 20 mM Hepes-Tris (pH 7.2)] with 4×15-s pulses in a food processor. Tissue is completely homogenized in 3 volumes of buffer A at high speed (2×30-s) in a Waring Blender. The total homogenate is centrifuged for 30 min at 2600 × g (4000 rpm) in a GSA-Sorvall rotor. The supernatant is preserved, and the pellet is rehomogenized in 3 volumes of buffer A and centrifuged as before. The combined 2600 × g supernatants are centrifuged at 10000 × g (8000 rpm) in the GSA-Sorvall, and the resulting supernatant is discarded. The 10000 × g pellets are resuspended and briefly homogenized in 0.6 M KCl, 5 mM Na-PIPES (pH 6.8) with two strokes of a motor-driven Teflon/glass homogenizer, followed by incubation on ice for 1 h. Salt-treated microsomes are sedimented at 90000 × g (32000 rpm) in a Beckman 35 rotor and resuspended in 10% w/w sucrose, 0.4 M KCl, 5 mM Na-PIPES (pH 6.8). This material is layered onto discontinuous sucrose gradients (5 ml 20%, 8 ml 30%, 6 ml 35%, 5 ml 40%) containing 0.4 M KCl, 5 mM Na-PIPES (pH 6.8) and is centrifuged overnight (18 h) at 26000 rpm in a Beckman SW.27 rotor. Fractions are collected from the sucrose interfaces by aspiration with a Pasteur pipette, diluted with ice-cold glass-distilled water, and pelleted at 90000 × g (32000 rpm) in a Beckman 35 rotor. Pelleted membranes are suspended in 0.3 M sucrose, 0.1 M KCl, 5 mM Na-PIPES (pH 6.8) and frozen in liquid nitrogen until use.

Chemicals

Phosphatidylethanolamine (PE) and phosphatidylserine (PS) were from Avanti Polar Lipids (Birmingham, Ala). n-Decane was purchased from Aldrich (Milwaukee, Wis). (+)-(methyl-^{3}H)PN200–110 (71 Ci/mmol) was from New England Nuclear (Boston, Mass). (–)Bay K8644, (+)Bay K8644, nifedipine, (–)nimodipine, (+)nimodipine, and racemic nimodipine were a gift from Dr. S. L. Hamilton and Dr. A. M. Brown at Baylor College of Medicine; S207–180 and PN200–100 were a gift from Dr. Ruegg at Sandoz. Racemic Bay K8644 was purchased from Calbiochem (La Jolla, Calif).

Results

Activation and Inhibition of T-Tubule Calcium Channels by DHPs

In all previous studies, we have used racemic Bay K8644 and found that micromolar levels are required to induce openings. With the pure agonist enantiomer (–)Bay K8644, openings are elicited at nanomolar concentrations. Figure 1 shows representative traces at holding potential (HP) 0 mV of rabbit T-tubule calcium channels activated with increasing concentrations of (–)Bay K8644. All traces are from the same experiment. Current carrier is 100 mM barium present on the internal side; thus, current flow is in the outward direction (E_{Ba} <–100 mV, E_{Na} = 0 mV).

Fig. 1. (Continued)Bay K8644 activation of calcium channels. Barium current at HP 0 mV through T-tubule calcium channels at the indicated concentrations of (–)Bay K8644 added to the trans (external) side. No openings were seen in the absence of the agonist. Channel insertion was monitored with the lowest concentration (10 nM). Each set of records at a given concentration is representative of 150 s of total recorded time at that concentration. Data are from a single experiment repeated three times. Sampling rate was 1 ms per point, and filter corner frequency was 0.1 KHz

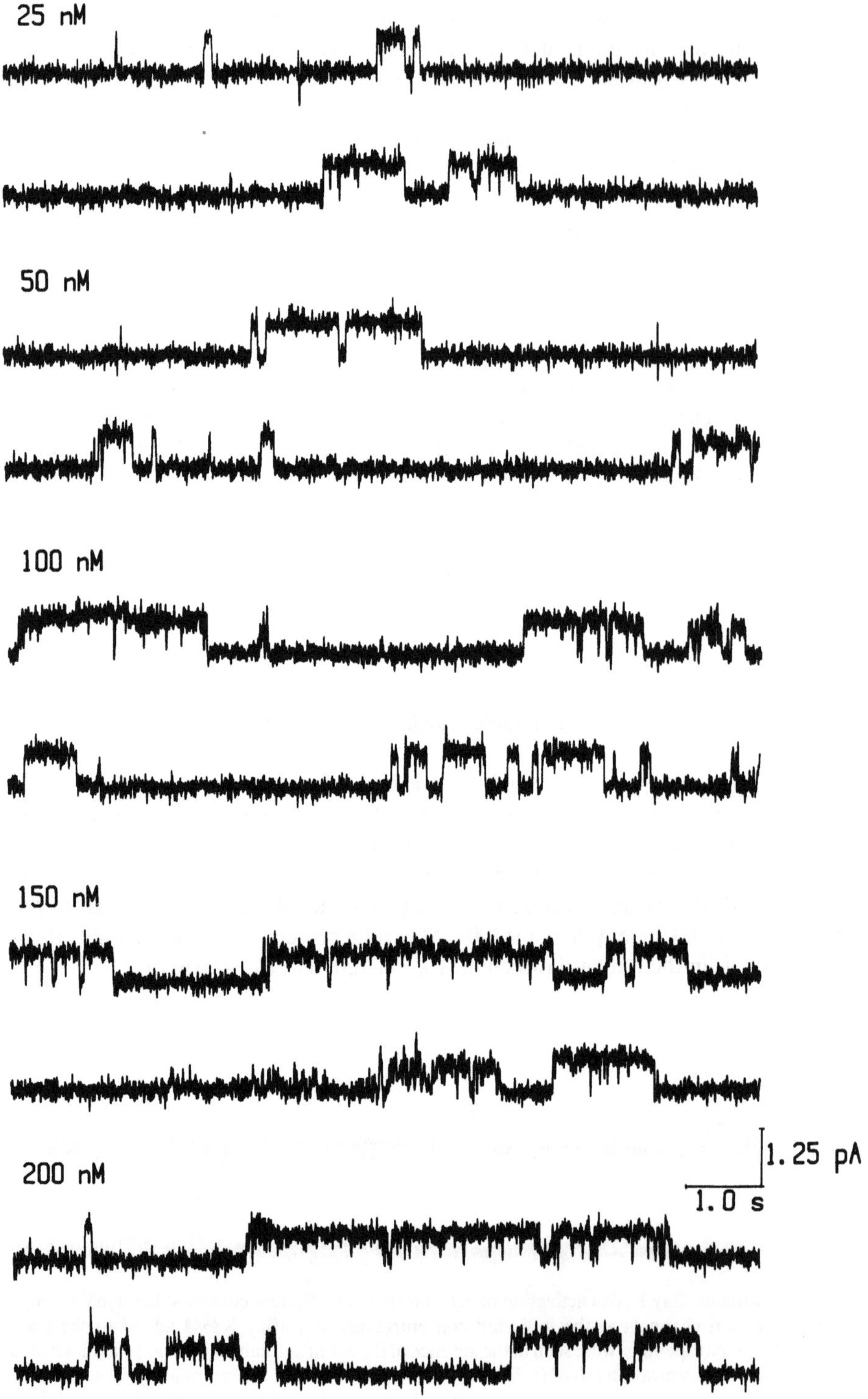

25 nM

50 nM

100 nM

150 nM

1.25 pA
1.0 s

200 nM

Fig. 1. (Continued)

Unlike in rat T-tubules (Affolter and Coronado 1985), we found no measurable spontaneous activity in the rabbit preparation in the absence of agonist. Threshold for activation by the pure enantiomer was approx. 10 nM DHP. At this concentration, however, we did not observe the long open events characteristic of Bay K-activated channels. At 25 nM and above, frequency of openings, channel lifetime, and fraction of time spent open increased with concentration, more in line with our measurements using the racemic compound. Over the concentration range tested, we observed no changes in channel conductance of the kind reported by Lacerda and Brown (1987) in heart cells. The distribution of observable lifetimes was biexponential, with both time constants being a function of drug concentration (Fig. 2). A fast time constant, denominated *tau fast*, is predominant at 10 nM and 25 nM, while a second component, *slow tau*, becomes more apparent at a higher concentration. Fast tau and slow tau reached a plateau at approx. 300 nM, with limiting values 60 ms and 400 ms, respectively. The midpoint (ED_{50}) for the concentration dependence of the fast and slow tau was 80 nM and 68 nM DHP, respectively. This affinity is in the same range as the IC_{50} of (−)Bay K8644 necessary to displace bound (^{3}H)PN200–110 in ligand-binding experiments (Table 1). Although lifetimes of open channels had a tendency to saturate with increasing concentration of (−)Bay K8644, a dose-response curve of open probability vs. concentration did not saturate (not shown). The experimental *np* product, that is, the number of channels times the open probability per channel, increased with concentration up to 250 μM (−)Bay K8644.

Stationary activation by agonist DHPs permitted us to establish a baseline or control activity at a constant HP of 0 mV and to inhibit this activity with antagonist DHPs in a dose-dependent manner. In all experiments described below, channel activity was measured as the fraction of time that one or more channels spent open averaged over 140–180 s of recording time. Surprisingly, inhibition was dependent on whether DHPs were added to the internal or external solutions. This is shown in Fig. 3 for the case of the antagonist DHP PN200–110. At 150 nM, cis-added PN200–110 had virtually no effect, while a lower concentration added trans blocked most openings. Dose-response curves constructed separately for cis and trans additions of DHPs are shown in Fig. 4 for PN200–110, nitrendipine, (−)nimodipine, and (+)Bay K8644. Each point in each dose-response curve represents 144 s of monitored activity. For all DHPs tested, the apparent affinity of inhibition was higher when drugs were added to the trans-external solution than when added to the cis-internal solution. In the most extreme case, that of PN200–110, there is an almost parallel shift of the dose-response curve by 1.2 orders of magnitude towards lower affinity when the DHP is added internally. Thus, the trans IC_{50} was 10 nM and the cis IC_{50} was 160 nM. In the case of (−)nimodipine, the shift is less, approx. 0.3 orders of magnitude.

As shown in Table 1, sidedness was observed with various degrees in all DHPs tested except in the case of the quaternary DHP S207–180. In most cases, ED_{50}s for cis and trans blockade differed by about 0.8–1.2 orders of magnitude, with (+)Bay K8644 showing the highest sidedness (12 μM cis vs 0.3 μM trans). All entries in Table 1 have a standard deviation less than 0.1 times the reported mean (approx. 10% error). Except for nitrendipine, all trans-external ED_{50}s are in the submicromolar range, with PN200–110 displaying the highest channel-drug affinity (ED_{50} = 10 nM). That the external end of the channel faces the trans solution and the internal end the cis solution was confirmed with the use of the charged phenylalkylamines, D575 and

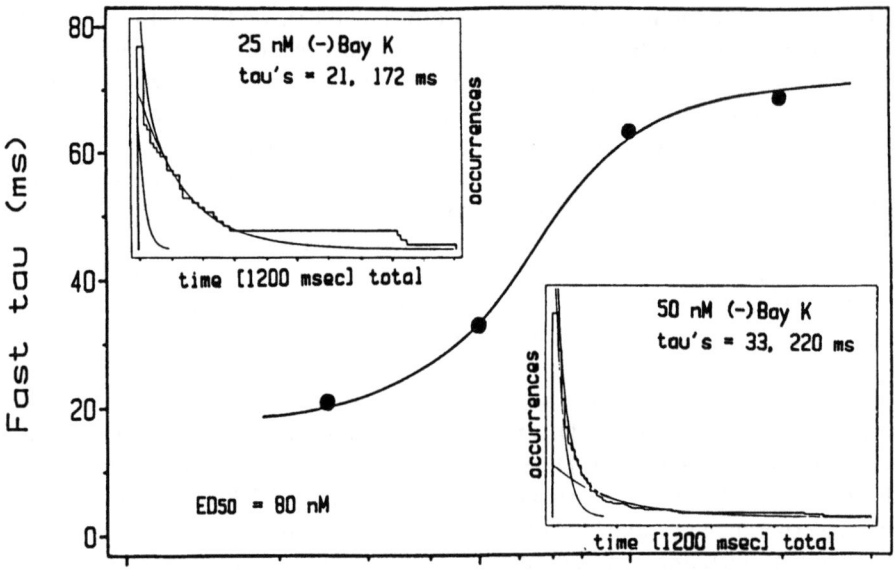
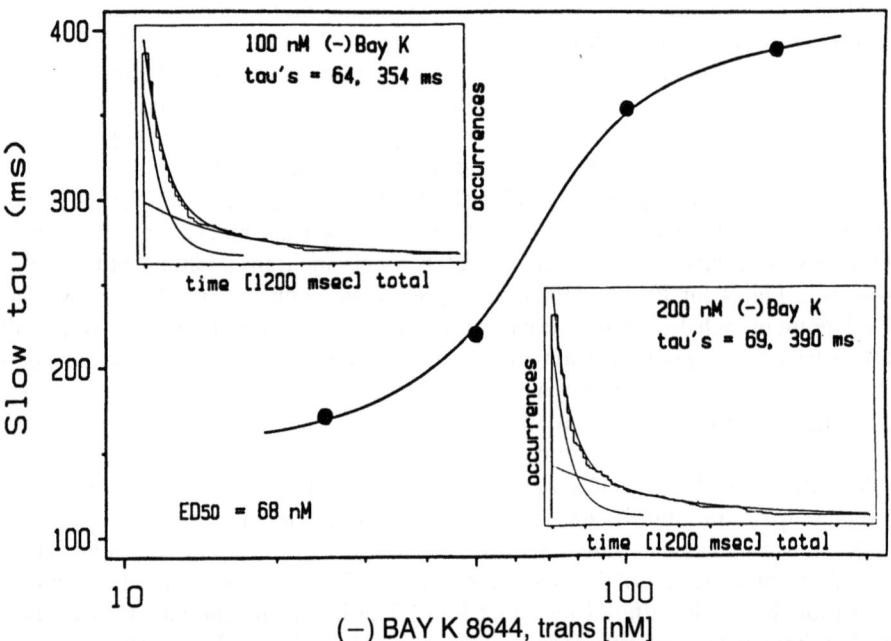

Fig. 2. Slow and fast lifetimes of the calcium channel are functions of (−)Bay K8644 concentration. Open time histograms and fitted exponentials are shown *(insets)* at the indicated concentrations of (−)Bay K8644. Numbers of openings were 48, 133, 162, and 191 for 25, 50, 100, and 200 n*M*, respectively (events <2 ms were excluded). *Main curves (top* and *bottom)* represent a fit of the concentration dependence of fast tau and slow tau to a single binding site isotherm with K$_d$s 80 nM and 68 n*M*, respectively

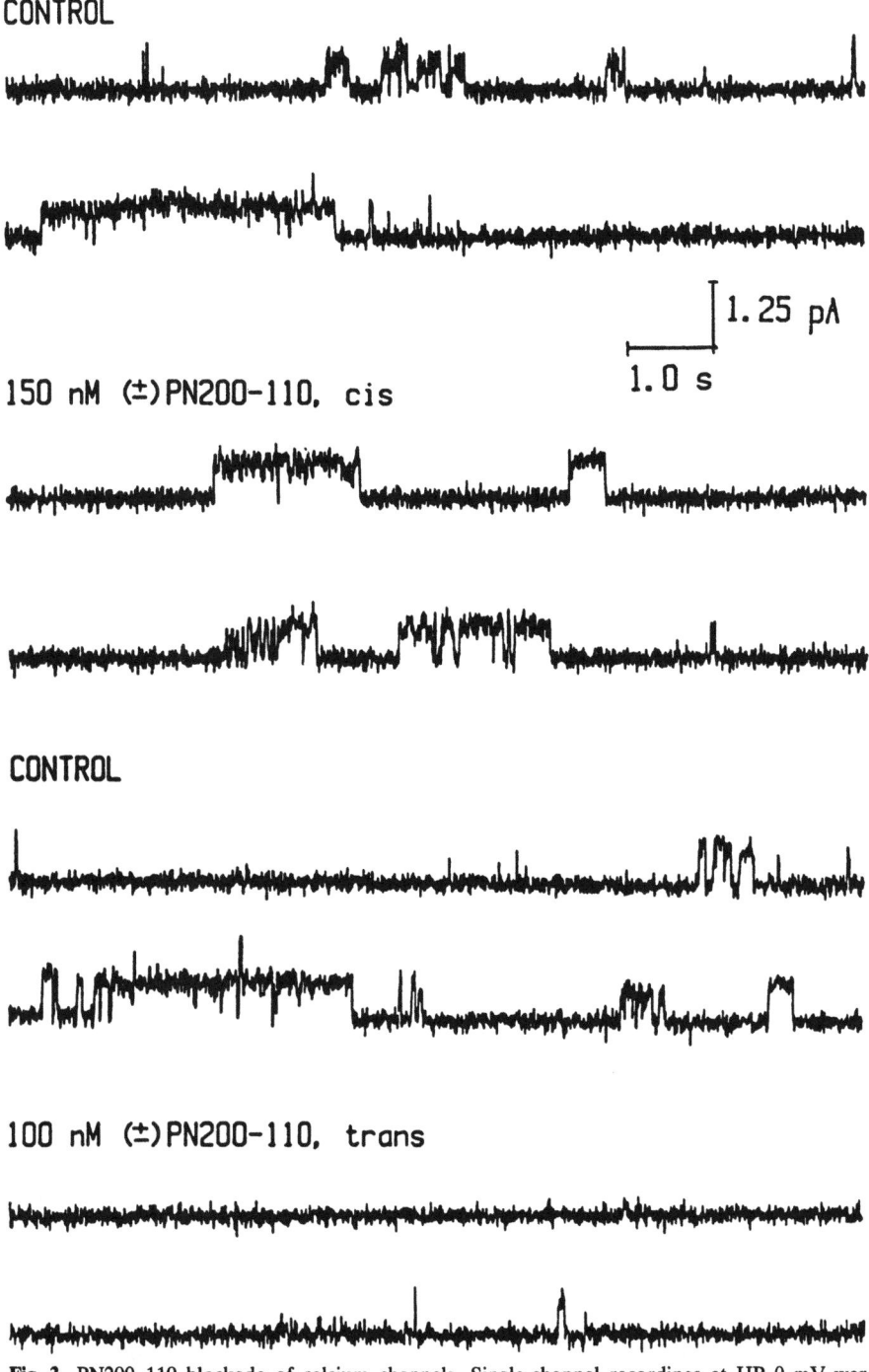

Fig. 3. PN200–110 blockade of calcium channels. Single-channel recordings at HP 0 mV were obtained using cis 1 μM racemic Bay K8644 as agonist. Fraction of open time (po) was 0.14 before and 0.12 after cis addition of 150 nM PN200–110. For trans side addition, po = 0.16 in control and po = 0.0048 after 100 nM PN200–110

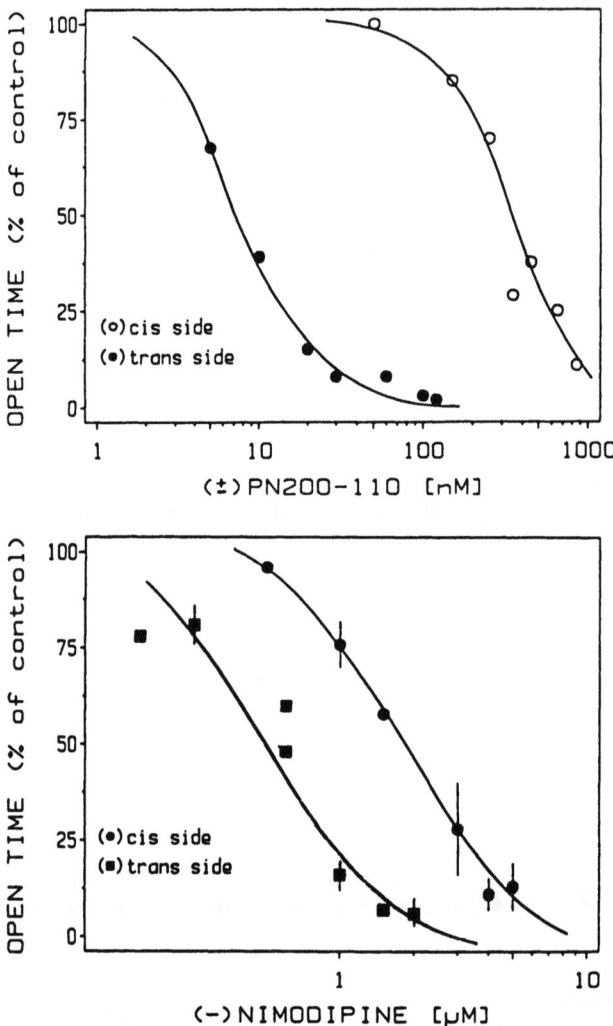

Fig. 4. Dose response of DHPs on cis (internal) and trans (external) sides of the calcium channel. Dose-response curves obtained at steady state and 0 mV holding potential are shown for the antagonist DHPs, racemic PN200-110, racemic nitrendipine, (−)nimodipine, and (+)Bay K8644. Inhibition of fraction open time (po) was scored over 150 s at each concentration. Cis and trans titrations were performed in separate experiments. Vesicles were always added to the cis chamber in order to keep channel orientation constant: cis internal and trans external. Racemic Bay K8644 (1 µM) was added to the cis chamber to elicit stationary activity. Po control was scored over 180–300 s prior to drug addition and was used to normalize data. Standard deviations are for two or three experiments

D890. As shown previously in rat (Affolter and Coronado 1986), D575 and D890 blocked channels when present in the cis side (side of protein addition and head-stage connection) with no effects up to a trans concentration of 50 µM. Clearly, the cis and trans IC$_{50}$s of the charged compounds constitute an excellent indicator of sidedness in the vitro system. Hescheler et al. (1982) described the same effects of D890 in

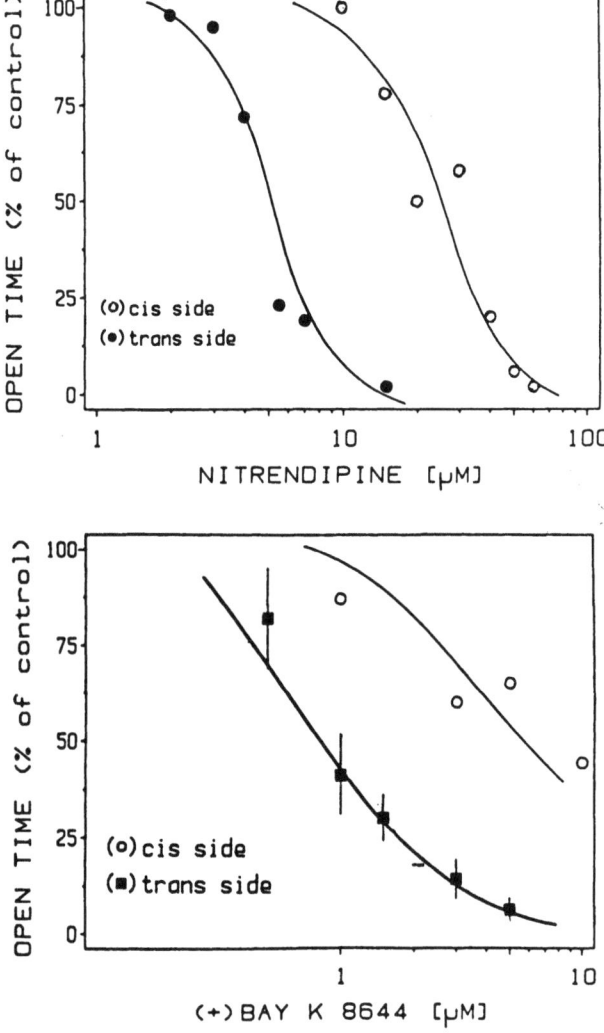

Fig. 4

ventricular heart cells, where blocking of calcium currents was only seen after injection of the compound into cells. Evidently, the quaternary ammonium in D890 prevents the compound from freely permeating the membrane and reaching its internally located blocking site. In our case, the cis location of the blocking sites for D890 and D575 is further supported by the observation that D600 and verapamil, two closely related tertiary amines, are indeed effective on either side (see Table 1).

In order to compare single-channel pharmacology with ligand-binding affinities of DHPs under identical conditions, we first studied in Fig. 5 the effect of $BaCl_2$ and Bay K8644 on binding of (^{3}H)PN200–110 to rabbit T-tubules. In Fig. 5, we also show inhibition of specific binding by cold PN200–110, nimodipine, nifedipine, and nitrendipine. Both barium and Bay K8644 are components of the recording system and were always present in single-channel experiments. Thus, they were also present along with the radioligand in binding experiments. In 50 mM NaCl and contaminant levels of

Table 1. Inhibition constants for calcium channels and PN200-110 binding in skeletal muscle T-tubules

Drug	Calcium channel		DHP receptor	
	Cis	Trans	Solution A	Solution D
Bay K8644	–	–	7.1	–
(−)Bay K8644	–	7.2	7.0	6.3
(+)Bay K8644	4.9	6.5	8.2	6.6
Nimodipine	5.1	6.2	8.0	6.5
(−)Nimodipine	5.7	6.8	8.4	7.4
(+)Nimodipine	5.0	6.4	8.1	6.5
Nitrendipine	4.7	5.7	8.7	7.4
S207-180	6.3	6.4	7.4	6.2
Nifedipine	5.7	6.5	8.3	7.0
PN200-110	6.8	8.0	9.0	8.0
Diltiazem	4.5	5.7	–	NI
Verapamil	5.8	6.0	–	NI
D600	4.6	6.4	–	–
D890	4.8	NI	–	–
D575	5.0	NI	–	–

All entries correspond to ED_{50}s expressed as $-\log_{10}[M]$. NI corresponds to experiments in which no significant inhibition was seen (< 20% of control) at the highest concentration tested. Channel IC_{50}s were performed at HP 0 mV in 1 µM racemic Bay K8644 (cis), 50 mM NaCl (cis, trans), 100 mM $BaCl_2$ (cis), 10 mM Herpes-Tris (pH 7.2 cis and trans). Cis-side and trans-side titrations are from separate experiments. Racemic Bay K was not present in single-channel and binding assays when (−)Bay K8644 was tested. Binding IC_{50}s were performed in solution A (50 mM NaCl, pH 7.2) or solution D (50 mM NaCl, 100 mM $BaCl_2$, 1 µM racemic Bay K8644). In all cases, standard deviation of entries was < 10% of the reported mean.

divalents, a Scatchard analysis of (^{3}H)PN200-110 binding (not shown) revealed an average B_{max} of 40 pmol/mg and a K_d of 0.8 nM, which is in excellent agreement with the binding capacity of DHPs reported by Fosset et al. (1983) and the binding affinities reported by Borsotto et al. (1984) in the same preparation. Binding parameters were obtained from least-squares regression of data from three T-tubule preparations. All curves in Fig. 5 have been scaled using the specific binding of 0.2 nM (^{3}H)PN200-110 as control. At that drug and receptor concentration, receptor occupancy was no more than 25%. Hence, ED_{50}s for inhibition of (^{3}H) PN200-110 binding follow closely the apparent affinity of the inhibitor DHP (Cheng and Prusoff 1973). Figure 5 also shows specific binding of (^{3}H)PN200-110 and displacement by cold PN200-110 in 50 mM NaCl. The ED_{50} of the displacement was 1 nM, and the apparent Hill coefficient was 1.03. When the same experiment was done in the presence of 1 µM racemic Bay K8644 (Fig. 5), the inhibition curve was displaced to the right with an apparent K_d of 10 nM and an apparent Hill coefficient of 1.05. The absolute specific binding in the presence of 1 µM Bay K8644 was approx. 30% of that seen in the absence of the agonist since Bay K8644 itself displaces (^{3}H)PN200-110 with an ED_{50} of approx. 100 nM (shown in Table 1).

The effect of 0.1 M BaCl$_2$ on binding parameters of (^{3}H)PN200-110 is shown in Fig. 5 in the absence or presence of 1 μM Bay K8644. Increasing concentration of BaCl$_2$ in the millimolar range inhibited (^{3}H)PN200-110 binding, which is again in agreement with the results obtained by Fosset et al. (1983) and Galizzi et al. (1986) in the same preparation. In our study, 0.1 M barium decreased the B$_{max}$ of (^{3}H)PN200-110 binding by approx. 60% (average B$_{max}$ = 16 pmol/mg) without a major change in affinity (average K$_d$ = 0.92 nM). However, slight changes in affinity were observed when other DHPs were used, as in the case of nimodipine and nifedipine (Fig. 5). Figure 5 shows that in addition to decreasing the number of available sites, barium induced an unexpected change in the apparent Hill coefficient of the displacement reaction, from close to unity in 50 mM sodium to 0.4-0.6 in 50 mM sodium plus 100 mM barium. The change in steepness of inhibition induced by barium can be clearly seen in each of the four displacement curves in Fig. 5, and in fact, it was present for all DHPs of Table 1. Even more peculiar is the fact that no change in Hill coefficient was observed when displacement was done in barium but in the presence of 1 μM Bay K8644 (compare open and filled squares in Fig. 5). We interpret this change in steepness as a manifestation of two interacting DHP sites which are either unmasked or induced by high barium. Evidence for two DHP sites is discussed later. However, it is sufficient to notice that the change in Hill coefficient is related to DHP occupancy since it is not observed when a large number of receptors are occupied by Bay K8644 (compare empty and filled squares in Fig. 5).

Correlation of ED$_{50}$s, in single-channel and binding experiments, is shown in Figs. 6, 7. *Cis* addition of DHPs (Fig. 6) and *trans* addition of DHPs (Fig. 7) were plotted separately, while binding data are the same for both Figures. The dotted line in both graphs has a slope of unity. All x, y entries are the same as those given in Table 1. The slope (n) and correlation coefficient (r) for the set of x, y pairs was n = 0.45, r = 0.85 in Fig. 6, and n = 0.8, r = 0.87 in Fig. 7. When nitrendipine data are omitted (because they show a notorious lack of correlation in both cases), the values become n = 0.55, r = 0.89 in Fig. 6, and n = 0.9, r = 0.92 in Fig. 7. As a whole, we conclude that the correlation holds quantitatively for trans-external – added DHPs, but not for cis-internal additions of the same. Surprisingly, the only DHP that shows equal effects on both sides and has a 1:1 correspondence in binding and channel experiments is the permanently charged DHP S207-180. In log units, S207-180 shows ED$_{50}$s 6.3, 6.4, and 6.2 for cis, trans, and binding affinities respectively (Table 1). In fact, S207-180 is the only DHP that in cis addition experiments (Fig. 6) is close to the dotted line. All other DHPs in Fig. 6 are below the dotted line, implying that affinities on the cis side of the channel are lower than binding affinities. In the most severe cases, there is an approx. 50-fold lower channel affinity for cis-added (+)Bay K8644 and (−)nimodipine, and a 500-fold lower channel affinity for cis-added nitrendipine (rightmost points in Fig. 6). In contrast, trans-added DHPs and binding data in Fig. 7 show a remarkable agreement, with the sole exception of nitrendipine. For trans-added nitrendipine the discrepancy is 50-fold, not 500-fold as for cis-added drug.

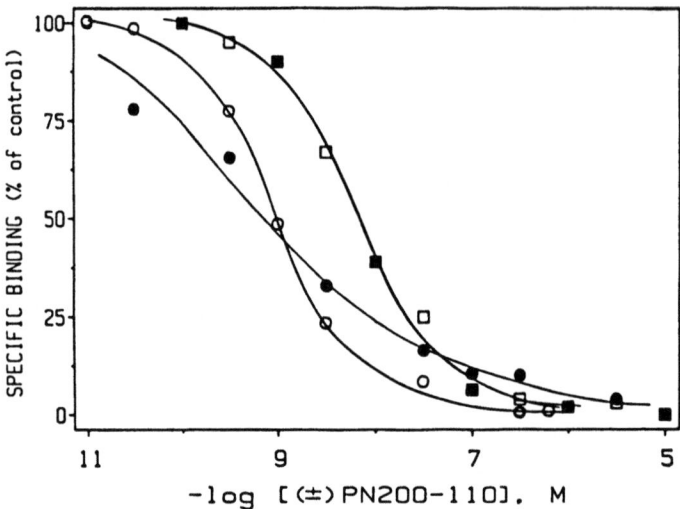

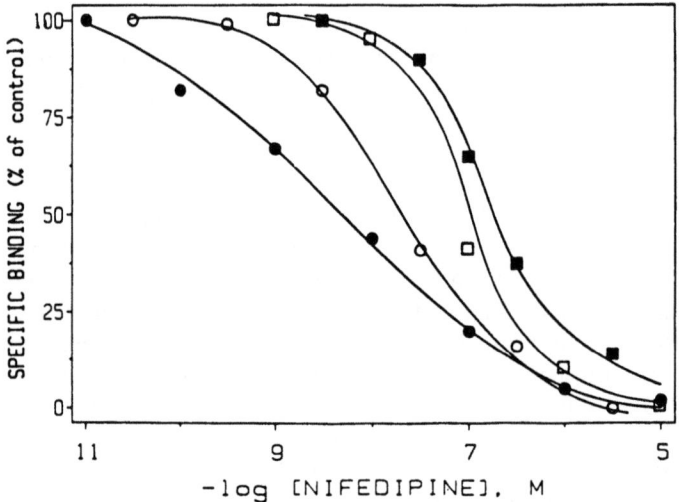

Fig. 5. Displacement of PN200–110 binding by DHPs under different conditions. (^{3}H)PN200–110 (0.2 nM) was incubated with 10–30 μg rabbit T-tubule vesicles and the indicated concentrations of racemic PN200–110, (+)nimodipine, nifedipine, and racemic nitrendipine. Incubation solutions were A 50 mM NaCl, 10 mM Hepes-Tris (pH 7.2) *(empty circles);* B 100 mM BaCl$_2$, 50 mM NaCl, 10 mM Hepes-Tris (pH 7.2) *(filled circles);* C and A plus 1 μM Bay K8644 *(empty squares);* D and B plus 1 μM Bay K8644 *(filled squares).* Specific binding was defined as cpm retained in the absence of displacer minus cpm retained in the presence of 1 μM PN200–110. This number was normalized to 100% and, in absolute terms, was approx. 3000 cpm for solution A and 2000 cpm for solution B. In experiments using Bay K8644 (solutions C, D), only a third of these values were retained. Larger signal-to-background ratios were obtained when using solutions C, D by increasing two fold the amount of protein and at the same time doubling the volume of incubation so that free (^{3}H)PN200–110 was kept constant

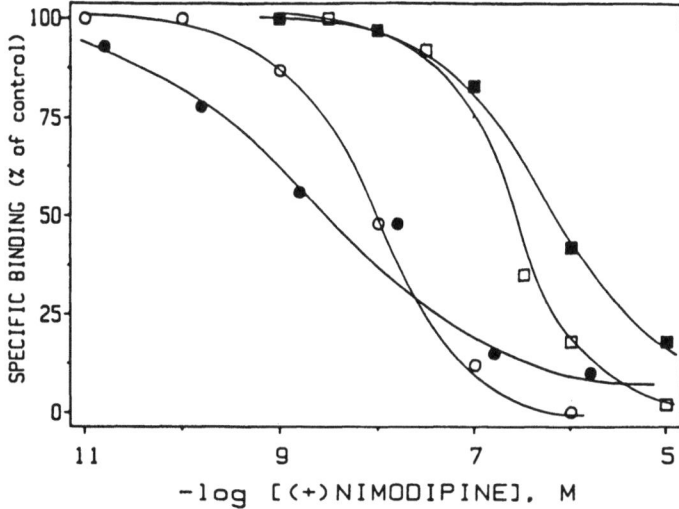

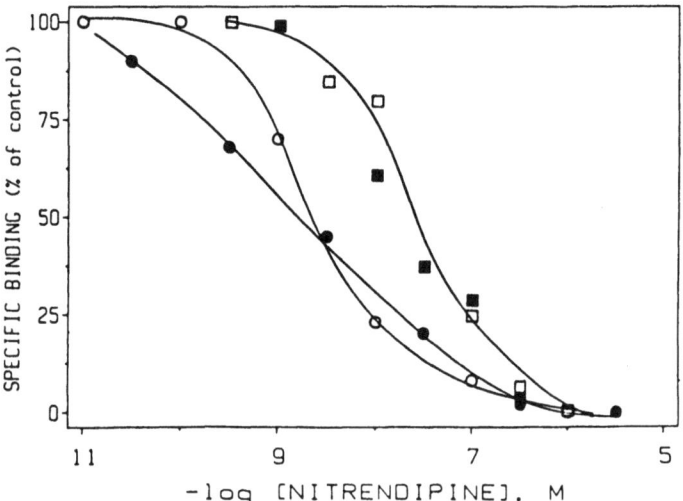

Fig. 5

Discussion

In this contribution, we have outlined for the first time the dihydropyridine pharmacology of the reconstituted calcium channel of skeletal muscle and correlated ED_{50}s for single channels and DHP receptors. Planar bilayers are particularly suited for this study since the technique permits easy access to both sides of the membrane where the channel is embedded, and it ensures a vectorial insertion of the protein. Thus, one can mimic "cytosolic" or "extracellular" manipulations in a reproducible manner. Unlike in single-cell studies, where DHPs are usually equilibrated with the

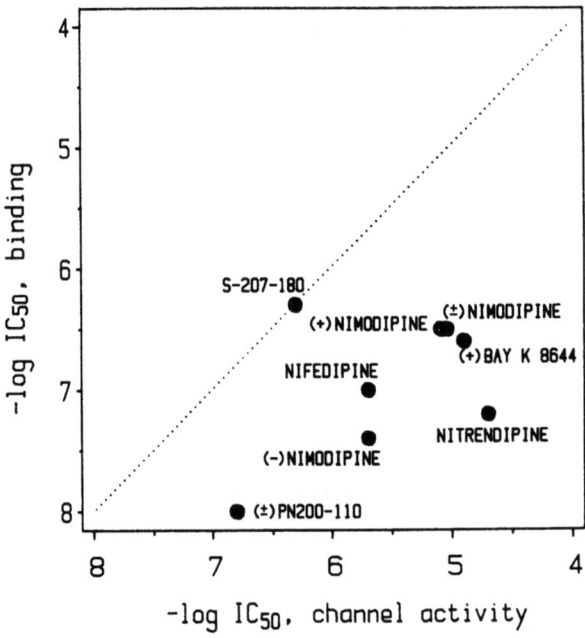

Fig. 6. Correlation between ligand-binding affinities and cis DHP blockade of calcium channels. Ligand-binding IC$_{50}$s of DHPs obtained by displacement of (^{3}H)PN200–110 *(y-axis)* are compared with single-channel IC$_{50}$s obtained by inhibition of fraction open time using cis-added DHPs *(x-axis)*. In both sets, experiments were performed at HP 0 mV in the presence of 1 μM racemic Bay K8644, 0.1 M barium. *Dotted line* has a a slope equal to 1

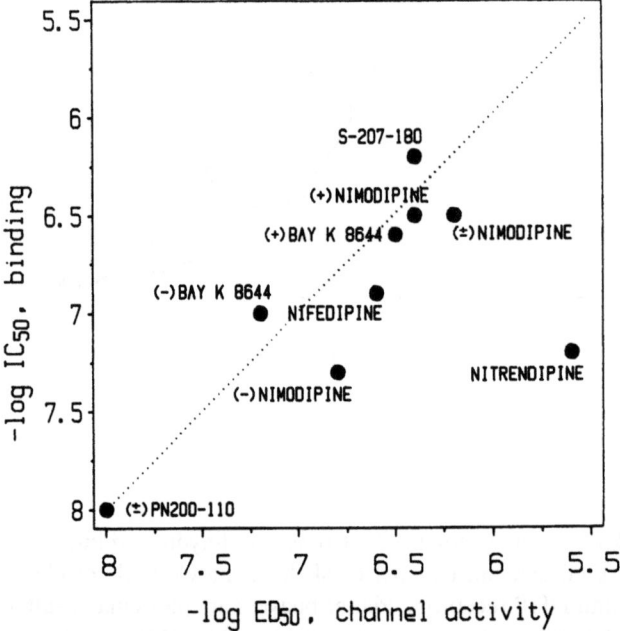

Fig. 7. Correlation between binding affinities and trans DHP blockade of calcium channels. Ligand-binding IC$_{50}$s plotted in Fig. 6 are compared with ED$_{50}$s for activation [(−)Bay K8644] and inhibition (all other DHPs) of fraction open time using trans-added DHPs *(x-axis)*. As channel activation by (−)Bay K8644 did not require the presence of racemic Bay K8644, its single-channel ED$_{50}$ was compared with ligand-binding IC$_{50}$ in the absence of racemic compound. *Dotted line* has a slope equal to 1. Note that the x, y scales differ from those in Fig. 6

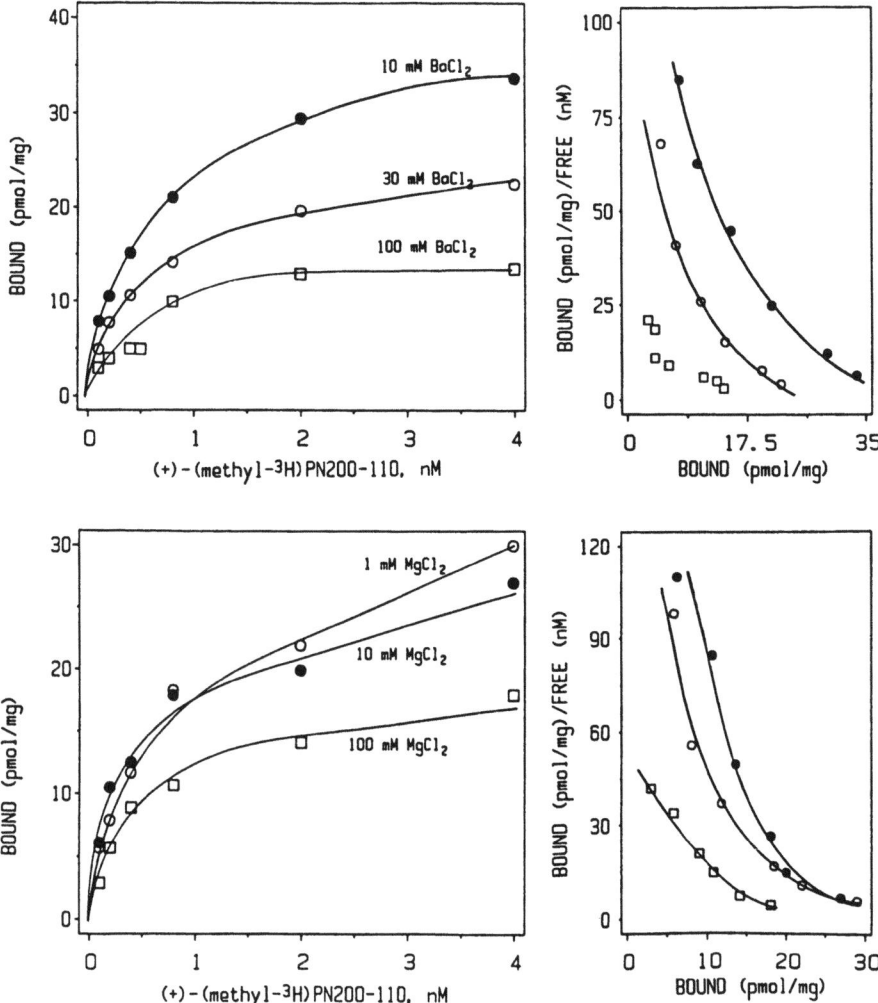

Fig. 8. Effects of barium and magnesium on binding of PN200–110 to T-tubule vesicles of skeletal muscle. Binding isotherms of (^{3}H)PN200–110 were obtained in 50 mM NaCl, 10 mM Tris-HCl (pH 7.2) and the indicated concentrations of BaCl$_2$ and MgCl$_2$. *Top:* Specific binding of the DHP *(left)* shows a substantial decrease in B$_{max}$ as BaCl$_2$ is raised from 10 to 100 mM. Scatchard plots *(right)* in the presence of barium could not be fit with a straight line, suggesting at least two binding sites. In the absence of barium, K$_d$ and B$_{max}$ were 0.36 nM and 40 pmol/mg, respectively. At 30 mM BaCl$_2$, we distinguished two limiting K$_d$s of 0.23 nM and 1.45 nM, respectively. *Bottom:* Multiple receptor sites are more apparent in millimolar magnesium. Binding curves *(left)* in 1 mM and 10 mM do not fully saturate, and Scatchard plots *(right)* are clearly nonlinear

extracellular solution, we here tested effects on cytosolic and extracellular sides and found significant differences in apparent affinities. This was surprising because the lipid solubility of DHPs is extremely high, and there was no previous indication that diffusion of DHPs in membranes was in any way restricted (Goll et al. 1984; Rhodes et al. 1985). Except for S207–180, we find that all DHPs have significantly lower cis than trans affinities. Moreover, the ranking orders of DHPs are different on both sides.

Potencies on the *cis* side are in the sequence PN200–110 > S207–180 > nifedipine = (−)nimodipine > (+)nimodipine = nimodipine > (+)Bay K8644 > nitrendipine, while on the *trans* side the sequence is PN200–110 > (−)nimodipine > nifedipine = S207–180 = (+)nimodipine = (+)Bay K8644 > nimodipine > nitrendipine. In both sequences, the affinity for PN200–110 is the highest. This is in agreement with binding data in T-tubules (see Table 1) and also in agreement with single-channel data in cardiac DHP channels (Hamilton et al. 1987). On both sides also, (−)nimodipine is more effective than (+)nimodipine, and the agonist action of (−)Bay K8644 has higher potency than that of the antagonist (+)Bay K8644. Again, these two results are in agreement with findings in heart calcium channels (Hamilton et al. 1987). Therefore, drugs added to either solution appear to gain access to a bona fide DHP site, although there are recognizable chemical differences between cis and trans effects and only trans IC_{50}s correlate well with receptor-binding IC_{50}s.

We have considered the possibility that cis and trans effects of DHPs reflect differences in accessibility of DHPs to a receptor site that is in equilibrium with the external bulk solution. In such a case, cis-added DHPs would have to negotiate their passage thermodynamically across the membrane to reach the binding site. The lower cis affinity could then be explained by a combination of factors, including differences in partition coefficients, ligand asymmetry, lateral diffusion, interfacial distributions, and in general, by parameters that would determine the diffusion rate of DHPs across the membrane. We consider that access limitations are rather unlikely. In general, the onset of DHPs was almost instantaneous in all cases and did not change significantly with time; that is, IC_{50}s did not depend to any great extent on the time of exposure of channels to DHPs, as would be the case if access rates were the limiting factor. Moreover, chemically identical DHPs, such as pairs of enantiomers, appear to have different stereoselectivities on both sides. For example, (−)nimodipine has a 25-fold difference between trans and cis IC_{50}s, while (+)nimodipine has a difference of only 12-fold. The ratio of affinities (+)nimodipine/(−)nimodipine is 2.5 on the trans side and 4.8 on the cis side. Because both steroisomers are expected to have identical diffusional properties, it is unlikely that the cis and trans differences reflect a difference in the rate of drug passage across the membrane.

Rather than an accessibility difference, we favor the hypothesis that the sidedness of DHPs reflects the presence of a second DHP site on the calcium channel. Besides the nimodipine data described above, there are significant differences in cis and trans effects of (−)Bay K8644 which suggest two chemically different DHP-binding moieties on opposite sides of the membrane. The pure enantiomer agonist increases channel lifetime on the trans side (Fig. 2), but when present on the cis side it *decreases* lifetime (not shown). Further ligand-binding evidence for two sites is provided by the decrease in apparent Hill coefficient or negative cooperativity seen in the presence of high barium (Fig. 5). The existence of high- and low-affinity DHP sites has been considered in numerous studies. Janis et al. (1984a, b) suggested this possibility on the basis of a nonlinear Scatchard plot and a change in Hill coefficient for Bay K8644 binding to heart and brain membranes. Hamilton et al. (1987) and Williams et al. (1985) suggested the same on the basis of the voltage dependence of DHPs, while Kokubun et al. (1987) noticed two allosterically coupled sites in studies with pure enantiomers of the DHP 202–791.

A direct confirmation of at least two binding sites in the DHP receptor of skeletal muscle is provided by the Scatchard plots of Fig. 8. High concentrations of barium or magnesium reduced the binding capacity of (^{3}H)PN200–110. At moderate concentrations, the Scatchard relationship is nonlinear, with at least two binding components. In the absence of added divalent, or above 0.1 M barium or magnesium, the two components fuse into an apparently single binding component. Interestingly, curvilinear Scatchard plots for (^{3}H)PN200–110 binding were not seen in calcium (not shown), which is the divalent of choice in most binding studies. Previous reports indicate that densities of DHP receptors increase with calcium concentration in brain (Gould et al. 1982) and decrease in skeletal muscle (Fosset et al. 1983), in both cases without changes in affinities. Phenylalkylamine receptors in skeletal muscle also decrease with increasing concentration of divalents (Galizzi et al. 1984). To our knowledge, our observation of multiple high-affinity binding sites for DHPs in barium and magnesium solutions has not been reported previously. As such, it confirms that under our single-channel recording conditions, the DHP receptor can be occupied by at least two DHPs. At the same time, the pharmacological sidedness of DHPs suggests that the highest affinity site, tracked by (^{3}H)PN200–110 binding, is external, while the lower affinity site is internal.

Finally, the correlation of receptor binding and single-channel effects of DHPs that we have clearly established strongly suggests that DHP receptor and the calcium channel are closely related and probably physically coupled. This conclusion stems from our finding that when all sources of discrepancies between DHP-receptor and calcium channel measurements are removed, the agreement between DHP effects on radioligand binding and channel inhibition is outstanding. In fact, the agreement is even much better than that reported for cardiac channels (Su et al. 1985). Only one compound, nitrendipine, has notoriously low affinity in our single-channel studies. At present, we have no clues as to why this is the case. Separate evidence to indicate that the T-tubule calcium channel and its DHP receptor remain tightly coupled during purification is being provided by the functional effects of DHP-receptor monoclonal antibodies on planar bilayer recordings (Vilven et al. 1988). The fact that the channel and the DHP receptor remain coupled in an in vitro system will permit further elucidation of the biochemical nature of the interaction of the calcium channel with its drug receptor. If the receptor site is embedded in the core of the phospholipid phase and DHPs reach this site due to their high partition coefficient, affinity constants for DHPs should be affected by changes in the composition of the lipid bilayer surrounding the channel. If, on the contrary, DHP blockade occurs through a site in direct equilibrium with the aqueous phases or a site insulated from the bilayer lipid phase (Coronado and Affolter 1986), affinity constants will be considerably independent of the physical state of the bilayer membrane. All relevant manipulations to test these alternatives (such as membrane thickness, phosholipid composition, fluidity, and surface charge) can be easily implemented in planar bilayers. Thus, they can now be applied to the study of the route of access of DHPs to functional sites in the DHP-receptor – calcium-channel complex.

References

Affolter H, Coronado R (1985) Agonist Bay-K8644 and CGP-28392 open calcium channels reconstituted from skeletal muscle transverse tubules. Biophys J 48:341–347

Affolter H, Coronado R (1986) Sidedness of reconstituted calcium channels from muscle transverse tubules as determined by D600 and D890 blockade. Biophys J 49:767–771

Bean BP (1984) Nitrendipine block of cardiac calcium channels: high affinity binding to the inactivated state. Proc Natl Acad Sci USA 81:6388–6392

Borsotto M, Barhanin J, Norman RI, Lazdunski M (1984) Purification of the dihydropyridine receptor of the voltage-dependent Ca channel from skeletal muscle transverse tubules using (+)[^{3}H]PN200–110. Biochim Biophys Res Commun 122:1357–1366

Cheng YC, Prusoff WH (1973) Relationship between the inhibition constant (Ki) and the concentration of inhibitor which causes 50 per cent inhibition (I_{50}) of an enzymatic reaction. Biochem Pharmacol 22:3099–3108

Coronado R, Affolter H (1986a) Characterization of dihydropyridine sensitive calcium channels from skeletal muscle transverse tubules. In: Miller C (ed) Ion channel reconstitution. Plenum, NY, pp 483–505

Coronado R, Affolter H (1986b) Insulation of the conductance pathway of skeletal muscle transverse tubules from the surface charge of bilayer phospholipid. J Gen Physiol 87:933–953

Coronado R (1987) Planar bilayer reconstitution of calcium channels: Lipid effects on single channel kinetics. Circ Res 61 (Supp II): 46–52

Coronado R, Smith JS (1987) Monovalent ion current through single calcium channels of skeletal muscle transverse tubules. Biophys J 51:497–502

Curtis B, Catterall WA (1986) Reconstitution of the voltage-sensitive calcium channel purified from skeletal muscle transverse tubules. Biochemistry 25:3077–3083

Donaldson SF, Kerrick WGL (1975) Characterization of the effects of Mg, Ca, on Sr-activated tension generation of skinned skeletal muscle fibers. J Gen Physiol 66:427–444

Ehrlich BE, Schen CR, Garcia ML, Kaczorowski GJ (1986) Incorporation of calcium channels from cardiac sarcolemma vesicles into planar bilayers. Proc Natl Acad Sci USA 83:193–197

Fosset M, Jaimovich E, Delpont E, Lazdunski M (1983) [^{3}H]Nitrendipine receptors in skeletal muscle. Properties and preferential location in tranverse tubules. J Biol Chem 258:6086–6092

Galizzi JP, Fosset M, Lazdunski M (1984) Properties of the receptor for the Ca channel blocker verapamil in transverse-tubule membranes of skeletal muscle. Stereospecificity, effects of Ca, and other inorganic cations, evidence for categories of sites and effects of nucleoside triphosphates. Eur J Biochem 144:211–215

Galizzi JP, Borsotto M, Barhanin J, Fosset M, Lazdunski M (1986) Characterization and photoaffinity labeling of receptor sites for the Ca channel inhibitors d-cis-Diltiazem, Bepridil, Desmethoxyverapamil, and (+)-PN 200–110 in skeletal muscle transverse tubule membranes. J Biol Chem 261:1393–1397

Goll A, Ferry D, Glossmann H (1984) Target size analysis and molecular properties of Ca channels labelled with [^{3}H]Verapamil. Eur J Biochem 141:177–186

Gould RJ, Murphy KMM, Snyder SH (1982) [^{3}H]Nitrendipine-labeled calcium channels discriminate inorganic calcium agonists and antagonists. Proc Natl Acad Sci USA 79:3656–3660

Green FJ, Farmer BB, Wiseman GL, Jose MJ, Watanabe AM (1985) Effect of membrane depolarization on binding of [^{3}H]Nitrendipine to rat cardiac myocytes. Circ Res 56:576–585

Hamilton SL, Yatani A, Brush K, Schwartz A, Brown AM (1987) A comparison between the binding and electrophysiological effects of dihydropyridines on cardiac membranes. Molec Pharmacol 31:221–231

Hescheler J, Pelzer D, Trube G, Trautwein W (1982) Does the organic Ca channel blocker D600 act from the inside or outside on the cardiac cell membrane? Pflugers Arch 393:287–291

Janis RA, Triggle DJ (1984) 1,4-Dihydropyridine Ca channel antagonists and activators: A comparison of binding characteristics with pharmacology. Drug Develop Res 4:257–274

Janis RA, Rampe D, Sarmiento JG, Triggle DJ (1984a) Specific binding of a calcium channel activator, [^{3}H]Bay K 8644, to membranes from cardiac muscle and brain. Biochem Biophys Res Comm 121:317–323

Janis RA, Sarmiento JG, Maurer SC, Bolger GT, Triggle DJ (1984b) Characteristics of the binding of [³H]Nitrendipine to rabbit ventricular membranes: Modification by other Ca channel antagonists and by the Ca channel agonist Bay K 8644. J Pharmacol Exp Ther 231:8–15

Janis RA, Bellemann P, Sarmiento JG, Triggle DJ (1985) The dihydropyridine receptor. In Bayer-Symposium IX. Cardiovascular effects of dihydropyridine-type calcium antagonists and agonists. Springer, Berlin Heidelberg New York, pp 140–155

Kokubun S, Prod'hom B, Becker C, Prozig H, Reuter H (1987) Studies on Ca channels in intact cardiac cells: Voltage-dependent effects and cooperative interactions of dihydropyridine enantiomers. Molec Pharmacol 30:571–584

Lacerda AE, Brown AM (1987) Dihydropyridine gating of calcium channels. Biophys J 51:262a

Ma J, Coronado R (1988) Heterogeneity of conductance states in calcium channels of skeletal muscle. Biophys J 53:387–395

Rosenberg RL, Hess P, Reeves J, Smilowitz H, Tsien RW (1986) Calicum channels in planar bilayers: new insights into the mechanism of permeation and gating. Science 231:1564–1566

Rhodes DG, Sarmiento JG, Herbette LG (1985) Kinetics of binding of membrane-active drugs to receptor sites, diffusion-limited rates for a membrane bilayer approach of 1,4-dihydropyridine calcium channel antagonists to their active sites. Molec Pharmacol 27:612–623

Sanguinetti MC, Kass RS (1984) Voltage-dependent block of calcium channel current in the calf cardiac Purkinje fiber by dihydropyridine calcium channel antagonists. Circ Res 55:336–349

Schwartz LM, McCleskey EW, Almers W (1985) Dihydropyridine receptors are voltage-dependent but most are not functional calcium channels. Nature 314:747–751

Su CM, Yousif FB, Triggle DJ, Janis RA (1985) Structure-function relationships of 1,4-dihydropyridines: Ligand and receptor properties. In Bayer-Symposium IX. Cardiovascular effects of dihydropyridine-type calcium antagonists and agonists. Springer, Berlin Heidelberg New York, pp 104–110

Tsien RW (1983) Calicum channels in excitable cell membranes. Annu Rev Physiol 45:341–358

Valdivia H, Coronado R (1988) Pharmacological profile of skeletal muscle calcium channels in planar lipid bilayers. Biophys J 53:555a

Vilven J, Leung AT, Imagawa T, Sharp AH, Campbell KP, Coronado R (1988) Interaction of calcium channels of skeletal muscle with monoclonal antibodies specific for its dihydropyridine receptor. Biophys J 53:665a

Williams JS, Grupp IL, Grupp G, Vaghy PL, Dumont L, Schwartz A, Yatani A, Hamilton S, Brown AM (1985) Profile of the oppositely acting enantiomers of the dihydropyridine 202–791 in cardiac preparations: Receptor binding, electrophysiological and pharmacological studies. Biochim Biophys Res Commun 131:13–21

Expression of mRNA Encoding Rat Brain Ca^{2+} Channels in Xenopus Oocytes

H. A. Lester, T. P. Snutch, J. P. Leonard, J. Nargeot, and N. Davidson*

Divisions of Biology and of Chemistry and Chemical Engineering, California Institute of Technology, Pasadena, CA 91125 USA

Introduction

Our laboratory is seeking a mechanistic picture of the action of the voltage-dependent, Ca^{2+}-selective ion channels that partially underlie the electrical excitability of membranes. Full explanations will ultimately require pictures of such molecules at atomic resolution (as obtained with X-ray crystallography) in all their states (closed, open, inactivated, etc.). Such structural information is not yet available for any channel.

Despite the lack of structural information, we believe that useful insights will be obtained with an approach that combines nucleic acid molecular biology with membrane biophysics. We employ *Xenopus* oocytes, as pioneered by Gurdon et al. (1971) and later, for electrophysiological studies of ion channels, by Barnard, Miledi, and their colleagues (Barnard et al. 1982; Miledi et al. 1982a; Miledi and Sumikawa 1982; Gundersen et al. 1983b). Although other expression systems are being developed for membrane channels (Claudio et al. 1987), the oocyte is now the most common, robust, and best characterized system. One may identify three major goals for experiments in which channels are expressed by RNA injections in oocytes:

1. *Reconstitution.* Oocyte injection provides a nucleic acid-based method to reconstitute membrane chemical and electrical excitability. In some cases, one expresses in oocytes to confirm the cloning of all subunits of a channel or a receptor (Mishina et al. 1985; White et al. 1985; Noda et al. 1986b; Goldin et al. 1986; Timpe et al. 1988; Iverson et al. 1988); in other cases, one reconstitutes to exploit the specific electrophysiological or biochemical advantages of the oocyte (Dascal et al. 1986b); in yet other cases, one reconstitutes to find out the number and size of the RNA components necessary for a response (Krafte et al. 1988; Lubbert et al. 1987a); in still other cases, the reconstitution gives otherwise unattainable information about hormonal regulation of mRNA levels (Boyle et al. 1987; Oron et al. 1987).

---

* This research was sponsored by grants from the National Institutes of Health (GM-10991 and GM-29836), and by postdoctoral fellowships from the American Heart Association to T. P. S. J. P. L. and J. N. thanks Laboratoire Servier for a travel grant.

2. *cDNA Cloning.* Once a reconstituted excitability assay is established, one uses this assay for screening a cDNA library and thus isolating a clone that encodes a channel, receptor, or transport molecule. Relevant recent examples concern the serotonin $5HT_{1C}$ receptor from choroid plexus (Lubbert et al. 1987b), the substance-K receptor from bovine stomach (Masu et al. 1987), and the Na-glucose transporter from intestine (Hediger et al. 1987b).

3. *Structure-Function Relations.* With complete clones for all subunits of the desired protein in hand, one employs bacteriophage promoters to synthesize RNA in vitro. Subunits are mixed to form interspecific hybrid proteins, and the clones are mutated to test specific hypotheses about amino acids important in the function. The studies from the Kyoto-Göttingen group and from our own group are revealing facts about agonist binding, drug blockade, and permeation at acetylcholine-receptor proteins (Sakmann et al. 1985; Mishina et al. 1986; Imoto et al. 1986; White et al.; Yoshii et al. 1987b).

At present our experiments on Ca^{2+} channels are still at stage 1. Thus, although full-length cDNA clones have been described that encode the dihydropyridine receptor of skeletal muslce (see Tanabe et al. 1987, and contribution by Schwartz et al., this volume), which is presumably a component of a voltage-dependent Ca^{2+} channel, no oocyte expression studies have been reported with RNA synthesized in vitro from these clones. This contribution describes further progress toward defining the components of brain mRNA that encode voltage-dependent Ca^{2+} channels.

Experimental Procedures

mRNA Isolation and Size Fractionation. Total cellular RNA was isolated from the brains of 14- to 16-day-old rats using a modification of the lithium chloride/urea procedure of Auffray and Rougeon (1980). Poly(A)+ RNA was isolated by a single binding to either oligo(dT)-cellulose (Collaborative Research, type 3) or poly(U)-Sepharose (Pharmacia, type 4B). A sucrose-gradient method was employed to fractionate RNA according to molecular length. Poly(A)+ RNA was heat denatured at 70 °C for 3 min, cooled on ice, and then layered onto a 6%–20% sucrose gradient containing 15 mM PIPES (pH 6.5), 5 mM EDTA, 0.25% Sarkosyl. Gradients were centrifuged in an SW 27.1 rotor at 24000 rpm for 18 h at 4 °C. Approximately 0.45-ml fractions were collected, and the RNA recovered by ethanol precipitation. Individual fractions were pooled into sets of three adjacent fractions, and the RNA was reprecipitated. The RNA from each pool was dissolved in H_2O to give a final concentration equivalent to 4.5 µg unfractionated poly(A)+ RNA per microliter. These samples were used for RNA blots and oocyte injections.

Oocyte Isolation, Injection, and Electrophysiology. Xenopus laevis were anesthetized in 0.15% tricaine (Sigma), and several ovarian lobes were excised. Ovarian tissue and follicular cells were removed by treatment with collagenase (2 mg/ml; Sigma, type 1A) in Ca^{2+}-free saline [82.5 mM NaCl, 2 mM KCl, 1 mM $MgCl_2$, 5mM Hepes (pH 7.4)]. Stage V and VI oocytes (Dumont 1972) were selected and allowed to recover overnight in ND96 [96 mM NaCl, 2 mM KCl, 1.8 mM CaCl$_2$, 1 mM MgCl$_2$, 5 mM Hepes (pH 7.4), 2.5 mM pyruvate, penicillin (100 U/ml), streptomycin (100 μg/ml). Oocytes were injected with 50–70 nl RNA solution by positive displacement, using a 10-μl micropipette (Drummond Scientific Co., Broomall, Pa).

RNA Transfer and Hybridization. RNA was denatured with 2.2 M formaldehyde, 50% formamide, 5 mM sodium acetate, 1 mM EDTA, 20 mM MOPS (pH 7.0) at 65 °C for 15 min and then cooled on ice. Samples were separated through a 1.1% agarose gel containing 2.2 M formaldehyde for 8 h at 3.5 V/cm. After electrophoresis, RNA

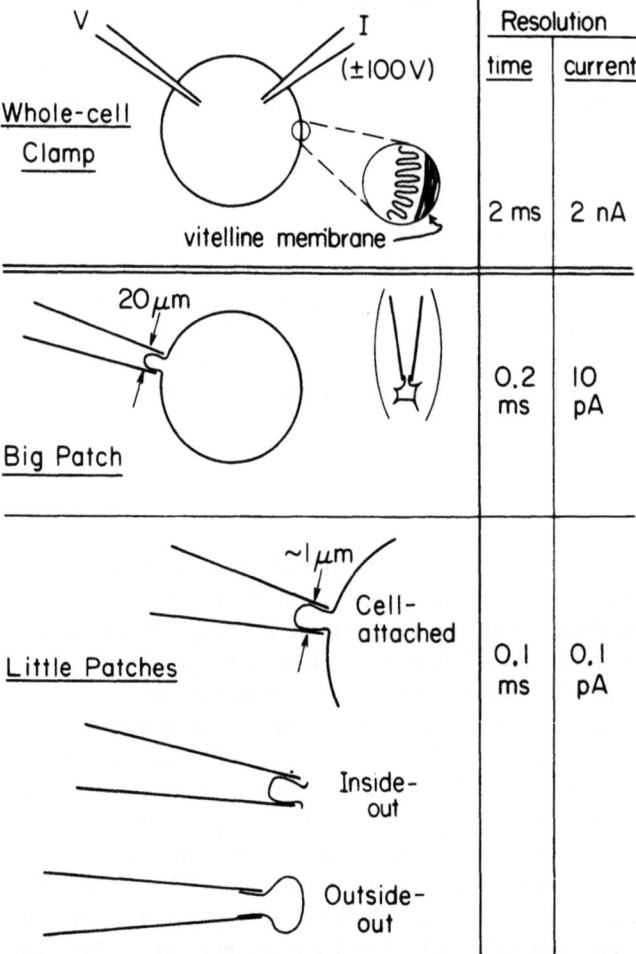

Fig. 1. Electrophysiological measurements on *Xenopus* oocytes. The *top panel* shows a standard two-microelectrode voltage clamp arrangement, suitable for recording macroscopic currents. The two *lower panels* show gigaohm-seal patch recording configurations that are possible after the vitelline layer has been stripped from contact with the plasma membrane. The "big patch" (Leonard et al. 1986a; Stuhmer et al. 1987) is useful for macroscopic recordings, at high temporal resolution, of channels present at densities too low for robust single-channel recordings. The standard "little patch" configurations are shown in the *bottom panel*. The temporal resolution and the smallest resolvable current are shown for each technique

was transferred overnight by capillary blot in 20 X SSPE (3.6 M NaCl, 0.2 M NaH_2PO_4, 20 mM EDTA) to Hybond-N (Amersham). The size distribution and recovery of mRNA from sucrose-gradient pools was determined by hybridization of Northern blot filters to a ^{32}P-labeled poly(dT) probe. Preparation of the probe and hybridization were performed as previously described (Goldin et al. 1986).

Electrophysiology. Macroscopic currents (see Fig. 1) were measured with a two-microelectrode voltage-clamp circuit (Dagan Instruments, Minneapolis, Minn) on oocytes incubated in ND96 (Dascal et al. 1986a) for 2–5 days after injection.

Microelectrodes were filled with 3 M KCl and had resistances of 0.5–2 MΩ. Ba currents were detected in 40 mM $Ba(OH)_2$, 50 mM NaOH, 2 mM KOH, 5 mM Hepes (pH 7.4) with methanesulphonic acid. To block Na currents, tetrodotoxin (TTX) was added to a final concentration of 1 μM. Other currents were detected in ND96 saline without added antibiotics and pyruvate. Pulse protocols and data analysis were as previously described (Leonard et al. 1987a).

Single channel currents were recorded on oocytes defolliculated by exposure to hypertonic solution, followed by manual stripping of the vitelline membrane. The currents were measured in the outside-out configuration (Fig. 1). The bath solution contained 70 mM $BaCl_2$, 10 mM Hepes (pH 7.4), and 1 μM TTX. The pipette was filled with 90 mM KCl, 10 mM NaCl, and 10 mM Hepes (pH 7.4).

Figure 2 presents the types of voltage-clamp tests used to detect Ca^{2+} channels in injected oocytes. A convenient but indirect measure of Ca^{2+} currents is given by the endogenous Cl channels activated by intracellular Ca^{2+} ions that flow through the RNA-induced Ca^{2+} channels. For more direct measurements, we have eliminated the Ca^{2+} from the extracellular solutions. The channels are then monitored by measuring the currents carried by Ba, which substitutes poorly if at all for Ca^{2+}, in two complicating phenomena: (a) activation of the Cl current and (b) inactivation of the Ca^{2+} channels themselves.

Macroscopic currents. The expressed currents have the following characteristics (Leonard et al. 1987a). Observable currents were present at test potentials more positive than about +30 mV. The current-voltage relation peaked at about +15 mV. The currents inactivated rather slowly, by about two-thirds after several seconds. The currents were insensitive to dihydropyridines – both antagonists such as nifedipine and agonists such as Bay K8644 – even at concentrations up to 20 μM. They were also insensitive to verapamil (10 μM). They were blocked by Cd^{2+} with an apparent dissociation constant of 6 μM and by Ni at about 500 μM, but they were not blocked by ω-conotoxin.

Single-Channel Currents. In single-channel experiments (Fig. 3), the Ba currents were recorded in outside-out patches. The openings appeared throughout the course of a 50-ms depolarization and summed to form a maintained macroscopic current (not shown) similar to the macroscopic currents recorded. The current-voltage relation (not shown) revealed a single-channel conductance of 12–15 pS.

RNA Fractionation. Oligo-dT blotting (Goldin et al. 1986; Krafte et al. 1988) allows the conclusion that sucrose-gradient fractionation produces a clear separation

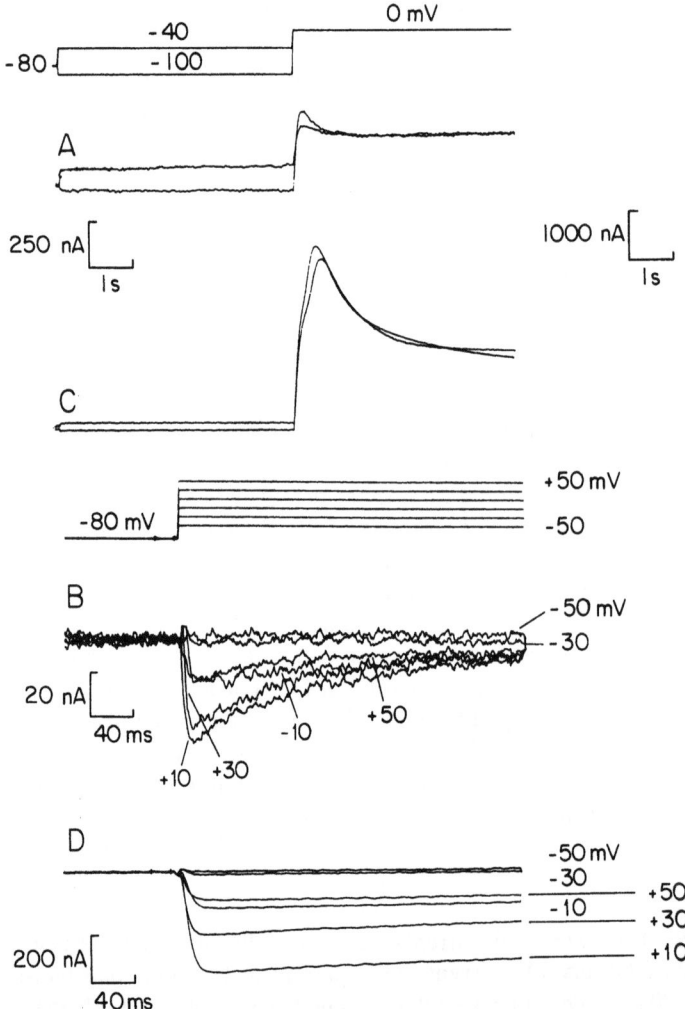

Fig. 2A–D. A comparison of Ca^{2+}-activated Cl currents *(left)* and Ba currents *(right)* from a noninjected *(A, B)* and an mRNA-injected *(C, D)* oocyte. Ca^{2+}-activated Cl current, elicited by depolarization to 0 mV from a holding potential of −100 or −40 mV in normal saline, was larger for mRNA-injected oocytes *(C)*, and it was less inactivated by holding at −40 mV than was the current from noninjected oocytes *(A)*. Ba current, in Ba-methane sulphonate saline, which was isolated by subtraction of Cd-insensitive currents, was considerably larger in mRNA-injected oocytes *(D)* than in noninjected cells *(B)*. Note the relatively rapid decay of Ba currents in noninjected oocytes. Note also differences in *calibration bars* for each panel. (From Leonard et al. 1987a)

between mRNAs of relatively large size (6–10 kb) and those of relatively small size (1–3 kb). As shown in Table 1, both of these size fractions induced voltage-dependent Ba currents in oocytes. Furthermore, the characteristics of the currents induced by fractionated RNA were similar to those induced by total RNA with regard to voltage dependence, inactivation, phorbol ester stimulation, and Cd blockade. Figure 4 gives an example of currents induced by the low-molecular-weight (MW) fraction.

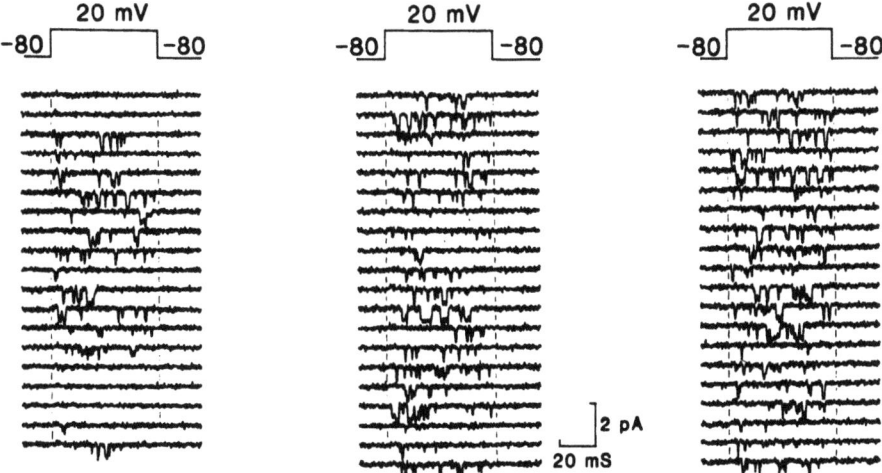

Fig. 3. Consecutive traces from an outside-out patch in a high-Ba solution from an oocyte injected several days previously with brain poly(A)+ RNA. Most of the traces show single-channel currents. See text for details

Table 1. RNA size fractionation of Ba and Na currents

Size fraction (kb)	I_{Na} (nA)	I_{Ba} (nA)	No. of oocytes
6–10	479 ± 219	361 ± 60	5
1–3	0	132 ± 13	15
Noninjected	0	22 ± 2	5

Amplitudes of voltage-dependent currents in oocytes injected with the various fractions are expressed as means ± SEM

Discussion

The characteristics of our induced Ba currents are in general agreement with those reported for Ca^{2+} fluxes in brain synaptosomes by Nachshen and Blaustein (1979) and by Nachshen (1985). The characteristics also agree in several respects with those reported for the N-type Ca^{2+} channels (Nowycky et al. 1985). There is pleasing agreement with the idea that N-type channels are primarily responsible for the Ca^{2+} influx that leads to transmitter release at brain synapses (Hirning et al. 1988). Perhaps the most serious point of disagreement is the insensitivity of our currents to ω-conotoxin (McCleskey et al. 1987).

The fractionation data point up several differences between the expression of brain Na and Ca channels in oocytes. First, the distribution of mRNA inducing Ca channels is much broader than that for Na channels. The low-MW RNA, for instance, yields detectable Ba currents, but no detectable Na currents. This observation is particularly

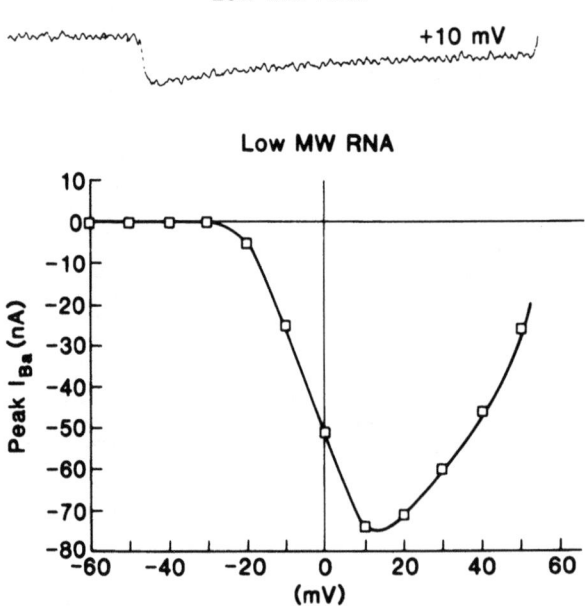

Fig. 4. A Ba-current waveform *(left)* for a voltage step to +10 mV from a holding potential of −80 mV in an oocyte injected several days previously with brain poly(A)+ RNA. The inactivation time course is quite similar to that observed with unfractionated RNA. The current-voltage relation *(right)* is also indistinguishable from that with unfractionated RNA

interesting in view of the fact that all known Na channels (Noda et al. 1986a) and the putative Ca^{2+}-channel subunit (Tanabe et al. 1987) are so large that their mRNAs have coding regions in excess of 5.5 kb. Clearly a shorter coding region is inducing currents in oocytes injected with rat brain mRNA. Although other interpretations are possible, the simplest is that some Ca^{2+} channels can be homo- or heterooligomers of subunits that correspond to only one or two of the four homology units in the known Na or Ca channels. Thus, the novel Ca-channel proteins might have the same organization as the K-channel proteins recently described from the *Shaker* lokus of *Drosophila* (Schwarz et al. 1988; Kamb et al. 1988) and expressed in oocytes (Timpe et al. 1988; Iverson et al. 1988).

A second point of difference concerns the apparent sufficiency of a single mRNA size fraction; that is, the pharmacology and kinetics of the induced Ba currents do not differ between total mRNA, the high-MW fraction, and the low-MW fraction. In contrast, normal Na currents are not induced in oocytes by RNA encoding the ~200-kd α-subunit alone; other proteins – encoded by RNA in the size range 2–4 kb – must also be present for normal kinetic and voltage-sensitive properties of the channel. Krafte et al. (1988) raise several possibilities about the nature of the additional protein(s): briefly, (a) mammalian brain and skeletal muscle Na channels contain one or two small subunits which could be necessary for normal function, (b) the additional protein(s) could be required for the very extensive and unusual glycosylations, and (c) the additional protein(s) could play another role in assembly. A similar but less completely described situation may apply to the expression of brain K currents in oocytes (Rudy et al. 1987).

There remain many questions about the molecular biology of neuronal Ca^{2+} channels and their expression in oocytes. For instance, dihydropyridine-sensitive high-threshold (L-type) and low-threshold transient (T-type) Ca^{2+} channels have often been described in the brain, yet these are not detected in the brain RNA-injected oocytes despite the fact that heart RNA induces dihydropyridine-sensitive channels in oocytes (Dascal et al. 1986a). As noted in the Introduction, further progress will call for cloning and expression of brain Ca^{2+} channels.

References

Auffray C, Rougeon F (1980) Purification of mouse immunoglobulin heavy-chain messenger RNAs from total myeloma tumor RNA. Eur J Biochem 107:303–324

Barnard EA, Miledi R, Sumikawa K (1982a) Translation of exogenous messenger RNA coding for nicotinic acetylcholine receptors produces functional receptors in *Xenopus* oocytes. Proc R Soc Lond B 215:241

Boyle MB, Azhderian NJ, MacLusky NJ, Naftolin F, Kaczmarek LK (1987) *Xenopus* oocytes injected with rat uterine RNA express very slowly activating potassium currents. Science 235:1221–1224

Claudio T, Green WN, Hartman DS, Hayden D, Paulson HL, Sigworth FJ, Sine SM, Swedlund A (1987) Genetic reconstitution of functional acetylcholine receptor channels in mouse fibroblasts. Science 238:1688–1694

Dascal N, Snutch TP, Lubbert H, Davidson N, Lester HA (1986a). Expression and modulation of voltage-gated calcium channels after RNA injection in Xenopus oocytes. Science 231:1147–1150

Dascal N, Ifune C, Hopkins R, Snutch TP, Lubbert H, Davidson N, Lester HA (1986b) Involvement of a GTP-binding protein in mediation of serotonin and acetylcholine responses in *Xenopus* oocytes injected with rat brain messenger RNA. Mol Brain Res 1:201–209

Dumont JN (1972) Oogenesis in *Xenopus laevis* (Daudin). I. Stages of oocyte development in laboratory maintained animals. J Morphol 136:153–180

Goldin AL, Dowsett A, Lubbert H, Snutch TP, Dunn R, Lester HA, Catterall WA, Davidson N (1986) Expression in *Xenopus* oocytes of the α subunit of the rat brain Na channel is sufficient for function. Proc Natl Acad Sci 83:7503–7507

Gundersen CB, Miledi R, Parker I (1983b) Voltage-operated channels induced by foreign messenger RNA in *Xenopus* oocytes. Proc R Soc Lond B 220:131–140

Gurdon JB, Lane CD, Woodland HR, Marbaix G (1971) Use of frog eggs and oocytes for the study of messenger RNA and its translation in living cell. Nature 233:177–182

Hediger M, Coady MJ, Ikeda TS, Wright EM (1987b) Expression cloning and cDNA sequencing of the Na^+/glucose co-transporter. Nature 330:379–381

Hirning LD, Fox AP, McCleskey EW, Olivera BM, Thayer SA, Miller RJ, Tsien RW (1988) Dominant role of N-type Ca^{2+} channels in evoked release of norepinephrine from sympathetic neurons. Science 239:57–61

Imoto K, Methfessel C, Sakamann B, Mishina M, Mori Y, Konno T, Fukuda K, Kurasaki M, Bujo H, Fujita Y, Numa S (1986) Location of a delta-subunit region determining ion-transport through the acetylcholine-receptor channel. Nature 324:670–674

Iverson L, Rudy B, Davidson N, Lester HA, Tanouye MA (1988) Expression of A-type potassium channels from Shaker cDNAs. Proc Natl Acad Sci USA in press.

Kamb A, Tseng-Crank J, Tanouye MA (1988) Multiple products of the Drosophila *Shaker* gene contribute to potassium channel diversity. Neuron in press.

Krafte DA, Snutch TP, Leonard JP, Davidson N, Lester HA (1988) Evidence for the involvement of more than one mRNA species in controlling the inactivation process of rat and rabbit brain Na channels expressed in *Xenopus* oocytes. J Neurosci in press

Leonard JP, Snutch TP, Lubbert H, Davidson N, Lester HA (1986a) Macroscopic Na currents with gigaohm seals on mRNA-injected *Xenopus* oocytes. Biophys J 49:386a

Leonard JP, Nargeot J, Snutch TP, Davidson N, Lester HA (1987) Ca channels induced in *Xenopus* oocytes by rat brain mRNA. J Neurosci 7:875–881

Lubbert H, Snutch TP, Dascal N, Lester HA, Davidson N (1987a) Rat brain $5HT_{1C}$ receptors are encoded by a 5–6 kbase mRNA size class and are functionally expressed in injected *Xenopus* oocytes. J Neurosci 7:1159–1165

Lubbert H, Hoffman B, Snutch TP, van Dyke T, Levine AJ, Hartig PR, Lester HA, Davidson N (1987b) cDNA cloning of a serotonin 5HT$_{1C}$ receptor by using electrophysiological assays of mRNA injected *Xenopus* oocytes. Proc Natl Acad Sci USA 84:4332–4336

Masu Y, Nakayama K, Tamaki K, Harada Y, Kuno M, Nakanishi S (1987) cDNA cloning of bovine substance-K receptor through oocyte expression system. Nature 329:836–838

McCleskey EW, Fox AP, Feldmann DH, Cruz LJ, Olivera BM, Yoshikami D (1987) x-Conotoxin: direct and persistent blockade of specific types of calcium channels in neurons but not muscle. Proc Natl Acad Sci USA 84:4327–4331

Miledi R, Sumikawa K (1982) Synthesis of cat muscle acetylcholine receptors by *Xenopus* oocytes. Biomedical Res 3:390–399

Miledi R, Parker I, Sumikawa K (1982a) Synthesis of chick brain GABA receptors by frog oocytes. Proc R Soc Lond B 216:509–515

Mishina M, Tobimatsu T, Imoto K, Tanaka KI, Fujita Y, Fukuda K, Kurasaki M, Takahashi H, Morimoto Y, Hirose T, Inayama S, Takahashi T, Kuno M, Numa S (1985) Localization of functional regions of acetylcholine receptor α-subunit by site-directed mutagenesis. Nature 313:364–369

Mishina M, Takai T, Imoto K, Noda M, Takahashi T, Numa S, Methfessel C, Sakmann B (1986) Molecular distinction between fetal and adult forms of muscle acetylcholine receptor. Nature 321:406–411

Nachshen DA, Blaustein MP (1979) The effects of some organic "calcium antagonists" on calcium influx in presynaptic nerve terminals. Mol Pharmacol 16:579–586

Nachshen DA (1985) The early time course of potassium stimulated calcium uptake in presynaptic nerve terminals isolated from the rat brain. J Physiology 361:251–268

Noda M, Ikeda T, Kayano T, Suzuki H, Takeshima H, Kurasaki M, Takahashi H, Numa S (1986) Existence of distinct sodium channel messenger RNAs in rat brain. Nature 320:188–192

Noda M, Ikeda T, Kayano T, Suzuki H, Takeshima H, Kuno M, Numa S (1986) Expression of functional sodium channels from cloned cDNA. Nature 322:826–828

Nowycky MC, Fox AP, Tsien RW (1985b) Three types of neuronal calcium channel with different calcium agonist sensitivity. Nature 316:440–443

Oron Y, Straub RE, Trautman P, Gershengorn MC (1987) Decreased TRH receptor mRNA activity precedes homologous downregulation: assay in oocytes. Science 238:1406–1408

Rudy B, Hoger JH, Davidson N, Lester HA (1987) Expression of different "A" currents in *Xenopus* oocytes injected with total or fractionated rat brain mRNA. Soc Neurosci Abst 13:178

Sakmann B, Methfessel C, Mishina M, Takahashi T, Takai T, Kurasaki M, Fukuda K, Numa S (1985) Role of acetylcholine receptor subunits in gating of the channel. Nature 318:538–543

Schwarz TL, Tempel BL, Papazian DM, Jan YN, Jan LY (1988) Multiple potassium-channel components are produced by alternative splicing at the *Shaker* locus in *Drosophila*. Nature 331:137–342

Stuhmer W, Methfessel C, Sakmann B, Noda M, Numa S (1987) Patch clamp characterization of sodium channels expressed from rat brain cDNA. Eur Biophys J 14:131–138

Tanabe T, Takeshima H, Mikami A, Flockerzi V, Takahashi H, Kangawa K, Kojima H, Matsuo H, Hirose T, Numa S (1987) Primary structure of the receptor for calcium channel blockers from skeletal muscle. Nature 328:313–318

Timpe LC, Schwarz TL, Tempel BL, Papazian DM, Jan YN, Jan LY (1988) Expression of functional potassium channels from *Shaker* cDNA in *Xenopus* oocytes. Nature 331:143–145

White MM, Mixter-Mayne K, Lester HA, Davidson N (1985) Mouse-*Torpedo* hybrid acetylcholine receptors: functional homology does not equal sequence homology. Proc Natl Acad Sci USA 82:4852–4856

Yoshii K, Yu L, Mixter-Mayne K, Davidson N, Lester HA (1987b) Equilibrium properties of mouse-*Torpedo* acetylcholine receptor hybrids expressed in *Xenopus* oocytes. J Gen Physiol 90:553–573

3 Calcium Channels in the Cardiovascular and Endocrine System

Ca^{2+} Pathways Mediating Agonist-Activated Contraction of Vascular Smooth Muscle and EDRF Release from Endothelium

N.J. Lodge and C. van Breemen

Department of Pharmacology, University of Miami School of Medicine, Miami, Florida 33101, USA

Vascular Smooth Muscle

Depletion of Ca^{2+} from the extracellular space surrounding vascular smooth muscle cells, or inhibition of Ca^{2+} influx into these cells by La^{3+}, completely blocks high potassium depolarization induced force development and abbreviates that brought about by agonists such as norepinephrine (see, for example, van Breemen et al. 1980). These well-known observations establish that force development in vascular smooth muscle is regulated by Ca^{2+}-dependent mechanisms, regardless of the type of myofilament interaction involved (see Rembold and Murphy 1986; Rasmussen et al. 1987).

Depolarization-Induced Force Development

Depolarization of vascular smooth muscle is known to increase the influx of extracellular Ca^{2+} into the cytosol (van Breemen et al. 1980) via voltage-gated Ca^{2+} channels (Bean et al. 1986; Friedman et al. 1986; Benham et al. 1987; Yatani et al. 1987). Moreover, a causal relationship between Ca^{2+} influx and force development has been established by the parallel block of force and $^{45}Ca^{2+}$ flux by Ca^{2+} antagonists (van Breemen et al. 1982). Considerable evidence suggests that this depolarization-induced Ca^{2+} influx into vascular smooth muscle cells is mediated by two distinct types of voltage-gated Ca^{2+} channels: one which is activated by relatively weak depolarizations but inactivates relatively rapidly (T or fast type) and another which is activated only be relatively strong depolarizations and inactivates relatively slowly (L or slow type). Both of these two channel types are blocked by La^{3+} (Friedman et al. 1986; Bean et al. 1986), Cd^{2+} (Sturek and Hermsmeyer 1986; Yatani et al. 1987), and Co^{2+} (Loirand et al. 1986; Yatani et al. 1987), but they may be distinguished by their differing thresholds for activation, voltage dependence of inactivation, sensitivity to dihydropyridines, half-time for current decay, relative permeabilities to Ba^{2+}, and single-channel conductance (for references see above).

Dacquet et al. (1987) have suggested that phasic K^+-induced tension may depend on the activation of both fast- and slow-type Ca^{2+} channels while maintained force may depend on Ca^{2+} influx through slow (dihydropyridine-sensitive) type Ca^{2+} channels. Such a postulate is consistent with the dependence of maintained K^+-induced force development on Ca^{2+} influx through a Ca^{2+} entry pathway sensitive to block by dihydropyridines and other organic Ca^{2+} antagonists (van Breemen et al. 1982).

Agonist-Induced Force Development

The Ca^{2+} required for agonist-induced force development may be derived from the release of Ca^{2+} from the sarcoplasmic reticulum (SR) in conjunction with Ca^{2+} influx from the extracellular space (Bond et al. 1984; Leijten and van Breemen 1984; Morgan and Morgan 1984) – the relative importance of the two processes depending on the nature and concentration of the agonist employed (van Breemen et al. 1980; Cauvin et al. 1985). However, the supply of Ca^{2+} from the SR, even with maximally effective concentrations of agonist, is only transient (van Breemen et al. 1981) necessitating a continued influx of extracellular Ca^{2+} for maintained force development. The anticipated agonist-induced Ca^{2+} influx has been measured routinely in vascular smooth muscle using $^{45}Ca^{2+}$ techniques in whole tissue (van Breemen et al. 1980) and in cultured cell monolayers (Wallnöfer et al. 1987; Zschauer et al. 1987).

Agonists including norepinephrine, 5-hydroxytryptamine, angiotensin, and histamine cause membrane depolarization in certain types of vascular smooth muscle (Keatinge 1979; Cauvin et al. 1985; Neild and Kotecha 1987) which, if of sufficient magnitude, would initiate Ca^{2+} influx through voltage-gated Ca^{2+} channels. However, agonist-induced force development may also occur with little or no depolarization, as well as in vessels continuously depolarized by high K^+ solution (Su et al. 1964; Somlyo and Somlyo, 1968; Haeusler 1978; Keatinge 1979; Bolton et al. 1984). For example, in the rabbit aorta, maintained NE (10^{-5} M norepinephrine)-induced contractions and $^{45}Ca^{2+}$ uptake proceed in the virtual absence of membrane depolarization (Cauvin et al. 1985). Moreover, the enhanced $^{45}Ca^{2+}$ uptake measured in the presence of NE was found to be relatively insensitive (in comparison to the high K^+ stimulated $^{45}Ca^{2+}$ uptake) to organic calcium antagonists (Fig. 1; Cauvin et al. 1983) yet still sensitive to inhibition by La^{3+} (Lukeman 1987). These observations, combined with the finding that the unidirectional $^{45}Ca^{2+}$ uptake stimulated by simultaneous application of maximally effective concentrations of NE and high K^+ was equal to the sum of that stimulated by each alone (i.e., an additive effect was demonstrated), led to the postulate that NE-stimulated Ca^{2+} uptake in the rabbit aorta was facilitated by a pathway distinct from that mediating the high K^+-induced Ca^{2+} uptake. This type of data led to the hypothesis that the vascular smooth muscle sarcolemma contains two distinct Ca^{2+} entry pathways: one activated by membrane depolarization and the other by agonist interaction with its receptor (van Breemen et al. 1979). A similar hypothesis was also proposed by Bolton (1979).

Direct electrophysiological evidence now exists for one type of receptor-gated channel in vascular smooth muscle (Benham and Tsien 1987a). This ATP-activated channel may represent an example of a channel gated directly by ligand, such as that activated by acetylcholine at the neuromuscular junction (Pepper et al. 1982), although coupling through a tightly associated transducing mechanism has not been ruled out. The ATP-gated channel exhibits a permeability ratio $P_{Ca}:P_{Na}$ of approximately 3 at near-physiological ion concentrations, is insensitive to block by nifedipine (5×10^{-6} M) and Cd^{2+} (0.5 mM), and can be activated at the cells resting membrane

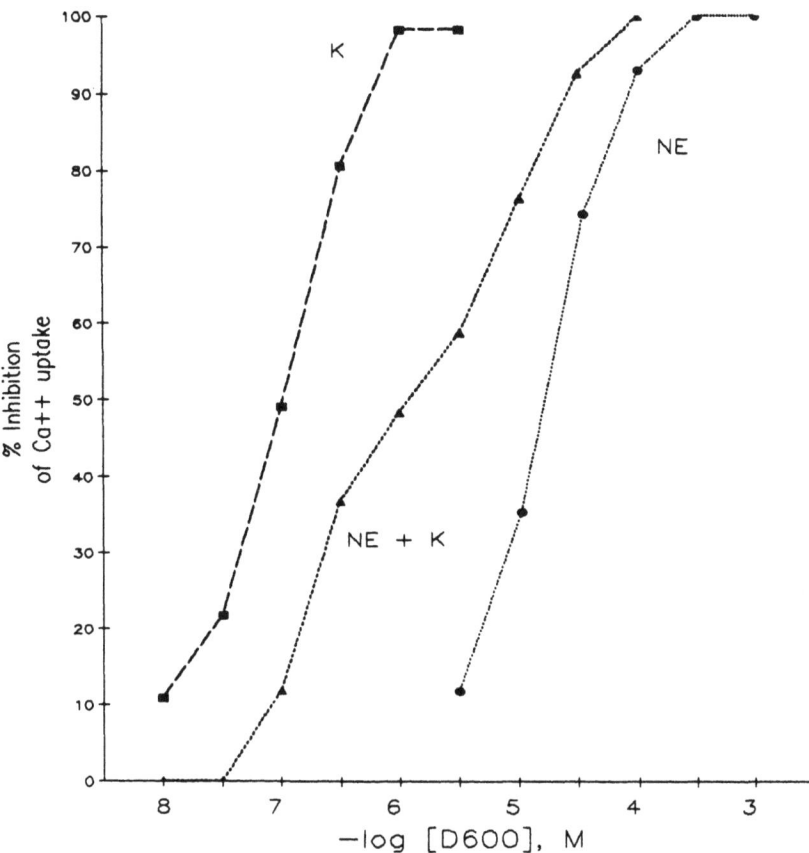

Fig. 1. The inhibition of 80 mM K$^+$ (K), 10^{-5}M norepinephrine (*NE*), and NE + K$^+$ stimulated ^{45}Ca^{2+} uptakes by increasing concentrations of D600. Stimulated ^{45}Ca^{2+} uptake refers to that above resting (basal) ^{45}Ca^{2+} uptake. The magnitude of the NE + K$^+$ stimulated ^{45}Ca^{2+} uptake was approximately equivalent to the sum of that stimulated by NE and K$^+$ alone (i.e., additivity was demonstrated). Note that K$^+$ stimulated ^{45}Ca^{2+} uptake was far more sensitive to D600 inhibition than NE-stimulated ^{45}Ca^{2+} uptake. The sensitivity of NE + K$^+$ stimulated ^{45}Ca^{2+} uptake to D600 was intermediate between the two individual stimulants and covered a broader concentration range, consistent with the hypothesis that NE and K$^+$ stimulate ^{45}Ca^{2+} uptake via different mechanisms. (Redrawn from Lukeman 1987)

potential. Electrophysiological experiments have yet to provide evidence for a Ca^{2+}-permeable channel directly gated by norepinephrine (Bean et al. 1986; Droogmans et al. 1987). However, it is noteworthy that a Ca^{2+}-permeable channel activated by elevated intracellular Ca^{2+} has recently been described in vascular smooth muscle (Sturek et al. 1988). Such a channel might be opened in response to an agonist-induced release of Ca^{2+} from the SR.

The only example we are aware of to date of indirect coupling between stimulant and ion-channel gating in vascular smooth muscle is the channel activated by elevated intracellular Ca^{2+} (Sturek et al. 1988). However, it is likely that other transducing mechanisms linking receptor occupation to channel opening will be found. For example, agonist-receptor interaction is linked to ion-channel opening in a variety of muscle and nonmuscle tissues through guanine nucleotide binding proteins (Brown and Birnbaumer 1988). Other candidates for indirect coupling include the products of agonist-activated phospholipase C hydrolysis of phosphotidylinositol 4,5-bisphosphate. Inositol 1,4,5-trisphosphate (IP_3) and inositol 1,3,4,5-tetrakisphosphate have been proposed as mediators of increased sarcolemmal Ca^{2+} permeability in sea urchin eggs and T-lymphocytes, respectively (Irvine and Moore 1986; Kuno and Gardner 1987).

Vascular smooth muscle has the capacity to respond to a wide variety of agonists, many of which stimulate $^{45}Ca^{2+}$ entry via a pathway which is relatively insensitive to organic calcium antagonists. Individual agonists such as angiotensin II and vasopressin (Zschauer et al. 1987), as well as norepinephrine, histamine, and serotonin (van Breemen et al. 1985), have all been shown to induce $^{45}Ca^{2+}$ influx which is additive to that stimulated by high K^+ but not with that stimulated by the other agonists. This suggests that these agonists share the same Ca^{2+}-influx pathway and raises the interesting possibility that multiple receptors may be coupled to a single Ca^{2+}-influx pathway.

Agonists may also affect Ca^{2+} influx by modulating voltage-gated Ca^{2+} channels. Benham and Tsien (1987b) have reported that norepinephrine, acting through non-α-, non-β-adrenergic receptors increases slow type current but not the fast type current in the rabbit ear artery. Endothelin (Yanagisawa et al. 1988), a potent constrictor of vascular smooth muscle, has also been shown to augment slow type Ca^{2+} current in vascular smooth muscle cells (see accompanying contributions, this volume). In contrast, α-adrenoceptor stimulation of the rabbit ear artery has been found to reduce peak Ca^{2+} current (Droogmans et al. 1987) and to inhibit slow type current (but stimulate fast-type current) in the portal vein (Pacaud et al. 1987). The α-adrenergic inhibition of slow current is hypothesized to be mediated by Ca^{2+} release from the SR (Pacaud et al. 1987). Ca^{2+}-dependent inactivation of Ca^{2+} current has recently been directly demonstrated in the rabbit portal vein by Ohya et al. 1988.

Role of the Sarcoplasmic Reticulum

Application of agonists such as norepinephrine or angiotensin to vascular smooth muscle has been shown to generate IP_3 from phosphatidylinositol 4,5-bisphosphate (Nabika et al. 1985; Hashimoto et al. 1986) while exogenously applied IP_3 has been demonstrated to release Ca^{2+} from the SR of permeabilized smooth muscle cells (Suematsu et al. 1984; Hashimoto et al. 1985; Somlyo et al. 1985; Yamamoto and van Breemen 1985). These observations implicate IP_3 as a second messenger coupling receptor occupation to SR Ca^{2+} release in vascular smooth muscle. Ca^{2+} may also be released from the SR via the process of Ca^{2+}-induced Ca^{2+} release (Saida 1981, 1982; Saida and van Breemen 1984), which may itself be triggered by the IP_3-induced Ca^{2+} release, thereby amplifying the rate of Ca^{2+} discharge from the SR (van Breemen and Saida, in press). Agonist dissociation from its receptor terminates Ca^{2+} release from the SR, presumably as a result of reduced IP_3 formation and the action of an endogenous IP$_3$-specific phosphatase (Walker et al. 1987). The SR may then promote relaxation by sequestering Ca^{2+} from the cytosol.

In resting vascular smooth muscle, or during stimulation with high K$^+$ or agonists which do not discharge the SR (or do so only to a small degree), a fraction of the Ca^{2+} entering the cell is thought to be sequestered by the SR (presumably the subsarcolemmal portion) which acts as a buffer against changes in cytosolic Ca^{2+}. This model is depicted schematically in Fig. 2. Evidence for this hypothesis comes from the following observations. (a) Stimulation by either high K$^+$ solution or the calcium-channel agonist Bay K8644 increases the Ca^{2+} content of the SR, as assessed by an increase in the ^{45}Ca^{2+} content of the caffeine-sensitive intracellular store (Leijten and van Breemen 1984; Hwang and van Breemen 1985). (b) Substances believed to stimulate the rate of SR Ca^{2+} sequestration (dibutyryl cAMP, forskolin) increase the amount of ^{45}Ca^{2+} influx required to generate a given level of force (Hwang and van Breemen, 1987). (c) Prior agonist-induced discharge of Ca^{2+} from the SR slows the rate of force development generated by subsequently applied high K$^+$ even though stimulated ^{45}Ca^{2+} influx proceeds normally (Loutzenhiser and van Breemen 1983). (d) The rate of Ca^{2+} entry into the cell determines the ability of a given amount of Ca^{2+} to elicit

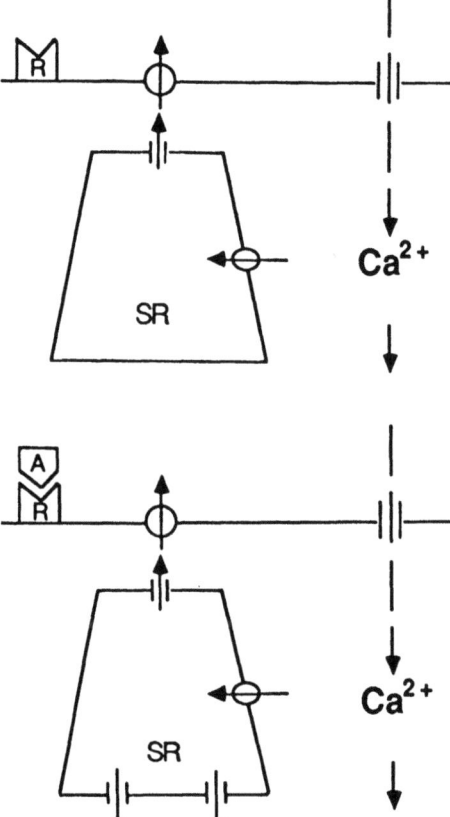

Fig. 2. Diagrammatic representation of the buffer barrier hypothesis. In the *upper panel* the receptor (*R*) located on the sarcolemmal membrane is unoccupied, and the sarcoplasmic reticulum (*SR*) functions to sequester influxing Ca^{2+}. This situation would occur in the resting and high K$^+$ stimulated tissue. The same mechanisms would also be operative with agonists which do not discharge the SR or do so to only a small degree. The Ca^{2+}-buffering efficiency of the SR is augmented by substances which stimulate the SR Ca^{2+} pumps, thereby enhancing the rate of Ca^{2+} accumulation by this organelle. In the *lower panel* receptor occupation by agonist (*A*) has discharged the SR Ca^{2+} by opening channels in the SR membrane. In this situation (e.g., during norepinephrine activation), the SR can no longer efficiently buffer influxing Ca^{2+}

force (van Breemen, 1977). It is proposed therefore that, depending on the nature of the stimulant, the SR may function either to supply additional Ca^{2+} to the cytosol or to buffer changes in cytosolic Ca^{2+} caused by influxing extracellular Ca^{2+}.

Endothelial Cells

The endothelial cells of blood vessels can modulate vascular smooth muscle tone by releasing constricting or relaxing substances in response to a variety of chemical or physical stimuli (Vanhoutte et al. 1986; Harder 1987). Endothelin, a potent endothelium-derived constricting factor, has recently been identified (Yanagisawa et al. 1988; see above) in addition to the relatively well-characterized endothelium-derived relaxing factor (EDRF) which is thought to be nitric oxide (Palmer et al. 1987). While the mechanism(s) involved in the release of endothelium-derived substances is yet to be fully determined, it is known that EDRF release in response to certain agonists is a Ca^{2+}-dependent process (Singer and Peach 1982; Furchgott 1984; Long and Stone 1985; Colden-Standfield et al. 1987). The following sections will consider the potential sources of Ca^{2+} for agonist-induced release of EDRF from endothelial cells.

Lack of Depolarization-Induced Ca^{2+} Influx

Membrane depolarization brought about by elevated K^+ failed to increase $^{45}Ca^{2+}$ uptake into endothelial cells (Johns et al. 1987) in agreement with electrophysiological experiments which were unable to detect inward current in response to depolarization, even with a high divalent cation concentration or with the Ca^{2+}-channel agonist Bay K8644 in the extracellular medium (Colden-Stanfield et al. 1987; Johns et al. 1987; Takeda et al. 1987). These data indicate that voltage-gated Ca^{2+} channels are absent from the sarcolemma of endothelial cells; a postulate also supported by the finding that organic Ca^{2+}-channel antagonists are usually ineffective at inhibiting EDRF release (Schoeffer and Miller 1986; Colden-Stanfield et al. 1987; Jayakody et al. 1987; Takeda et al. 1987).

Agonist-Induced Ca^{2+} Influx

The observations that depletion of extracellular Ca^{2+} diminished both EDRF release (see above) and agonist-induced increases in cytosolic Ca^{2+} (Hallam and Pearson, 1986; Colden-Stanfield et al. 1987; Morgan-Boyd et al. 1987) suggest that an agonist-induced Ca^{2+} influx pathway is present in the endothelial cell sarcolemma. Direct support for this supposition comes from the finding that $^{45}Ca^{2+}$ influx into endothelial cells is increased by the agonists thrombin and bradykinin (Johns et al. 1987; Johns and van Breemen, unpublished observations). The application of either agonist has also been shown to generate an inward current in these cells, although only in a small fraction of those cells tested. This current displayed a reversal potential close to 0 mV and is believed to be carried by nonselective cation channels (Johns et al. 1987).

Furthermore, single-channel Ba^{2+} currents have been recorded in cell-attached membrane patches when bradykinin was included in the patch pipette solution. These channels, which were not observed in the absence of agonist, exhibited an extrapolated reversal potential (zero current) of approximately 0 mV and a slope conductance of approximately 20 pS. As with the whole-cell currents, these channels were only observed in a small fraction of the endothelial cells tested (Lodge et al. 1988).

Endothelial cells also respond to mechanical stimuli such as stretch and shear/flow. The channels opened by stretch are relatively nonselective and may admit Ca^{2+} (Lansman et al. 1987), while shear stress activates a K^+ dependent current (Oleson et al. 1988).

Ca^{2+} release from the Endoplasmic Reticulum

Agonist-induced Ca^{2+} release from a nonmitochondrial intracellular site (presumably the endoplasmic reticulum) has been demonstrated in endothelial cells by the finding that agonists could still increase $^{45}Ca^{2+}$ efflux (Johns et al. 1987; Lodge et al. 1988) and fura-2 or Quin-2 fluorescence (Hallam and Pearson 1986; Luckhoff and Busse 1986; Colden-Stanfield et al. 1987; Morgan-Boyd et al. 1987) in the absence of extracellular Ca^{2+}.

Activation of endothelial cells by bradykinin is associated with an increase in the concentration of cytosolic IP_3 (Derian and Moscowitz 1986; Lambert et al. 1986). This finding, coupled with the observation that exogenously applied IP_3 releases Ca^{2+} from permeabilized endothelial cells (Lodge et al. 1988), supports the postulate that agonist-receptor interaction at the sarcolemma is coupled to Ca^{2+} release from the endoplasmic reticulum by IP_3.

References

Bean BP, Sturek M, Pugg A, Hermsmeyer K (1986) Calcium channels in muscle cells isolated from rat mesenteric arteries: modulation by dihydropyridine drugs. Circ Res 59:229–235

Benham CD, Hess P, Tsien RW (1987) Two types of calcium channels in single smooth muscle cells from rabbit ear artery studied with whole-cell and single channel recordings. Circ. Res 61(suppl I):I10–I16

Benham CD, Tsien RW (1987a) A novel receptor-operated Ca^{2+}-permeable channel activated by ATP in smooth muscle. Nature 328:275–278

Benham CD, Tsien RW (1987b) Noradrenaline increases L-type calcium current in smooth muscle cells of rabbit ear artery independently of α- and β-receptors. J Physiol 390:98P

Bolton TB (1979) Mechanisms of action of transmitter and other substances on smooth muscle. Physiol Revs 59:606–718

Bolton TB, Lang RJ, Takewaki T (1984) Mechanisms of action of noradrenaline and carbachol on smooth muscle of guinea-pig anterior mesenteric artery. J Physiol 371:289–304

Bond M, Kitazawa T, Somlyo AV (1984) Release and recylcing of calcium by the sarcoplasmic reticulum in guinea pig portal vein smooth muscle. J Physiol Lond 355:677–695

Brown AM, Birnbaumer L (1988) Direct G protein gating of ion channels. Am J Physiol 254:H401–H410

Cauvin C, Loutzenhiser R, van Breemen C (1983) Mechanisms of calcium antagonist-induced vasodilation. Ann Rev Pharmacol Toxicol 23:373–374

Cauvin C, Lukeman S, Cameron J, Hwang O, van Breemen C (1985) Differences in norepinephrine activation and diltiazem inhibition of Ca^{2+} channels in isolated rabbit aorta and mesenteric resistance vessels. Circ Res 56:822–828

Colden-Stanfield M, Schilling WP, Ritchie AK, Eskin SG, Navarro LT, Kunze DL (1987) Bradykinin-induced increases in cytosolic calcium and ionic currents in cultured bovine aortic endothelial cells. Circ Res 61:632–640

Dacquet C, Loirand G, Mironneau C, Mirronneau J, Pacaud P (1987) Spironoactone inhibition of contraction and calcium channels in rat portal vein. Br J Pharmacol 92:535–544

Derian CK, Moskowitz MA (1986) Polyphosphoinositide hydrolysis in endothelial cells and carotid artery segments. J Biol Chem 261(8):3831–3837

Droogmans G, Declerck I, Casteels R (1987) Effect of adrenergic agonists on Ca^{2+} channel currents in single vascular smooth muscle cells. Pflügers Arch 409:7–12

Droogmans G, Raeymaekers L, Casteels R (1977) Electro- and pharmacomechanical coupling in the smooth muscle cells of the rabbit ear artery. J Gen Physiol 70:129–148

Friedman ME, Suarez-Kurtz G, Kaczorowski GJ, Katz GM, Reuben JP (1986) Two calcium currents in a smooth muscle cell line. Am J Cardiol 49:507–510

Furchgott RF (1984) The role of endothelium in the responses of vascular smooth muscle to drugs. Ann Rev Pharmacol Toxicol 24:175–197

Haeusler G (1978) Relationship between noradrenaline-induced depolarization and contraction in vascular smooth muscle. Blood Vessels 15:46–54

Hallam TJ, Pearson JD (1986) Exogenous ATP raises cytoplasmic free calcium in fura-2 loaded piglet aortic endothelial cells. FEBS 207:95–99

Harder DR (1987) Pressure-induced myogenic activation of cat cerebral arteries is dependent on intact endothelium. Circ Res 60:102–107

Hashimoto T, Hirata M, Ieoh T, Kanamura Y, Kuriyama H (1986) Inositol 1,4,5-trisphosphate activates pharmacomechanical coupling in smooth muscle of the rabbit mesenteric artery. J Physiol 370:605–618

Hwang KS, van Breemen C (1985) Effects of the Ca agonist Bay K8644 on ^{45}Ca influx and net Ca uptake into rabbit aortic smooth muscle. Eur J Pharmacol 116:299–305

Hwang KS, van Breemen C (1987) Effect of db-cAMP and forskolin on the ^{45}Ca influx, net Ca uptake and tension in rabbit aortic smooth muscle. Eur J Pharmacol 134:155–162

Irvine RF, Moore RM (1986) Micro-injection of inositol 1,3,4,5-tetrakisphosphate activates sea urchin eggs by a mechanism dependent on external Ca^{2+}. Biochem J 240:917–920

Jayakody RL, Kappagoda CT, Senartne MPJ, Sreeharan N (1987) Absence of effect of calcium antagonists on endothelium-dependent relaxation in rabbit aorta. Br J Pharmacol 91:155–164

Johns A, Lategan TW, Lodge NJ, Ryan US, van Breemen C, Adams DJ (1987) Calcium entry through receptor-operated channels in bovine pulmonary artery endothelial cells. Tissue and Cell 19:733–745

Keatinge WR (1979) Blood-vessels. Br Med Bull 35:249–254

Kuno M, Gardner P (1987) Ion channels activated by inositol 1,4,5-trisphosphate in plasma membrane of human T-lymphocytes. Nature 236:301–304

Lambert TL, Kent RS, Whorton AR (1986) Bradykinin stimulation of inositol polyphosphate production in porcine aortic endothelial cells. J Biol Chem 261:15288–15293

Lansman JB, Hallam TJ, Rink TJ (1987) Single stretch-activated ion channels in vascular endothelial cells as mechanatransducers? Nature 325:811–813

Leijten PAA, van Breemen C (1984) The effects of caffeine on the noradrenaline-sensitive calcium store in rabbit aorta. J Physiol 357:327–329

Lodge NJ, Adams DJ, Johns A, Ryan US, van Breemen C (1988) Calcium activation of endothelial cells. In: Proceedings of the second international symposium on resistance arteries. Halpern W (ed) Perinatology Press, Ithaca, NY

Loirand G, Pacaud P, Mironneau C, Mironneau J (1986) Evidence for two distinct calcium channels in rat vascular smooth muscle cells in short-term primary culture. Pflügers Arch 407:566–568

Long CJ, Stone TW (1985) The release of endothelium-derived relaxant factor is calcium dependent. Blood Vessels 22:205–208

Loutzenhiser R, van Breemen C (1983) The influence of receptor occupation on Ca^{2+} influx-mediated vascular smooth muscle contraction. Circ Res 53(Suppl 4):97–103

Luckhoff A, Busse R (1986) Increased free calcium in endothelial cells under stimulation with adenine nucleotides. J Cell Physiol 126:414–420

Lukeman DS (1987) A pharmalogic study of calcium influx pathways in rabbit aortic smooth muscle. Ph D Thesis, University of Miami, Miami

Morgan-Boyd R, Stewart JM, Savrek RJ, Hassid A (1987) Effects of bradykinin and angiotensin II on intracellular Ca^{2+} dynamics in endothelial cells Am J Physiol 253:C588–C598

Morgan JP, Morgan KG (1984) Stimulus-specific patterns of intracellular calcium levels in smooth muscle of ferret portal vein. J Physiol 351:116–122

Nabika T, Velletri PA, Lovenberg W, Beaven MA (1985) Increase in cytosolic calcium and phosphoinositide metabolism induced by angiotensin II and [arg] vasopressin in vascular smooth muscle cells. J Biol Chem 260:4661–4670

Neild TO, Kotecha N (1987) Relation between membrane potential and contractile force in smooth muscle of the rat tail artery during stimulation by norepinephrine, 5-hydroxytryptamine, and potassium. Circ Res 60:791–795

Ohya Y, Kitamura K, Kuriyama H (1988) Regulation of calcium current by intracellular calcium in smooth muscle cells of rabbit porta vein. Circ Res 62:375–383

Oleson S-P, Clapham DE, Davies PF (1988) Haemodynamic shear stress activates a K^+ current in vascular endothelial cells. Nature 331:168–170

Pacaud P, Loirand G, Mironneau C, Mironneau J (1987) Opposing effects of noradrenaline on the two classes of voltage-dependent calcium channels of single vascular smooth muscle cells in short-term primary culture. Pflügers Arch 410:557–559

Palmer RMJ, Ferrige AG, Moncada S (1987) Nitric oxide release accounts for the biological activity of endothelium-derived relaxing factor. Nature 327:524–526

Pepper K, Bradley RJ, Dreyer F (1982) The acetylcholine receptor at the neuromuscular junction. Physiological Rev. 62:1271–1340

Rasmussen H, Takuwa Y, Park S (1987) Protein kinase C in the regulation of smooth muscle contraction. FASEB J 1:177–185

Rembold CM, Murphy RA (1986) Myoplasmic calcium, myosin phosphorylation and regulation of the cross bridge cycling in swine arterial smooth muscle. Circ Res 58:803–815

Saida K (1981) Ca^{2+}- and depolarization-induced Ca^{2+} release in skinned smooth muscle fibers. Biochem Res 2:453–455

Saida K (1982) Intracellular Ca^{2+} release in skinned smooth muscle. J Gen Physiol 80:191–202

Saida K, van Breemen C (1984) Cyclic AMP modulation of adrenoceptor-mediated arterial smooth muscle contraction. J Gen Physiol 84:307–318

Schoeffter P, Miller RC (1986) Role of sodium-calcium exchange and effects of calcium entry blockers on endothelial-mediated responses in rat isolated aorta. Mol Pharmacol 30:53–57

Singer HA, Peach MH (1982) Calcium- and endothelial-mediated vascular smooth muscle relaxation in rabbit aorta. Hypertension 4(Suppl. 2):II19–25

Somlyo AV, Somlyo AP (1968) Electromechanical and pharmaconmechanical coupling in vascular smooth muscle. J Pharmacol Exp Ther 159:129–145

Sturek M, Thayer SA, Miller RJ (1988) Intracellular Ca release activates Ca-permeable ion channels in coronary artery smooth muscle cells. Biophys J 53:561 abstract

Sturek M, Hermsmeyer K (1986) Calcium and sodium channels in spontaneously contracting vascular muscle cells. Science 233:475–478

Su C, Bevan JA, Ursillo RC (1964) Electrical quiescence of pulmonary artery smooth muscle during sympathomimetic stimulation. Circ Res 15:20–27

Takeda K, Schini V, Stoechkel H (1987) Voltage-activated potassium but not calcium currents in cultured bovine aortic endothelial cells. Pflügers Arch 410:385–393

van Breemen C (1977) Calcium requirement for activation of intact aortic smooth muscle. J Physiol 272:317–329

van Breemen C, Aaronson P, Loutzenhiser R (1979) Na^+, Ca^{2+} interactions in mammalian smooth muscle. Pharmacol Rev 30(2):167–208

van Breemen C, Aaronson P, Loutzenhiser R, Meisheri K (1980) Ca^{2+} movements in smooth muscle. In: Zelis R, Schroeder JJ (eds) Symposium on calcium, calcium antagonists and cardiovascular disease. Chest 78:157S–165S van Breemen C, Hwang K, Loutzenhiser R, Lukeman S, Yamamoto H (1985) Ca entry into vascular smooth muscle. In: Cardiovascular effects of dihydropyridine-type calcium antagonists and agonists. 58–71. Springer Verlag, Berlin Heidelberg

van Breemen C, Loutzenhiser R, Mangel A, Meisheri K (1981) Ca^{2+} compartments in vascular smooth muscle. In: Varga E, Kövér A, Kovács (eds) Molecular and cellular aspects of muscle function. Advanced Physiological Sciences (vol 5)

van Breemen C, Mangel A, Fahim M, Meisheri K (1982) Selectivity of calcium antagonistic action in vascular smooth muscle. Am J Cardiol 49:507–510

van Breemen C, Saida K (1988) Cellular mechanisms regulating smooth muscle contraction. Ann Rev Physiol. In press

Vanhoutte PM, Rubanyi GM, Miller VM, Houston DS (1986) Modulation of vascular smooth muscle contraction by the endothelium. Ann Rev Physiol 48:307–320

Walker JW, Somlyo AV, Goldman YE, Somlyo AP, Trentham DR (1987) Kinetics of smooth and skeletal muscle activation by laser pulse photolysis of caged inositol 1,4,5-trisphosphate. Nature 327:249–252

Wallnöfer A, Cauvin C, Rüegg UT (1987) Vasopressin increases $^{45}Ca^{2+}$ influx in rat aortic smooth muscle cells. Biochem Biophys Res Comm 148:273–278

Yanagisawa M, Kierihara H, Kimura S, Tombe Y, Kobayashi M, Mitsui Y, Yazaki Y, Goto K, Masaki T (1988) A novel potent vasoconstrictor peptide produced by vascular endothelial cells. Nature 332:411–415

Yatani A, Seidel CL, Allen J, Brown AM (1987) Whole-cell and single-channel calcium currents of isolated smooth muscle cells from saphenous vein. Circ Res 60:523–533

Zschauer A, Scott-Burden T, Bühler FR, van Breemen C (1987) Vasopressor peptides and depolarization stimulated Ca^{2+} entry in cultured vascular smooth muscle. Biochem Biophys Res Comm 148:225–231

Calcium Channels and the Heart*

J. S. Dillon, X. H. Gu, and W. G. Nayler

Department of Medicine, University of Melbourne, Austin Hospital, Heidelberg, Victoria, 3084 Australia

Introduction

Calcium ions play a key role in muscle contraction. In the resting state the cytosolic Ca^{2+} is maintained at a concentration of around 10^{-7} M, but upon excitation this level increases by a factor of 100 or more. In cardiac muscle this increase involves an influx of Ca^{2+} through the voltage-sensitive Ca^{2+}-selective channels which traverse the sarcolemma and a secondary release of intracellularily stored Ca^{2+} (Fabiato and Fabiato 1979). The structure, biochemistry, ionic selectivity and voltage regulation of these trans-sarcolemmal channels have been the subject of previous papers in this symposium, as has their isolation and reconstitution.

The α_1 subunit of the Ca^{2+} channels contain specific high-affinity binding sites for the drugs which modulate their Ca^{2+}-conducting activity. These drugs are the Ca^{2+} antagonists, and the binding sites with which they interact function as "receptors." This paper deals primarily with some of the conditions which modulate the functioning of these binding sites in the heart, particularly with respect to their density (B_{max}) and to their affinity (K_D) for the calcium antagonists. Our own experiments have concentrated on the effects of ischaemia, oxygen deprivation, and chemically induced energy depletion. Before presenting the results of these experiments, however, it may be of interest to summarize some of the other conditions which modulate the density and affinity of these sites. These conditions include hypertension, cardiomyopathy, alcoholism, morphine, chronic calcium antagonist therapy and ontogeny.

Hypertension

Several studies (Chatelain et al. 1984; Sharma et al. 1986) have provided evidence of an increase in density, without any change in the affinity, of the cardiac dihydropyridine (DHP) binding sites in aged hypertensive rats. Chatelain et al. (1984), for example, reported a density of 96 fmol/mg protein in cardiac membranes obtained from 24-week-old normotensive Wistar Kyoto (WKY) rats, compared with 137 fmol/mg protein in cardiac membranes obtained from spontaneously hypertensive (SH) rats of the same age. At 9 weeks of age, however, there appeared to be no difference.

---

* These investigations were supported by the National Health and Medical Research Council, and the National Heart Foundation of Australia.

Ishii et al. (1986) also found a significant increase in the density of DHP binding sites in the hearts of mature SH rats, relative to age-matched WKY rats (187 fmol/mg protein in the WKY, and 218 fmol/mg protein in the SH rats). Although these studies provide evidence of an increase in DHP binding site density with no change in affinity in hearts of mature SH rats, they do not indicate whether the increase is a direct effect of the hypertension or a secondary consequence of the accompanying hypertrophy.

Cardiomyopathy

Cardiomyopathic hamsters have also been shown to exhibit an increase in the number of DHP binding sites which are present in cardiac membrane fragments (Wagner et al. 1986). This effect is not peculiar to DHP recognition sites, however, because there is a similar increase in the number of D888-binding sites. However, in both cases the change in density occurs without any change in affinity (Wagner et al. 1986).

Alcohol

The effect of chronic alcoholism on cardiac calcium-antagonist binding site density and affinity has not yet been established, but in the brain the density (B_{max}) increases by more than 90% (Messing et al. 1986) after only 25 days. Again, the change in density occurs without any change in affinity.

Morphine

Also here, data for heart is not yet available, but in the brain, 3-day morphine therapy (75-mg pellet given subcutaneously) causes a 60% increase in density without any change in affinity (Ramkumar and El-Fakahany 1984).

Chronic Calcium-Antagonist Therapy

Chronic therapy with high doses of nifedipine causes a decrease in cardiac DHP binding site density (Hawthorn et al. 1987), suggesting that down-regulation may take place. High-dose levels of the agonist Bay K8644 have also been shown to cause a decrease in density, and low-dose levels an increase (Gengo et al. 1988). In general, therefore, there are conditions which mediate changes in the density of the high-affinity calcium-antagonist binding sites. The fact that these changes occur in the absence of a change in affinity may indicate that the receptors are capable of being translocated into an intracellular pool, in which case their pattern of behaviour mimics that of other membrane-located receptors.

Ontongeny

This provides us with another example of a situation in which the density of cardiac calcium-antagonist (DHP) binding sites increases without a change in affinity. An increase occurs during the first few days of post-natal development, but 1 or 2 weeks of post-natal development are required for the B_{max} of these high-affinity binding sites to reach asymptote (Erman et al. 1983; Marangos et al. 1984; Renaud et al. 1984). This increase in density is not associated with the conversion of calcium antagonist-insensitive channels to calcium antagonist-sensitive L channels, but it may be associated with the activation of a cyclic AMP-mediated process and a protein kinase C mediated pathway (Trautwein et al. 1987).

The results which have been outlined in the previous paragraphs indicate that the number of high-affinity calcium-antagonist binding sites present in cardiac membrane preparations varies in accordance with the pathological state of the tissue, its age and its exposure to certain drugs – including morphine and alcohol. Other conditions have also been shown to affect the density of these sites. Ischaemia, for example, causes a reduction in density without any change in affinity or selectivity (Nayler et al. 1985; Dillon and Nayler 1987; Gu et al. 1988). For example, within 30 min of global ischaemia at 37°C, the density of high-affinity DHP binding sites in rat cardiac membranes decreases from the control level of 210.4 ± 7.3 to 133.3 ± 20.8 fmol/mg protein. This is a loss of 36%, and it occurs without any change in affinity or selectivity (Gu et al. 1988). We have now extended these earlier observations by undertaking experiments aimed at establishing whether this ischaemia-induced reduction in the density of DHP-binding sites is (a) reversible, (b) prevented by hypothermic conditions, (c) mimicked by hypoxia, or (d) mimicked by chemically induced high-energy phosphate depletion (produced by dinitrophenol-induced inhibition of oxidative phosphorylation, with and without iodoacetate-induced inhibition of glycolysis).

Methods

Adult male Sprague-Dawley rats (250–300 g) were used for these studies. They were maintained on a standard diet, and tail cuff measurements showed them to be normotensive, with a systolic blood pressure of 135 ± 4 mmHg (mean ± SEM of 24 measurements, measured on the third occlusion).

Perfusion

The rats were anaesthetized with diethylether-O_2 mixture and heparinized (Nayler et al. 1985). The hearts were quickly excised, arrested by immersion into ice-cold modified Krebs-Henseleit (K–H) buffer, trimmed of extraneous fat and tissue and then perfused in the Langendorff mode at a mean coronary flow of 10–12 ml/min. The K–H buffer, which was gassed with 95% O_2 + 5% Co_2, was delivered to the aortic cannula at 37°C and contained (in mM): NaCl, 119.0; KCl, 4.6; NaHCO$_3$, 25.0; KH$_2$Po$_4$, 1.2; MgSO$_4$, 1.2; CaCl$_2$, 1.3; and glucose, 11.0.

Ischaemia Series

After 15 min perfusion as described above the hearts were randomly divided into two groups: one perfused at 37°C, the other at 22°C.

In group 1, perfusion was under normothermic conditions. Heart in this series were maintained at 37°C by means of heated, water-filled jackets. They were randomly divided into three sub-groups:

Aerobically perfused: Here K–H perfusion was continued as described for up to 75 min. Left ventricular wall temperature was maintained at 37°C.

Ischaemic: These hearts were perfused for another 15 min at 37°C and then made globally ischaemic by totally occluding coronary flow for up to 60 min. Left ventricular wall temperature was maintained at 37°C, irrespective of flow.

Reperfused: Hearts in this group were made globally ischaemic for 30 or 60 min and then reperfused with K–H buffer at 37°C for 15 min.

In group 2, perfusion was under hypothermic conditions. The temperature of the hearts (measured in the left ventricular wall) in this series was rapidly dropped to 22°C by changing the heated water-filled jackets to jackets filled with water maintained at 22°C and switching from K–H buffer at 37°C to buffer already cooled to 22°C. After 15 min perfusion at 22°C the hearts were randomly divided into two sub-groups:

Aerobically perfused. Here perfusion was continued at 22°C for up to 60 min.

Ischaemic. Here the hearts were made globally ischaemic for up to 60 min as described above, but at 22°C.

Hypoxia Series

After 15 min perfusion with K–H buffer the hearts in this series were randomly divided into three groups:

Aerobic perfusion. These hearts were perfused for another 15 min with K–H buffer and then for either 30 or 60 min with glucose-free K–H buffer (22 mM sucrose was added to the glucose-free K–H to maintain tonicity).

Hypoxic perfusion. These hearts were perfused for another 15 min with K–H buffer and then with glucose-free hypoxic buffer (gassed with 95% O_2 + 5% CO_2).

Reoxygenation series. These hearts were perfused with glucose-free hypoxic buffer for 30 or 60 min and then for 15 min with K–H buffer containing glucose and gassed with 95% O_2 + 5% CO_2.

High-Energy Phosphate Depletion Series

Hearts here were divided into two groups, treated, respectively in the following ways:

Uncoupling of oxidative phosphorylation. After 30 min aerobic perfusion with glucose-containing K–H buffer, hearts in this series were perfused for another 30 or 60 min in the presence or absence of 0.1 mM 2,4 dinitrophenol (DNP), with and without glucose.

Uncoupling of oxidative phosphorylation and inhibition of glycolysis. Hearts in this series were perfused for 30 or 60 min with glucose-free K–H buffer containing 0.1 mM DNP and 1 mM iodoacetic acid.

Measurement of 1,4 Dihydropyridine Ca^{2+} Antagonist Binding Sites

Cardiac Membrane Isolation

After the required period of perfusion, ischaemia and reperfusion, or hypoxia and reoxygenation, the hearts were immersed into ice-cold homogenizing medium containing 20 mM $NaHCO_3$, and 0.1 mM phenylmethylsulphonyl fluoride (PMSF). Cardiac membranes were isolated using the method described by Glossmann and Ferry (1985). The ventricles were trimmed, minced into small pieces (< 1 mm^3), and then homogenized in 8 ml homogenizing medium/g wet weight tissue, using two 15-s bursts of an Ultra Turrax homogenizer at three-quarters maximal speed. The homogenate was diluted 1:20 (wet weight:homogenizing volume) and then centrifuged at 1500 g for 15 min at 4°C in a Sorvall RC 2–B centrifuge, using a fixed-angle SS34 rotor. The supernatant was centrifuged for 15 min at 48000 g and the resultant pellet suspended in ice-cold 50 mM Tris buffer (containing 0.1 mM PMSF, pH 7.4) and recentrifuged. This procedure was repeated twice. The pellet from the final spin was suspended in 50 mM Tris buffer containing 0.1 mM PMSF (pH 7.4) to provide a final protein concentration of 0.8–1.2 mg protein/ml, and stored in liquid N_2. Protein was assayed as described by Lowry et al. (1951), using bovine serum albumin as standard.

Binding studies

$(+)[^3H]$PN200–110 binding was monitored as described by Glossmann and Ferry (1985) and was performed in duplicate at 25°C using a protein concentration of 0.1–0.5 mg/ml in a final volume of 0.25 ml. The incubation buffer contained 50 mM Tris and 0.1 mM PMSF (pH 7.4). For saturation binding, 0.015–0.7 nM $(+)[^3H]$PN200–110 was used. Non-specific binding was defined by adding 1 µM Bay K8644. After 60 min incubation, bound and free $(+)[^3H]$PN200–110 were separated by vacuum filtration, after diluting with 3.5 ml ice-cold 50 mM Tris buffer containing 6.6% polyethyleneglycol (PEG; pH 7.4). Filtration was across Whatman GF/C filters and was followed by two additional washes with Tris-PEG buffer. The radioactivity of the filters was counted (40% efficiency) in Filter Count Scintillant (Packard, Illinois, USA), using a Packard Tricarb spectrometer. Binding selectivity was characterized by using enantiomers of the DHPs: $(+)$PN200–110 (10^{-13}–10^{-8} M); $(-)$PN200–110 (10^{-12}–3×10^{-7} M); $(-)$Bay K8644 (10^{-12}–3×10^{-7} M); $(+)$Bay K8644 (10^{-11}–10^{-6} M); or $(-)$D600 (gallopamil); (10^{-10}–10^{-5} M) or the α_1 adrencoceptor antagonist, prazosin (10^{-10}–10^{-5} M), to displace bound $(+)[^3H]$PN200–110. D-*cis* diltiazem (10^{-10}–10^{-5} M) stimulation of $(+)[^3H]$PN200–110 binding was monitored at 37°C (Ferry et al. 1983).

Because of the photolability of the DHPs, binding studies were performed under a sodium lamp.

Data Analysis

Initial estimates of equilibrium binding parameters (K_D and B_{max}) were obtained from Scatchard, Hill and Hofstee analysis, using the EDBA programme (McPherson

1983). A file was then produced, and the data analysed with the aid of a weighted, non-linear, least-squares computer curve-fitting programme (Munson and Rodbard 1980) to obtain final parameter estimates. K_D is defined as the concentration of ligand required to occupy 50% of the binding sites. B_{max} is the density of binding sites.

The binding data was subjected to analysis of variance followed by a modified Student's t test (with the Bonferroni adjustment of the t statistic) for multiple comparison (Wallenstein et al. 1980). The level of significance was taken at $p < 0.05$.

Reagents

$(+)[^3H]$PN200–110 (specific activity 71 Ci/mmol) was obtained from Amersham International (UK). $(+)$PN200–110, $(-)$PN200–110, $(+)$Bay K8644, and $(-)$Bay K8644 were obtained as gifts from Sandoz Ltd. (Switzerland) and Bayer AG (FRG). $(-)$D600 was a gift from Knoll AG (FRG) and D-*cis* diltiazem from Tanabe Laboratories (Japan). Prazosin was obtained from Pfizer Ltd. (Australia). All other reagents were from Sigma Chemical Company (St. Louis, Missouri, USA).

Results

The yields of membrane fragments obtained from the homogenized aerobically perfused, globally ischaemic and reperfused, and hypoxic and reoxygenated hearts used here were not significantly different from one another. It is unlikely, therefore, that any change in K_D or B_{max} obtained for membranes prepared from hearts subjected to the different experimental conditions can be attributed to the preferential selection of a particular population of membrane fragments.

$(+)[^3H]$PN200–110 Binding to Membranes Isolated from Freshly Excised Hearts

Membranes isolated from non-perfused hearts contained a single population of high-affinity $(+)[^3H]$PN200–110 binding sites, with a K_D of 0.035 ± 0.005 nM, a B_{max} of 230.3 ± 15.5 fmol/mg protein, and a Hill coefficient centering around unity. These results are similar to those obtained by Lee et al. (1984) for rat cardiac membranes.

$(+)[^3H]$PN200–110 Binding to Membranes Isolated from Hearts Perfused Under Aerobic Conditions

Normothermic Conditions

Membranes isolated from hearts perfused aerobically at 37°C retained a single population of high-affinity $(+)[^3H]$PN200–110 binding sites. Table 1 shows that the K_D and B_{max} of these sites remained constant during normothermic perfusion (up to 90 min). The Hill coefficients remained centered around unity (Table 1), indicating

Table 1. $(+)[^3H]PN200-110$ binding to membranes from aerobically perfused, ischaemic and reperfused rat hearts

	Normothermia (37°C)			Hypothermia (22°C)		
	K_D (nM)	B_{max} (fmol/mg protein)	Hill coefficient	K_D (nM)	B_{max} (fmol/mg protein)	Hill coefficient
10-min ischaemia series						
Aerobic perfusion	0.043 ± 0.003	248.5 ± 31.9	0.991 ± 0.006	0.040 ± 0.002	274.0 ± 25.0	0.991 ± 0.005
Ischaemia	0.039 ± 0.002	250.2 ± 27.3	0.978 ± 0.012	0.040 ± 0.002	297.8 ± 23.6	0.993 ± 0.003
20-min ischaemia series						
Aerobic perfusion	0.040 ± 0.002	219.2 ± 9.4	0.982 ± 0.023			
Ischaemia	0.048 ± 0.004	224.3 ± 8.2	0.964 ± 0.012			
30-min ischaemia series						
Aerobic perfusion	0.042 ± 0.003	210.4 ± 7.3	0.980 ± 0.008	0.046 ± 0.002	247.5 ± 14.5	0.987 ± 0.004
Ischaemia	0.045 ± 0.001	133.3[a] ± 20.8	0.982 ± 0.012	0.048 ± 0.002	296.7 ± 15.5	0.997 ± 0.005
Reperfusion	0.058[a] ± 0.004	190.8 ± 7.3	0.974 ± 0.008			
60-min ischaemia series						
Aerobic perfusion	0.045 ± 0.004	219.3 ± 20.3	0.983 ± 0.008	0.046 ± 0.002	190.8[b] ± 7.7	0.981 ± 0.003
Ischaemia	0.047 ± 0.003	147.5[a] ± 7.8	0.965 ± 0.006	0.044 ± 0.002	269.5 ± 13.1	0.986 ± 0.003
Reperfusion	0.046 ± 0.002	138.5[a] ± 11.4	0.975 ± 0.007			

Each result is mean ± SEM of six to eight separate experiments. Reperfusion was for 15 min.
[a] $p < 0.05$ relative to the appropriate aerobic control at 39°C.
[b] $p < 0.05$ relative to 10 min aerobic reperfusion at 22°C.

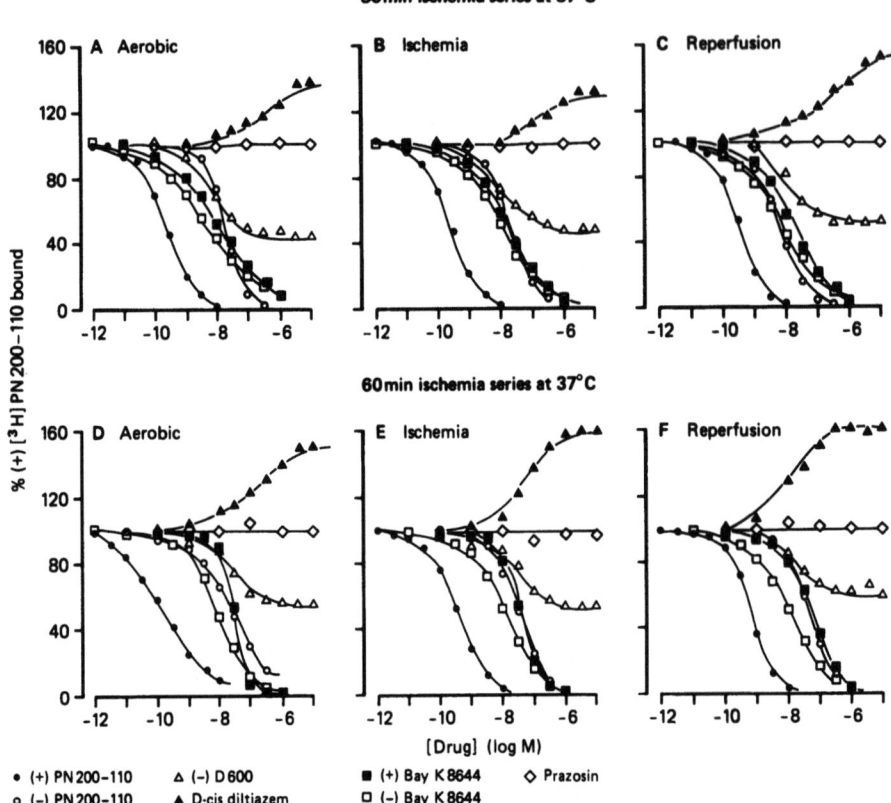

Fig. 1A–F. Allosteric and stereoselective interaction of (+)PN200–110 (●), (−)PN200–110 (○), (+)Bay K8644 (■), (−)Bay K8644 (□), D600 (△), D-*cis* diltiazem (▲), and prazosin (◊), with (+)[^{3}H]PN200–110 binding sites in cardiac membranes isolated after aerobic perfusion for 30 (**A**) or 60 (**D**) min, ischaemia for 30 (**B**) or 60 (**E**) min, and after 15 min post-ischaemic reperfusion after 30 (**C**) or 60 (**F**) min of ischaemia. Each curve is the representative of three separate experiments. The hearts were maintained at 37°C throughout

the continued presence of a single population of (+)[^{3}H]PN200–110 binding sites. The selectivity of these sites was consistent with that described by other investigators (Lee et al., 1984; Glossmann et al., 1985), with (+)PN200–110 > (−)Bay K8644 > (−)PN200–110 > (+)Bay K8644 >> (−)D600 in displacing bound (+)[^{3}H]PN200–110 (Fig. 1A, D). D-*cis* diltiazem stimulated (+)[^{3}H]PN200–110 binding, whereas prazosin had no effect (Fig. 1A, D).

Hypothermic Conditions

A single population of (+)[^{3}H]PN200–110 binding sites was also identified in cardiac membranes isolated from hearts which had been perfused aerobically for up to 60 min

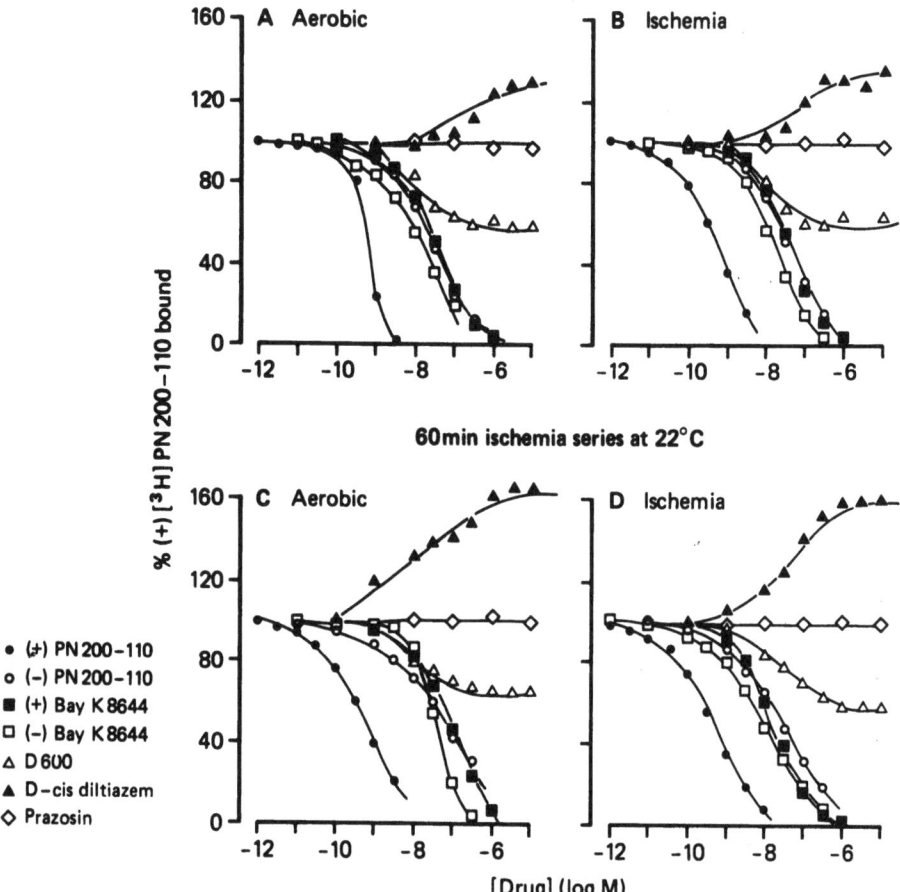

Fig. 2A–D. Allosteric and stereoselective interaction of (+)PN200–110 (●), (−)PN200–110 (○), (+)Bay K8644 (■), (−)Bay K8644 (□), (−)D600 (△), D-*cis* diltiazem (▲), and prazosin (◇), with (+)[^{3}H]PN200–110 binding sites after 30 (**A**) or 60 (**C**) min aerobic perfusion or 30 (**B**) or 60 min (**D**) ischemia at 22°C

at 22°C, instead of 37°C. The K_D was similar to that obtained after comparable periods of perfusion at 37°C (Table 1), and stereoselectivity was maintained, with D-*cis* diltiazem continuing to stimulate (+)[^{3}H]PN200–110 binding (Fig. 2A, C). There was, however, a gradual decline in B_{max} (Fig. 3A), which became significant ($p < 0.05$) after 60 min perfusion (Fig. 3A; Table 1).

These results establish that DHP binding sites survive prolonged periods of aerobic perfusion at either 37°C or 22°C, without any change in affinity or selectivity. At 22°C, but not at 37°C, there is a small decline in density.

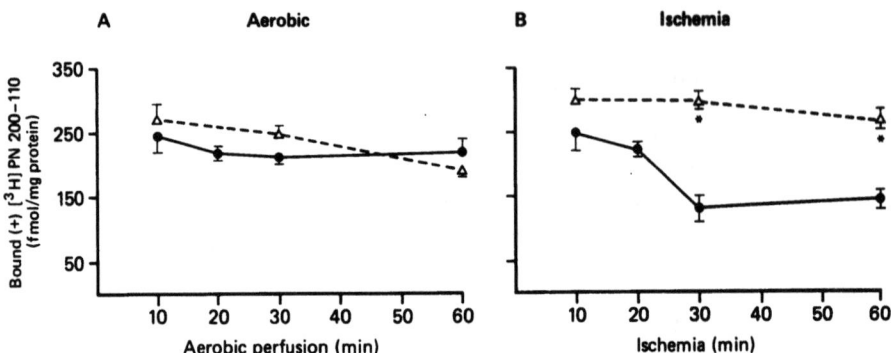

Fig. 3A,B. Effect of up to 60 min aerobic perfusion at either 37°C (●) or 22°C (△) **(A)** or 60 min global ischaemia at 37°C (●) or 22°C (△) **(B)** on the density of $[+][^3H]PN200-110$ binding in rat cardiac sarcolemmal fragments. Cardiac membrane fragments were isolated after the indicated periods of aerobic perfusion or ischaemia, after the initial 30 min aerobic perfusion as described in text. Minutes of aerobic perfusion refers to perfusion after the initial 30 min perfusion. *Asterisks* indicate significant difference ($p < 0.05$) between the effect of normothermic and hypothermic ischaemia on the B_{max} of DHP binding sites

$(+)[^3H]PN200-110$ Binding to Membranes Isolated from Ischaemic Hearts

Normothermic Conditions

Figure 3B and Table 1 show that 10 or 20 min global ischaemia at 37°C has no effect on the K_D or B_{max} of DHP binding sites, compared to the relevant aerobic controls. However, in membranes isolated from hearts which had been globally ischaemic at 37°C for 30 min, there was a significant ($p < 0.05$) reduction (36%) in density (Fig. 3B) without any change in affinity (Table 1), relative to the appropriate aerobic controls. Longer periods of ischaemia (up to 60 min) caused no further decrease in the density (Fig. 3B; Table 1) or affinity. Selectivity was also maintained, with $(+)PN200-110 > (-)Bay K8644 > (-)PN200-110 = (+)Bay K8644 \gg (-)D600$ in displacing bound $(+)[^3H]PN200-110$. Prazosin, again, did not displace bound $(+)[^3H]PN200-110$, whereas D-*cis* dilitiazem stimulated it (Fig. 1B, E).

Hypothermic Conditions

Figure 3A and Table 1 show that continued *aerobic perfusion* at 22°C causes a gradual but progressive decline in B_{max} without any change in affinity or selectivity of DHP binding sites. By contrast, continued *global ischaemia* at 22°C for up to 60 min did not cause any change in density (Fig. 3B), affinity (Table 1) or selectivity (Fig. 2B, D).

In summary, therefore, although prolonged periods of normothermic ischaemia reduce the density of DHP binding sites (without changing their affinity or selectivity) no such change occurs during ischaemia at 22°C.

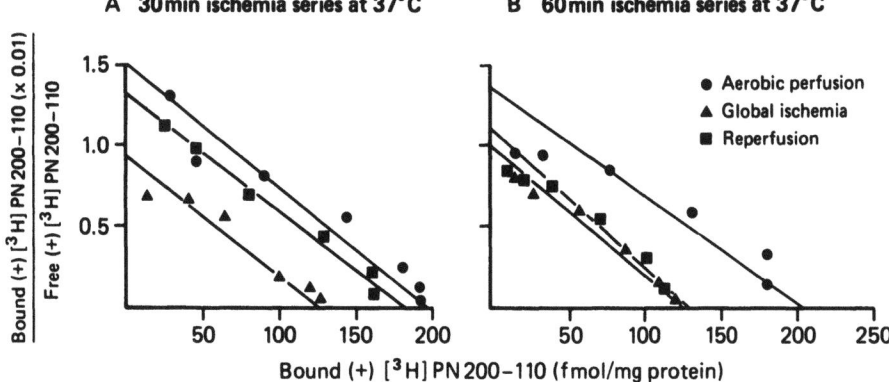

Fig. 4A, B. Scatchard curves of $(+)[^3H]PN200$–110 binding to membranes isolated from aerobically perfused hearts (see text) and after 30 (**A**) and 60 (**B**) min ischaemia followed by 15 min reperfusion at 37°C. The binding parameters for these curves are shown in Table 1. *Ordinate*, bound/free $[^3H]PN200$–110; abscissa, $(+)[^3H]PN200$–110 (fmol/mg protein). Similar estimates were obtained from six separate experiments, using duplicate estimate for each point

$(+)[^3H]PN200$–110 Binding to Membranes Isolated from Reperfused Hearts

Cardiac membranes isolated from hearts which were reperfused for 15 min after 30 or 60 min ischaemia at 37°C contained a single population of DHP binding sites. Analysis of the Scatchard plots revealed that reperfusion after 30 min ischaemia resulted in the B_{max} returning to a level close to that obtained for aerobically perfused hearts (Fig. 4A). At the same time, and although selectivity was maintained (Fig. 1C), there was a small increase in K_D ($p < 0.05$) relative to the values obtained after aerobic perfusion (Table 1).

Reperfusion after 60 min ischaemia did not result in the B_{max} returning to the levels obtained for aerobically perfused hearts (Fig. 4B; Table 1). Under these conditions there was no change in selectivity or affinity, and the Hill coefficient remained centered around unity.

In summary, these results show that post-ischaemic reperfusion reverses the ischaemia-induced reduction in the density of DHP binding sites after 30 but not after 60 min normothermic ischaemia at 37°C. Irrespective of the duration of the ischaemic episode, reperfusion had no effect on the selectivity of sites, but reperfusion after 30 min (but not 60 min) ischaemia at 37°C resulted in a small reduction in affinity.

Effect of Hypoxia and Reoxygenation

Glucose-free Aerobic Perfusion

Membranes prepared from hearts which had been aerobically perfused with glucose-free K–H buffer for 30 or 60 min retained a single population of high-affinity

Table 2. $(+)[^3H]PN200-110$ binding to membranes isolated from aerobically perfused, hypoxic and reoxygenated rat hearts

	30-min series			60-min series		
	K_D (nM)	B_{max} (fmol/mg protein)	Hill coefficient	K_D (nM)	B_{max} (fmol/mg protein)	Hill coefficient
Aerobic perfusion (with glucose)	0.042 ± 0.003	210.4 ± 7.3	0.980 ± 0.008	0.045 ± 0.004	219.3 ± 20.3	0.983 ± 0.008
Aerobic perfusion (without glucose)	0.044 ± 0.002	194.0 ± 15.7	0.973 ± 0.009	0.078[a] ± 0.005	215.7 ± 16.0	0.993 ± 0.004
Hypoxia	0.048 ± 0.002	181.7 ± 6.9	0.979 ± 0.007	0.080[a] ± 0.005	224.4 ± 5.8	0.998 ± 0.003
Reoxygenation	0.058 ± 0.002	169.8 ± 12.4	0.986 ± 0.005	0.077[a] ± 0.005	172.3[b,c] ± 9.6	0.993 ± 0.004

Each result is a mean ± SEM of four to six separate experiments.
[a] $p < 0.05$ relative to the appropriate group of 30-min series under the same conditions.
[b] $p < 0.05$ relative to the appropriate aerobic perfusion groups with glucose.
[c] $p < 0.05$ relative to the aerobic controls (without glucose) and hypoxia.

$(+)[^3H]PN200-110$ binding sites. B_{max} remained comparable with that found for hearts which had been perfused with glucose containing K-H buffer, but after 60 min there was a small ($p < 0.05$) decline in affinity (Table 2). Glucose-free aerobic perfusion had no effect on selectivity, with $(+)PN200-110 > (-)Bay K8644 > (-)PN200-110 = (+)Bay K8644 >> (-)D600$ in displacing $(+)[^3H]PN200-110$ (Fig. 5). Hence, DHP binding sites survive glucose-free aerobic perfusion.

Glucose-free Hypoxic Perfusion

Table 2 and Fig. 6A, B show that neither 30 nor 60 min hypoxia has any effect on the B_{max} of these binding sites. The selectivity (Fig. 5) of these sites was maintained, and, as with the aerobically perfused hearts, there was a small ($p < 0.05$) decrease in the affinity (Table 2) after 60 but not 30 min of hypoxia.

Hypoxia, therefore, does not mimic the effect of ischaemia on the density of cardiac DHP binding sites.

Reoxygenation

After 30 min hypoxia reoxygenation had no effect on the affinity or density (Table 2; Fig. 6A) of DHP sites. However, reoxygenation after 60 min of hypoxia caused a

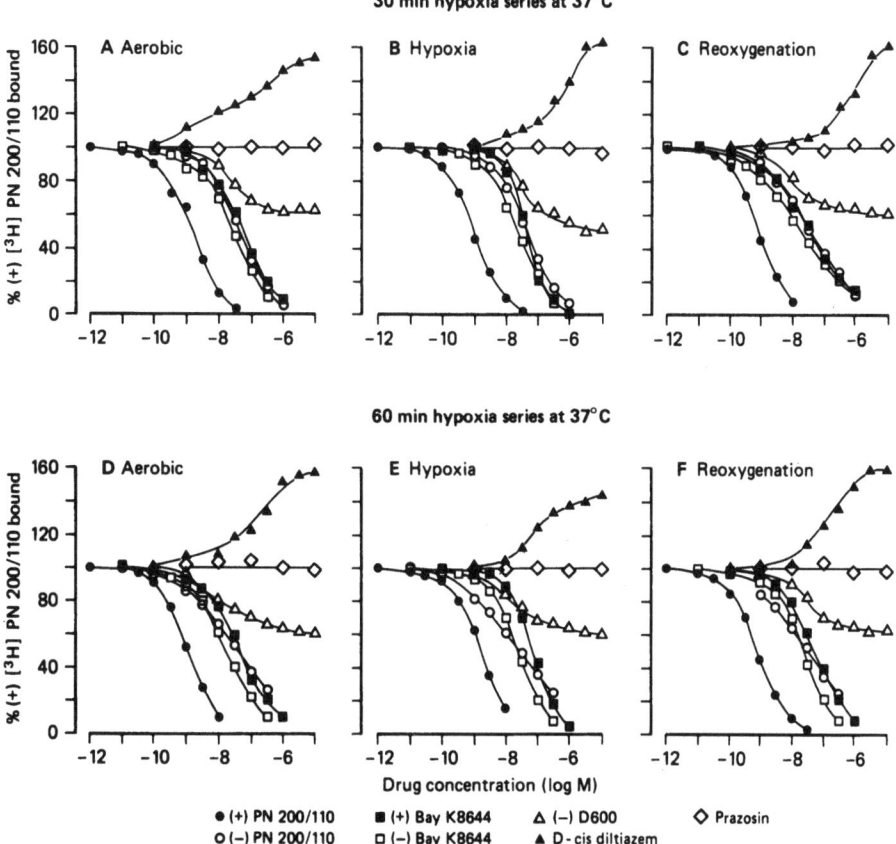

Fig. 5A–F. Allosteric and stereoselective interaction of (+)PN200-110 (●), (−)PN200-110 (○), (+)Bay K8644 (■), (−)Bay K8644 (□), (−)D600 (▲), D-cis diltiazem (▲), and prazosin (◊), with (+)[^{3}H]PN200-110 binding sites in cardiac membranes isolated after glucose-free aerobic perfusion for 30 **(A)** or 60 **(D)** min, hypoxia for 30 **(B)** or 60 **(E)** min and after 15 min reoxygenation after 30 **(C)** or 60 **(F)** min of hypoxia. Each curve is the representative of three separate experiments

significant decrease in B_{max}, with values falling from 224.4 ± 5.8 at the end of the hypoxic episode to 172,7 ± 9.6 fmol/mg protein after 15 min reoxygenation (Table 2; Fig. 6B). Selectivity (Fig. 5) was unchanged.

Effect of High-energy Phosphate Depletion

Perfusing hearts with 0.1 mM DNP, a concentration which is sufficient to cause a rapid and sustained decrease in tissue adenosine triphosphate and creatine phosphate (results not shown) failed to decrease the density of DHP binding sites (Fig. 7). This same response was observed when the period of DNP perfusion was extended to 60

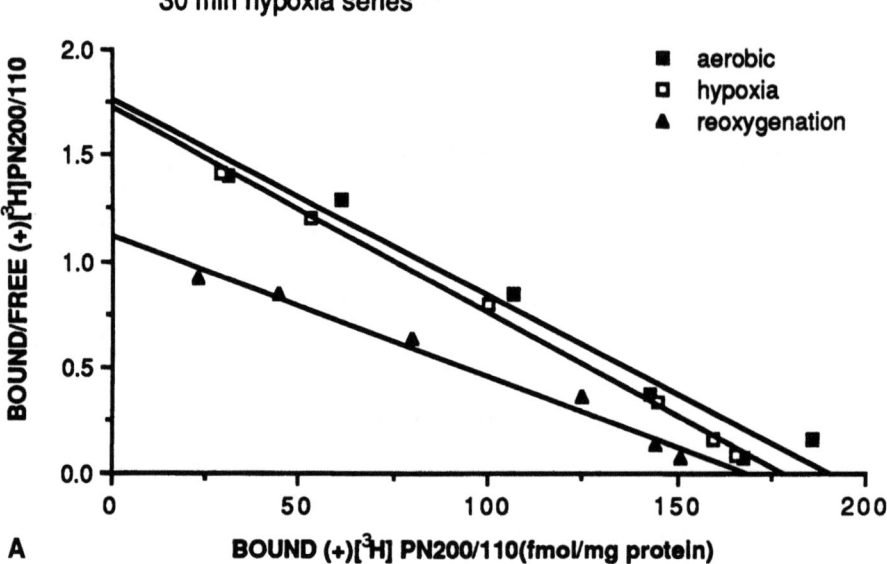

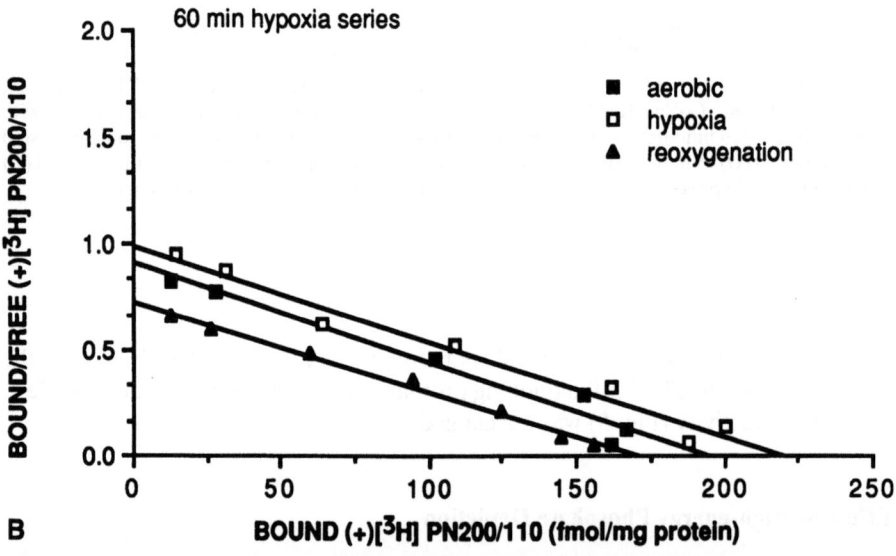

Fig. 6A, B. Scatchard curves of $(+)[^3H]PN200$–110 binding to membranes isolated from glucose-free aerobically perfused (■) hearts (see text) and after 30 **(A)** and 60 **(B)** min hypoxia (□) followed by 15 min reoxygenation (▲) at 37°C. The binding parameters for these curves are shown in Table 2. Similar estimates were obtained from six separate experiments, using duplicate estimate for each point

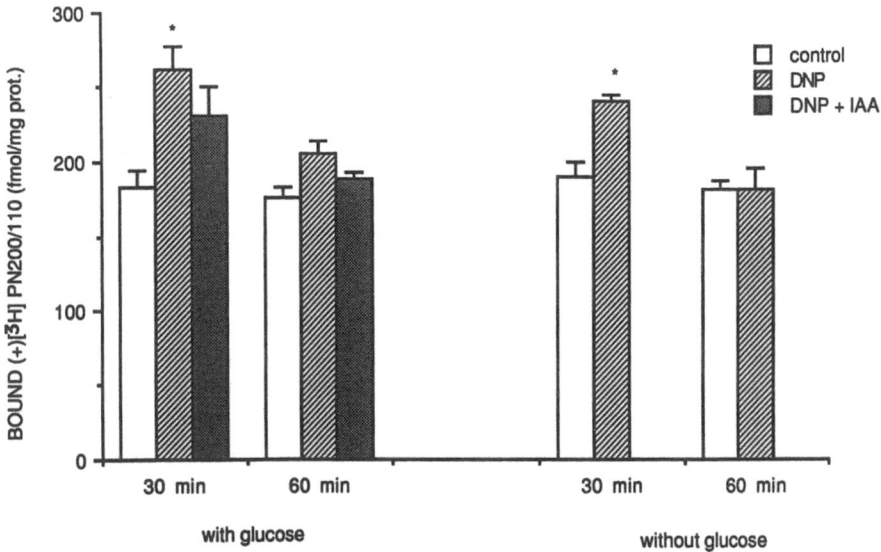

Fig. 7. Histogram of (+)[^{3}H]PN200–110 binding site density (fmol/mg protein) monitored in cardiac membranes isolated from hearts that have perfused aerobically with K–H buffer (control) (□), with K–H buffer containing 0.1 mM 2,4 dinitrophenol (DNP) (▨), and with K–H buffer containing 0.1 mM DNP plus 1 mM iodoacetic acid (IAA) (■) for either 30 or 60 min at 37°C. Estimates were obtained from six separate experiments. *Asterisks* indicate significant difference ($p < 0.05$) in density of DHP binding sites after perfusion with DNP containing K–H buffer relative to control

min, and even when the glycolytic inhibitor, iodoacetic acid was added (Fig. 7). The selectivity of binding sites was again maintained, with (+)PN200–110 > (−)Bay K8644 > (−)PN200–110 = (+)Bay K8644 >> (−)D600 in displacing bound (+)[^{3}H]PN200–110, and D-*cis* diltiazem stimulating the binding.

Control studies showed that when DNP (10^{-10}–10^{-3} M) was added directly to isolated cardiac membrane fragments, it did not interfere with DHP binding, nor did it alter the selectivity of binding sites.

Discussion

These results show that ischaemia-induced reduction in the density of DHP-binding sites is rapidly reversed upon reperfusion, after short but not after long periods of ischaemia, and that it is attenuated by hypothermic conditions. The results also show that neither hypoxia nor chemically induced high-energy phosphate depletion causes a similar reduction in DHP binding site density, and that reoxygenation after a prolonged period of hypoxia decreases the number of binding sites which are available for DHP binding. On the basis of these results it can be concluded that energy depletion alone cannot account for the ischaemia-induced reduction in the density of these binding sites, and that reoxygenation has an effect which is quite different from that of reperfusion.

The rapid recovery of receptor binding site density during post-ischaemic reperfusion after short but not long periods of ischaemia may be interpreted as meaning that the ischaemia-induced reduction in the density of these sites involves internalization of the receptors, and not their degradation. The de novo synthesis of new receptors is an improbable explanation for their recovery upon reperfusion, because the half-life of these binding sites is measured in hours, not minutes. Internalization is not the only possible explanation, however. The insertion of the binding sites into the channel complex may have been modified, or the ability of the ligand to approach the receptor may have been affected. Whatever the cause, however, the affinity of the receptor was unchanged.

Since ischaemia, but neither hypoxia nor energy depletion, causes a reduction in DHP binding site density, some other consequence of ischaemia must be involved. Oxygen deprivation has been ruled out, because hypoxia failed to produce a similar response. Ischaemic hearts become severely acidotic, and it is therefore possible that proton accumulation is responsible, in part at least, for the decrease in binding site density.

Other investigators (Mukherjee et al. 1982; Maisel et al. 1985; Vatner et al. 1988) have shown that ischaemia externalizes cardiac β-adrenoceptors. Their experiments were performed on dog hearts, however, and since an ischaemia-induced externalization of α-adrenoceptors (Corr and Crafford 1981) has been found to be species dependent, with no externalization occurring in rat hearts (Dillon et al. 1988), we do not know whether the internalization of DHP binding sites which has been described here occurs in other species. However, if it is a general phenomenon, and since binding sites are associated with the L-type voltage-sensitive Ca^{2+} channels, then it could be argued that it represents a homeostatic response designed to limit Ca^{2+} overload and preserve the energy reserves of the myocardium – by reducing cardiac work.

References

Chatelain P, Demol D, Roba J (1984) Comparison of [^{3}H] nitrendipine binding to heart membranes of normotensive and spontaneously hypertensive rats. J Cardiovasc Pharm 6:220–223

Corr PB, Witkowski FX, Sobel BE (1978) Mechanisms contributing to malignant dysrhythmias induced by ischemia in the cat. J Clin Invest 61:109–119

Dillon JS, Nayler WG (1987) [^{3}H]-verapamil binding to rat cardiac sarcolemmal membrane fragments: an effect of ischaemia. Br J Pharm 90:99–109

Dillon JS, Gu XH, Nayler WG (1988) Alpha$_1$ adrencoceptors in the ischemic myocardium. J Mol Cell Cardiol, in press

Erman RD, Yamamura HI, Roseke WR (1983) The ontogeny of specific binding sites for the calcium channel antagonist nitrendipine, in mouse heart and brain. Brain Res 278:327–331

Fabiato A, Fabiato F (1979) Calcium and cardiac excitation – contraction coupling. Ann Rev Physiol 41:473–484

Ferry DR, Goll A, Glossmann H (1983) Differential labelling of potative skeletal muscle calcium channels by [^{3}H]nitrendipine [^{3}H]nimidipine and [^{3}H]PN 200–110. Naunyn Schmied Arch Pharm 323:276–277

Gengo P, Skattebol A, Moran JF, Gallant S, Hawthorn M, Triggle DJ (1988). Regulation by chronic drug administration of neuronal and cardiac calcium channel, beta-adrenoceptor and muscarinic receptor levels. Biochem Pharm 37:627–634

Glossmann H, Ferry DR (1985) Assay for calcium channels. Methods Enzymol 109:513–510

Gu XH, Dillon JS, Nayler WG (1988) Dihydropyridine binding sites in aerobically perfused, ischemic and reperfused rat hearts: effect of temperature and time. J Cardiovasc Pharm. In press

Hawthorn M, Skattebol A, Gengo P, Rampe D, Moran JF, Wei XY, Triggle DJ (1987) Intl Symposium on Calcium Antagonists. NY CITY Feb 10–13 (Abst)

Ishii K, Kano T, Ando J, Yoshida H (1986) Binding of [^{3}H]-nitrendipine to cardiac and cerebral membranes from normotensive and renal, deoxycorticosterone/NaCl and spontaneously hypertensive rats. European J Pharm 123:271–278

Lee HR, Roeske WR, Yamamura HI (1984) High affinity specific [^{3}H](+)PN200-110 binding to dihydropyridine receptors associated with calcium channels in rat cerebral cortex and heart. Life Sci 35:721–732

Lowry OH, Rosebrough NJ, Farr AL, Randall RJ (1951) Protein measurement with Folin phenol reagent. J Biol Chem 193:265–275

Maisel AS, Motulsky HJ, Insel PA (1985) Externalization of β-adrenergic receptors promoted by myocardial ischemia. Science 230:183–186

Marangos PJ, Sperelakis N, Patel J (1984) Ontogeny of calcium antagonist binding sites in chick brain and heart. J Neurochem 42:1338–1342

McPherson GA (1983) A practical computer-based approach to the analysis of radioligand binding experiments. Comput Prog Biomed 17:107–114

Messing RO, Carfenter CL, Diamond I, Greenberg DA (1986) Ethanol regulates calcium channels in clonal neural cells. Proc Nat Acad Science 83:6213–6215

Muhkerjee A, Wong TM, Buja LM, Lefkowitz RK, Willerson JT (1979) Beta adrenergic and muscarinic cholinergic receptors in canine myocardium. J Clin Invest 64:1423–1428

Munson PJ, Rodbard D (1980) Ligand: a versatile computerized approach for the characterization of ligand binding system. Anal Biochem 107:220–239

Nayler WG, Dillon JS, Elz JS, McKelvie M (1985) An effect of ischemia on myocardial dihydropyridine binding sites. European J Pharm 115:81–89

Ramkumar V, El-Fakahany EE (1984) Increase in [^{3}H] nitrendipine binding sites in the brain in morphine tolerant mice. European J Pharm 102:371–372

Renaud JF, Kazazoiglou T, Schmid A, Romey G, Lazdunski M (1984) Differentiation of receptor sites for [^{3}H] nitrendipine in chick hearts and physiological relation to slow Ca^{2+} channel and to excitation contraction coupling. European J Biochem 139:673–681

Sharma RV, Butters R, Bhalla RC (1986) Alterations in the plasma membrane properties of the myocardium of spontaneously hypertensive rats. Hypertension 8:583–591

Trautwein W, Cavalié A, Flockerzi V, Hofmann F, Pelzer D (1987) Modulation of calcium channel by phosphorylation in guinea pig ventricular cells and phospholipid bilayer membranes. Circ Res 61 (Suppl 1) 17–23

Vatner DE, Knight DR, Shen Y-T, Thomas JX Jnr, Homey CJ, Vatner SF (1988) One hour of myocardial ischaemia in conscious dogs increases β-adrenergic receptors but decreases adenylate cyclase activity. J Mol Cell Cardiol 20:75–82

Wagner JA, Reynolds IJ, Weisman HF, Dudeck P, Weisfeldt ML, Snyder S (1986) Calcium antagonist receptors in cardiomyopathic hamster: selective increases in heart, muscle and brain. Science 232:515–518

Wallenstein S, Zucker CI, Fleiss JL (1980) Some statistical methods useful in circulation research. Circ Res 47:1–9

Role of Calcium Channels in Cardiac Arrhythmias

E. Carmeliet

Laboratory of Physiology, University of Leuven, Campus Gasthuisberg, Herestraat, 3000 Leuven, Belgium

In heart muscle three different calcium channels can be distinguished, of which two are located in the plasma membrane, the L- and the T-type calcium channels, and one in the sarcoplasmic reticulum membrane. All three may be involved in the genesis of different arrhythmias.

L- and T-Type Calcium Channels in the Plasma Membrane

Before discussing the role of the two types of plasmalemma calcium channels in cardiac arrhythmias, it is of interest to recall some of their basic characteristics. As shown by a number of authors [1, 30, 33], the channels differ in threshold, time course of inactivation, and sensitivity to blocking agents and neurotransmitters. The T-type channel has a low threshold (−70 mV), is rapidly inactivated (transient, or "T-type"), is preferentially blocked by Ni (100 μM) and amiloride [39], is not blocked by verapamil or nifedipine, and is not sensitive to β-receptor stimulation ([1, 20, 42], but see [30] for a different result). The L-type channel has a high threshold (positive to −40 mV), its inactivation is slow (long-lasting, or "L-type"), it is blocked by Cd (20–50 μM) and by substances belonging to the verapamil, dihydropyridine, and diltiazem groups, and it is highly sensitive to β-receptor modulation. Both channel types can be inhibited by flunarizine [43].

L-Type Calcium Channel and Arrhythmias

Under physiological conditions, activation of the L-type channel is responsible for the upstroke of the action potential in the sinoatrial (SA) and atrioventricular (AV) nodes. Slow conduction of the impulse through the AV node is explained by the slow rate of depolarization. In atrial and ventricular cells the upstroke is caused by the fast sodium current, but activation of the L-type calcium channel is responsible for a substantial calcium influx. Under pathological conditions, activation of the L-type channel can be involved in the genesis of arrhythmias which may be either of the pacemaker or the reentry type. Pacemaker activity comparable to SA-node activity has been described in Purkinje fibers isolated from an infarcted area [19, 27]. It appears as early after potentials or steady-state oscillations at the plateau level. Calcium-mediated action potentials, because of their slow conduction velocity, may be responsible for reentry phenomena in the AV node as such, in the AV node in conjunction with an abnormal muscular pathway between auricles and ventricles, and in plain atrial or ventricular muscle cells depolarized by exposure to ischemic conditions.

The role of calcium current in junctional or supraventricular tachycardias and the efficiency of calcium-channel blockers in the treatment of this type of arrhythmias are well established [26]. In all these cases the AV node is directly involved. In atrial flutter and fibrillation calcium-channel antagonists are able to slow and regularize the ventricular response, and sometimes to correct atrial flutter. The slowing effect can be explained by an effect on the node. Regularization suggests a more direct effect on atrial fibers. This is not surprising since L-type calcium channels are present in atrial fibers; in diseased state atrial muscle is known to be partly depolarized, so that calcium-mediated action potentials can be generated.

The role of L-type calcium channels in ventricular arrhythmias, and especially during early ischemia, is less sure. Variable effects have been obtained by using the classic calcium-channel antagonists in different experimental arrhythmia models [7]. It is not clear to what extent positive results are due to indirect metabolic effects, secondary to vasodilatation, or to decrease in mechanical load. Measurements of conduction velocity in ischemic hearts and in hearts perfused with high potassium- and catecholamine-containing solutions have shown that conduction velocity in ischemic hearts is too high to be explained by activation of L-type calcium channels [25]. This type of experiment, however, does not rule out a possible contribution of the T-type calcium-channel activity. In the light of the recent distinction between these two types a reevaluation of this problem seems indicated.

T-Type Calcium Channel and Arrhythmias

While the upstroke of the action potential in the SA node is entirely due to activation of the L-type calcium channel, activation of the T-type channel is responsible for the later part of the diastolic depolarization [20, 32]. Its role in causing diastolic depolarization in subsidiary pacemakers in the atrium may even be more important, but direct experimental evidence is still lacking. In plain atrial and ventricular cells calcium influx through the T-type channel probably plays an important role in eliciting calcium release from the sarcoplasmic reticulum.

Since the activation voltage range of the sodium current and T-type calcium current overlap, both currents can participate in causing slowly propagating action potentials and eventually reentry arrhythmias when cells are partly depolarized. The absence of an effect on arrhythmias by dihydropyridines or verapamil is no definitive criterion to exclude a contribution of calcium channels in general, because the T-type calcium channel is not sensitive to these drugs [33, 43]. More specific blockers of the T-type channel should thus be used to determine the relative contribution of sodium and calcium channels.

During recent years the hypothesis has been formulated that reperfusion arrhythmias are causally related to calcium overload of the cell [17, 28]. How such calcium overload is caused is only partially understood. An increase of intracellular sodium and consequent calcium influx via Na–Ca exchange is one possibility. Depolarization of the cell and activation of T-type calcium channels is another possibility. In this respect it is worthwhile to mention that Ni and amiloride, but not nifedipine are able to block this extra calcium influx [34]. This result does not allow the distinguishing between Na–Ca exchange and channel activation, but eliminates any large contribution of the L-type channel activity.

The Sarcoplasmic Reticulum Calcium Channel

The sarcoplasmic reticulum represents an intracellular store of calcium ions, which releases its content during depolarization of the plasma membrane and in such a way provides the link between excitation and contraction [13]. The calcium channel responsible for this release is activated by calcium ions (μM) and ATP (mM); it is blocked by Mg (mM) and ruthenium red (μM) [5, 35, 37]. According to Fabiato [13], activation is more determined by the rate of calcium change than by the concentration itself; high concentrations of calcium result in inactivation. The selectivity of the channel for calcium over Tris is about 8. The channel shows a high affinity for ryanodine; its effect differs according to the experimental condition, and it can lock the channel in its activated or closed state [15, 29].

Under pathological conditions of calcium overload (digitalis intoxication, ischemia, dilatation) the release of Ca through these calcium channels is disturbed. It not only occurs during an action potential but also following a depolarization (triggered activity) and even in the resting state [14]. Under such conditions of excessive calcium release a transient inward current (i_{TI}) is activated in the plasma membrane, resulting in after-depolarizations and, if sufficient in amplitude, extrasystolic action potentials and arrhythmia [18, 23]. The nature of the transient inward current is complex and probably due to activation of the Na–Ca exchange system [16] and of a nonspecific cation channel [2]. The central role of an excessive oscillatory calcium release is well established and is based on the results of a number of experiments. Simultaneous measurement of tension and potential (or current) have shown that after-contractions and delayed after-potentials (or transient inward current) appear in parallel [45]. Both disappear when the cell is dialyzed with a solution containing a buffering concentration of EGTA or following exposure to ryanodine [3, 38]. Treatment with ryanodine also prevents digitalis and reperfusion arrhythmias [21, 41] and is able to reduce the frequency of subsidiary pacemakers in dilated atria [12].

Arrhythmias due to oscillatory calcium release are more likely to occur during *reperfusion* than during *severe ischemia*, as suggested by the following observations. First, reperfusion is accompanied by a net influx of calcium ions, exacerbating the calcium overload already present during ischemia [34]. Sodium-proton exchange plays a major role in causing this extra calcium influx, as shown by the inhibitory effect of amiloride (a blocker of Na–H exchange) on calcium uptake and reperfusion arrhythmias [9, 40]. A transient reperfusion with acidotic solution also improves the mechanical recovery [24]. Second, delayed after depolarizations, elicited by digitalis intoxication or blockade of the Na–K pump, are markedly depressed under conditions of simulated ischemia [4, 8, 10]. This effect is not due to the well-known hypoxia-induced increase of potassium conductance [22, 44] but to an inhibitory effect of ischemia on the transient inward current. Voltage clamp experiments have shown that

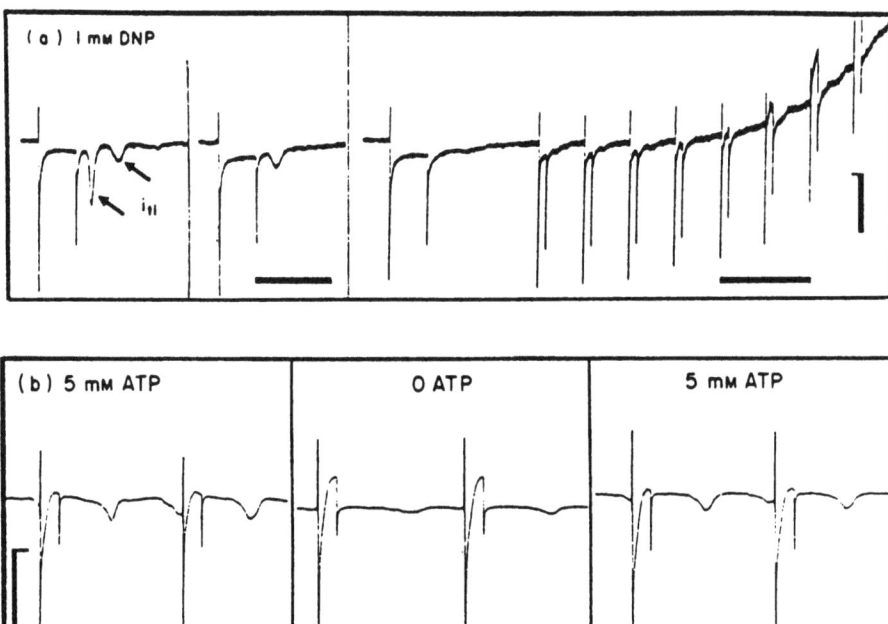

Fig. 1a, b. a The effect of DNP on the transient inward current in guinea pig ventricular myocyte. The currents were measured using a chopped clamp technique with a conventional microelectrode. The transient inward current was evoked by applying 500-ms steps from -55 to 0 mV in Tyrode's solution containing 0.54 mM K and 5.4 mM Ca. *Left panel* shows the current in low potassium, high calcium solution; *middle panel*, the current is shown 45 s after application of 1 mM DNP; *right panel* shows the evolution of the currents 2 min after the application of DNP. The time base was changed after the first clamp step in the left panel. The time calibration represents 1 s for the first three clamp steps and 6 s for the slower speed. The current calibration is 0.5 nÅ with the *horizontal tick* mark representing zero current. **b** The effect of perfusion with ATP-free solution on the transient inward current in a guinea pig ventricular myocyte using the whole cell clamp technique. *Left*, the current in response to a voltage step from -55 to 0 mV, in a cell exposed to 0.54 mM K, 5.4 mM Ca solution and perfused using a pipette containing a solution with 5 mM ATP; *middle*, current recorded 10 min after the pipette was perfused with ATP-free medium. A strong reduction of i_{TI} is seen, accompanied by an increase in outward current during the clamp step. The calcium current is not reduced; *right*, reversibility after reapplication of ATP in the pipette. The time calibration is 1 s; the current calibration is 0.5 nÅ. (Modified from [6])

i_{TI} disappears before the outward current increases (Fig. 1a). More directly it was demonstrated that i_{TI} requires the presence of a critical concentration of ATP and is suppressed in the absence of ATP ([6]; Fig. 1b). The mechanism underlying this effect has not been elucidated. The following three possibilities remain open: (a) ATP is required for a normal function of the Na–Ca exchange in the plasma membrane [11]; (b) ATP acts as fuel for the Ca-ATPase in the sarcoplasmic reticulum; or (c) ATP is necessary for activation of the calcium channel in the sarcoplasmic reticulum [5, 35, 37].

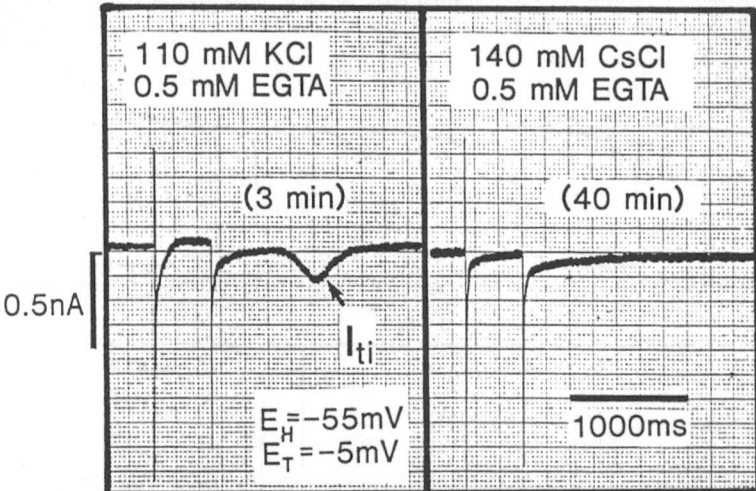

Fig. 2. Effect of intracellular K or Cs on the transient inward current in guinea pig ventricular myocyte. Whole cell recording. The suction pipette contained 110 mM KCl, 0.5 mM EGTA *(left panel)* or 140 mM CsCl, 0.5 mM EGTA. Appearance of transient inward current after 3 min when the cell was superfused with a solution containing 0.54 mM K, 5.4 mM Ca *(left)*. In the presence of intracellular Cs no transient inward current was observed even after 40 min of superfusion with a low potassium, high calcium solution *(right)*. (Unpublished observation by W. A. Coetzee)

An important component of the calcium release mechanism may be a simultaneous *movement of potassium ions into* the sarcoplasmic reticulum. In the absence of simultaneous efflux of anions or influx of counter cations, the outward movement of calcium ions should create an electrical gradient that inhibits the release itself. Some experimental results are suggestive for the existence of a counter potassium K-transport. (a) In endoplasmic reticulum from brain [36] and hepatocytes [31] ^{86}Rb influx was found to accompany calcium release induced by inositol trisphosphate (IP$_3$). Both are blocked by tetraethylammonium (TEA) and 9-TEA, known as potassium-channel blockers. (b) When a cardiac cell is dialyzed with a KCl solution and superfused with a solution low in K (to block the active Na–K pump), it develops transient inward currents; it never does so when KCl in the pipette is substituted by CsCl (Fig. 2); cesium ions are known to block different types of potassium channels.

Conclusion

For the future it may be important to develop drugs that bind selectively to the T-type calcium channel of the plasma membrane and the calcium and potassium channels of the sarcoplasmic reticulum. Drugs interfering with the sarcoplasmic reticulum channels may not only be important in the treatment of arrhythmias but also for the correction of inotropic disturbances.

References

1. Bean BP (1985) Two types of calcium channel in canine atrial cells. Differences in kinetics, selectivity and pharmacology. J Gen Physiol 86:1–30
2. Cannell MB, Lederer WJ (1986) The arrhythmogenic current i_{TI} in the absence of electrogenic sodium-calcium exchange in sheep cardiac Purkinje fibres. J Physiol 374:201–219
3. Cannell MB, Vaughan-Jones RD, Lederer WJ (1985) Ryanodine block of calcium oscillations in heart muscle and the sodium-tension relationship. Federation Proc 44:2964–2969
4. Carbonin P, DiGennaro M, Valle R, Weisz AM (1981) Inhibitory effect of anoxia on reperfusion- and digitalis-induced ventricular tachyarrhythmias. Am J Physiol 240:H730–H737
5. Chamberlain BK, Volpe P, Fleischer S (1984) Calcium-induced calcium release from purified cardiac sarcoplasmic reticulum vesicles. J Biol Chem 259:7540–7546
6. Coetzee W, Biermans G, Callewaert G, Vereecke J, Opie L, Carmeliet E (1988) The effect of inhibition of mitochondrial energy metabolism on the transient inward current of isolated guinea-pig ventricular myocytes. J Mol Cell Cardiol 20, in press
7. Coetzee WA, Dennis SC, Opie LH, Muller CA (1987) Calcium channel blockers and early ischemic ventricular arrhythmias: electrophysiological versus anti-ischemic effects. J Mol Cell Cardiol (Suppl) 19:77–97
8. Coetzee WA, Opie LH (1987) Effects of components of ischemia and metabolic inhibition on delayed afterdepolarizations in guinea pig papillary muscle. Circ Res 61:157–165
9. Dennis SC, Coetzee WA, Siko L, Opie LH (1987) Reperfusion arrhythmias: evidence for a role for Na^+/H^+ exchange and an antiarrhythmic effect of amiloride. Circulation 76: suppl. IV–56
10. DiGennaro M, Vassalle M, Iacono G, Pahor M, Bernabel R, Carbonin PU (1986) On the mechanism by which hypoxia eliminates digitalis-induced tachyarrhythmias. Eur Heart J 7:341–352
11. DiPolo R (1977) Effect of ATP on the calcium efflux in dialyzed squid giant axons. J Gen Physiol 64:503–517
12. Escande D, Corabœuf E, Planché C (1987) Abnormal pacemaking is modulated by sarcoplasmic reticulum in partially-depolarized myocardium from dilated right atria in humans. J Mol Cell Cardiol 19:231–241
13. Fabiato A (1983) Calcium-induced release of calcium from the cardiac sarcoplasmic reticulum. Am J Physiol 245:C1–C14
14. Fabiato A (1985) Spontaneous versus triggered contractions of "calcium-tolerant" cardiac cells from the adult rat ventricle. Basic Res Cardiol 80: Suppl 2, 83–88
15. Fabiato A (1985) Effects of ryanodine in skinned cardiac cells. Federation Proc 44:2970–2976
16. Fedida D, Noble D, Rankin AC, Spindler AJ (1987) The arrhythmogenic transient inward current i_{TI} and related contraction in isolated guinea-pig ventricular myocytes. J Physiol 392:523–542
17. Ferrier GR, Moffat MP, Lukas A (1985) Possible mechanisms of ventricular arrhythmias elicited by ischemia followed by reperfusion. Circ Res 56:184–194
18. Ferrier GR, Saunders JH, Mendez C (1973) A cellular mechanism for the generation of ventricular arrhythmias by acetylstrophanthidin. Circ Res 32:600–609
19. Friedman PL, Stewart JR, Fenoglio JJ, Wit AL (1973) Survival of subendocardial Purkinje fibers after extensive myocardial infarction in dogs: In vitro and in vivo correlations Circ Res 33:597–611
20. Hagiwara N, Irisawa H, Kameyama M (1988) Contribution of two types of calcium currents to the pacemaker potentials of rabbit sino-atrial node cells. J Physiol 395:233–253
21. Hajdu S, Leonard E (1961) Action of ryanodine on mammalian cardiac muscle: effects on contractility and reversal of digitalis-induced ventricular arrhythmias. Circ Res 9:1291–1298
22. Isenberg G, Vereecke J, Van Der Heyden G, Carmeliet E (1983) The shortening of the action potential by DNP in guinea-pig ventricular myocytes is mediated by an increase of a time-independent K conductance. Pflügers Arch 396:251–259
23. Kass RS, Tsien RW, Weingart R (1978) Ionic basis of transient inward current induced by strophanthidin in cardiac Purkinje fibers. J Physiol 281:209–226
24. Kitakaze M, Weisfeldt ML, Marban E (1987) Acidosis during reperfusion prevents myocardial stunning in ferret hearts. Circulation 76: Suppl IV–57

25. Kléber AG, Janse MJ, Wilms-Schopmann FJG, Wilde AAM, Coronel R (1986) Changes in conduction velocity during acute ischemia in ventricular myocardium of the isolated porcine heart. Circulation 73:189–198
26. Krikler DM, Rowland E (1985) The clinical value of calcium antagonists in the treatment of supraventricular arrhythmias. In: Zipes DP, Jalife J (ed) Cardiac electrophysiology and arrhythmias. Grune & Stratton, Orlando, p 567
27. Lazzara R, El Sherif N, Scherlag B (1974) Electrophysiological properties of canine Purkinje cells in one day-old myocardial infarction. Circ Res 33:722–734
28. Lukas A, Ferrier GR (1986) Interaction of ischemia and reperfusion with subtoxic concentrations of acetylstrophanthidin in isolated cardiac ventricular tissues: effects on mechanisms of arrhythmia. J Mol Cell Cardiol 18:1143–1156
29. Meissner G (1986) Ryanodine activation and inhibition of the Ca^{2+} release channel of sarcoplasmic reticulum. J Biol Chem 261:6300–6306
30. Mitra R, Morad M (1986) Two types of calcium channels in guinea pig ventricular myocytes. Proc Natl Acad Sci USA 83:5340–5344
31. Muallem S, Schoeffield M, Pandol S, Sachs G (1985) Inositol triphosphate modification of ion transport in rough endoplasmic reticulum. Proc Natl Acad Sci USA 82:4433–4437
32. Nilius B (1986) Possible functional significance of a novel type of cardiac Ca channel. Biochim Biophys Acta 45, Suppl K37–45
33. Nilius B, Hess P, Lansman JB, Tsien RW (1985) A novel type of cardiac calcium channel in ventricular cells. Nature 316:443–446
34. Poole-Wilson PA, Harding DP, Bourdillon PDV, Tones MA (1984) Calcium out of control. J Mol Cell Cardiol 16:175–187
35. Rousseau E, Smith JS, Henderson JS, Meissner G (1986) Single channel and $^{45}Ca^{2+}$ flux measurements of the cardiac sarcoplasmic reticulum calcium channel. Biophys J 50:1009–1014
36. Shah J, Pant HC (1988) Potassium-channel blockers inhibit inositol trisphosphate-induced calcium release in the microsomal fractions isolated from the rat brain. Biochem J 250:617–620
37. Smith JS, Coronado R, Meissner G (1985) Sarcoplasmic reticulum contains adenine nucleotide-activated calcium channels. Nature 316:446–449
38. Sutko JL, Kenyon JL (1983) Ryanodine modification of cardiac muscle responses to potassium-free solutions. J Gen Physiol 82:385–404
39. Tang C-M, Morad M (1988) Amiloride selectively blocks the low threshold (T) calcium channel. Biophys J 53:22a
40. Tani M, Neely JR (1987) Roles of H^+ and Na^+/H^+ exchange in Ca^{++} uptake and recovery of function in ischemic rat heart. Circulation 70 Suppl IV-56.
41. Thandroyen FT, McCarthy J, Burton KP, Opie LH (1988) Ryanodine and caffeine prevent ventricular arrhythmias during acute myocardial ischemia and reperfusion in rat heart. Circ Res 62:306–314
42. Tytgat J, Nilius B, Vereecke J, Carmeliet E (1988) The T-type Ca channel in guinea-pig ventricular myocytes is insensitive to isoproterenol. Pflüg Arch, in press
43. Tytgat J, Vereecke J, Carmeliet E (1988) Differential effects of verapamil and flunarizine on cardiac L-type and T-type Ca channels. Naunyn Schmied Arch Pharm, in press
44. Vleugels A, Vereecke J, Carmeliet E (1980) Ionic currents underlying hypoxia in voltage-clamped cat ventricular muscle. Circ Res 47:501–508
45. Wier WG, Hess P (1984) Excitation-contraction coupling in cardiac Purkinje fibers. J Gen Physiol 83:395–415

General Anaesthetics and Antiarrhythmics Antagonize the Ca^{2+} Current in Single Frog Cardiac Cells

P. Scamps, K. Mongo, A. Undrovinas, and G. Vassort

Laboratoire de Physiologie Céllulaire Cardiaque, Inserm U-241, Université Paris-Sud, Bât. 443, 91405 Orsay, France

The depressant effect of general anaesthetics such as halothane and ethanol in the nervous system is well documented. It is also clear that high doses of ethanol depress the maximum velocity of the upstroke, the total amplitude and duration of the action potential, and the force of contraction of cardiac muscle. Also, many agents with a local anaesthetic effect on nerves have a class 1 antiarrhythmic action on myocardium, which is associated with a negative inotropic effect. In both tissues, such agents reduce the maximum rate of depolarization as a consequence of inhibition of the fast sodium channel. The resting and rate-dependent depression of this channel has been carefully characterized, and the sodium-channel blockade by these agents has been modelled in the modulated receptor hypothesis (Hondeghem and Katzung 1984) and the guarded receptor hypothesis (Starmer and Grant 1985).

Class 1 antiarrhythmic agents also affect other electrical properties of cardiac membranes. At therapeutic concentrations, they depress the spontaneous diastolic depolarization, increase the threshold of excitability and reduce the conduction velocity. However, the exact mechanism whereby these agents exert these effects are not clearly understood, and it is probable that in cardiac tissues they affect more than one ionic channel (Carmeliet 1984). Recognition of the importance of slow action potentials and slow conduction in the genesis of arrhythmias has been stressed (Cranefield and Wit 1975). The inflow of Ca^{2+} ions during the cardiac action potential is a major determinant for the onset and magnitude of the concomitant contraction. Consequently, it could be anticipated that the action of class 1 antiarrhythmics on the Ca^{2+} current produces, at least in part, the negative inotropic effect generally reported for these agents.

In the present study, we compared the effects of four class 1 antiarrhythmic agents (quinidine, flecainide, ethmozin and ethacizin) and four straight-chain alcohols (ethanol, hexanol, octanol and dodecanol) on the kinetic parameters of the patch-clamp Ca^{2+} current (I_{Ca}) in isolated frog ventricular heart cells, using whole-cell recording. Special attention was paid to the effect of stimulus frequency since this factor is essential to demonstrate that drug-induced depression is more severe at high driving rates.

Materials, and Methods

Experiments were performed on single ventricular cells isolated from frog *(Rana esculenta)* heart. Cells were dissociated enzymatically and used within 36 h. The experimental arrangements for whole-cell patch-clamp recording and superfusion of the cells have been previously described (Fischmeister and Hartzell 1986). Briefly, for routine monitoring of I_{Ca}, cells were depolarized every 8 s from -80 mV holding potential (HP) to 0 mV for 200 ms. To measure I_{Ca} accurately, with no contamination of other ionic currents, the cells were bathed in potassium-free, 20-mM-cesium Ringer's solution, containing 88.4 mM NaCl, 20 mM CsCl, 22.9 mM $NaHCO_3$, 0.6 mM NaH_2PO_4, 1.8 mM $CaCl_2$, 1.8 mM $MgCl_2$, 5 mM D-glucose, 5 mM sodium pyruvate and 0.3 µM tetrodotoxin. The standard internal solution in the patch electrode (1–3 MΩ resistance) contained 120 mM CsCl, 5 mM K_2EGTA, 4 mM $MgCl_2$, 5 mM Na_2PCr, 3 mM Na_2ATP and 0.4 mM Na_2GTP, adjusted to pH 7.1 with KOH. Solutions were applied to the exterior of the cell by placing the cell at the opening of a 250 µm-inner-diameter capillary tube (flow rate 10 µl/min). All external solutions were gassed with 95% O_2/5% CO_2 (pH = 7.4), and experiments were carried out at room temperature (19.5 °C–22.5 °C). At various times during the course of the experiment, appropriate voltage-clamp protocols were applied to determine I_{Ca} characteristics.

Results

Quinidine and Flecainide Block of the I_{Ca}. Figure 1 illustrates the relative potency of quinidine and flecainide as inhibitors of the pure L-type I_{Ca} in the frog ventricle (Argibay et al. 1988), elicited by 200-ms depolarizing pulses to 0 mV from HP of -80 mV every 8 s. A 50% decrease of I_{Ca} amplitude was achieved with 10 µM quinidine

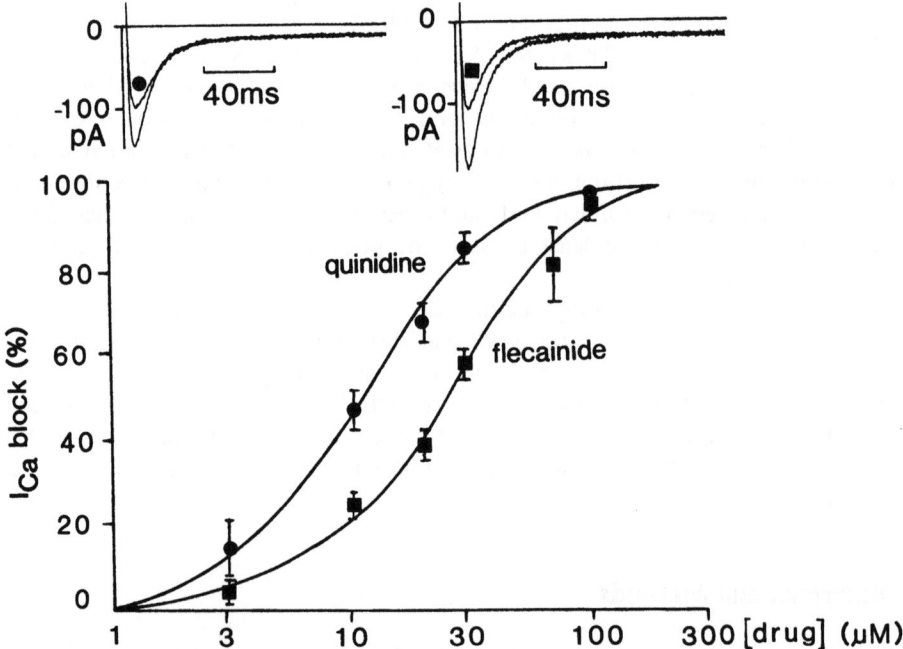

Fig. 1. Dose dependent block by quinidine and flecainide of the peak Ca^{2+} current *(I_{Ca})* elicited by membrane depolarizations to 0 mV. The upper traces are current recordings obtained on two different single cells in control and in the presence of quinidine (15 µM) or flecainide (30 µM)

and 20 µM flecainide. In both cases, we noticed that the Hill coefficient for the dose-response relationship was close to 2; however, we have no interpretation for this observation. To check the selectivity of these two antiarrhythmics, we also tested their blocking effect on the sodium current I_{Na} elicited in frog ventricular cells. Concentrations of quinidine (3 µM) and flecainide (10 µM) which induced about 10% decrease of I_{Ca} also decreased peak I_{Na} by 50 ± 7% ($n = 2$) and 94 ± 3% ($n = 2$) respectively. When the stimulation frequency was increased, lower concentrations were sufficient to induce the same reduction in I_{Na}.

In control conditions, a positive staircase of I_{Ca} was often observed if the stimulation frequency was increased up to 2 Hz (Argibay et al. 1988). This positive effect of frequency was suppressed by both drugs. However, no significant additive block or use-dependent block was observed. In other words, whatever the stimulation frequency, a given dose-dependent inhibition was achieved at first depolarization following quinidine or flecainide application. Reversibility was also complete on the first step after wash out of the drugs.

To assess whether the degree of block was linked to the degree of activation of the Ca^{2+} channels, current-voltage relationships were determined in control and in the presence of the drugs. At 30 µM, both drugs generally induced a 5- to 10-mV hyperpolarizing shift in the potential at which I_{Ca} achieved maximal amplitude. If one plots the ratio of I_{Ca} control/I_{Ca} in the presence of drug vs membrane potential, it appears that the degree of block increases linearly with increasing depolarizations (not shown). Such a shift was also observed in the availability curve: this could be attributed to a stronger affinity of the drug for the inactivated state of the channel (Hondeghem and Katzung 1984), or to a retention of the drug by closed inactivation gates (Starmer and Grant 1985). Since applying prehyperpolarizations did not remove the inhibition induced by the drugs, it appears that quinidine and flecainide mainly produce a tonic block, probably due to preferential binding to the rested channels.

Ethmozin and Ethacizin Block of I_{Ca}. Under control conditions at a stimulation frequency of 0.125 Hz, the two phenothiazine derivatives ethmozin and ethacizin seem to have different effects: ethmozin did not affect I_{Ca}, while ethacizin decreased I_{Ca} in a rest-block and use-dependent manner. Indeed, at the basal stimulation frequency (0.125 Hz), a very high concentration of ethmozin (70 µM) only slighly reduced I_{Ca}. With increasing stimulation frequency, I_{Ca} inhibition was more marked, being nearly complete at 2.5 Hz. However, the inactivation rate of I_{Ca} was not altered. The effects of 10 µM ethacizin were evaluated on another cell, using the same range of imposed stimulation frequencies. The first depolarization applied after switching to the ethacizin-containing solution decreased I_{Ca} markedly. It decreased further with repeated depolarizations at 0.125 Hz. The inhibitory effect was amplified at a higher stimulation frequency (1.6 Hz).

Figure 2 shows the dose-response curves for both the tonic and the use-dependent components of inhibition, at steady state, determined at three stimulation frequencies. It demonstrates that both drugs inhibited I_{Ca} primarily in a use-dependent manner. The apparent weak use-dependent inhibition observed with ethacizin at the highest stimulation frequencies (1 and 2 Hz) is a consequence of the very large tonic block.

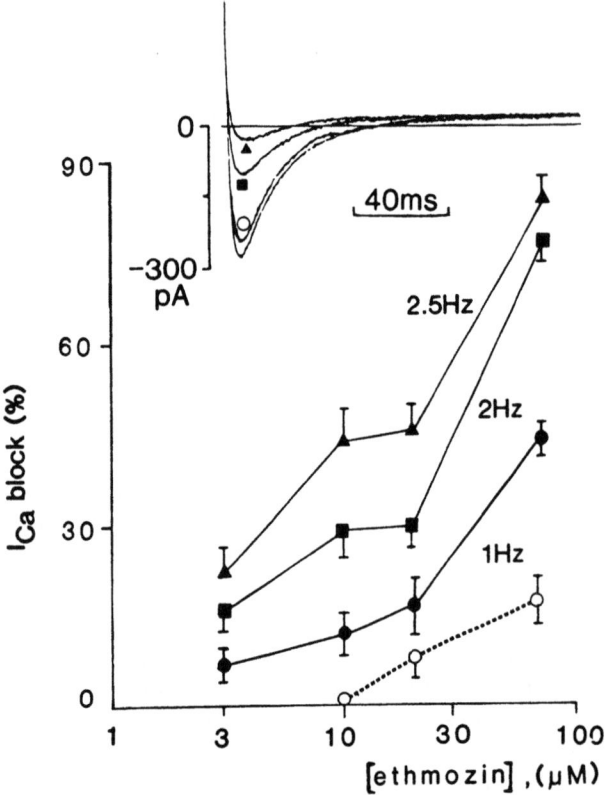

Fig. 2. Dose- and frequency-dependent block by ethmozin *(left)* and ethacizin *(right)* of the peak Ca^{2+} current (I_{Ca}) elicited by membrane depolarizations to 0 mV. The *dotted lines* represent the amount of rest block. The current traces (upper part) were obtained in control and in the presence of ethmozin (70 µM) or ethacizin (10 µM) during steady state at the indicated frequencies.

The inhibitory effects of these agents on I_{Ca} and I_{Na} were compared at a stimulation frequency of 0.125 Hz. Ethmozin (3 µM), which did not induce any I_{Ca} block, decreased I_{Na} to $25 \pm 7\%$ ($n = 2$) of control amplitude. Similarly, 1 µM ethacizin, which induced about 10% I_{Ca} block, produced at $42 \pm 6\%$ ($n = 2$) reduction in I_{Na}.

The current-voltage relationships determined at a stimulation frequency of 0.125 Hz in the presence of 10 µM ethmozin or ethacizin exhibited hyperpolarizing shift of up to 10 mV compared with control. At this frequency, ethmozin produced little rest or use-dependent block. This shift was more evident at 2.5 Hz (not shown). Availability of I_{Ca} was evaluated in the presence of the two compounds. In both cases, the curves were slightly shifted towards hyperpolarized potentials in the range of −50 to +10-mV prepulse potentials. As for quinidine and flecainide, we investigated whether this shift was indicative of an affinity of these drugs for the inactivated state of the channel. The effects of these drugs were also independent of the prepulse duration, and, moreover, the inhibition was not removed by hyperpolarization.

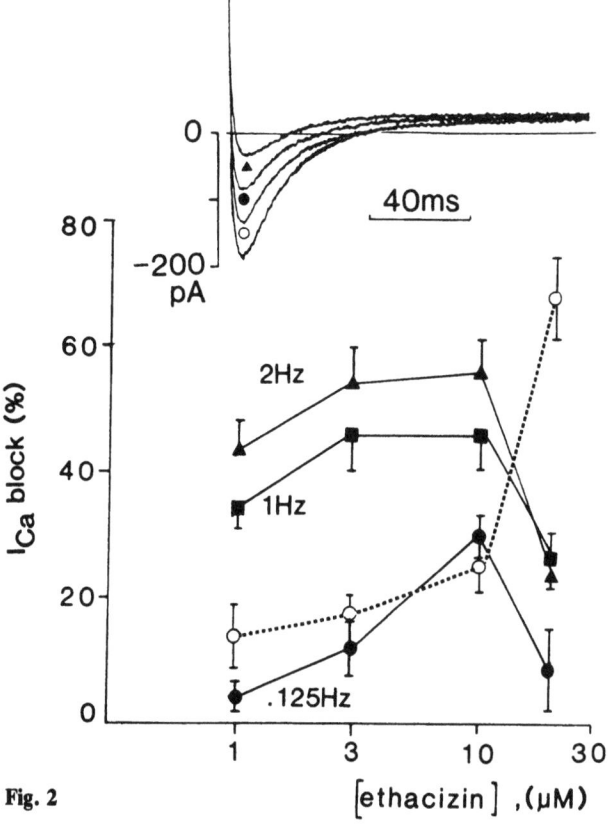

Fig. 2

Alteration in I_{Ca} by Straight-Chain Alcohols. At relatively high concentrations (100–200 mM), ethanol decreased I_{Ca}. However, this effect was mostly transient since after 3–5 min the current reached a new steady state, sometimes close to the control level. At low concentrations (10–20 mM), ethanol induced a slight increase in I_{Ca}. Neither this increase nor the subsequent recovery observed at high concentrations was prevented by the addition of propranolol (1 μM). These effects on peak I_{Ca} occured rapidly and were fully reversible.

It is well known that the efficiency of alcohol to slow down nerve action potential propagation increases with the number of carbon atoms in the chain. This is directly in line with the oil-water partition coefficient of the molecule (Haydon et al. 1983; Vassort et al. 1986) The inhibitory effects on I_{Ca} of hexanol, octanol and dodecanol were qualitatively very similar to the results obtained with ethanol, including the biphasic effect. Figure 3 summarizes the results recorded at maximal inhibition and provides evidence that 0.5 μM dodecanol was enough to inhibit 50% of I_{Ca}. Only the alcohol concentrations which induced 90% or more I_{Ca} inhibition were partially reversible. Rather than membrane disorder, this partial reversibility could be related to the strong contracture of the cell which occurred in such conditions.

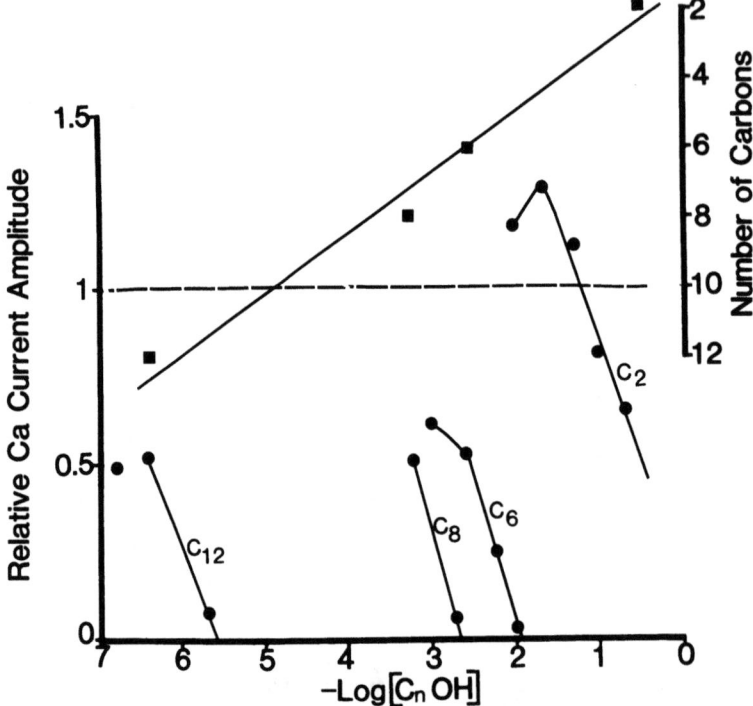

Fig. 3. Inhibition of the peak Ca^{2+} current elicited by depolarizations to 0 mV is a function of the dose and the length of the straight-chain alcohols. Their relative potency expressed as the log ED_{50} appears to be a linear function of the number of carbon atoms (C_n) in the chain, as is their oil/water partition coefficient

In the presence of alcohols, I_{Ca} kinetics were not significantly altered. There was also no evidence of voltage-dependent inhibition and shift in the availability curve except with dodecanol: in the presence of dodecanol, the observations were very similar to those in the presence of antiarrhythmics.

Discussion

Arrhythmia is a multifactorial phenomenon. One of the expected primary actions of antiarrhythmic agents is blocking of impulse propagation by inhibition of the fast I_{Na} of well-polarized cardiac tissues. However, the availability of the fast I_{Na} in depolarized tissues is reduced. In this range of membrane potential (above -65 mV) during ischaemia, premature beats or membrane oscillations related to the transient inward current, the ascending phase of the action potential is markedly dependent on inward I_{Ca}. Thus, general anaesthetics and antiarrhythmics can also exert their antiarrhythmic action by inhibiting this current. In this context, it has been shown that frog atrial and mammalian ventricular heart muscles develop slow oscillatory potential in response to a depolarizing current, which strongly depends on external Ca^{2+} concentration and is antagonized by quinidine (Ducouret 1976; Grant and Katzung 1976).

These effects on Ca^{2+} movements could occur together with inhibition of the I_{Na} since the range of active concentrations is rather similar. It should be stressed that the antiarrhythmic agents used in this study are more potent than some so-called Ca^{2+} antagonists in inhibiting the I_{Ca} [ED_{50} of diltiazem 5×10^{-5} µM at 0.33 Hz, ED_{50} of D-600 5×10^{-6} µM at 0.33 Hz in frog atrial cells (Uehara and Hume 1985)]. Only the dihydropyridines are more potent, having effects in the 10^{-7} M range. However, a more detailed comparison would be required to investigate the effects of each compound under the same experimental conditions, to allow more insight into the mechanisms of drug-channel interaction and to define another Ca^{2+}-antagonist binding site. At present, we know that the biological effects are poorly correlated with the electronic characteristics (pK_a values of quinidine, flecainide, ethmozin and ethacizin are 8.6, 9.4, 6.7 and 10.4 respectively) and with the molecular weight (324, 414, 461 and 450 respectively). We have more reason to expect a better relationship with the lipophilic character of the drugs since membrane channels are thought to represent hydrophilic (aqueous) pores.

Under our experimental conditions, ethmozin and ethacizin induced rest and use-dependent block, while quinidine and flecainide only had a rest-block effect. The latter observation could be a consequence of a very fast block rate: the association of quinidine and flecainide with the open channel would be complete during one (200-ms) pulse. Similarly, egress of the compounds might also occur during the first pulse on return to control solution. A use-dependent inhibition of I_{Ca} by quinidine has been recently evidenced in canine ventricular myocytes under experimental conditions which should favour its occurence, namely a holding potential at -40 mV (Salata and Wasserstrom 1988); this is in agreement with the much slower recovery that we observed at less negative holding membrane potentials. These observations, at least with ethmozin and ethacizin, are fully comparable with those recently reported with organic Ca^{2+}-channel antagonists. It was demonstrated that all three types (D-600, diltiazem and nifedipine) induced use-dependent inhibition (McDonald et al. 1984; Uehara and Hume 1985); thus, the modulated receptor hypothesis as originally developed for local anaesthetic interaction with sodium channels (Hondeghem and Katzung 1984) was proposed to be applicable. A more recent, simpler model, the guarded receptor hypothesis, characterized by ligand binding to periodically accessible receptors (Starmer and Grant 1985), may also account for the effects of local anaesthetics on sodium channels and for the above results obtained with ethacizin on Ca^{2+} channels (Starmer et al., to be published).

The way in which general anaesthetics alter membrane properties remains a matter of controversy. The most popular theories, involving lipids as the primary targets, are supported by the correlation of the anaesthetics' potency with the non-aqueous/aqueous partition coefficient. In these cases, adsorption of hydrocarbons into the interior of a lipophilic region would cause expansion in a direction normal to the membrane surface. This expansion, together with alteration in membrane fluidity, could modulate the kinetics of membrane-bound proteins due to the intramembranous location of sarcolemmal ion channels mediating ion flux and the transmembrane nature of ligand-receptor coupling.

There is now evidence accumulating which suggests that the most likely target sites of anaesthesia could be hydrophobic pockets of some proteins. Recent experiments with aequorin (Neering and Fryer 1986) or luciferase (Franks and Lieb 1985) demonstrate that the potency of inhibition by the homologous series of n-alcohols was in the same order as the oil/water partition coefficients of the alcohols. Similarly, alcohols alter the Ca^{2+} sensitivity of contractile proteins in cardiac skinned fibers according to their chain length (Ventura-Clapier and Vassort, unpublished work). It is worth noting that in line with the reported effects of short-chain alcohols on peak light intensity of aequorin (Neering and Fryer 1986) and on adenylate cyclase activity (Chatelain et al. 1986), ethanol increased peak I_{Ca} at 10 and 20 mM, while larger concentrations decreased I_{Ca} to below control values. In the present study, there was a continuous, direct relationship between alcohol chain length and potency for decreasing peak I_{Ca}, which corresponds roughly with the oil/water partition coefficient of the alcohols (Fig. 3). However, this is not sufficient to define more precisely the target alteration of membrane phospholipid fluidity or hydrophobic site of the Ca^{2+}-channel protein.

In conclusion, the above-mentioned alterations of peak I_{Ca} by general anaesthetics and antiarrhythmics could be a consequence of non-specific interaction of drugs with the lipid bilayer, as well as voltage-controlled accessibility to more specific Ca^{2+}-antagonist binding sites controlling the opening probability and the number of single Ca^{2+} channels. The data from both biochemical and electrophysiological experiments demonstrate at least six different drug-binding sites to the sodium channel, including one for type 1 antiarrhythmic agents (Sheldon et al. 1987). One can anticipate that besides verapamil, diltiazem and dihydropyridines, several other Ca^{2+} antagonists will soon be associated with a specific binding site.

References

Argibay J, Fischmeister R, Hartzell HC (1988) Inactivation, reactivation and pacing dependence of calcium current in frog cardiocytes: correlation with current density. J Physiol (Lond) (in press)

Carmeliet E (1984) Selectivity of antiarrhythmic drugs and ionic channels: a historical overview. A. NY Acad Sci 427:1–15

Chatelain P, Robberecht P, Waelbroeck M, Camus JC, Christophe J (1986) Modulation by n-alkanols of rat cardiac adenylate cyclase activity. J Membrane Biol 93: 23–32

Cranefield PF, Wit AL (1979) Cardiac arrhythmias. Annu Rev Physiol 41:459–472

Ducouret P (1976) The effect of quinidine on membrane electrical activity in frog auricular fibres studied by current and voltage clamp. Br J Pharmacol 57: 163–184

Fischmeister R, Hartzell HC (1986) Mechanism of action of acetylcholine on calcium current in single cells from frog ventricle. J Physiol (Lond) 376:183–202

Franks NP, Lieb WR (1985) Mapping of general anesthetic target sites provides a molecular basis for cutoff effects. Nature 316:349–351

Grant AO, Katzung BG (1976) The effects of quinidine and verapamil on electrically induced automaticity in the ventricular myocardium of guinea pig. J Pharmacol Exp Ther 196:407–419

Haydon DA, Urban BW (1983) The action of alcohols and other non-ionic surface active substances on the sodium current of the squid giant axon. J Physiol (Lond) 341:411–427

Hondeghem LM, Katzung BG (1984) Antiarrhythmic agents: the modulated receptor mechanism of action of sodium and calcium channel-blocking drugs. Ann Rev Pharmacol Toxicol 24:387–423

McDonald TF, Pelzer D, Trautwein W (1984) Cat ventricular muscle treated with D600: characteristics of calcium channel block and unblock. J Physiol (Lond) 352:217–241

Neering LR, Fryer MW (1986) The effect of alcohols on aequorin luminescence. Biochem Biophys Acta 882:39–43

Salata JJ, Wasserstrom JA (1988) Effects of quinidine on action potentials and ionic currents in isolated canine ventricular myocytes. Circul Res 62:324–337

Sheldon RS, Cannon NJ, Duff HJ (1987) A receptor for type I antiarrhythmic drugs associated with rat cardiac sodium channels. Circul Res 61:492–497

Starmer CF, Grant AO (1985) Phasic ion channel blockade. A kinetic model and parameter estimation procedure. Mol Pharmacol 28:348–356

Uehara A, Hume JR (1985) Interactions of organic calcium channel antgonists with calcium channels in single frog atrial cells. J Gen Physiol 85:621–647

Vassort G, Whittembury J, Mullins LJ (1986) Increases in internal Ca^{2+} and decreases in internal H^+ are induced by general anesthetics in squid axons. Biophys J 50: 11–19

The Calcium Channel and Vascular Injury

S. Kazda

Institute of Pharmacology, Bayer AG, D-5600 Wuppertal 1, FRG

Introduction

The specific calcium antagonists have well-defined mechanisms of action and have become tools for identification of the pathogenetic role of calcium ions in a variety of diseases. Hypertension is a typical example. By using verapamil, nifedipine, or diltiazem it was possible to show that increased transmembrane calcium availability in the small resistance vessels is the cause of elevated vascular tone in experimental or human hypertension (Aoki et al. 1976; Lederballe-Pedersen 1981). Moreover, the calcium content in vascular wall of spontaneously hypertensive rats with established hypertension progressively increases with increasing age without additional increase in blood pressure. This "secondary" increase in calcium concentration even in the large vessels may also be prevented by specific calcium antagonists (Fig. 1; Flecken-

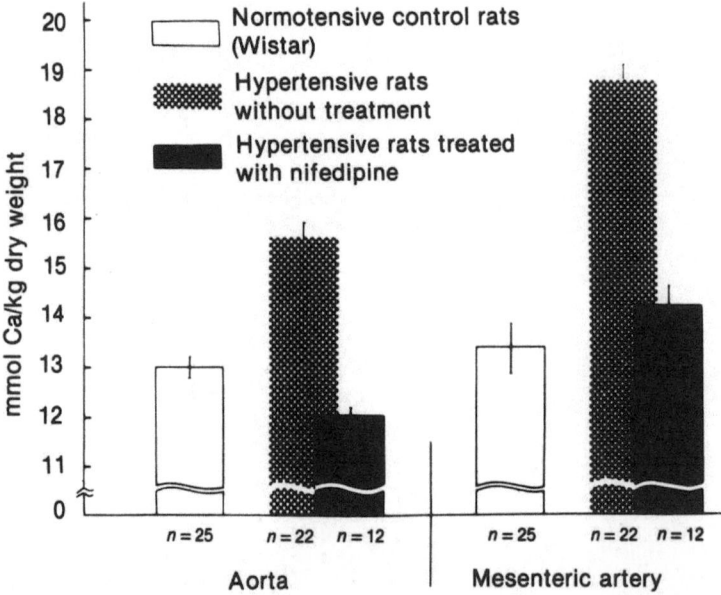

Fig. 1. Chronic treatment of spontaneously hypertensive rats with nifedipine during 5 months prevents the increase of calcium content of the arterial walls (aorta, mesenteric artery). (From Fleckenstein et al. 1983, with permission)

stein et al. 1983) showing that this "calcium overload" is also mediated by the vascular calcium channel. In advanced hypertension the vascular calcium overload results in fibrinoid necrosis and sclerotic changes of the vessel wall and in parenchymatous damage (Fleckenstein et al. 1983; Kazda et al. 1983).

In the present paper we suggest: (a) that high arterial blood pressure per se is not the cause of calcium-dependent tissue damage, (b) that some additional factors are involved in the excessive enhancement of calcium influx in advanced or accelerated hypertension, and (c) that excess parathormone may be one of the calcium-promoting factors resulting in tissue damage in hypertensive disease.

Salt-Induced Hypertension and Tissue Damage

Dahl rats are a suitable model for studying malignant hypertension and vascular damage. Fulminant hypertension in the salt-loaded (S) rat results in necrotizing vasculopathy with preferential localization in the kidney (Jaffé et al. 1970). Our histological investigations of Dahl rats revealed fibrinoid intimal degeneration, medial hyperplasia, and periarteritis, predominantly in the afferent glomerular arteries of the hypertensive S rats on a high-salt diet, leading to collapse and degeneration of the glomeruli (Fig. 2; Luckhaus et al. 1982). Electron-microscopic studies performed in our experimental series by J. Staubesand, Freiburg, revealed huge calcium deposits in the wall of the renal arteries of hypertensive Dahl rats (Fig. 3). No

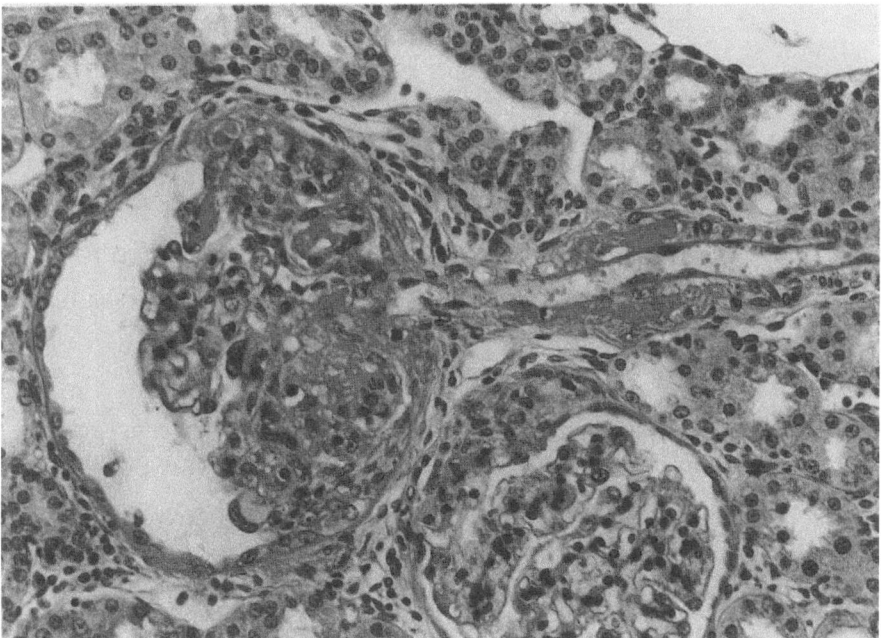

Fig. 2. Fibrinoid impregnation of the vas afferens and glomerular collapse in hypertensive salt-loaded S Dahl rat. (From Luckhaus et al. 1982)

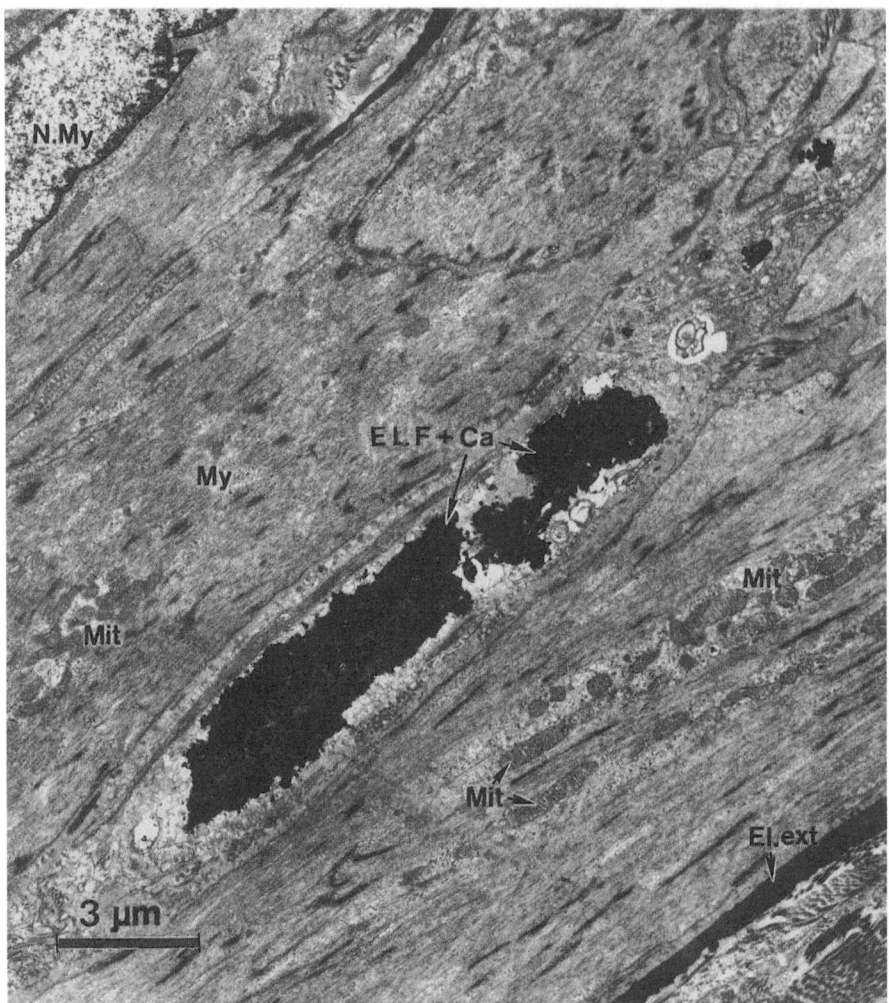

Fig. 3. Calcium deposits (Ca) in the wall of renal artery of salt-loaded S Dahl rat. Electron microscopy using pyroantimonate technique (J. Staubesand, personal communication)
My = Myocyte; N.My = nucleus of the myocyte; Mit = mitochondria; El. F= elastic fibres; El. ext = lamina elastica externa

pathological findings were detected in S rats that remained normotensive on a low-sodium diet. Salt-loaded Dahl salt-resistant (R) rats also remained morphologically intact.

Prevention by Calcium Antagonists

Treatment of Dahl S rats on a high-salt diet with nifedipine or nitrendipine prevented an increase in blood pressure and reduced cardiac hypertrophy and mortality (Fig. 4).

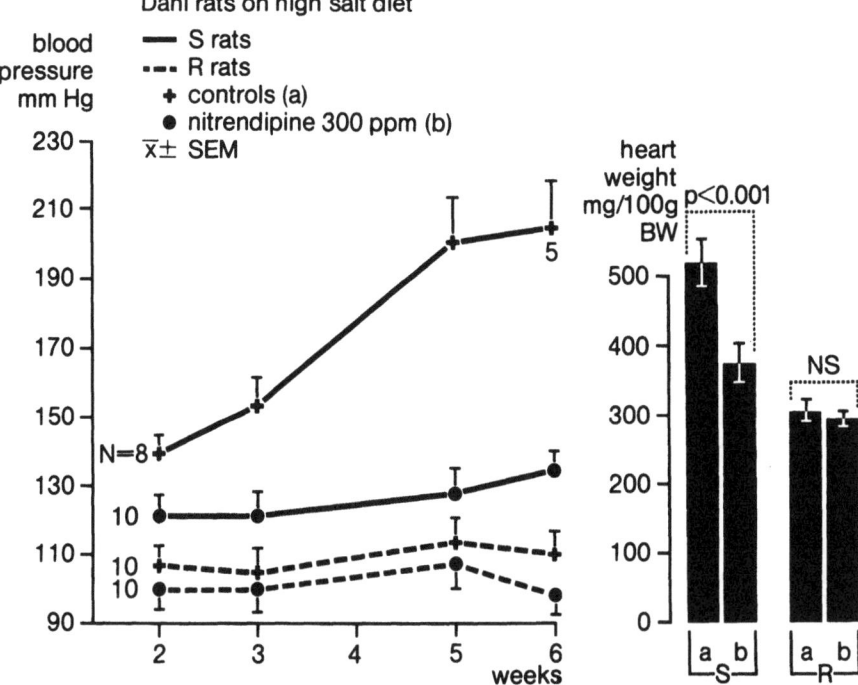

Fig. 4. Prevention of fulminant hypertension, mortality, and cardiac hypertrophy by chronic treatment with nitrendipine in salt-loaded sensitive (S) Dahl rats. No effect of nitrendipine treatment in salt-resistant (R) Dahl rats

Moreover, no morphological vascular changes were recorded in S rats after chronic treatment with calcium antagonists (Luckhaus et al. 1982). Also, the subcellular calcium deposits were absent in the renal arteries of nifedipine-treated rats.

Aggravation by the Calcium Agonist

In the Dahl S rat chronic treatment with Bay K8644 accelerated the development of salt-induced hypertension, precipitated mortality, and aggravated vascular lesions. In S rats on a low-salt diet (0.4% NaCl) which normally remain normotensive for a long period, Bay K8644 produced only a transient slight increase in blood pressure. At the end of the experiments, some of these agonist-treated low-salt rats had a high serum creatinine concentration and high plasma renin activity. These animals developed renovascular morphological changes which had never previously been observed in S rats on a low-salt diet (Kazda et al. 1986). This finding is the first hint that the vascular damage in S rats does not necessarily depend on the systemic blood pressure.

Salt-Accelerated Tissue Damage in Genetic Hypertension

Stroke-prone (SP) rats were originally derived from the Okamoto strain of spontaneously hypertensive rats (SHR; Okamoto et al. 1974). At the age of 10–12 months they spontaneously develop cerebrovascular lesions and brain infarction. In our experiments with adult male SP-rats aged 5 months the dietary salt load (8% NaCl) did not produce any additional increase in blood pressure. It did, however, drastically accelerate the mortality rate and produced severe cerebro- and renovascular lesions.

Blood Pressure Independent Tissue Protection

Effect of Nimodipine. Simultaneous treatment with the calcium antagonist nimodipine resulted in a dramatic reduction in mortality and vascular lesions in salt-loaded adult SHR SP. In this experiment all control rats died within 9 weeks as a result of the high-salt diet. Histological investigations revealed severe cerebro- and renovascular lesions. More than half the nimodipine-treated rats survived for 24 weeks. Nimodipine is a calcium antagonist derivative which has only a weak peripheral vasodilator effect. In this experiment the high blood pressure was not decreased by nimodipine (Fig. 5). In contrast, in the very ill controls, the blood pressure dropped ante finem; on average it was lower than that of nimodipine-treated animals (Kazda et al. 1982). These experiments on SHR SP show that, rather than preventing high blood pressure, nimodipine primarily prevented the deterioration of hypertensive disease, the vascular damage, and the mortality. These findings support our assumption that the vascular damage in hypertensive or hypertension-prone rats is not entirely dependent on the systemic blood pressure. In addition to high blood

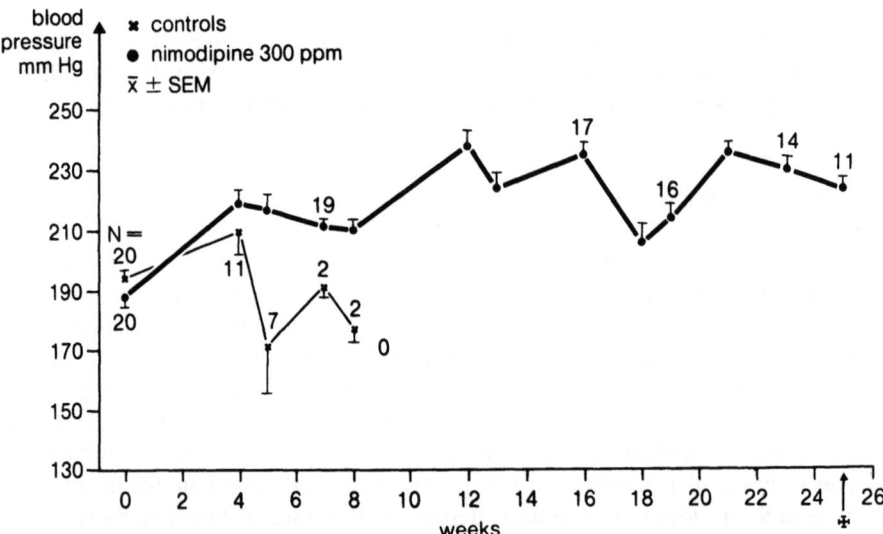

Fig. 5. Chronic treatment with nimodipine (•) dramatically increased the survival time of salt-loaded stroke-prone spontaneously hypertensive rats (8% NaCl) compared with that of control rats (X) without decreasing high blood pressure

pressure, other factors are presumably involved in the vascular damage induced by calcium overload.

Effect of Parathyroidectomy. In desoxycorticosterone acetate (DOCA)-salt hypertension, renal and cardiac lesions have been reported to be largely reduced by surgical parathyroidectomy without influencing the high blood pressure (Nickerson and Conran 1981). In DOCA-salt hypertensive rats parathyroidectomy had a similar, blood pressure independent protective effect to that of nimodipine in our experiments with SHR SP. This similarity prompted us to compare the effect of nimodipine with that of parathyroidectomy in salt-loaded SHR SP. Parathyroidectomy was performed in one of three groups of 12-week-old male SHR SP; two groups, each of 18 rats, were sham-operated. After 17 days all three groups received a diet containing 8% NaCl, and one group of sham-operated animals was treated with nimodipine added to the high salt diet (Kazda et al. 1986).

All control animals died in the course for 4 weeks on a high-salt diet (blood pressure 254 ± 9 mmHg). After parathyroidectomy all rats developed tetany after the introduction of the high-salt diet, and more than half died in the early weeks. Six animals recovered rather quickly and survived the observation period of 17 weeks but had extremely high blood pressure (276 ± 6 mmHg). All the nimodipine-treated rats developed well until the end of observation (17 weeks) despite persistently high values of blood pressure (259 ± 4 mmHg). Morphological lesions typical of hypertensive vascular disease in these rats were severe in all control rats but completely absent in the surviving parathyroidectomy group and in all nimodipine-treated rats.

In general, there were some similarities in the effects of parathyroidectomy and nimodipine in these experiments. In contrast with controls, both nimodipine-treated rats and some in the parathyroidectomy group survived for a long period but had extremely high blood pressure. Obviously, high blood pressure alone is not the cause of stroke and high mortality in salt-loaded SHR SP. It is conceivable that the high salt intake produces fatal tissue damage by inducing a harmful calcium overload. The life-saving effect of nimodipine, as mimicked by parathyroidectomy, may be due to inhibition of an excessive increase of cellular calcium concentration

Salt-Independent Calcium Overload in Stroke-Prone Rats

Seven-month-old male SHR SP were divided into three groups: in one group bilateral parathyroidectomy was performed, whereas the other groups were sham-operated. On one of these groups a continuous nimodipine treatment was started at that age; the third group served as a control. All animals, fed a standard rat chow containing less than 1% NaCl were observed for 14 weeks.

Nearly half the parathyroidectomized rats developed fatal tetany in the first weeks after surgery; the surviving rats developed well until the end of observation. In the last days of observation (10.5 months of age) all the control rats showed neurological symptoms typical for stroke (high irritability, transitory convulsions, hemiplegia or paraplegia; one animal died in clonic convulsions). None of the nimodipine-treated or the parathyroidectomized rats showed any neurological symptoms until the end of observation. The final blood pressure of the nimodipine-treated rats was identical

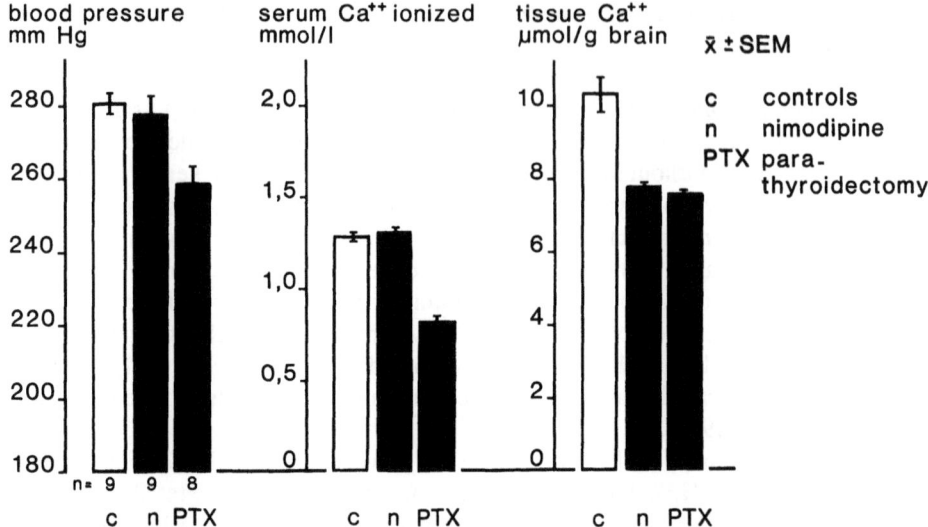

Fig. 6. Systolic blood pressure, serum Ca^{2+} concentration and calcium content in the brain of 42-week-old stroke-prone spontaneously hypertensive male rats on a normal salt diet. Determinations were performed 14 weeks after parathyroidectomy (*PTX*) or introduction of nimodipine treatment. (Experimental data of M. Grunt, not published)

with that of the control rats; a small but insignificant decrease was measured at the end of observation in parathyroidectomized rats (Fig. 6).

Since all the controls showed symptoms of stroke in week 14, the experiments were terminated. Brains were removed and homogenized, and the calcium content was measured by atomic absorption. In addition, the calcium content of brains from adult normotensive Wistar-Kyoto rats as well as from young SHR SP was investigated. The brain calcium content of the old control SHR SP was more than 50% higher than that of young SHR SP or adult Wistar-Kyoto rats. The calcium content of brains from both nimodipine-treated and parathyroidectomized rats was only slightly higher than that of young SHR SP at prehypertensive age but much lower than that of diseased old SHR SP (Fig. 6).

The results of this experiment confirm our previous observation that tissue damage in severe hypertensive disease is not primarily caused by high blood pressure. The protective effect of parathyroidectomy suggests that the activity of the parathyroid gland may be activated, and that an excess of parathormone may be at least partly responsible for the calcium overload and tissue damage in advanced hypertension. This hypothesis explains why nimodipine has a similar effect to removal of the parathyroid gland. It prevents the harmful calcium overload, even in the presence of long-standing, extremely high hypertension.

Conclusions

An excessive increase in cellular calcium content is the crucial step in cellular damage in advanced hypertension. In a susceptible rat strain (Dahl S rats) hypertension and tissue damage, induced by a high-salt diet, are preventable by calcium-antagonistic dihydropyridine derivatives. Moreover, the enhancement of calcium influx by a dihydropyridine calcium agonist (Bay K8644) results in tissue damage in normotensive S rats on a low-salt diet.

There is some evidence that an increase in the number of dihydropyridine receptors and related voltage-sensitive calcium channels may be involved in the pathogenesis of calcium overload and tissue damage. In Syrian cardiomyopathic hamster the cardiac lesions associated with calcium overload can be prevented by calcium antagonists. Numbers of DHP-receptor binding sites in heart, brain, skeletal muscle, and smooth muscle were markedly increased in parallel to the increased uptake of ^{45}Ca into tissues in these animals (Wagner et al. 1986). Little is known about changes of calcium channel in hypertensive vascular disease. Recently, Garthoff and Bellemann (1987) found that high-salt diet markedly increased density of nitrendipine binding sites in myocardial membranes of SHR SP. In our experiments a dietary salt load of SHR SP resulted in brain lesions. Tissue damage was largely prevented, and survival was increased by nimodipine without decreasing high blood pressure.

A similar protective effect, independent of blood pressure, was achieved by bilateral parathyroidectomy in salt-loaded SHR SP. Parathormone is known to stimulate calcium uptake into isolated kidney and heart cells and increases the intracellular calcium concentration resulting in an early cellular death (Bork 1968; Bogin et al. 1981). Administration of parathyroid extract to uremic dogs produced a significant rise in brain calcium concentration which was not related to the plasma calcium concentration. In acutely uremic dogs calcium content in the grey and white matter of the brain was increased, and this increment was prevented by parathyroidectomy (Arieff and Massry 1974).

In our experiments the natural appearance of stroke in SHR SP (without any additional salt load) was correlated with the increase in calcium content in brain tissue. Nimodipine prevented spontaneous strokes and the increase in brain calcium content without affecting the high blood pressure in SHR SP. Parathyroidectomy also mimicks the protective and life-saving effect of nimodipine.

Presumably, some calcium promoting factors are activated in advanced hypertension, resulting in cellular damage. Excess parathormone may be one of these factors.

References

1. Aoki K, Yamashita K, Suzuki A, Takikawa K, Hotta K (1976) Uptake of calcium ions by sarcoplasmic reticulum from heart and arterial smooth muscle in the spontaneous hypertensive rat (SHR). Clin Exp Pharmacol Physiol 3:27–30
2. Arieff AI, Massry SG (1974) Calcium metabolism of brain in acute renal failure. J Clin Invest 53:387–392
3. Bogin E, Massry SG, Harary J (1981) Effect of parathyroid hormone on rat heart cells. J Clin Invest 67:1215–1227
4. Bork AB (1968) Effects of purified parathyroid hormone on the calcium metabolism of monkey kidney cells. Endocrinology 83:1312--1322
5. Fleckenstein A, Frey M, von Witzleben H (1983) Vascular calcium overload – a pathogenic factor in arteriosclerosis and its neutralization by calcium antagonists. In: Kaltenbach M, Neufeld HN (eds) 5th International Adalat Symposium Excerpta Medica. Amsterdam-Oxford-Princeton pp 36–54

6. Garthoff B, Bellemann P (1987) Effects of salt loading and nitrendipine on dihydropyridine receptors in hypertensive rats. J Cardiovasc Pharmacol 10 (suppl 10):S36–S38
7. Jaffé S, Sutherland LE, Barker DV, Dahl LK (1970) Effect of chronic salt ingestion: morphologic findings in kidneys of rats with differing susceptibilities of hypertension. Arch Pathol 90:1–16
8. Kazda S, Garthoff B, Luckhaus G, Nash G (1982) Prevention of cerebrovascular lesions and mortality in stroke-prone spontaneously hypertensive rats by the calcium antagonist nimodipine. In: Godfraind T, Albertini A, Paoletti R (eds) Calcium modulators. Elsevier Biomedical Press, New York, pp155–167
9. Kazda S, Garthoff B, Luckhaus G (1983) Calcium antagonists in hypertensive disease: experimental evidence for a new therapeutic concept. Postgraduate Med J 59 (suppl 2):78–83
10. Kazda S, Garthoff B, Hirth C, Preis W, Stasch JP (1986) Parathyroidectomy mimicks the protective effect of the calcium antagonist nimodipine in salt-loaded stroke-prone spontaneously hypertensive rats. J Hypertension 4 (suppl 3):S482–S485
11. Kazda S, Garthoff B, Luckhaus G (1986) Calcium and malignant hypertension in animal experiments: effects of experimental manipulation of calcium influx. Am J Nephrol 6 (suppl 1): 145–150
12. Lederballe-Pedersen O (1981) Calcium blockade as a therapeutic principle in arterial hypertension. Acta Pharmacol Toxicol (Copenh) 49 (suppl II):1–31
13. Luckhaus G, Garthoff B, Kazda S (1982) Prevention of hypertensive vasculopathy by nifedipine in salt-loaded Dahl rats. Arzneim Forsch (Drug Res) 32:1421–1425
14. Nickerson PA, Conran RM (1981) Parathyroidectomy ameliorates vascular lesions induced by deoxycorticosterone in rat. Am J Pathol 105:185–190
15. Okamoto K, Yamori Y, Nagaoka A (1974) Establishment of the stroke-prone spontaneously hypertensive rat (SHR). Circ Res 34,35 (suppl 1):143–153
16. Wagner JA, Reynolds IJ, Weisman HF, Dudeck P, Weisfeldt ML, Snyder SH (1986) Calcium antagonist receptors in cardiomyopathic hamster: selective increases in heart, muscle, brain. Science 232:515–518

Endocrine Effects of Calcium Channel Agonists and Antagonists in a Pituitary Cell System*

P. M. Hinkle, R. N. Day, and J. J. Enyeart

Department of Pharmacology and the Cancer Center,
University of Rochester School of Medicine and Dentistry, Rochester, NY 14642, USA

Introduction

The endocrine regulation of the anterior pituitary gland is complex. Hypothalamic and circulating hormones control the rate of secretion of each of the adenohypophyseal hormones and also the rate of de novo hormone biosynthesis. A widely used model system for studying regulation of secretory endocrine cells is the GH_3 and related cell lines, collectively referred to as GH cells. GH cells are clonal lines derived from a rat pituitary tumor and have the characteristics of a somatomammotrope, i. e., they synthesize and secrete both growth hormone and prolactin. GH cells respond to many of the stimuli that regulate anterior pituitary function in vivo and are a particularly attractive model because they afford the opportunity to study control of both the release and the synthesis of prolactin and growth hormone. Although this chapter summarizes work from this laboratory, the molecular mechanisms of regulation of GH cell function have been studied by many investigators and are the subject of a number of recent reviews (Gourdji et al. 1981; Gershengorn, 1986; Hinkle, 1988).

Release of hormone from the cell occurs via two pathways, constitutive and regulated (Kelly, 1985). Prolactin released via the constitutive pathway is discharged from the cell shortly after it is synthesized, and the rate of secretion depends only on the rate of synthesis. Prolactin released via the regulated pathway is sorted and packaged into dense secretory granules. Exocytosis occurs in response to stimulatory signals such as a rise in intracellular free calcium ion, cyclic AMP, or protein kinase C activity. Prolactin and growth hormone released via this pathway have been stored during granule maturation and hence are "older" in metabolic labeling experiments than hormone released constitutively (Statchura, 1983). Hormone release via the regulated pathway is subject to hormonal inhibition as well as hormonal stimulation.

Calcium serves as a potent stimulus to the release of hormones stored in secretory granules from the GH cell line, as it does for almost all endocrine and exocrine cells. A rise in cytoplasmic free calcium ion could occur, in principle, by increased influx of extracellular calcium through channels, by release of calcium from intracellular calcium-rich stores (organelles or calcium-binding proteins), or by decreased extrusion of cytoplasmic calcium outside the cell or to other organelles. There has been a great deal of progress in elucidating the mechanisms underlying the tight control of

---

* This research was supported in part by NIH grant AM 19974, Cancer Center Core Research Grant 11 198, and NIH training grants AM 07 092 and NS 07 184 (JJE) and GM 07 102 (RND).

cytoplasmic calcium ion concentration. Much less is understood about how the rise in calcium causes the release of hormone from secretory granules.

GH cells possess at least two types of voltage-sensitive calcium channels (VSCCs). One class, like the L-type channels described in other cells, is typified by activation at strongly depolarized potentials, slow inactivation, and sensitivity to dihydropyridines (DHPs) (Nowycky et al. 1985; Matteson and Armstrong 1986; Miller 1987; Cohen and McCarthy 1987). GH cells maintain a resting potential of about −40 to −50 mV and fire spontaneous, mainly calcium action potentials (Taraskevich and Douglas 1980; Ozawa and Kimura 1979).

Response to Depolarization

Depolarization with KCl leads to an immediate burst of secretion of prolactin and growth hormone from normal and pituitary tumor cells. The increase in secretion can be observed within 5 s, and the initial exocytosis is largely complete by 1 min (Aizawa and Hinkle 1985a, b). The relative increases in prolactin and growth hormone are virtually identical. Secretion occurs as if both hormones were stored in the same granules, but this has never been demonstrated and is in fact not predicted, based on what has been found in normal somatomammotropes, which store the two hormones in different granules of the same cell (Hashimoto et al. 1987).

A typical secretory pattern in response to repeated pulses with a depolarizing concentration of potassium is shown in Fig. 1. The burst of release elicited by depolarization occurs after no measurable delay, and the secretory rate quickly returns toward normal. A transient pattern of release is also found if high potassium is present continuously (Aizawa and Hinkle 1985b). Cells respond to repeated challenges with high potassium with continued bursts of secretion. There is little apparent desensitization, although the size of the response gradually declines as intracellular hormone stores are depleted. Because about 1 h is required from the time that translation of prolactin and growth hormone mRNAs begins and the time that newly synthesized hormone first appears in the medium, new synthesis does not contribute at all to the responses shown in Fig. 1. Other experiments have demonstrated that transient depolarization does not change the rate of hormone synthesis.

Pharmacology of GH-Cell Calcium Channels

Effects of Calcium-Channel Antagonists

The effect of drugs on VSCCs has been assessed in four ways: inhibition of depolarization-induced hormone secretion, inhibition of $^{45}Ca^{2+}$ influx, blockade of calcium current measured in patch clamp experiments, and blockade of the rise in intracellular free calcium measured with intracellularly trapped dyes.

In the absence of a depolarizing stimulus, organic calcium-channel blockers cause only minimal inhibition of prolactin secretion from GH cells over 10 min. Although the cells are firing spontaneous calcium action potentials, the amount of calcium entering the resting cell over a few minutes is apparently not large. When high

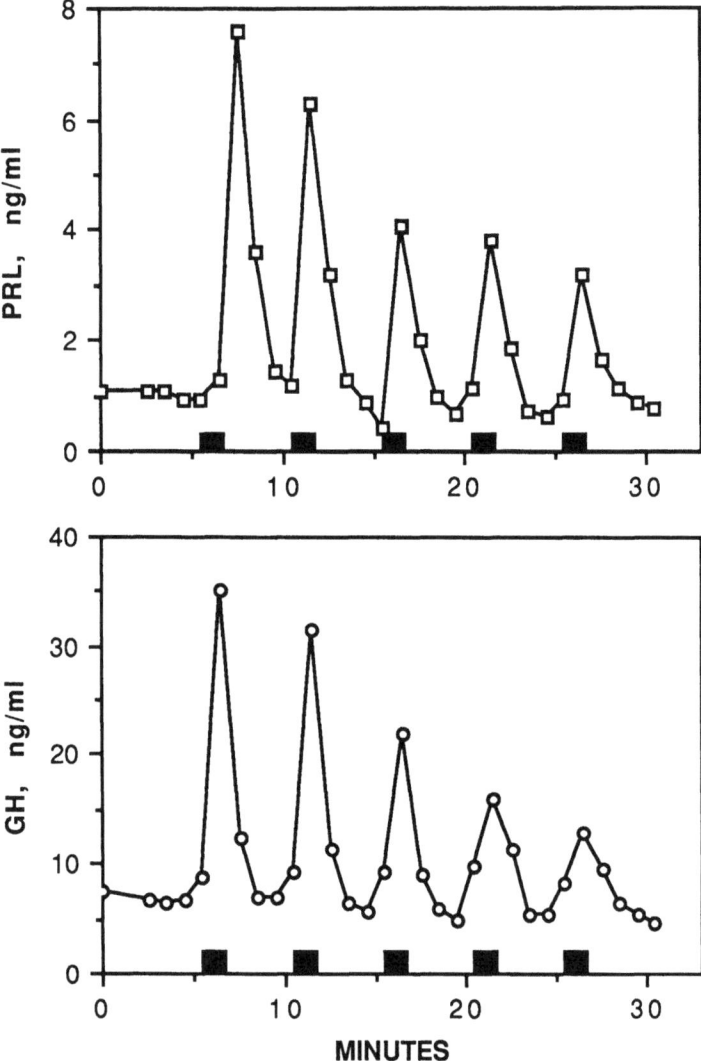

Fig. 1. Secretory response to depolarization. GH cells in a perifusion system were treated with 50 mM KCl at the times shown by the *solid bars*. *PRL*, Prolactin. (From [2])

potassium is added to depolarize cells, there is a prompt three fold increase in $^{45}Ca^{2+}$ uptake (Fig. 2A) and a five fold increase in prolactin release (Fig. 2B). The rise in cytoplasmic free calcium ion concentration can be quantitated with intracellularly trapped dyes; Figure 3 shows a typical example where $[Ca^{2+}]_i$ increased approximately four fold with 50 mM KCl. The depolarization-induced increase in $^{45}Ca^{2+}$ uptake was almost completely blocked by nitrendipine ($IC_{50} = 2$ nM) as was the increase in hormone secretion. The fact that DHP antagonists completely block calcium influx and secretion in these experiments suggests that the entry of calcium

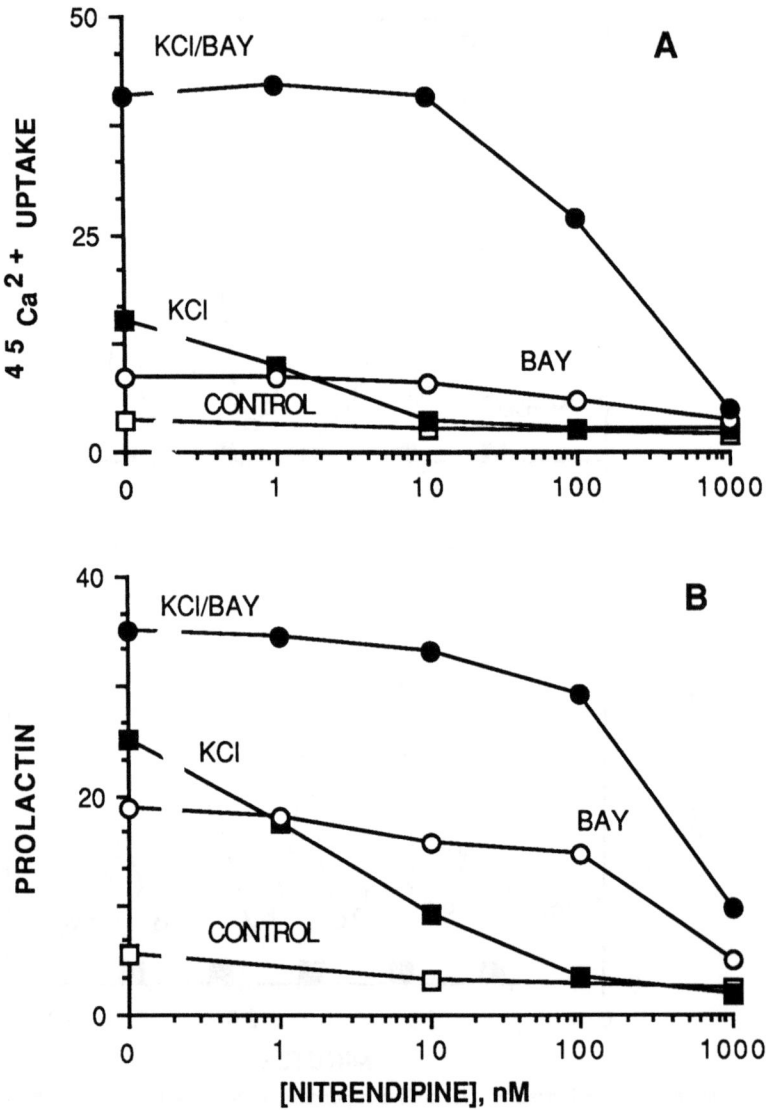

Fig. 2A, B. Drug effects on $^{45}Ca^{2+}$ uptake and prolactin secretion. GH cells were incubated for 10 min with nitrendipine and (□) no additional drug, (■) 54 mM KCl, (○) 2 μM Bay K8644, or (●) KCl + Bay K8644. **A** Calcium uptake is shown as cpm/dish × 10^{-3}. **B** Prolactin secretion is shown as ng/dish per 10 min. Errors averaged less than 5%. (From [8])

through L-type VSCCs is the prime contributor to the rise in cytoplasmic calcium stimulated by depolarization over 10 min. The increase in intracellular free calcium elicited by depolarization with KCl was likewise prevented if DHP antagonists were added first (Fig. 3). $[Ca^{2+}]_i$ also returned towards the resting value promptly if a channel blocker was added after depolarization.

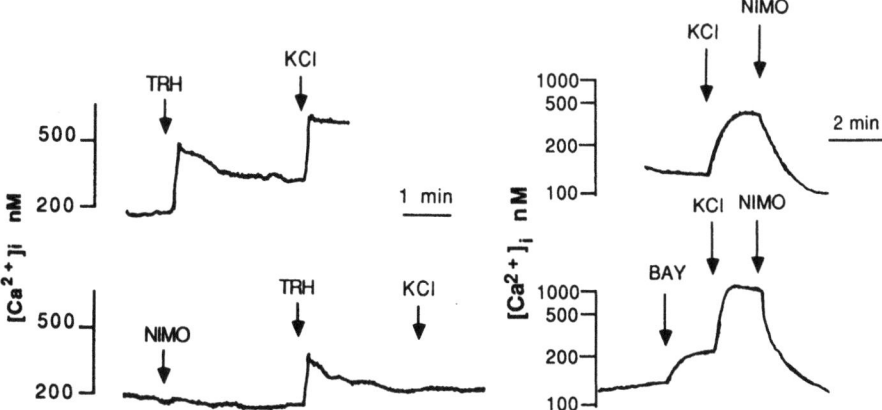

Fig. 3. Effects of depolarization, thyrotropin-releasing hormone *(TRH)* and dihydropyridines on $[Ca^{2+}]_i$. *Left panels*, at the times indicated by the *arrows* 100 nM nimodipine *(NIMO)*, 100 nM TRH, or 50 mM KCl were added to Fura2-loaded GH cells in Hepes-buffered Hank's balanced salt solution maintained at 37°C (from [5].) *Right panels, arrows* show addition of 1 μM Bay K8644 *(BAY)*, 50 mM KCl, and 100 nM nimodipine to cells loaded with Quin-2, essentially as described in [3]

The activation of VSCCs has also been measured using the whole cell variation of the patch clamp technique. Calcium currents were studied in near isolation by blocking sodium currents with tetrodotoxin and potassium currents with cesium. Curve 1 of Fig. 4A shows the activation of calcium influx by a 300-ms 50-mV depolarizing pulse from a resting potential of −40 mV. The slow inward calcium current was virtually abolished when the preparation was treated with the DHP-channel blocker nisoldipine (Curve 2).

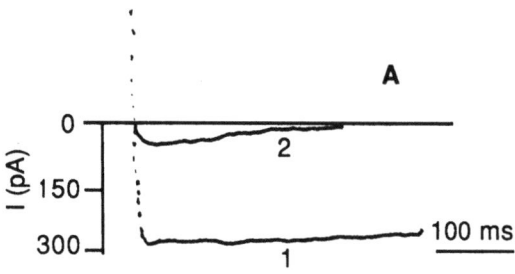

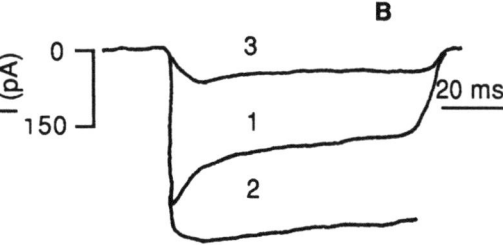

Fig. 4A, B. Electrophysiological effects of dihydropyridines in GH cells. Two experiments are shown in which a GH cell was subjected to a 300-ms (**A**) or 100-ms (**B**) 50 mV-depolarizing pulse from a holding potential of −40 mV. **A** Curve *1*, control; curve *2*, 90 s after addition of 100 nM nisoldipine (from [8]). **B** Curve *1*, control; curve **2**, 500 nM Bay K8644; curve *3*, 500 nM Bay K8644 + 500 nM nimodipine (from [9])

Effects of Calcium-Channel Agonists

The DHP calcium-channel agonist Bay K8644 prolongs the mean open time of channels (Nowycky et al. 1985). When added to pituitary cells, Bay K8644 causes an increase in $^{45}Ca^{2+}$ influx through voltage-sensitive channels, as monitored with intracellular calcium sensing dyes, hormone secretion, and calcium currents. In a patch clamp experiment using GH cells, Bay K8644 was shown to increase the amplitude of the slow calcium current, and addition of the more potent antagonist nimodipine together with Bay K8644 virtually abolished the effect (Fig. 4B).

Bay K8644 and high K^+ stimulate calcium influx in a synergistic fashion, as might be expected (Fig. 2A). However, the effects of Bay K8644 and high potassium on secretion are less than additive (Fig. 2B). This is most simply interpreted as evidence that the calcium-sensitive secretory mechanisms are saturated at calcium concentrations below those attainable with Bay K8644 and high K^+. Typically, depolarization alone causes a three to four fold rise in intracellular calcium while Bay K8644 alone causes a two fold rise. In the experiment shown in Fig. 3, depolarization and Bay K8644 together increased $[Ca^{2+}]_i$ from a resting value of 120 to 1200 nM. The increase in intracellular calcium elicited by hormonal stimuli such as hypothalamic peptides is invariably smaller than that attained with sustained depolarization. It is reasonable that the cellular response mechanism would activate and saturate over a calcium range attainable with physiological regulators.

The actions of Bay K8644 and DHP-channel antagonists appear to be strictly competitive in the pharmacological sense. The dose-response curves for Bay K8644 stimulation of prolactin secretion are shifted to the right by increasing concentrations of nimodipine; for example, the ED_{50} for Bay K8644 shifted from 200 nM (no antagonist) to > 1000 nM (200 nM nimodipine) (Enyeart and Hinkle 1984). The converse is also true (e.g., Fig. 2), namely, the dose-response for nitrendipine inhibition of KCl-stimulated secretion and calcium influx is shifted to the right by almost two orders of magnitude (2 to 100 nM) by 2 μM Bay K8644. The amplitude of the secretory response evoked by Bay K8644 increases with increasing extracellular calcium concentration. The results indicate that Bay K8644 and DHP antagonists probably bind to the same site on the calcium channel. The DHP antagonists are considerably more potent, on a molar basis, than Bay K8644.

Available data suggest that the VSCCs on normal and tumor cells are substantially the same. For example, the DHP-channel blockers completely prevent depolarization-induced secretion of prolactin from cells isolated from normal pituitary glands as well as from pituitary tumor cells (Enyeart et al. 1985). Similarly, Bay K8644 increases prolactin secretion from normal mammotropes (Enyeart et al. 1986) (Fig. 5).

Unconventional Calcium-Channel Blockers

Using the techniques described above, we have studied the activity of both classical organic channel blockers and of a number of drugs not generally considered as calcium-channel antagonists (Enyeart et al. 1987). These data are summarized in Table 1. Surprisingly, several of the nonconventional drugs are potent blockers of L-type calcium channels. Pimozide, an antipsychotic dopamine-receptor antagonist,

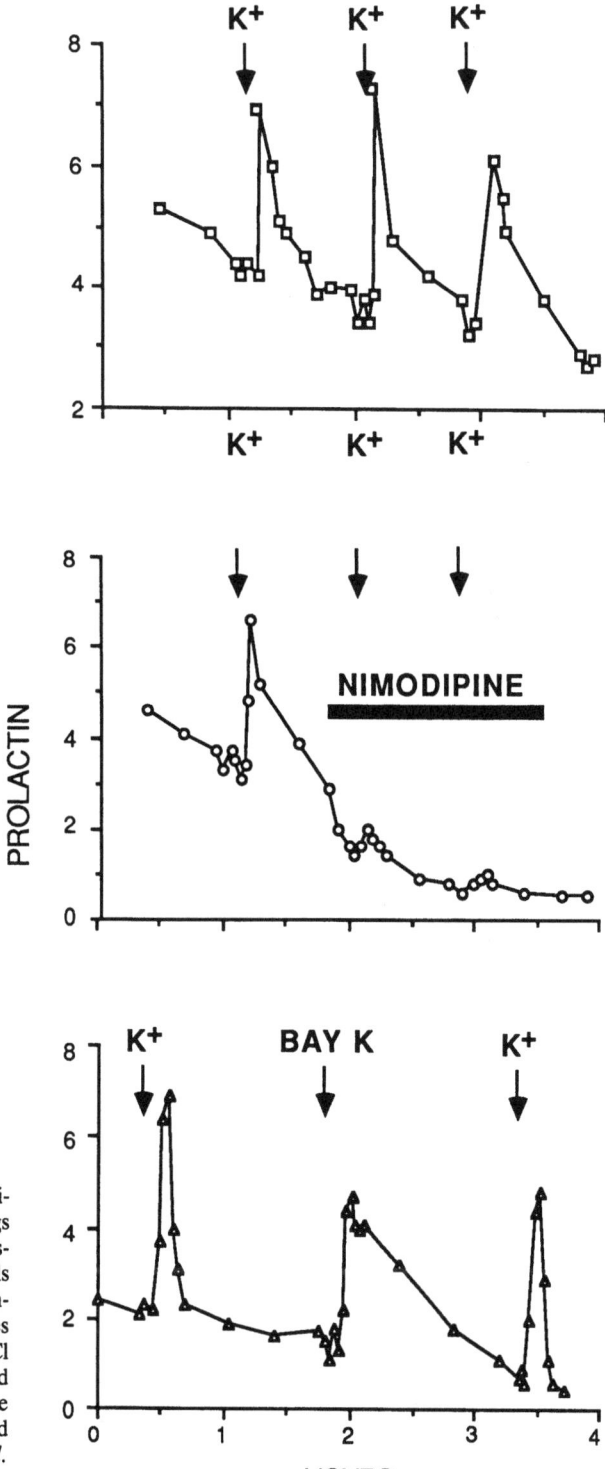

Fig. 5. Effects of voltage-sensitive calcium channel active drugs on normal pituitary cells. Dispersed rat anterior pituitary cells were perifused with serum-containing medium. At the times shown by the *arrows* 60 mM KCl or 500 nM Bay K8644 were added for 2.5 min; 50 nM nimodipine was present during the time noted by the *bar* in the *middle panel*. (From [6, 7])

Table 1. Drug inhibition of calcium influx and prolactin secretion

Drug	$^{45}Ca^{2+}$ uptake	PRL secretion
Nitrendipine	2.5	3.0
Nimodipine		1.8
Nifedipine		6.0
Nisoldipine		1.8
Verapamil	1000	1000
Diltiazem	2500	3500
Cinnarizine		1000
Pimozide	80	100
Haloperidol	3000	5000
Calmidazolium	400	600
Veratridine	1000	1000

The half-maximally effective concentration of drugs for inhibition of high potassium induced $^{45}Ca^{2+}$ uptake and prolactin (PRL) secretion were measured in experiments such as that shown in Fig. 2. Values are nM. (From [6, 8])

blocked both high K$^+$-induced $^{45}Ca^{2+}$ uptake and prolactin secretion and also calcium currents measured in whole cell patch clamp experiments at nanomolar concentrations. Pimozide was not acting via a dopamine receptor, since GH cells lack functional dopamine binding sites (Faure et al. 1980). Calmidazolium, a widely used calmodulin antagonist (VanBelle, 1981), and the sodium-channel activator veratridine likewise blocked VSCCs. These drugs were acting at the calcium channel because their effects on calcium currents could be shown directly in patch clamp experiments and were antagonized by Bay K8644. Veratridine effects were not taking place via the sodium channel since they were not prevented by tetrodotoxin, and other more potent sodium-channel activators such as batrachatoxin were not effective calcium-channel antagonists. The finding that a number of drugs act at similar concentrations on several important membrane regulatory molecules suggests structural similarities, something that is confirmed by the homologies in the amino acid sequences of the sodium and calcium channels (Tanabe et al. 1987). The results also indicate the need for caution when using pharmacologic agents as specific blockers of key regulatory proteins.

Calcium Channels and Cadmium Toxicity

Depolarization-induced calcium influx and secretion can be blocked by inorganic ions as well as organic channel blockers. Cadmium is a potent channel blocker, with an IC$_{50}$ for inhibiting calcium uptake to GH cells of less than 10 μM. Surprisingly, cadmium toxicity can be largely prevented with channel blockers such as nimodipine, verapamil, and diltiazem at concentrations that block calcium influx (Hinkle et al. 1987). Cadmium toxicity is enhanced by Bay K8644 and high potassium. Direct $^{109}Cd^{2+}$ uptake experiments have established that cadmium uptake is diminished by VSCC blockade, accounting for the protective effects of channel blockers. The $^{109}Cd^{2+}$ flux studies confirm what has been well established by electrophysiological means, namely, that the uptake of cadmium through voltage-sensitive channels is very low relative to that of calcium (< 1%). However, cadmium is toxic at extremely low intracellular concentrations, and entry via L-type channels is the major route of uptake by the GH cell. While further work is required to demonstrate whether cadmium uptake through voltage-sensitive channels is important in vivo, the results raise the possibility that calcium-channel blockers can be used to minimize cadmium uptake following acute exposure.

Effects on Prolactin Synthesis

The role of calcium in prolactin gene activation is not well understood and is quite controversial. Depolarization with high potassium and the divalent cation ionophore A23187 do not increase the rate of prolactin gene transcription measured in nuclear run-on assays 45 min after drug treatment (Murdoch et al. 1985). This suggests that changes in intracellular calcium do not have an immediate effect on the rate of prolactin mRNA formation. On the other hand, when cells are maintained in low calcium medium and calcium is readded, there is up to a 150-fold increase in prolactin mRNA after a 6- to 9-h delay (White et al. 1981). Overall transcriptional activity does not increase under these conditions. These apparently contradictory findings may simply reflect the fact that changes in cytoplasmic calcium regulate prolactin gene transcription, but only after a lag of several hours.

We have found that the rate of de novo prolactin synthesis is substantially altered by long-term incubation with VSCC antagonists or agonists (Enyeart et al. 1987b). For example, cells incubated for 8 days with 200 nM nimodipine had rates of prolactin synthesis only 16% of control, whereas cells incubated for 8 days with 300 nM Bay K8644 made prolactin at twice the control rate (Table 2). Since GH cells are firing spontaneous calcium action potentials, it is likely that drugs affect resting calcium levels over a period of hours to days by changing the amount of calcium entering during each action potential. The data are all consistent with the idea that there is some intracellular calcium-sensing mechanism controlling the rate of prolactin synthesis by GH cells. It is uncertain whether this mechanism involves a calcium-dependent protein kinase, calmodulin-sensitive enzyme, or a more direct process such as a calcium-sensitive transcription factor. To date, the region at the 5' end of the prolactin gene responsible for the calcium sensitivity has not been mapped.

One hypothesis to explain the effect of a number of drugs on hormone release and synthesis is that the cell can sense the intracellular hormone concentration. Depletion of intracellular hormone stores might "turn on" hormone synthesis. This cannot be

Table 2. Selective effect of calcium channel active drugs on prolactin synthesis

	Protein	Prolactin	Growth hormone
Control	635 ± 20	11.6 ± 0.3	44.4 ± 2.2
Bay K8644	670 ± 30	26.0 ± 1.6	30.0 ± 2.0
Nimodipine	670 ± 10	1.84 ± 0.2	34.0 ± 1.9

Prolactin secreted by GH cells over 8 days was measured in cells incubated with no drug, 300 nM Bay K8644, or 200 nM nimodipine. Values are µg. (From [9])

the entire explanation for the effects observed with VSCC agonists and antagonists. Channel blockers inhibit prolactin and growth release and cause the expected transient rise in intracellular stores, but this does not lead to diminished synthesis measured 72 h later. Persistent blockade of release does decrease prolactin synthesis but does not affect growth hormone (Enyeart et al. 1987). Conversely, treatment with Bay K8644, which increases prolactin release and therefore transiently depletes intracellular stores, does not lead to increased de novo synthesis unless it is present continually. While the cell may be able to respond to changes in prolactin content, these must be sustained, and the cell does not respond to changes in intracellular growth hormone with altered synthesis.

Role of Calcium Channels in Hormonal Regulation

Prolactin secretion by GH cells is stimulated by a variety of peptide hormones and transmitters acting by different mechanisms (Gourdji et al. 1981; Gershengorn 1986; Hinkle 1988). Hormones affecting the release of prolactin act via receptors on the cell surface and their action is almost immediate. The hormones stimulating prolactin release may be broadly divided into those that stimulate polyphosphoinositide breakdown and those that stimulate adenylate cyclase. For both classes, the signal emanating from the interaction of a peptide with a limited number of receptors is transduceed via a G protein to an enzyme (adenylate cyclase or phospholipase) which gives rise to increased concentrations of an intracellular second messenger. In the case of hormones such as vasoactive intestinal peptide (VIP), receptor activation stimulates cAMP formation via Gs, the stimulatory G protein. In the case of the hypothalamic tripeptide, thyrotropin-releasing hormone (TRH), receptor activation stimulates hydrolysis of phosphatidylinositol(4,5)biphosphate to give inositol trisphosphate, thought to release intracellular calcium from endoplasmic reticulum, and diacylglycerol, thought to activate protein kinase C. The TRH receptor is coupled to phospholipid breakdown via an as yet unidentified G protein, which is not pertussis toxin sensitive so it is unlikely to be either Gi or Go.

TRH causes an increase in intracellular free calcium measured with intracellularly trapped dyes (reviewed in Albert and Tashjian 1984; Gershengorn 1986; Hinkle 1988). Calcium increases quickly to a peak, but this maximum, or "first-phase" spike, is never as high as the calcium spike caused by depolarization. Calcium elevation persists for several minutes in the continued presence of TRH. The important question of whether the calcium rise in response to TRH occurs via increased calcium entry through voltage-gated calcium channels has been approached in two ways. One is to remove extracellular calcium. This generally leads to a rapid decrease in $[Ca^{2+}]_i$ by itself and reduces but never abolishes the TRH-induced increase in calcium. A second is to employ calcium-channel blockers. One such experiment, showing that the TRH-induced rise in calcium is only slightly reduced by a concentration of nimodipine adequate to block VSCC completely, is shown in Fig. 3. The results for hormone secretion are similar: VSCC blockers do not prevent the immediate burst of prolactin release stimulated by TRH. In the experiment shown in Table 3, TRH and high K^+ produced similar increases in prolactin secretion, but the depolarization-induced response was prevented by nimodipine while the TRH response was not.

Table. 3. Effect of calcium-channel blockers on prolactin secretion induced by 54 mM KCl and 100 nM TRH

	Control	TRH	KCl
No addition	7.4	22.5	30.0
10 nM Nimodipine	6.9	21.6	9.6
200 nM Nimodipine	7.3	16.0	0.3

Secretion was measured over 20 min. Errors were within ± 5%. (From [6])

Over a longer time frame, channel blockers both reduce calcium concentrations in the absence of any stimuli and reduce the TRH response. These results support the concept, which is now widely accepted, that the initial rise in calcium is due to inositol triphosphate induced intracellular calcium redistribution. TRH causes an increase in spontaneous action potential frequency (Ozawa and Kimura 1979; Taraskevich and Douglas 1980). It follows that VSCC blockers will, over several minutes or longer, reduce the calcium rise (often referred to as "second phase") elicited by TRH, which is in part due to increased influx of calcium during action potentials.

Summary and Conclusions

The rate of secretion of prolactin and growth hormone from regulated secretory granules is controlled by the intracellular calcium ion concentration. DHP-channel blockers completely prevent secretion stimulated by depolarization at nanomolar concentrations, whereas agonists enhance the response. Prolactin synthesis appears to be regulated by intracellular calcium ion concentrations, and voltage-sensitive channel-active drugs selectively alter transcription of the prolactin gene. The molecular mechanisms by which calcium controls secretory granule fusion and lysis and prolactin gene activity remain to be elucidated.

Acknowledgements. The authors are grateful to Ann Marie Zavacki and Devin Coppola for excellent technical assistance.

References

Aizawa T, Hinkle PM (1985a) TRH rapidly stimulates a biphasic secretion of PRL and GH in GH_4C_1 rat pituitary tumor cells. Endocrinology 116:73–82

Aizawa T, Hinkle PM (1985b) Differential effects of TRH, vasoactive intestinal peptide, phorbol ester and depolarization in GH_4C_1 rat pituitary cells. Endocrinology 116:909–919

Albert PR and Tashjian AH Jr (1984) Thyrotropin-releasing hormone induced spike and plateau in cytosolic free Ca^{2+} concentrations in pituitary cells. J Biol Chem 259:5827–5832

Cohen CJ, McCarthy RT (1987) Nimodipine block of calcium channels in rat anterior pituitary cells. J Physiol (Lond) 387:195–225

Day RN (1987) Inhibitory regulation of prolactin synthesis and release. PhD Thesis, University of Rochester, Rochester NY

Enyeart JJ, Aizawa T, Hinkle PM (1985) Dihydropyridine Ca^{2+} antagonists: potent inhibitors of secretion from normal and transformed pituitary cells. Am J Physiol 248 (Cell Physiol) 250:C510–519

Enyeart JJ, Aizawa T, Hinkle PM (1986) Interaction of the dihydropyridine Ca^{2+} agonist Bay K8644 with normal and transformed pituitary cells. Am J Physiol 250 (Cell Physiol 17) 250:C95–102

Enyeart JJ, Sheu S-S, Hinkle PM (1987a) Pituitary Ca^{2+} channels: blockade by conventional and novel Ca^{a+} antagonists. Am J Physiol 253 (Cell Physiol) 250:C162–170

Enyeart JJ, Sheu S-S, Hinkle PM (1987b) Dihydropyridine modulators of voltage sensitive Ca^{2+} channels specifically regulate prolactin production by GH_4C_1 pituitary tumor cells. J Biol Chem 262:3154–3159

Enyeart JJ, Hinkle PM (1984) The calcium agonist Bay K8644 stimulates secretion from a pituitary cell line. Biochem Biophys Res Commun 122:991–996

Faure N, Cronin MJ, Martial JA, Weiner RI (1980) Decreased responsiveness of GH_3 cells to the dopaminergic inhibition of prolactin. Endocrinology 107:1022–1026

Gershengorn MC (1986) Mechanism of the thyrotropin-releasing hormone stimulation of pituitary hormone secretion. Annu Rev Physiol 48:515–526

Gourdji D, Tougard C, Tixier-Vidal A (1981) Clonal prolactin strains as a tool in neuroendocrinology. In Ganong WF, Martini L (eds) Frontiers in neuroendocrinology vol 7. Raven Press, NY, pp 317

Hashimoto S, Fumagalli G, Zanini A, Meldolesi J (1987) Sorting of three secretory proteins to distinct secretory granules in acidophilic cells of cow anterior pituitary. J Cell Biol 105:1579–1586

Hinkle PM (1988) Regulation of prolactin synthesis and secretion by peptide hormones. In Negro-Vilar A, Conn PM (eds) Peptide hormones: effects and mechanisms of action. CRC Press Inc, Boca Raton, FL, in press

Hinkle PM, Kinsella PA, Osterhoudt KC (1987) Cadmium uptake and toxicity via voltage-sensitive calcium channels. J Biol Chem 262:16333–16337

Kelly RB (1985) Pathways of protein secretion in eukaryotes. Science 230:25–32

Matteson DR, Armstrong CM (1986) Properties of two types of calcium channels in clonal pituitary cells. J Gen Physiol 87:161–182

Miller RJ (1987) Multiple calcium channels and neuronal function. Science 235:46–52

Murdoch GH, Waterman M, Evans RM, Rosenfeld MG (1985) Molecular mechanisms of phorbol ester, thyrotropin-releasing hormone, and growth factor stimulation of prolactin gene transcription. J Biol Chem 260:11852–11858

Nowycky MC, Fox AP, Tsien RW (1985) Three types of neuronal calcium channels with different Ca^{2+} agonist sensitivity, Nature Lond. 316:440–443

Ozawa S, Kimura N (1979) Membrane potential changes caused by thyrotropin-releasing hormone in the clonal GH_3 cells and their relationship to secretion of pituitary hormones. Proc Natl Acad Sci USA 76:6017–6021

Statchura M (1982) Sequestration of an early-release pool of growth hormone and prolactin in GH_3 rat pituitary tumor. Endocrinology 111:1769–1777

Tanabe T, Takeshima H, Mikami A, Flockerzi V, Takahashi H, Kangawa K, Kojima M, Matsuo H, Hirose T, Numa S (1987) Primary structure of the receptor for calcium channel blockers from skeletal muscle. Nature 328:313–318

Tarakevich PS, Douglas WW (1980) Electrical behavior in a line of anterior pituitary cells (GH cells) and the influence of the hypothalamic peptide, thyrotropin releasing factor. Neuroscience 5:421–431

VanBelle HR (1981) R24571 A potent inhibitor of calmodulin-activated enzymes. Cell Calcium 2:483–494

White BA, Bauerle LR, Bancroft FC (1981) Calcium specifically stimulates prolactin synthesis and messenger RNA sequences in GH_3 cells. J Biol Chem 256:5942–5945

L-Type Calcium Channels and Adrenomedullary Secretion*

C. R. Artalejo, M. G. López, C. F. Castillo, M. A. Moro, and A. G. García

Departamento de Farmacología, Facultad de Medicina, Universidad Autónoma de Madrid, Arzobispo Morcillo, 4, 28029 Madrid, Spain

Introduction

On the basis that the calcium removal from solutions perfusing cat adrenal glands abolished acetylcholine-evoked catecholamine release, and that acetylcholine caused some increase in membrane permeability to common species of ions including Ca, particularly at the muscle end-plate (Del Castillo and Katz, 1956), Douglas and Rubin (1961) coined the expression "stimulus-secretion coupling" to define the role of external Ca entering chromaffin cells as the necessary event triggering the catecholamine ejection process. This was extrapolated from the earlier concept of "excitation-contraction coupling" coined by Shandow in 1952 to stress the role of calcium ions in muscle contraction. Although the stimulus-secretion coupling concept emerged from experiments performed in adrenomedullary chromaffin cells, it was soon extended to many other secretory systems, including various neuronal cell types (Douglas, 1968) where we now know that secretion occurs by exocytosis, a process in which the contents of small intracellular vesicles are released after fusion of the vesicle membrane with the plasmalemma.

Although this is a calcium-dependent process, little is known about the sites at which these ions act, as well as the biochemical steps leading to membrane fusion and secretion. However, for a number of reasons it seems clear that the site of action of Ca is intracellularly located. (a) Secretion is increased after microinjection of Ca into pre-synaptic nerve terminals of the squid giant synapse (Miledi 1973). (b) Calcium ionophores promote secretion from a variety of cells, including chromaffin cells (García et al. 1975; Carvalho et al., 1982). (c) Cytosolic calcium concentrations rise after acetylcholine stimulation of chromaffin cells (Knight and Kesteven, 1983). (d) Ca evokes secretion in electropermeabilized chromaffin cells (Baker and Knight, 1978), and (e) radioactive calcium uptake into chromaffin cells precedes in few seconds, and closely parallels, catecholamine release (Artalejo et al. 1986).

---

* This research was supported by grants from Fundación Ramón Areces, CAICYT (numbers 0626/81 and PB86–0119) and FISS, Spain.

Access of External Calcium Ions to the Cytosolic Secretory Machinery

The question then arises as to how external calcium ions gain the chromaffin cell interior during its physiological stimulation by acetylcholine released from splanchnic nerve terminals. Although there has been some dispute as to whether the acetylcholine-associated ionophore contributes in an important manner to calcium entry (see reviews by García et al. 1984a, 1986, 1987; and by Artalejo and García 1988), recent evidence from our laboratory suggests that during exocytosis, external Ca gains access to the intracellular secretory machinery primarily via potential-sensitive calcium channels; these channels would be activated by the depolarization signal that results when Na enters the cell through the acetylcholine-receptor ionophore. This assertion is based on experiments showing that a highly potent dihydropyridine (DHP) calcium-channel blocker, PN200-110, blocks nicotinic-evoked catecholamine release from perfused cat adrenals only in the presence of sodium ions; the blockade exhibits a high degree of stereoselectivity between the (+) and (−) enantiomers that is lost in the absence of sodium ions (Cárdenas et al. 1988).

Chromaffin Cell Calcium Channels

Various techniques, including intracellular recordings of membrane and action potentials, radiolabelled calcium fluxes and monitoring of intracellular free calcium concentrations, have provided data suggesting that adrenomedullary chromaffin cells contain in their plasma membrane specific voltage-dependent calcium channels.

Electrophysiological Evidence

By recording the mean value of the transmembrane potential, acetylcholine or high potassium was shown to depolarize cultured chromaffin cells of gerbil (Douglas and Kanno, 1967; Douglas et al. 1967) and rat adrenals (Biales et al. 1976; Brandt et al. 1976; Kidokoro et al. 1982). In rat chromaffin cells, a voltage-dependent calcium conductance increase was seen as a regenerative action potential; it was present after sodium removal and blocked by Co, suggesting that calcium channels were carrying such current (Brandt et al. 1976).

Using more direct patch-clamp techniques, D600 and cobalt-sensitive calcium currents have been described in bovine adrenomedullary chromaffin cells (Fenwick et al. 1982; Hoshi et al. 1984; Kim and Neher, 1988), suggesting the presence of specific voltage-dependent calcium channels in their plasmalemma.

Evidence from Studies of Calcium, Strontium and Barium Fluxes

A second, quite direct piece of evidence favouring the presence in the chromaffin cell of calcium channels comes from the fact that both nicotinic stimulation and high potassium markedly increase the rate of ^{45}Ca uptake into chromaffin cells, either maintained in situ in the intact adrenal gland (Douglas and Poisner, 1962; Wakade et al. 1986; Fonteriz et al. 1987; Borges et al. 1987; Artalejo et al. 1988) or isolated in suspensions or primary cultures (Holz et al. 1982; Kilpatrick et al. 1982; García et al. 1984c, Wada et al. 1983, 1985; Artalejo et al. 1986, 1987). Calcium uptake into potassium-depolarized chromaffin cells was strongly inhibited by the inorganic calcium channel blockers La, Co and Mg with IC_{50} values of 300 nM, 80 µM and 8 mM, respectively. In addition, the increase of the calcium gradient enhances calcium uptake, and cells take up ^{90}Sr and ^{133}Ba also, with relative permeabilities Ba > Sr > Ca (Artalejo et al. 1987). These constitute adequate criteria to identifiy a putative calcium channel (Hagiwara and Byerly, 1981; Tsien, 1983).

Properties of Chromaffin Cell Calcium Channels

The kinetics of bovine adrenal chromaffin cell calcium currents were studied in detail by Neher's group (Fenwick et al. 1982). Following a depolarizing step, chromaffin cell calcium channels tend to open in bursts, separated by long silent periods. Both currents and unit conductances are easier to study when Ca is replaced as the major cation by Ba; this is logical since in the presence of 2.5 mM external Ca, single channel calcium current is only a fraction of a pÅ. As in other cell types, the peak value of the current occurred when the membrane was completely depolarized; the apparent reversal potential for the calcium current was estimated to be close to 60 mV; since the theoretical equilibrium potential for calcium currents is greater, the low reversal potential obtained in chromaffin cells must be due to permeation of the calcium channel by an intracellular cation, perhaps K. As in other secretory cells, calcium currents flow for hundreds of milliseconds or even seconds and show little inactivation.

Because of the slow inactivation of calcium currents, we thought that it could be possible to analyse the kinetics of calcium channels by means other than electrophysiological techniques, using fast ionic flux studies in the range of few seconds. Therefore, we designed experiments to perform measurements of ^{45}Ca, ^{90}Sr and ^{133}Ba fluxes in bovine adrenal chromaffin cells to study the kinetic properties of these channels at much shorter times (1–10 s) than in previous studies (Artalejo et al. 1987). We demonstrated that calcium uptake into cells exposed to 59 mM K was linear only during the first 5 s of depolarization, and that the initial rate of calcium entry varied from 0.06 fmol/cell per second at 0.125 mM external Ca to 2.85 fmol/cell per second at 7.5 mM Ca. While calcium uptake inactivated quickly, strontium uptake inactivated later, and barium uptake was linear for the entire depolarization period (2–60 s) – just opposite to the cell permeability sequence for the three cations, which was higher for Ba, followed by Sr and lastly by Ca. This agrees with the fact that chromaffin cell calcium currents are better seen with Ba, as discussed above.

Cytosolic Calcium Levels, and not Voltage, Control the Inactivation of Chromaffin Cell Calcium Channels

Neither voltage changes nor calcium ions passing through the channels seemed to cause their inactivation. Experiments using the cell-permeable chelator Quin-2 to lower intracellular calcium concentrations, or the ionophore A23187 to increase them, support the view that as first suggested in *Paramecium* and *Aplysia* neurons (Brehm and Eckert, 1978; Tillotson, 1979) intracellular calcium accumulation during depolarization, rather than voltage, causes inactivation of chromaffin cell calcium channels (Artalejo et al. 1987). This contrasts with the views of Baker and Rink (1975) and Schiavone and Kirpekar (1982) who, using a rather indirect approach, concluded that the decline of the rates of catecholamine release during prolonged stimulation with high potassium was attributable to voltage-dependent inactivation of calcium channels.

Expected Cytosolic Calcium Levels Early After Potassium Depolarization of Chromaffin Cells

From the tangential slope of the curve of calcium uptake during the first 5 s of depolarization with 59 mM K of bovine adrenal chromaffin cells, we calculated an initial rate of calcium entry of 0.8 fmol/cell per second, which corresponds to 58 pmol/cm^2 surface per second (Artalejo et al. 1987). Assuming a density of channels of 10/μm^2 (Fenwick et al. 1982; García et al. 1984c; Baker and Knight, 1984), 33 400 ions/channel per second entered each cell, giving an expected cytosolic calcium concentration in 1 s of 340 μM. This is several orders of magnitude greater than the free calcium concentration (around 1 μM) that produces 50% activation of catecholamine release in electropermeabilized chromaffin cells (Baker and Knight, 1978). The calculated Ca$_i$ is also much higher than the submicromolar to few micromolar levels of Ca$_i$ measured with the fluorescent probe Quin-2 (Knight and Kesteven, 1983; Burgoyne and Cheek, 1985; Kao and Schnyder, 1986) or aequorin (Cobbold et al. 1987) upon exposure of chromaffin cells to the same depolarizing stimuli (acetylcholine, high potassium) that promoted a marked calcium uptake into these cells. This suggests either that the calcium probes are underestimating the real early Ca$_i$ transients taking place during stimulation, or that powerful Ca$_i$ sequestering systems reduce almost instantaneously Ca entering the cell to submicromolar concentrations. The latter conclusions seem to be correct in the light that saturation of calcium uptake was not visible at times less than 5 s depolarization (Artalejo et al. 1987).

In any case, it is likely that just in the area beneath the plasmalemma, where the calcium channels are, Ca$_i$ must reach much higher levels than those estimated with fluorescent or light-sensitive probes; since the exocytotic sites are probably located near zones where calcium channels are in higher densities (active zones equivalent to those seen in nerve terminals for exocytotic transmitter release?; Zucker and Landó, 1986) it is likely that these high levels of localized cytosolic Ca are physiologically relevant to trigger the secretory event.

Calcium Channel Modulators and Adrenal Catecholamine Secretion

Full characterization of chromaffin cell calcium channels requires the availability of specific blockers or agonists that might affect calcium currents, calcium fluxes and the hypothetical function that they mediate – in our case, catecholamine secretion. There are at least four chemical classes of calcium-channel blockers (1,4-DHP, benzothiazepine, dibenzylalkylamine and piperazine derivatives) and one class of calcium-channel activators of the DHP type. They all affect calcium-channel function and muscle contraction of cardiovascular tissues (Fleckenstein, 1985); however, neuronal and various endocrine secretory cells exhibit a different degree of sensitivity to these drugs. In spite of this, and provided that certain experimental conditions are met, the chromaffin cell displays an exquisite sensitivity to them, particularly to 1,4-DHP derivatives.

Calcium Channel Activators

The so-called calcium channel agonists or activators represent a particularly exciting recent pharmacological development. Small modifications to the nifedipine molecule produced a derivative, Bay K8644 (methyl-1,4-dihydro-2,6-dimethyl-3-nitro-4-(2-trifluoromethylphenyl)-pyridine-5-carboxylate) that in contrast to the calcium-channel blocking agents, stimulated cardiac and vascular smooth muscle contractility (Schramm et al. 1983). Other DHPs with interesting agonist properties have been recently synthesized by Ciba-Geigy (CGP 28392) and by Sandoz (202–791). During the past 2 years, several features on the actions of novel calcium-channel agonists have emerged from experiments performed in our laboratory on perfused cat adrenal glands and isolated bovine adrenal chromaffin cells (García et al. 1984a, b; 1987; Montiel et al. 1984; Borges et al. 1986; Sala et al. 1986; Artalejo and García, 1986; Ladona et al. 1987; Fonteriz et al. 1987).

Dihydropyridine Agonists do not Affect the Basal Rates of Spontaneous Catecholamine Release

In contrast to the ionophores A23187 (García et al. 1975) or ionomycin (Carvalho et al. 1982; Ceña et al. 1983), neither Bay K8644, CGP 28392 or (+) Sandoz 202–791 modified the basal outputs of catecholamines in perfused adrenal glands or cultured chromaffin cells. DHP agonists thus do not seem to act on resting calcium channels.

Dihydropyridine Calcium-Channel Activators Potentiate the Secretory Response to Weak but not Strong Cell Depolarizations

This is a very reproducible experimental finding in both the perfused cat adrenal gland and cultured bovine chromaffin cells. In the latter system, Bay K8644 potentiated markedly [^{3}H]noradrenaline release and ^{45}Ca uptake at low (17.7 mM) but not at high (59 mM) potassium depolarizations. When using pulses of 17.7 or 35 mM K as depolarizing stimuli, the three calcium-channel activators enhanced catecholamine release in a concentration-dependent manner; however, Bay K8644 was about 100-fold more potent than CGP 28392 and (+) Sandoz 202–791.

It seems plausible to conclude from these experiments that the number of calcium channels that are opened at a given stimulus strength is not changed by Bay K8644; it is likely that, as in the heart (Hess et al. 1984), the drug is enhancing the mean open time of individual channels and decreasing their closing time, thereby promoting more calcium entry, especially at low depolarization strengths.

Selective Effects of Bay K8644 on Several Secretagogues

Several ionic manipulations (deprivation of external divalent cations followed by calcium reintroduction) or drugs (nicotine, ouabain, methacholine) are known to favour catecholamine release from cat adrenal glands through different mechanisms. Bay K8644 potentiated markedly the secretory effects of those secretagogues that are likely to cause cell depolarization (nicotine, ouabain, divalent cation deprivation, high potassium) but not those acting through muscarinic- or nicotinic-receptor associated ionophores or activation of Na–Ca exchange system (by potassium deprivation). It therefore seems that the drug acts only upon depolarizing stimuli that selectively alter the rate of activation of voltage-sensitive calcium channels.

The Effects of Bay K8644 on Secretion Depend on the Divalent Permeant Cation Used

These experiment were performed with other divalent cations (Sr and Ba) that can substitute for Ca in releasing catecholamines from cat adrenal glands. The relative conductances and permeability ratios of calcium channels for Ca, Sr and Ba are known to be quite different in several neuronal cell types (Hagiwara and Byerly, 1981; Tsien, 1983; Artalejo et al., 1987); in fact, barium currents are larger than strontium and calcium currents in the bovine chromaffin cell (Fenwick et al. 1982). In the perfused cat adrenal gland, Bay K8644 markedly potentiated potassium-evoked secretion when Ca was the permeant cation; in the presence of Sr, the potentiation observed was modest or absent (Sala et al. 1986) and with Ba it was absent. These results are compatible with Bay K8644 acting selectively on calcium channels; obviously, the drug should potentiate least the secretory response in the presence of the cation (Ba) that moves best through the open calcium channel and do not cause its inactivation.

Bay K8644 does not Act Intracellularly

This assumption is based on two sets of experiments. In one, we observed that Bay K8644 did not modify the rates of catecholamine release evoked by A23187, a secretory response that by-passes calcium channels (Artalejo and García, 1986). If the drug were acting intracellularly, it should also potentiate the ionophore-evoked response. A second set of experiments demonstrated that Bay K8644 does not modify the rates of secretion evoked by Ca from cultured bovine adrenal chromaffin cells permeabilized with digitonin (Ladona et al. 1987).

A Common Site of Action for Dihydropyridine Agonists and Antagonists?

At high concentrations (around 10 μM), Bay K8644 becomes itself an "antagonist", since its potentiating effects on secretion substantially decrease. This is also observed with the pure enantiomer (+)Sandoz 202–791. This raises the question as to whether a common site of action for DHP-agonist and -antagonist drugs exists. Although Bay K8644 displaces the binding of [^{3}H]nitrendipine to bovine adrenomedullary membranes (García et al. 1984c), its potentiating effects on secretion are antagonized by nifedipine (Montiel et al. 1984) or nitrendipine (Borges et al. 1986) in a dubious competitive manner.

The Effects of Dihydropyridine Agonists are Dependent on Calcium Concentrations

The potentiating effects of Bay K8644 are best seen at low depolarizing stimuli and at low extracellular calcium concentrations. However, provided that (Ca_o) is maintained low (0.25 mM or less), the drug will potentiate catecholamine release even with 59 or 118 mM K. In other words, Bay K8644 seems to increase greatly the ability of external Ca to gain access to the intracellular secretory machinery.

Chirality of Calcium-Channel Activators

The (+)enantiomer of the compound Sandoz 202–791 potentiates secretion while the (−)enantiomer behaves as a potent inhibitor (Fonteriz et al. 1987). Something similar occurs with Bay K8644, the (−)enantiomer being a potent activator and the (+)enantiomer a weak activator but pure blocker of secretion (unpublished results). It is curious that pure enantiomers also lose their potentiating effects at concentrations around 10 μM. This might be due to cross-contamination of the (+)enantiomer of Sandoz 202–791 with the (−)enantiomer or to desensitization of the DHP receptor when exposed for prolonged periods to high concentrations of the agonist. In any case, these disparate actions of two enantiomers are extremely unusual, and, as demonstrated here, tenuous changes (i. e. only in the configuration of two substances) might produce different and even opposite actions.

Inorganic Calcium-Channel Blockers

Inorganic calcium channel blockers Cd (Holz et al. 1982), Co (Kidokoro and Ritchie, 1980), La (Borowitz, 1972), Mg (Holz et al. 1982; Wada et al. 1985) and Mn (Wakade, 1981) have been sporadically shown to inhibit depolarization-evoked adrenomedullary catecholamine release. More recently, we have studied in more detail the effects of various inorganic calcium-channel blockers on potassium-evoked secretion from perfused cat adrenal glands and shown the following order of potencies: Zn > Cd > La > Ni > Co > Mn > Mg. While Zn had an IC_{50} of 51 μM, Mg had an IC_{50} of 980 μM. Since La, Co and Mg potently block potassium-evoked calcium uptake into chromaffin cells (Artalejo et al., 1987), it seems that the secretory actions of these cations are related to their well-known blocking effects on calcium channels.

Organic Calcium-Channel Blockers

A growing number of so-called calcium antagonists or calcium-channel blockers are known to interfere with calcium currents and various physiological functions in different excitable cells (Fleckenstein, 1985). They constitute a heterogeneous group with disparate chemical structures (1,4-DHP, benzylalkylamine, benzothiazepine and piperazine derivatives), pharmacological effects and clinical profiles.

One of the main objections to the use of these drugs to characterize calcium channels is their limited selectivity and their efficacy to interfere with functions in certain neurones, skeletal muscle and secretory cells. Therefore, verapamil, the first calcium-entry blocker reported potently to depress cardiac tissues (Fleckenstein, 1964), inhibited catecholamine release in response to acetylcholine or high potassium only at concentrations several orders of magnitude greater than those required to affect heart or smooth muscle tissues (Pinto and Trifaró, 1976). In addition, verapamil also blocks sodium channels in synaptosomes (Norris et al. 1983), has muscarinic cholinoceptor (El-Fakahany and Richelson, 1983) and α_2 adrenoceptor affinity (Galzin and Langer, 1983) and also inhibits the calcium conductance of acetylcholine endplate channels (Bregestovski et al. 1979).

The more recent organic calcium antagonists of the DHP type seem to be more potent and specific in PC12 (Takahashi and Ogura, 1983) and bovine adrenal chromaffin cells (Ceña et al. 1983) since they inhibited secretion in the nM concentration range. In contrast, 0.5 mM verapamil was required to block by 75% high potassium-evoked catecholamine release from rat adrenal glands (Wakade, 1981) and 10 µM verapamil or D600 to inhibit acetylcholine-evoked secretion in bovine adrenal chromaffin cells (Lemaire et al. 1981).

In view of the lack of data that could allow the definition of how calcium-channel blockers of the various known chemical groups affect potassium-evoked adrenomedullary secretion, we recently performed a comprehensive study using 16 drugs belonging to four major classes of organic calcium-channel blockers (Gandia et al., 1987). Their relative order of potencies to block potassium-evoked secretion from perfused cat adrenal glands were (+)PN200–110 > nisoldipine > nitrendipine > (±)PN200–110 > nicardipine > (-)Sandoz 202–791 > nimodipine > nifedipine > (-)PN200–110 > niludipine > verapamil > diltiazem > flunarizine > amlodipine. By far the most potent were 1,4-DHP derivatives [IC_{50} for (+)PN200–110 was 0.84 nM], followed by verapamil, diltiazem and lastly by flunarizine (IC_{50} = 2980 nM). Substantial differences were also observed between drugs of the same group; for instance, niludipine was 100-fold less potent than (+)PN200–110, and nitrendipine or nicardipine was 10- to 20-fold more potent than nifedipine, the prototype drug in cardiovascular tissues (IC_{50} = 38 nM).

Evidence for Separate Binding Sites and Mechanisms of Action on Chromaffin Cell Calcium Channels

Radioligand binding studies have generated a model for three separate, but allosterically linked, binding sites for DHPs, phenylalkylamines and benzothiazepines (Triggle and Janis, 1987); this model has only limited and controversial functional support in non-secretory tissues (DePover et al. 1983; Spedding, 1983; Yousif and Triggle, 1985). We took advantage of the well-known voltage-dependent effects of some of these drugs on calcium currents (Lee and Tsien, 1983; Sanguinetti and Kass, 1984; Bean, 1984; Hueara and Hume, 1985; Gurney et al. 1985; Sanguinetti et al. 1986; Rane et al. 1987) to design functional experiments in which the blocking effects of these drugs on cat adrenomedullary secretion could be studied as a function of the membrane potential.

For this purpose, experiments similar to those used by electrophysiologists when measuring calcium-current inactivation by conditioned pre-pulses (Fenwick et al. 1982) were designed. They had various original features of previous experiments from our laboratory performed to study the inhibitory effects of calcium-channel blockers on potassium-evoked secretion (Ceña et al. 1983; García et al. 1984c; Ladona et al. 1986; Gandia et al. 1987). First, tissues were impregnated with the drugs under hyperpolarizing (1.2K0Ca) or depolarizing (118K0Ca) conditions with the intention of manipulating the affinities of the drugs for their receptors as a function of the membrane potential. Secondly, secretion was triggered by a brief calcium test pulse applied to adrenals previously perfused with calcium-free solutions containing low (1.2 mM) or high (118 mM) potassium; in these conditions, drugs attach to their receptors with variable affinities depending on the membrane potential. And, thirdly, the duration of the calcium pulses was very short (10 s) in order to decrease maximally the opportunity of drugs to change their equilibrium receptor binding during the pre-test 10-min perfusion period. For instance, the pre-hyperpolarized glands were exposed for 10 s to Ca plus 118 mM K to trigger secretion; this pulse should be as short as possible, ideally 1 s, because, as Sanguinetti and Kass (1984) have demonstrated, the onset of DHPs blockade of cardiac calcium currents is completed in about 10 s. It is clear then that in our earlier experiments (Ceña et al. 1983; Gandia et al. 1987) we were really measuring drug effects mainly in depolarized cells, since the test secretory pulse lasted over 60 s, and the pre-perfusion period with drugs was performed only in resting (5.9 mM K) conditions. It is therefore important to emphasize that experiments to study the effects of calcium-channel modulators on secretion of neurotransmitters and hormones should consider the strong voltage dependence of these drugs.

Table 1 shows that the various categories of calcium-channel blockers exhibit different degrees of voltage dependence in inhibiting calcium-evoked catecholamine

Table 1. Voltage-dependent effects of five calcium-channel blockers on adrenomedullary secretion

Drug	N	IC_{50} 1.2K	IC_{50} 118K	$\dfrac{IC_{50}\ 1.2K}{IC_{50}\ 118K}$
Nitrendipine	10	214	0.99	215
Cinnarizine	8	189	5.23	36
Diltiazem	11	2040	106	19
Verapamil	7	1770	217	8
Cadmium	8	1150	1510	0.76

Catecholamine release from perfused cat adrenal glands was evoked by a test pulse of 118 mM K, 0.5 mM Ca^{2+} for 10 s, after pre-treatment with hyperpolarizing (1.2K0Ca) or depolarizing (118K0Ca) solutions for 10 min. IC_{50}, drug concentration inhibiting by 50% the initial control release (values expressed in nM); N, number of paired experiments

release. These experiments show that inhibition by various calcium-channel blockers of adrenomedullary catecholamine release evoked by brief pulses of Ca is strongly dependent on whether the drugs impregnated the glands in hyperpolarizing or depolarizing conditions. The secretory responses in hyperpolarized glands were very resistant to blockade by the four drugs tested, while those obtained in depolarized glands were highly sensitive to the drugs. While nitrendipine exhibited a strong dependence on the membrane potential to attach to its receptor and block secretion, the inorganic calcium-channel blocker Cd had no voltage dependence at all. The organic calcium-channel blockers cinnarizine, diltiazem and verapamil showed intermediate degrees of voltage dependence.

The question then arises as to whether the observed secretory patterns might be correlated with events taking place in membrane calcium channels. The fact that Cd, a potent inorganic blocker of L-type calcium channels (Nowicky et al. 1985) indistinctly inhibited both secretory responses strongly indicates that this is so. However, the more conspicuous evidence favouring the correlation between both events comes from the fact that nitrendipine binds with much higher affinity to cat adrenomedullary tissues, and blocks much more efficaciously ^{45}Ca uptake into and catecholamine release from pre-depolarized as compared to pre-hyperpolarized tissues (Artalejo et al. 1988). It is therefore plausible to explain the present results on the basis of voltage dependence of nitrendipine, cinnarizine, diltiazem and verapamil to bind to their receptors on L-type calcium channels.

Voltage-dependent calcium channels are dynamic molecular entities that undergo constant transitions between resting, open and inactivated states (Hess et al. 1984). Since there is a close parallelism between calcium uptake and secretion (Artalejo et al. 1986), and sustained depolarization in the absence of Ca (for instance, 10 min in 118K0Ca) causes little inactivation of chromaffin cell calcium channels (Sala et al. 1986; Artalejo et al. 1987), it seems that nitrendipine, cinnarizine, diltiazem and verapamil preferentially bind to their receptors when calcium channels are in their open state. This conclusion is consistent with the interpretation of the mechanism of action of DHPs based on electrical recordings of calcium currents, showing that the drugs display an increased affinity for the L-type channel as the holding potential of the cell is increased (Hueara and Hume, 1985; Cohen and McCarthy, 1987); however, it differs from other reports suggesting that nitrendipine tightly binds to the inactivated state of the channel (Bean, 1984; Sanguinetti and Kass, 1984).

The differences in potencies to block secretion in hyperpolarized or depolarized glands are quite notorious. In other words, voltage-dependent blocking effects are very marked for DHPs, followed by cinnarizine and diltiazem, scarce for verapamil and non-existent for the inorganic calcium-channel blocker Cd. This suggests different sites and mechanisms of action of the five agents; it is possible that these sites are allosterically coupled, as suggested by radioligand binding studies in isolated membranes (Triggle and Janis, 1987). In any case, these different degrees of voltage dependence might help to explain why verapamil and diltiazem are much more cardioactive compounds than DHPs, or why cinnarizine might have some selectivity for certain vascular beds, since these tissues contain cells with different resting membrane potential. What is clear from these data is that, at clinically relevant concentrations, DHPs might interfere with adrenomedullary secretion, helping in this manner the vasodilating effects of the drugs in treating hypertensive patients, through mutual and complementary mechanisms acting on vascular smooth muscle and chromaffin cells.

Coupling of L-Type Calcium Channels to the Adrenomedullary Secretion Process

Various experimental designs have attempted to demonstrate the coupling between calcium channels and secretory activities of chromaffin cells. For instance, in intact perfused rat adrenal glands (Ishikawa and Kanno, 1978), the chromaffin cell membrane potential fell progressively as the potassium concentration rose; there was an almost linear relationship between the logarithm of the potassium concentration, the membrane potential and the rate of adrenaline release, supporting the view that depolarization induced by excess K causes the influx of Ca through calcium channels, which then initiates the exocytotic event.

On the other hand, chemical clamping of the membrane potential by raising the external potassium concentration causes a quick enhancement of calcium uptake into, and catecholamine release from, cultured bovine adrenal chromaffin cells; this is followed by their rapid decline to almost resting levels in spite of continued depolarization (Artalejo et al. 1986). Both the increase and the inactivation of calcium uptake precede by 5–10 s and closely parallel the changes in secretion. Since the two processes are delayed in a parallel manner in the presence of Sr or Ba, it seems clear that the activity of calcium channels tightly modulates the kinetics of secretion.

In a third, more direct approach, we have taken advantage of the voltage-dependent effects of nitrendipine to design experiments trying to correlate its tissue binding with its effects on calcium uptake into and secretion from intact adrenomedullary tissue of perfused cat adrenal glands. Until now, equating DHP receptors to voltage-gated calcium channels has been hampered by poor correlations between functional and binding parameters. For instance, PN200–110 can bind to intact sartorii muscle without causing block of calcium currents (Schwartz et al. 1985), and the block of cardiac calcium currents (Lee and Tsien, 1983) requires 100- to 1000-fold higher concentrations of DHPs than expected from K_D values obtained from binding studies (Bolger et al. 1982). On the other hand, it is controversial whether transverse tubular charge movements occurring during skeletal muscle contraction are coupled (Rios and Brum 1987) or not (Walsh et al. 1987) to DHP-sensitive calcium channels; also, in neural tissues, calcium uptake and transmitter release are either insensitive (Danielle et al. 1983; Ogura and Kahashi, 1984) or weakly sensitive (Turner and Goldin, 1985) to DHPs. These discrepancies have remained unresolved because in intact cells or tissues, where the pharmacological effects of DHPs must be tested, it is difficult to measure radiolabelled DHP binding.

In recent experiments we have shown that when [^{3}H]nitrendipine binding is enhanced in adrenomedullary cells, its blocking effects on ^{45}Ca uptake into and catecholamine release from those cells are concomitantly increased; the reverse occurs in hyperpolarized glands, where a poor binding of the DHP is accompanied by a weak effect on both functional parameters (Artalejo et al. 1988) (Table 2).

The above results obtained in the intact adrenal gland suggest that membrane potential must be considered when we attempt to compare DHP binding data with

Table 2. Correlation between [^{3}H]nitrendipine binding, ^{45}Ca uptake into and catecholamine release from perfused cat adrenomedullary tissues

	Hyperpolarized gland (1.2K0Ca)	Depolarized gland (118K0Ca)
[^{3}H]Nitrendipine binding (fmol/mg protein of intact adrenomedullary tissue)	2 ± 0.4	50 ± 6
^{45}Ca uptake (pmol/mg protein of intact adrenomedullary tissue)	195 ± 4	10 ± 1
Catecholamine release (ng/Ca pulse)	808 ± 102	139 ± 20

Glands were perfused at 37°C with hyperpolarizing (1.2 mM K in the absence of Ca) or depolarizing solutions (118 nM K in the absence of Ca) in the presence of 10 nM [^{3}H]nitrendipine. Ca uptake and catecholamine release were studied by applying 10-s pulses of 0.5 mM Ca in 118 mM K containing 16 uCi/ml ^{45}Ca. Data are means ± SE of six paired experiments

biological effects. Receptor binding studies performed in membranes identify channels that are presumably in inactivated or open states, since the potential difference across the membrane is lost. The consistent, close correlation between nitrendipine binding and its blocking effects on calcium uptake and catecholamine secretion, can be taken as strong direct evidence supporting the belief that a DHP receptor is indeed coupled to a (probably) L-type calcium channel to control its kinetics and, through it, the availability of external calcium ions to the internal secretory machinery. This conclusion contrasts with the recent suggestion that in rat sympathetic neurones, which share with chromaffin cells a common embriological origin, N-type DHP-resistant calcium channels – and not L-type calcium channels – seem to control depolarization-evoked release of noradrenaline (Hirning et al. 1988).

Acknowledgements. The generous, continuing supply of various DHP calcium-channel blockers and activators by Prof. F. Hoffmeister, Dr. E. Truscheit and Dr. F. Seuter (Bayer), and by Dr. R.P. Hof (Sandoz) have made possible the research presented here; we deeply acknowledge their kindness. We thank Mrs M. C. Molinos for typing the manuscript.

References

Artalejo CR, Bader MF, Aunis D, García AG (1986) Inactivation of the early calcium uptake and noradrenaline release evoked by potassium in cultured chromaffin cells. Biochem Biophys Res Commun 134:1–7

Artalejo CR, García AG (1986) Effects of Bay-K-8644 on cat adrenal catecholamine secretory responses to A23187 or ouabain. Br J Pharmacol 88:757–765

Artalejo CR, García AG (1988) Regulation of a dihydropyridine-sensitive calcium channel and its relation to the chromaffin cell secretory response. In: Dahlström A (ed) Proceedings of the 6th International Catecholamine Symposium, Alan R. Liss, Inc., New York, In press

Artalejo CR, García AG, Aunis D (1987) Chromaffin cell calcium channel kinetics measured isotopically through fast calcium, strontium, and barium fluxes. J Biol Chem 262:915–926

Artalejo CR, López MG, Moro MA, Castillo CF, Pascual R, García AG (1988) Voltage-dependence of nitrendipine provides direct evidence for dihydropyridine receptor coupling to calcium channels in intact cat adrenals. Biochem Biophys Res Commun, In press

Baker PF, Knight DE (1978) Calcium-dependent exocytosis in bovine adrenal medullary cells with leaky plasma membranes. Nature 276:620–622

Baker PF, Knight DE (1984) Calcium control of exocytosis in bovine adrenal medullary cells. TINS 7:120–126

Baker PF, Rink TJ (1975) Catecholamine release from bovine adrenal medulla in response to maintained depolarization. J Physiol 253:593–620

Bean B (1984) Nitrendipine block of cardiac calcium channels: high affinity binding to the inactivated state. Proc Natl Acad Sci USA 81: 6388–6392

Biales B, Dichter M, Tischler A (1976) Electrical excitability of adrenal chromaffin cells. J Physiol 262:743–753

Bolger GT, Gengo PJ, Luchowski EM, Siegel H, Triggle DJ, Janis RA (1982) High affinity binding of a calcium channel antagonist to smooth and cardiac muscle. Biochem Biophys Res Commun 104:1604–1609

Borges R, Ballesta JJ, García AG (1987) M_2 muscarinoceptor-associated ionophore at the cat adrenal medulla. Biochem Biophys Res Commun 144:965–972

Borges R, Sala F, García AG (1986) Continuous monitoring of catecholamine release from perfused cat adrenals. J Neurosci Methods 16:289–300

Borowitz JL (1972) Effect of lanthanum on catecholamine release from adrenal medulla. Life Sci Part I 11:959–964

Brandt BL, Hagiwara S, Kidokoro Y, Miyazaki S (1976) Action potentials in the rat chromaffin cells and effects of acetylcholine. J Physiol 263:417–439

Bregestovski PD, Miledi R, Parker I (1979) Calcium conductance of acetylcholine-induced endplate channels. Nature 279:638–639

Brehm P, Eckert R (1978) Calcium entry leads to inactivation of calcium channel in *Paramecium*. Science 202:1203–1206

Burgoyne RD, Cheek TR (1985) Is the transient nature of the secretory response of chromaffin cells due to inactivation of calcium channels? FEBS Lett 182:115–118

Cárdenas AM, Montiel C, Artalejo AR, Sánchez-García P, García AG (1988) Sodium-dependent inhibition by PN200-110 enantiomers of nicotinic adrenal catecholamine release. Br J Pharmacol, In press

Carvalho MH, Prat JC, García AG, Kirpekar SM (1982) Ionomycin stimulates secretion of catecholamines from cat adrenal gland and spleen. Am J Physiol 242:E137–E145

Ceña V, Nicolás GP, Sánchez-García P, Kirpekar SM, García AG (1983) Pharmacological dissection of receptor-associated and voltage-sensitive ionic channels involved in catecholamine release. Neuroscience 10:1455–1462

Cobbold PH, Cheek TR, Cuthbertson KSR, Burgoyne RD (1987) Calcium transients in single adrenal chromaffin cells detected with aequorin. FEBS Lett 211:44–48

Cohen CJ, McCarthy RT (1987) Nimodipine block of calcium channels in rat anterior pituitary cells. J Physiol 387:195–225

Daniell LC, Barr EM, Leslie SW (1983) $^{45}Ca^{2+}$ uptake into rat whole brain synaptosomes unaltered by dihydropyridine calcium antagonists. J Neurochem 41:1455–1459

Del Castillo J, Katz B (1956) Biophysical aspects of neuromuscular transmission. Progr Biophys 6:121–170

DePover A, Grupp IL, Grupp G, Schwartz A (1983) Diltiazem potentiates the negative inotropic action of nimodipine in heart. Biochem Biophys Res Commun 114:922–929

Douglas WW (1968) Stimulus-secretion coupling; the concept and clues from chromaffin and other cells. Br J Pharmacol 34:451–474

Douglas WW, Kanno T (1967) The effect of amethocaine on acetylcholine induced depolarization and catecholamine secretion in the adrenal chromaffin cell. Br J Pharmacol 30:612–619

Douglas WW, Kanno T, Sampson SR (1967) Effects of acetylcholine and other medullary secretagogues and antagonists on the membrane potential of adrenal chromaffin cells: an analysis employing techniques of tissue culture. J Physiol 118:107–120

Douglas WW, Poisner AM (1962) On the mode of action of acetylcholine in evoking adrenal medullary secretion: increased uptake of calcium during the secretory response. J Physiol 162:385-392

Douglas WW, Rubin RP (1961) The role of calcium in the secretory response of the adrenal medulla to acetylcholine. J Physiol 159:40-57

Fenwick EM, Marty A, Neher E (1982) Sodium and calcium channels in bovine chromaffin cells. J Physiol 331:599-635

El-Fakahany E, Richelson R (1983) Effect of some calcium antagonists on muscarinic receptor-mediated cyclic GMP formation. J Neurochem 40:705-710

Fleckenstein A (1964) Die Bedeutung der energiereichen Phosphate für Kontraktilität und Tonus des Myokards. Verh Dtsch Ges Inn Med 70:81-99

Fleckenstein A (1985) Calcium antagonists and calcium agonists: fundamental criteria and classification. In: Fleckenstein A, Van Breemen C, Gross R, Hoffmeister F (eds) Bayer-Symposium IX Cardiovascular effects of dihydropyridine-type calcium antagonists and agonists. Springer-Verlag, Berlin pp 3-31

Fonteriz RI, Gandia L, López MG, Artalejo CR, García AG (1987) Dihydropyridine chirality at the chromaffin cell calcium channel. Brain Res 408:359-362

Galzin AM, Langer SZ (1983) Presynaptic alpha$_2$ adrenoceptor antagonism by verapamil but not by diltiazem in rabbit hypothalamic slices. Br J Pharmacol 78:571-577

Gandia L, López MG, Fonteríz RI, Artalejo CR, García AG (1987) Relative sensitivities of chromaffin cell calcium channels to organic and inorganic calcium antagonists. Neurosci Lett 77:333-338

García AG, Artalejo CR, Borges R, Reig JA, Sala F (1986) Pharmacological properties of the chromaffin cell calcium channel. In: Atwater I, Rojas E, Soria B (eds) Biophysics of the pancreatic B-cell. Plenum, New York, pp 139-157

García AG, Ceña V, Frias F (1984a) Pharmacological dissection of ionic channels involved in catecholamine release from the chromaffin cell. Actualités Chimie Therapeutique Lavoisier, Paris, 11:165-185

García AG, Kirpekar SM, Prat JC (1975) A calcium ionophore stimulating the secretion of catecholamines from the cat adrenal. J Physiol 244:253-262

García AG, Sala F, Ladona MG, Ceña V, Montiel C (1984b) Analysis of the catecholamine secretory process by using a novel dihydropyridine calcium "agonist" and potassium or calcium gradients. In: Vizi ES, Magyar K (eds) Regulation of transmitter function. Elsevier, Amsterdam, pp 51-63

García AG, Sala F, Reig JA, Viniegra S, Frias J, Fonteriz RI, Gandia L (1984c) Dihydropyridine Bay-K-8644 activates chromaffin cell calcium channels. Nature 308:69-71

García AG, Sala F, Ceña V, Montiel C, Ladona MG (1987) Modulation by calcium of the kinetics of the chromaffin cell secretory response. In: Rosenheck K, Lelkes PI (eds) Stimulus-secretion coupling in chromaffin cells. CRC Press, Boca Raton, Florida, pp 97-115

Gurney AM, Nerbonne JM, Lester HA (1985) Photoinduced removal of nifedipine reveals mechanisms of calcium antagonist action on single heart cells. J Gen Physiol 85:353-379

Hagiwara S, Byerly L (1981) Calcium channel. Ann Rev Neurosci 4:69-125

Hess P, Lansman JB, Tsien RW (1984) Different modes of Ca channels gating behaviours favored by dihydropyridine Ca agonists and antagonists. Nature 311:538-544

Hirning LD, Fox AP, McCleskey EW, Olivera BM, Thayer SA, Miller RJ, Tsien RW (1988) Dominant role of N-type Ca^{2+} channels in evoked release of norepinephrine from sympathetic neurons. Science 239:57-61

Holz RW, Senter RA, Frye RA (1982) Relationship between Ca^{2+} uptake and catecholamine secretion in primary dissociated cultures of adrenal medulla. J Neurochem 39:635-646

Hoshi T, Rothlein J, Smith SJ (1984) Facilitation of Ca^{2+}-channel currents in bovine adrenal chromaffin cells. Proc Natl Acad Sci USA 81:5871-5875

Hueara A, Hume JR (1985) Interactions of organic calcium channels antagonists with calcium channels in single frog atrial cells. J Gen Physiol 85:621-647

Ishikawa K, Kanno T (1978) Influences of extracellular calcium and potassium concentrations on adrenaline release and membrane potential in the perfused adrenal medulla of the rat. Jap J Physiol 28:275-289

Kao LS, Schneider AS (1986) Calcium mobilization and catecholamine secretion in adrenal chromaffin cells. A Quin-2 fluorescence study. J Biol Chem 261:4881-4888

Kidokoro Y, Miyazaki S, Ozawa S (1982) Acetylcholine-induced membrane depolarization and potential fluctuations in the rat adrenal chromaffin cell. J Physiol 324:203–220

Kidokoro Y, Ritchie AK (1980) Chromaffin cell action potential and their possible role in adrenaline secretion from rat adrenal medulla. J Physiol 307:199–216

Kilpatrick DL, Slepetis RJ, Corcoran JJ, Kirshner N (1982) Calcium uptake and catecholamine secretion by cultured bovine adrenal medulla cells. J Neurochem 38:427–435

Kim YI, Neher E (1988) IgG from patients with Lambert-Eaton syndrome blocks voltage-dependent calcium channels. Science 239:405–408

Knight DE, Kesteven NT (1983) Evoked transient intracellular free Ca^{2+} changes and secretion in isolated bovine adrenal medullary cells. Proc R Soc Lond B 218:177–179

Ladona MG, Aunis D, Gandia L, García AG (1987) Dihydropyridine modulation of the chromaffin cell secretory response. J Neurochem 48:483–490

Lee KS, Tsien RW (1983) Mechanism of calcium channels blockade by verapamil, D600, diltiazem and nitrendipine in single dialysed heart cells. Nature 302:790–794

Lemaire S, Derome T, Tseng R, Mercier P, Lemaire I (1981) Distinct regulations by calcium of cyclic GMP levels and catecholamine secretion in isolated bovine adrenal chromaffin cells. Metabolism 30:462–468

Miledi R (1973) Transmitter release induced by injection of calcium ions into nerve terminals. Proc R Soc Lond B 183:421–425

Montiel C, Artalejo AR, García AG (1984) Effects of the novel dihydropyridine Bay-K-8644 on adrenomedullary catecholamine release evoked by calcium reintroduction. Biochem Biophys Res Commun 120:851–857

Norris PJ, Dhaliwal DK, Druce DP, Bradford HF (1983) The suppression of stimulus-evoked release of amino acid neurotransmitters from synaptosomes by verapamil. J Neurochem 40:514–521

Nowycky MC, Fox A, Tsien RW (1985) Three types of neuronal calcium channels with different calcium agonist sensitivity. Nature 316:440–443

Ogura A, Kahashi M (1984) Differential effect of a dihydropyridine derivative to Ca^{2+} entry pathways in neuronal preparations. Brain Res 301:323–330

Pinto JEB, Trifaró JM (1976) The different effects of D-600 (methoxy-verapamil) on the release of adrenal catecholamines induced by acetylcholine, high potassium or sodium deprivation. Br J Pharmacol 57:127–132

Rane SG, Holz IV GG, Dunlap K (1987) Dihydropyridine inhibition of neuronal calcium current and substance P release. Pflügers Archiv Eur J Physiol 409:361–366

Rios E, Brum G (1987) Involvement of dihydropyridine receptors in excitation-contraction coupling in skeletal muscle. Nature 325:717–720

Sala F, Fonteriz RI, Borges R, García AG (1986) Inactivation of potassium-evoked adrenomedullary catecholamine release in the presence of calcium, strontium or Bay-K-8644. FEBS Lett 196:34–38

Sandow A (1952) Excitation-contraction coupling in muscular response. Yale J Biol Med 25:176–201

Sanguinetti MC, Kass RS (1984) Voltage-dependent block of calcium channel current in the calf Purkinje fiber by dihydropyridine calcium channel antagonists. Circ Res 55:336–348

Sanguinetti MC, Krafte DS, Kass RS (1986) Voltage dependent modulation of Ca channel current in heart cells by Bay-K-8644. J Gen Physiol 88:369–392

Schiavone MT, Kirpekar SM (1982) Inactivation of secretory responses to potassium and nicotine in the cat adrenal medulla. J Pharmacol Exp Ther 223:743–749

Schramm M, Thomas G, Towart R, Franckowiak G (1983) Novel dihydropyridines with positive inotropic action through activation of Ca^{2+} channels. Nature 303:535–537

Schwartz LM, McCleskey EW, Almers W (1985) Dihydropyridine receptors in muscle are voltage-dependent but most are not functional calcium channels. Nature 314:747–751

Spedding M (1983) Functional interactions of calcium-antagonists in K^+-depolarized smooth muscle. Br J Pharmacol 80:485–488

Takahashi M, Ogura A (1983) Dihydropyridines as potent calcium channels blockers in neuronal cells. FEBS Lett 152:191–194

Tillotson D (1979) Inactivation of Ca conductance dependent on entry of Ca ions in molluscan neurons. Proc Natl Acad Sci USA 76:1497–1500

Triggle DJ, Janis RA (1987) Calcium channel ligands. Ann Rev Pharmacol Toxicol 27:347–369

Tsien RW (1983) Calcium channels in excitable cell membranes. Ann Rev Physiol 45:341–358

Turner TJ, Goldin SM (1985) Calcium channels in rat brain synaptosomes: identification and pharmacological characterization. High affinity blockade by organic Ca^{2+} channel blockers. J Neurosci 5:841–849

Wada A, Takara H, Izumi F, Kobayashi H, Yanagihara N (1985) Influx of ^{22}Na through acetylcholine receptor-associated Na channels: relationship between ^{22}Na influx, ^{45}Ca influx and secretion of catecholamines in cultured bovine adrenal medulla cells. Neuroscience 15:283–292

Wada A, Yanagihara N, Izumi F, Sakurai S, Kobayashi H (1983) Trifluoperazine inhibits $^{45}Ca^{2+}$ uptake and catecholamine secretion and synthesis in adrenal medullary cells. J Neurochem 40:481–486

Wakade AR (1981) Studies on secretion of catecholamines evoked by actylcholine or transmural stimulation of the rat adrenal gland. J Physiol 313:463–480

Wakade AR, Malhotra RK, Wakade TD (1986) Phorbol ester facilitates ^{45}Ca accumulation and catecholamine secretion by nicotine and excess K^+ but not by muscarine in rat adrenal medulla. Nature 321:698–700

Walsh KB, Bryant SH, Schwartz A (1987) Suppression of charge movement by calcium antagonists is not related to calcium channel block. Pflügers Arch 409:217–219

Yousif F, Triggle DJ (1985) Functional interactions between organic calcium channel antagonists in smooth muscle. Can J Physiol Pharmacol 63:193–195

Zucker RS, Landó L (1986) Mechanism of transmitter release: voltage hypothesis and calcium hypothesis. Science 231:574–578

Regulation of Signal Transduction by G Proteins in Exocrine Pancreas Cells

I. Schulz, S. Schnefel, and R. Schäfer

Max-Planck-Institut für Biophysik, Kennedyallee 70, 6000 Frankfurt am Main 70, FRG

Introduction

The second messenger for hormone-induced Ca^{2+} release is inositol 1,4,5-triphosphate (IP_3) [14]. Following binding of an agonist to its receptor, phospholipase C (PLC) is activated and phosphatidylinositol 4,5-bisphosphate is broken down to IP_3 and diacylglycerol (Fig. 1). While IP_3 releases Ca^{2+} from a nonmitochondrial compartment, which is most likely the endoplasmatic reticulum [15], diacylglycerol activates protein kinase C which in many cells leads to the final cell response by kinase C mediated phosphorylation of target proteins [9]. IP_3 can be metabolized by dephosphorylation to inositol 1,4-bisphosphate (IP_2) or by phosphorylation to inositol 1,3,4,5-tetrakisphosphate (IP_4), which is supposed to be involved in Ca^{2+} influx into the cell, the mechanism of which is yet not quite clear. The two molecules IP_4 and IP_3 seem to act together to control Ca^{2+} influx [5, 8]. A current model is based on the hypothesis that Ca^{2+} enters the cell through an IP_3-sensitive Ca^{2+} pool in a manner similar to that proposed by Putney [10, 11], and that IP_4 modulates Ca^{2+} entry into that Ca^{2+} store [8]. Thus, the Ca^{2+} pool can be filled from the outside of the cell, and Ca^{2+} influx takes place only if the pool is emptied due to IP_3-induced Ca^{2+} release. IP_4 is dephosphorylated to inositol 1,3,4-trisphosphate of which a second messenger function is not yet known. Evidence suggests that in receptor-mediated activation of PLC GTP-binding proteins (G proteins) are involved [3]. Similarly to other systems such as the adenylylcyclase [6], activation of PLC can be influenced by bacterial toxins such as cholera toxin (CT) and pertussis toxin (PT) and by the weakly hydrolyzable GTP analog GTPγS [6]. In order to characterize G proteins involved in receptor-mediated activation of PLC we have measured IP_3 production and the effects of bacterial toxins on it. Furthermore, bacterial toxin-induced ADP-ribosylation and photolabeling of plasma membranes with GTP-γ-azidoanilide in response to hormones has been studied.

Methods

Acinar cells were isolated from rat pancreas by means of collagenase treatment and were purified in an albumin gradient [13]. Cells were permeabilized by washing them in a nominally Ca^{2+}-free buffer [13], and IP_3 production was measured in [^3H]*myo*-inositol prelabeled cells by separation of inositol phosphates on Dowex anion

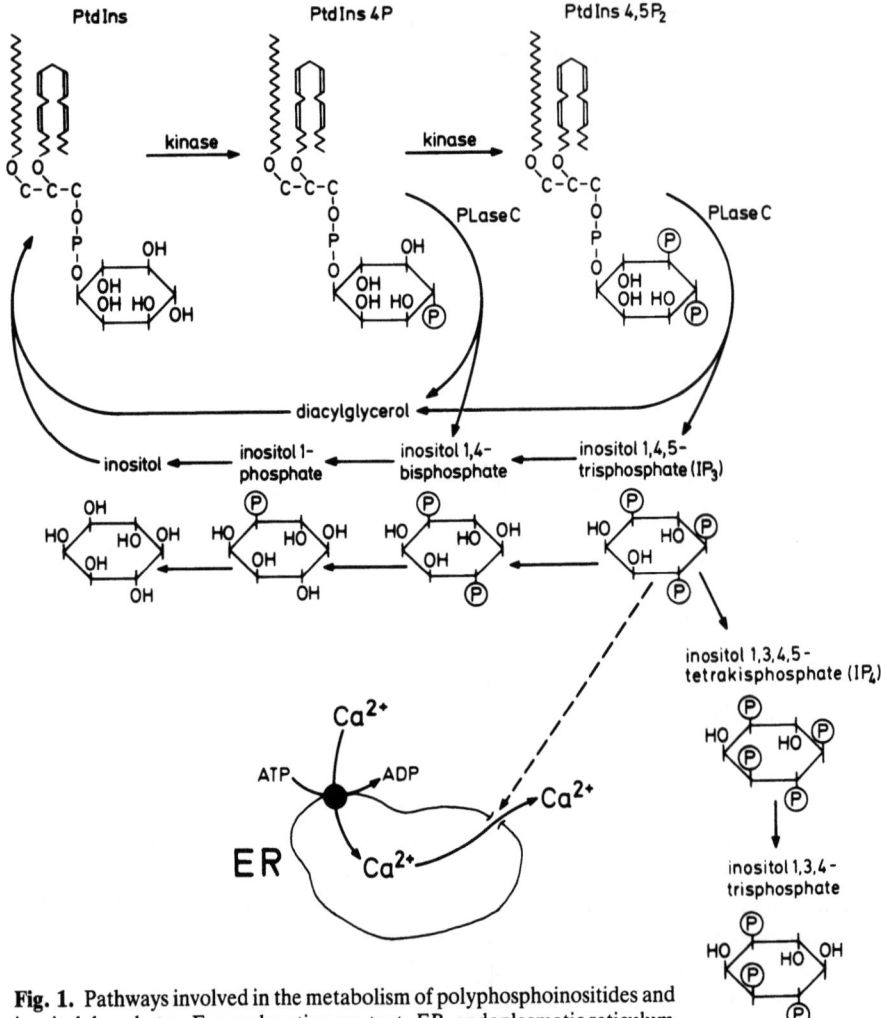

Fig. 1. Pathways involved in the metabolism of polyphosphoinositides and inositolphosphates. For explanation see text. *ER*, endoplasmatic reticulum

exchange columns [16]. Plasma membranes were prepared from isolated acinar cells as described previously [1] using a MgCl$_2$ precipitation method. ADP-ribosylation was performed on isolated plasma membranes in the presence of [^{32}P]NAD$^+$ and of activated CT or PT as described previously [12]. Photolabeling with [α-^{32}P]GTP-γ-azidoanilide (specific activity, 265 Ci/mmol) was performed on isolated plasma membranes in the presence of the GTP photolabel and different hormones. Photolysis was carried out by irradation at 350 nm for 7 min. The photolyzed membranes were subjected to sodium dodecyl sulfate (SDS) gel electrophoresis. Incorporated activity into proteins was visualized by autoradiography on Kodak films.

Effect of Cholera Toxin on Cholecystokinin-, Carbachol-, and GTPγ-Induced IP$_3$ Production

Pretreatment of permeabilized acinar cells with preactivated CT (40 µg/ml) in the presence of NAD$^+$ did not affect basal phosphoinositide metabolism. However, cholecystokinin-octapeptide (CCK-OP) induced IP$_3$ production following 30 min of incubation of the cells was inhibited by 49% when preactivated CT was present in the incubation medium. Consequently, also CCK-OP-induced Ca^{2+} release was diminished (Fig. 2, Table 1). Similarly, GTPγS-induced IP$_3$ production was inhibited by CT by 75% following 25 min of incubation (see Table 1). In contrast, carbachol

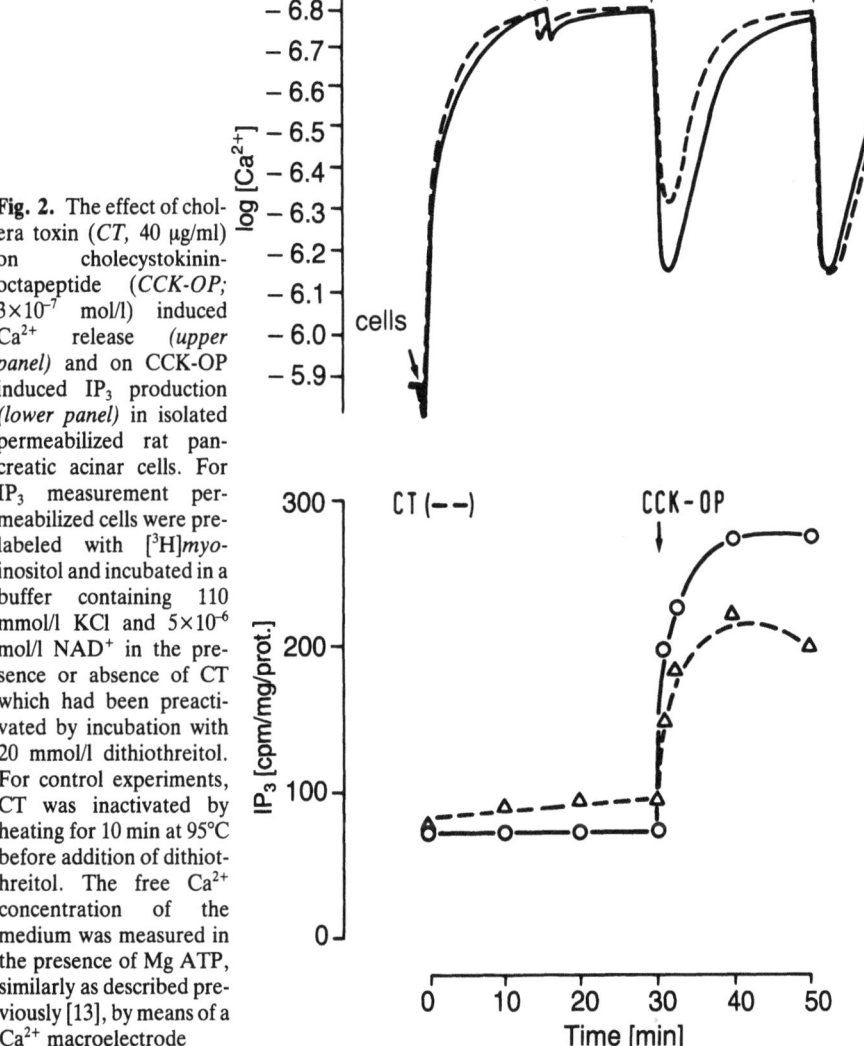

Fig. 2. The effect of cholera toxin (*CT*, 40 µg/ml) on cholecystokinin-octapeptide (*CCK-OP*; 3×10^{-7} mol/l) induced Ca^{2+} release *(upper panel)* and on CCK-OP induced IP$_3$ production *(lower panel)* in isolated permeabilized rat pancreatic acinar cells. For IP$_3$ measurement permeabilized cells were prelabeled with [^{3}H]*myo*-inositol and incubated in a buffer containing 110 mmol/l KCl and 5×10^{-6} mol/l NAD$^+$ in the presence or absence of CT which had been preactivated by incubation with 20 mmol/l dithiothreitol. For control experiments, CT was inactivated by heating for 10 min at 95°C before addition of dithiothreitol. The free Ca^{2+} concentration of the medium was measured in the presence of Mg ATP, similarly as described previously [13], by means of a Ca^{2+} macroelectrode

Table 1. Effect on cholera toxin on basal, cholecystokinin-octapeptide-, carbachol-, and GTPγS-induced IP_3 production in isolated permeabilized cells

	No addition	CCK	CCh	GTPγS
Control	100 ± 18 (10)	342 ± 74 (7)	187 ± 11 (3)	516 ± 161 (3)
with CT	107 ± 6 (10)	218 ± 26 (7)	196 ± 37 (3)	204 ± 38 (3)

Permeabilized cells were prelabeled with [^{3}H]*myo*-inositol and incubated under standard conditions in the presence or absence of cholera toxin (CT, 40 μg/ml). Cholecystokinin-octapeptide (CCK-OP; 3×10^{-7} mol/l, carbachol (CCh; 5×10^{-5} mol/l), and GTPγS (10^{-5} mol/l) were added after 30 min of preincubation. IP_3 production during 30 min of preincubation with no addition, 100% = 105 ± 18 cpm/μg protein. IP_3 production 10 min after addition of CCK or CCh shows means in percentage of the control ± SE. GTPγS-induced IP_3 production was measured 25 min after addition of GTPγS, at which time it was maximal (means in percentage of the control ± SE). The number of separate experiments is given in parentheses. (From [12])

(CCh) induced IP_3 production was unaffected by pretreatment of the cells with CT (see Table 1). In three experiments, preactivated PT (20–2000 μg/ml) in the presence of NAD$^+$ had no effect on basal or on CCK-OP- or CCh-induced IP_3 production (not shown).

Does Stimulation of Adenylylcyclase Mediate the CT Effect on CCK-OP-Induced Activation of Phospholipase C?

CT stimulates cAMP production in pancreatic acinar cells [12] presumably by ADP-ribosylation of the G_S protein of the adenylylcyclase system. In order to determine whether inhibition of PLC stimulation was due either to a direct effect of the activated G_S protein on PLC or to elevation of cAMP levels following stimulation of adenylylcyclase by CT, we investigated the effect of 8-bromo-cAMP, a poorly hydrolyzable cAMP analog, as well as of vasoactive intestinal polypeptide (VIP), a stimulatory hormone for adenylylcyclase, on CCK-OP-induced IP_3 production. The inclusion of 8-bromo-cAMP in the incubation buffer prior to the addition of permeabilized cells had no effect on either basal or CCK-OP-induced IP_3 production, whether it was given 30 min or 1 min before addition of CCK-OP to the incubation medium (not shown). Although cAMP production induced by VIP in pancreatic acinar cells was similar to that induced by CT, VIP had no effect on CCK-OP-induced IP_3 production (not shown).

Effect of Agonists on CT-Dependent ADP-Ribosylation of 40-kd Protein

When pancreatic acinar cell membranes were treated with CT and [^{32}P]NAD$^+$, four proteins with molecular masses of 50, 48, 45, and 40 kd were ADP-ribosylated (Fig. 3). Addition of GTPγS to the incubation medium markedly stimulated the ADP-ribosylation of the 50-, 48-, and 45-kd proteins, presumably corresponding to

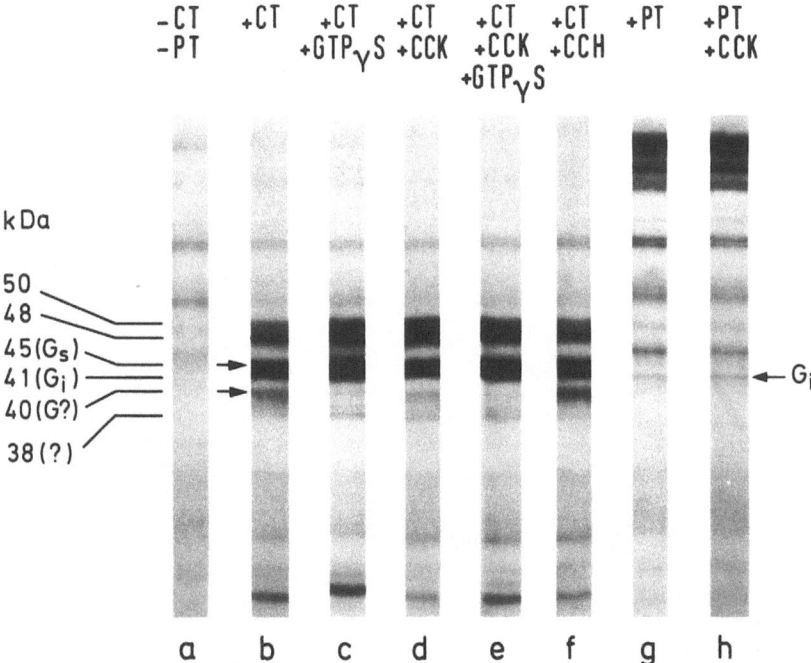

Fig. 3. Autoradiography of an SDS gel illustrating cholera toxin (*CT;* 40 µg/ml) or pertussis toxin (*PT;* 5 µg/ml) catalyzed ADP-ribosylation in the presence of [^{32}P]NAD$^+$ of proteins from pancreatic plasma membranes in the absence or presence of GTPγS (10^{-5} mol/l), of cholecystokinin-octapeptide (*CCK-OP;* 3×10^{-7} mol/l), or of carbachol (*CCh,* 10^{-6} mol/l). (From [12])

the α-subunits of the stimulatory G protein of the adenylylcyclase (G_s) (see Fig. 3). In contrast, GTPγS (see Fig. 3) and GTP (not shown) inhibited the ADP-ribosylation of the 40-kd protein. Addition of CCK-OP, which stimulates PLC, also inhibited the CT-dependent ADP-ribosylation of the 40-kd protein, but was without any effect on the labeling of other protein bands. Simultaneous addition of GTPγS and CCK-OP also inhibited CT-induced ADP-ribosylation (see Fig. 3). However, CCh, which stimulates PLC (see Table 1) had no effect on the CT-dependent ADP-ribosylation of the 40-kd protein. CCK-OP had no effect on the PT-dependent ADP-ribosylation of the 41-kd protein, which presumably corresponds to the α-subunit of the G_i protein.

Effect of Carbachol and Cholecystokinin-Octapeptide on Photolabeling of a 40-kd G Protein from Pancreatic Plasma Membranes With [α^{32}P]GTP-γ-azidoanilide

Photoactivation of [α^{32}P]GTP-γ-azidoanilide in the presence of pancreatic plasma membranes resulted in covalent incorporation of the label predominantly into a protein of 40-kd, as demonstrated by separation of the photolyzed proteins by SDS gel

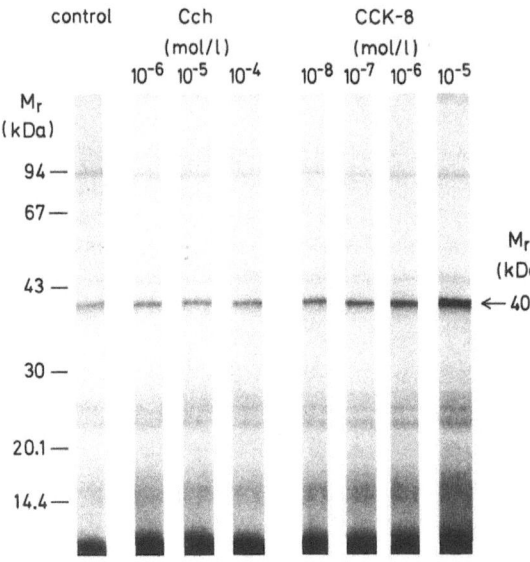

Fig. 4. Autoradiography of an SDS gel demonstrating photolabeling of proteins from pancreatic plasma membranes with [α^{32}P]GTP-γ-azidoanilide (2.5×10^{-7} mol/l) in the absence or presence of carbachol *(CCh)* or cholecystokinin-octapeptide *(CCK-8)*. Incubation of plasma membranes with the label and subsequent photolysis resulted in incorporation of the label predominantly into a protein of 40 kd and to less extent into proteins of 94, 45, 38, 30, 27, and 24 kd

electrophoresis and subsequent autoradiography (Fig. 4). The photoaffinity reagent was also incorporated, although to less extent, into proteins banding at 94, 45, 38, 30, 27, and 24 kd. The label was specifically displaced by the weakly hydrolyzable GTP analogs GTPγS and Gpp(NH)p when added before photolysis (not shown). Incubation of pancreatic plasma membranes with GTP-γ-azidoanilide in the presence of CCK-OP and subsequent photoactivation resulted in increased covalent incorporation of the photolabel into the 40-kd GTP-binding protein (see Fig. 4). Addition of the acetylcholine (ACh) analog CCh, which activates PLC in pancreatic acinar cells (see Table 1), caused only a small increase in the photolabeling of the 40-kd protein even at a high concentration of 10^{-4} mol/l (see Fig. 4). Photoincorporation of GTP-γ-azidoanilide was not increased in the presence of vasoactive intestinal polypeptide at concentrations at which adenylylcyclase activity was fully stimulated in pancreatic acinar cells (not shown).

Effect of Cholera Toxin-Catalyzed ADP-ribosylation of Plasma Membranes on Photolabeling with GTPγS Azidoanilide

Treatment of permeabilized acinar cells with CT (40 μg/ml) inhibited CCK-OP-induced IP_3 production but not that induced by CCh (see Table 1). We have therefore investigated the influence of CT-catalyzed ADP-ribosylation of plasma membranes, isolated from pancreatic acinar cells, on photolabeling of the 40-kd protein with [α^{32}P]GTP-γ-azidoanilide. ADP-ribosylation of the plasma membranes by CT reduced the covalent incorporation of GTP-γ-azidoanilide into the 40-kd protein band by about 30% as compared to nonribosylated plasma membranes (Fig. 5). This reduction in the covalent incorporation of the GTP photolabel into the 40-kd binding

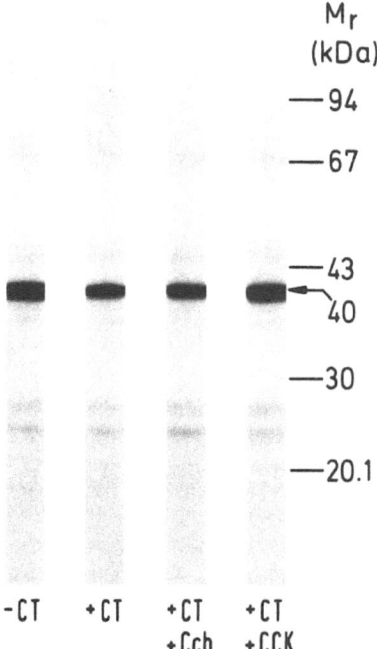

Fig. 5. Autoradiography of an SDS gel illustrating the effect of cholera toxin *(CT;* 40 µg/ml) catalyzed ADP-ribosylation on photolabeling of a 40-kd protein with [α^{32}P]GTP-γ-azidoanilide in the absence and presence of carbachol *(CCh;* 10^{-6} mol/l) or of cholecystokinin-octapeptide *(CCK;* 3×10^{-7} mol/l)

protein was totally reversed if the CT-induced ADP-ribosylation was carried out in the presence of CCK-OP (5×10^{-7} mol/l). The addition of CCh at a concentration of 10^{-5} mol/l did not restore photolabeling of the 40-kd protein to the control value, and only a small increase in the photoincorporation of the GTP photolabel could be achieved.

Discussion

These studies in isolated permeabilized pancreatic acinar cells and isolated plasma membranes show that ACh and CCK receptors functionally couple by different G proteins to PLC in pancreatic acinar cells. Neither of them is a substrate for PT, and only the CCK receptor-coupling, not the ACh receptor-coupling, G-protein is ADP-ribosylated by CT with consequent inhibition of PLC.

Furthermore, photolabeling of a 40-kd protein by the GTP analog [α^{32}P]GTP-γ-azidoanilide is influenced by CCK-OP but not by CCh. Whereas binding of the label was enhanced in the presence of CCK-OP, the ACh analog CCh had no effect on GTP binding (see Fig. 4). Similarly, in the presence of CCK-OP, but not of CCh, binding of the GTP-photoaffinity label into ADP-ribosylated 40-kd protein was reduced (see Fig. 5). These experiments indicate that the GTP-binding protein of 40-kd is a good candidate for the coupling of the CCK-receptor to PLC. CT either inhibits coupling of CCK receptors to PLC by ADP-ribosylating an activating 40-kd protein, or it inhibits PLC activity by activating an inhibitory G protein of 40 kd. In the latter case the mechanism of how CT should act on this inhibitory G protein could be similar as that

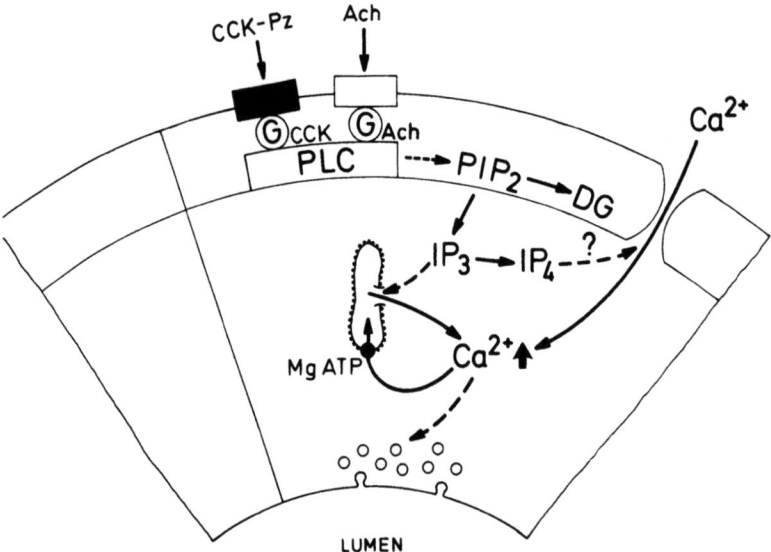

Fig. 6. Model for hormone-induced signal transduction in the acinar cell of the exocrine pancreas. Two different GTP-binding proteins, G_{CCK} and G_{Ach}, functionally couple cholecystokinin and muscarinic acetylcholine receptors, respectively, to phospholipase C *(PLC)*. For further explanation see text

of G_S activation of adenylylcyclase. Activation of G_S by CT leads to dissociation of $α_S$ from G_S and inactivation of GTPase activity, with consequent permanent activation of adenylylcyclase. In the case of an inhibitory G protein that couples PLC, one would have to assume that dissociation of α-GTP from the inhibitory PLC coupling G protein would lead to inactivation of PLC. Since we do not know yet whether both activating and inhibitory G proteins are involved in the regulation of PLC activity, the mechanisms of how GT inhibits CCK-induced PLC stimulation remains unknown.

At present we can distinguish three different classes of G proteins, which are involved in receptor-mediated regulation of phospholipase C. These proteins can be differentiated by their different apparent sensitivities to bacterial toxins. They are either inactivated by PT (G_i-like proteins), or they are inactivated by CT, or they are insensitive to both of them [2, 4, 7]. The molecular mass of 40 kd of the protein supposed to couple CCK receptors to PLC is very similar to that of Gi-like proteins. Isolation of this protein and determination of its amino acid sequence will be needed to clarify whether or not this 40-kd protein can be attributed to the G_i-like family of G proteins [4]. The model shown in Fig. 6 summarizes the data obtained: receptor-mediated activation of specific G proteins leads to stimulation of PLC activity and production of IP_3 by hydrolysis of phosphatidylinositol 4,5-bisphosphate. IP_3 induces Ca^{2+} release from a nonmitochondrial Ca^{2+} pool, which is endoplasmic reticulum (ER) or a subcompartment of it. The mechanism of Ca^{2+} release most likely involves opening of a Ca^{2+} channel in the ER membrane by IP_3. Phosphorylation of IP_3 leads to generation of IP_4, which is likely to be involved in Ca^{2+} influx into the cell. At rest Ca^{2+} reuptake by a Ca^{2+} ATPase into Ca^{2+} pools takes place.

References

1. Bayerdörffer E, Eckhardt L, Haase W, Schulz I (1985) Electrogenic calcium transport in plasma membrane of rat pancreatic acinar cells. J Membr Biol 84:45–60
2. Birnbaumer L, Codina J, Mattera R, Yatani A, Scherer N, Toro M-J, Brown AM (1987) Signal transduction by G proteins. Kidney Int 32:14–37
3. Cockcroft S, Gomperts BD (1985) Role of guanine nucleotide binding protein in the activation of polyphosphoinositide phosphodiesterase. Nature 314:534–536
4. Graziano MP, Gilman AG (1987) Guanine nucleotide-binding regulatory proteins: mediators of transmembrane signaling. Trends Pharmacol Sci 8:478–481
5. Irvine RF, Moor RM (1986) Micro-injection of inositol-1,3,4,5-tetrakisphosphate activates sea urchin eggs by a mechanism dependent on external Ca^{2+}. Biochem J 240:917–920
6. Levitzki A (1984) Receptor to effector coupling in the receptor-dependent adenylate cyclase system. J Recept Res 4:399–409
7. Lo WWY, Hughes J (1987) Receptor-phosphoinositidase C coupling. Multiple G-proteins? FEBS Lett 224:1–3
8. Morris AP, Gallacher DV, Irvine RF, Petersen OH (1987) Synergism of inositol triphosphate and tetrakisphosphate in activating Ca^{2+}-dependent K^+ channels. Nature 330:653–657
9. Nishizuka Y (1983) Phospholipid degradation and signal translation for protein phosphorylation. Trends Biochem Sci 8:13–16
10. Putney JW Jr (1986) A model for receptor-regulated calcium entry. Cell Calcium 7:1–12
11. Putney JW Jr (1987) Formation and actions of calcium-mobilizing messenger, inositol 1,4,5-trisphosphate. Am J Physiol 252:149–G157
12. Schnefel S, Banfic H, Eckhardt L, Schultz G, Schulz I (1988) Acetylcholine and cholecystokinin receptors functionally couple by different G-proteins to phospholipase C in pancreatic acinar cells. FEBS Lett 230:125–130
13. Streb H, Schulz I (1983) Regulation of cytosolic free Ca^{2+} concentration in acinar cells of rat pancreas. Am J Physiol 245:G347–G357
14. Streb H, Irvine RF, Berridge MJ, Schulz I (1983) Release of Ca^{2+} from a nonmitochondrial intracellular store in pancreatic acinar cells by inositol-1,4,5-triphosphate. Nature 306:67–69
15. Streb H, Bayerdörffer E, Haase W, Irvine RF, Schulz I (1984) Effect of inositol-1,4,5-trisphosphate on isolated subcellular fractions of rat pancreas. J Membr Biol 81:241–253
16. Streb H, Heslop JP, Irvine RF, Schulz I, Berridge MJ (1985) Relationship between secretagogue-induced Ca^{2+} release and inositol polyphosphate production in permeabilized pancreatic acinar cells. J Biol Chem 260:7309–7315

Calcium Channel Blockers and Renin Secretion*

P. C. Churchill

Department of Physiology, Wayne State University School of Medicine, 540 East Canfield, Detroit, Michigan 48201, USA

Introduction

The renin-angiotensin system plays a central role in salt and water balance and in the regulation of arterial blood pressure. The level of activity of this system is determined primarily by the rate at which the granulated juxtaglomerular cells (JG cells) secrete renin into the blood. Physiologically, renin secretory rate is controlled by a number of first messengers: afferent arteriolar transmural pressure or some function of it, such as stretch (the baroreceptor mechanism); solute transport in the macula densa segment of the nephron (the macula densa mechanism); catecholamines released from the renal nerves and the adrenal medulla (the β-adrenergic mechanism); extracellular concentrations of many organic and inorganic substances including angiotensin II, vasopressin, K^+, and Mg^{2+} [24, 36, 68]. In addition to these physiological first messengers, a number of pharmacological agents affect renin secretion [36]. This review concerns the renin secretory effects of calcium-channel blockers.

It is an accepted principle of cellular biology that first messengers act by affecting the intracellular concentrations of only a few second messengers. There is evidence to suggest that intracellular free ionic calcium (Ca_i^{2+}), cyclic AMP, and cyclic GMP are second messengers in renin secretion [13, 14, 28, 29, 31, 38].

The evidence that Ca_i^{2+} is an inhibitory second messenger (JG cell Ca_i^{2+} is inversely related to secretory activity) has been reviewed recently [13, 14, 28, 29], and only some of the pieces of evidence are presented, but in sufficient detail to explain two seeming paradoxes. First, an inverse relationship between JG cell Ca_i^{2+} and renin secretion might suggest the existence of inverse relationship between extracellular Ca^{2+} (Ca_e^{2+}) and renin secretion. However, changes in Ca_e^{2+} do not produce parallel, or even directionally similar, changes in Ca_i^{2+}. Second, if increased Ca_i^{2+} inhibits renin secretion, then calcium-channel blockers might be expected to stimulate renin secretion. Sometimes they do, but sometimes they do not [14]. However, calcium-channel blockers "specifically" antagonize only Ca^{2+} influx through calcium channels which are gated or opened by membrane depolarization [27, 30, 58]. Possibly JG cells are not always depolarized; if their calcium channels are not open, then calcium-channel blockers have no open channels to block.

---

* This research was supported by NIH grant HL-24880.

Review articles are cited extensively, since space limitations preclude citing the original relevant literature.

Calcium as an Inhibitory Second Messenger

Extracellular and Intracellular Calcium

Very few data on JG cell Ca_i^{2+} are available. Recently, Kurtz et al. [38] used the Quin-2 fluorescence method of measurement, and the results of one experiment indicate a concentration of approximately 3×10^{-7} M. A priori, JG cell Ca_i^{2+} must be approximately 10^{-7} M in the steady state [13], since these cells contain mitochondria, and since Ca_i^{2+} must be about 10^{-7} M in order for oxidative phosphorylation to occur [9]. Moreover, Ca_i^{2+} measured in other cells ranges between 5×10^{-8} and 3.5×10^{-7} M in the non-calcium-activated state to about 1.5×10^{-6} M in the calcium-activated state [57, 58, 60]. A priori, again, Ca_i^{2+} must range between these values, since this concentration range is consistent with the calcium-binding constants of the calcium-modulated proteins which mediate the effects of Ca_i^{2+} on cellular responses, such as calmodulin [57, 58]. If JG cell Ca_i^{2+} is approximately 10^{-7} M, an enormous electrochemical gradient for Ca^{2+} influx exists, since Ca_e^{2+} is approximately 10^{-3} M, and since JG cell membranes are electrically polarized, intracellular fluid some 60–70 mV negative with respect to extracellular fluid [8, 26].

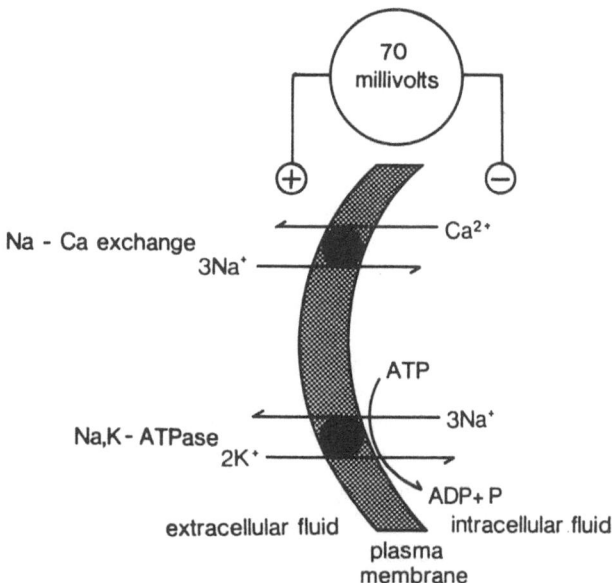

Fig. 1. A hypothetical JG cell membrane. The large electrochemical gradient favoring Ca^{2+} influx proves that efflux pathways must exist, even if intracellular Ca^{2+} can be buffered by mitochondrial uptake, by sequestration in organelles such as endoplasmic reticulum, and by calcium-binding molecules such as proteins and nucleotides

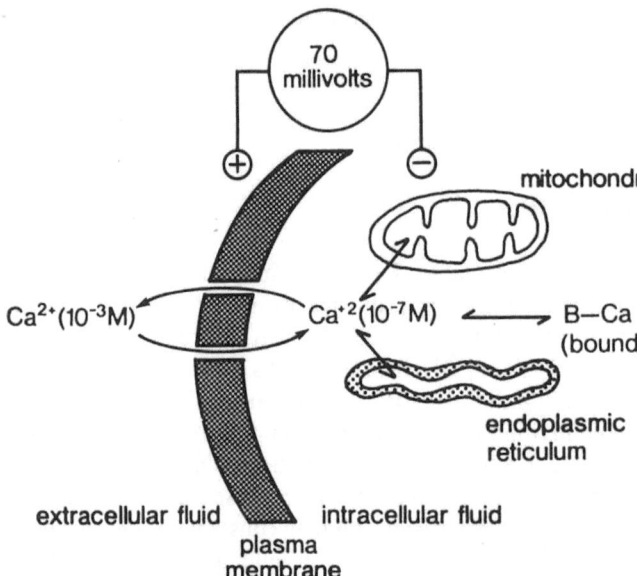

Fig. 2. A hypothetical JG cell membrane with Na–Ca exchange. Na^+ influx down its electrochemical gradient is coupled with Ca^{2+} efflux against its electrochemical gradient. Transmembrane Na^+ and K^+ concentration gradients, and ultimately the transmembrane potential, are maintained by primary active Na^+ and K^+ transport (Na,K-ATPase)

How can the JG cell, shown in Fig. 1, maintain such a low Ca_i^{2+} in the face of such a large electrochemical gradient for Ca^{2+} influx? Even if the cell membrane is relatively impermeable to Ca^{2+}, and even if Ca_i^{2+} is buffered by mitochondrial uptake through Ca^{2+} binding to intracellular molecules such as calmodulin and through Ca^{2+} sequestration in organelles such as the endoplasmic reticulum [3, 9, 57, 58, 60, 67], in the final analysis there must be specific Ca^{2+} extrusion mechanisms, since cell membranes are not completely impermeable to Ca^{2+}, and since no buffering mechanism has infinite capacity. Only two cellular mechanisms for Ca^{2+} extrusion are known to exist [3, 5, 57, 58, 60, 67]: Na–Ca exchange, which is ultimately dependent on the active transport of Na^+ and K^+ (Fig. 2), and primary active calcium transport (Fig. 3). A priori, it seems reasonable to infer that one or both of these mechanisms must contribute to regulating JG cell Ca_i^{2+}. The pharmacological evidence which supports this inference, and which simultaneously supports the hypothesis that Ca_i^{2+} is an inhibitory second messenger in renin secretion, is presented in the following subsection.

It should be stressed that an inverse relationship between Ca_i^{2+} and renin secretion (Ca^{2+} acting as a second messenger) does not imply an inverse relationship between Ca_e^{2+} and renin secretion (Ca^{2+} acting as a first messenger). The implicit assumption is incorrect. Indeed, changes in Ca_e^{2+} can induce *opposite* changes in Ca_i^{2+}, due to the membrane-stabilizing property of Ca_e^{2+}. Increasing Ca_e^{2+} decreases cell membrane Ca^{2+} permeability [56, 58, 60, 67], and despite the enormous electrochemical gradient, this can actually decrease Ca^{2+} influx. Conversely, decreasing Ca_e^{2+} destabilizes and depolarizes cell membranes, and depending upon the transmembrane concentration gradient, Ca^{2+} influx and Ca_i^{2+} can increase as Ca_e^{2+} decreases [56]. For this reason, one must be cautious in interpreting the results of in vitro experiments in which cells were exposed to a nominally "calcium-free" physiological saline solu-

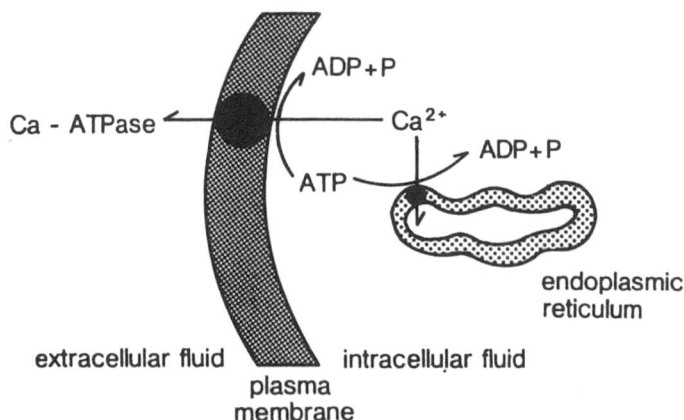

Fig. 3. A hypothetical JG cell membrane with primary active (ATP-dependent) Ca^{2+} transport. Calcium-activated ATPase activity, associated with the cell membrane and the membranes of organelles such as endoplasmic reticulum, is the biochemical correlate of primary active Ca^{2+} efflux and non mitochondrial sequestration

tion. A gradient of indeterminant magnitude for Ca^{2+} influx still exists, since Ca_i^{2+} is submicromolar but "calcium-free" solutions without chelators invariably contain 1–10 $\mu M\ Ca^{2+}$, depending upon the quality of the water, the quality of the reagents, and the amount of Ca^{2+} leached from the glassware used for preparation and storage of solutions [56]. It is not surprising, therefore, that "calcium-free" extracellular fluid has been observed to inhibit, to have no effect on, and to stimulate renin secretion in in vitro preparations [13]. Similarly, increased plasma Ca^{2+} has been observed to inhibit, to have no effect on, and to stimulate renin secretion and/or increase plasma renin concentration in intact animals [36]. Since the effects of Ca_e^{2+} on Ca_i^{2+} are unpredictable, except when the transmembrane concentration gradient is actually reversed or the cell membrane Ca^{2+} permeability is increased, these observations cannot be taken as evidence for or against a second messenger role of Ca_i^{2+}.

Adding a calcium-binding substance to a nominally calcium-free physiological saline solution reverses the normal transmembrane Ca^{2+} concentration gradient which favors Ca^{2+} influx, and ethylaminediaminetetraacetate (EDTA) and ethyleneglycol-bis(β-amino ethyl ether)N,N'-tetraacetate (EGTA) stimulate renin release in several in vitro preparations, including rat glomeruli, rat renal cortical slices, pig renal cortical slices, and isolated perfused kidneys [13, 28, 29]. Conceivably, some of the instances in which "calcium-free" media stimulated renin release [28, 29] could be attributed to the calcium-binding properties of albumin, since rather high concentrations of albumin were used in these experiments. In theory, chelator-stimulated renin "release" could be attributed to the leakage of intracellular renin across permeabilized cell membranes, rather than to stimulated renin "secretion" per se. However, EGTA-stimulated renin release is unrelated to the release of another intracellular protein – lactate dehydrogenase [10, 46]. Therefore, the observation that reversal of the transmembrane Ca^{2+} concentration gradient, which cannot fail to decrease JG cell Ca_i^{2+}, stimulates renin secretion can be taken as one piece of evidence that Ca_i^{2+} is an inhibitory second messenger.

The renin secretory effects of La^{3+}, Mg^{2+}, and calcium ionophores add to this evidence [13, 28, 29]. La^{3+} blocks transmembrane Ca^{2+} fluxes [56, 58, 60], and La^{3+} blocks the stimulatory effect of chelators on renin secretion. Mg_e^{2+} acts as a membrane stabilizer, reducing cell membrane Ca^{2+} permeability [56, 58, 60]; in the presence of a normal Ca_e^{2+}, increased Mg_e^{2+} stimulates renin secretion in vivo, in isolated perfused rat kidneys, and in rat renal cortical cell suspensions. Calcium ionophores facilitate passive Ca^{2+} fluxes in either direction across cell membranes. Although there are some discrepancies, calcium-ionophores appear either to inhibit or to stimulate renin secretion in vitro, depending upon whether the Ca^{2+} concentration gradient favors influx or efflux, respectively. The in vivo renin secretory effects of calcium ionophores and of calcium chelators are nearly impossible to interpret; the concentrations required to increase or decrease Ca_i^{2+} in JG cells would do so in *all* cells, and this would almost certainly be lethal.

Collectively, the renin secretory effects of calcium chelators, calcium ionophores, Mg^{2+}, and La^{3+} are consistent with the hypothesis that Ca_i^{2+} is an inhibitory second messenger. It does not follow that Ca_e^{2+} is a physiological first messenger, either inhibitory or stimulatory. On the other hand, Mg_e^{2+} could be considered a stimulatory first messenger, but if so, its importance in the physiological regulation of renin secretion remains to be determined.

Calcium Efflux and Sequestration

The Sodium-Calcium Exchange Mechanism

Na–Ca exchange is shown in Fig. 2 as a carrier that binds Na^+ and Ca^{2+} in a 3:1 ratio. In the presence of a normal transmembrane electrochemical gradient for Na^+ influx, Na^+ influx down its gradient is coupled with Ca^{2+} efflux against its gradient [3, 5]. Transmembrane Na^+ and K^+ concentration gradients, and ultimately the transmembrane potential, are maintained by primary active Na^+ and K^+ transport (Na,K-ATPase [63]). Accordingly, Na,K-ATPase inhibitors should inhibit renin secretion (increased Na_i^+, decreased Na^+ gradient, decreased Ca^{2+} efflux, and increased Ca_i^{2+}), and agents that promote Ca^{2+} efflux (calcium chelators) should block the inhibitory effect. Conversely, Na,K-ATPase stimulators should stimulate renin secretion (decreased Na_i^+, increased Na^+ gradient, increased Ca^{2+} efflux, and decreased Ca_i^{2+}), and agents that increase Ca_i^{2+} should block the stimulatory effect.

Ouabain, vanadate, and potassium-free extracellular fluid inhibit Na,K-ATPase activity, and all three have inhibitory effects on renin secretion in several in vitro preparations [13]. Ouabain and vanadate inhibit renin secretion in vivo as well [13]. Although hypokalemia is not associated with inhibited renin secretion in vivo [36], plasma K_e^+ cannot be lowered sufficiently to block Na,K-ATPase, since Na–K transport is fully activated by $K_e^+ < 2$ mM [63]. In any case, the inhibitory effects of ouabain, vanadate, and potassium-free extracellular fluid on in vitro renin secretion are mediated by increased Ca_i^{2+}, since they can be blocked by reversing the transmembrane Ca^{2+} gradient with calcium chelators [13].

Na,K-ATPase is activated by increasing K_e^+ from 0 to approximately 2 mM [63], and can be stimulated further by phenytoin and by β-adrenergic agonists via increased cyclic AMP [55, 61]. Increasing K_e^+ from 0 to 2 mM stimulates renin secretion in vitro, and both phenytoin and β-adrenergic agonists stimulate renin secretion in vitro and in vivo [13, 24, 28, 29, 36]. Moreover, ouabain blocks the stimulatory effects on renin secretion of increasing K_e^+ [13], of phenytoin [13], of isoproterenol [13, 28, 29], and of dibutyryl cyclic AMP [13].

Primary Active Calcium Transport

Ca-ATPase (Fig. 3) is the biochemical correlate of primary active calcium efflux and nonmitochondrial calcium sequestration [3, 9, 57, 58, 67]. Based on this model, one would predict that Ca-ATPase inhibitors and stimulators would inhibit and stimulate renin secretion, respectively. As mentioned above, vanadate inhibits Na,K-ATPase activity. However, vanadate also inhibits Ca-ATPase activity, and either action produces an increase in Ca_i^{2+} that can be blocked by calcium chelators. Therefore, its inhibitory effect on renin secretion, and the blockade of its inhibitory effect by calcium chelators, can be taken as evidence that Ca_i^{2+} is an inhibitory second messenger, but it cannot be taken as evidence that vanadate increases Ca_i^{2+} by a specific single action. Similarly, as mentioned above, there is evidence for cyclic AMP-stimulated Na,K-ATPase activity, which increases Ca^{2+} efflux via Na–Ca exchange. However, there is also evidence for cyclic AMP-stimulated Ca-ATPase activity, which increases active Ca^{2+} efflux and sequestration [57, 67]. Whichever mechanism is involved, Ca_i^{2+} decreases, and two pieces of evidence indicate that this cyclic AMP-dependent decrease in Ca_i^{2+}, attributable to increased efflux, mediates the stimulatory effect of β-adrenergic agonists on renin secretion. First, La^{3+} blocks Ca^{2+} efflux [56, 58, 60], and La^{3+} blocks the stimulatory effect of isoproterenol on renin secretion [42]. Second, several manipulations and substances antagonize and/or block isoproterenol-stimulated and dibutyryl cyclic AMP-stimulated renin secretion, and the only known effect that they all have in common is to increase Ca_i^{2+}. These include α-adrenergic agonists, angiotensin II, vasopressin, ouabain, vanadate, potassium-free extracellular fluid, and potassium depolarization [13, 28, 29].

Intracellular Calcium Sequestration

Variations in the rates of calcium sequestration and mobilization can affect Ca_i^{2+}, and therefore potentially affect renin secretion. Consistently, TMB-8 (8-(N,N-diethylamino) octyl 3,4,5-trimethoxy-benzoate) antagonizes calcium mobilization from intracellular sequestration sites, and there are several reports that TMB-8 stimulates renin secretion in vitro [29].

In summary, the above observations are consistent with Ca_i^{2+} being an inhibitory second messenger in renin secretion. Further, they suggest that JG cell Ca_i^{2+} can be affected by the rates of two Ca^{2+} efflux mechanisms (Na–Ca exchange and primary active calcium transport) and the rates of intracellular calcium sequestration and mobilization.

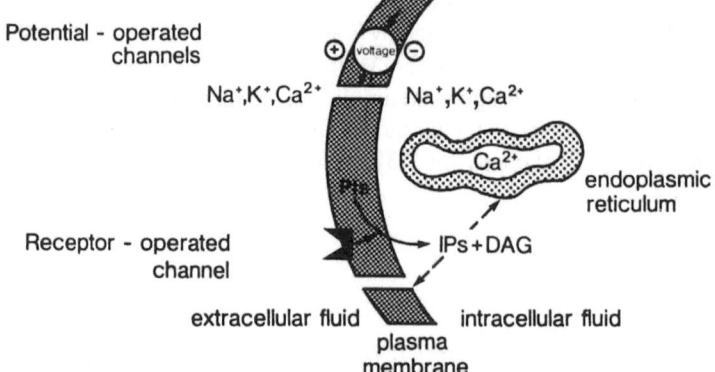

Fig. 4. A hypothetical JG cell membrane with potential-operated and receptor-operated channels. Potential-operated channels for Na$^+$, K$^+$, and Ca^{2+} have been characterized in many cells. Potential-operated calcium channels are opened by membrane depolarization (and by Bay K8644, a channel agonist); calcium-channel blockers (nifedipine, verapamil) antagonize depolarization-induced Ca^{2+} influx. Similarly, tetrodotoxin antagonizes Na$^+$ influx through potential-operated Na$^+$ channels. In contrast, receptor-operated channels are activated when an agonist occupies its receptor on the cell membrane; this usually leads to stimulation of phospholipase activity, hydrolysis of phosphoinositides *(PIs)* associated with the membrane, and the production of inositol phosphates *(IPs)*, which increase Ca^{2+} influx and/or mobilize Ca^{2+} from intracellular sequestration sites, and diacylglycerol *(DAG)*. Receptor antagonists block agonist-induced changes in ion permeability, but, in general, agonist-induced ion fluxes are resistant to the effects of channel blockers (e.g., tetrodotoxin, nifedipine). Probably angiotensin II and vasopressin (and possibly A$_1$-adenosine receptor agonists and α-adrenergic agonists) activate JG cell receptor-operated channels

Calcium Influx and Mobilization

Potential-Operated Calcium Channels

Potential-operated, or voltage-dependent, ion channels for Na$^+$, K$^+$, and Ca^{2+} have been described and characterized in contractile and secretory cells [6, 27, 30, 56, 58, 60]. This observation, taken together with the observation that the renin-secretion JG cells are derived from smooth muscle, suggests that depolarization of the JG cell might activate potential operated calcium channels and increase Ca^{2+} influx, thereby increasing Ca$_i^{2+}$ and decreasing renin secretion (Fig. 4). Although few electrophysiological data are available, the membrane potential of JG cells approximates a potassium-diffusion potential [8, 26], and increasing K$_e^+$ causes a progressive depolarization [26]. Potassium depolarization (raising K$_e^+$ to 50–60 mM) inhibits renin secretion in many in vitro preparations, including isolated perfused rat kidneys and renal cortical slices from rats, rabbits, dogs, and pigs ([13, 14, 17, 28, 29]; Fig. 5). Several observations demonstrate that the inhibitory effect is mediated by increased Ca^{2+} influx through potential-operated calcium channels, resulting in increased Ca$_i^{2+}$. Calcium chelation reverses the transmembrane Ca^{2+} gradient, and it abolishes the inhibitory effect. Moreover, the inhibitory effect is antagonized, in concentration-dependent manner, by four different organic calcium-channel blockers: nifedipine,

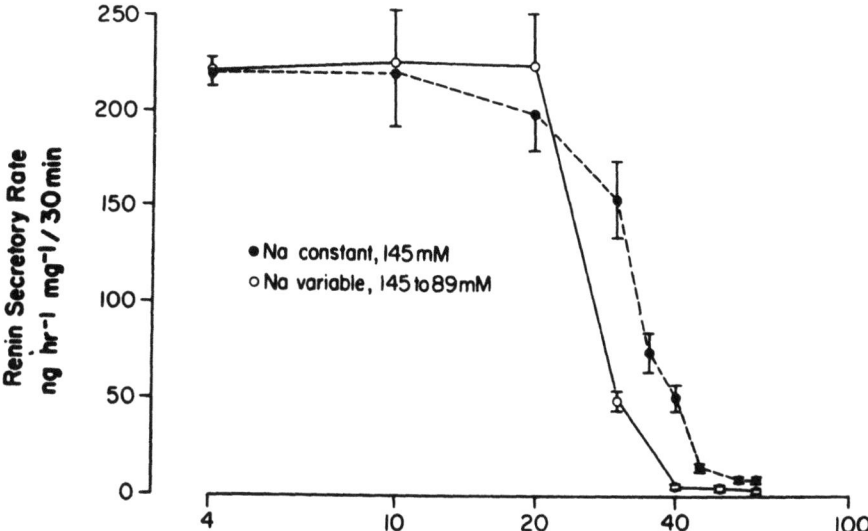

Fig. 5. Renin secretory rates of rat kidney slices as a function of extracellular K^+ concentration. *Open circles,* NaCl was reduced as KCl was increased, such that osmolality and Cl^- were held constant. *Filled circles,* KCl was added to the normal medium, such that Na^+ was constant but osmolality and Cl^- increased with increasing K^+. In either case, renin secretion was virtually abolished by $K^+ > 40$ mM. (From [17])

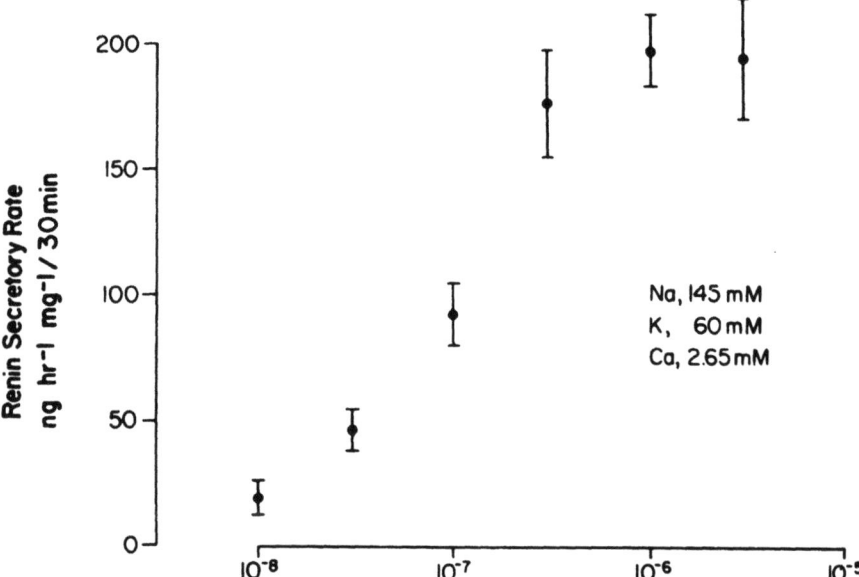

Fig. 6. Renin secretory rates of rat renal cortical slices incubated in potassium-depolarizing media, as a function of methoxyverapamil concentration. Potassium depolarization (60 mM K^+) nearly abolishes secretion, and methoxyverapamil antagonizes the inhibitory effect of potassium depolarization in a concentration-dependent manner. The IC_{50} is approximately 0.1 μM, and 0.,5 μM restores secretory rate to the level found in nondepolarized cells. (From [17])

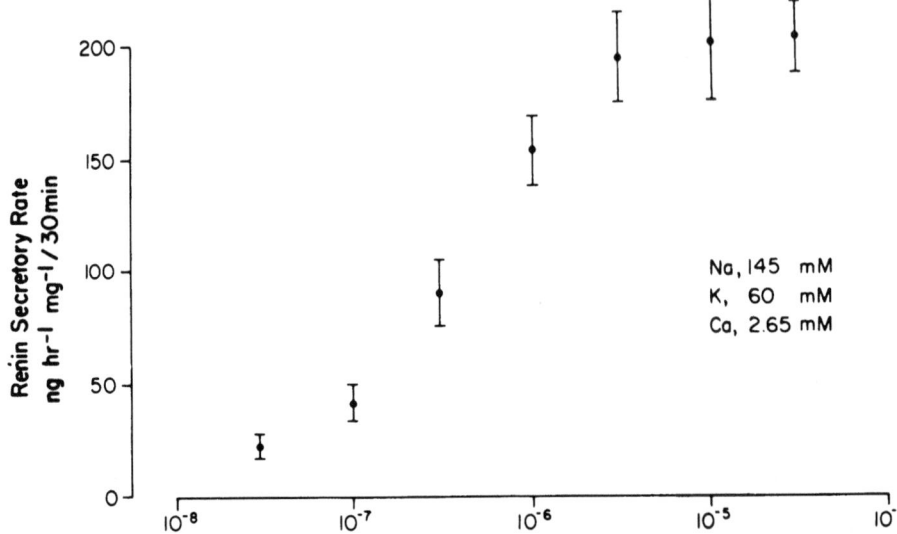

Fig. 7. Renin secretory rates of rat renal cortical slices incubated in potassium-depolarizing media, as a function of verapamil concentration. Potassium depolarization (60 mM K$^+$) nearly abolishes secretion, and verapamil antagonizes the inhibitory effect of potassium depolarization in a concentration-dependent manner. The IC$_{50}$ is approximately 0.5 μM, and 1.0 μM restores secretory rate to the level found in nondepolarized cells. (From [17])

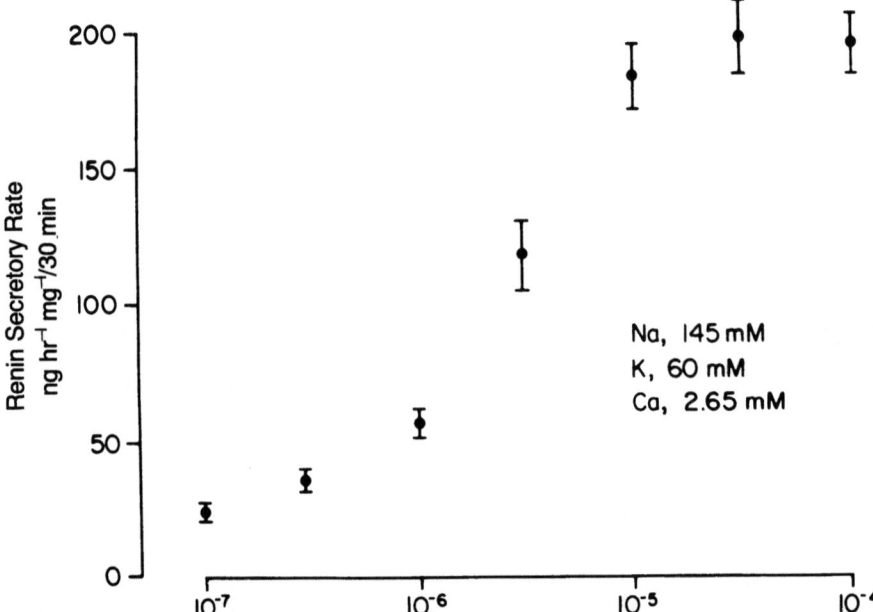

Fig. 8. Renin secretory rates of rat renal cortical slices incubated in potassium-depolarizing media, as a function of diltiazem concentration. Potassium depolarization (60 mM K$^+$) nearly abolishes secretion, and diltiazem antagonizes the inhibitory effect of potassium depolarization in a concentration-dependent manner. The IC$_{50}$ is approximately 5 μM, and 30 μM restores secretory rate to the level found in nondepolarized cells. (From [20])

methyoxyverapamil (Fig. 6), verapamil (Fig. 7), and diltiazem (Fig. 8). The order of potency (nifedipine > methoxyverapamil > verapamil > diltiazem) and the IC_{50} values are similar to those that have been reported for antagonizing potential-operated calcium channels that have been characterized electrophysiologically. Lowering Ca_e^{2+} potentiates the ability of calcium-channel blockers to antagonize the effect of potassium depolarization (Fig. 9), which is consistent with competitive action of Ca_e^{2+} and these drugs at potential-operated calcium channels. Finally, Bay K8644, a calcium-channel agonist, inhibits renin secretion, and the inhibitory effect is antagonized by calcium-channel blockers [29]. Collectively, even in the absence of electrophysiological data, these observations constitute unequivocal pharmacological evidence that JG cells have potential-operated calcium channels, and that Ca_i^{2+} is an inhibitory second messenger.

It follows that first messengers could affect renin secretion by altering JG cell membrane potential – inhibitory first messengers by depolarizing and increasing Ca^{2+} influx, stimulatory first messengers by hyperpolarizing and decreasing Ca^{2+} influx. Indeed, the hypothesis that JG cell membrane potential-controlled renin secretion was originally advanced, by Fishman [26], more than a decade ago. Subsequently, Fray and Park and their coworkers [28, 29] hypothesized that Ca^{2+} influx is the link between membrane potential and renin secretion. As attractive as this unifying hypothesis is, and despite the recent evidence that can be marshalled in its support [8], it ignores other mechanisms of altering JG cell Ca_i^{2+}, such as Na–Ca exchange and primary active calcium transport (vide supra) and receptor-mediated mobilization of sequestered calcium (vide infra). Moreover, there is evidence that the stimulatory

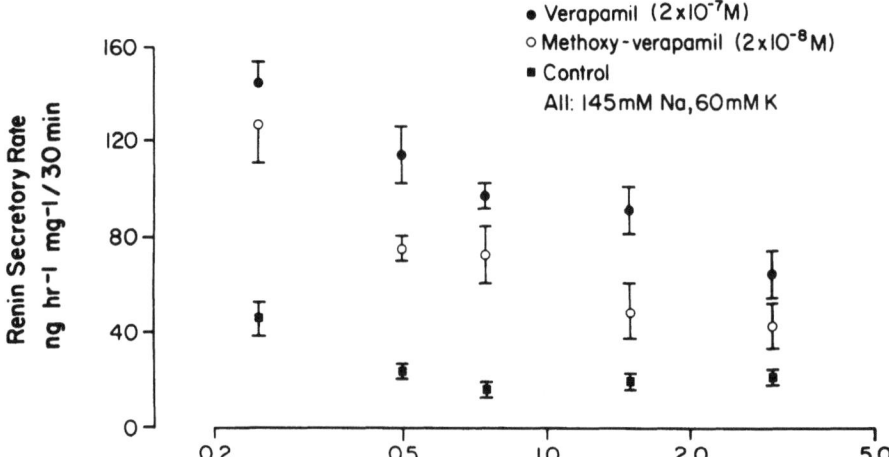

Fig. 9. Renin secretory rates of rat renal cortical slices incubated in potassium-depolarizing media. Potassium depolarization (60 mM K$^+$) nearly abolishes secretion. Reducing extracellular Ca^{2+} from normal to one-tenth normal (0.25 mM) has little effect because there is still an enormous electrochemical gradient for Ca^{2+} influx (Ca_i^{2+} is submicromolar). Low concentrations of verapamil and methoxyverapamil also have little effect when extracellular Ca^{2+} is normal; however, as extracellular Ca^{2+} is lowered, these low concentrations of calcium-channel blockers become increasingly more effective in antagonizing the inhibitory effect of potassium depolarization, as if they and extracellular Ca^{2+} "competed" for the channels. (From [17])

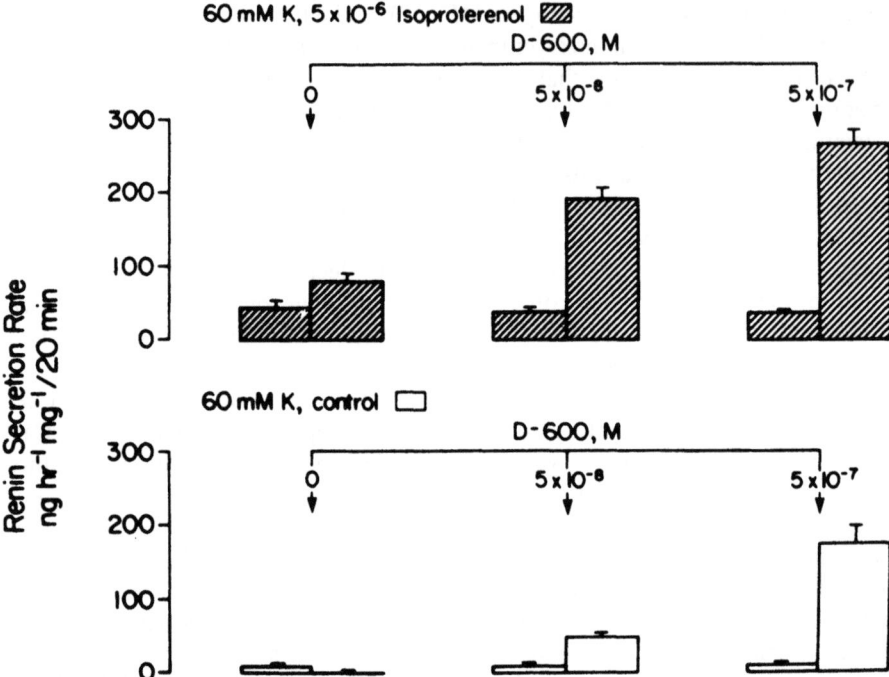

Fig. 10. Renin secretory rates of rat renal cortical slices incubated in potassium-depolarizing media. Each pair of columns represents the rates during two consecutive incubation periods (with or without additions to the incubation media between the periods) in a given experiment. Potassium depolarization (60 mM K$^+$) nearly abolishes secretion; secretory rate is approximately 130 (same units) in the presence of 4 mM K$^+$. (Note that slices were incubated for 20-min periods in these experiments, and for 30-min periods in the experiments shown in Fig. 5–9). *Lower panel*, methoxyverapamil (D-600) was added between the incubation periods, and it antagonized the inhibitory effect of potassium depolarization; *upper panel*, all incubation media contained isoproterenol. By comparing upper and lower panels, it can be seen that isoproterenol stimulated renin secretion, and that methoxyverapamil potentiated this stimulation, despite complete depolarization. Therefore, the stimulatory effect of isoproterenol cannot be attributed to membrane hyperpolarization. (From [16])

effect of β-adrenergic agonists ([16]; Fig. 10) and the inhibitory effects of angiotensin II [11, 20] and vasopressin [12, 20] are completely independent of membrane potential [Fig. 11 and 12].

Ironically, even if increased plasma K$^+$ inhibits renin secretion in vivo [36], the effect is probably not mediated by direct potassium-depolarization of the JG cells, for the following reasons. First, increasing plasma K$^+$ by as little as 0.5 mM inhibits renin secretion in vivo [69], and increases in K$_e^+$ of this magnitude have no appreciable effect on JG cell membrane potential [26]. Second, and more definitively, the inhibitory effect of increased plasma K$^+$ is observed in filtering kidneys, but not in nonfiltering kidneys [62]. Since direct potassium depolarization of Jg cells would occur independently of filtration, this latter finding suggests that the inhibitory effect is mediated instead by a tubular mechanism, that is, by a K$^+$-induced increase in macula densa NaCl load.

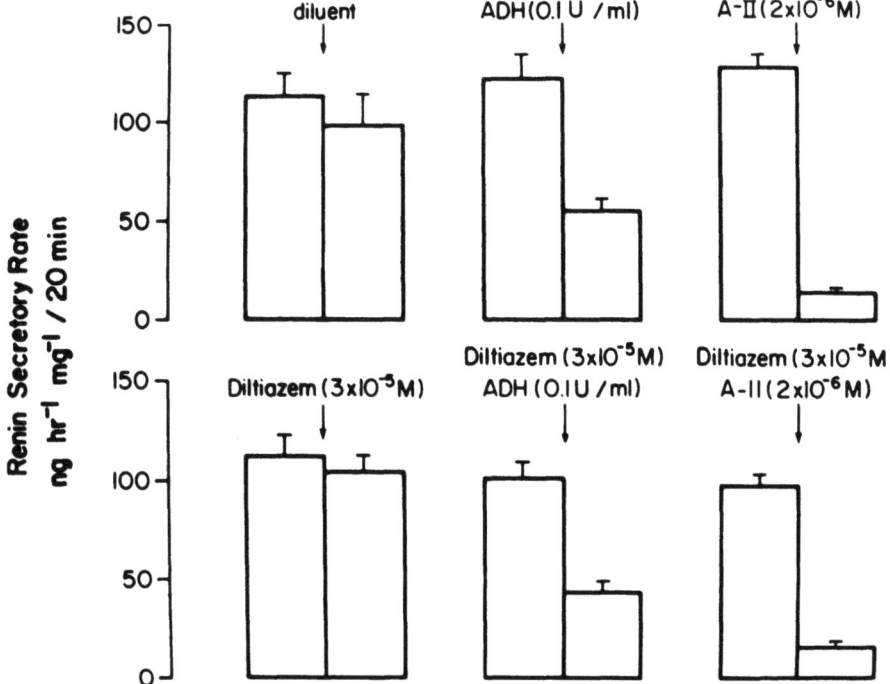

Fig. 11. Renin secretory rates of rat renal cortical slices incubated in normal media (4 mM KCl). Each pair of colums represents the rates during two consecutive incubation periods in a given experiment, with or without additions to the incubation media between the periods. Note that the slices were incubated for 20-min periods in these experiments and for 30-min periods in the experiments of Fig. 5–9. Antidiuretic hormone *(ADH)* and angiotensin II *(A–II)* inhibit renin secretion *(upper panel)*, and the inhibitory effects are not blocked by the simultaneous addition of diltiazem *(lower panel)*. Diltiazem by itself has no effect on renin secretion *(left pair of columns in lower panel)*. (From [20])

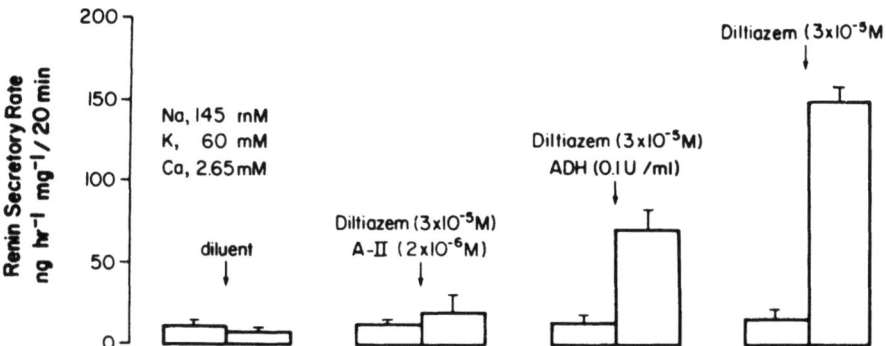

Fig. 12. Renin secretory rates of rat renal cortical slices incubated in potassium-depolarizing media. Each pair of columns represents the rates during two consecutive incubation periods in a given experiment, with or without additions to the incubation media between the periods. Potassium depolarization (60 mM K^+) nearly abolishes secretion; compare secretory rates with those in Fig. 11. Diltiazem (30 μM) restores secretory rate to the level found in nondepolarized cells *(two right columns versus two left columns)*. If angiotensin II *(A–II)* or antidiuretic hormone *(ADH)* are added along with 30 μM diltiazem, secretory rate is considerably less than the level found in nondepolarized cells. Therefore, both peptides have inhibitory effects on renin secretion which are not susceptible to antagonism by calcium-channel blockers and which cannot be attributed to membrane depolarization, since cells were already depolarized with 60 mM K^+. (From [20])

In summary, JG cells have potential-operated calcium channels, and depolarization-induced Ca^{2+} influx inhibits renin secretion. However, some first messengers act independently of changes in JG cell membrane potential. What first messenger does act by changing JG cell membrane potential? A reasonable answer is presented below ("Physiological Regulation of Renin Secretion").

Receptor-Operated Ion Channels

In general, "receptor-operated ion channels" are local increases in ion permeability that result from the interaction of an agonist with its specific receptor on the cell membrane ([6, 56, 58, 60]; Fig. 4). Agonist-induced increases in ion permeability are not initiated by membrane depolarization, but they can produce either depolarization or hyperpolarization depending upon the ion selectivity of the channel (e.g., neurotransmitter-induced excitatory and inhibitory postsynaptic potentials). In particular, receptor-operated calcium channels [6] appear to be agonist-induced increases in Ca_i^{2+} that are attributable to phosphoinositide hydrolysis and the production of inositol phosphates, which increase Ca^{2+} influx and/or mobilize calcium from intracellular sites of sequestration, and of diacylglycerol, which, together with the increased Ca_i^{2+}, activates protein kinase C [51]. Since such receptor-induced increases in Ca_i^{2+} are not dependent upon membrane depolarization, they are relatively resistant to the effects of calcium-channel blockers, as evidenced by the observation that the IC_{50} values for antagonizing vascular smooth muscle contractions produced by potassium depolarization versus α-adrenergic agonists and angiotensin II generally differ by several orders of magnitude [58].

The observation that secretory and vascular smooth muscle cells have receptor-operated calcium channels [6, 27, 30, 51, 56, 58], taken with the observation that the renin-secreting JG cells are derived from vascular smooth muscle, suggests that JG cells might have receptor-operated calcium channels. Indeed, Peart and coworkers [54] pointed out many years ago that substances which increase vascular smooth muscle contractility, presumably by receptor-induced increases in Ca_i^{2+}, tend to inhibit renin secretion. Four examples of this will be considered next: angiotensin II, vasopressin, A_1-adenosine receptor agonists, and α-adrenergic agonists.

It is well-known that angiotensin II and vasopressin inhibit renin secretion, both in vivo and in vitro [13, 24, 28, 29, 36, 54, 68]. It is very likely that the inhibitory effects are attributable to activation of receptor-operated calcium channels, increased phospholipid metabolism [40], and an increase in Ca_i^{2+} that is due to mobilization from intracellular sites of sequestration. Although calcium chelation blocks the inhibitory effects, blockade is not instantaneous, which is consistent with the concept that mobilization, rather than influx, of Ca^{2+} is involved [11, 12]. Furthermore, although both peptides can depolarize JG cells [8], their inhibitory effects on renin secretion occur independently of membrane potential ([11, 12, 20]; Fig. 11, 12).

It is well-known that adenosine can both increase renovascular resistance and decrease renin secretion [15, 53, 64]. Pharmacological studies have shown that both responses are produced by the A_1 subclass of adenosine receptors [15, 19, 49]. Furthermore, both cellular responses are mediated by increased Ca_i^{2+}, although the mechanisms for increasing Ca_i^{2+} appear to differ in vascular smooth muscle cells and JG cells [15]. Calcium-channel blockers antagonize adenosine-induced renal vasoconstriction, suggesting depolarization-induced Ca^{2+} influx as the mechanism. In contrast, calcium-channel blockers do not antagonize adenosine-induced inhibition of renin secretion. The inhibitory effect is mediated by increased Ca_i^{2+}, however, since it can be blocked by calcium chelation [15, 19].

Finally, it is well-known that α-adrenergic agonists can both increase renovascular resistance and inhibit renin secretion, particularly in in vitro preparations [13, 28, 29, 36]. The inhibitory effects on renin secretion are mediated by increased Ca_i^{2+} since they are antagonized and/or blocked by Mn^{2+} (an inorganic "calcium-channel blocker" [58, 60], by calcium chelation, and by organic calcium-channel blockers. Collectively, these observations suggest that either α-adrenergic agonists activate potential-operated calcium channels directly (that is, the drugs are "calcium-channel agonists" like Bay K8644), or α-adrenergic agonists activate some kind of receptor-operated ion channel which induces depolarization, and the depolarization in turn opens potential-operated calcium channels. The recent findings that norepinephrine depolarizes JG cells [8] and increases the calcium permeability of their plasma membranes [39, 40] are consistent with either possibility.

In summary, agonists of four distinct receptors inhibit renin secretion by mechanisms dependent upon increased Ca_i^{2+}. There is some uncertainty and controversy concerning the mechanisms by which Ca_i^{2+} increases in response to these agonists [13, 14, 28, 29], but this should not detract from the fact that the inhibitory effects on renin secretion have been shown, in many ways, to be calcium-dependent, and this strengthens the evidence that Ca_i^{2+} is an inhibitory second messenger.

Intracellular Calcium Receptors

Collectively, the above observations indicate that JG cell Ca_i^{2+} and therefore renin secretion can be altered by all of the following: calcium chelators, calcium ionophores, inorganic ions which block Ca^{2+} flux or act as calcium antagonists (Mg^{2+}, Mn^{2+}, La^{3+}), inhibitors and stimulators of Na–Ca exchange (Na,K-ATPase inhibitors and stimulators), inhibitors and stimulators of primary active calcium transport and sequestration (Ca-ATPase inhibitors and stimulators), substances which antagonize the mobilization of sequestered calcium (TMB-8), substances which act on receptors to release or mobilize calcium (angiotensin II, vasopressin, and possible adenosine), substances which depolarize the membrane and thereby open potential-operated calcium channels (α-adrenergic agonists), and agonists (Bay K8644) and antagonists (nifedipine, methoxyverapamil, verapamil, diltiazem) of potential-operated calcium channels. These observations constitute very strong evidence that Ca^{2+} is an inhibitory second messenger.

This appears to be one of only a very few exceptions to the rule that Ca^{2+} is a stimulatory second messenger in secretory cells [51, 56–58, 60]. Obviously, some step must be reversed in the sequence leading from a change in Ca_i^{2+} to changes in the secretory activities of JG cells versus, for example, adrenal medullary chromaffin cells. If both JG and chromaffin cells release their stored secretory products by a process that involves exocytosis, then increased Ca_i^{2+} must change the concentration/

effect of some intermediate in precisely the opposite directions in JG and chromaffin cells. So far the intermediate remains elusive.

Regardless of whether Ca_i^{2+} is an inhibitory or a stimulatory second messenger in a particular cell, some intracellular entity must act as a "calcium receptor", and calmodulin is a likely candidate for this role. Further, if Ca_i^{2+} is an inhibitory second messenger in renin secretion, and if calmodulin is the receptor, then antagonists of calcium-calmodulin activity should stimulate renin secretion. Several investigators have shown that calcium-calmodulin antagonists stimulate basal renin secretion and antagonize the effects of several inhibitors of renin secretion, including angiotensin II and potassium depolarization [13, 14, 28, 29]. However, these observations cannot be taken as evidence of calcium-calmodulin involvement since several commonly used calcium-calmodulin antagonists (trifluoperazine, calmidazolium, chlorpromazine) block Ca^{2+} influx, expecially depolarization-induced Ca^{2+} influx [37]. This effect can account for their stimulatory effects on basal secretion, and their antagonism of the effects of some inhibitors, particularly potassium depolarization. In other words, the renin secretory effects of these drugs can be explained equally well by either their putative intracellular effect of binding to and antagonizing the calcium-calmodulin complex, or their effect on the cell membrane of antagonizing Ca^{2+} influx, independently of any direct interaction with intracellular calcium-calmodulin. If the latter, then their renin secretory effects provide no evidence that calmodulin is involved in the renin secretory process. Indeed, although they do not interpret their results in this manner, some investigators have shown that trifluoperazine is as effective as verapamil in blocking ^{45}Ca influx into cultured JG cells [39]. Therefore, calmodulin could be the intracellular receptor, but there is no evidence that it is.

The calcium activated phosphilipid-dependent protein kinase C is also an intracellular calcium-receptor, and activation of this enzyme is believed to be involved in the secretory process in some cells [51]. Tumor-promoting phorbol esters, such as 12-O-tetradecanolylphorbol 13-acetate (TPA), activate protein kinase C in two ways: by substituting for diacylglycerol and by increasing the Ca^{2+} affinity such that ambient Ca_i^{2+} is sufficient. If calcium-activation of protein kinase C mediates the inhibitory effect of increased Ca_i^{2+} in JG cells, then TPA should inhibit renin secretion. It does [18, 40], but since phorbol esters have many effects on cells in addition to activation of protein kinase C, this cannot be taken as conclusive evidence of a role of protein kinase C.

It must be admitted that the sequence of events by which increases and decreases in JG cell Ca_i^{2+} lead to decreases and increases in renin secretion, respectively, remain to be fully elucidated. However, the same statement also applies to the role of Ca_i^{2+} in virtually all secretory cells [60].

Physiological Regulation of Renin Secretion

Intrarenal Mechanisms

Baroreceptor Mechanism

It is well established that renin secretion is physiologically controlled, at least in part, by an intrarenal baroreceptor mechanism. The secretory activity of JG cells, which are found primarily in the media of the afferent arteriole, is influenced by physical changes in the afferent arteriole. Most experimental observations are consistent with the hypothesis that renin secretion is inversely related to stretch of the afferent arteriole: increased renal perfusion pressure increases stretch, and increased stretch somehow inhibits renin secretion; conversely, decreased renal perfusion pressure, or increased renal interstitial fluid pressure, decreases stretch, and decreased stretch somehow stimulates renin secretion [24, 28, 36, 68].

Fray and coworkers [28, 29] have marshalled evidence to support the hypothesis that voltage-dependent changes in Ca^{2+} influx underlie the baroreceptor mechanism for controlling renin secretion: increased stretch leads to depolarization, followed by increased Ca^{2+} influx, which then leads to increased JG cell Ca_i^{2+} and inhibited renin secretion. Conversely, decreased stretch leads to hyperpolarization (or to less depolarization), followed by decreased Ca^{2+} influx, leading to decreased JG cell Ca_i^{2+} and stimulated renin secretion (Fig. 13). In brief, the evidence is: (a) increased stretch inhibits renin secretion of isolated JG cells [28, 29]; (b) mechanical distortion and stretch depolarize isolated JG cells [26]; (c) potassium depolarization inhibits renin secretion [13, 14]; and (d) calcium-channel blockers antagonize the inhibitory effects of depolarization (vide supra) and of high renal perfusion pressure [28, 29] on renin secretion.

It is interesting that afferent arteriolar vascular smooth muscle cells in general also exhibit the phenomenon of stretch-induced depolarization, followed by increased contraction [32]. This response to stretch could obviously account for renal autoregulatory changes in afferent arteriolar resistance: increased stretch leads to depolarization and contraction (increased resistance) whereas decreased stretch leads to hyperpolarization and vasodilation (decreased resistance). Consistently, calcium-channel blockers abolish renal autoregulation [28, 29].

Macula Densa Mechanism

It is well established that renin secretion is physiologically controlled, at least in part, by an intrarenal "chemoreceptor" mechanism which is sensitive to some function of tubular fluid NaCl in the macula densa segment of the nephron. Most experimental observations are consistent with Vander's [68] original hypothesis that renin secretion is inversely related to the amount of Na^+ actively reabsorbed from tubular fluid in the macula densa segment [24, 36]. (It may be true that Na^+ and Cl^- are cotransported in this segment, but the active reabsorption of both Na^+ and Cl^- is ultimately dependent upon Na,K-ATPase activity [52], which is stimulated by increased Na_i^+ [63].) Thus, in the absence of transport inhibitors such as furosemide increased load of NaCl delivered to the macula densa leads to increased transport and inhibited renin secretion, whereas decreased NaCl load leads to decreased transport and stimulated renin secretion.

How the macula densa cells in the early distal tubule transmit information concerning NaCl transport to the renin-secreting JG cells is unknown. Some granulated cells are quite distant from any macula densa cells. A plausible hypothesis, originally advanced by Tagawa and Vander [66], is as follows: as macula densa active transport increases, ATP hydrolysis and adenosine production increase; adenosine, released from the macula densa cells, inhibits the secretory activity of nearby JG cells [15, 53, 64].

Extrarenal Mechanisms

In addition to the two completely intrarenal mechanisms, a whole host of extrarenal mechanisms is involved in the physiological control of renin secretion [24, 36, 68]. Among the most important, as assessed by the effects of their blockade, are (a) the β-adrenergic stimulation of renin secretion, by increased activity of the renal nerves and increased plasma concentrations of catecholamines (β-blockers almost invariably inhibit renin secretion in vivo), and (b) the negative-feedback inhibition of renin secretion by plasma angiotensin II (receptor antagonists and converting enzyme inhibitors almost invariably stimulate renin secretion in vivo). Both cyclic AMP and Ca_i^{2+} are probably involved in the β-adrenergic response; a cyclic AMP-induced decrease in Ca_i^{2+} could stimulate renin secretion. A receptor-induced increase in Ca_i^{2+} probably mediates the negative-feedback effect of angiotensin II. If depolarization-induced Ca^{2+} influx mediated the inhibitory effect of angiotensin, as has been concluded by some investigators [8, 28, 29, 31, 38–40] despite evidence to the contrary ([11, 12, 20]; Fig. 10), then calcium channel blockers would interrupt the negative-feedback loop and stimulate renin secretion as effectively as converting enzyme inhibitors do – but they do not [14].

Renin-Secretory Effects of Calcium Channel Blockers In Vivo

There is strong pharmacological evidence for the existence of JG cell potential-operated calcium channels (vide supra): potassium depolarization and calcium-channel agonists produce inhibitory effects on renin secretion; the inhibitory effects are antagonized by calcium-channel blockers; the order of potency of calcium-channel blockers is "correct", and increasing Ca_e^{2+} antagonizes their effects. However, there are very few indications that potential-operated calcium channels play a significant role in the physiological regulation of renin secretion. If they did, then since Ca^{2+} is an inhibitory second messenger, these drugs should invariably stimulate renin secretory rate and increase plasma renin concentration. Although stimulatory effects have been reported [1, 2, 4, 21, 33, 34, 41, 43–45, 47, 59, 65, 70], no effects [22, 25, 33] and even inhibitory effects [7, 59] have also been reported.

A hypothetical explanation of these diverse observations can be advanced. In vivo, the secretory activity of JG cells is simultaneously affected by numerous signals –

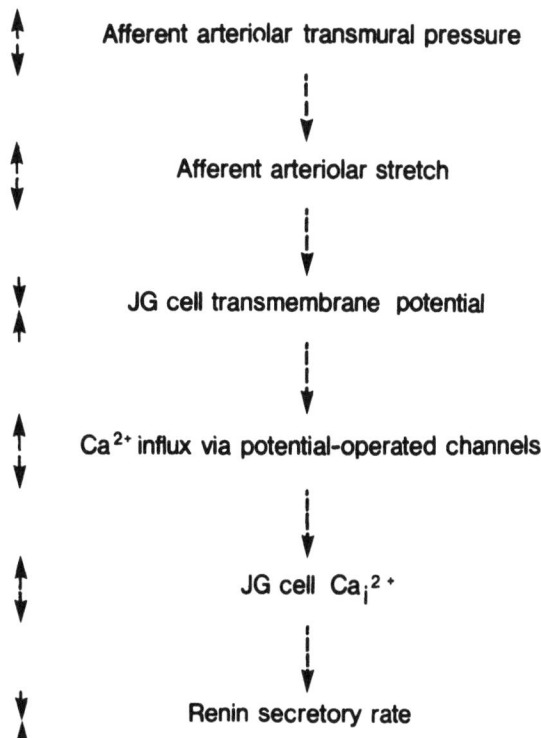

Fig. 13. The putative mechanism of operation of the "intrarenal baroreceptor mechanism" for controlling renin secretion. Increased renal perfusion pressure stretches the afferent arterioles, thereby depolarizing the JG cells which are located therein, and a depolarization-induced influx of Ca^{2+} leads to inhibition of renin secretion. Calcium-channel blockers block the inhibitory effects on renin secretion of increased renal perfusion pressure, increased stretch, and potassium depolarization (see text)

afferent arteriolar stretch, NaCl reabsorption in the macula densa segment, the activity of the renal sympathetic nerves, and the plasma concentrations of both inhibitors (e.g., angiotensin II and vasopressin) and stimulators (e.g., catecholamines). There is evidence that baroreceptor-mediated changes in renin secretion are produced by depolarization-induced Ca^{2+} influx (Fig. 13), but perhaps the baroreceptor mechanism is the only controlling mechanism operative in vivo which is susceptible to calcium-channel blockade. If so, one would predict that calcium-channel blockers would stimulate renin secretion (by blocking the inhibitory effect of the prevailing renal perfusion pressure) but only if calcium-channel blockers do not produce some other effect that counteracts the stimulation. In fact, calcium-channel blockers have at least two effects which could mask this putative stimulatory effect. First, they antagonize neurotransmitter release and catecholamine release from the adrenal medulla [60]. Such actions could inhibit renin secretion, by reducing any stimulatory input to the JG cells from the renal adrenergic nerves and circulating catecholamines. Second, calcium-channel blockers frequently induce natriuresis [1, 2, 7, 23, 25, 34, 35, 48, 50, 59]. Therefore, they could increase NaCl load in the macula densa segment and thereby increase the intensity of the inhibitory input to the JG cells from the macula densa cells. In this context, adenosine is the putative chemical signal between macula densa and JG cells, and calcium-channel blockers do not antagonize its inhibitory effect on renin secretion (vide supra). The results of one study support this interpretation: verapamil induces natriuresis and inhibits renin secretion in filtering dog kidneys, but stimulates renin secretion in nonfiltering dog kidneys in which macula densa influences on renin secretion are either excluded or unchanging [59].

Summary and Conclusions

There is evidence that cyclic AMP and Ca^{2+} are stimulatory and inhibitory second messengers in renin secretion, respectively. Some first messengers alter cyclic AMP by altering the rates of its synthesis or destruction (by affecting adenylate cyclase and phosphodiesterase activities). Other first messengers, and probably the second messenger cyclic AMP itself, alter Ca_i^{2+} by altering the rates of Ca^{2+} efflux and sequestration (Na–Ca exchange, Ca-ATPase activity) or the rates of Ca^{2+} influx and mobilization (receptor-operated and potential-operated calcium channels). There is evidence that pressure-induced changes in renin secretion are mediated by depolarization-induced Ca^{2+} influx (the baroreceptor mechanism), and on this basis, calcium-channel blockers should stimulate renin secretion. Although stimulation is occasionally observed, other effects of these drugs must sometimes mask or counteract this stimulatory effect.

Acknowledgements. If it hadn't been for my wife Monique, I wouldn't have been able to draw many of my conclusions, since she designed and did most of the experiments. However, she wouldn't have been able to do the experiments, nor I to draw conclusions, without support from the National Institutes of Health (HL-24880). Therefore, both of us are grateful to the NIH for support.

References

1. Abe Y, Komori T, Miura K, Takada T, Imanishi M, Okahara T, Yamamoto K (1983) Effects of the calcium antagonist nicardipine on renal function and renin release in dogs. J Cardiovasc Pharmacol 5:254–259
2. Abe Y, Yukimura T, Iwao H, Mori N, Okahara T, Yamamoto K (1983) Effects of EDTA and verapamil on renin release in dogs. Jpn J Pharmacol 33:627–633
3. Baker PF (1976) The regulation of intracellular calcium. In: Duncan DJ (ed) Calcium in biological systems. Cambridge, Cambridge University Press, p 67
4. Blackshear JL, Orlandi C, Williams GH, Hollenberg NK (1986) The renal response to diltiazem and nifedipine: comparison with nitroprusside. J Cardiovasc Pharmacol 8:37–43
5. Blaustein MP (1974) The interrelationship between sodium and calcium fluxes across cell membrane. Rev Physiol Biochem Pharmacol 70:33–82
6. Bolton TB (1979) Mechanisms of action of transmitters and other substances on smooth muscle. Physiol Rev 59:606–718
7. Brown B, Churchill P (1983) Renal effects of methoxyverapamil in anesthetized rats. J Pharmacol Exp Ther 225:372–377
8. Buhrle CP, Nobiling R, Taugner R (1985) Intracellular recordings from renin-positive cells of the afferent glomerular arteriole. Am J Physiol 249:F272–F281
9. Carafoli E, Crompton M (1976) Calcium ions and mitochondria. In: Duncan DJ (ed) Calcium in biological systems. Cambridge, Cambridge University Press, p 89
10. Churchill PC (1979) Possible mechanism of the inhibitory effect of ouabain on renin secretion from rat renal cortical slices. J Physiol (London) 294:123–134

11. Churchill PC (1980) Effect of D-600 on inhibition of in vitro renin release in the rat by high extracellular potassium and angiotensin II. J Physiol (London) 304:449–458
12. Churchill PC (1981) Calcium dependency of the inhibitory effect of antidiuretic hormone on in vitro renin secretion in rats. J Physiol (London) 315:21–30
13. Churchill PC (1985) Second messengers in renin secretion. Am J Physiol 249:F175–F184
14. Churchill PC (1987) Calcium channel antagonists and renin secretion. Am J Nephrol 7 (Suppl 1):32–38
15. Churchill PC, Bidani AK (1988) Adenosine and renal function. In: Williams M (ed) The adenosine receptor. New York, Humana Press, in press
16. Churchill PC, Churchill MC (1980) Biphasic effect of extracellular [K] on isoproterenol-stimulated renin secretion from rat kidney slices. J Pharmacol Exp Ther 214:541–545
17. Churchill PC, Churcill MC (1982) Ca-dependence of the inhibitory effect of K-depolarization on renin secretion from rat kidney slices. Arch Int Pharmacodyn Ther 258:300–312
18. Churchill PC, Churchill MC (1984) 12-O-Tetradecanoylphorbol 13-acetate inhibits renin secretion of rat renal cortical slices. J Hypertension 2 (Suppl 1):25–28
19. Churchill PC, Churchill MC (1985) A_1 and A_2 adenosine receptor activation inhibits and stimulates renin secretion of rat renal cortical slices. J Pharmacol Exp Ther 232:589–594
20. Churchill PC, McDonald FD, Churchill MC (1981) Effect of diltiazem, a calcium antagonist, on renin secretion from rat kidney slices. Life Sci 29:383–389
21. Corea L, Miele N, Bentivoglio M, Boschetti E, Agabiti-Rosei E, Muiesan G (1979) Acute and chronic effects of nifedipine on plasma renin activity and plasma adrenaline and noradrenaline in controls and hypertensive patients. Clin Sci 57:115s–117s
22. Cruz-Soto MA, Benabe JE, Lopez-Novoa JM, Martinez-Maldonado M (1982) Renal Na^+–K^+-ATPase in renin release. Am J Physiol 243:F598–F603
23. Cruz-Soto M, Benabe JE, Lopez-Novoa JM, Martinez-Maldonado M (1984) Na^+–K^+-ATPase inhibitors and renin release: relationship to calcium. Am J Physiol 247:F650–F655
24. Davis JO, Freeman RH (1976) Mechanisms regulating renin release. Physiol Rev 56:1–56
25. Dietz JR, Davis JO, Freeman RH, Villarreal D, Echtenkamp SF (1983) Effects of intrarenal infusion of calcium entry blockers in anesthetized dogs. Hypertension 5:482–488
26. Fishman MC (1976) Membrane potential of juxtaglomerular cells. Nature 260:542–544
27. Fleckenstein A (1981) Fundamental actions of calcium antagonists on myocardial and cardiac pacemaker cell membranes. In: Weiss GB (ed) New perspectives on calcium antagonists. Bethesda, American Physiological Society, p 59
28. Fray JCS, Lush DJ, Park CS (1986) Interrelationship of blood flow, juxtaglomerular cells, and hypertension: role of physical equilibrium and Ca. Am J Physiol 251:F643–F662
29. Fray JCS, Park CS, Valentine AND (1987) Calcium and the control of renin secretion. Endocrine Rev 8:53–93
30. Godfraind T (1981) Mechanisms of action of calcium entry blockers. Federation Proc 40:2866–2871
31. Hackenthal E, Schwertschlag U, Taugner R (1983) Cellular mechanisms of renin release. Clin Exp Hyper Theory & Practice A5(7&8): 975–993
32. Harder DR, Gilbert R, Lombard JH (1987) Vascular smooth muscle cell depolarization and activation in renal arteries on elevation of transmural pressure. Am J Physiol 253:F778–F781
33. Hiramatsu K, Yamagishi F, Kubota T, Yamada T (1982) Acute effects of the calcium antagonist, nifedipine, on blood pressure, pulse rate, and the renin-angiotensin-aldosterone system in patients with essential hypertension. Am Heart J 104:1346–1350
34. Imagawa J, Kurosawa H, Satoh S (1986) Effects of nifedipine on renin release and renal function in anesthetized dogs. J Cardiovasc Pharmacol 8:636–640
35. Johns EJ (1985) The influence of diltiazem and nifedipine on renal function in the rat. Br J Pharmacol 84:707–713
36. Keeton TK, Campbell WB (1980) The pharmacologic alteration of renin release. Pharmacol Rev 32:81–227
37. Klockner U, Isenberg G (1987) Calmodulin antagonists depress calcium and potassium currents in ventricular and vascular myocytes. Am J Physiol 253:H1601–H1611
38. Kurtz A, Bruna RD, Pfeilschifter J, Taugner R, Bauer C (1986) Atrial natriuretic peptide inhibits renin release from juxtaglomerular cells by a cGMP-mediated process. Proc Natl Acad Sci USA 83:4769–4773

39. Kurtz A, Pfeilschifter J, Bauer C (1984) Is renin secretion governed by the calcium permeability of the juxtaglomerular cell membrane? Biochem Biophys Res Commun 124:359–366
40. Kurtz A, Pfeilschifter J, Hutter A, Buhrle C, Nobiling R, Taugner R, Hackenthal E, Bauer C (1986) Role of protein kinase C in inhibition of renin release caused by vasoconstrictors. Am J Physiol 250:C563–C571
41. Lederballe Peterson O (1983) Calcium blockade in arterial hypertension. Hypertension 5:1174–1179
42. Logan AG, Tenyi I, Quesada T, Peart WS, Breathnach AS, Martin BGH (1975) Blockade on renin release by lanthanum. Clin Sci Mol Med 48:31s–32s
43. Lopez-Novoa JM, Garcia JC, Cruz-Soto MA, Benabe JE, Martinez-Maldonado M (1982) Effect of sodium orthovanadate on renal renin secretion in vivo. J Pharmacol Exp Ther 222:447–451
44. Loutzenhiser R, Epstein M, Horton C, Hamburger R (1985) Nitrendipine-induced stimulation of renin release by the isolated perfused rat kidney. Proc Soc Exp Biol Med 180:133–136
45. Luft FC, Aronoff GR, Sloan RS, Fineberg NS, Weinberger MH (1985) Calcium channel blockade with nitrendipine. Effects on sodium homeostasis, the renin-angiotensin system, and the sympathetic nervous system in humans. Hypertension 7:438–442
46. Lyons HJ (1980) Studies on the mechanism of renin release from rat kidney slices: calcium, sodium and metabolic inhibition. J Physiol (London) 304:99–108
47. Macias-Nunez JF, Garcia-Iglesias C, Santos JC, Sanz E, Lopez-Novoa JM (1985) Influence of plasma renin content, intrarenal angiotensin II, captopril, and calcium channel blockers on the vasoconstriction and renin release promoted by adenosine in the kidney. J Lab Clin Med 1065:562–567
48. Marre M, Misumi J, Raemsch K-D, Corvol P, Menard J (1982) Diuretic and natriuretic effects of nifedipine on isolated perfused rat kidneys. J Pharmacol Exp Ther 223:263–270
49. Murray RD, Churchill PC (1985) The concentration-dependency of the renal vascular and renin secretory responses to adenosine receptor agonists. J Pharmacol Exp Ther 232:189–193
50. Narita H, Nagao T, Yabana H, Yamaguchi S (1983) Hypotensive and diuretic actions of diltiazem in spontaneously hypertensive and Wistar Kyoto Rats. J Pharmacol Exp Ther 227:472–477
51. Nishizuka Y (1984) Turnover of inositol phospholipids and signal transduction. Science 225:1365–1369
52. O'Grady SM, Palfrey HC, Field M (1987) Characteristics and functions of Na–K–Cl cotransport in epithelial tissues. Am J Physiol 253:C177–C192
53. Osswald H (1984) The role of adenosine in the regulation of glomerular filtration rate and renin secretion. Trends Pharmacol Sci 5:94–97
54. Peart WS (1978) Intra-renal factors in renin release. Contrib Nephrol 12:5–15
55. Phillis JW, Wu PH (1981) Catecholamines and the sodium pump in excitable cells. Prog Neurobiol 17:141–184
56. Putney JW Jr, Askari A (1978) Modification of membrane function by drugs. In: Andreoli TE, Hoffman JF, Fanestil DD (eds) Physiology of membrane disorders. New York, Plenum, p 417
57. Rasmussen H, Barrett PQ (1984) Calcium messenger system: an integrated view. Physiol Rev 64:938–984
58. Rosenberger L, Triggle DJ (1978) Calcium, calcium translocation, and specific calcium antagonists. In: Weiss GB (ed) Calcium in drug action. New York, Plenum, p 3
59. Roy MW, Guthrie GP Jr, Holladay FP, Kotchen TA (1983) Effects of verapamil on renin and aldosterone in the dog and rat. Am J Physiol 245:E410–E416
60. Rubin RP (1982) Calcium and cellular secretion. New York, Plenum
61. Schwartz A, Lindenmayer GE, Allen JC (1975) The sodium-potassium adenosine triphosphatase: pharmacological, physiological and biochemical aspects. Pharmacol Rev 27:3–134
62. Shade RE, Davis JO, Johnson JA, Witty RT (1972) Effects of arterial infusion of sodium and potassium on renin secretion in the dog. Circulation Res 31:719–727
63. Skou JC (1975) The $(Na^+ + K^+)$ activated enzyme system and its relationship to transport of sodium and potassium. Q Rev Biophys 4:401–434
64. Spielman WS, Thompson CI (1982) A proposed role for adenosine in the regulation of renal hemodynamics and renin release. Am J Physiol 242:F423–F435
65. Sundet WD, Wang BC, Hakumako MOK, Goetz KL (1984) Cardiovascular and renin responses to vanadate in the conscious dog: attenuation after calcium channel blockade. Proc Soc Exp Biol Med 175:185–190

66. Tagawa H, Vander AJ (1970) Effects of adenosine compounds on renal function and renin secretion in dogs. Circulation Res 26:327–338
67. VanBreemen C, Aaronson P, Loutzenhiser R (1979) Sodium-calcium interactions in mammalian smooth muscle. Pharmacol Rev 30:167–208
68. Vander AJ (1967) Control of renin release. Physiol Rev 47:359–382
69. Vander AJ (1970) Direct effects of potassium on renin secretion and renal function. Am J Physiol 219:445–449
70. Waeber B, Nussberger J, Brunner HR (1985) Does renin determine the blood pressure response to calcium entry blockers? Hypertension 7:223–227

Action of Parathyroid Hormone on Calcium Transport in Rat Brain Synaptosomes is Independent of cAMP

C. L. Fraser[1] and P. Sarnacki[2]

[1] Department of Medicine, Divisions of Nephrology and Geriatrics,
  Veterans Administration Medical Center, 4150 Clement Street, San Francisco, CA 94121, USA
[2] University of California at San Francisco, San Francisco, California, USA

Introduction

We previously investigated the effects of uremia on calcium transport in synaptosomes from rat brains and found that both the Na–Ca exchanger and the ATP-dependent calcium pump were increased by uremia [1]. We also observed that parathyroidectomy of these rats before they were made acutely uremic resulted in correction of these transport defects [2]. Additionally, the administration of parathyroid hormone (PTH) to parathyroidectomized rats resulted in a return of the calcium transport abnormalities. In nonuremic rats, the absence or presence of PTH also affected calcium transport in a manner similar to that which we observed in uremia, in that parathyroidectomy and PTH administration decreased and increased calcium transport respectively [3]. Based on these observations we decided to investigate the mechanism of action of PTH on calcium transport in synaptosomes in order to determine whether its action was cAMP-mediated, as has been described in tissues such as bone [4] and kidney [5, 6, 7].

Methods

Isolation of Synaptosomes

Synaptosomes were isolated from 200-g male Sprague-Dawley rats as previously described [8]. In brief, the rats were decapitated and their cerebral cortex removed and immediately placed in 10 ml ice-cold isolation medium containing 320 mM sucrose, 0.2 mM K-EDTA, 5 mM Tris-HCl, pH 7.4, at 0–4°C. The brains were chopped finely with scissors, washed with the isolation medium, and homogenized in a glass Dounce homogenizer (clearance 1 mm). The crude synaptosomal-mitochondrial pellet was obtained after several centrifugations at 1300 g for 3 min and a final spin at 18000 g for 10 min. The purified synaptosomal preparation was then obtained by differential centrifugation on a discontinuous Ficoll gradient for 60 min [8]. The synaptosomal protein is diluted in isolation medium to a final concentration of 8–10 mg/ml and used for transport studies.

Calcium Transport Assay

Uptake Assay

Calcium uptake studies were carried out by the Na–Ca exchange mechanism as previously described [1]. A 0.5-ml aliquot of synaptosomal protein (approximately 5 mg protein) was brought up to 2 ml in a preequilibration medium (140 mM NaCl, 1 mM MgSO$_4$, 10 mM glucose, 5 mM Hepes-Tris, pH 7.4) and allowed to incubate for 10 min at 37°C. The suspension was then spun at 20000 g for 5 min and the resulting pellet was resuspended in the same medium and spun at the same speed for 5 min. At the end of the second spin, the final pellet was resuspended in 400 μl of the preequilibration medium and kept on ice (0–4°C) until transport studies were commenced.

Uptake was initiated by adding 5 μl synaptosomal suspension (approximately 50 μg protein) to 95 μl of an external medium consisting of 140 mM KCl, 1 mM MgSO$_4$, 10 mM glucose, 10 μM CaCl$_2$, 0.1 μCi ^{45}Ca^{2+} (40000 cpm/20 μl), 5 mM Hepes-Tris, pH 7.4, at 25°C. Depending on the compound being studied, either 100 ng/ml bovine parathyroid extract, 10^{-7} M 1–34 bovine PTH (bPTH), 10^{-7} M 1–84 bPTH, 10^{-5} M 8-bromo-cAMP, 10^{-5} M dibut-cAMP, or 10^{-5} M forskolin was added to the external medium before transport study was begun. After the desired period of incubation, uptake was terminated by adding to the uptake medium 2 ml ice-cold 150 mM KCl solution (stop solution). The mixture was immediately vacuum-filtered through a 0.45-μm pore size cellulose acetate filter and washed twice with 2 ml of the cold stop solution as previously described [1]. The filters were dissolved in phase combine scintillant (PCS) and counted by a Packard counter (model 2000 CA). Measurements at each time point were made in triplicate and the value obtained at time zero, which we attributed to binding of calcium to protein and filter, was subtracted from each of the points of observation [1, 8]. As described previously, nonspecific binding to the filter and synaptosomal protein was less than 0.05% [1].

Efflux Assay

For calcium efflux studies the vesicles are loaded by the Na–Ca exchanger for 10 min as discussed above. At the end of this period, the loaded vesicles are spun at 20000 g for 10 min and resuspended to a protein concentration of 10 mg/ml in the external medium [140 mM KCl, 1 mM MgSO$_4$, 10 mM glucose, 10 μM Cacl$_2$, 0.1 μCi^{45}Ca^{2+} (40000 cpm/20 μl), 5 mM Hepes-Tris, pH 7.4] and kept on ice until efflux studies were commenced. Efflux was promptly started by diluting 5 μl of the calcium-loaded vesicles in 95 μl of a calcium-free efflux medium consisting of 140 mM NaCl, 1 mM MgSO$_4$, 10 mM glucose, 5 mM Hepes-Tris, pH 7.4, at 25°C with appropriate additions of either 10^{-7} M 1–34 bPTH, 10^{-7} M 1–84 bPTH, 10^{-6} M 8–bromo-cAMP, 10^{-6} M dibut-cAMP or 10^{-5} M forskolin. At the appropriate period of efflux the reaction was terminated by adding 2 ml ice-cold stop solution to 100 μl of the transport mixture [9]. The mixture was then filtered as above, washed twice with the stop solution, and counted. The initial time point was obtained at the instant prior to starting the efflux studies, and this was taken as the maximum value of calcium where comparison was made over time.

Purification of Canine Renal Cortical Plasma Membranes

Highly purified canine renal cortical plasma membranes (CRCPM) were isolated from freshly removed canine kidney at 0–4°C. Cortical tissue was dissected free from medullary tissues, cut into small pieces with scissors, and washed in 0.9% NaCl, 1 mM EDTA, 5 mM Tris-HCl, pH 7.5. The medium was decanted and the tissue was disrupted in 3 vol (relative to wet weight) ice-cold 0.25 M sucrose, 1 mM EDTA, 5 mM Tris-HCl, pH 7.5 (SET buffer), using two or three short bursts (< 5 s) of a Polytron homogenizer (Brinkman Instruments, Westbury, NY). Tissue was further homogenized using 10 strokes of a motor-driven loose Teflon pestle. After centrifugation at 1475 g for 10 min, the pellets were washed once with SET buffer then resuspended in 1 vol (relative to initial wet weight) 2 M sucrose, 1 mM EDTA, 5 mM Tris-HCl, pH 7.5. The suspension was spun at 13300 g for 10 min and the supernatant was decanted and diluted eight fold with 1 mM EDTA, 5 mM Tris-HCl, pH 7.5. This suspension was centrifuged at 20000 g for 15 min and the white upper layer of the pellet was gently removed, resuspended in SET buffer, and respun. This process of centrifugation and resuspension was repeated three times to yield a homogeneous white fluffy pellet [10].

Guanyl Nucleotide-Amplified Adenylate Cyclase Assay

Adenylate cyclase activity was measured by the conversion of α-[^{32}P]ATP to [^{32}P]cAMP. Standard incubations were carried out in a final volume of 0.1 ml containing 25 mM Tris-HCl, pH 7.5, 2 mM MgCl$_2$, 0.1% BSA, 0.1 mM α-[^{32}P]ATP (100–300 cpm/pmol), an ATP-regenerating system consisting of 30 µg creatine phosphokinase and 10 mM creatine phosphate, and 1 mM 1-methyl,3-isobutylxanthine, a potent inhibitor of cyclic nucleotide phosphodiesterase [11]. The hydrolysis-resistant GTP analog, 100 µM Gpp(NH)p, was also added to augment the sensitivity of adenylate cyclase to PTH [10]. Appropriate concentrations of either 1–34 bPTH, 1–84 bPTH, forskolin, NaF or isoproterenol were also added, depending on experimental protocol. Incubation tubes were made of borosilicate glass and were pretreated with dichlorodimethylsilane (Aldrich Chemical Co., Milwaukee, Wis.) to reduce adsorption of PTH. Incubations were started by adding either CRCPM or synaptosomes in SET buffer or buffer alone for assay blank. After 30 min incubation at 30°C, enzyme activity was determined by the addition of 0.1 ml of a solution containing 10 mM unlabeled ATP, 2% sodium dodecyl sulfate (vol/vol), 50 mM Tris-HCl at pH 7.5, 1 mM [^{3}H]cAMP (30000 cpm). Assay tubes were immediately placed in a boiling water bath for 3 min and [^{32}P]cAMP was isolated by a chromatographic procedure [10, 12]. For studies with hypotonically lysed synaptosomal vesicles, lysis was achieved with 20 mM Tris-HCl as previously described [9].

Materials

Highly purified bovine PTH tetratriacontapeptide (1–34 bPTH, 1–84 bPTH, 3–34 bPTH), α-[^{32}P]ATP, [^{3}H]cAMP, and Gpp(NH)p were all generous gifts from Dr.

Claude Arnoud (Endocrine Division, VA Medical Center, San Francisco, Calif). ATP, cAMP, forskolin, 1-methyl,3-isobutylxanthine, creatine phosphate, creatine phosphokinase, and NaF were obtained from Sigma Chemical Co. (St. Louis, Mo.). $^{45}Ca^{2+}$ was obtained from New England Nuclear (Boston, Mass.). All other chemicals were of reagent grade and were obtained from Sigma Chemical Co. (St. Louis, Mo.).

Results

Calcium uptake by the Na–Ca exchanger was first performed as previously described and observations were made between 3 and 60 s [1]. With either 100 ng/ml parathyroid extract or 10^{-7} M 1–34 bPTH, uptake was significantly increased ($p < 0.001$) from 0.31 ± 0.01 to 0.44 ± 0.01 and 0.48 ± 0.01 nmol/mg protein respectively at 5 s (Fig. 1). This stimulation by PTH resulted in an increase in uptake of 42% with the extract and 55% with 1–34 bPTH. In the presence of PTH, uptake was significantly increased only up to 15 s; thereafter no difference was observed between PTH and the control group (Fig. 1). This suggest that the ability of PTH to stimulate calcium uptake in synaptosomes is transient and may be related to a finite amount of available substrate (ATP) within the vesicles. In contrast to 1–34 bPTH and the extract, 3–34 bPTH did not increase calcium uptake above control values during the period of observation. To determine whether calcium uptake was also stimulated by cAMP, we next performed uptake studies in the presence of 10^{-6} M of either 8–bromo-cAMP or dibut-cAMP for

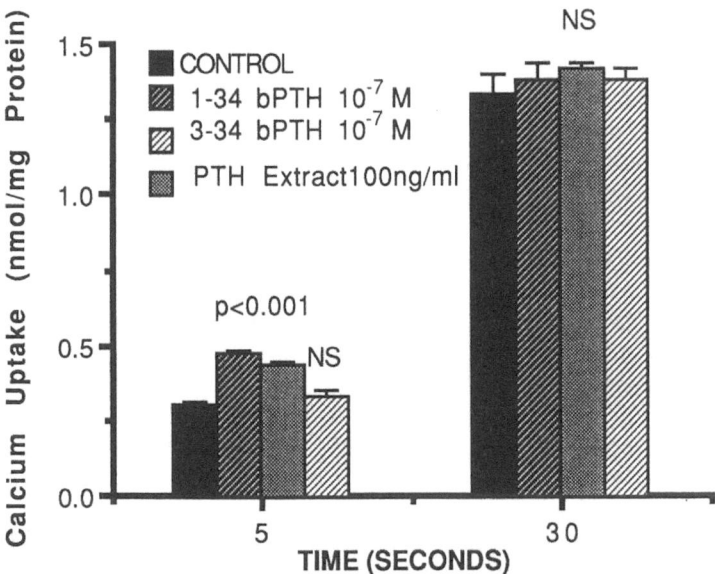

Fig. 1. The effects of three different PTH molecules on calcium uptake in synaptosomes. The presence of either 100 ng/ml parathyroid extract or 10^{-7} M 1–34 bPTH significantly increased calcium uptake above control at 5 s ($p < 0.001$), but 10^{-7} M 3–34 bPTH had no stimulatory effects on uptake. At 30 s uptake with all three groups was not different from control. Data are expressed as mean ± SE of eight experiments

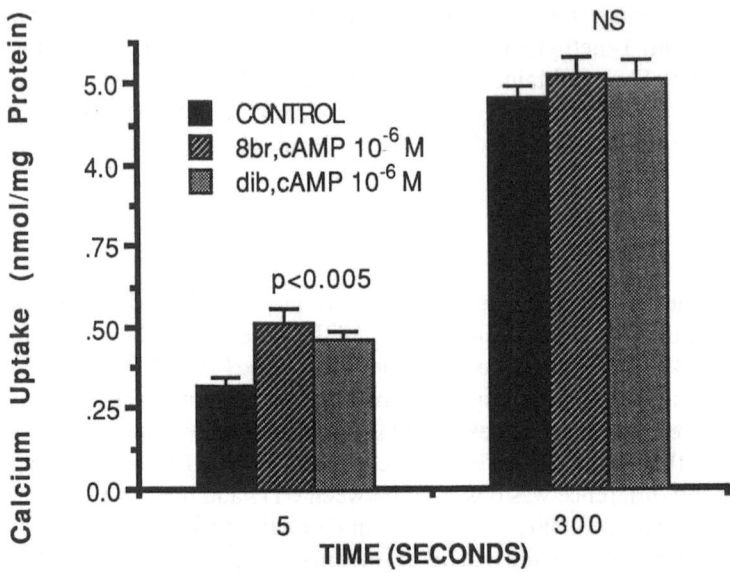

Fig. 2. The effects of 8-bromo-cAMP and dibut-cAMP on calcium uptake. In the presence of cAMP, uptake is significantly greater than in control vesicles at 5 s. No difference in uptake was observed between the groups at 5 min, by which time maximum uptake was achieved. Data are expressed as mean ± SE of 10 experiments

up to 5 min (Fig. 2). Between 3 and 30 s, both cAMP molecules significantly ($p < 0.005$) increased calcium uptake in a manner similar to PTH. At 5 s 8-bromo-cAMP significantly increased uptake by 63% from 0.32 ± 0.02 to 0.52 ± 0.03 nmol/mg protein, and dibut-cAMP increased uptake by 44% above control. However, by 5 min, calcium uptake was no longer stimulated by cAMP and the values in all three groups were not significantly different from each other (Fig. 2). Based on these data, it appears that PTH stimulates calcium transport in synaptosomes by a cAMP-mediated action. To clarify whether this was the case we next investigated the role of forskolin on calcium transport, in order to determine whether direct stimulation of the adenylate cyclase catalytic subunit by forskolin would increase synaptosomal calcium transport, as was observed with PTH [13].

Calcium uptake was assessed over time (3 s to 10 min) in the presence of 10^{-5} M forskolin. As shown in Fig. 3, forskolin significantly increased uptake ($p < 0.005$) above control by approximately 40%. Since forskolin activates the cyclase system by binding to the catalytic subunit, it becomes apparent that the native intrasynaptosomal ATP is sufficient to produce a sustained uptake by PTH.

We next evaluated calcium efflux in the presence of 10^{-7} M 1–34 bPTH and found that efflux was significantly increased ($p < 0.005$) by 29%–50% between 3 and 10 s (Fig. 4). Similarly, 10^{-7} M 1–84 bPTH significantly increased calcium efflux by 32%–73% at the same time interval. With 10^{-5} M forskolin, efflux was increased above control by 42%–120%, again suggesting the possible dependene of PTH stimulation of calcium transport on cAMP. To determine whether this is so, we then

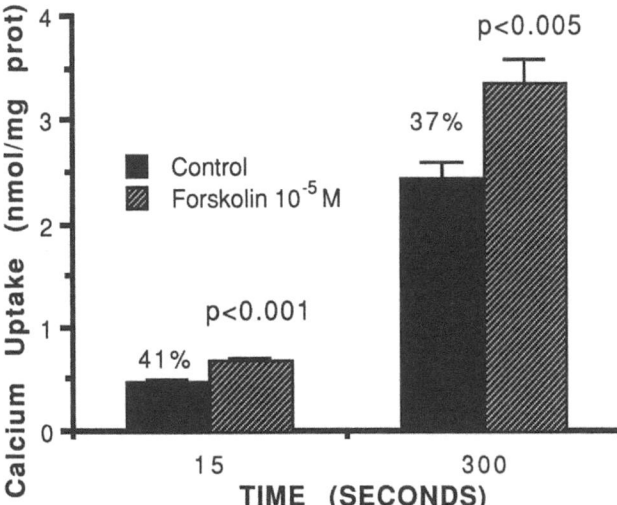

Fig. 3. Forskolin (10^{-5} M), activator of the catalytic subunit of adenylate cyclase, increased calcium uptake by 41% and 37% at 15 s and 5 min respectively. Data are expressed as mean ± SE of nine experiments

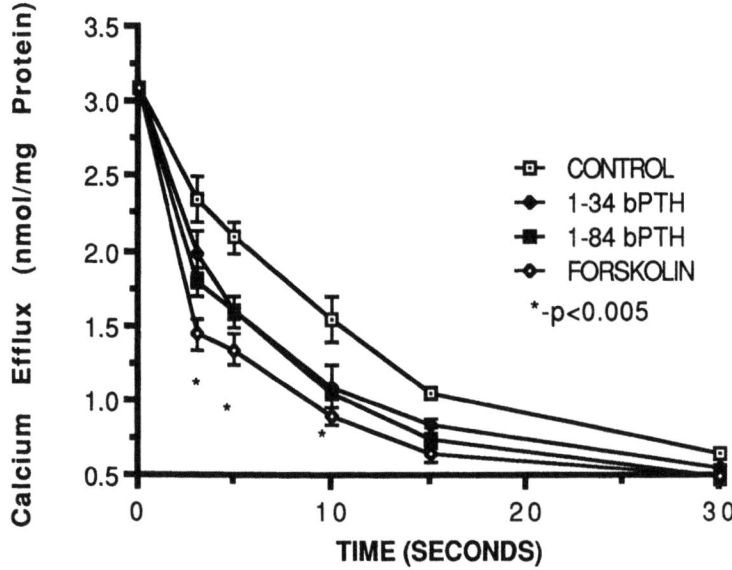

Fig. 4. Calcium efflux from synaptosomes in the presence of 10^{-7} M of either 1–34 bPTH or 1–84 bPTH. In the presence of PTH calcium efflux is significantly greater ($p < 0.005$) than in control vesicles between 3 and 10 s. Data are expressed as mean ± SE of eight experiments

measured cAMP production rates in synaptosomes in the presence of either PTH or the membrane adenylate cyclase activators forskolin and NaF.

cAMP production was evaluated during the conversion of α-[^{32}P]ATP to α-[^{32}P]cAMP in the presence of both an ATP-regenerating system and an inhibitor of cyclic nucleotide phosphodiesterase. The purpose of these additions was to provide

Table 1. Adenylate cyclase activity measured as a function of increasing concentration of cyclase activators

Agent	Concentration	Adenylate cyclase activity (pmol cAMP mg protein^{-1} 30 min^{-1})	
		CRCPM	Synaptosomes
No. of experiments		5	5
1–34 bPTH (ng/ml)	0	278 ± 11	911 ± 51
	30	763 ± 25 (2.7)	–
	300	1230 ± 36 (4.4)	953 ± 57
	100 000	1411 ± 40 (5.1)	922 ± 44
1–84 bPTH (ng/ml)	0	250 ± 15	1300 ± 98
	10	519 ± 22 (2.1)	1357 ± 34
	500	1036 ± 45 (4.1)	1290 ± 75
Forskolin (mM)	0	239 ± 34	911 ± 51
	0.1	3330 ± 145 (14)	7466 ± 420 (8.2)
NaF (mM)	0	239 ± 34	911 ± 51
	100	6669 ± 310 (28)	3562 ± 248 (4)

Adenylate cyclase activity was measured in the presence of 1 mM IBMX, 10 mM creatine phosphate, and 30 µg creatine phosphokinase. Numbers in parentheses indicate the factor by which cyclase activity increased above control. Data are expressed as mean ± SE, and measurements at each time point were made in triplicate

enough substrate to maintain an adequate concentration of ATP during the production of cAMP, and also to prevent the breakdown of cAMP once it has been formed. Studies were performed simultaneously in CRCPM and in synaptosomes. The CRCPM were used as control vesicles. Table 1 shows that the addition of increasing concentrations of 1–34 bPTH resulted in a respective increase in adenylate cyclase activity in CRCPM. However, no increase in cyclase activity was observed in synaptosomes with the same concentrations of 1–34 bPTH. Similarly, with either 10 or 500 ng/ml of 1–84 bPTH, cyclase activity increased from 250 ± 15 to 519 ± 22 and to 1036 ± 45 pmol cAMP/mg protein respectively in CRCPM but was not affected in synaptosomes (1300 ± 98 to 1357 ± 34 and 1290 ± 75 pmol cAMP/mg protein respectively). To determine whether either the catalytic or the guanine nucleotide regulatory subunits were operative in synaptosomes, we first carried out experiments in the presence of forskolin and found that cyclase activity was increased in CRCPM from 239 ± 34 to 3330 ± 145 pmol cAMP/mg protein and in synaptosomes from 911 ± 51 to 7466 ± 420 pmol cAMP/mg protein. Similarly, NaF also significantly increased cyclase activity manifold in both groups. These results suggest that both the catalytic and the regulatory subunits of the adenylate cyclase system are intact and functional in synaptosomes, and that the mechanism of action of PTH on calcium transport in synaptosomes is most likely independent of cAMP.

To investigate whether the PTH effect on calcium transport was indeed independent of cAMP, it was necessary for us to first determine whether an intact adenylate cyclase system exists in synaptosomes. To achieve this, we evaluated cyclase activity in the presence of isoproterenol, a β-adrenergic agonist which is a known adenylate cyclase activator [13]. Table 2 shows that 10 mM isoproterenol resulted in a 90%

Table 2. Ability of 1–34 bPTH and isoproterenol to activate adenylate cyclase activity in synaptosomes and CRCPM

Agent	Concentration	Adenylate cyclase activity (pmol cAMP mg protein^{-1} 30 min^{-1})	
		CRCPM	Synaptosomes
1–34 bPTH (ng/ml)	0	242 ± 8	1080 ± 41
	30	3357 ± 41 (14)	1198 ± 37
	1000	4363 ± 57 (18)	1148 ± 16
Isoproterenol (µM)	0	253 ± 14	1116 ± 21
	100	390 ± 21 (1.5)	1307 ± 7 (1.2)
	10000	367 ± 7 (1.5)	2099 ± 114 (1.9)

Activity was measured in the presence of GTP, 1 mM IBMX, 10 mM creatine phosphate, and 30 µg creatine phosphokinase. Numbers in parentheses indicate the factor by which cyclase activity increased due to the agents. Data are expressed as mean ± SE

increase in cyclase activity in synaptosomes and a 50% increase in CRCPM. These observations suggest that there is an intact adenylate cyclase system in synaptosomes that is at least as active as that in CRCPM. Therefore, the inability of PTH to stimulate cAMP production in synaptosomes appears to be due not to the absence of an intact adenylate cyclase system, but to the inability of PTH to activate this system in synaptosomes.

Discussion

The studies described above show that PTH increases both the uptake and the efflux of calcium from rat brain synaptosomes in a manner similar to either cAMP or forskolin. Based on these findings, it was initially felt that the mechanism of action of PTH on calcium transport in synaptosomes was cAMP-dependent. It was also observed that parathyroid extract and 1–34 bPTH increased synaptosomal calcium transport, while 3–34 bPTH had no such effect (Fig. 1). This suggested that the amino acids in positions 1 and 2 of the PTH molecule are necessary for PTH to stimulate calcium transport in synaptosomes. Thus, as a result of these observations we used 1–34 bPTH and 1–84 bPTH in all subsequent studies.

As mentioned above, our initial observations led us to believe that the mechanism of action of PTH on calcium transport was cAMP-dependent. However, the inability of PTH to generate cAMP in synaptosomes suggested to us that the relationship between PTH and synaptosomal calcium transport was more complicated than we initially thought. This complexity became more intriguing when PTH increased cyclase activity in CRCPM five-fold in simultaneously run experiments (Table 1). Our aim, then, was to determine whether there was an intact adenylate cyclase system in synaptosomes, and if so whether its function could be demonstrated by other known activators of the adenylate cyclase system. Our first task was to measure cAMP production in synaptosomes and CRCPM in the presence of either forskolin or NaF, and then to compare the results in these two membrane preparations. In both preparations, forskolin as well as NaF increased cyclase activity manifold (Table 1), suggesting that synaptosomes contain both a catalytic and the guanine nucleotide regulatory subunits similar to those in CRCPM. Additionally, isoproterenol, a β-adrenergic agonist which exerts many of its functions with cAMP as its messenger, increased cyclase activity by 50% in CRCPM and 90% in synaptosomes. This suggested that, similarly to CRCPM, synaptosomes contain an intact adenylate cyclase system which can be stimulated by cyclase activators other than PTH.

As with our observations in synaptosomes, studies in other tissues have shown that PTH can produce cellular changes which are independent of the adenylate cyclase system. In a very recent publication, Atkinson et al. showed that PTH stimulation of mitosis in rat thymic lymphocytes was also independent of cAMP [14]. This study showed that although PTH significantly increased calcium uptake in thymocytes following both $^{45}Ca^{2+}$ and Quin 2 fluorescence, it was unable to activate thymocyte membrane adenylate cyclase, which was responsive to other activators. In another study it was shown that PTH increased intracellular calcium in proximal tubular cells of the mammalian kidney independently of the cyclic nucleotides [15]. Also in the opossum kidney cell line, which exhibits proximal tubular characteristics, it was shown that PTH increased cytoplasmic calcium and produced a dose-dependent stimulation of inositol triphosphate and diacylglycerol which were not mimicked by cAMP [16]. Similar observations were also made in the isolated papillary muscle from rat heart, where PTH action on papillary muscle contraction was found to be independent of cAMP [17]. The ionotropic action of PTH on papillary muscle was shown to be mediated by the release of endogenous norepinephrine from the myocardium, and both the release of norepinephrine and muscle contraction to be independent of adenylate cyclase activation [17]. Thus, it is evident that the mechanism of action of PTH on cellular functions is quite variable and does not necessarily involve the adenylate cyclase system.

References

1. Fraser CL, Sarnacki P, Arieff AI (1985) Calcium transport abnormality in uremic rat brain synaptosomes. J Clin Invest 76:1789–1795
2. Fraser CL, Arieff AI (1986) Abnormalities of transport in synaptosomes from uremic rat brain: role of parathyroid hormone. J Gen Physiol 88:24 (Abstract)
3. Fraser CL, Sarnacki P (1988) Parathyroid hormone mediates changes in calcium transport in uremic rat brain synaptosomes. Am J Physiol 23: (June)
4. Chase LR, Fedack SA, Aurback GD (1969) Activation of skeletal adenyl cyclase by parathyroid hormone in vitro. Endocrinology 84:761–768
5. Teitelbaum AP, Nissenson RA, Arnaud CD (1982) Coupling of the canine renal parathyroid hormone receptor to adenylate cyclase: modulation by guanyl nucleotides and N-ethylmaleimide. Endocrinology 111:1524–1533
6. Nissenson RA, Arnaud CD (1979) Properties of the parathyroid hormone receptor-adenylate cyclase system in chicken renal plasma membranes. J Biol Chem 254:1469–1475
7. Segre GV, Rosenblatt M, Reiner BL, Mahaffey JE, Potts JT (1979) Characterization of parathyroid hormone receptors in canine renal cortical plasma membranes using a radioiodinated sulfur-free hormone analogue. J Biol Chem 254:6980–6986
8. Fraser CL, Sarnacki P, Arieff AI (1985) Abnormal sodium transport in synaptosomes from brains of uremic rats. J Clin Invest 75:2014–2023
9. Gill DL, Grollman EF, Kohn LD (1981) Calcium transport mechanisms in membrane vesicles from guinea pig brain synaptosomes. J Biol Chem 256:184–192

10. Nissenson RA, Abbot SR, Teitelbaum AP, Clark OH, Arnaud CD (1981) Endogenous biologically active human parathyroid hormone: measurement by a guanyl nucleotide-amplified renal adenylate cyclase assay. J Clin Endocrinol Metab 52:840–846
11. Beavo JA, Crofford OB, Hardman JG, Sutherland EW, Newman EV (1970) Effects of xanthine derivatives on lipolysis and on adenosine 3',5'-monophosphate phosphodiesterase activity. Mol Pharmacol 6:597–603
12. Salomon Y, Londos C, Rodbell M (1974) A highly sensitive adenylate cyclase assay. Anal Biochem 58:541–548
13. Seamon KB, Daly JW (1983) Forskolin, cyclic AMP and cellular physiology. TIPS 120–123
14. Atkinson MJ, Hesch RD, Cade C, Wadwah M, Perris AD (1987) Parathyroid hormone stimulation of mitosis in rat thymic lymphocytes is independent of cAMP. J Bone and Mineral Research 2 (4):303–309
15. Hruska KA, Gligorsky M, Scoble J, Tsutsumi M, Westbrook S, Moskowitz D (1986) Effects of parathyroid hormone on cytosolic calcium in renal proximal tubular cultures. Am J Physiol 20:F188–F198
16. Hruska KA, Moskowitz D, Esbrit P, Civitelli R, Westbrook S, Huskey M (1987) Stimulation of inositol triphosphate and diacylglycerol production in renal tubular cells by parathyroid hormone. J Clin Invest 79:230–239
17. Katoh Y, Klein KL, Kaplan RA, Sanborn WG, Kurokawa K (1981) Parathyroid hormone has a positive inotropic action in the rat. Endocrinology 109:2252–2254

Calcium and the Mediation of Tubuloglomerular Feedbacks Signals*

P. D. Bell, M. Franco[1], and M. Higdon

Nephrology Research and Training Center, Departments of Physiology and Biophysics and of Medicine, University of Alabama at Birmingham, Birmingham, AL 35294, USA
[1] Visiting scientist from the Instituto de Cardiologia I. Chavez, Mexico City, Mexico

Introduction

The process of urine formation by the mammalian kidney is under the general control of two metabolically distinct processes. First, there is the filtration of fluid across the glomerular capillaries and into Bowman's capsule, the initial portion of the tubular network. The magnitude of filtrate formation is controlled by the hemodynamic forces that act across the glomerular capillary wall. The second process is the modification of this fluid as it flows along the tubular network. Tubular fluid is modified through a complex series of reabsorptive and secretory mechanisms that extensively alter both the volume and composition of the fluid. Although these two events appear to be completely separate, an intrarenal mechanism exists that couples the process of filtration with the tubular reabsorptive events. This mechanism is called tubuloglomerular feedback.

Morphological Bases for the Feedback Mechanism

The tubuloglomerular feedback mechanism is based on the anatomical connection that exists between the distal tubule and the glomerulus of each individual nephron unit [3]. As shown in Fig. 1, the thick ascending limb of the loop of Henle emerges from the medullary region of the kidney and forms an intimate association with its own glomerulus and vascular pole. At the site of contact between the glomerulus and the thick ascending limb, the tubular cells that face the glomerulus are specialized. These cells are referred to as macula densa cells. The cells on the opposite side of the tubule and those that are located on either side of the macula densa resemble typical thick ascending limb cells [31]. Figure 2 is an electron micrograph of the macula densa cells and associated structures of the juxtaglomerular apparatus. As can be seen, the cells of the macula densa rest on a thick and irregular basement membrane that abuts against layers of diffuse extraglomerular mesangial cells. These extraglomerular mesangial cells are in continuity with the intraglomerular mesangial cells and also with the smooth muscle cells of the afferent and efferent arterioles. From the figure, it is readily apparent that the cells of the macula densa differ markedly from the surround-

---

* This work was supported by grants from the NIH (AM 32032) and from the American Heart Association, of which P. D. Bell is an Established Investigator.

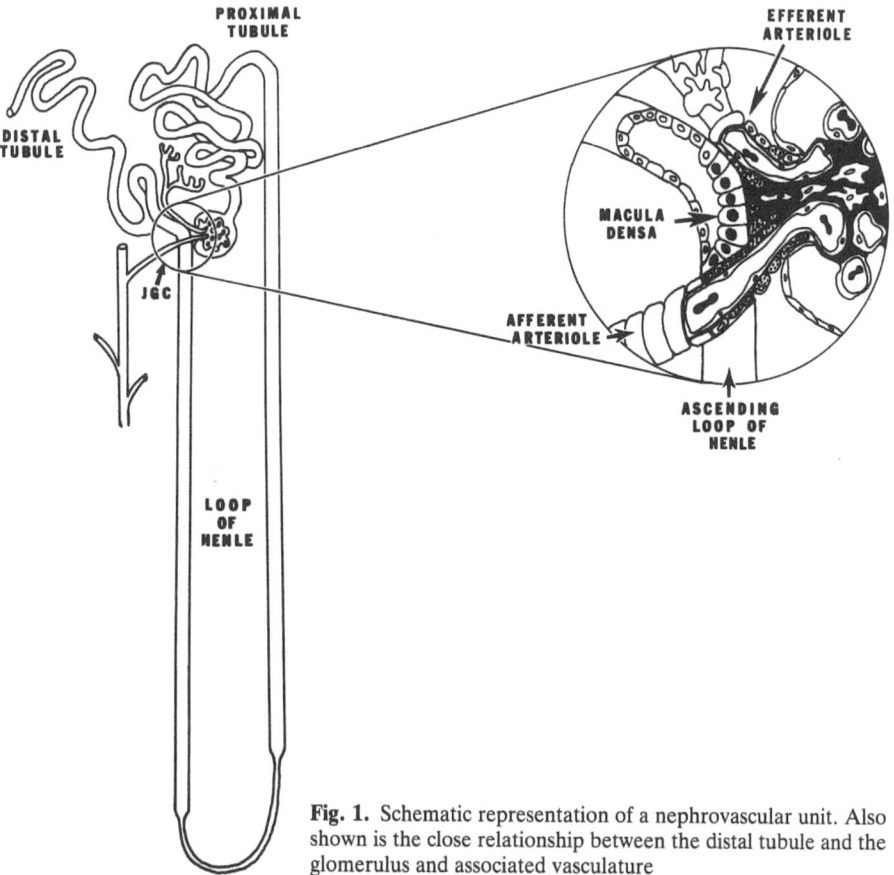

Fig. 1. Schematic representation of a nephrovascular unit. Also shown is the close relationship between the distal tubule and the glomerulus and associated vasculature

ing tubular cells. Macula densa cells are much taller than the surrounding thick ascending limb cells, possess a large prominent nucleus, numerous mitochondria, and endoplasmic reticulum, and lack apically located Tamm-Horsfall glycoprotein [19]. Also, in most renal cells, the basolateral membrane forms inwardly directed infoldings between which are located rows of mitochondria. These infoldings serve to amplify the surface area of the basolateral membrane and along with the adjacent mitochondria increase transport capability. In the macula densa cells, no such infoldings exist; rather, there are protrusions of the basolateral membrane through the basement membrane and into the area of the extraglomerular mesangial cells. In agreement with this observation, morphological studies [18] on isolated macula densa basement membrane demonstrate that this basement membrane is unique in the renal epithelium in that it contains numerous large irregular holes. Protrusions of the basal membranes of macula densa cells extend into these basement membrane holes and apparently are in contact with extraglomerular mesangial cells. However, it is unclear whether or not specialized areas of contact exist between macula densa and extraglomerular mesangial cells; no gap junctions have been observed between the two cell types [3, 50].

The extraglomerular mesangial cells consist of long spindle-shaped cells with numerous cytoplasmic processes. In addition, there are extensive regions of extracel-

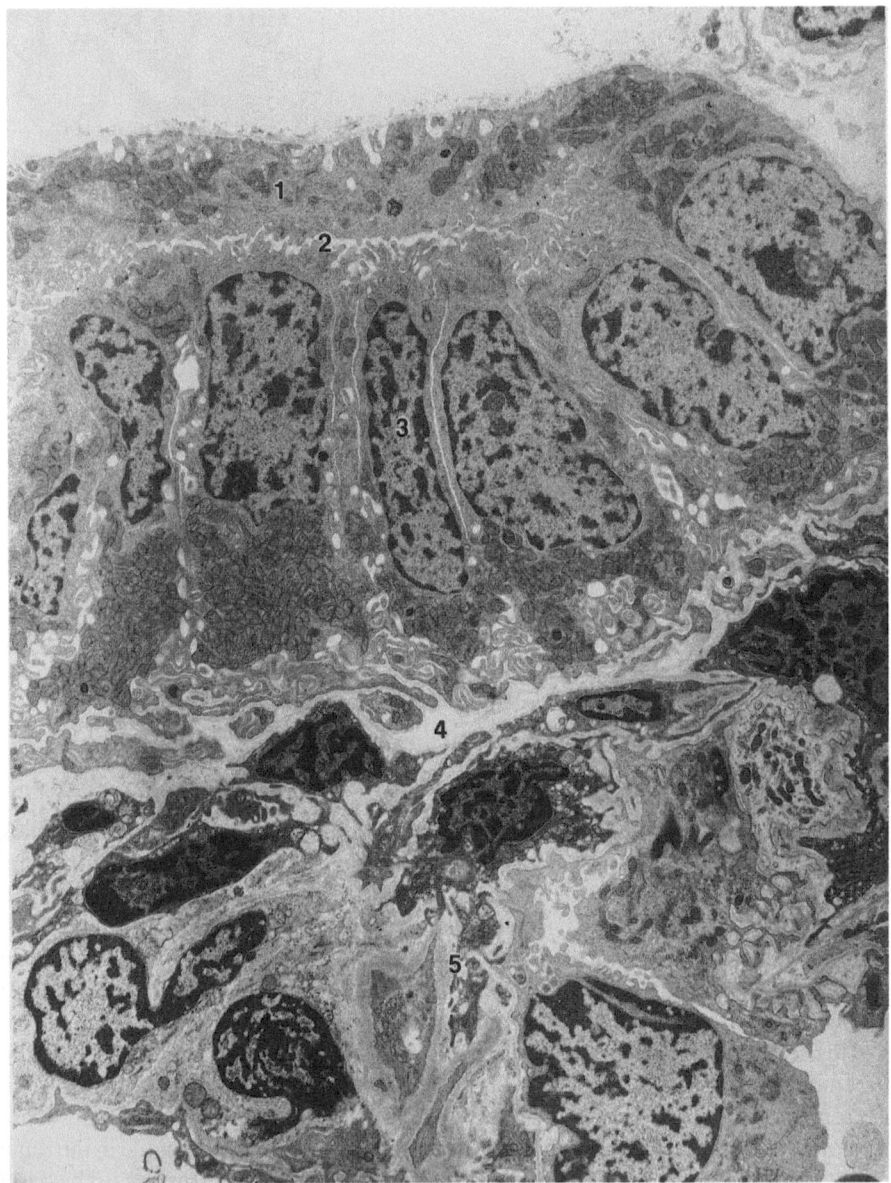

Fig. 2. Electron micrograph of the macula densa – juxtaglomerular apparatus. *1* Cortical thick ascending limb cells; *2* collapsed lumen of tubule; *3* macula densa cells; *4* basement membrane between macula densa cells and extraglomerular mesangial cells; *5* extraglomerular mesangial cells

lular matrix-like material occupying the spaces between these cells. In contrast to macula densa–extraglomerular mesangial contact points, numerous gap junctions have been found between extraglomerular mesangial cells. In addition, there is morphological continuity between the extraglomerular mesangial cells, the intraglomerular mesangial cells and the smooth muscle cells of the afferent and efferent arterioles. Furthermore, this entire area displays gap junctions, suggesting that it may act as a functional syncytium [27, 50].

The cells of the macula densa appear to be functionally and biochemically distinct from the surrounding thick ascending limb cells and indeed from other renal epithelial cells. The Golgi apparatus is located near the base of the macula densa cells as compared to the more apical location of the Golgi apparatus found in all other renal tubular cells. Macula densa cells contain low levels of Krebs cycle enzymes, high levels of glucose-6-phosphate dehydrogenase activity, high RNA synthetic activity, and stain positive for carbonic anhydrase [3]. In a recent study, Schnermann and Marver [45] found that Na-K ATPase activity in the macula densa was some 50 times lower than in the surrounding thick ascending limb cells. This suggests that the macula densa cells may have a very limited transepithelial transport capacity.

Another morphological and functional characteristic of the macula densa cells concerns the water permeability of this small cell plaque. As mentioned earlier, macula densa cells lack Tamm-Horsfall protein, which is present on the apical membranes of the surrounding thick ascending limb cells. It has been suggested that Tamm-Horsfall protein may be responsible for the water impermeability of the thick ascending limb [49]. In morphological studies [32, 44], remarkable large spaces have been found between macula densa cells. Since the luminal fluid at the macula densa is normally hypotonic (see below), this suggests that the cells of the macula densa may be permeable to water: the dilated spaces between macula densa cells would then be due to water flow across the macula densa plaque. This suggestion was supported, in other morphological studies [32], by the finding that physiological conditions which increase luminal fluid osmolality at the macula densa lead to the diminution in width of lateral spaces bbetween macula densa cells. In recent physiological stdies, Kirk and coworkers [34] used an isolated perfused thick ascending limb – macula densa preparation in combination with differential interference contrast miscroscopy. They found that the width of spaces between macula densa cells increased as the luminal fluid was changed from isotonic to hypotonic values. No change in the spaces between adjacent thick ascending limb cells were observed. These studies indicate that the macula densa plaque may be a small water-permeable group of cells located within the water-impermeable thick ascending limb. We will now consider the basic characteristics of the tubuloglomerular feedback mechanism and discuss current knowledge of the cellular mechanisms responsible for the feedback signal transmission process.

Characteristics of the Feedback Mechanism

The initial step in the activation of the tubuloglomerular feedback mechanism resides in the transport characteristics of the thick ascending limb of the loop of Henle. Since this tubular segment is virtually impermeable to water, active transepithelial transport of NaCl results in the dilution of tubular fluid. Under normal conditions, tubular fluid

NaCl concentration and osmolality are approximately one-third of the plasma values [4, 7, 9, 13]. However, the degree of hypotonicity is dependent on the flow rate through the loop of Henle. At low flow rates, maximum transport gradients can be achieved and therefore the NaCl concentration and osmolality reaching the macula densa is low. As flow rate is increased, less NaCl is reabsorbed per unit of flow and therefore the NaCl concentration and osmolality reaching the macula densa is increased. Over the physiological range, there is a linear relationship between flow rate and tubular fluid NaCl concentration and osmolality (Fig. 3). This flow rate-dependent change in luminal fluid composition, serves as the initial stimulus for the activation of the tubuloglomerular feedback mechanism.

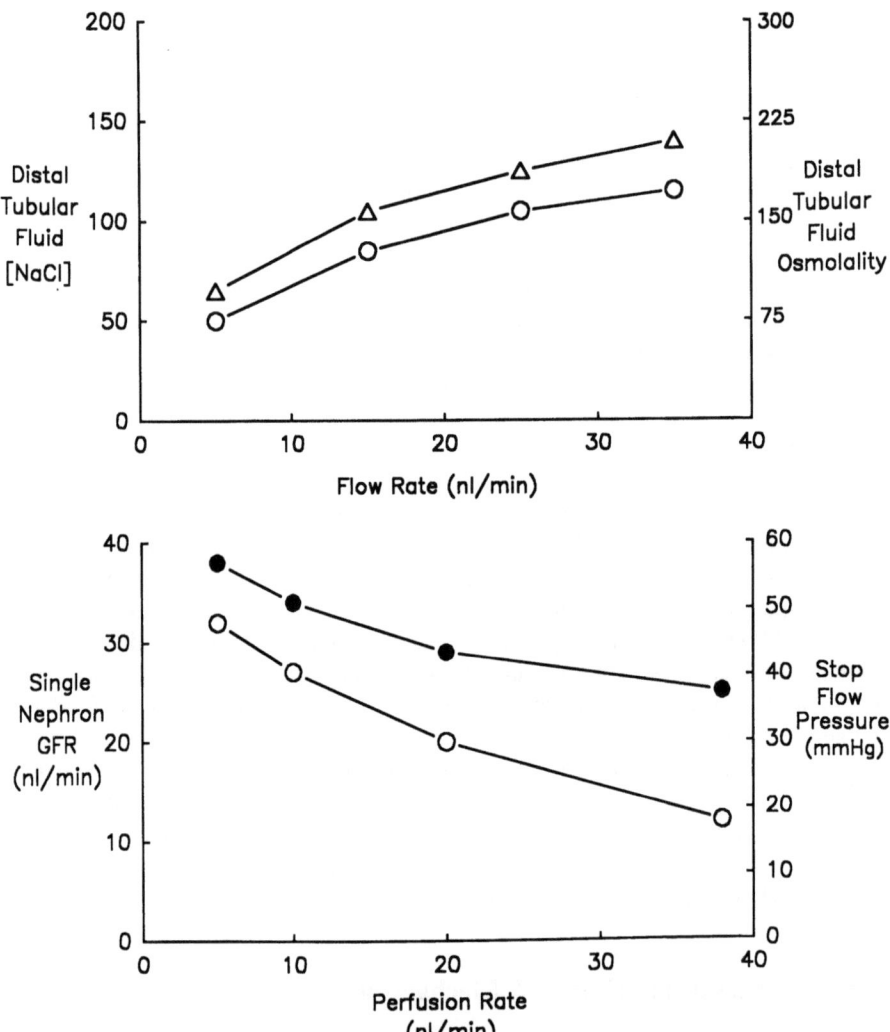

Fig. 3. *Top:* the relationship between tubular flow rate (nl/min) and distal tubular fluid NaCl concentration and osmolality. *Bottom:* the relationship between perfusion rate, single-nephron glomerular filtration rate and glomerular capillary pressure. In most studies stop flow pressure is used as an index of glomerular capillary pressure

It is generally accepted that a flow rate-dependent change in luminal fluid NaCl concentration and osmolality is detected by the macula densa cells and that signals are transmitted to the extraglomerular mesangial cells and on to the smooth muscle cells [13, 19]. An increase in luminal fluid NaCl concentration and osmolality is associated with vasoconstriction and decreases in glomerular capillary pressure and glomerular filtration rate. Conversely, decreases in luminal fluid sodium chloride concentration and osmolality are associated with vasodilation and increased glomerular capillary pressure and glomerular filtration rate. The relationships between luminal fluid compositional changes and single-nephron glomerular filtration rate (SNGFR) and glomerular capillary pressure or stop flow pressure (SFP) are also shown in Figure 3 [15, 42]. It is readily apparent from this relationship that the tubuloglomerular feedback mechanism can exert a powerful influence on glomerular hemodynamics. It is generally thought that under conditions where glomerular filtration rate is elevated, flow rate into the loop of Henle is increased, resulting in higher distal tubular fluid NaCl concentration and osmolality [9]. Macula densa cells detect this increase in luminal fluid composition, which results in feedback-mediated decreases in glomerular filtration rate. Likewise, when glomerular filtration is depressed, tubular flow rate decreases, resulting in lower values for luminal fluid NaCl concentration and osmolality at the macula densa. This would lead to the transmission of vasodilator signals and increases in glomerular filtration rate. Also, the suggestion has been made that the tubuloglomerular feedback mechanism tends to regulate of at least guard against large changes in transepithelial transport in tubular segments proximal to the macula densa [19]. According to this scheme, if transepithelial transport of salt and water were inappropriately depressed, flow to the macula densa would increase, inducing feedback-mediated decreases in glomerular filtration and ultimately decreasing filtered load to the tubule. Thus, the tubuloglomerular feedback mechanism serves an important role in coupling the hemodynamic mechanisms that control the process of filtrate formation with renal transport processes that are responsible for the volume and composition of the final renal excretory product.

Over the last few years, a number of studies have been conducted in an effort to elucidate the mechanisms involved in the transmission of feedback signals. Many of these studies have utilized in vivo micropuncture techniques. With this approach it is possible to study the tubuloglomerular mechanism at the single nephron level. It is not possible to directly study macula densa cells, since these cells are inaccessible to micropuncture procedures. It has also recently been possible to isolate that portion of the thick ascending limb containing the macula densa cells and attached glomeruli. Using techniques for the perfusion of isolated single renal tubules, studies are now under way that directly examine functional properties of the macula densa.

Luminal Mechanisms in Feedback Signal Transmission

The first step in the transmission of feedback signals involves an alteration in luminal fluid composition [4, 17, 20, 47, 52]. A number of studies have sought to determine the component(s) of tubular fluid detected by the macula densa cells. Most of these studies have used microperfusion techniques to alter the luminal environment at the macula densa in an effort to determine the effects of electrolyte or solute substitutions on the transmission of feedback signals. At the present time, the results of these studies are equivocal. Early studies by Schnermann and coworkers [47] suggested that the macula densa cells specifically detected changes in luminal fluid chloride concentration. On the other hand, studies in our laboratory [4, 13] suggested that the macula densa cells may not specifically detect a single ion but that the detection process might be more generally related to the osmolality of the luminal fluid. As the issue now stands, neither of these views appears to be entirely correct. Clearly, feedback responses can be obtained in the presence of either luminal fluid sodium or chloride and the mechanism is not limited to just chloride. However, conflicting results have been obtained using nonelectrolyte solutions, suggesting that the detection process may be more complex than a simple osmotic effect across the luminal membrane of the macula densa cells. Thus at the present time, the nature of the luminal stimulus that activates the feedback signal transmission process remains unclear. In addition, the cellular mechanisms involved in this detection process remain to be elucidated.

Role of Cytosolic Calcium in Feedback Signal Transmission

In recent years, we have obtained evidence that the macula densa cells may detect changes in luminal fluid composition through a cytosolic calcium system. The first evidence for a role of cytosolic calcium in feedback signal transmission was obtained using the calcium ionophore A23187. Using in vivo microperfusion techniques [12], A23187 was added to a hypotonic solution that alone did not activate the transmission of feedback signals. Perfusion with this solution into the distal tubule resulted in feedback-mediated vasoconstriction which was of similar magnitude to the responses obtained with an isotonic Ringer's solution. Thus, the calcium ionophore which increases luminal membrane calcium permeability, and therefore elevates cytosolic calcium concentration, mimicked the effects of increased luminal fluid composition. This suggested that the macula densa cells may detect changes in luminal fluid composition through changes in cytosolic calcium concentration. A summary of these and other studies discussed in this paper are presented in Fig. 4.

In other studies, we examined the mechanism for the increase in macula densa cytosolic calcium concentration during the transmission of feedback signals [14]. We found that removal of luminal calcium, addition of a calcium chelator (EGTA), or addition of a calcium channel blocker did not inhibit the feedback responses obtained with an isotonic Ringer's solution. Therefore, elevations in cytosolic calcium concentration were not the result of calcium entry from luminal fluid into the macula densa cells. In other studies [14], TMB-8, an intracellular calcium antagonist, [36] was used to block intracellular calcium release. TMB-8 was effective in blocking the feedback responses obtained with an isotonic Ringer's solution. This suggests that one component of calcium activation during the transmission of feedback signals involves calcium mobilization from intracellular storage sites. TMB-8 inhibits calcium mobilization but does not block the actions of calcium, since the inhibitory effects of TMB-8 could be overcome by the addition of a calcium ionophore.

In addition, it is possible that calcium entry across the basolateral membrane may also participate in the regulation of macula densa cytosolic calcium concentration or in the activation of these cells during the transmission of feedback signals. This sugges-

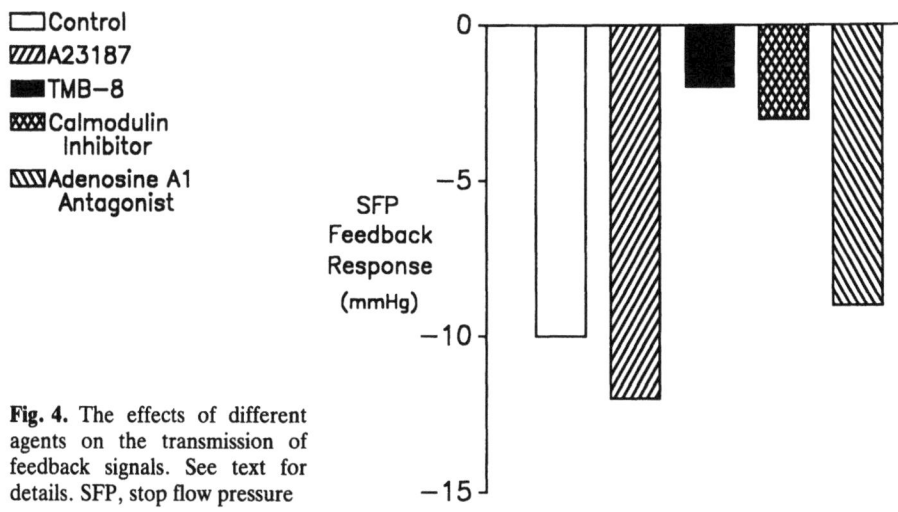

Fig. 4. The effects of different agents on the transmission of feedback signals. See text for details. SFP, stop flow pressure

tion is supported by recent studies performed in the isolated perfused thick ascending limb – macula densa preparation [11]. In these studies, the basolateral membrane potential of macula densa cells was measured using microelectrodes. We found that increases in luminal fluid NaCl concentration from 45 to 150 mM resulted in the depolarization of the basolateral membrane potential. Conceivably, this may result in an increase in calcium entry into these cells. Thus, activation of macula densa cytosolic calcium during the transmission of feedback signals may occur through both calcium mobilization and calcium entry across the basolateral membranes. It is possible, and in fact even probable, that the cellular mechanisms which initiate calcium mobilization and calcium entry may differ. This may help explain some of the confusion and apparent inconsistencies in studies attempting to define the component of tubular fluid that initiates feedback signals. For instance, it is possible that increased luminal fluid NaCl concentration may be capable of increasing macula densa cytosolic calcium concentration through basolateral membrane depolarization and influx of calcium into these cells. On the other hand, increased luminal ionic strength or osmolality may be capable of inducing calcium mobilization. Thus, there may be several different mechanisms that can alter macula densa cytosolic calcium concentration.

The recent development of the isolated perfused thick ascending limb – macula densa preparation also has provided the means of directly testing for the existence of a macula densa cytosolic calcium system that responds to changes in luminal fluid composition [10]. This has recently been accomplished by combining the isolated perfused tubule technique with fluorescent microscopy and the use of Fura 2, a recently developed calcium-sensitive fluorescent probe. Fura 2 fluorescence was measured at excitation wavelengths of 340 and 380 nm and emission wavelength of 510 nm using a photometer-diaphragm system that restricted the field so that photon counting could be performed on only those cells of interest. In preliminary studies, [10] we confirmed that luminal application of a calcium ionophore, ionomyocin, increased cytosolic calcium concentration in cells of the macula densa five- to sixfold.

In addition, we have tested the effects of increases in luminal fluid NaCl concentration on macula densa cytosolic calcium concentration. When the luminal fluid NaCl concentration was increased from 45 to 150 mM, the directly measured macula densa cytosolic calcium concentration increased fom 50 to 150 nM. In contrast, similar measurements in surrounding thick ascending limb cells revealed that thick ascending limb cytosolic calcium averaged approximately 100 nM and did not change significantly as luminal fluid NaCl concentration was increased from 45 to 150 mM. Thus, these results directly support the existence of a macula densa cytosolic calcium system that is responsive to alterations in luminal fluid composition.

Ca^{2+}-Calmodulin System

It is well known that calmodulin is an important component in calcium-mediated events [16, 23, 25, 43]. In addition, calmodulin plays an important role in regulating cytosolic calcium concentration through the calcium-dependent activation of calcium transport systems that result in either calcium uptake into intracellular sequestering sites or extrusion of calcium from cells. This later role for calmodulin may be important in the transmission of feedback signals. In previous studies we found that intraluminal administration of putative calmodulin antagonists such as trifluoperazine and calmidazolium [30] did not inhibit the transmission of feedback signals, suggesting that calmodulin does not play a direct role in the transmission of feedback signals [6]. However, these agents did alter the pattern of feedback responsiveness. Under normal conditions, termination of intraluminal perfusion with an isotonic Ringer's solution results in a rapid (< 60 s) return of SFP to control values. After administration of the calmodulin antagonist, the time required for the return of SFP to control values was prolonged. This indicates that, in the presence of calmodulin blockade, vasoconstriction was prolonged. Thus, in the transmission of feedback signals, calmodulin may serve in the regulation of cytosolic calcium concentration. The prolonged vasoconstrictive response in the presence of calmodulin inhibition may simply reflect a diminished ability of the macula densa cells to rapidly lower cytosolic calcium concentration.

cAMP-Ca^{2+} Interactions

Previous studies have shown that cAMP may serve as an important negative modulator of feedback responses. We have found that several agents which elevate intracellular cAMP concentration markedly inhibit feedback responses [5]. This includes the phosphodiesterase inhibitors, DcAMP (dibut-cAMP), and forskolin, an adenylate cyclase stimulator [48]. In addition, the inhibition of feedback responses by these agents appears to be calcium-dependent, since concurrent administration of a calcium ionophore can overcome the inhibition of feedback responses obtained under conditions of elevated intracellular cAMP. It is therefore possible that cAMP inhibits the transmission of feedback signals by acting directly upon the macula densa cytosolic calcium system. In this regard, studies in other sytems have shown that cAMP can directly stimulate calcium sequestration or calcium extrusion from cells

[26]. It is interesting to speculate that cAMP may be one means of regulating the sensitivity of the feedback mechanism. Studies have demonstrated that the ability of the feedback mechanisms to induce vasoconstriction is altered under a variety of physiological and pathophysiological conditions [2]. Receptor-activated cAMP production by the macula densa cells may be one means by which feedback sensitivity is altered.

Adenosine A1 Receptor Activation of the Feedback Mechanism

Another system that may play an important role in the transmission of feedback signals is adenosine. Recent studies have shown that adenosine, and activation of adenosine receptors, plays an important role as a local regulator of epithelial cell function [1, 21, 22, 35]. At the present time two membrane-bound receptors for adenosine (A1, A2) have been identified [24]. The adenosine A1 receptor causes a decrease in cell cAMP levels and also a direct increase in cytosolic calcium concentration. The A2 receptor stimulates the production of cAMP. In previous work, Osswald and coworkers [39–41, 46] suggested that adenosine may help mediate feedback responses, since systemic administration of theophylline, an adenosine receptor antagonist, blocked feedback responses. Also, dipyridamole, an agent that prevents reuptake of adenosine and therefore increases extracellular adenosine concentration, enhanced feedback responses. In an effort to directly evaluate the effects of adenosine and adenosine receptors in the transmission of feedback signal transmission, we have administered adenosine receptor agonist and antagonists directly into the distal tubule [28]. In recent studies [28] we have found that intraluminal perfusion of the adenosine receptor agonist N^6-cyclopentyl adenosine resulted in feedback-mediated decreases in SFP. The feedback responses obtained with this agent were similar to those obtained with an isotonic Ringer's solution. This effect could be blocked by coinfusion of an adenosine receptor antagonist. These results are intriguing because they suggest that adenosine may participate in the transmission of feedback signals not by directly acting on the smooth muscle cells but rather by acting at the macula densa. Indeed, it is possible that adenosine A1 receptors may be located on the luminal membrane of the macula densa cells and that the effect of activation of adenosine receptors may occur through the macula densa cytosolic calcium system. In this regard, we have recently found that luminal administration of an adenosine A1 receptor agonist produces a 200% increase in cortical thick ascending limb cytosolic calcium concentration which is associated with an inhibition of transepithelial transport as assessed by a decrease in transepithelial potential difference [8]. This effect was specifically related to the intraluminal administration of this agent, since the adenosine A1 receptor agonist was ineffective when added to the bathing solution. Thus, it is possible that in the macula densa, adenosine A1 receptors may play an important role in the transmission of feedback signals. Other studies have shown that adenosine production is increased under conditions of enhanced transepithelial transport rate [33]. Accordingly, an elevation in luminal fluid NaCl concentration may increase adenosine production by the cortical thick ascending limb and possibly by the macula densa cells. This may result in an increase in luminal fluid adenosine concentration and the activation of macula densa adenosine A1 receptors. Thus, there may be two pathways for the initiation of feedback signal transmission, one involving changes in luminal fluid NaCl concentration or osmolality and another involving the activation of adenosine A1 receptors. This second pathway may help explain certain observations which have suggested that an endogenous substance can activate the feedback mechanism independent of luminal fluid NaCl concentration and osmolality [53].

Transmission of Signals to the Vasculature

The mechanism by which the macula densa cells transmit information to the extraglomerular mesangial cells and on to the smooth muscle cells is not known at this time. As mentioned earlier, gap junctions have not been detected betweeen the macula densa cells and the extraglomerular mesangial cells. Feedback signal transmission via the production and release of a chemical mediator by the macula densa cells has been proposed, but the nature of this chemical mediator has not been identified. In a previous study [29], we found that intraluminal administration of arachidonic acid produced feedback responses. However, subsequent inhibitor studies failed to identify which, if any, of the arachidonic acid metabolites might be mediating this effect. Thus, it is possible that some unidentified product of arachidonic acid may serve in the transmission of feedback signals. Alternatively, recent work has shown that arachidonic acid can directly mobilize cytosolic calcium [51], and it is possible that the enhanced feedback response with arachidonic acid is a consequence of mobilization of cytosolic calcium in the macula densa.

Mechanism for Feedback-Mediated Vasoconstriction

The transmission of feedback signals ultimately results in the activation of glomerular contractile cells. Under normal conditions, changes in afferent arteriole resistance can fully account for feedback-mediated alterations in renal hemodynamics. Previous studies have shown that the increases in afferent arteriolar tone induced by the tubuloglomerular feedback mechanism are dependent upon calcium channels. Systemic [38] or local peritubular capillary [37] administration of calcium-channel blockers completely abolishes feedback responses. Thus, the final step in the transmission of feedback signals involves calcium-channel activation and changes in vascular tone.

In summary, the tubuloglomerular feedback mechanism is an important intrarenal control mechanism for the regulation of renal hemodynamics. It is also a rather novel and perhaps unique means of controlling vascular resistance. Cytosolic calcium appears to play an important role in the transmission of feedback signals at two distinct sites (Fig. 5). First, the macula densa cells appear to detect changes in luminal fluid composition through the activation of a cytosolic calcium system. In this system, calcium is mobilized from intracellular storage sites; however, calcium may also enter across the basolateral membrane, since increased luminal fluid NaCl concentration is associated with basolateral membrane depolarization. Second, the transmission of feedback signals results in contraction of the smooth muscle cells of the afferent arteriole. This response appears to occur through the activation of voltage-dependent calcium channels that are sensitive to calcium-channel blockade.

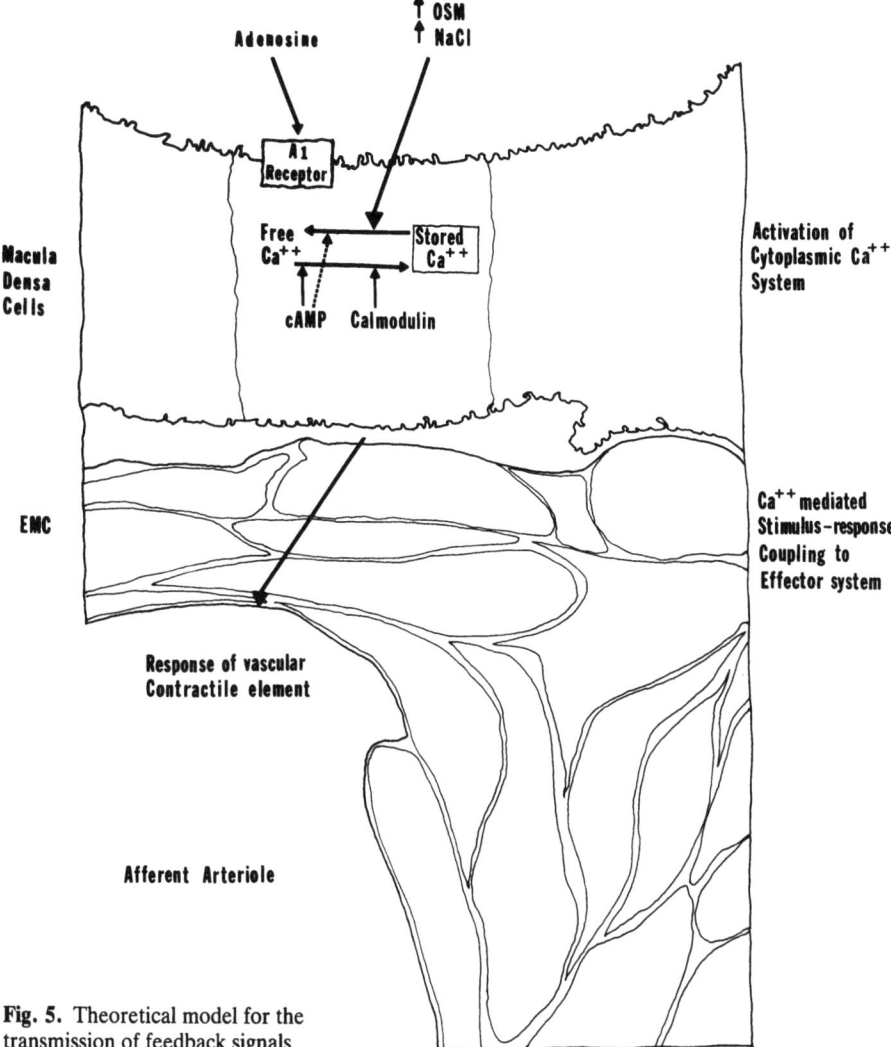

Fig. 5. Theoretical model for the transmission of feedback signals

Acknowledgements. We would like to extent our appreciation for the help provided by previous collaborators.

References

1. Arend LJ, Thompson CI, Spielman WS (1985) Dipyridamole decreases glomerular filtration in the sodium-depleted dog. Evidence for mediation by intrarenal adenosine. Circ Res 56:242–251
2. Arendshors WI (1987) Altered reactivity of tubuloglomerular feedback. Ann Rev Physiol 49:295–317
3. Barajas L, Salido EC, Powers KV (1988) Anatomical basis of the tubuloglomerular feedback mechanism: the juxtaglomerular apparatus. In: Persson AEG, Boberg U (eds) The juxtaglomerular apparatus. Elsevier Science Publishers New York pp 7–26

4. Bell PD (1982) Luminal and cellular mediation of tubuloglomerular feedback responses. Kidney Int 22:S97–S103 Suppl 12
5. Bell PD (1985) Cyclic AMP-calcium interaction in the transmission of tubuloglomerular feedback signals. Kidney Int 28:728–732
6. Bell PD (1986) Tubuloglomerular feedback responses in the rat during calmodulin inhibition. Am J Physiol 250:F715–719
7. Bell PD, Franco-Guevara M, Abrahamson DR, Lapointe JY, Cardial J (1988) Cellular mechanisms for tubuloglomerular feedback signaling. In: Persson AEG, Boberg U (eds) The juxtaglomerular apparatus. Elsevier Science Publishers New York pp 63–77
8. Bell PD, Franco M, Higdon M (1988) Adenosine A1 agonist increases cytosolic calcium concentration in the isolated perfused cortical thick ascending limb. Fed Proc 72:3681
9. Bell PD, Franco M, Navar LG (1987) Calcium as a mediator of tubuloglomerular feedback. Ann Rev Physiol 49:275–293
10. Bell PD, Krause A, Franco M (1988) Macula densa cytosolic calcium concentration during changes in luminal fluid osmolality. Kidney Int 33:150
11. Bell PD, Lapointe J-Y, Cardinal J (submitted) Direct measurement of the basolateral membrane potential from cells of the macula densa. Am J Physiol
12. Bell PD, Navar LG (1982) Cytoplasmic calcium in the mediation of macula densa tubuloglomerular feedback responses. Science 215:670–673
13. Bell PD, Navar LG (1982) Macula densa feedback control of glomerular filtration: role of cytosolic calcium. Miner Electrolyte Metab 8:61–77
14. Bell PD, Reddington M (1983) Intracellular calcium in the transmission of tubuloglomerular feedback signals. Am J Physiol 245:F295–F302
15. Bell PD, Reddington M, Ploth D, Navar LG (1984) Tubuloglomerular feedback-mediated decreases in glomerular pressure in Munich-Wistar rats. Am J Physiol 247:F877–880
16. Berridge MG (1984) Cellular control through interactions between cyclic nucleotides and calcium. Adv Cylic Nucleotide Protein Phosphor Res 17:329–335
17. Boberg U, Wright FS (1988) Distal NaCl concentration as signal for the tubuloglomerular feedback mechanism. In: Persson AEG, Boberg U (eds) The juxtaglomerular apparatus. Elsevier Science publishers New York, pp 89–119
18. Bonsib SM (1987) The macula densa tubular basement membrane: a plaque of specialized basement membrane. Proc Xth Internatl Cong of Nephr p 315
19. Briggs JP, Schnermann J (1987) The tubuloglomerular feedback mechanisms: functional and biochemical aspects. Ann Rev Physiol 49:00–00
20. Briggs JP, Schubert G, Schnermann J (1982) Further evidence for an inverse relationship between macula densa NaCl concentration and filtration rate. Pflügers Arch 392:372–378
21. Brines ML, Forrest JN Jr (1987) Autoradiographic localization of A1 adenosine receptors to tubules in the red medulla and papilla of the rat kidney. Kidney Int 33:109
22. Bruns RF, Daly SH, Snyder SH (1983) Adenosine receptor binding: structure-activity analysis generates extremely potent xanthine antagonist. Proc Natl Acad Sci USA 80:2077–2080
23. Cheung WJ (1980) Calmodulin plays a pivotal role in cellular regulation. Science 207:19–27
24. Daly JW (1982) Adenosine receptors: target for future drugs. J Med Chem 25:197–207
25. Exton JH (1985) Role of calcium and phosphoinositides in the actions of certain hormones and neurotransmitters. J Clin Invest 75:1753–1757
26. Feinstein MB, Egan JJ, Sha'afi RI, White J (1985) The cytoplasmic concentration of free calcium in platelets is controlled by stimulators of cyclic AMP production (PDG_2, PGE_1, forskolin). Biochem Biophys Res Commun 113:598–604
27. Forssmann WG, Taugner R (1977) Studies on the juxtaglomerular apparatus in Typania with special references to intercellular contacts. Cell Tissue Res 177:291–305
28. Franco M, Bell PD, Navar LG (submitted) Effect of Adenosine A1 analog on tubuloglomerular feedback mechanism. Am J Physiol
29. Franco M, Bell PD, Navar LG (in press) Evaluation of prostaglandins as mediators of tubuloglomerular feedback. Am J Physiol
30. Gietzen K, Wuthrich A, Bader H (1981) A new powerful inhibitor of red blood cell Ca^{++}-transport ATPase and of calmodulin-regulated function. Biochem Biophys Res Commun 101:418–425

31. Kaissling B, Peter S, Kriz W (1977) The transition of the thick ascending limb of Henle's loop into the distal convoluted tubule in the nephron of the rat kidney. Cell Tiss Res 182:111–118
32. Kaissling B, Kriz W (1982) Variability of intercellular spaces between macula densa cells: a transmission electron microscopic study in rabbits and rats. Kidney Int 22:S9–S17 (Suppl 12)
33. Kelley GG, Assar OS, Forrest JN Jr (1987) Adenosine released during hormonal stimulation is a feedback inhibitor of chloride transport in the shark rectal gland. Kidney Int 33:123
34. Kirk KL, Bell PD, Barfuss DW, Ribadenerira M (1985) Direct visualization of the isolated and perfused macula densa. Am J Physiol 248:F890–894
35. Lear SP, Silva P, Epstein FH (1985) Adenosine and PGE2 modulate transport by isolated thick ascending limb. Clin Res 33:586
36. Misbahuddin M, Isosaki M, Houchi J, Oka M (1985) Muscarinic receptor-mediated increase in cytosolic Ca^{2+} in isolated bovine adrenal medullary cells. Effects of TMB-8 and phorbol ester TPA. FEBS Lett 190:25–28
37. Mitchell KD, Navar LG (1988) Single nephron responses to peritubular capillary infusion of verapamil. Fed Proc 2:5675
38. Muller-Suur R, Gutsche HU, Schurek HJ (1976) Acute and reversible inhibition of tubuloglomerular feedback mediated afferent vasoconstriction by the calcium antagonist verapamil. Curr Probl Clin Biochem 6:291–298
39. Osswald H (1984) The role of adenosine in the regulation of glomerular filtration rate and renin secretion. Trends in pharmacol Sci 5:94–97
40. Osswald H, Hermes H, Nabakowsky G (1982) Role of adenosine in the transmission of tubuloglomerular feedback. Kidney Int 22 (Suppl 12):136–S142
41. Osswald H, Nabakowsky G, Hermes H (1980) Adenosine as a possible mediator of metabolic control of glomerular filtration rate. Int J Biochem 12:263–267
42. Persson AEG, Gushwa LC, Blantz RC (1984) Feedback pressure-flow responses in normal and angiotensin-prostaglandin-blocked rats. Am J Physiol 247:F925–931
43. Rasmussen H, Barret PQ (1984) Calcium messenger system: an integrated view. Physiol Rev 64:938–987
44. Schnabel E, Kriz W (1984) Morphometric studies of the extraglomerular mesangial cell field in volume expanded and volume depleted rats. Anat Histol Embryol 170:217–222
45. Schnermann J, Marver D (1986) ATPase activity in macula densa cells of the rabbit kidney. Pflügers Arch 407:82–86
46. Schnermann J, Osswald H, Hermle M (1977) Inhibitory effect of methylxanthines on feedback control of glomerular filtration rate in the rat kidney. Pflügers Arch 369:39–48
47. Schnermann J, Ploth EW, Hermle M (1976) Activation of tubuloglomerular feedback by chloride transport. Pflügers Arch 362:229–240
48. Seamon KB, Daley JW (1981) A unique diterpene activator of cyclic AMP-generating systems. J Cyclic Nucleotide Res 224:201–224
49. Sikri K, Forster CL, MacHugh N, Marshall RD (1980) Localization of Tamm-Horsfall glycoprotein in the human kidney using immuno-fluorescence and immunoelectron microscopical techniques. J Anat 132:597–605
50. Taugner R, Schiller A, Kaissling B, Kriz W (1979) Gap junctions coupling between JGA and the glomerular tuft. Cell Tissue Res 186:279–285
51. Wolfe BA, Turk J, Sherman WR, McDaniel ML (1986) Intracellular Ca^{2+} mobilization by arachidonic acid. Comparison with myo-inositol, 4,5,6-triphosphate in isolated pancreatic islets. J Biol Chem 261:3501–3511
52. Wright FS, Schnermann J (1977) Interference with feedback control of glomerular filtration rate by furosemide, triflocin, and cyanide. J Clin Invers 53:1695–1708
53. Wunderlich PF, Brunner FP, Davis JM, Haberle DA, Tholen H, Thiel G (1980) Feedback activation in rat nephrons by sera from patients with acute renal failure. Kidney Int 17:497–506

Inhibition of Parathyroid Hormone Secretion by Calcium: The Role of Calcium Channels

L. A. Fitzpatrick[1] and H. Chin[2]

[1] Department of Medicine, Division of Endocrinology, Univesity of Texas Health Science Center, 7703 Floyd Curl Drive, San Antonio, TX 78284, USA
[2] Laboratory of Biochemical Genetics, National Heart, Lung, and Blood Institute, National Institutes of Health, Bethesda, MD 20892, USA

Calcium controls an array of biological processes, including cell growth, cytodifferentiation, and muscle contraction, as well as neurotransmitter and hormone release [1]. The mechanisms involved in these functions are incompletely understood. Extracellular calcium is tightly regulated by the effects of parathyroid hormone on its target tissues, and the secretion of parathyroid hormone itself is responsive to extracellular calcium concentrations [2]. We have studied the role of calcium channels in regulating the response of parathyroid hormone release to calcium.

Inhibition of Parathyroid Hormone Release by Extracellular Calcium

Calcium is the major physiological regulator of parathyroid hormone (PTH) synthesis and secretion. Unlike the case for most secretory cells, a rise in extracellular calcium inhibits PTH release and a fall in extracellular calcium levels stimulates PTH release [2]. The mechanism by which calcium controls PTH secretion has yet to be elucidated. The entry of calcium into parathyroid cells is dependent upon functional membrane glycoproteins called calcium channels, and the transfer of the extracellular calcium signal to intracellular sites via these channels is mediated by a guanine nucleotide regulatory protein [3–5]. In 1942 Patt and Luckhardt [6] were the first to gather evidence indicating that circulating calcium ion is an important physiological regulator of secretion. Further evidence that serum calcium concentration regulates PTH release was provided by the studies of Habener et al. [7], who demonstrated an inverse relationship between extracellular calcium and PTH release in slices of bovine parathyroid tissue. Development of dispersed parathyroid cell systems demonstrated this relationship in quantitative terms [8]. Brown [9] described a sigmoidal relationship between extracellular calcium and PTH secretion in dispersed human and bovine parathyroid glands.

Intracellular Regulation of Parathyroid Hormone Release

PTH release appears to be regulated by a number of intracellular second messengers. In many secretory cells, factors that modify cell secretion act through changes in intracellular 3',5'-cyclic adenosine monophosphate (cAMP). In the parathyroid cell, cAMP mediates β-adrenergic stimulation of the parathyroid cell [2]. However, quantitative changes in cAMP do not account for inhibition of PTH secretion by calcium, suggesting that other intracellular messengers are involved [3, 10].

An intracellular messenger of particular importance is the calcium ion. In secretory cells, calcium is a key regulator of stimulus-secretion coupling, such that an elevation in cytosolic (intracellular) calcium precedes secretion [1]. Classical stimulus-secretion coupling is paradoxically reversed in the parathyroid cell: increasing cytosolic calcium concentrations are associated with inhibition of PTH release. Recent work with the calcium indicator quin2 has directly correlated extracellular calcium measurements with cytosolic calcium levels in the parathyroid cell [11], further establishing this paradoxical relationship.

Several hypotheses have been proposed to account for the link between extracellular calcium, calcium fluxes, and calcium-mediated intracellular events in the secretory process. Cytosolic calcium levels are maintained in the 100–200 nM range in spite of millimolar concentrations of extracellular calcium [12]. This great differential between extracellular and intracellular calcium levels is tightly regulated both by calcium entry into the cell and by the storage of calcium in cell organelles. Stimulation of secretory cells is associated with release of intracellular calcium stores and/or opening of channels in the plasma membrane. Both mechanisms may account for the rise in intracellular calcium in response to an increase in extracellular calcium. The recent development of calcium-sensing fluorescent indicators has suggested that there is an initial transient rise in intracellular calcium followed by a sustained "steady-state" increase in intracellular calcium levels in the parathyroid cell [13].

Role of G Proteins. Guanine nucleotide regulatory proteins (G proteins) transmit extracellular signals to the cell interior in a variety of cell systems [14]. As membrane-bound constituents, G proteins function as signal transducers, coupling receptors to effectors. The family of G proteins includes two proteins that are linked to stimulation (G_s) or inhibition (G_i) of adenylate cyclase. Since the discovery of these two G proteins, other G proteins linked to effector systems besides adenylate cyclase have been discovered. The recognition that G proteins may serve as signal transducers to other hormone pathways such as phosphotidylinositol hydrolysis [15] and ion channels [16] has enlarged greatly the fundamental biological importance of G proteins. With respect to noncyclase pathways, the newly discovered G proteins appear to be G_i-like, in that they are substrates for an inhibitory interaction with pertussis toxin. Pertussis toxin catalyzes ADP ribosylation and inactivation of G_i and of certain other guanine nucleotide regulatory proteins [14].

Our recent studies have explored the potential role of a G protein(s) in the regulation of PTH secretion by calcium. Treatment of dispersed bovine parathyroid cells with pertussis toxin abolishes inhibition of PTH secretion by calcium (Fig. 1). The ability of pertussis toxin to abolish the actions of extracellular calcium suggests that a G protein(s) is involved in coupling the extracellular calcium signal to suppression of PTH secretion. Furthermore, the action of pertussis toxin was not associated with any change in cAMP concentration, suggesting that the pathway utilized by calcium and mediated by a G protein(s) is independent of cAMP. Oetting et al. [17] confirmed that G proteins may modulate PTH release in permeabilized parathyroid cells.

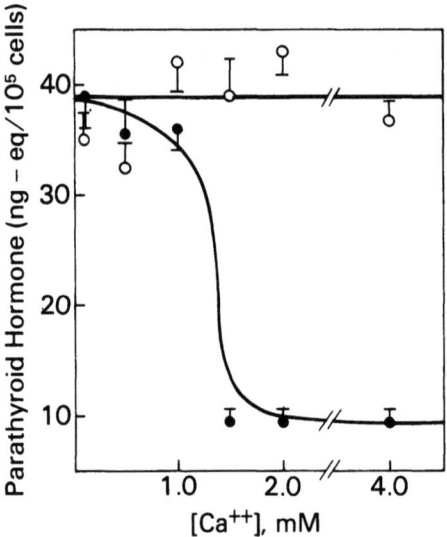

Fig. 1. Inhibition of parathyroid hormone (PTH) release by calcium. Dispersed bovine parathyroid cells were suspended in MEM (minimal essential medium) buffer and incubated with or without pertussis toxin (0.5 µg/ml) for 4 h. Cells were washed free of toxin and transferred to media containing varying calcium concentrations (0.25–4.0 mM). Cells were incubated at 37°C for 90 min in a shaking water bath, and PTH release was determined in cell supernatant by midregion specific radioimmunoassay. PTH release was maximal at low calcium concentrations regardless of pertussis toxin treatment. Calcium added to control cells *(closed circles)* caused a marked inhibition of PTH release. Calcium did not inhibit hormone release in cells treated with pertussins toxin *(open circles)*

Evidence for the presence of a G protein(s) in the parathyroid cell was documented by directly demonstrating the substrate for pertussis toxin in the ADP ribosylation reaction (Fig. 2). A 42-kDa protein is ADP-ribosylated by activated pertussis toxin in parathyroid cell membranes incubated with ^{32}P-labeled NAD. A 42-kDa protein consistent with the α-subunit of G_i is present after ADP ribosylation, suggesting that a G protein(s) is involved in calcium-controlled secretion in the parathyroid cell.

The inhibitory action of calcium on hormone release from parathyroid cells is virtually unique among secretory cells [2, 3]. We have shown that inhibition of PTH

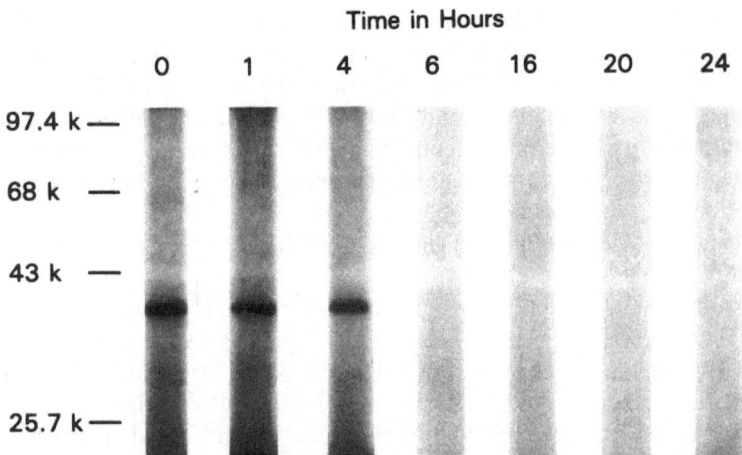

Fig. 2. ADP ribosylation of bovine parathyroid cell membranes. Membranes were prepared from bovine parathyroid cells incubated with pertussis toxin (0.5 µg/ml) for the times indicated *(top)*. After ADP ribosylation with ^{32}P-labeled NAD, membranes were fractionated by SDS polyacrylamide gel electrophoresis, and autoradiography was performed. ADP ribosylation increasing with time in the intact cell appears as a progressive reduction in labeling of a 42-kDa protein in parathyroid cell membranes

release by calcium is blocked by pertussis toxin, suggesting that a G protein(s) is responsible for transfer of extracellular calcium signal to intracellular sites affecting PTH release.

Pertussis toxin utilizes several G protein(s) as substrates, and further characterization has been accomplished. Polyclonal antibodies raised against the subunits of G proteins indicate that abundant G_i is present in parathyroid cells. There is little if any G_o (a G protein linked to neuronal calcium channels [16]) present. Further biochemical studies will be necessary to confirm this hypothesis.

Site of G Protein regulation. In classical receptor studies, G proteins couple stimulatory or inhibitory hormone receptors to effectors. Calcium does not bind to a classical receptor per se, suggesting that the role of G proteins may not be to couple receptor-effector function in this context. The divalent ionophore A23187 helped to define the role of G proteins in mediating control of PTH secretion. A23187 is a lipid-soluble antibiotic that raises intracellular calcium by permitting entry of extracellular calcium [18]. Parathyroid cells incubated with pertussis toxin which inactivates G proteins show an increase in PTH release over control cells at 2.0 mM calcium (Fig. 3). When the same experiment is performed in the presence of A23187, pertussis toxin is unable to reverse the inhibitory action of calcium. A23187 inhibits PTH release regardless of pertussis toxin treatment and bypasses the site of G protein action.

Calcium Channels

Pharmacological Studies. Calcium channels are large glycoproteins that allow controlled entry of extracellular calcium into a cell [19]. The availability of agents that bind to calcium channels makes it possible to study the role of calcium channels in the regulation of parathyroid cell secretion. Calcium-channel antagonists block the slow

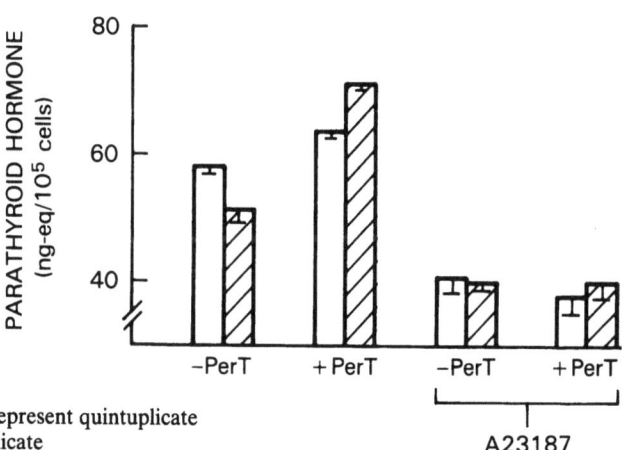

Fig. 3. The effect of A23187 on parathyroid cells treated with pertussis toxin. Parathyroid cells were prepared and incubated with *(+PerT)* or without *(−PerT)* pertussis toxin. Parathyroid hormone (PTH) release was measured at two calcium concentrations: 1.0 mM *(open bars)* and 2.0 mM *(hatched bars)*. A23187 (100 nM) was added to control and pertussis toxin-treated cells before incubation for 90 min at 37 °C. PTH release was determined on cell supernatants by radioimmunoassay. Data represent quintuplicate samples each analyzed in duplicate

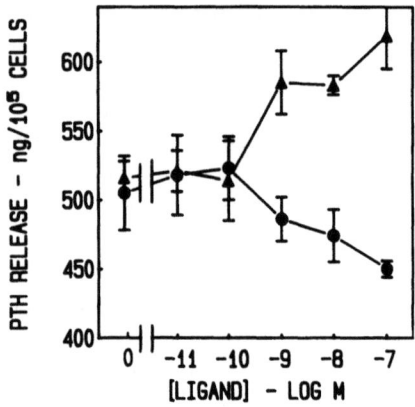

Fig. 4. The effect of calcium-channel agonist (+)202–791 and antagonist (−)202–791 on parathyroid hormone *(PTH)* release. Dispersed bovine parathyroid cells were incubated with (+)202–791 *(circles)* or (−)202–791 *(triangles)* at the concentrations indicated. Extracellular calcium concentration was 0.25 mM. Cell supernatants were separated by centrifugation for PTH determination by radioimmunoassay. Results represent the mean ± SEM of five samples each assayed in duplicate (From [4])

inward current of calcium into cells [20]. Conflicting evidence has been presented on the effects of calcium-channel antagonists on parathyroid cell secretion [21–24]. These studies have respectively suggested that calcium-channel antagonists stimulate, inhibit, or have no effect on PTH release. Previous studies with these compounds were often hampered because the compounds were available as racemic mixtures. Resolution of the racemates into their enantiomeric forms has led to a clearer understanding of the role of calcium-channel agents in PTH release. Our work suggests that the (+)form of the calcium-channel-specific compound 202–791 [isopropyl-4-(2,1,3-benzoxadiazol-4-yl)-1,4-dihydro-2,6-dimethyl-5-nitro-3-pyridinecarboxylate] is an agonist, whereas the pure (−)enantiomer has antagonist properties. The calcium-channel agonist (+)202–791 inhibited PTH release by 40%; the enantiomer (−)202–791 stimulated PTH release.(Fig. 4).

Calcium inhibits PTH release in a dose-dependent manner (Fig. 5). Addition of the calcium-channel antagonist (−)202–791 shifts the calcium inhibition curve to the right, implying that the calcium-channel blocker inhibits access of calcium to the channel, and that such access is functionally required for calcium regulation of PTH secretion. The calcium-channel agonist mimics the effect of calcium and inhibits secretion even at low calcium concentrations.

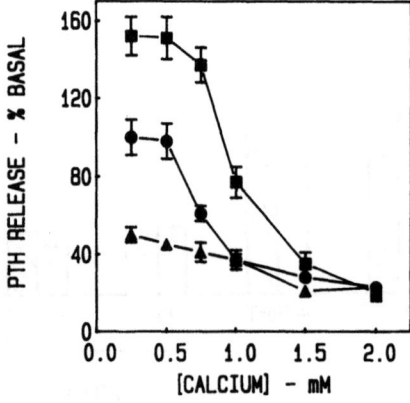

Fig. 5. The effect of calcium-channel antagonist and agonist at varying calcium concentrations in bovine parathyroid cells. Dispersed bovine parathyroid cells were incubated with calcium *(circles)*, 5×10^{-8} M (−)202–791 *(squares)*, or (+)202–791 *(triangles)*. Parathyroid hormone *(PTH)* release was determined by midregion specific radioimmunoassay. Data represent the mean ± SEM of five samples, assayed in duplicate and repeated in triplicate. (From [4])

Table 1. Effect of calcium-channel agonist and antagonist on cAMP accumulation

Concentration (M)	cAMP (pmol/10^5 cells)	
	(+)202–791	(−)202–791
Basal	1.03 ± 0.04	0.87 ± 0.06
10^{-11}	0.96 ± 0.02	0.77 ± 0.05
10^{-10}	1.08 ± 0.05	0.83 ± 0.05
10^{-9}	1.07 ± 0.07	0.86 ± 0.02
10^{-8}	1.01 ± 0.05	0.72 ± 0.05
10^{-7}	0.95 ± 0.06	0.81 ± 0.04
10^{-6}	0.90 ± 0.04	0.06 ± 0.09

Dispersed bovine parathyroid cells were prepared as described in the test. Cells were incubated with the calcium-channel agonist (+)202–791 or antagonist (−)202–791. Intracellular cAMP was extracted with 1-propanol and assayed by automated radioimmunoassay according to the method of Harper and Brooker [40]

We have compared several of the classical calcium-channel antagonists/agonists with optically pure enantiomers of the dihydropyridine family. Calcium-channel agonists [(+)202–791, Bay K8644] which open calcium channels mimic the action of calcium and inhibit secretion in parathyroid cells. Inhibition by (+)202–791 was greater than that of the agonist Bay K8644.

The calcium-channel agonist (+)202–791 and antagonist (−)202–791 were incubated with parathyroid cells, and intracellular cAMP was extracted. No significant differences, from basal values were noted with the addition of the agonist or antagonist (Table 1).

The role of cAMP as an intracellular messenger has been proposed as part of the calcium-channel complex in pituitary cells [25]. We could find no effect on intracellular cAMP accumulation by the calcium-channel agents in our pilot experiments. Inhibition of cAMP accumulation is difficult to measure because the inhibited levels are close to the detection limit of our automated assay. Despite raising the levels of cAMP with isoproterenol, a known stimulator of cAMP in the parathyroid cell [2], we were still unable to show inhibition of cAMP by calcium-channel agonists. This does not completely rule out the possibility that cAMP may be involved in the mediation of inhibition of PTH release by calcium, but under the conditions of our experiments we cannot provide evidence for this possibility.

We have previously demonstrated that inhibition of PTH secretion by calcium is blocked by treatment of parathyroid cells with pertussis toxin, and that there is concomitant ADP ribosylation of a 42-kDa protein in cell membranes (Fig. 2). Pertussis toxin treatment of parathyroid cells also blocks the PTH inhibition by the calcium-channel agonists (+)202–791 and Bay K8644 (data not shown). The ability of pertussis toxin to block the inhibition of PTH release by both calcium and calcium-channel agonist suggests a significant interaction between calcium channels and a G protein(s). This observation takes our initial data one step further by providing an operational localization of the G protein to a calcium channel.

Studies with Calcium-Channel Antibodies. Voltage-sensitive calcium channels facilitate the influx of calcium through the cell membrane and regulate hormone secretion, neurotransmitter release, and muscle contraction [20]. The calcium channel of particular interest to this project is sensitive to dihydropyridine compounds: the so-called L-type calcium channel. Agonists of this class promote the open state of the channel, and antagonists block the channel. We have studied the calcium channel in parathyroid cells by raising polyclonal, monospecific mouse antibodies against highly purified preparations of the α-subunits of voltage-sensitive calcium channels from rat transverse tubules (T-tubules). The effects of these antibodies on parathyroid cell secretion were tested.

Membrane fractions enriched in T-tubule vesicles were prepared from rat skeletal muscle. Calcium channels were purified from the rat T-tubule membranes as described by Curtis and Catterall [26] with modifications [5]. Female BALB/c mice were immunized intraperitoneally with purified calcium-channel subunits, and antisera obtained. Bovine parathyroid cells were dispersed with collagenase and DNase as previously described [3]. After incubation of the cells with the antibodies, cell supernatants were assayed for PTH by midregion specific radioimmunoassay [27].

Mouse antisera were examined for their effects on PTH release at two calcium concentrations, 0.5 mM (which normally stimulates PTH secretion) and 2.0 mM (which normally inhibits PTH secretion). Three different mouse antisera (MC-2, MC-3, MC-4) inhibited PTH release at a concentration of calcium normally associated with stimulated PTH release. One rabbit antiserum stimulated PTH release when incubated with parathyroid cells (HK-6). Normal BALB/c mouse serum at the same dilution (1 : 1000) and same calcium concentration did not alter PTH release (Fig. 6).

The effect of antiserum MC-4 was detected at dilutions as low as 1 : 10000 (Fig. 7). MC-4 antiserum diluted from 1 : 100000 to 1 : 1000000 had little or no effect on PTH release. Affinity-purified MC-4 was also tested in serial dilutions and effectively inhibited PTH release at 1 : 50000 dilution.

To test the specificity of the MC-4 antibody, antibody was preabsorbed with the purified α-subunit of rat T-tubule calcium-channel proteins, which had been immobilized on a nitrocellulose sheet. The preabsorbed antiserum was no longer

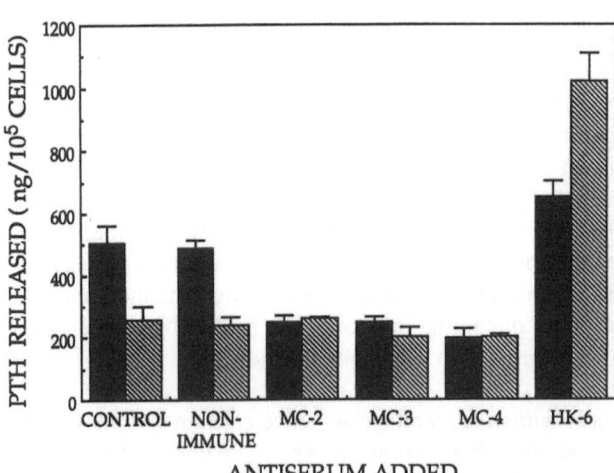

Fig. 6. Effect of polyclonal antiserum to the α-subunit of calcium channels on parathyroid hormone *(PTH)* secretion. Dispersed bovine parathyroid cells were incubated with polyclonal monospecific mouse antiserum *(MC-2, MC-3, MC-4)*, with rabbit antiserum *(HK-6)*, or with normal BALB/c mouse antiserum *(NON-IMMUNE)* at 1:1000 dilution. PTH secretion in control cells (no antiserum added) was suppressed by 50% at 2.0 mM calcium *(hatched bars)* as compared with cells in 0.5 mM calcium *(dark bars)*. Data are expressed as the mean ± SEM of duplicate experiments each analyzed in duplicate

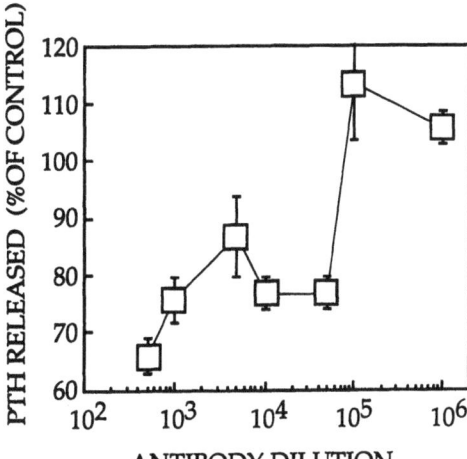

Fig. 7. Concentration response of MC-4 antiserum in terms of parathyroid cell secretion. Antiserum MC-4 was added to cells in medium containing 0.5 mM calcium, and parathyroid hormone *(PTH)* levels were determined on cell supernatants by radioimmunoassay. Data are expressed as a percentage of PTH release from control cells (no MC-4 antiserum added)

Table 2. Effect of preabsorption of antiserum on PTH release

Antiserum Added[a]	% PTH release [Ca^{2+}] mM	
	0.5	2.0
None	100 ± 6	42 ± 3
MC-4	75 ± 3	51 ± 2
MC-4 preabsorbed	100 ± 4	40 ± 3

MC-4 antiserum was absorbed for 24 h with purified α-subunits of the calcium channel which had been immobilized on nitrocellulose sheets. Bovine parathyroid cells were dispersed and incubated without MC-4, with MC-4, or with preabsorbed MC-4 for 60 min at 37°C. Midregion PTH was determined in cell supernatants by radioimmunoassays after separation by centrifugation. Data are expressed as the mean ± SEM of triplicate experiments. Triplicate samples from each experiment were analyzed in duplicate
[a] At 1:10000 dilution

effective as an inhibitor of low calcium-stimulated PTH release (Table 2). Preabsorbed antiserum was similarly ineffective at 1 : 50000 and 1 : 100000 dilutions.

The agonistlike properties of MC-4 antiserum suggest that this antiserum facilitates the open state of the calcium channel, allowing influx of extracellular calcium and inhibition of PTH secretion. We tested this hypothesis by measuring ^{45}Ca uptake in parathyroid cells. Dispersed bovine parathyroid cells were incubated to equilibrium with radiolabeled calcium. At time zero, MC-4 antiserum or normal BALB/c serum was added to the cells. Cells were filtered under vacuum at the times indicated, and radioactivity was determined by liquid scintillation spectrometry. MC-4 antiserum caused a marked increase in uptake of ^{45}Ca (Fig. 8). Normal BALB/c mouse did not cause an increase in ^{45}Ca uptake (data not shown).

G Protein Mediation of Calcium-Channel Regulation. The next set of studies has provided further support for localizing G protein interaction to the calcium channel. Dispersed bovine parathyroid cells were incubated with pertussis toxin for 4 h and

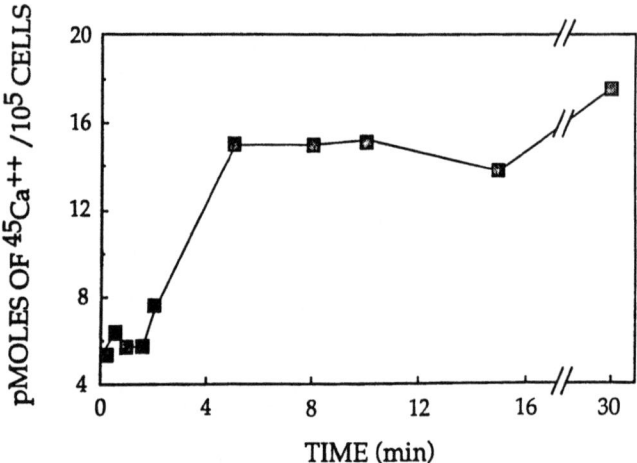

Fig. 8. Uptake of ^{45}Ca in parathyroid cells. Dispersed parathyroid cells were incubated with ^{45}Ca to exquilibrium (30 min), and MC-4 antiserum (1:10000 dilution) was added at time zero. Data represent the mean ± SEM of quintuplicate determinations

Table 3. Interaction of pertussis toxin and MC-4 antiserum

Addition(s)	PTH secreted (ng/10^5 cells)
None	1066 ± 38.8
MC-4	678 ± 27.0
MC-4 + pertussins toxin	942 ± 98.2

Bovine parathyroid cells were dispersed and incubated with or without pertussis toxin (0.1 µg/ml) for 4 h at room temperature. Cells were washed, and MC-4 antiserum was added at 1:10000 dilution. Cells were incubated at 37°C for 60 min, and PTH was determined in cell supernatants by radioimmunoassay. Each value represents the mean ± SEM of triplicate samples assayed in duplicate; one representative of three experiments is shown.

assessed for PTH release in conjunction with MC-4 antibody. In three experiments, pertussis toxin reversed the MC-4 – mediated inhibition of PTH release (Table 3).

Our study suggests that the dihydropyridine-sensitive calcium channels in parathyroid cells are regulated via G proteins. Further experiments and biochemical evaluation will be necessary to confirm this hypothesis and define more precisely the pertussis toxin-sensitive substrate involved in calcium-channel regulation.

Biochemical Evidence. Membrane proteins from bovine parathyroid cells were subjected to SDS-PAGE and electrophoretically transferred to nitrocellulose sheets. The latter were incubated overnight with affinity-purified mouse antibody to the α-subunit(s) of calcium-channel proteins or mouse antiserum absorbed against the α-subunit proteins. A 150-kDa protein is present, and no change in molecular weight is noted in reducing vs nonreducing conditions on immunoblot of the parathyroid cell membranes [5].

The 150-kDa protein does not bind with WGA (wheat germ agglutinin) peroxidase, suggesting that this protein is not a glycoprotein and is similar to the α_1-subunit of skeletal muscle calcium channel. The α_1-subunit contains sites for specific photoaffinity labeling by dihydropyridine derivatives [28] and is phosphorylated by cAMP-dependent protein kinase [29]. This suggests that the 150-kDa protein present in parathyroid tissue is similar to the α_1-subunit of calcium channels found in skeletal muscle.

Discussion

Although control of PTH secretion represents one of the first calcium-regulated physiological systems recognized [6], the biochemical mechanisms involved have yet to be elucidated. The parathyroid cell is unusual in that increasing concentrations of extracellular calcium inhibit PTH release. We have evaluated the role of calcium channels and guanine nucleotide regulatory proteins in the control of parathyroid cell secretion.

Pertussis toxin ADP-ribosylates and inactivates several G proteins. Treatment of parathyroid cells with pertussis toxin prevents the inhibition of PTH secretion by calcium. The result implies that G protein is involved in coupling the extracellular calcium signal to suppression of PTH secretion. A23187 causes suppression of PTH release irrespective of pertussis toxin treatment and thus bypasses the function linked to the guanine nucleotide regulatory protein.

The availability of agents that interact with calcium channels makes it possible to study the functions of these channels in diverse control processes regulated by calcium. Many electrically excitable cells contain voltage-sensitive calcium channels [20]. Calcium-channel antagonists, by blocking calcium channels, inhibit insulin release in pancreatic cells [30], GnRH-induced release of LH in anterior pituitary cells [31], and TRH-induced release of TSH in GH_4C_1 pituitary cells [32, 33]. In adrenal glomerulosa cells, calcium-channel antagonists block the aldosterone response to angiotensin II and potassium ions [34]. The calcium-channel agonists raise levels of intracellular calcium and stimulate release of insulin from pancreatic tissue [35].

In the parathyroid cell, the calcium-channel agonist (−)202−791 mimics the action of calcium and strikingly inhibits secretion. The calcium-channel antagonist (+)202−791 stimulates the release of PTH. The calcium-channel agonist and antagonist did not affect intracellular cAMP levels in this cell. Treatment of the parathyroid cell with pertussis toxin blocked the inhibition of PTH release by the calcium-channel agonist.

We further evaluated the role of calcium channels in PTH release by testing the effects of polyclonal antiserum directed against purified α-subunits of calcium channels from rat skeletal muscle T tubules. Antibodies to the α-subunits of dihydropyridine-sensitive calcium channels were found to function as channel agonists and inhibited secretion in parathyroid cells. One antiserum studied in detail caused a marked increase in uptake of ^{45}Ca.

Previous reports have suggested the involvement of G proteins in the regulation of ion channels [16]. In AtT-20/D16−16 pituitary cells, a G protein mediates somatostatin-induced inhibition of voltage-dependent calcium currents [36]. A GTP-binding protein regulates voltage-sensitive calcium channels in NG108–15 neuroblastoma-glioma hybrid cells [16]. Incubation of parathyroid cells with pertussis toxin blocked most of the inhibition by the antiserum. This suggests that a pertussis-sensitive G protein mediates the activation of calcium channels in the parathyroid cell.

Recent reports indicate that the α-subunit of rabbit skeletal muscle calcium channels consists of α_1- and α_2-subunits, two polypeptides that are distinguishable by their different biochemical properties. The α_1-subunit (170–175 kDa) is not a glycoprotein, and the electrophoretic mobility of this subunit remains unchanged after disulfide bonds are reduced. The α_1-subunit of skeletal muscle calcium channels contains specific binding sites for dihydropyridine [37, 38] and phenylalkylamine [39] derivatives and is phosphorylated by cAMP-dependent protein kinase [29]. Depolarization of GH_3 pituitary cells results in phosphorylation and activation of calcium channels. The α_2-subunit (143 kDa) is heavily glycosylated and linked by disulfide bonds to a protein of approximately 25 kDa. We suggest that the 150 kDa protein in parathyroid cell membranes is similar to the α_1-subunit of the skeletal muscle calcium channel.

Summary

The parathyroid cell is unusual in that increasing concentrations of extracellular calcium inhibit the release of hormone. This inhibition of parathyroid hormone release by calcium is mediated by a pertussis toxin-sensitive guanine nucleotide regulatory protein [3]. The effect of calcium-channel agonists and antagonists implies that calcium channels are involved in the regulation of PTH secretion. Calcium-channel agonists which open calcium channels allow influx of extracellular calcium and inhibition of PTH release [4].

Polyclonal, monospecific mouse antibodies against highly purified preparations of α-subunits of voltage-sensitive calcium channels from rat skeletal muscle T tubules were prepared, and the effects of these antibodies on PTH cell secretion were tested. Three of the antibodies inhibited PTH release; one antiserum stimulated release. The results suggest that antibodies bind to α-subunits of voltage-sensitive calcium channels and activate the channels, allowing influx of calcium, which in turn leads to an inhibition of PTH secretion. In addition, we suggest that the function of the dihydropyridine-sensitive channel in parathyroid cells is linked to a pertussis toxin-sensitive guanine nucleotide regulatory protein.

References

1. Rasmussen H (1986) The calcium messenger system. N Engl J Med 314:1094–1101
2. Brown EM (1982) PTH secretion *in vivo* and *in vitro*. Regulation by calcium and other secretagogues. Mineral Electrolyte Metab 8:130–150
3. Fitzpatrick LA, Brandi M-L, Aurbach GD (1986) Calcium controlled secretion is effected through a guanine nucleotide regulatory protein in parathyroid cells. Endocrinology 119:2700–2703
4. Fitzpatrick LA, Brandi ML, Aurbach GD (1986) Control of PTH secretion is mediated through calcium channels and is blocked by pertussis toxin treatment of parathyroid cells. Biochem Biophys Res Commun 138:960–965

5. Fitzpatrick LA, Chin H, Nirenberg M, Aurbach GD (1988) Antibodies to an α subunit of skeletal muscle calcium channels regulate parathyroid cell secretion. Proc Natl Acad Sci USA 85:2115–2119
6. Patt HM, Luckhardt AB (1942) Relationship of a low blood calcium to parathyroid secretion. Endocrinology 31:384
7. Habener JF, Kemper B, Potts JT Jr (1975) Calcium-dependent intracellular degradation of parathyroid hormone. A possible mechanism for the regulation of hormone stores. Endocrinology 97:431–441
8. Brown EM, Hurwitz S, Aurbach GD (1976) Preparation of viable isolated bovine parathyroid cells. Endocrinology 99:1582
9. Brown EM (1983) Four-parameter model of the sigmoidal relationship between parathyroid hormone release and extracellular calcium concentration in normal and abnormal parathyroid tissue. J Clin Endocrinol Metab 56:572–581
10. LeBoff MS, Chang J, Henry M, Brandoin D, Swiston L, Brown EM (1986) Role of cytosolic calcium in the control of cAMP content by calcium in bovine parathyroid cells. Molec Cell Endocrinology 45:127–135
11. Shoback DM, Thatcher J, Lemobruno R, Brown EM (1984) Relationship between parathyroid hormone secretion and cytosolic calcium concentration in dispersed bovine parathyroid cells. Proc Natl Acad Sci USA 81:3113–3117
12. Sawyer DW, Sullivan JA, Mandell GL (1985) Intracellular free calcium localization in neutrophils during phagocytosis. Science 230:663–666
13. Nemeth EF, Scarpa A (1987) Rapid mobilization of cellular Ca^{2+} in bovine parathyroid cells evoked by extracellular divalent cations. Evidence for a cell surface calcium receptor. J Biol Chem 262:5188–5196
14. Spiegel AM (1987) Review: Signal transduction by guanine nucleotide binding proteins. Mol Cell Endocrinol 49:1–16
15. Verghese MW, Smith CD, Snyderman R (1984) Potential role for a guanine nucleotide regulatory protein in chemoattractant receptor mediated polyphosphoinositide metabolism, Ca^{++} mobilization and cellular responses by leukocytes. Biochem Biophys Res Commun 127:414
16. Hescheler J, Rosenthal W, Trautwein W, Schultz G (1987) The GTP-binding protein, G_O, regulates neuronal calcium channels. Nature 325:445–447
17. Oetting M, LeBoff M, Swiston L, Preston J, Brown E (1986) Guanine nucleotides are potent secretagogues in permeabilized parathyroid cells. FEBS Lett 208:99–104
18. Reed PW, Hardy HA (1972) A23187: a divalent cation ionophore. J Biol Chem 247:6170
19. Stanfield PR (1986) Voltage-dependent calcium channels of excitable membranes. Br Med Bull 42:359–367
20. Glossmann H, Ferry DR, Goll A, Striessnig J, Zernig G (1985) Calcium channel and calcium channel drugs: recent biochemical and biophysical findings. Arzneim-Forsch/Drug Res 35:1917–1935
21. Larsson R, Akerstrom G, Gylfe E, Johansson H, Ljunghall S, Rastad J, Wallfelt C (1985) Paradoxical effects of K^+ and D-600 on parathyroid hormone secretion and cytoplasmic Ca^{2+} in normal bovine and pathological human parathyroid cells. Biochim Biophys Acta 847:263–269
22. Cooper CW, Borosky SA, Farrell PE, Steinsland OD (1986) Effects of the calcium channel activator BAY K8644 on in vitro secretion of calcitonin and parathyroid hormone. Endocrinology 118:545–549
23. Wallace J, Pintado E, Scarpa A (1983) Parathyroid hormone secretion in the absence of extracellular free Ca^{2+} and transmembrane Ca^{2+} influx. FEBS Lett 151:83–88
24. Hove K, Sand O (1981) Evidence for a function of calcium influx in the stimulation of hormone release from the parathyroid gland in the goat. Acta Physiol Scand 113:37–43
25. Gunning R (1987) Increased numbers of ion channels promoted by an intracellular second messenger. Science 235:80–82
26. Curtis BM, Catterall WA (1985) Phosphorylation of the calcium antagonist receptor of the voltage-sensitive calcium channel by cAMP-dependent protein kinase. Proc Natl Acad Sci USA 82:2528–2532
27. Sharp ME, Marx SJ (1985) Radioimmunoassay for the middle region of human parathyroid hormone: comparison of two radioiodinated synthetic peptides. Clin Chim Acta 145:59–66

28. Takahashi M, Catterall WA (1987) Indentification of an alpha subunit of dihydropyridine-sensitive brain calcium channels. Science 236:88–91
29. Armstrong D, Eckert R (1987) Voltage-activated calcium channels that must be phosphorylated to respond to membrane depolarization. Proc Natl Acad Sci USA 84:2518–2522
30. Malaisse-Lagae F, Mathias PCF, Malaisse WJ (1984) Gating and blocking of calcium channels by dihydropyridines in the pancreatic B-cell. Biochem Biophys Res Commun 123:1062–1068
31. Chang JP, McCoy EE, Graeter J, Tasaka K, Catt KJ (1986) Participation of voltage-dependent calcium channels in the action of gonadotropin-releasing hormone, J Biol Chem 261:9105–9018
32. Enyeart JJ, Aizawa T, Hinkle PM (1985) Dihydropyridine Ca^{2+} antagonists: potent inhibitors of secretion from normal and transformed pituitary cells. Am J Physiol 248 (5 Pt 1):C510–C519
33. Tan K-N, Tashjian AH Jr (1984) Voltage-dependent calcium channels in pituitary cells in culture. J Biol Chem 259:427–434
34. Aguilera G, Catt KJ (1986) Participation of voltage-dependent calcium channels in the regulation of adrenal glomerulosa function by angiotensin II and potassium. Endocrinology 118:112–118
35. Kojima I, Kojima K, Rasmussen H (1985) Characteristics of angiotensin II–, K^+- and ACTH-induced calcium influx in adrenal glomerulosa cells. Evidence that angiotensin II, K^+, and ACTH may open a common calcium channel. J Biol Chem 260:9171–9176
36. Lewis DL, Weight FF, Luini A (1986) A guanine nucleotide-binding protein mediates the inhibition of voltage-dependent calcium current by somatostatin in a pituitary cell line. Proc Natl Acad Sci USA 83:9035–9039
37. Sharp AH, Imagawa T, Leung AT, Campbell KP (1987) Indentification and characterization of the dihydropyridine-binding subunit of the skeletal muscle dihydropyridine receptor. J Biol Chem 262:12309–12315
38. Takahashi M, Seagar MJ, Jones JF, Reber BF, Catterall WA (1987) Subunit structure of dihydropyridine-sensitive calcium channels from skeletal muscle. Proc Natl Acad Sci USA 84:5478–5482
39. Striessnig J, Moosburger K, Goll A, Ferry DR, Glossmann H (1986) Stereoselective photoaffinity labelling of the purified 1,4-dihydropyridine receptor of the voltage-dependent calcium channel. Eur J Biochem 161:603–609
40. Harper JF, Brooker G (1975) Fentomol senitive radioimmunoassay for cyclic AMP and cyclic GMP after 2'o-acetylation by acetic anhydride in aqueous solution. J Cyclic Nucleotide Res 1-207–210

4 Neuropharmacology of Calcium Channels

The Neuropharmacology of Ca^{2+} Channels*

R. J. Miller, D. A. Ewald, S. N. Murphy, T. M. Perney, I. J. Reynolds, S. A. Thayer, and M. W. Walker

Department of Pharmacological and Physiological Sciences, University of Chicago, Chicago, IL 60637, USA

Introduction

Ion channels that allow the influx of Ca^{2+} into neurons or other cells have been traditionally divided into two groups: voltage-sensitive Ca^{2+} channels (VSCC) and receptor-operated channels (ROC). Ca^{2+} influx via these and other pathways is one method by which an increase in the intracellular free Ca^{2+} concentration, [Ca^{2+}]$_i$, can be achieved. The second method is through release of Ca^{2+} from intracellular bound stores. Such increases in [Ca^{2+}]$_i$ act as signals that trigger a large number of important neuronal responses. These include the release of neurotransmitters, the regulation of certain ion channels, modification of the cytoskeleton, and the production of long-term changes in the efficacy of synaptic transmission such as those associated with long-term potentiation (LTP). The division of ion channels into VSCC and ROC is somewhat artificial, however, as VSCC can certainly be modulated by receptor-mediated events and some ROC can function in a voltage-dependent fashion under physiological conditions. In this chapter we shall focus on two examples. The first concerns the inhibitory modulation of VSCC in dorsal root ganglion (DRG) neurons by a variety of neurotransmitters such as neuropeptide Y (NPY). The second concerns the regulation of Ca^{2+} entry into central neurons by the excitatory amino acid neurotransmitter glutamate. This latter process involves the activation of a number of glutamate receptor subtypes and a number of ion channels.

Neuropeptide Y

Neuropeptide Y (NPY) is a 36-amino-acid peptide which is very widely distributed in both the central and peripheral nervous systems (Allen et al. 1983; Chronwall et al. 1983; Ekblad et al. 1984). It is often found colocalized with other neurotransmitters such as norepinephrine (Ekblad et al. 1984; Everitt et al. 1984). NPY exhibits extremely close sequence homologies to the gut peptide PYY and quite close homologies to the pancreatic peptide (PP) family of hormones (Tatemoto et al. 1982). NPY has been shown to have pre- and postsynaptic actions at several neuroeffector junctions (Wahlestedt et al. 1985). In particular, at sympathetic synapses, NPY has

---

* This work was supported by PHS grants DA-02121 and MH-40165 and by grants from Miles Inc. and Marion Labs.

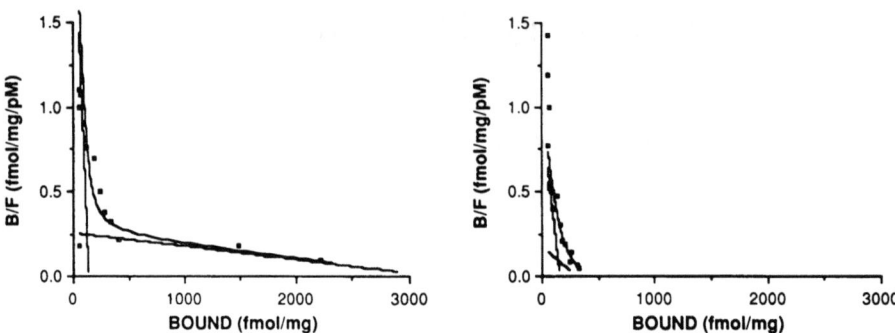

Fig. 1. Binding of ^{125}I-NPY to membranes prepared from rat DRG neurons grown in vitro. *Lower panel* indicates that inclusion of Gpp(NH)p reduces binding to the lower-affinity binding site

been shown to potently block the evoked release of norepinephrine (Serfozo et al. 1986). Furthermore, we have demonstrated that NPY blocks the evoked release of substance P from rat DRG neurons in vitro (Walker et al. 1988). We attempted to elucidate the mechanism by which NPY produced inhibition of evoked neurotransmitter release.

Figure 1 shows that ^{125}I-NPY labels two high-affinity binding sites in cultured DRG cells. Binding to one of these sites is reduced when Gpp(NH)p, a nonhydrolyzable analog of GTP, is inclued in the binding assay. This result indicates that one group of NPY receptors is coupled to a G protein. NPY could potentially reduce the release of substance P by several mechanisms. The most straightforward of these would be the direct inhibition of Ca^{2+} entry into neurons. Indeed, we observed that NPY was able to potently reduce the Ca^{2+} current (I_{Ca}) measured in DRG cells using the whole-cell voltage-clamp procedure (Fig. 2). It is interesting to note that NPY reduced both the total I_{Ca} evoked from negative holding potentials and also the sustained Ca^{2+} current evoked from more positive holding potentials. This implies that NPY probably inhibits both the L-type VSCC that produce the sustained DRG I_{Ca} and also the N-type VSCC that produce the major portion of the inactivating I_{Ca} (Miller 1987; Nowycky et al. 1985). Clearly the inhibition of Ca^{2+} influx produced by NPY is likely to be responsible for its observed ability to inhibit neurotransmitter release.

Another question of interest is the mechanism by which NPY produces inhibition of the Ca^{2+} current. We found that the effects of NPY were completely abolished following treatment of DRG cells with pertussis toxin (Fig. 2). This result also suggests that a G protein mediates the effects of NPY and that this G protein is a pertussis toxin substrate. We attempted to elucidate the identity of the G protein involved by introducing the purified α subunits of a number of pertussis toxin substrates isolated from bovine brain into pertussis toxin-treated DRG cells. The α subunit of 39-KD pertussis toxin substrate previously designated G_o (Gilman 1987; Sternweis and Robishaw 1984) reconstituted the NPY induced inhibition of the DRG I_{Ca} in a concentration- and time-dependent manner (Fig. 3). (Ewald et al. 1988). The α subunit of a 41-KD G protein (G_1i) produced a modest amount of reconstitution, but the α subunit of a 40-KD G protein (G_2i) was completely ineffective. As G_o is found in

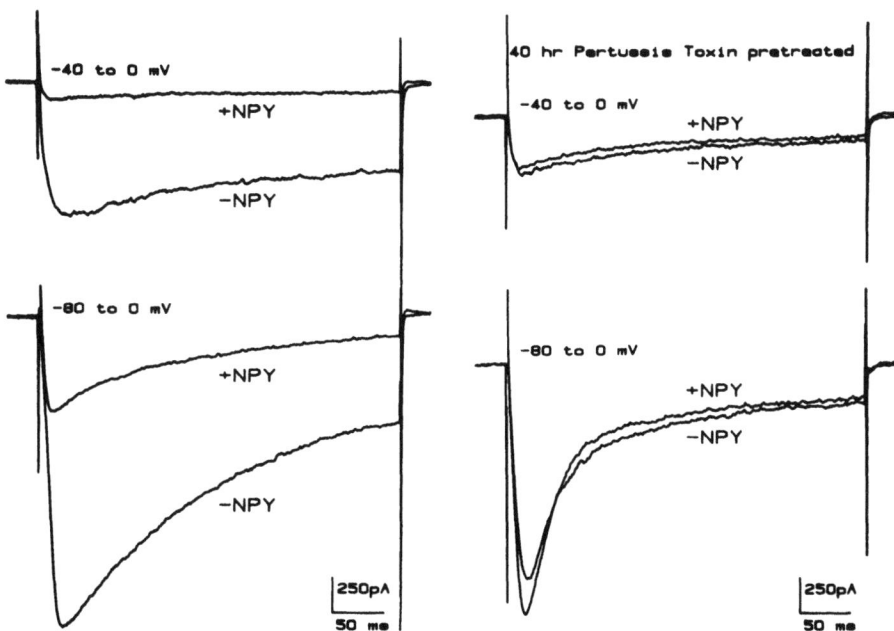

Fig. 2. Effect of NPY ($10^{-7}M$) on the I_{Ca} in rat DRG neurons. *Left panel:* the I_{Ca} evoked from two different holding potentials is effectively reduced by NPY. *Right panel:* the effects of NPY are completely blocked by treating cells with pertussis toxin

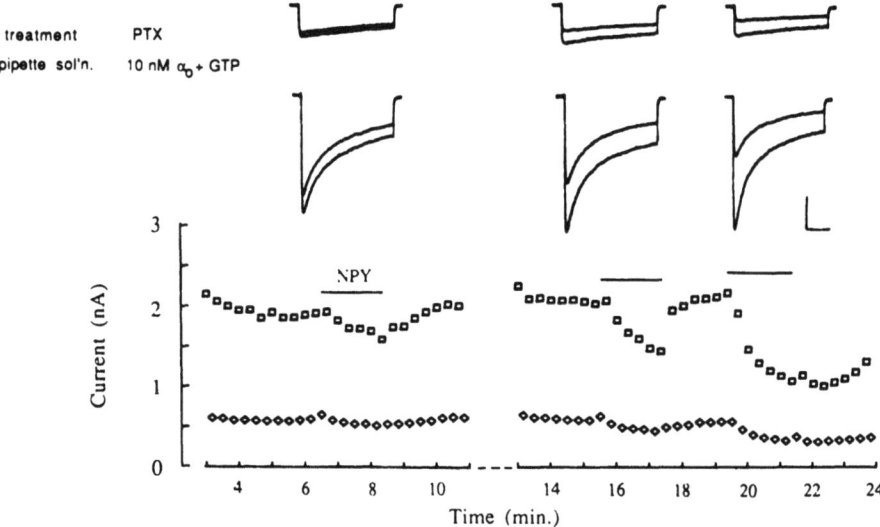

Fig. 3. Reconstitution of NPY inhibition of the I_{Ca} in DRG cells. The purified α subunit of G_o has been introduced into this pertussis toxin *(PTX)*-treated DRG neuron by means of a patch pipette. The NPY-mediated inhibition of the I_{Ca} returns in a time-dependent manner

neurons, including DRG cells, in particularly high concentrations (Asano et al. 1988), these results imply that it may be the G protein that normally couples NPY receptors to VSCC in DRG cells. An interesting comparison can be made between the effects of NPY and similar effects we have observed with bradykinin (BK). BK also blocks the DRG I_{Ca} in a pertussis toxin-sensitive fashion. We attempted to reconstitute this inhibitory effect with the same G proteins used in our experiments with NPY. In this case we found that all three G proteins were effective. Indeed, the 40-KD protein was almost as effective as the 39-KG protein. This result implies that the specificity observed with NPY is probably related to the NPY receptor/G protein interaction rather than coupling between the G protein and the VSCC.

We also asked whether the NPY receptors and their associated G proteins were directly coupled to the VSCC, as in the case of activation of cardiac atrial K^+ channels by muscarinic agonists, or whether a diffusible second messenger is involved. It has been observed that the I_{Ca} of chick DRG neurons can be blocked by phorbol esters, implicating the enzyme protein kinase C (PKC) in coupling receptors to VSCC in DRG cells (Rane and Dunlap 1986). We tested this hypothesis by examining the effects of NPY in cells which contained no PKC following down-regulation of the enzyme by chronic phorbol ester treatment (Ewald et al. 1988). Under these circumstances we observed that there was some reduction in the ability of both NPY and BK to reduce the sustained I_{Ca} evoked from a holding potential of −40 mV. However, both NPY and BK retained their full inhibitory effect on the transient portion of the I_{Ca}. Thus, the major portion of the inhibitory effects of NPY and BK do not seem to involve PKC and may be due to direct coupling of VSCC to receptors. However, there may be a role for PKC specifically in the inhibitory modulation of L-type VSCC.

With these conclusions in mind it was also of interest to examine the effects of NPY on DRG cell phospholipid metabolism. Again, we compared these effects with those of BK. We found that NPY stimulated the synthesis of inositol triphosphate (IP_3) and

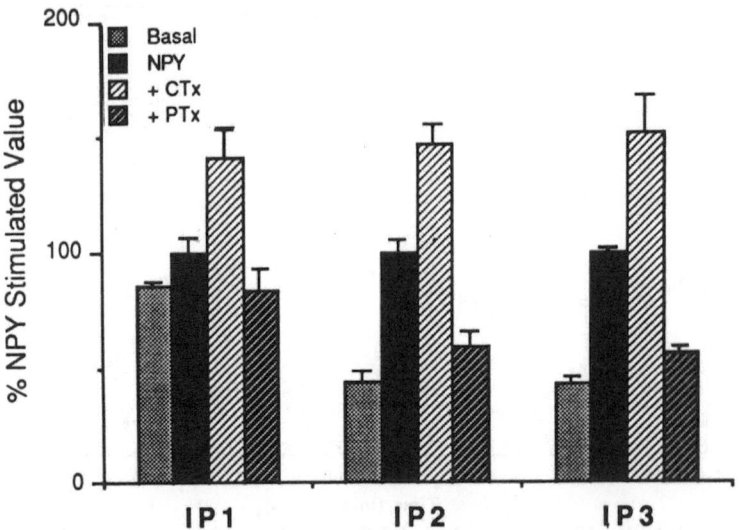

Fig. 4. NPY (10^{-7} *M*) stimulates inositol triphosphate *(IP$_3$)* production by cultured rat DRG neurons. Note effects of pertussis toxin *(PTx)* and cholera toxin *(CTx)*

diacylglycerol (DAG). The stimulation of IP$_3$ synthesis was completely blocked by pertussis toxin (Fig. 4). NPY had no effect on the release of arachidonic acid from labeled neurons. BK also stimulated IP$_3$ and DAG synthesis and in addition stimulated the release of arachidonic acid. In contrast to the effects of NPY, the ability of BK to stimulate IP$_3$ synthesis was unaffected by pertussis toxin. Thus, both NPY and BK produced DAG, which is consistent with the proposal that PKC mediates some of their effects on the I$_{Ca}$. Furthermore the results with BK support the notion that BK receptors can interact with a variety of G proteins in these cells whereas NPY receptors seem to exhibit a much higher degree of specificity.

Excitatory Amino Acids

VSCC are also important for controlling the influx of Ca^{2+} into neurons of the central nervous system (CNS) (Miller 1987). In addition, however, at least one other important pathway has been defined. This second pathway is regulated by the excitatory amino acid neurotransmitter glutamate. As in peripheral neurons the entry of Ca^{2+} into neurons of the CNS has many important functions. Of particular interest is its involvement in the generation of forms of synaptic plasticity such as LTP (Abrams and Kandel 1988). In addition, it has also become clear that prolonged abnormally high [Ca^{2+}]$_i$ can be toxic to neurons (Choi 1987). It has been shown that large amounts of glutamate are released from cells in the brain during periods of cerebral ischemia (Sanchez-Prieto and Gonzales 1988). This leads to an abnormally large uptake of Ca^{2+} by neurons and their subsequent death. Such a process is obviously of great relevance for studies of brain disorders involving neurodegeneration. How exactly is amino acid-induced Ca^{2+} uptake produced? Glutamate can act at a number of different types of receptor which directly gate a variety of ion channels (Mayer and Westbrook 1987). Each subtype of glutamate receptor is typified by the action of certain agonists and antagonists. The glutamate analog N-methyl-D-aspartic acid (NMDA) is the archetypal agonist at one receptor type. Figure 5 demonstrates that

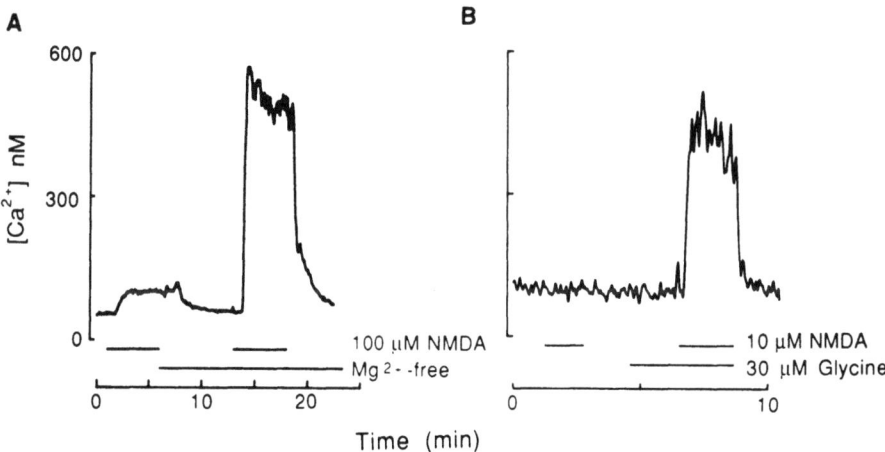

Fig. 5. NMDA stimulates Ca^{2+} infux into striatal neurons. *Left panel:* effect of NMDA in the absence and presence of Mg^{2+}. *Right panel:* synergistic effects of NMDA and glycine

NMDA can produce a large influx of Ca^{2+} into CNS neurons. There are several key properties associated with this response. The first is its complete blockade by physiological concentrations of Mg^{2+}. However, the Mg^{2+}-induced block is highly voltage-dependent and is relieved when cells are depolarized (Nowak et al. 1984). Zn^{2+} also block the effects of NMDA but in a non-voltage-dependent fashion, suggesting a different site of action (Westbrook and Mayer 1987). A second important influence on the NMDA receptor appears to be the amino acid glycine (Johnson and Ascher 1987). Figure 5 shows that whereas neither NMDA nor glycine added alone produce an increase in $[Ca^{2+}]_i$ when they are added together a very large effect is seen. The physiological significance of this powerful influence of glycine is not yet clear. Thus the concentration of glycine in the extracellular space is about 1 μM, which would be high enough to produce maximal augmentation. The response produced by NMDA under the appropriate conditions is not blocked by manipulations designed to block neuronal VSCC. Thus it appears that NMDA gates a unique Ca^{2+}-permeable ionophore that is distinct from any of the known VSCC. This conclusion is supported by electrophysiological studies (MacDermott et al. 1986).

The NMDA receptor is further distinguished by the effects of a group of specific antagonists. Competitive antagonists such as AP5 and CPP[+] block the binding of the agonist by direct competition at its recognition site (Foster and Fagg 1984). However phencyclidine (PCP) and related agents block responses to NMDA in a voltage- and use-dependent fashion (MacDonald et al. 1987). Thus the binding site for PCP and for MK801 is probably within the ion channel that is directly gated by NMDA. The relative effects of these various modulatory influences on the NMDA receptor can be seen in the experiment illustrated in Figure 6. Cortical membranes were treated with a combination of glutamate and glycine to allow "opening" of the NMDA-linked ionophore. ^{3}H-MK801 was then allowed to bind to its receptor (Reynolds and Miller

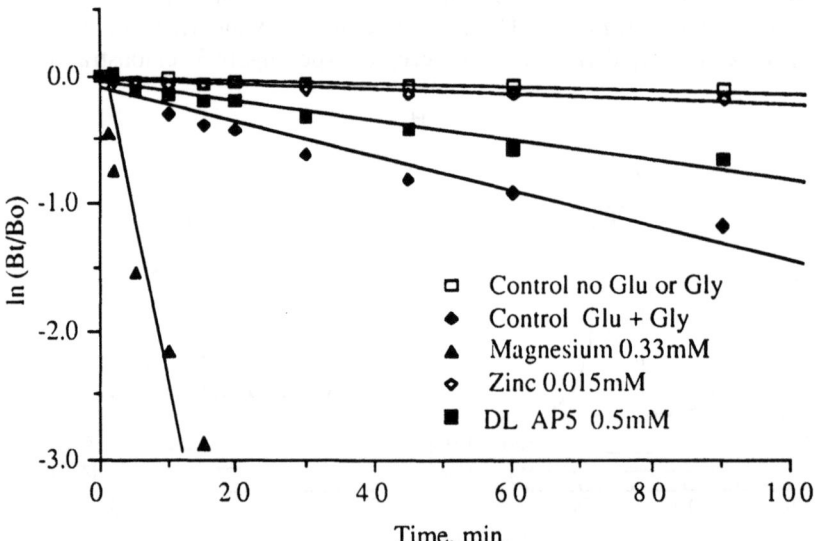

Fig. 6. Dissociation of ^{3}H-MK801 from rat cortical membranes. Effects of various modulators of NMDA receptor function

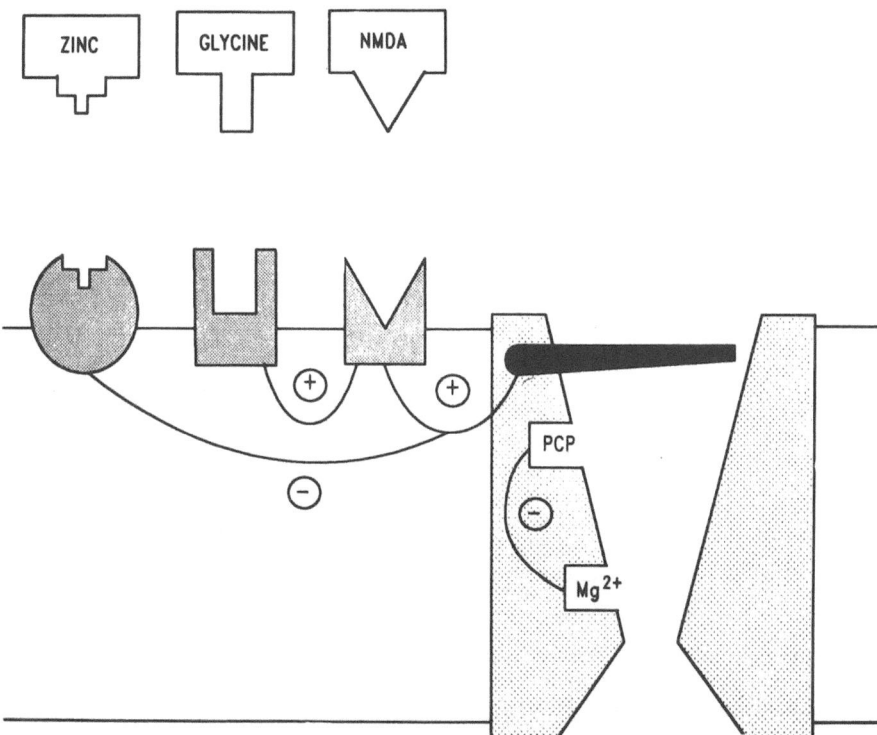

Fig. 7. Proposed model for the interaction of various modulators with the NMDA receptor

1988). Following the establishment of equilibrium, the dissociation of the drug was examined. Several effects are clear. Drugs that act at the same site as MK801 (e.g. PCP) have no effect on the drug dissociation rate. Agents that lead to the dissociation of agonist either by competitive interaction (e.g. AP5) or by noncompetitive mechanisms (e.g. Zn^{2+}) greatly reduce the drug dissociation rate. Presumably when the agonist dissociates the channel closes and the drug is trapped. Mg^{2+}, on the other hand, greatly accelerates the rate of ^{3}H-MK801 dissociation. Mg^{2+} also acts within the ionophore, and presumably the binding sites for Mg^{2+} and MK801 are allosterically coupled. The results allow us to suggest a model of the NMDA receptor as shown in Fig. 7.

In addition to NMDA, other archetypal glutamate agonists such as kainate are also extremely neurotoxic. Figure 8 shows that kainate also produces a large increase in $[Ca^{2+}]_i$. As in the case of NMDA, this increase is due to Ca^{2+} influx, as it is absent in Ca^{2+}-free medium. However, the response produced by kainate clearly has different properties from that produced by NMDA. For example, it is not blocked by Mg^{2+}. However, when neuronal VSCC are blocked by a combination of voltage-dependent inactivation and nitrendipine, the effects of 100 μM kainate are greatly reduced. It is known that normally kainate activates an ionophore that is rather selective for Na^+ (Mayer and Westbrook 1987). Thus kainate depolarizes neurons and this in turn leads

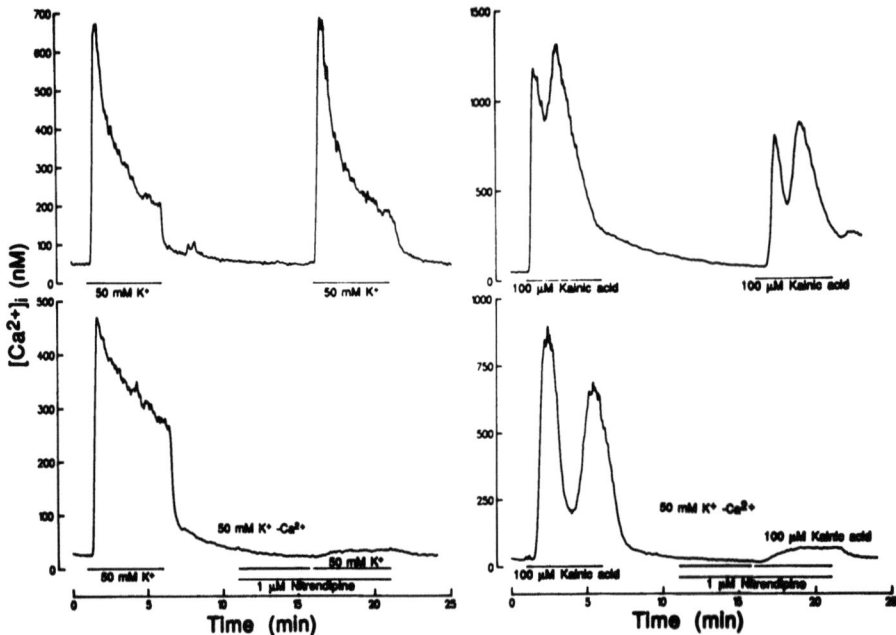

Fig. 8. Effect of Ca^{2+}-channel blockade on kainate-induced Ca^{2+} influx into striatal neurons. *Left panels:* the effects of K^+ depolarization before and after blockade of Ca^{2+} channels by a combination of depolarization and nitrendipine. *Right panels:* such treatment also substantially blocks the effects of kainate

to the opening of VSCC and Ca^{2+} entry via this route. Interestingly, however, at higher kainate concentrations a further phenomenon can be observed. Kainate-induced increases in $[Ca^{2x}]_i$ become progressively more resistant to the blockade of VSCC. Thus high kainate concentrations appear to activate a second pathway for Ca^{2+} entry in addition to VSCC. Additional studies reveal that this secondary pathway is not the NMDA receptor and probably represents the kainate-gated ionophore itself. When glutamate is released during periods of cerebral ischemia, for example, it can induce Ca^{2+} entry into CNS neurons by a variety of pathways. All of these must be considered when designing novel anti-ischemic agents.

References

Abrams TW and Kandel ER (1988) Is contiguity detection in classical conditioning a system or cellular property? Learning in *Aplysia* suggests a possible molecular site. Trends in Neurosci 11:128–135

Allen YS, Adrian TE, Allen JM, Tatemoto K, Crow TJ, Bloom SR and Polak JM (1983) Neuropeptide Y distribution in the rat brain. Science 221:877–879

Asano T, Semba R, Kamiya N, Ogasawara N and Kato K (1988) G_o, a GTP binding protein: immunological and immunohistochemical localization in the rat. J Neurochem 50:1164–1169

Choi DW (1987) Ionic dependence of glutamate neurotoxicity. J Neurosci 7:369–379

Chronwall BM, DiMaggio DA, Massari VJ, Pickel VM, Ruggiero DA and O'Donohue TL (1985) The anatomy of neuropeptide Y containing neurons in the rat brain. Neurosci 15: 1159–1181

Ekblad E, Hakanson R and Sundler F (1984) VIP & PH1 coexist with NPY like peptides in intramural nerves of the small intestine. Reg Pep 10:47–58

Everitt BJ, Hokfelt T, Terenius L, Tatemoto K, Mutt V and Goldstein M (1984) Differential coexistence of neuropeptide Y (NPY) with catecholamines in the central nervous system of the rat. Neurosci 11:443–462

Ewald DA, Sternweis PC and Miller RJ (1988) G_o induced coupling of NPY receptors to calcium channels in sensory neurons. Proc Natl Acad Sci (USA) (in press)

Ewald DA, Matthies HJG, Perney TM, Walker MW and Miller RJ (1988) The effect of down regulation of protein kinase C on the inhibitory modulation of dorsal root ganglion neuron Ca^{2+} currents by neuropeptide Y. J Neurosci (in press)

Foster AC and Fagg GE (1984) Acidic amino acid binding sites in mammalian neuronal membranes: their characteristics and relationship to synaptic receptors. Brain Res Rev 7:103–164

Gilman AG (1987) G proteins: transducers of receptor generated signals. Ann Rev Biochem 56:615–649

Johnson JW and Ascher P (1987) Glycine protentiates the NMDA response in cultured mouse brain neurons. Nature 325:529–531

MacDermott A, Mayer ML, Westbrook GL, Smith SJ and Barker JC (1986) NMDA receptor activation increases cytoplasmic Ca^{2+} concentration in cultured spinal cord neurons. Nature 321:519–522

MacDonald JF, Miljkovic Z and Pennefather P (1987) Use dependent block of excitatory amino acid currents in cultured neurons by ketamine. J Neurophysiol 58:251–266

Mayer ML and Westbrook GL (1987) The physiology of excitatory amino acids in the vertebrate central nervous system. Prog Neurobiol 28:197–276

Miller RJ (1987) Multiple calcium channels and neuronal function. Science 235:46–52

Nowak C, Bregestowski P, Ascher P, Herbet A and Prochiantz A (1984) Magnesium gates glutamate activated channels in mouse central neurons. Nature 307:462–465

Nowyck MC, Fox AP and Tsien RW (1985) Three types of neuronal calcium channels with different calcium agonist sensitivity. Nature 316:440–443

Rane SG and Dunlap K (1986) Kinase C activator 1,2-oleylacetylglycerol attenuates voltage dependent calcium current in sensory neurons. Proc Natl Acad Sci (USA) 83:184–188

Reynolds IJ and Miller RJ (1988) Multiple sites for the regulation of the N-methyl-D-aspartate receptor. Molec Pharmacol (in press)

Sanchez-Prieto J and Gonzalez P (1988) Occurrence of a large Ca^{2+} independent release of glutamate during anoxia in isolated nerve terminals (synaptosomes). J Neurochem 50:1322–1324

Serfozo P, Bartfai T and Vizi ES (1986) Presynaptic effects of neuropeptide Y on ^{3}H-noradrenaline and ^{3}H-acetylcholine release. Reg Pep 16:117–123

Sternweis PC and Robishaw JD (1984) Isolation of two proteins with high affinity for guanine nucleotides from membranes of bovine brain. J Biol Chem 259:13806–13813

Tatemoto K, Carlquist M and Mutt V (1982) Neuropeptide Y: a novel brain peptide with structural similarities to peptide YY and pancreatic polypeptide. Reg Pep 13:317–328

Wahlestedt C, Yanaihara N and Hakanson R (1985) Evidence for different pre- and postjunctional receptors for neuropeptide Y and related peptides. Reg Pep 13:317–328

Walker MW, Ewald DA, Perney TM and Miller RJ (1988) Neuropeptide Y modulates neurotransmitter release and Ca^{2+} currents in rat sensory neurons. J Neurosci (in press)

Westbrook GL and Mayer ML (1987) Micromolar concentrations of Zn^{2+} antagonize NMDA and GABA responses of hippocampal neurons. Nature 328:640–643

Kinetic Characteristics of Different Calcium Channels in the Neuronal Membrane

P. G. Kostyuk[1], Ya. M. Shuba[1], A. N. Savchenko[1], and V. I. Teslenko[2]

[1] Bogomoletz Institute of Physiology, Bogomoletzstr., 4, 252601 GSP, Kiev 24, USSR,
[2] Institute for Theoretical Physics, Ukrainian Academy of Sciences, Kiev, USSR

Introduction

It is clear now that the calcium conductance of neuronal membrane is based on the activity of several types of calcium channels. Originally two types of calcium channels were discovered in the somatic membrane of DRG neurons of rat (Veselovsky and Fedulova 1983). As they are different in their operational potential range, these calcium channels can be identified as low- and high-threshold. Channels of the first type become active at membrane potentials between -60 and -30 mV and reveal a clear voltage- and time-dependent inactivation (LTI channels, or T type according to Tsien's nomenclature). The second type required more positive potentials for activation (up to 0 mV) and their inactivation was much less pronounced. Subsequently this finding was confirmed in a number of investigations in which properties of both types of channels were studied in detail (Carbone and Lux, 1984a, b, 1987a, b; Fedulova et al., 1985; Kostyuk et al., 1986). However, it has been shown recently with sensory neurons of vertebrates that high-threshold calcium channels are also nonhomogeneous and can be separated into two classes – inactivating and noninactivating (HTI, HTN; N and L according to Tsien) (Nowycky et al., 1985b; Kostyuk et al., 1987; Fox et al., 1987a, b).

It is obvious that in parallel with differences in potential dependence, selectivity, pharmacological sensitivity, etc., the three types of calcium channels may differ in their kinetic properties, in particular in activation kinetics. However, the latter question is still insufficiently studied. One of the reasons for this is the absence of a kinetic model for calcium-channel activation which could describe experimental data observed on both macroscopic and single-channel current levels. Indeed, as was shown in our early experiments the activation of both low- and high-threshold calcium currents in snail and mammalian neurons can be well described by a modified Hodgkin-Huxley model using a square power of th m variable (Kostyuk et al., 1977; Fedulova et al., 1981, 1985). These data have been confirmed in other laboratories. The model suggests the presence of two independent, equally charged gating particles, whose transition along the channel axis can be described by first-order kinetic equations with potential-dependent rate constants. This model fits well to the data about stationary characteristics and kinetics of asymmetric displacement currents accompanying the activation of calcium channels which may represent the movement of the suggested gating particles (Kostyuk et al., 1981). However, spectrum analysis of high-threshold calcium-current fluctuations and direct analysis of single calcium channel activity have shown the insufficiency of this simple model. The data obtained let to the assumption of the presence of a fast potential-independent kinetic stage that precedes the open state of the channel (Krishtal et al., 1980; Fenwick et al., 1982; Hagiwara and Byerly, 1983; Brown et al., 1982; Hagiwara and Ohmori, 1983; Brown et al., 1984; Shuba and Savchenko, 1985). The two approaches to the understanding of kinetic mechanisms of calcium channels are still little related to one another. Recently, therefore, we tried to describe the activation of different types of single calcium channels using a four-state model for their gating mechanisms

$$R \underset{\beta}{\overset{2\alpha}{\rightleftarrows}} C \underset{2\beta}{\overset{\alpha}{\rightleftarrows}} A \underset{b}{\overset{a}{\rightleftarrows}} O \qquad (1)$$

which is a natural generalization of both approaches. In this model α and β represent potential-dependent rate constants for the transitions from resting state R into intermediate closed C and activated A states; a and b are those for fast transitions between activated A and open O states (a, b $\gg$ α, β).

To prove the validity of this model, different preparations which possess specific types of calcium channels were used. Such approach allows the activity of single calcium channels of the appropriate type to be analyzed without fear that the results will be distorted by the activity of other types of channels. The influence of permeant ion species on the activation kinetics of calcium channels was also examined in the context of the proposed model.

Channel Type Separation

Experiments in our group were performed on several preparations using the patch-clamp technique (Hamill et al., 1981). Pharmacological separation of three types of calcium channels and kinetic characterization of HTI channels were made on cultured dorsal root ganglion (DRG) neurons from 12- to 14-day-old mouse embryos. The procedure for cell cultivation was described in detail by Skibo and Koval (1984). Most cells were used after 5 days in culture. The activation kinetics of LTI channels was investigated on neuroblastoma cells of clone N1E-115, and the pheochromocytoma cell line PC-12 was used for the description of HTN channels.

After formation of a gigaseal between the membrane of cultured DRG neurons of mouse embryos and the tip of the recording pipette in cell-attached configuration, inward current pulses could be recorded in about 80% of the patches investigated. As the pipette solution was composed so as to suppress the activity of all types of ion channels except calcium ones, these pulses reflected the activity of unitary calcium channels. The character of the activity observed varied essentially from one patch to another. The current pulses in different patches could be seen in different potential ranges both for cells placed in normal Ringer's solution and for those after zeroing V_r by high external K^+. They differed also in their unitary amplitude and kinetic features. Detailed examination of experimental current records has shown that all these observations result from the functioning of three distinct types of calcium channels.

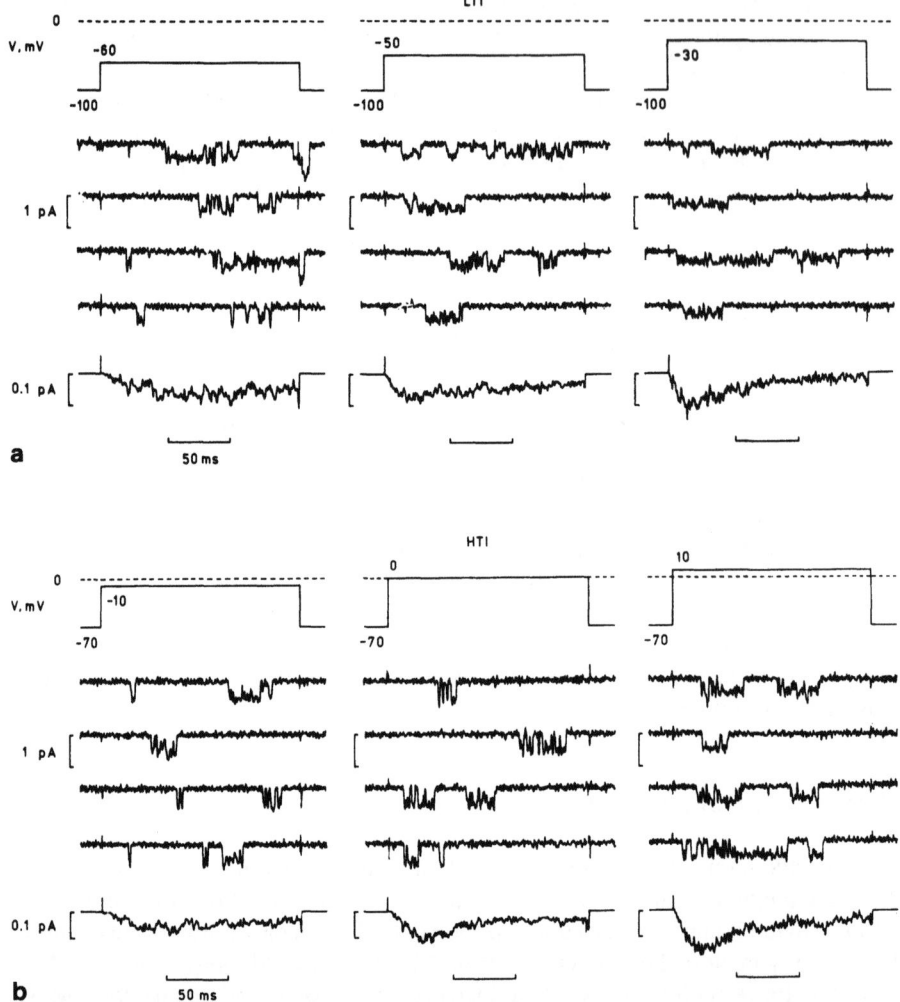

Fig. 1a-c. Three types of single calcium channel activity of the membrane of mouse sensory neurons. Original current records (obtained in cell-attached configuration) in response to four successive depolarizations at each potential (shown at top) for **a** LTI, **b** HTI, and **c** HTN channels. 60 mM Sr^{2+} in the pipette. *Lower traces:* corresponding averages of idealized current records (V$_r$ −15 mV taken into account). Data from different cells

Figure 1 presents examples of the activity of all these types of calcium channels with 60 mM Sr^{2+} as permeant cation. LTI (type T) channels with the lowest unitary conductance (5.7 pS with 60 mM Sr^{2+} or 7.2 pS with 60 mM Ba^{2+}) were activated in the potential range between −60 and −20 mV and revealed a clear potential- and time-dependent inactivation (Fig. 1a). They preserved their activity for a long time in excised membrane patches and were insensitive to the dihydropyridine calcium channel agonist Bay K8644. Corresponding whole-cell current could be decreased by 40% with 25 µM D-600.

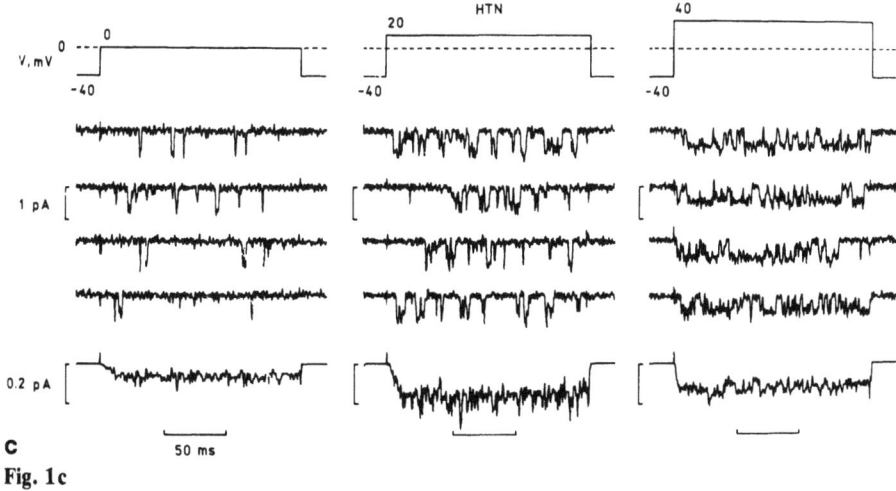

Fig. 1c

HTI (type N) calcium channels had a somewhat higher unitary conductance (7 pS and 11.4 pS for Sr^{2+} and Ba^{2+} respectively at 60 mM) and required more positive depolarizations to be activated (Fig. 1b). They also revealed potential- and time-dependent inactivation. HTI calcium channels were much more sensitive to intracellular metabolic support. D-600 inhibited the corresponding HTI whole-cell current by about 10%.

HTN (type L) channels were the most conducting ones – 9 pS and 18.4 pS for 60 mM Sr^{2+} or 60 mM Ba^{2+} (Fig. 1c). The main characteristic feature of HTN channels was a practically complete absence of inactivation. They required still more positive potentials to be activated (0 – 40 mV) and their functioning was strongly metabolically dependent. Whole-cell HTN current could be slightly enhanced by 10 μM Bay K8644 due to some prolongation of channel mean open time. However, the effect of Bay K8644 was much less pronounced than that reported for cardiac cells. HTN whole-cell current could be almost completely blocked by 25 μM D-600.

The three types of calcium channels described revealed different potential dependence and absolute values of mean channel open time. Open-time histograms for LTI channels can be fitted satisfactorily by a single exponent; mean open time for 60 mM SR^{2+} as a charge carrier decreased from 2 to 0.8 ms with an increase in depolarization from −60 to −30 mV. Open-time distribution for HTI channels at all test potentials was also single exponential with mean open time ~ 1.4 ms under the same conditions. This value remained practically unchanged with an increase in membrane potential. At the same time, with Sr^{2+} as a charge carrier the mean open time of HTN channels increased with an increase of depolarization from 0.57 ms at 0 mV to 1.67 ms at 40 mV. However, it should be pointed out that open-time histograms for HTN channels could hardly be fitted with single exponentials. Better results could be obtained using double exponential fitting.

Separation of HTI and HTN channels at single-channel level is much simpler using Ba^{2+} or Sr^{2+} as a charge carrier instead of Ca^{2+}. This is mainly due to more pronounced differences in unitary current amplitudes in barium or strontium solutions than in calcium solutions. According to our observations, the probability of obtaining functional HTI channel increases depending on permeant ion species in the sequence Ca > Sr > Ba. For HTN channels the sequence seems to be the opposite.

HTI Channels

As mentioned above, in conditions with Ca^{2+} ions as natural charge carriers HTI channels are the most frequent in single-channel experiments. That is why the model shown in Eq. 1 was the first to be applied to the analysis of the channels of this type.

Figure 2 presents histograms for HTI channel lifetimes in different states and their approximations by single or double-exponential functions using the least-square method. The open-time histograms (Fig. 2a), as already mentioned, were monoexponential for all test potentials with only slightly potential-dependent mean open time $\tau_{op} \approx 1.2$ ms (Fig. 3a). The closed-time histograms had a more complicated form (Fig. 2b). Most events came in the initial part (0–2 ms) followed by a low-amplitude "tail" lasting up to 100 ms. Such distribution was observed at all test potentials. The "fast"

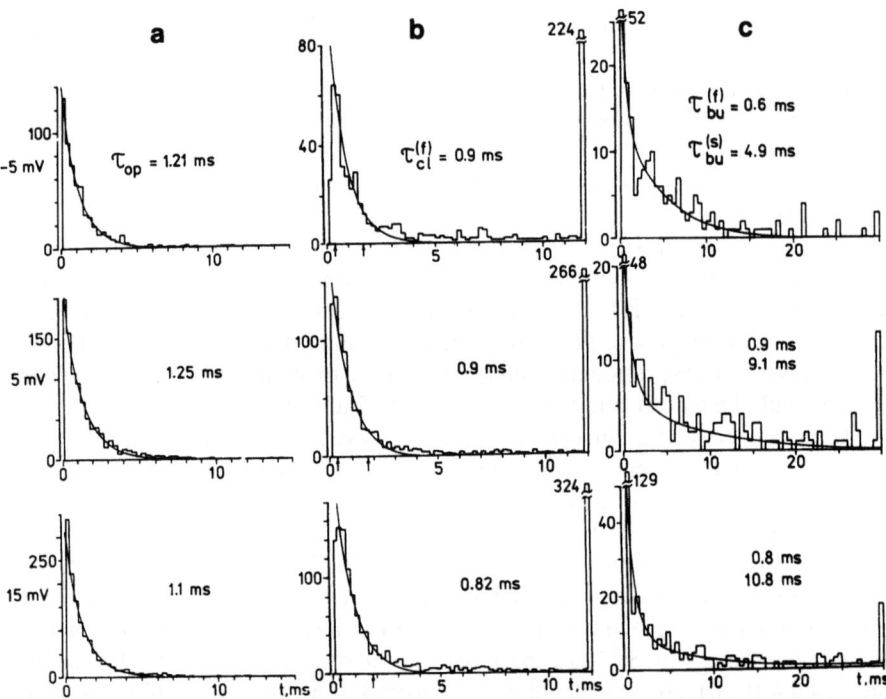

Fig. 2a–c. Temporal characteristics of functioning of single HTI calcium channels. **a** Open-time, **b** closed-time, and **c** burst-length histograms at three membrane potentials (indicated on *left*) and their approximations by single (**a, b**) and biexponential (**c**) curves. *Arrows* (**b**) indicate a part of the histogram used for approximation. The corresponding time constants are indicated. Binwidth **a** 0.3 ms; **b** 0.2 ms; **c** 0.5 ms. The last bin of each histogram contains all remaining events up to 100 ms. The figures near each last bin in **b** and each first bin in **c** indicate the number of events containing in these bins

component of this distribution could be successfully fitted by a single exponential function with time constant $\tau_{cl}^{(f)} = 0.9$ ms. As can be seen from Fig. 3b, the latter was practically independent of test potential.

The open-time histogram (Fig. 2a) and the "fast" component of the closed-time histogram (Fig. 2b) evidently represented potential independent conformational transitions of the channel between closed activated and open states (transitions $A_b \overset{a}{\rightleftharpoons} O$) which produce its "flickering" during bursts. These transitions do not involve transmembrane charge movement but only local transformations of the channel gating mechanism.

The small contribution of the "slow" tail to the overall closed-time histogram indicates that the number of gaps during the burst is significantly higher than the number of interburst closed times. As such channel behavior was observed at all test potentials, we used a simple procedure for the identification of bursts, choosing a characteristic time interval which would include all intraburst channel closures. Proceeding from the experimentally obtained closed-time histograms, the interval could be $0 \leq t \leq 3\, \tau_{cl}^{(f)}$ (Magleby and Pallotta, 1983). During identification of bursts, all channel closures which fell under this interval were considered as intraburst gaps and were neglected. Figure 2c presents histograms for burst durations obtained in such a way. They could be fitted by two exponentials – a "fast", practically potential-

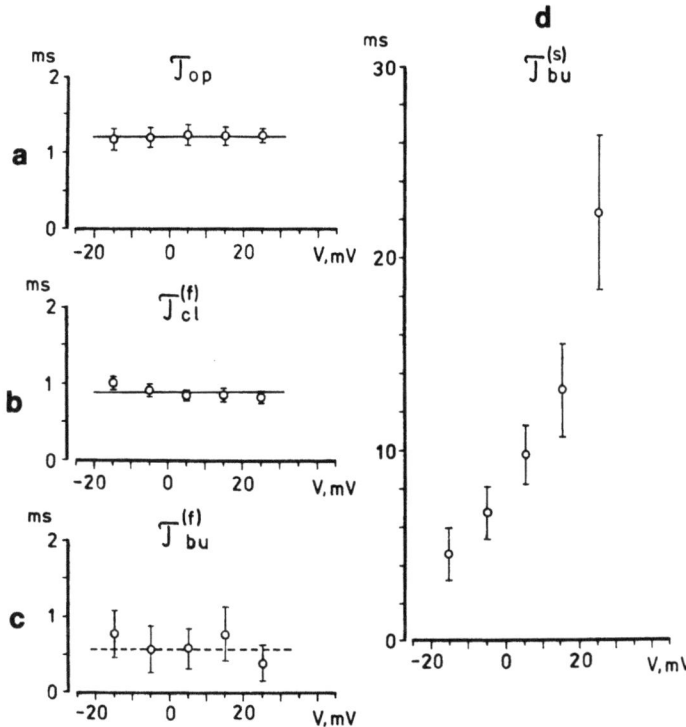

Fig. 3a–d. Potential dependence of temporal parameters of functioning of single HTI calcium channels. **a** Mean open time (τ_{op}); **b** mean time of intraburst gaps ($\tau_{cl}^{(f)}$); **c** "fast" component of burst length ($\tau_{bu}^{(f)}$); **d** "slow" component of burst length ($\tau_{bu}^{(s)}$). *Solid lines* (**a**, **b**), corresponding mean values for all test potentials; *dashed line* (**c**) result of calculation according to the expressions in Eq. 3

independent exponential with time constant $\tau_{bu}^{(f)}$ of about 0.6 ms (Fig. 3c), and a "slow" exponential with time constant $\tau_{bu}^{(s)}$ increasing from 4 ms at -15 mV to 20 ms at 25 mV (Fig. 3d). The "fast" component corresponds to short single openings of the channel which are not united into bursts, while the "slow" one reflects the distribution of real bursts. Physically, the "fast" component represents the process of reaching an equilibrium between activated (A) and open (O) states of the channel, and the "slow" one the exponential decay of the total state AO = A + O.

Theoretical analysis of the stochastic properties of a four-state model (Eq. 1) using an approximation characteristic for calcium channels

$$a, b \gg \alpha, \beta \qquad (2)$$

shows that the measured time constants τ_{op}, $\tau_{cl}^{(f)}$, $\tau_{bu}^{(f)}$, and $\tau_{bu}^{(s)}$, which characterize single-channel behavior, are associated with rate constants a, b, and β by the following expressions (Shuba and Teslenko, 1987):

$$\tau_{op} = 1/b; \quad \tau_{cl}^{(f)} = 1/a; \quad \tau_{bu}^{(f)} = (a+b)^{-1}; \quad \tau_{bu}^{(s)} = (a+b)/2b\beta \qquad (3)$$

Thus, using the experimentally obtained values τ_{op}, $\tau_{cl}^{(f)}$ and $\tau_{bu}^{(s)}$ and the theoretical expressions in Eq. 3 one can determine the rate constants a, b, and β (see Table 1).

Table 1. Mean values (ms) of the constants a^{-1}, b^{-1}, α^{-1}, β^{-1}, τ_s, and τ_{vs} for different potentials

Potential (mV)	a^{-1}	b^{-1}	α^{-1}	β^{-1}	τ_s	τ_{vs}
-15	1.0	1.14	25.0	3.6	2.92	122.0
-5	0.9	1.16	14.0	5.6	2.74	35.7
5	0.82	1.24	11.3	8.2	2.6	22.0
15	0.86	1.22	6.6	11.1	2.27	9.6
25	0.82	1.24	5.4	18.9	2.17	6.7

To identify α, the waiting-time histograms for the first channel opening were analyzed. These histograms, in the approximation shown in Eq. 2, should be described in the proposed model by the difference of two exponentials with time constants τ_s and τ_{vs} related to α and β as follows (Shuba and Teslenko, 1987):

$$\frac{1}{\tau_s, \tau_{vs}} = \frac{1}{2}\left\{3\alpha + \beta \pm [\alpha^2 + 6\alpha\beta + \beta^2]^{1/2}\right\}; \quad \frac{1}{\alpha} = (2\tau_s \tau_{vs})^{1/2} \qquad (4)$$

As can be seen from Fig. 4, the waiting-time histograms had a maximum which shifted to shorter time intervals with increased depolarization. The amplitude and position of the maximum are very sensitive to the changes in α and β values which makes these histograms very convenient for such identification. The approximation of the waiting-time histograms by the difference of two exponentials is also presented in Fig. 4. The α values for each test potential were determined using least-square deviations between theoretical curves and experimental histograms, and the β values independently from the already known values of τ_{op}, $\tau_{cl}^{(f)}$, and $\tau_{bu}^{(s)}$ according to Eq. 3. The kinetic constants obtained are summarized in Table 1.

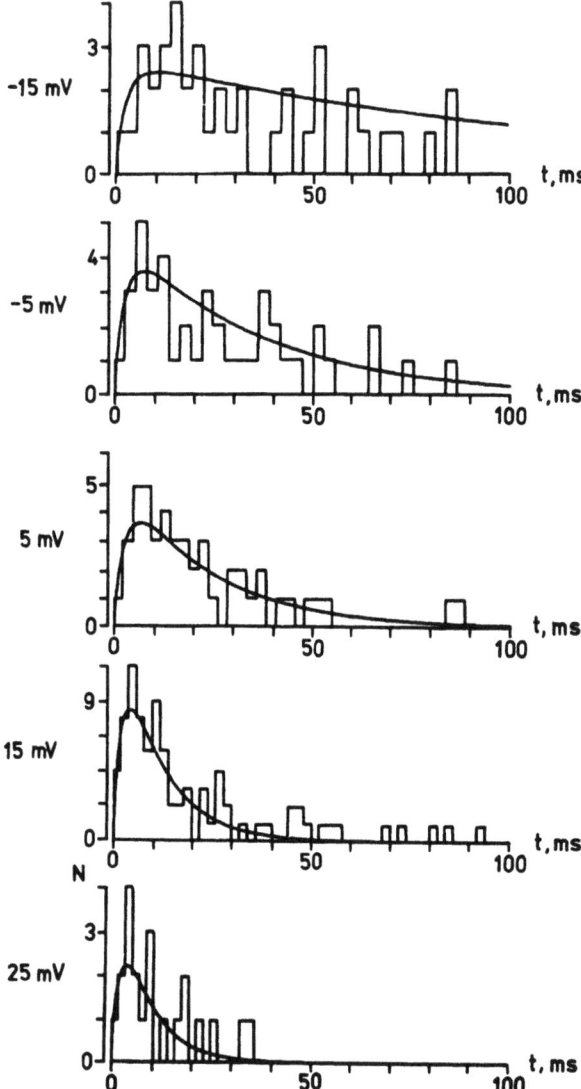

Fig. 4. Waiting-time histograms at different membrane potentials (indicated on *left*) and their approximations by the difference of two exponentials. See text time for constants. Binwidth 2.8, 2.8, 2.4, 2.0, and 1.8 ms for test potentials −15, −5, 5, 15 and 25 mV respectively

Thus, the complete set of rate constants was obtained determining the behavior of single calcium channels in the framework of the model in Eq. 1. To test the adequacy of this description of the kinetic characteristics of the calcium channel, we compared the experimentally obtained overall closed-time distributions with theoretical predictions according to which these distributions can be described by the sum of three exponentials with time constants $\tau_{cl}^{(f)}$, τ_s, and τ_{vs}. The latter two – "slow" and "very slow" are the same as for waiting-time histograms. The amplitudes of corresponding exponential components relative to the amplitude of "fast" components A_s and A_{vs} in the approximation in Eq. 2 can be calculated according to the following expressions (Shuba and Teslenko, 1987):

$$A_s = \frac{2[\beta \tau_{cl}^{(f)}]^2 \tau_{vs}}{(\tau_{vs}-\tau_s)(2\alpha \tau_{vs}-1)}; \quad A_{vs} = \frac{2[\beta \tau_{cl}^{(f)}]^2 \tau_s}{(\tau_{vs}-\tau_s)(1-2\alpha \tau_s)} \quad (5)$$

The results of this comparison are presented in Fig. 5a. Parameters of "slow" and "very slow" exponential components were determined using the expressions in Eqs. 4 and 5, while the time constant $\tau_{cl}^{(f)}$ and the inversely proportional amplitude of the "fast" component of the closed time distribution ($A_{cl}^{(f)} \sim 1/\tau_{cl}^{(f)}$) (Shuba and Teslenko, 1987) were determined using least-square deviations between experimental and theoretical histograms. The figure shows satisfactory agreement between the theoretical and experimental curves, thus supporting the applicability of the proposed model for description of the kinetic behavior of single HTI calcium channels. The calculated values of $\tau_{cl}^{(f)}$ for test potentials −5, 5, and 15 mV were 1.12, 0.98, and 1.00 ms

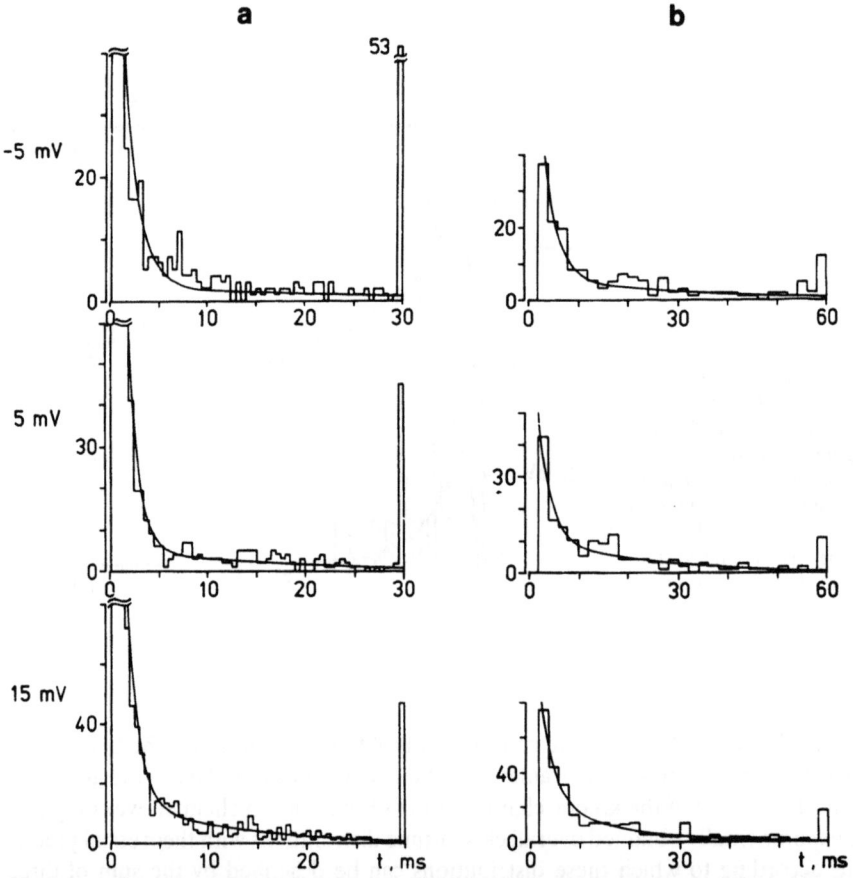

Fig. 5a, b. Closed time (**a**) and interburst closed time (**b**) histograms at three membrane potentials (indicated on *left*). **a** Binwidth 0.5 ms; *continuous lines*, results of data fitting with theoretical PDFs. **b** Binwidth 2 ms; *continuous lines*, results of biexponential fitting using least-squares procedure. Corresponding time constants given in text. Designations as in Fig. 2

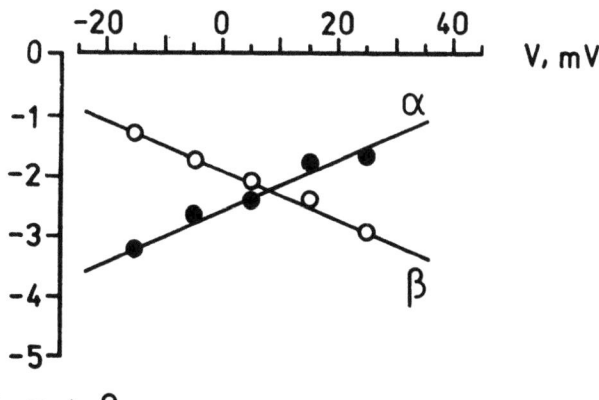

Fig. 6. Potential dependence of rate constants α and β

respectively, close to those obtained from direct fitting of the "fast" component of the closed time distribution by the least-squares procedure (see Fig. 2b); this indicates that the error made during burst identification was not high.

Figure 5b presents the histograms of interburst time intervals and the results of their approximation by two exponents using the least-squares procedure. The values of $\tau_{cl}^{(s)}$ and $\tau_{cl}^{(vs)}$ (in ms) obtained for test potentials −5, 5, and 15 mV respectively were as follows: 2.9 and 37.2, 2.5 and 22.5, 2.2 and 12.8. These values of $\tau_{cl}^{(s)}$ and $\tau_{cl}^{(vs)}$ are very close to the time constants τ_s and τ_{vs} obtained independently from the waiting-time histograms (see Fig. 4 and Table 1). Amplitude ratios A_s and A_{vs}, calculated from the expressions in Eq. 5, and the corresponding values $A_{s,\ exp}$ and $A_{vs,\ exp}$, obtained experimentally as the ratios of the amplitudes of two exponential components of the interburst closed-time distribution (Fig. 5b) to the amplitude of the "fast" exponential component of total closed-time distribution (Fig. 2b) (taking into account the difference of bin size of the corresponding histograms), had nearly the same numerical values.

The presented determination of the rate constants α and β responsible for the initial two stages of channel activation seems to be quite rigorous. Therefore, it would be desirable to evaluate the degree of their potential dependence. Figure 6 presents, in a semilogarithmic scale, the mean values of α and β at five different test potentials. They fall on straight lines having semilogarithmic slopes of about +1.1 and −1.1. Thus, the potential dependence of α (V) and β (V) rate constants is, as expected, exponential in type. The logarithm of the relation α (V) / β (V) has a slope of 2.2, representing the effective valency of a postulated single-gating *m* particle in case of its transmembrane transition during channel activation. Similar values (2.4−2.9) have been obtained in the analysis of the asymmetric displacement currents related to calcium-channel activation in nerve cell membranes (Kostyuk et al., 1981).

LTI Channels

To investigate the influence of permeant ion species on activation kinetics of calcium channels, neuroblastoma cells of clone N1E-115 were used. These cells possess calcium channels which are rather similar in their properties to LTI channels of the membrane of DRG neurons. This similarity is evidenced by the following features: (a) Macroscopic whole-cell currents through the calcium channels of neuroblastoma cells have activation and inactivation kinetics similar to LTI currents in the neuronal membrane. As shown on the whole-cell level, the activation kinetics of the LTI calcium current of DRG neurons of rat can be satisfactorily described using a square power of the m variable in a modification of the Hodgkin-Huxley model (Fedulova et al., 1985). (b) Calcium channels in neuroblastoma cells demonstrate a selective sequence to divalent cations of alkaline earth metals – $Ca^{2+}: Sr^{2+}: Ba^{2+} = 1 : 1.12 : 0.94$ – which resembles that for LTI channels (Fedulova et al. 1985; Carbone and Lux, 1987a) and differs from the same sequence for calcium channels carrying the main high-threshold component of the overall calcium current – $Ca^{2+}: Sr^{2+}: Ba^{2+} = 1 : 1.3 : 1.8$ (Fedulova et al., 1985) in DRG neurons of vertebrates. (c) Influence of the same ions on the position of activation curves along the voltage axis for calcium channels in neuroblastoma cells and LTI channels of DRG neurons is similar. In their efficiency in shifting activation curves in depolarizing direction, these ions form the following sequence: $Ba^{2+} > Ca^{2+} > Sr^{2+}$. The sequence for high-threshold calcium

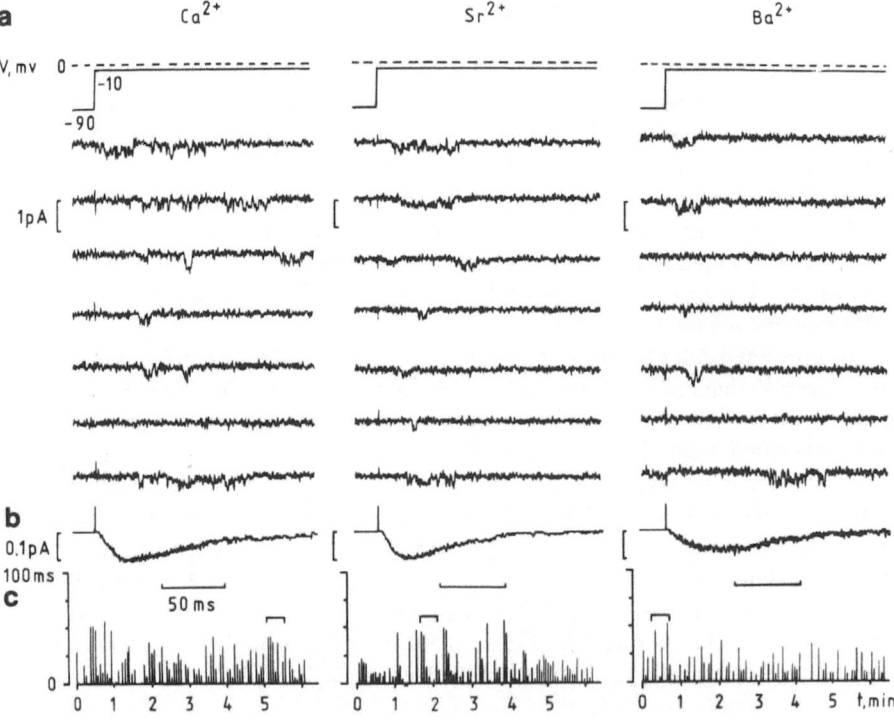

Fig. 7a–c. Activity of single LTI calcium channels in neuroblastoma cells. **a** Current records from cell-attached configuration for Ca^{2+}, Sr^{2+}, and Ba^{2+} (60 mM) as permeant cations. *Upper traces*, potential scales. V_r zeroed with high external K^+, $\Delta f = 1$ kHz. **b** Corresponding averages of idealized current records. **c** Total time which channel spent in open state at every consecutive depolarization of 400 ms duration (column height) plotted against time of recording. *Horizontal bars*, records plotted in **a**

channels in a large variety of preparations is practically opposite: $Ca^{2+} > Sr^{2+} > Ba^{2+}$.
(d) Functioning of calcium channels in neuroblastoma cells as well as of LTI channels of neuronal membrane is practically independent of intracellular metabolic support. Both of them are also insensitive to the dihydropyridine calcium-channel agonist Bay K8644.

The adequacy of the model in Eq. 1 for calcium channels of neuroblastoma cells was shown using the whole procedure of data fitting as described above for HTI calcium channels of DRG neurons of mouse embryos.

The traditional divalent permeant cations Ca^{2+}, Sr^{2+}, and Ba^{2+} (Fig. 7), as well as monovalent cations Na^+ and Li^+ (Fig. 8), were used as charge carriers through the calcium channels. The latter two were used after modification of channel selectivity by lowering of external Ca^{2+} concentration to $10^{-8}M$ (Kostyuk at al., 1983; Almers et al., 1984; Hess et al., 1986; Carbone and Lux, 1987b). In such concentration, Ca^{2+} ions themselves do not affect the modified calcium-channel kinetics (Kostyuk et al., 1983; Carbone and Lux, 1987b).

We found that the conductance of a calcium channel was practically identical with differently permeant divalent cations at 60 mM – 7.2 pS. However, modification of channel selectivity, allowing monovalent cations to pass, caused an essential increase of channel conductance. In parallel, a strong shift ($\approx$ 70 mV) of all potential-dependent characteristics in hyperpolarizing direction took place. For Na^+ at concen-

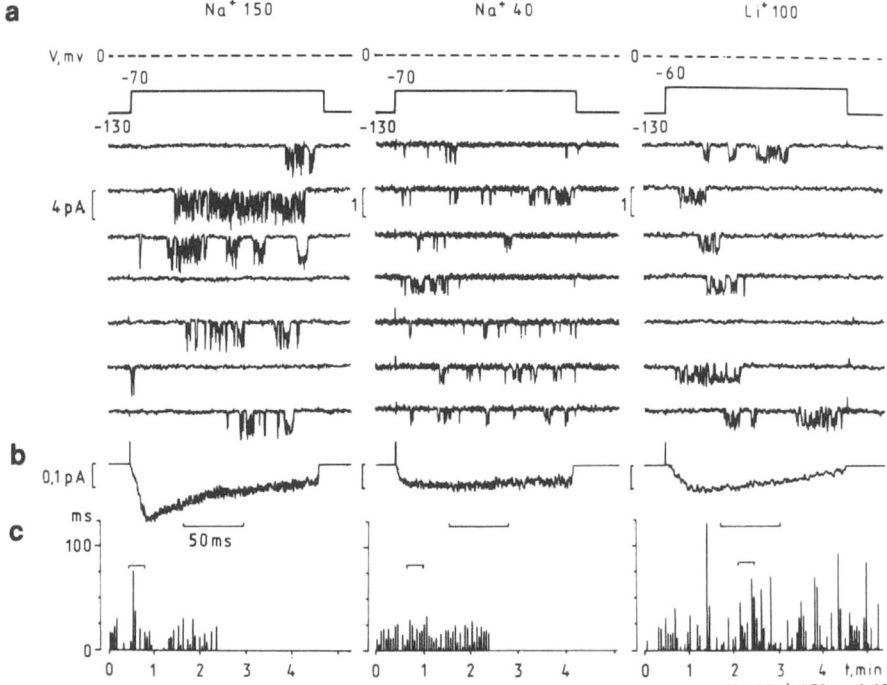

Fig. 8a–c. Activity of single modified LTI calcium channels in neuroblastoma cells. Na^+ 150 and 40 mM, Li^+ 100 mM as charge carriers. $\Delta f = 2$ kHz for Na^+ current, $\Delta f = 1$ kHz for Li^+ current. Recording conditions and designations as in Fig. 7

trations of 40, 60, 100, and 150 mM, conductance was 10, 18, 31, and 57 pS respectively. For Li$^+$ ions at 100mM the channel conductance was 11 pS. The modified calcium-channel permeability toward monovalent cations, measured according to the shift in reversal potential, was as follows: Na : K : Li = 1: 0.51 : 0.37.

Mean channel open times τ_{op} for Ca^{2+}, Sr^{2+}, and Ba^{2+} as charge carries had close absolute values and similar potential dependence – with increase of depolarization this decreased (Fig. 9a). As mentioned above, the same behavior of τ_{op} is characteristic for LTI calcium channels of DRG neurons. Modifikation of channel selectivity led to essential changes of τ_{op} which were dependent on both permeant monovalent ion species and its concentration (Fig. 9a). At high external sodium concentrations (> 100 mM) τ_{op} became potential-independent with essentially lower absolute values than with Ca^{2+}, Sr^{2+}, and Ba^{2+}. With decrease in sodium concentration, potential dependence became more and more prominent, becoming the strongest for Li$^+$ ions.

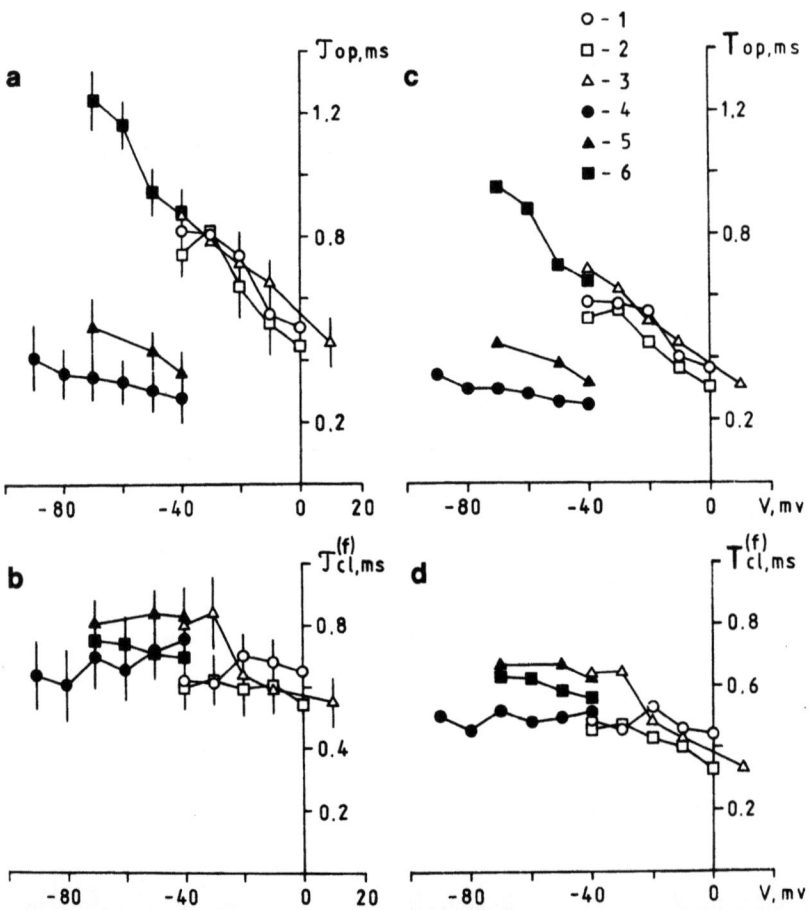

Fig. 9a–d. Potential dependence of temporal parameters for single LTI calcium channels in neuroblastoma cells. **a** Mean open time (τ_{op}), and **b** mean time of intraburst gaps ($\tau_{cl}^{(f)}$) for different charge carriers. **c, d** Corresponding data corrected for missed events (T$_{op}$, T$_{cl}^{(f)}$). Charge carriers: *1* Ca^{2+} 60 mM ; *2* Sr^{2+} 60 mM; *3* Ba^{2+} 60 mM; *4* Na$^+$ 100 mM; *5* Na$^+$ 40 mM; *6* Li$^+$ 100 mM

Simultaneously, an increase of τ_{op} absolute values took place (Fig. 9a). At the same time, mean duration of intraburst closed time $\tau_{cl}^{(f)}$ for different ions before and after modification of channel selectivity was nearly the same and practically potential-independent (Fig. 9b).

The same behavior of τ_{op} and $\tau_{cl}^{(f)}$ for differently permeant ions was preserved after correction of their numerical values on events missed due to the limited band width of the recording system. Indeed, we recorded Na$^+$ currents at $\Delta f = 2$ kHz, while Ca^{2+}, Sr^{2+}, Ba^{2+}, and Li$^+$ currents were obtained at $\Delta f = 1$ kHz (Figs. 7, 8). Using the data of Blatz and Magleby (1986), it is possible to show that in approximations a, b $\gg$ α, β and $t_d \ll \tau_{op}, \tau_{cl}^{(f)}$, which are true for calcium channels and our recording conditions, real mean channel open time T_{op} and real mean intraburst closed time $\tau_{cl}^{(f)}$ is related to measured times τ_{op} and $\tau_{cl}^{(f)}$ as follows:

$$\frac{1}{b} = T_{op} \approx \tau_{op}\left(1 - \frac{t_d}{\tau_{cl}^{(f)}}\right); \quad \frac{1}{a} = T_{cl}^{(f)} \approx \tau_{cl}^{(f)}\left(1 - \frac{t_d}{\tau_{op}}\right) \quad (6)$$

Results of such data correction are shown in Fig. 9c, d.

Apart from the changes in τ_{op} and $\tau_{cl}^{(f)}$, which describe the last kinetic stage in the model in Eq. 1 connected with local transformations of channel gating mechanism, permeant ion species affected also the first two stages which normally can be attributed to the transmembrane movement of two charged gating particles. Figure 10a shows potential dependence of mean burst duration of channel openings $\tau_{bu}^{(s)}$. The strongest potential dependence of $\tau_{bu}^{(s)}$ was observed for Ba^{2+} ions, the weakest for Ca^{2+} ions. Modification of channel selectivity did not cause essential changes in $\tau_{bu}^{(s)}$ for Na$^+$ current compared to Ba^{2+} current. Figure 10b presents, in a semilogarithmic scale, the mean values of rate constants α and β at different test potentials. Experimental data could be fitted by straight lines, indicating exponential dependence of α and β on membrane potential. The potential at which α and β lines cross one another

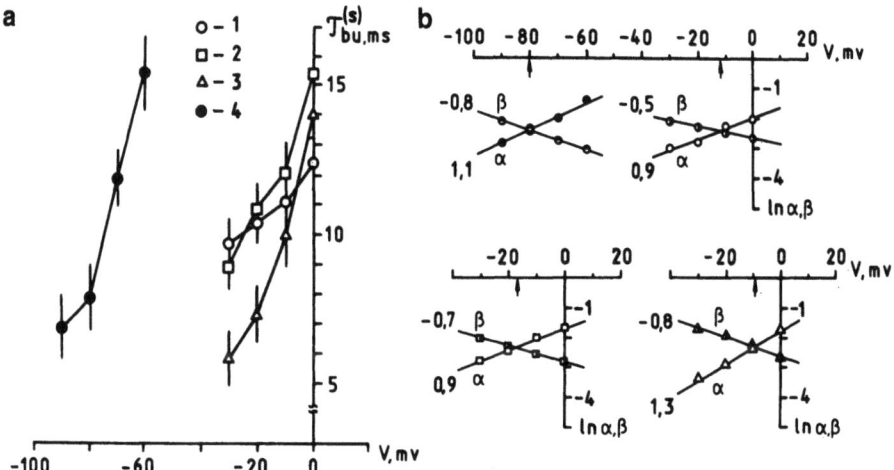

Fig. 10a–b. Potential dependence of **a** burst duration $\tau_{bu}^{(s)}$ and **b** rate constants α and β for single LTI calcium channels of neuroblastoma cells. Charge carriers: *1* Ca^{2+} 60 mM; *2* Sr^{2+} 60 mM; *3* Ba^{2+} 60 mM; *4* Na$^+$ 100 mM. Figures near each line in **b** show semilogarithmic slope for α and β

corresponds to 50% activation of the channels. Observed shifts in this potential for differently permeant ions (Fig. 10) are in good accord with similar observations on whole-cell currents. The semilogarithmic slopes of α and β for various charge carriers are different. As $\ln \alpha(V)/\beta(V)$ represents the effective valency of a postulated single-gating m particle, it is possible to conclude that this value is noticeably affected by permeant ion species.

It is known that permeant ions, when passing through the calcium channel, bind to some molecular group (probably carboxylic) located inside the channel. According to the data on whole-cell LTI calcium current of DRG neurons of chick, the dissociation constant of Ca^{2+} ions with this binding site is 3.3 mM (Carbone and Lux, 1987a). The value obtained in single-channel experiments on the same preparation was somewhat higher – 10.3 mM (Carbone and Lux, 1987b). It is possible to suggest that various divalent cations bound to the channel binding site are able to change the potential relief of the channel, thus affecting in a different way the intramembrane movement of charged gating particles.

HTN Channels

Kinetic behavior of HTN channels was studied on pheochromocytoma cell line PC-12. This cell line originates from mouse chromaffin cells of adrenal medulla (Green and Tischler, 1976) which have many similarities to sympathetic neurons and normally respond to Ca^{2+} entry by release of catecholamines.

Calcium channels of pheochromocytoma cells are practically identical to HTN channels of DRG neurons. We have found that their unitary conductance in 60 mM Ba^{2+} solution is 17 pS (18.4 pS for DRG neurons) and that they are sensitive to the dihydropyridine calcium-channel agonist Bay K8644. Excision of membrane patch led to rapid diminishing of channel activity. Nevertheless, as one can see from averaged currents (Fig. 11), inactivation of calcium channels is somewhat more prominent in pheochromocytoma cells than in DRG neurons. As can also be seen from Fig. 11, the calcium channel under investigation demonstrates the so-called mode 2 activity (Nowycky et al. 1985a) usual for HTN channels.

The main characteristic feature of this channel is, however, a biexponential distribution of its open times for all test potentials (Fig. 12). The time constant of the "fast" component $\tau_{op}^{(f)}$ decreased from 0.5 ms to 0.2 ms with a 30 mV increase in depolarization, while the "slow" one $\tau_{op}^{(s)}$ remained practically unchanged at ≈ 1.8 ms. Simultaneously with the increase of depolarization, the contribution of the "slow" component to the overall open-time histogram increased. Such behavior of open-time distributions can hardly be attributed to distortions due to poor time resolution. As was shown by Blatz and Magleby (1986), an additional component can appear only if more than 40%–50% of all channel openings and closures are lost, and its effective duration has to be less than two "dead times" – $2t_d$. In our experiments for 2 kHz bandwidth it should be less than 0.18 ms, which is not the case. Thus, biexponential distribution of open times reflects the presence of two conducting states with different life-times for HTN calcium channels of pheochromocytoma cells.

All other kinetic distributions for calcium channels under investigation were similar to those described already for LTI and HTI channels. Direct approximation of these

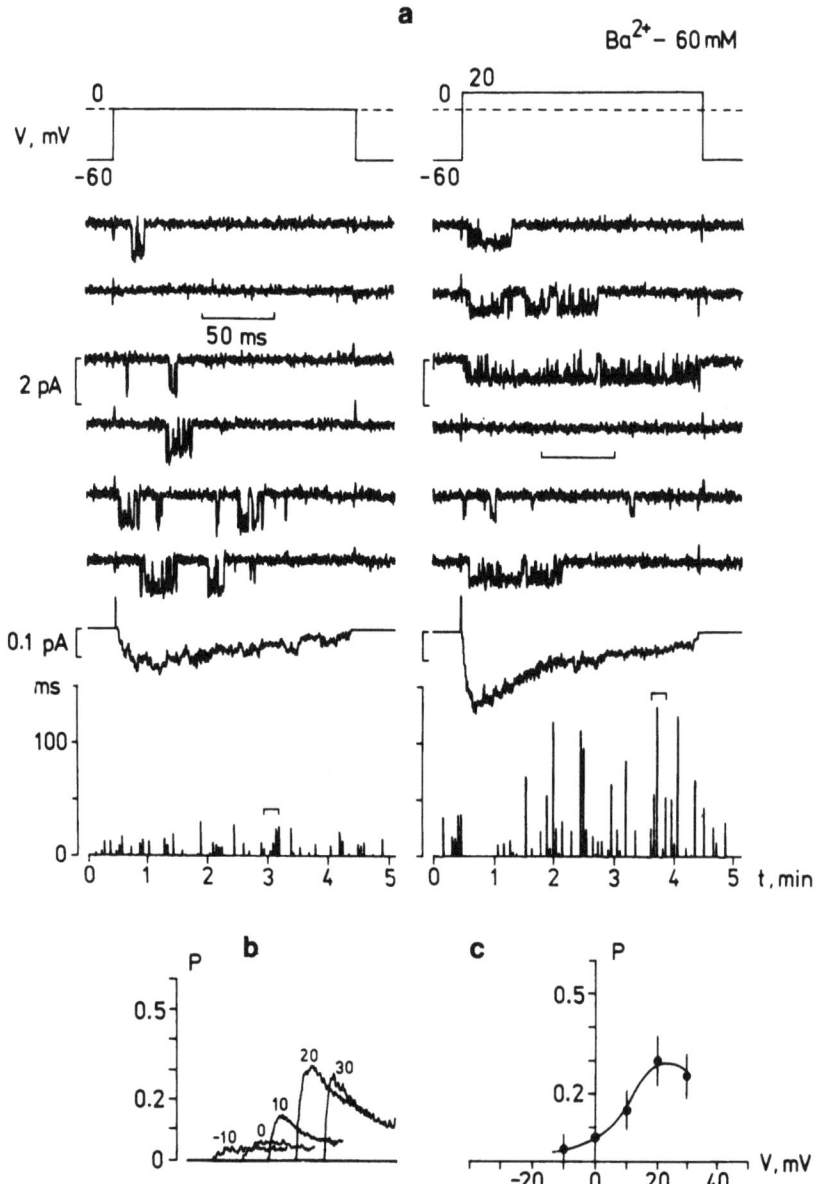

Fig. 11a–c. Activity of single HTN calcium channels of pheochromocytoma cells. **a** Original current traces and corresponding averages of idealized records obtained in cell-attached configuration for two membrane potentials (shown at top). Charge carrier: Ba^{2+} 60 mM. V_r zeroed with high external K^+. Δf = 2 kHz. *Lower traces:* changes in total channel open time (height of columns) in response to every consecutive depolarization of 160 ms duration during time of recording. *Horizontal bars:* records which are plotted. **b** The curves of probabilities of the channel to be open during 50 ms of depolarization at the indicated membrane potentials. **c** Activation curve for a single HTN calcium channel

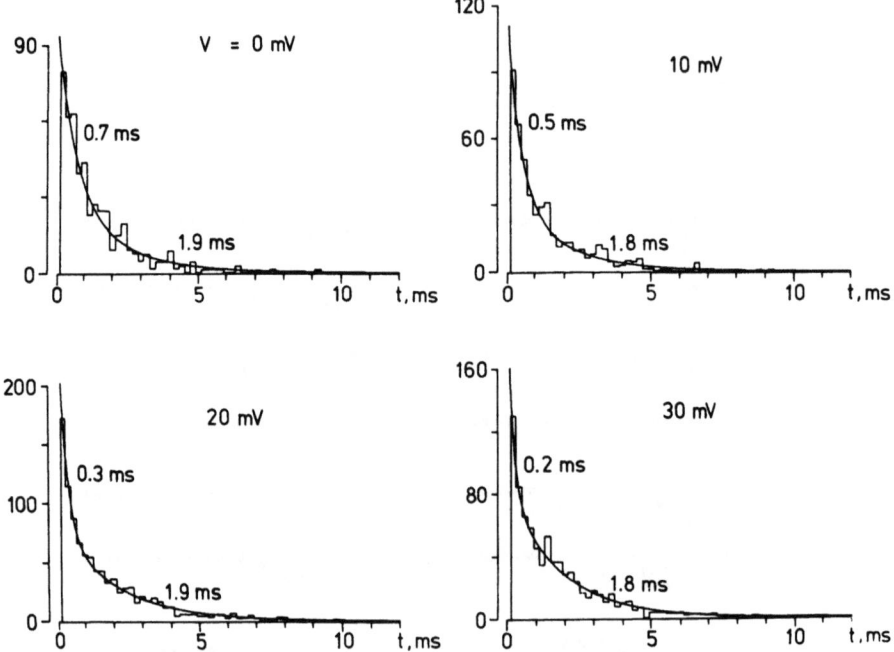

Fig. 12. Distributions of HTN channel open times for indicated membrane potentials and their fit with the sum of two exponentials. Corresponding time constants shown near each component of histograms. $\Delta f = 2$ kHz

data using the least-square procedure showed that burst-length distributions could be approximated by the sum of two exponentials, and two components of waiting-time histograms, as well as those of intraburst closed-time histograms, had similar time constants. This indicates that the two stages preceding the activated state of the channel in the model in Eq. 1 are similar in nature for HTN channels, as well as for HTI and LTI channels.

Figure 13 shows the effect of 10 μM Bay K8644 on distributions of open and closed times of calcium channels of pheochromocytoma cells for 1 kHz bandwidth. Time constants of both exponential components in the open-time histogram increased with Bay K8644 from 0.5 to 1.3 ms ($\tau_{op}^{(f)}$) and from 3.3 to 4.2 ms ($\tau_{op}^{(s)}$). A parallel decrease of $\tau_{cl}^{(f)}$ from 0.5 to 0.4 ms was observed. If we suggest the presence of two open states in the last kinetic stage on the model in Eq. 1,

$$A \underset{b_f}{\overset{a_f}{\rightleftharpoons}} O_f \qquad A \underset{b_s}{\overset{a_s}{\rightleftharpoons}} O_s \tag{7}$$

where O_f and O_s are the "short"- and "long"-living open states and a_f, b_f, a_s, and b_s are the rate constants, then correction of $\tau_{op}^{(f)}$ and $\tau_{op}^{(s)}$ for missed events according to the

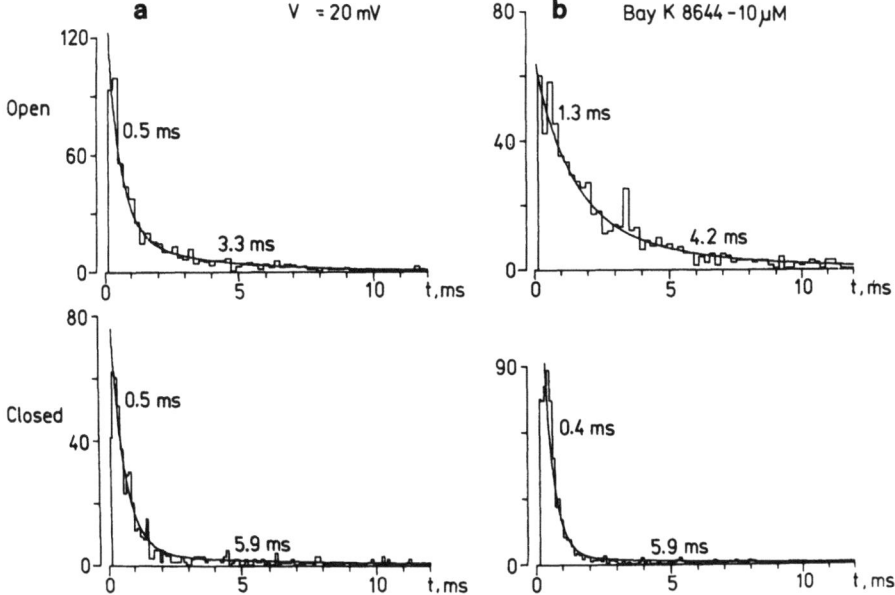

Fig. 13a–b. Action of Bay K 8644 (10 μM) on single HTN calcium channels of pheochromocytoma cells. Histograms of open and closed times **a** before and **b** after application of Bay K8644. $\Delta f = 1$ kHz. *Smooth lines:* biexponential fitting of experimental histograms. Corresponding time constants are shown near each exponential component

expressions in Eq. 6 shows that $T_{op}^{(f)} = 1/b_f$ and $T_{op}^{(s)} = 1/b_s$ increase correspondingly from 0.31 ms to 0.71 ms and from 2.1 ms to 2.3 ms respectively with Bay K8644. Thus, data correction shows that Bay K8644 affects mainly the "fast" component of open-time distribution, increasing its time constant by a factor of about 2.5, leaving the "slow" one practically unchanged.

If we accept the scheme in Eq. 7 for the last activation stage of HTN channels in the framework of the model in Eq. 1, the measured mean intraburst closed time $\tau_{cl}^{(f)}$ must theoretically be

$$\tau_{cl}^{(f)} = 1/(a_f + a_s)$$

Consequently, the observed decrease of $\tau_{cl}^{(f)}$ with Bay K8644 may be due to an increase of both a_f and a_s. As, according to our data, Bay K8644 alters only b_f, it is possible to assume that changes in $\tau_{cl}^{(f)}$ are also due to a_f and that Bay K8644 affects only the transitions leading to the short-lived open state O_f.

Comparison of Different Gating Mechanisms: Theoretical Considerations

Our findings obtained in different cell types are in qualitative agreement with the results of Nowycky et al. (1985b) obtained in chick embryonic sensory neurons. Some quantitative differences which appear mainly in unitary current amplitudes and corresponding channel conductances may be due to the use of different ions as charge carriers. It is rather interesting that the mean open time for LTI and HTN channels passing strontium current appeared to be potential-dependent; for LTI channels it decreased with depolarization increase, whereas for HTN channels the potential dependence was opposite – at higher depolarizations the mean open time increased. This difference can hardly be considered as an artifact due to insufficient resolution of channel activity; no potential dependence was observed for the mean open time of HTI channels, although their unitary current amplitude was practically within the same range as for LTI and HTN channels.

However, overestimation of the mean open time for HTN channels due to poor time resolution cannot be excluded. Indeed, the comparison of our data on single-channel activity with investigations of calcium "tail" currents in response to membrane repolarization demonstrates that the deactivation time constant for LTI current at low membrane potentials is of the order of 1–2 ms (Carbone and Lux, 1984a, 1987a; Veselovsky et al., 1985; Cota, 1986), whereas the predominant exponential component in the deactivation process for the overall high-threshold calcium current has in many preparations the time constant of only 0.1–0.4 ms (Fenwick et al., 1982; Brown et al., 1983; Veselovsky et al., 1985; Cota, 1986; Carbone and Lux, 1987a). As the deactivation time constant at high negative repolarization potentials reflects the mean channel open time (Shuba and Teslenko, 1987), there is good accord between these values for LTI channels. However, the mean open times for HTI and HTN channels do not coincide with the observed "fast" deactivation time constants for the overall high-threshold calcium current. As HTN channels seem to transfer the main portion of high-threshold calcium current and are, probably, the most frequent ones in many cells, HTN channels might be considered to be solely responsible for the "fast" tail currents. Thus, the discrepancy mentioned can be explained if we assume some distortions of the open-time distribution for this type of channels in single-channel experiments due to insufficient time resolution of the recording system. The evidence in favor of this was obtained from HTN calcium channels of pheochromocytoma cells demonstrating biexponential distribution of open times.

The presented analysis of the activity of calcium channels in the membrane of variously excitable cells confirms the possibility of an adequate description of the main experimental findings in the framework of a four-state sequential kinetic model. This model has solved all the difficulties which arose earlier during the analysis of a three-state model. An important feature of the new model is the description of the closed-time histogram by the sum of three (not two) exponential components. The first, "fast" component reflects the potential independent transition $A \xrightarrow{a} O$, while two slower ones – "slow" and "very slow" – reflect the potential-dependent transfer of charged gating particles. These three closed-time distribution components can be reliably detected in original current records, in which the "fast" one represents transient channel closing during a single burst of activity. Several bursts aggregate into clusters with a mean intracluster closed time of about 3 ms for HTI channels. This clustering can be especially well observed at low depolarizations. Finally, closed invervals between clusters form the third "very slow" component of the total shut times distribution. Similar relations between different closed times were observed by Cavalie et al. (1986) in their study of single calcium channels in the membrane of isolated cardiomyocytes.

It can be seen from Fig. 5A that although the theoretical probability density functions describe correctly the majority of events in the closed-time histogram (~ 96%) for HTI channels, a small proportion of them (~ 4%) form an "extraslow" tail that cannot be described by the model in Eq. 1. Most probably, this tail is due to inactivation of calcium channels. An indication of the presence of this process in the investigated test potential range can be found in the averaged current records shown in Fig. 1b: a definite decrease in current amplitude can be seen after the latter reaches its maximum value. The identification of the potential-dependent rate constants α and β was based in our calculations on the histograms of burst durations and waiting times to the first channel opening. These histograms are least affected by the inactivation process. Due to this fact we could maximally avoid distortions of the results describing activation of the calcium channel by superposition of inactivation. Extension of the model to include inactivated channel states seems premature because of insufficient experimental data about inactivation of single calcium channels.

We may relate the calculated rate constants $\alpha(V)$ and $\beta(V)$ to the time constants $\tau_m(V)$ and $\tau_{tail}(V)$ of corresponding macroscopic current relaxation obtained during membrane depolarization (m process) and repolarization (tail process). At large depolarizations the proposed model corresponds exactly to the modified Hodgkin-Huxley equations for calcium-channel activation using the square power of the m variable. Under these conditions $\beta(V) \ll \alpha(V) \ll a, b$ and the following relation holds: $\alpha^{-1}(V) \approx \tau_m(V)$. Consequently, when the test potential is shifted to more positive values, the value of $\tau_m(V)$ must decrease exponentially for HTI channels, for example from 6.7 ms at $V_t = 15$ mV to 2.8 ms at $V_t = 35$ mV (see Fig. 6, Table 1). Similar voltage-dependence of $\tau_m(V)$ was found by Kostyuk et al. (1977, 1981) during their study of the calcium-current kinetics in the membrane of snail neurons. During repolarization the time constant of the calcium tail current [$\tau_{tail}(V)$] must practically coincide with the lifetime of the single-channel burst [$\tau_{bu}^{(s)}(V)$]. At small repolarization potentials, when $\alpha(V) \ll \beta(V) \ll a, b$ the value of $\tau_{tail}(V) \approx (a + b)/2b\beta(V)$. Thus, during repolarization of the membrane to these potentials τ_{tail} values must decrease exponentially (see Fig. 6); however, at very high negative potentials, when $\alpha(V) \ll a, b \ll \beta(V)$ this decrease should be limited and τ_{tail} should theoretically become equal to the mean open lifetime of the channel $\tau_{tail} \approx b^{-1}$ (as $a \approx b$ for HTI channel). Qualitatively such a voltage dependence of $\tau_{tail}(V)$ has been recently observed by Veselovsky et al. (1985). A more detailed quantitative analysis of the voltage dependence of $\tau_{tail}(V)$ is now in progress.

Investigation of the influence of permeant ion species on the activation properties of calcium channels has shown that the mean channel open time τ_{op}, which characterizes the rate of channel conformational transition to the nearest closed state, is the most affected. This permits the supposition that the process of channel closure occurs in close vicinity to the selective filter region and may be due to its steric narrowing.

There are two factors determining channel permeability: the first is due to electrostatic interaction of permeant ions with anionic groups in the selective filter, and the second one relates to their steric correspondence (Hille 1975). The contribution of the first factor grows with increasing ion valency z_i and decreasing ion radius ϱ_i, while sterical limitations increase with increasing ϱ_i. For divalent cations ($\varrho_{Ca} = 1.04$ Å, ϱ_{Sr}

$= 1.2$ Å, $\varrho_{Ba} = 1.38$ Å), both factors can compensate one another, which may explain the similar conductance values g_i of calcium channels of neuroblastoma cells with respect to these ions.

Substitution of divalent by monovalent ion leads to a twofold decrease of z_i. Due to this the electrostatic limitations also decrease. Li$^+$ ions ($\varrho_{Li} = 0.68$ Å) may pass the channel in such conditions without steric difficulties. However, electrostatic interaction because of small radius should still be high. Passing of K$^+$ ions ($\varrho_K = 1.33$ Å) is determined mainly by steric factors. Correspondingly, the permeability of modified calcium channels towards these ions is of the same order and nearly twice as low as that for sodium ions ($\varrho_{Na} = 0.98$ Å).

Considering these data it is possible to suppose that the selective filter of calcium channels has the dimensions 3×4.8 Å and that only hydrated Li$^+$ ions with three water molecules can pass it without steric difficulties. Passing of all other permeant ions causes deformations of the steric region. Such deformations result in displacements of charged molecular groups of the selective filter at a distance S_i depending on ϱ_i. This leads to the formation of dipoles along the ion trajectory which exist during the time of dipole relaxation $\tau \approx 10^{-7}$ s (Burfoot and Taylor, 1979). The mean dipole moment d_i depends on the rate of ion transit through the channel ω_i as follows (Teslenko, 1985):

$$\omega_i = \frac{kT}{e} \cdot \frac{g_i}{Z_i^2 e}; \quad d_i = \frac{\omega_i \tau e S_i}{\omega_i \tau + 1}$$

where $e = 1.6 \cdot 10^{-19}$ K, $kT/e = 0.025$ V.

As, according to our assumption, $S_{Li} = 0$, then $d_{Li} = 0$. For divalent cations, $S_{Ca} = 0.2$ Å, $S_{Sr} = 0.4$ Å, and $S_{Ba} = 0.6$ Å, which, considering that $g_{Ca, Sr, Ba} = 7.2$ pS, gives $d_{Ca} = 0.01$ D, $d_{Sr} = 0.02$ D and $d_{Ba} = 0.03$ D. Such low dipole moments produced by divalent ion fluxes cannot essentially change the initial value of the dielectric constant ε_c of channel protein ($\varepsilon_c \approx 2-3$) and thus affect the potential dependence of channel transition from open to closed state compared with Li$^+$ ions.

For sodium ions the situation is rather different. At $S_{Na} = 2.8$ Å and $g_{Na} = 10-100$ pS (depending on concentration), $\omega_{Na} = 10^6 - 10^7$ s^{-1}, very close to the characteristic frequency of dipole relaxation $1/\tau \approx 10^7$ s^{-1}. As a result, at low sodium concentrations $d_{Na} \approx 0.1$ D and ε_c increases nearly twofold, which may exert only a small effect. However, at high sodium concentrations $d_{Na} \approx 0.5$ D, leading to the increase of the dielectric constant of the steric region of the channel to 20–30.

After channel transition from the open into the nearest closed state, the steric region of the selective filter becomes narrow and ion flux through the channel is no longer possible. This is why the values of time constant $\tau_{cl}^{(f)}$ characterizing this transition are practically independent of both the species and the concentration of the permeant ion. However, due to the different energies of the divalent ions bound to the molecular groups of the channel outer mouth (Kostyuk et al., 1983) and the selective filter, it is possible to suppose that these ions will affect to a varying extent the potential dependence of the first two stages in the model in Eq. 1. As is clearly seen from Fig. 10, variations of charge carriers lead to a shift of the half-activation potential of the channel and to changes in logarithmic slopes for rate constants α (V) and β (V). These effects can be explained if we suppose that divalent ions binding sites of the channel outer mouth are in close vicinity to the initial location of the gating particles in the resting state. In these conditions the energies of states R and C in the model in Eq. 1 will include electrostatic interaction of cation with gating charge which will depend on cation radius, affinity to the binding site, and valency. In favor of the significance of such interaction is the fact that the shift of activation curves for divalent charge carriers compared to monovalent ones is about 70 mV, considerably higher than the maximum possible shift due to the changes in surface potential V_s ($\sim$ 30–50 mV) (Kostyuk et al., 1983).

The discovery of the biexponential distribution of open times for HTN calcium channels of pheochromocytoma cells, which reflects the presence of two open states for these channels, is in good accord with previous observations on whole-cell calcium current in normal chromaffin cells (Fenwick et al., 1982). Indeed, as was shown by Fenwick et al. (1982), the relaxation of calcium current in response to membrane repolarization is biexponential. Two components were also found in the spectrum of current fluctuations. Time constants which correspond to the cut-off frequencies of the spectrum at membrane potential -12 mV were 0.25 ms and 2.48 ms, very close to the relaxation time constants 0.38 ms and 2.46 ms. As is seen from Fig. 11, about the same values are characteristic for open-time distributions of HTN channels in pheochromocytoma cells. Thus, the presence of two open states for these channels seems to be established. It is rather interesting that the calcium channel agonist Bay K8644 affects only the "fast" component of open-time distributions, increasing their characteristic time by a factor of nearly 2.5. This result differs from those described for cardiac cells (Hess et al., 1984) and chick sensory neurons (Nowycky et al., 1985a), where Bay K8644 induced the appearance of a "slow" component in open-time distribution.

References

Almers W, McCleskey EW, Palade PT (1984) A non-selective cation conductance in frog muscle membrane blocked by micromolar external cacium ions. J Physiol 353:565–583

Blatz AL, Magleby KL (1986) Correcting single channel data for missed events. Biophys J 49:967–980

Brown AM, Lux HD, Wilson DL (1984) Activation and inactivation of single calcium channels in snail neurons. J gen Physiol 83:751–769

Brown AM, Tsuda Y, Wilson DL (1983) A description of activation and conduction in calcium channels based on tail and turn-on current measurements in the snail. J Physiol 344:549–584

Burfoot JC, Taylor GW (1979) Polar dielectrics and their applications. The Macmillan Press LTD

Carbone E, Lux HD (1984a) A low voltage-activated calcium conductance in embryonic chick sensory neurones. Biophys J 46:413–418

Carbone E, Lux HD (1984b) A low voltage-activated, fully inactivating Ca channel in vertebrate sensory neurones. Nature 310:501–503

Carbone E, Lux HD (1987a) Kinetics and selectivity of a low voltage-activated calcium current in chick and rat sensory neurones. J Physiol 386:547–570

Carbone E, Lux HD (1987b) Single low-voltage-activated calcium channels in chick and rat sensory neurones. J Physiol 386:571–601

Cavalie A, Pelzer D, Trautwein W (1986) Fast and slow gating behaviour of single calcium channels in cardiac cells. Pflügers Arch ges Physiol 406:241–258

Colquhoun D, Sigworth FJ (1983) Fitting and statistical analysis of single channel records. In: Sakmann B, Neher E (eds) Single channel recording, Plenum Publishing Co., New York, p 191–263

Cota G (1986) Calcium channel currents in pars intermedia cells of the rat pituitary gland. J gen Physiol 88:83–105

Fedulova SA, Kostyuk PG, Veselovsky NS (1981) Calcium channels in the somatic membrane of the rat dorsal root ganglion neurons. Effect of cAMP. Brain Res 214:210–214

Fedulova SA, Kostyuk PG, Veselovsky NS (1985) Two types of calcium channels in the somatic membrane of new-born rat dorsal root ganglion neurons. J Physiol 359:431–446

Fenwick EM, Marty A, Neher E (1982) Sodium and Ca channels in bovine chromaffin cells. J Physiol 331:599–636

Fox AP, Nowycky MC, Tsien RW (1987a) Kinetic and pharmacological properties distinguishing three types of calcium currents in chick sensory neurones. J Physiol 394:149–172

Fox AP, Nowycky MC, Tsien RW (1987b) Single-channel recordings of three types of calcium channels in chick sensory neurones. J Physiol 394:173–200

Green LA, Tischler AS (1976) Establishment of a noradrenergic clonal line of a rat adrenal pheochromocytoma cells which respond to nerve growth factor. Proc natl Acad Sci USA 73:2424–2428

Hagiwara S, Byerly L (1983) The calcium channel. Trends in Neuroscience 6:189–193

Hagiwara S, Ohmori H (1983) Studies of single calcium channel currents in rat clonal pituitary cells. J Physiol 336:649–661

Hamill OP, Marty A, Neher E, Sakmann B, Sigworth FJ (1981) Improved patch-clamp techniques for high-resolution current recording from cells and cell-free membrane patches. Pflügers Arch ges Physiol 391:85–100

Hess P, Lansman JB, Tsien RW (1984) Different models of Ca channel gating behaviour favoured by dihydropyridine Ca agonists and antagonists. Nature 311:538–544

Hess P, Lansman JB, Tsien RW (1986) Calcium channel selectivity for divalent and monovalent cations. J gen Physiol 88:293–319

Hille B (1975) Ionic selectivity of Na and K channels of nerve membranes. In: Eisenman G (ed) Lipid bilayers and biological membranes: dynamic properties, Marcel Dekker Inc, New York Basel 255:323

Kostyuk PG, Krishtal OA, Pidoplichko VI (1981) Calcium inward current and related charge movements in the membrane of snail neurones. J Physiol 310:403–421

Kostyuk PG, Krishtal OA, Shakhovalov YA (1977) Separation of sodium and calcium currents in the somatic membrane of mollusc neurones. J Physiol 270:545–568

Kostyuk PG, Mironov SL, Shuba YM (1983) Two ion-selecting filters in the calcium channel of the somatic membrane of mollusc neurones. J Membrane Biol 76:83–93

Kostyuk PG, Shuba YM, Savchenko AN (1986) Single channels of low- and high-threshold calcium currents in the membrane of the mouse sensory neurones. Neurophysiology (Kiev) 18:412–416

Kostyuk PG, Shuba YM, Savchenko AN (1987) Three types of calcium channels in the membrane of mouse sensory neurons. Biol Membranes (Moscow) 4:366–373

Krishtal OA, Pidoplichko VI, Shakhovalov YA (1980) Properties of single calcium channels in the neuronal membrane. Bioelectrochemistry and Bioenergetics 7:195–207

Magleby KL, Pallotta BS (1983) Burst kinetics of single calcium-activated potassium channels in cultured rat muscle. J Physiol 344:605–624

Nowycky MC, Fox AP, Tsien RW (1985a) Long-opening mode of gating of neuronal calcium channels and its promotion by the dihydropyridine calcium agonist Bay K8644. Proc natl Acad Sci USA 82:2178–2182

Nowycky MC, Fox AP, Tsien RW (1985b) Three types of neuronal calcium channel with different calcium agonist sensitivity. Nature 316:440–443

Shuba YM, Savchenko AN (1985) Single calcium channels in rat dorsal root ganglion neurons. Neurophysiology (Kiev) 17:673–682

Shuba YM, Teslenko VI (1987) Kinetic model for activation of single calcium channels in mammalian sensory neurone membrane. Biol Membranes (Moscow) 4:315–329

Skibo GG, Koval LM (1984) Ultrastructural characteristics of synaptogenesis in monolayer cultures of spinal cord. Neurophysiology (Kiev) 16:336–343

Teslenko VI (1985) Thermodynamics of the stationary blockage of ionic channels in biological membrances. Biol Membranes (Moscow) 2:1162–1169

Veselovsky NS, Fedulova SA (1983) Two types of calcium channels in the somatic membrane of rat dorsal root ganglion neurones. Dokl Akad Nauk SSSR (Moscow) 268:747–756

Veselovsky NS, Kostyuk PG, Fedulova SA, Shirokov RE (1985) Deactivation of calcium currents in the somatic membrane of dorsal root ganglion neurons with removal of the membrane potential depolarization shift. Neurophysiology (Kiev) 17:682–691

Increased Calcium Currents in Rat Hippocampal Neurons During Aging

P. W. Landfield

Department of Physiology and Pharmacology, Bowman Gray School of Medicine, of Wake Forest University, 300 South Hawthorne Road, Winston-Salem, N.C. 27103, USA

This paper will review evidence from our laboratory which indicates that calcium currents in rat hippocampal neurons are increased with aging and that these alterations may play a key role in functional and morphological brain changes during aging. In addition, evidence will be reviewed to show that a form of calcium-dependent inactivation of calcium currents, which had been seen previously only in invertebrate neurons, is also present in mammalian hippocampal neurons. It is not clear whether changes in this inactivation mechanism are responsible for age-related changes in hippocampal calcium current (I_{Ca}), but this possibility is presently under active investigation.

Calcium Currents in the Hippocampus

Hippocampal pyramidal cells exhibit pronounced inward calcium currents, as evidenced by the large calcium spikes and currents that can be elicited in these cells following treatment with tetrodotoxin (TTX) (Schwartzkroin and Slawsky 1977; Wong and Prince 1981; Johnston et al. 1980; cf. review in Miller 1987). Along with cerebellar Purkinje cells (Llinas and Hess 1976) hippocampal pyramidal cells are among the few neuronal types that generate large calcium spikes. Calcium potentials have been widely implicated in seizure activity (for reviews, cf. Delgado-Escueta et al. 1986) and the prominent calcium influx found in hippocampal neurons may be one important reason that the hippocampus is a common site for active epileptic foci, and is among the structures with the lowest seizure thresholds.

Intracellular single-electrode voltage-clamp (SEVC) or whole-cell-clamp studies have indicated that voltage-dependent calcium currents are pronounced in hippocampal neurons (Johnston et al. 1980; Gray and Johnston 1987). Some of these currents have been found to be relatively non-inactivating (Brown and Griffith 1983).

There have been few studies of calcium channels within the context of the T, N, and L calcium channels found in other cell types (Nowycky et al. 1985; Miller 1987). However, several recent studies have reported that dissociated or cultured hippocampal neurons examined with patch-clamp or whole-cell-clamp techniques contain channels that appear to correspond in part to the T, N, and L channels (Gray and Johnston 1986; Bley et al. 1987). That is, T-like channels in these cells appear to exhibit rapid voltage-inactivation and to be characterized by low thresholds, whereas other channels exhibit L-like properties.

Nevertheless, studies conducted in dissociated or cultured cells must use techniques to retard both short- and long-term "rundown" of calcium conductance (Eckert et al. 1984; Miller 1987), so it is unclear whether these channel properties correspond to those found in intact tissues, or whether calcium current-dependent inactivation normally modulates these currents in vivo.

Calcium-Dependent Inactivation of Calcium Currents in Hippocampal Slice Neurons

Although hippocampal slice preparations are not suitable for patch-clamp analyses, they are accessible to intracellular SEVC techniques (Johnston et al. 1980; Brown and Johnston 1983; Halliwell 1983). In addition, they possess the important virtue of retaining many properties of intact tissues. Thus calcium spikes and calcium currents are prominent in these cells, in normal calcium concentrations (e. g. 2 mM), and there is little or no calcium current "rundown" over many hours of study.

Using this preparation to study calcium currents in cesium-loaded, TTX-treated hippocampal CA1 neurons, we recently found a form of apparently calcium-dependent inactivation of these currents (Pitler and Landfield 1987). In these cells, cesium blocks the slow calcium-dependent potassium conductance (Johnston et al. 1980; Brown and Johnston 1983), among other potassium currents, and this blockade was confirmed in each cell studied. By using repetitive stimulation design, moreover, it was possible to avoid the confounding effects of voltage-dependent potassium or other outward currents not blocked by cesium. For example, it was shown with hyperpolarizing commands that no outward currents flowed during the intervals between repetitive depolarizing command steps. Further, the inactivation was clearly calcium-dependent, occuring more rapidly with high calcium, and not developing in barium-substituted media (Fig. 1). The inactivation during repetitive pulses was seen for calcium spikes in current-clamp mode (Fig. 1) and for calcium currents below spike threshold in single-electrode voltage-clamp mode (Fig. 2) (Pitler and Landfield 1987).

Nevertheless, one alternative interpretation of the apparent inactivation, which could not be ruled out fully in those studies, was that a fast voltage- and calcium-dependent potassium current (I_C) was activated during each depolarizating command step; if this outward I_C increased with calcium accumulation, it might appear as if inward I_{Ca} were inactivating. An I_C that is blocked by tetraethylammonium (TEA) has been seen in vertebrate sympathetic neurons (Adams et al. 1982), and was recently described in hippocampal neurons (Lancaster and Adams 1986; Storm 1987).

Consequently, in recent studies, we examined inactivation of I_{Ca} before and after TEA application (Campbell et al. 1988). The efficacy of TEA block was assessed by its dramatic prolongation of the calcium spike. In SEVC mode, the degree of inactivation was similar before and after TEA application, indicating that apparent inactivation of I_{Ca} is not an artifact of I_C activation.

In addition, the baseline calcium current influx and degree of inactivation were both reduced by nimodipine (Fig. 3) (Campbell et at. 1988), a dihydropyridine (DHP) calcium channel antagonist. Since nimodipine and other DHP compounds act primarily on the L channels (cf. review in Scriabine 1987), it appears as if these relatively low

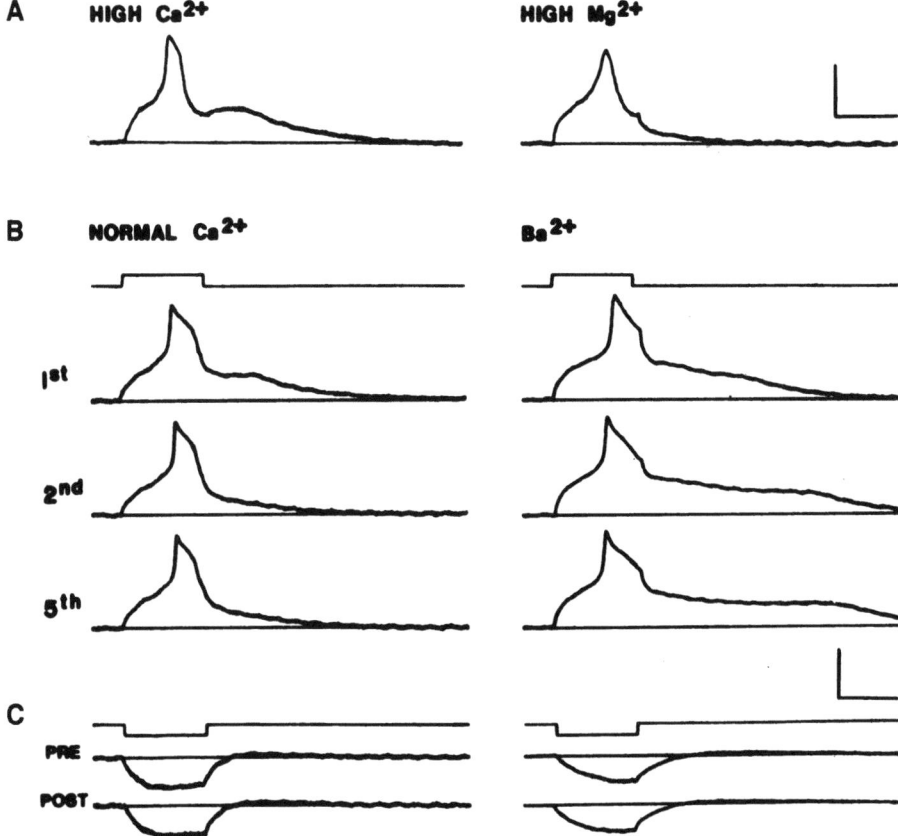

Fig. 1A–C. A. TTX-resistant spikes in cesium-loaded cells. *Left*, Spike from a cell bathed in a high-calcium medium; *right*, spike from a cell bathed in high magnesium-medium. *Bar*, 30 mV, 2 nA, 50 ms. **B** Current-clamp recordings of TTX-resistant responses from cesium-loaded cells bathed in either a normal-calcium medium or in a barium-substituted medium, during a 2-Hz train of five depolarizing pulses. Significant inactivation of the potential was observed only in calcium-containing medium. The top trace shows the same constant-current depolarizing pulse used to elicit each of the TTX-resistant spikes. Below, the 1st, 2nd, and 5th spikes of the train are shown. **C** *Top trace* is the same hyperpolarizing current pulse used to assess input conductance before and after the train of depolarizing pulses. *Bottom traces* show the voltage deflections of the membrane in response to the same intensity hyperpolarizing pulse both directly preceding *(PRE)* and 500 ms following *(POST)* the train of depolarizing pulses, for the cells shown in B. No drop in input resistance was seen 500 ms following the train, in either medium. *Bar* for **B** and **C** 30 mV, 2 nA, 50 ms. (Reprinted from Pitler and Landfield (1987)

threshold currents (activated by 20–40 mV steps from -60 mV) are carried by L-like channels in hippocampal slices. Alternatively, N- or T-like channels in hippocampus may be sensitive to nimodipine as well. Thus, moderately low-threshold, DHP-sensitive currents in the hippocampus appear to develop at membrane currents slightly more negative than calcium spike threshold (approximately -40 mV) and to be subject to calcium-dependent inactivation.

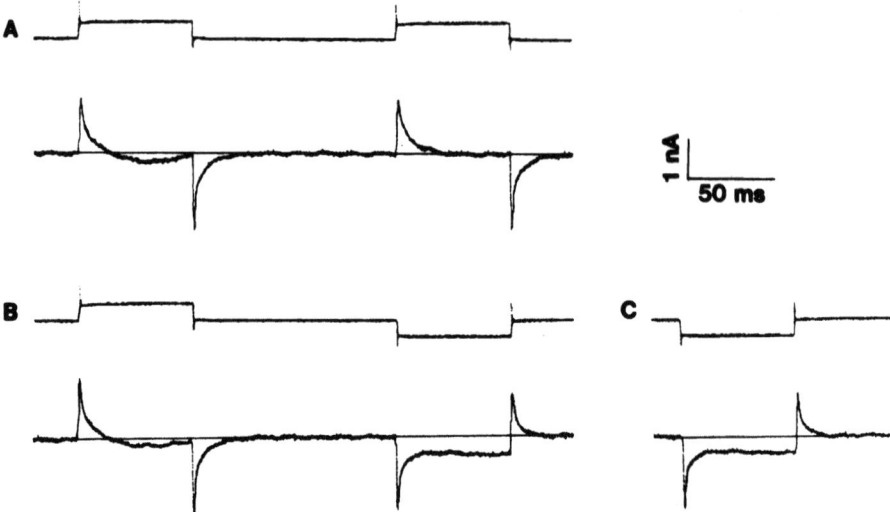

Fig. 2 A–C. Single-electrode voltage-clamp record showing inactivation of inward current during identical paired voltage steps **A** below calcium spike threshold (200 ms interpulse interval). *Upper record,* voltage; *lower record,* current. **B, C** Control experiments with hyperpolarizing steps show that no change in the control level (**B**) of leakage current was present 200 ms following the first depolarizing step (**C**). *Bar,* 50 mV, 1 nA, 50 ms. (Reprinted from Pitler and Landfield 1987)

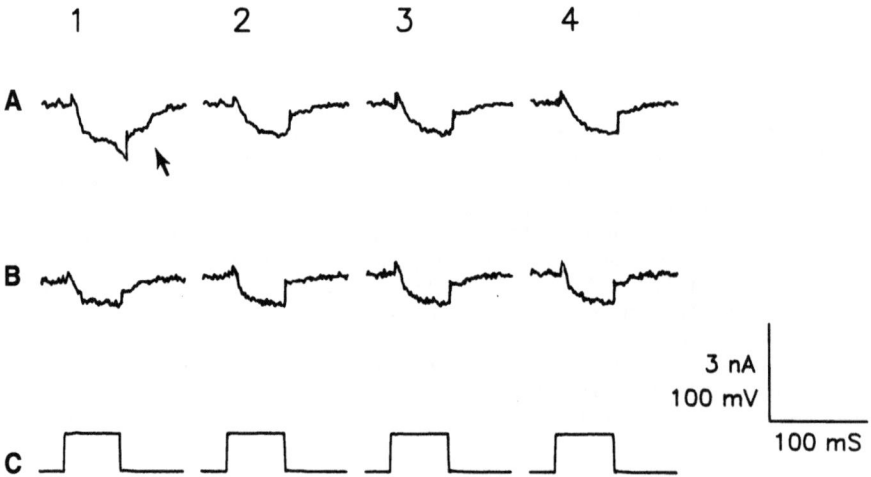

Fig. 3A–C. Single-electrode voltage-clamp measures of inward calcium currents during a 5-Hz train of depolarizing command steps in cesium-loaded, TTX- and TEA-treated CA1 pyramidal cells of hippocampal slices. Holding potential of approximately −60 mV. **A** The first four current responses to a train of 25-mV command steps are shown, prior to administration of nimodipine. Substantial inactivation of current is seen during the train, particularly for the prolonged tail current (*arrow*). **B** Following application of nimodipine to the same cell. The initial calcium current, including the tail current, is reduced in response to the first command step, and there is little inactivation during the train. **C** The first four voltage commands during the 5-Hz train for the current traces shown above. Current traces are corrected for leak current and capacitance by adding the current trace induced by an equal hyperpolarizing voltage step. (Data from Campbell et al. 1988)

Functional Role of Calcium Currents During Hippocampal Synaptic Transmission

Some of the first evidence indicating that calcium currents were abnormal in aged rat hippocampus arose indirectly from studies of frequency potentiation (FP) (growth of synaptic responses during repetitive synaptic activation) in the hippocampus of young and aged rats, both in intact animals (Landfield et al. 1978) and in slice (Landfield and Lynch 1977; Landfield et al. 1986) preparations. These findings led to further analyses of the basic properties of FP in the brain and these, in turn, revealed some unusual aspects of the calcium-dependent properties of central FP. By analogy with facilitation in peripheral nervous systems (Martin 1977), it appeared reasonable to think that hippocampal FP was also likely to be calcium-dependent. Therefore, it seemed to us that elevating extracellular magnesium, a calcium antagonist, might be able to strengthen FP by preventing transmitter depletion, as it can strengthen facilitation in peripheral systems. We examined the effects of high magnesium on FP in hippocampal slices (Landfield et al. 1986) and in intact animals (Landfield and Morgan 1984) and found that it did in fact improve FP of both the EPSP and of cell firing.

However, it became apparent that the effects of magnesium on synaptic transmission in the brain were more complex than in the peripheral nervous system. That is, in the hippocampus, FP was improved at only moderately elevated magnesium concentrations and magnesium/calcium ratios – levels which did not appear to alter synaptic transmission to single pulses (Landfield et al. 1986). By contrast, in the periphery, magnesium-dependent improvement of facilitation by prevention of depletion requires very high concentrations (e.g. 10 mM), and these substantially reduce baseline release to a single pulse (Martin 1977).

In addition, intracellular recording in hippocampus showed that part of the mechanism of the magnesium-dependent improvement of FP, particularly for postsynaptic spike generation, was due to partial blockade of an apparent synaptically-induced potassium-mediated hyperpolarization in the postsynaptic cell (Fig. 4). This blockade increased the probability of spike firing for a given EPSP amplitude. If a spike-induced hyperpolarization in the presynaptic terminals were also blocked by magnesium, then presynaptic spike failure might be less pronounced (e.g. Smith 1980), and EPSP potentiation might also be strengthened.

Moreover, the recent finding in hippocampus of calcium-dependent inactivation of calcium currents, as described above, provides another mechanism through which magnesium might reduce synaptic depression or strengthen synaptic potentiation during repetitive activation. That is, magnesium could slow the inactivation of calcium currents in presynaptic terminals by reducing intracellular calcium, thereby maintaining greater release. In fact, recent studies of synaptic vesicle redistribution during repetitive stimulation in the hippocampus indicate that vesicle depletion is not the only factor that underlies synaptic depression (Applegate and Landfield 1988) suggesting that calcium current inactivation may play a role in synaptic depression (cf. also Augustine and Eckert 1984).

Regardless of the number or nature of the multiple calcium-dependent mechanisms that govern synaptic and neuronal transmission in the hippocampus, however, it appears that high calcium tends to weaken and high magnesium to strengthen short-term synaptic potentiation by each of these mechanisms (Fig. 4). Thus, although FP is calcium-dependent, too much calcium appears, paradoxically, to impair this process.

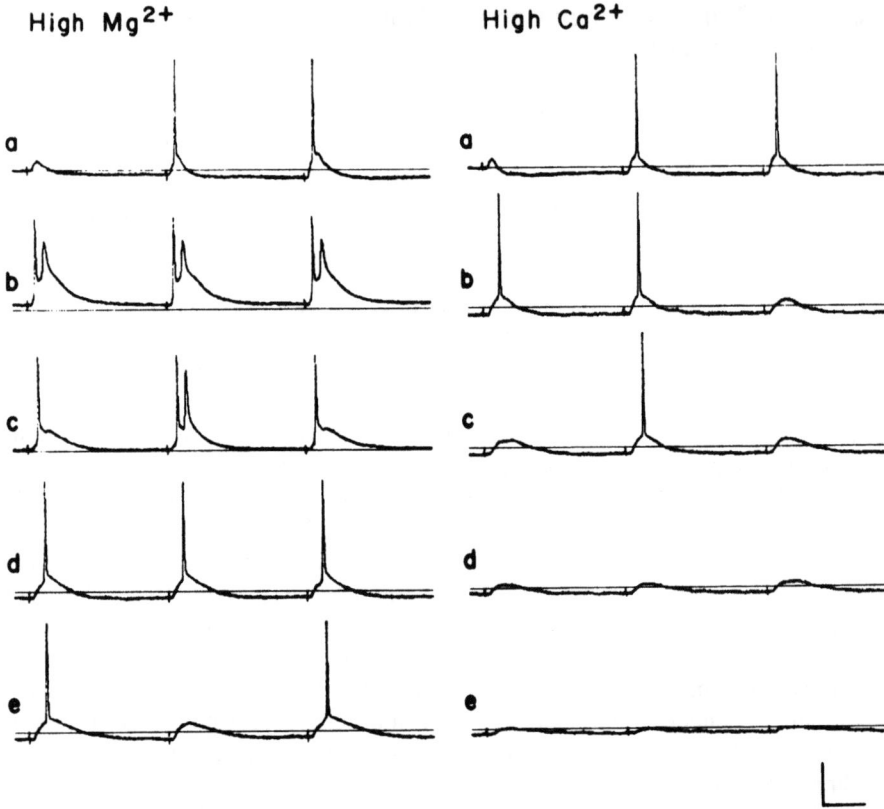

Fig. 4 a–e. Representative intracellular recordings from a CA1 cell maintained in high magnesium medium (2.6 mM Mg, 1.4 mM Ca) (*left*) and a comparable cell maintained in high-calcium medium (2.6 mM Ca, 1.4 mM Mg) (*right*). **a** Onset of 10-Hz stimulation (first response is the control EPSP set at 75% of spike threshold); **b** EPSPs at 15 s of 10-Hz stimulation; **c** EPSPs at 30 s of stimulation; **d** EPSPs at 1 min of stimulation; **e** EPSPs at 4 min of stimulation. *Calibration,* 30 mV and 30 ms. Control resting potential is indicated by the *solid horizontal line* across each figure. An initial hyperpolarization follows even the subthreshold control EPSP (**a**) and grows larger during repetitive stimulation. The hyperpolarization is reduced in high magnesium, leading to greater spike activation. Double spikes (**b**, *left*) and some spike inactivation are common at depolarized levels of membrane potential. Depression of the EPSP was pronounced by 4 min of 10-Hz stimulation in most high-calcium cells. (Reprinted from Landfield et al. 1986)

Evidence of Aging-Dependent Increases in Hippocampal Calcium Currents

In studies of neuronal function in aged animals, which were generally conducted concomitantly with basic studies of calcium dependent synaptic processes described above, we noted that magnesium substantially improved FP and, in addition, appeared to exert a stronger effect on FP in aged than in young hippocampus. Thus, the percent inprovement of FP by high magnesium was greater in aged hippocampal slices (Landfield et al. 1986) (Figs. 5, 6).

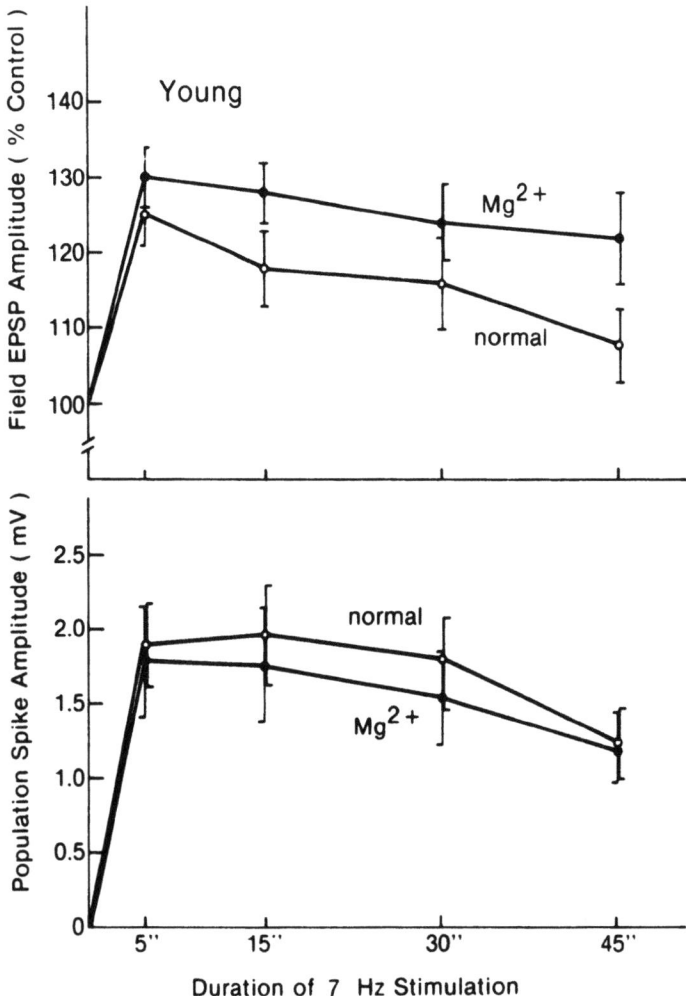

Fig. 5. Mean (±SE) of frequency potentiation for the field EPSP and population spike, as functions of magnesium/calcium ratios in the medium and 7-Hz stimulation duration in slices from young rats. *Filled circles*, high magnesium/calcium ratio (2:1); *open circles*, normal medium (1:1). (From Landfield et al. 1986)

We have suggested for some years that FP may be a critical mechanism for information processing and storage, and that it may serve to amplify and increase the retention of biologically meaningful material (Landfield et al. 1978; 1986). If this were the case, then one prediction would be that raising brain or CSF magnesium might improve learning and memory in aged rats. We tested this prediction by feeding rats a very high magnesium diet (2.0%) for 4 days. The literature indicates that chronic elevation of plasma magnesium will gradually affect CSF concentrations (e. g., Katzman 1975; Buck et al. 1979) and, in this study, the high-magnesium diet improved

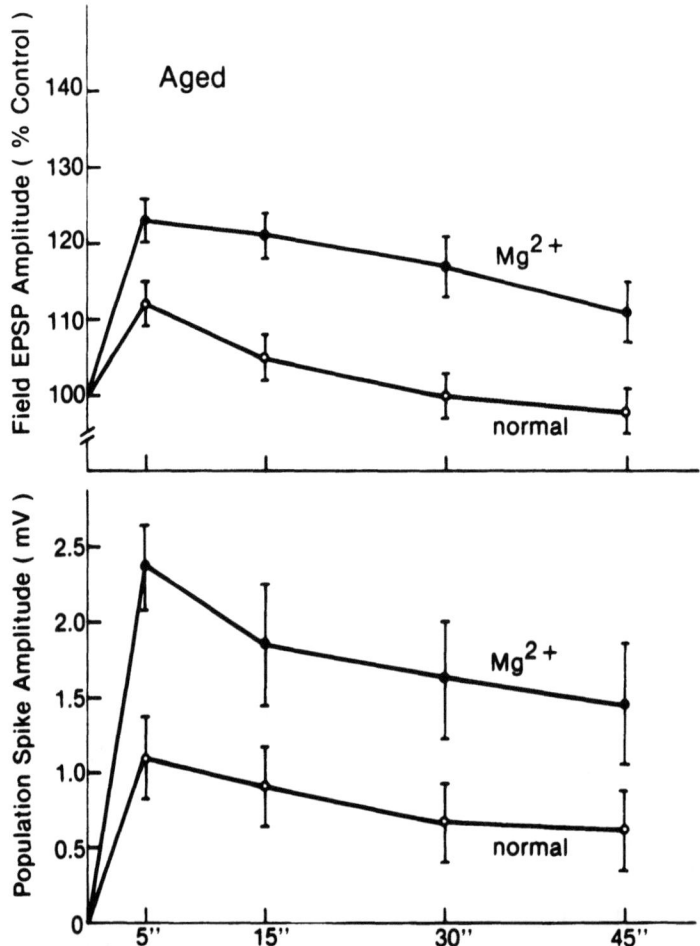

Fig. 6. Mean (±SE) of frequency potentiation for the field EPSP and population spike, as functions of magnesium/calcium ratios and 7-Hz stimulation duration in slices from aged rats. *Filled circles*, high magnesium/calcium ratio; *open circles*, normal medium. (From Landfield et al. 1986)

hippocampal FP in the intact animal (Landfield and Morgan 1984), as high magnesium improved FP in the slices (Landfield et al. 1986). With regard to the test of the hypothesis, we found that the high-magnesium diet did in fact improve acquisition of a maze reversal learning paradigm by aged rats, and increased their performance to the level found in young animals (Fig. 7). Moreover, although high magnesium also tended to improve performance in young animals, the effects were greater in aged rats (Landfield and Morgan 1984).

Although the more pronounced effects magnesium on both FP and behavior of aged rats could have been due to a "ceiling" effect in young animals, they also raised the possibility that calcium homeostasis differed in the brain neurons of young and aged rats. Consequently, we tested this possibility by measuring a process that is more

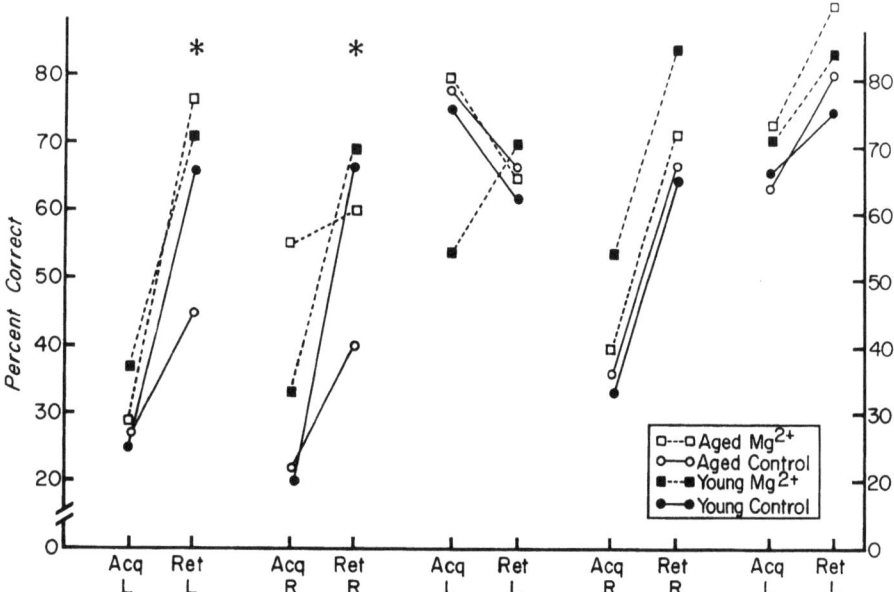

Fig. 7. Group mean percent correct choices for acquisition and retention of T-maze reversal learning, for each of five test days. Each set of three acquisition (*Acq*) trials of a new reversal was followed in 48–72 h by a set of three retention trials (*Ret*) to the same side. The three retention trials were then followed in 5 min by three acquisition trials of a new reversal to the opposite side (*L, R,* left and right). On the first two retention tests the aged magnesium-treated group performed significantly better than the aged control group, and comparably to the two young groups (*asterisks*). Data on running latencies showed similar results. (From Landfield and Morgan 1984)

directly and solely calcium-dependent than is synaptic transmission, namely, the slow calcium-dependent potassium-mediated afterhyperpolarization (AHP) (Alger and Nicoll 1980; Wong and Prince 1981; Johnston et al. 1980; Lancaster and Adams 1986; Storm 1987).

In quantitative studies in hippocampal slices, we found that the $AHP_{(Ca)}$ was prolonged substantially in neurons from aged animals (Landfield and Pitler 1984) (Fig. 8). However, it was not fully clear from this result alone whether calcium influx was increased, whether potassium efflux was increased, or whether calcium buffering and/or extrusion were reduced. Consequently, we studied calcium spikes in aged and young hippocampal neurons, and, in additional studies, found the calcium spikes to be prolonged in aged rat neurons (Landfield and Pitler 1987). In particular, the calcium spikes of aged neurons exhibited prolongation of the second depolarization plateau phase that follows the initial rapid spike. Recent studies also indicate that this second plateau phase can be reduced by nimodipine, suggesting that it is mediated by slowL-like calcium channels.

In summary, the data suggest that, at least in hippocampal cell somata, voltage-dependent calcium currents are increased with aging. Some of this increase seems to occur through prolonged activation of L-type channels, or perhaps through the

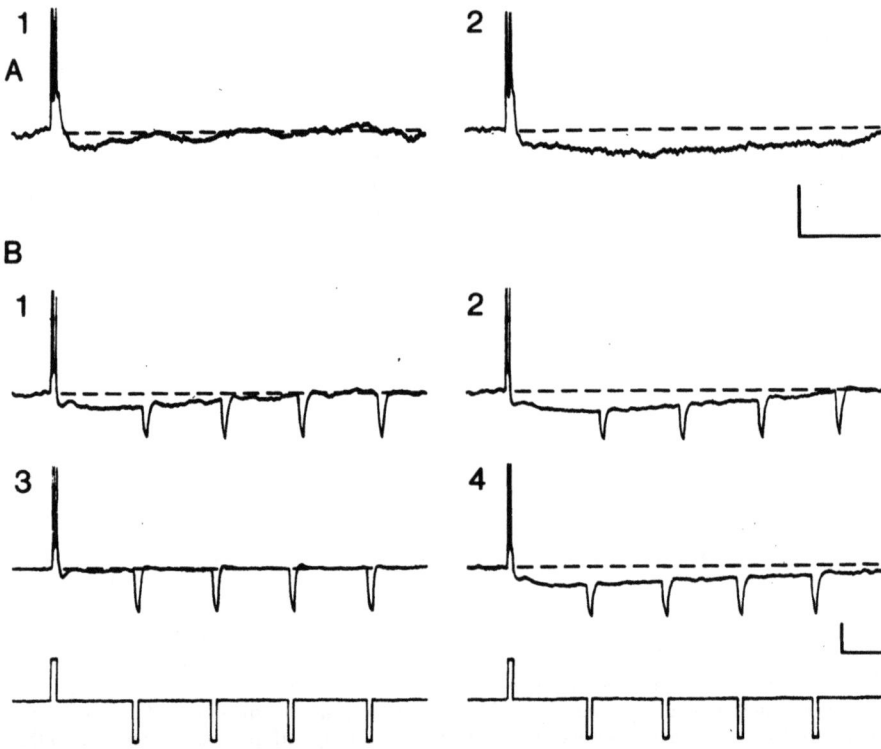

Fig. 8A, B. Intracellular current-induced bursts of action potentials and subsequent AHPs in CA1 neurons of hippocampal slices from young and aged rats. **A** AHPs following a 0.4-nA current-induced burst of two spikes. A_1 Cell from a young rat. A_2 Cell from an aged rat. **B** AHPs and concomitant conductance increases following a 0.4-nA current-induced burst of three spikes. B_1 Young rat cell in normal calcium. B_2 Aged rat cell in normal calcium. B_3 Young rat cell in low calcium, high magnesium. B_4 Young rat cell in high calcium, low magnesium. *Dashed lines* show resting potentials before the burst. At the *bottom* are shown the initial intracellular depolarizing current pulses used to induce a spike burst and the subsequent 2-Hz train of 0.4-nA hyperpolarizing pulses used to assess input conductance during the AHP, for cells shown in **B** *Calibrations* 10 mV, 250 ms (From Landfield and Pitler 1984)

presence of more L channels. In addition, synaptic plasticity is impaired in hippocampal slices of aged rats in a manner analogous to the impairment that can be induced in high-calcium media (e.g., impaired frequency potentiation). Therefore, a prolongation of calcium currents into various neuronal compartments might well account for the impairment of synaptic function, and perhaps of learning and memory, that is found in aged rats.

Implications for Human Brain Aging and Alzheimer's Disease

The human hippocampus is well established as a critical structure for the storage of new memories (Squire 1986; Olton 1983), and this structure is also among the most severely deteriorated in Alzheimer's disease (Wisniewski and Terry 1973). Further, even in "normal" aging, the hippocampus shows considerable morphological change (Tomlinson and Henderson 1976; Ball 1977), and humans exhibit significant memory deficits (Craik 1984). In humans, the hippocampus has a low seizure threshold, and is often the site of epileptic foci.

If the findings on increased calcium conductance in hippocampal neurons of aged rodents can be generalized to aged humans, then it seems possible that a variety of synaptic physiological and morphological changes in the hippocampus, and consequently, in the cognitive function, of aging humans might result from these changes. Neuronal calcium homeostasis appears to be altered by aging in a number of ways (cf. reviews in Khachaturian 1984; Gibson and Peterson 1987) and it has been suggested that increased membrane conductance to calcium is an important early step in age-related disturbances of calcium homeostasis (Landfield 1987; Landfield and Pitler 1984). Moreover, it appears highly conceivable that if these age-related membrane conductance changes were accelerated by an additional factor (e. g., impaired genomic control of calcium channels, virus- or aluminum-induced changes, etc.) then the full Alzheimer's disease syndrome might result. Elevated calcium influx is known to be cytotoxic in ischemia (Nayler et al. 1979; Schlaepfer and Hasler 1979), and it therefore seems readily conceivable that brain cell degeneration might also occur if calcium currents were increased significantly by aging and/or Alzheimer's disease.

References

Adams PR, Constanti A, Brown DA, Clark RB (1982) Intracellular Ca^{2+} activates a fast voltage-sensitive K^+ current in vertebrate sympathetic neurons. Nature (London), 296:746–749

Alger BE, Nicol RA (1980) Epileptiform burst afterhyperpolarization: clacium-dependent potassium potential in hippocampal CAl pyramidal cells. Science 210:1122–1144

Applegate MD, Landfield PW (1988) Synaptic vesicle redistribution during hippocampal frequency potentiation and depression in young and aged rats. J Neurosci 8:1096–1111

Augustine GJ, Eckert R (1984) Calcium-dependent inactivation of presynaptic calcium channels. Soc Neurosci Abstr 10:194

Ball MJ (1977) Neuronal loss, neurofibrillary tangles and granulovacuolar degeneration in the hippocampus with aging and dementia: A quantitative study. Acta Neuropatholog 37:11–118

Bley KR, Madison DV, Tsien RW (1987) Multiple types of calcium channels in hippocampal neurons: Characterization and localization. Soc Neurosci Abstr 13:1010

Brown DA, Griffith WH (1983) Persistent slow inward calcium current in voltage-clamped hippocampal neurones of the guinea pig. J Physiol (London) 337:303–320

Brown TH, Johnston D (1983) Voltage-clamp analysis of mossy fiber synaptic input to hippocampal neurons. J Neurophysiol 50:487–507

Buck DR, Mahoney AW, Hendricks DF (1979) Effects of cerebral intraventricular magnesium injections and a low magnesium diet on nonspecific excitability level, audiogenic seizure susceptibility and serotonin. Pharmacol Biochem Behav 10:487–491

Campbell LW, Hao SY, Landfield PW (1988) Calcium-dependent inactivation of calcium currents in hippocampal neurons: Effects of tetraethylammonium and nimodipine. Soc Neurosci Abstr 14:in press

Craik FIM (1984) Age differences in remembering. In: Squire LR, Butters N (eds) Neuropsychology of memory. Guilford Press. New York, pp 3–12

Delgado-Escueta AV, Ward AA, Woodbury DM, Porter RJ (eds) (1986) Basic mechanisms of the epilepsies: Molecular and cellular approaches. Raven, New York

Eckert R, Ewald D (1983) Inactivation of calcium conductance characterized by tail current measurements in neurones of Aplysia californica. J Physiol (London) 345:549–565

Gibson GE, Peterson C (1987) Calcium and the aging nervous system. Neurobiol Aging 8:329–344

Gray R, Johnston D (1986) Multiple types of calcium channels in acutely-exposed neurons from adult hippocampus. Biophy J 49:432a

Gray R, Johnston D (1987) Noradrenaline and β-adrenoceptor agonists increase activity of voltage-dependent calcium channels in hippocampal neurons. Nature 327:620–622

Halliwell JW (1983) Caesium loading reveals two distinct Ca-currents in voltage-clamped guinea-pig hippocampal neurones in vitro. J Physiol (London) 341:10–11

Johnston D, Hablitz JJ, Wilson WA (1980) Voltage clamp discloses slow inward current in hippocampal burst firing neurones. Nature 286:391–393

Katzman R (1975) Cerebrospinal fluid physiology: role of secretory and mediated transport systems. In: Tower DB (ed) The nervous system, Vol 1: The basic neurosciences. Raven Press, New York, pp 291–297

Khachaturian ZS (1984) Towards theories of brain aging. In: Kay D, Burrows GD (eds) Handbook of studies on psychiatry and old age. Elsevier, Amsterdam

Lancaster B, Adams PR (1986) Calcium-dependent current generating the afterhyperpolarization of hippocampal neurons. J Neurophysiol 55:1268–1282

Landfield PW (1987) "Increased calcium current" hypothesis of brain aging. Neurobiol Aging 8:346–347

Landfield PW, Lynch G (1977) Impaired monosynaptic potentiation in *in vitro* hippocampal slices from aged, memory-deficient rats. J Gerontol 32:523–533

Landfield PW, McGaugh JL, Lynch G (1978) Impaired synaptic potentiation processes in the hippocampus of aged, memory-deficient rats. Brain Res 150:85–101

Landfield PW, Morgan G (1984) Chronically elevating plasma Mg^{2+} improves hippocampal frequency potentiation and reversal learning in aged and young rats. Brain Res 322:167–171

Landfield PW, Pitler TA (1984) Prolonged Ca^{2+}-dependent afterhyperpolarizations in hippocampal neurons of aged rats. Science 226:1089–1092

Landfield PW, Pitler TA (1987) Calcium spike duration: prolongation in hippocampal neurons of aged rats. Soc Neurosci Abstr 13:718

Landfield PW, Pitler TA, Applegate MD (1986b) The effects of high Mg^{2+}-to Ca^{2+} ratios on frequency potentiation in hippocampal slices of young and aged rats. J Neurophysiol 56:797–811

Llinas R, Hess R (1976) Tetrodotoxin resistant dendritic spikes in avian Purkinje cells. Proc Natl Acad Sci USA 73:2520–2523

Martin AR (1977) Presynaptic mechanisms. In: Brookhart JM, Mountcastle VB (eds) Handbook of Physiology I: The nervous system. American Physiological Society, Bethesda, pp 329–355

Miller RJ (1987) Calcium channels in neurones. In: Venter JC, Triggle D (eds) Structure and physiology of the slow inward calcium channel. Alan R Liss, New York, p 161

Nayler WG, Poole-Wilson PA, Williams A (1979) Hypoxia and calcium. J Molec Cell Cardiol 11:683–706

Nowycky MC, Fox AP, Tsien RW (1985) Three types of neuronal calcium channel with different agonist sensitivity. Nature 316:440–443

Olton DS (1983) Memory functions and the hippocampus. In: Seifert W (ed) Neurobiology of the hippocampus, Academic Press, New York, pp 335–373

Pitler TA, Landfield PW (1987a) Probable Ca^{2+}-mediated inactivation of Ca^{2+} currents in mammalian brain neurons. Brain Res 410:147–153

Schlaepfer WW, Hasler MB (1979) Characterization of the calcium-induced disruption of neurofilaments in rat peripheral nerve. Brain Res 168:299–309

Schwartzkroin DA, Slawsky MA (1977) Probable calcium spikes in hippocampal neurones. Brain Res 135:157–161

Scriabine A (1987) Ca^{2+} channel ligands: Comparative pharmacology. In: Venter JC, Triggle D (eds) Structure and physiology of the slow inward calcium channel. Alan R. Liss, New York, p 51

Smith DO (1980) Mechanisms of action potential failure at sites of axon branching in the crayfish. J Physiol (Lond) 301:243–259

Squire LR (1986) Mechanisms of memory. Science 232:1612–1619

Storm J (1987) Action potential repolarization and a fast afterhyperpolarization in rat hippocampal pyramidal cells. J Physiol (Lond) 385:733–759

Tomlinson BE, Henderson G (1976) Some quantitative cerebral findings in normal and demented old people. In: Terry RD, Gershon S (eds) Neurobiology of aging. Raven Press, New York, pp 183–204

Wisniewski HM, Terry RD (1973) Morphology of the aging brain, human and animal. Prog Brain Res 40:167–186

Wong RKS, Prince DA (1981) After potential generation in hippocampal pyramidal cells. J Neurophysiol 45:87–97

Regulation of Brain 1,4-Dihydropyridine Receptors by Drug Treatment*

V. Ramkumar[1] and E. E. El-Fakahany[2]

[1] Department of Medicine, Duke University Medical Center, Box 3444, Durham, NC 27710, USA
[2] Department of Pharmacology and Toxicology, University of Maryland School of Pharmacy, 20 N. Pine Stret, Baltimore, MD 21201, USA

Introduction

The voltage-sensitive Ca^{2+} channel promotes depolarization-dependent entry of Ca^{2+} into cells in order to maintain multiple functions such as skeletal, smooth, and cardiac muscle contraction, glandular secretion, metabolic events, and the release of various neurotransmitters [11, 12, 35]. In the brain, three different types of voltage-sensitive Ca^{2+} channels have been described: the L, the N, and the T channels [19]. These various channel types differ in terms of their activation and inactivation kinetics [19] and in their sensitivity to 1,4-dihydropyridine Ca^{2+} agonists and antagonists [17, 19]. The use of radiolabeled 1,4-dihydropyridines has expanded our understanding of the properties of brain Ca^{2+} channels. Thus, it has been demonstrated that 1,4-dihydropyridine binding sites are differentially localized in the brain [7] Furthermore, these sites are coupled allosterically to two other discrete domains on the Ca^{2+} channel, namely the verapamil and the diltiazem binding sites [3].

Neuroleptic drugs are known to be high-affinity antagonists of dopamine (D_2) receptors and are believed to manifest their therapeutic effects against schizophrenia by interacting with these receptors in the brain. In addition, these drugs bind to the 1,4-dihydropyridine sites in the brain, albeit with lower affinities [6, 22]. It has been suggested that neuroleptic drugs interact specifically with the verapamil binding site on the Ca^{2+} channel [6]. For example, the diphenylbutylpiperidine, pimozide, interacts with the verapamil binding site with nanomolar affinity and is believed to mediate its unique therapeutic effects, in part, via such a site [6].

Chlorpromazine is a neuroleptic drug belonging to the phenothiazine class. While this drug interferes with Ca^{2+} channel functions [16] and the binding of radioligands to 1,4-dihydropyridine sites [22], a full characterization of the drug at the latter sites has not been undertaken. In the present study, therefore, the in vitro effects of chlorpromazine at the Ca^{2+} channel were studied to determine whether this drug, like pimozide, interacts primarily with the verapamil binding site. Prolonged administration of neuroleptics is associated with the syndrome of tardive dyskinesia [15]. The possibility that the Ca^{2+} channel might be involved in the manifestation of this syndrome prompted us to test the effects of administration of chlorpromazine, fluphenazine, and pimozide on 1,4-dihydropyridine binding sites in the brain. In addition, changes in brain 1,4-dihydropyridine receptors induced by the administra-

---

* This work was supported by a grant from the Miles Institute for Preclinical Pharmacology

tion of morphine were compared to the effects of the neuroleptics. The changes in the properties of 1,4-dihydropyridine binding sites induced by in vivo drug treatment described below underscore the plasticity of these sites in the brain.

In Vitro Effects of Chlorpromazine at the 1,4-Dihydropyridine Binding Sites

[^{3}H]Nimodipine binding in mouse brain exhibited saturability and high affinity. Transformation of saturation data yielded linear Rosenthal plots with B_{max} averaging 186.7 ± 6.8 fmol/mg protein (mean ± SEM, $n = 4$) (Fig. 1). Low concentrations of chlorpromazine (< 10 µM) produced no significant effects on these control binding parameters, while higher concentrations of the neuroleptic (10–30 µM) specifically decreased the B_{max} by 15.3% and 26.6% respectively without affecting the K_d (Fig. 1). At higher concentrations, chlorpromazine decreased the B_{max} and increased the K_d of [^{3}H]nimodipine (Fig. 1). Schild analysis of saturation plots obtained at different concentrations chlorpromazine yielded linear plots with slopes averaging 0.8 (significantly different from unity, $P < 0.02$). These findings are consistent with the notion that chlorpromazine interacts noncompetitively with the 1,4-dihydropyridine binding site in the brain.

Displacement experiments indicated that chlorpromazine inhibited the binding of [^{3}H]nimodipine (0.04 nM) with an IC_{50} of 34.3 ± 7.0 µM. However, a 10-fold elevation in radioligand concentration led to only a twofold increment in the IC_{50} of chlorpromazine, suggesting a deviation from the law of mass action (Fig. 2). This is readily apparent when the [^{3}H]nimodipine concentrations and the experimentally determined IC_{50} values are compared to the theoretical values expected for a competi-

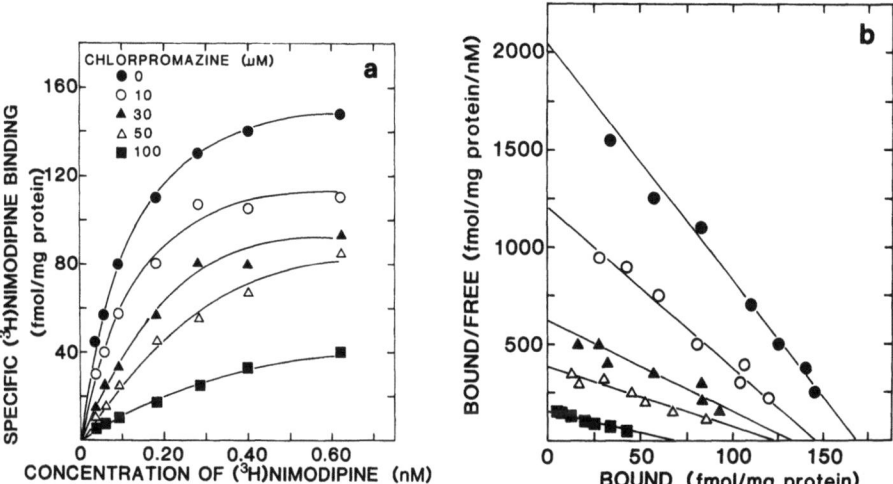

Fig. 1 a, b. Saturation curves of specific [^{3}H] nimodipine binding in mouse brain homogenates. **a** Experiments were conducted by incubating seven different concentrations of [^{3}H]nimodipine (0.04–0.06 nM) with 0.2 mg protein of rat brain membranes in the absence and presence of 0.1–100 µM chlorpromazine. Nonspecific binding was defined by 1 µM nifedipine. **b** Rosenthal plots of the data

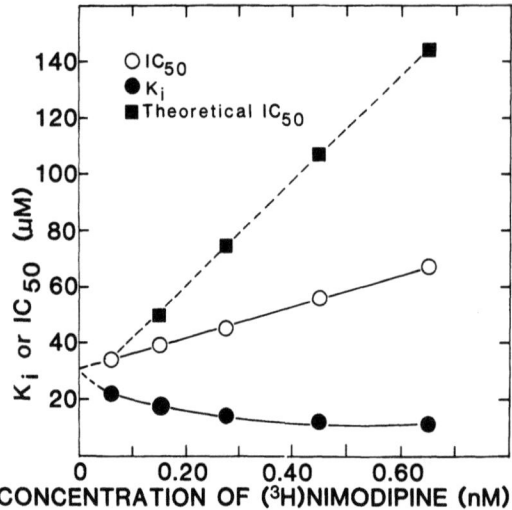

Fig. 2. The dependence of the IC_{50} and K_i values of chlorpromazine on the concentration of [^{3}H]nimodipine. K_i values were calculated from IC_{50} obtained from chlorpromazine displacement curves using a K_d value for [^{3}H]nimodipine of 0.12 nM according to the method of Cheng and Prusoff. A K_i value of 22.5 µM, calculated using the lowest ligand concentration (0.063 nM), was used to determine the theoretical IC_{50} relationship represented by the dashed line

tive interaction (Fig. 2). Furthermore, the gradual decrease in K_i values with increasing radioligand concentrations and the curvilinear relationship between these two parameters suggests nonconformity to a simple competitive interaction.

The ability of a drug to alter the rate of dissociation of a radioligand initiated by infinite dilution indicates a noncompetitive interaction between the drug and the primary radioligand binding site. Therefore, we tested the effects of chlorpromazine on the dissociation of [^{3}H]nimodipine induced by 50-fold dilution with buffer. In all cases, dissociation plots were exponential and monophasic (Fig. 3). Half-life was 22.3 ± 3.6 min ($n = 3$) in the absence of the neuroleptic, and 19.0 ± 0.8 min ($n = 3$) and

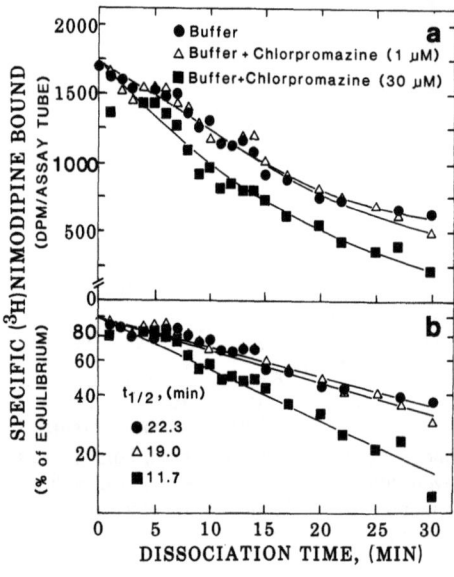

Fig. 3 a, b. Effect of chlorpromazine on the rate of dissociation of [^{3}H]nimodipine from brain homogenate. a Tissue was incubated with 0.25 nM of [^{3}H]nimodipine for 90 min and, following this, dissociation was initiated by a 50-fold dilution with 50 mM Tris buffer (pH 7.4) at 25 °C in the absence and presence of chlorpromazine. b The linear regression plot of the log of percent equilibrium binding (determined at time 0). Data are from one of three independent experiments, each performed in duplicate

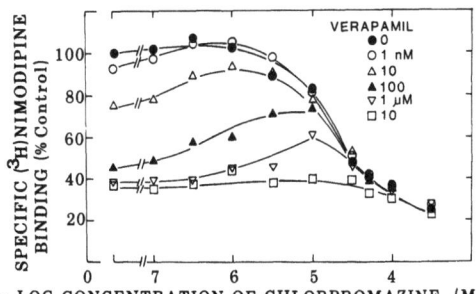

Fig. 4. Chlorpromazine-induced reversal of the inhibition of [^{3}H]nimodipine binding by verapamil. The 100% level represents binding in the absence of both chlorpromazine and verapamil. The data are from one of three similar experiments, each performed in duplicate

11.7 ± 0.6 min ($n = 3$) in the presence of 1 and 30 µM chlorpromazine respectively. Taken together, these data clearly demonstrate a noncompetitive interaction between chlorpromazine and the 1,4-dihydrophyridine binding site.

The verapamil and diltiazem sites represent discrete ligand binding domains comprising the Ca^{2+} channel [3]. In order to study indirectly the effects of chlorpromazine at these sites, the ability of this neuroleptic to affect the actions of verapamil and diltiazem was determined. Chlorpromazine shifted verapamil inhibition curves of [^{3}H]nimodipine binding to the right in a dose-dependent manner, with significant inhibition of verapamil's effects being observed at 1 µM of the neuroleptic and abolition of its effect evident at 30 µM. Thus, it appears that chlorpromazine interacts with the verapamil binding site with affinity one order of magnitude higher than that observed for the 1,4-dihydropyridine site (i.e., IC_{50} of 40 µM at this latter site). Furthermore, chlorpromazine stimulated [^{3}H]nimodipine binding in the presence of verapamil (Fig. 4). EC_{50} values for this effect ranged from 0.2 µM to 3 µM of chlorpromazine, depending on the concentration of verapamil used. In addition, diltiazem shifted chlorpromazine inhibition curves to the right in a dose-dependent fashion. Thus, the IC_{50} values of chlorpromazine were increased from 28 µM in the absence of diltiazem to 32, 55, and 107 µM in the presence of 1, 10, and 100 µM diltiazem respectively.

The effects of chlorpromazine on the rate of dissociation of [^{3}H]nimodipine in the absence and presence of verapamil and diltiazem were also tested. In these experiments, the dissociation reactions were allowed to proceed for 15 min, and the remaining specific [^{3}H]nimodipine binding was used to calculate the extent of dissociation. As shown in Fig. 5, chlorpromazine decreased the fraction of the bound radioligand remaining after 15 min in a concentration-dependent manner, reflecting an enhanced rate of dissociation induced by this drug. This inhibitory effect of the neuroleptic evident at 1 and 3 µM was significantly attenuated in the presence of 0.1 µM verapamil. However, diltiazem (10 µM) did not influence the ability of chlorpromazine to increase the dissociation of [^{3}H]nimodipine (Fig. 5).

Taken together, the present data provide evidence for the action of chlorpromazine at both the verapamil and diltiazem binding sites. Since these sites are coupled allosterically to the 1,4-dihydropyridine binding site, it is possible that chlorpromazine exerts its influence at the latter site by interacting primarily at the verapamil and the diltiazem binding sites. The higher affinity of chlorpromazine at these nondihydropyridine sites tends to support such a proposal.

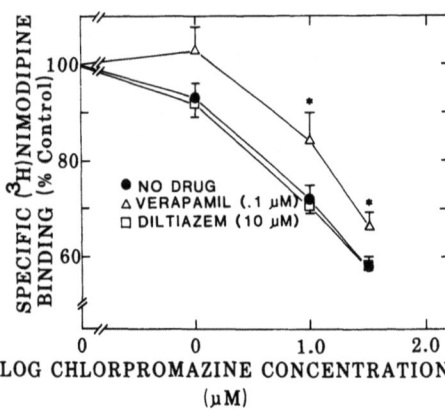

Fig. 5. Dissociation of [^{3}H]nimodipine by chlorpromazine in the presence of verapamil and diltiazem. Dissociation of the radioligand was initiated by a 50-fold dilution with buffer alone (no drug), or buffer containing 0.1 µM verapamil or 10 µM diltiazem. Specific binding remaining after 15 min was used to calculate the extent of dissociation. The 100% level of binding was set by corresponding control samples containing buffer, verapamil, or diltiazem without chlorpromazine. Each data point represents the mean ± SEM of six independent experiments, each performed in duplicate. *Asterisks* represent statistically significant inhibition of dissociation ($P < 0.05$)

In Vivo Effects of Neuroleptics

It has been shown previously that administration of the Ca^{2+} channel antagonists verapamil and nifedipine to rats decreases brain [^{3}H]nitrendipine binding sites and diminishes K$^+$-stimulated ^{45}Ca^{2+} uptake [20]. Our demonstration that chlorpromazine interacts allosterically with the 1,4-dihydropyridine binding sites prompted us to test the effects of prolonged administration of this drug and other neuroleptics on these sites in vivo. For these experiments mice were administered chlorpromazine hydrochloride (0.5 mg/g diet) and pimozide (0.0025 mg/g diet) orally and fluphenazine decanoate (16.6 mg/kg in polyethylene glycol) intramuscularly once every 2 weeks. Mice serving as controls were administered a drug-free diet (controls for chlorpromazine and pimozide) or polyethylene glycol (controls for fluphenazine).

The dose of chlorpromazine used in this study was one-half that used previously for chronic administration in mice [16]. The dose of pimozide used was estimated as the amount of this drug which produces an equivalent degree of inhibition of D$_2$ receptors as the above dose of chlorpromazine [27]. Furthermore, the dose of fluphenazine injections was equivalent to that needed to produce spontaneous orofacial dyskinesias in rat [36], a syndrome whose features are characteristics of tardive dyskinesia. Apart from sedation observed during the 1st week of treatment, no other overt side effects were observed subsequent to administration of chlorpromazine and fluphenazine. Similarly, clinical administration of the phenothiazines is generally associated with a rapid development of tolerance to sedation [2].

Figure 6a indicates that the administration of chlorpromazine resulted in a significant increase in [^{3}H]nimodipine binding affinity (1/K$_d$). Affinity constants averaged 6.35, 10.64, and 15.87 nM^{-1} for controls and following drug administration for 4 and 8 weeks respectively [24]. Upon withdrawal of chlorpromazine, binding affinites were 6.67, 1.98, and 7.80 nM^{-1} at 1, 2, and 3 weeks, respectively [24]. Under a similar treatment schedule, no changes in B$_{max}$ between control and treated brains were observed (Fig. 6b). In addition, verapamil competition curves of [^{3}H]nimodipine binding in control and treated brains were superimposable, implying no alterations in either the affinity or number of verapamil binding sites between the two groups.

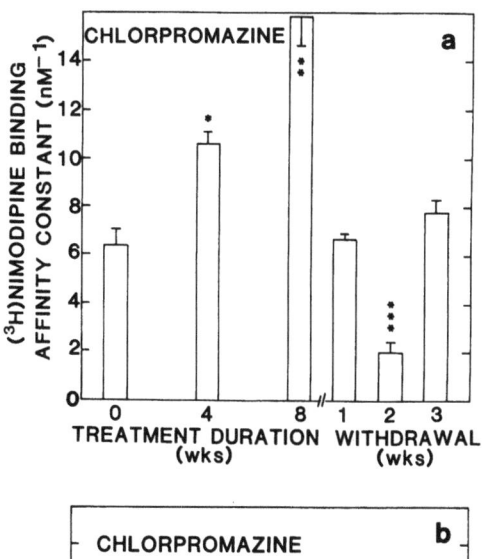

Fig. 6 a, b. Effect of in vivo chlorpromazine treatment and withdrawal on 1,4-dihydropyridine binding sites in mouse brain. Mice were treated with chlorpromazine (0.5 mg/g diet) for various periods as indicated. The figures indicate the effect of chlorpromazine on the affinity (**a**) and the B_{max} (**b**) of [^{3}H]nimodipine. The data are presented as the mean ± SEM of 3–11 brains. * $P < 0.05$; ** $P < 0.01$; *** $P < 0.001$

Similar to chlorpromazine, fluphenazine administration induced a time-dependent increase in the affinity of [^{3}H]nimodipine binding sites in the brain (Fig. 7a). For example, the affinity constants averaged 9.29, 7.55, and 14.06 nM^{-1} for controls and following treatment for 2 and 6 weeks respectively. Upon withdrawal of the drug, the affinities returned to control levels within the first week and remained relatively constant thereafter. Membranes prepared from brains of treated mice also showed a small but significant decrease (14%) in receptor number during the fluphenazine treatment period. The B_{max} determined from Rosenthal plots [28] averaged 189.2, 162.6, and 163.2 fmol/mg protein for controls and mice treated for 2 and 6 weeks respectively. These values also recovered rapidly to control levels following drug withdrawal (Fig. 7b).

In contrast to the other neuroleptics tested, the administration of pimozide produced no change in affinity apart from an initial decrease observed by the 2nd week of drug treatment (Fig. 8a). Subsequent drug administration resulted in no further

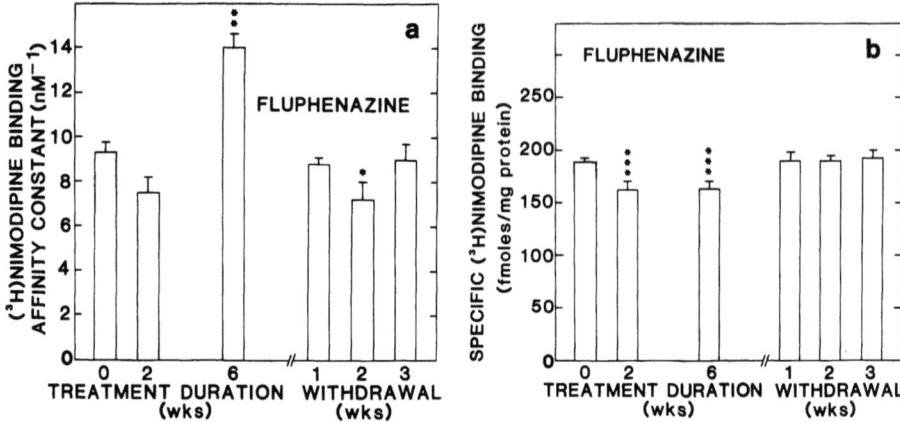

Fig. 7 a, b. Effect of in vivo fluphenazine treatment and withdrawal on 1,4-dihydropyridine binding sites in mouse brain. Mice were injected intramuscularly with fluphenazine decanoate (16.6 mg/kg) once every 2 weeks. Brain membranes were prepared at the indicated time and [^{3}H]nimodipine binding experiments were performed. The data are presented as the mean ± SEM of 3–11 brains for the affinity (**a**) and B_{max} (**b**) of the radioligand. See Fig. 6 for explanation of asterisks.

change in affinity from control values. In addition, administration of pimozide resulted in a small but significant decrease (14%–16%) in the number of [^{3}H]nimodipine binding sites in the brain, observed in the 2nd and 4th weeks of treatment. Receptor number following drug administration and subsequent withdrawal was unchanged from control values (Fig. 8b).

The finding that changes at the Ca^{2+} channel occur in the absence [36] or presence [31] of increases in dopamine receptor levels suggests that the Ca^{2+} channel is an important site for the manifestation of the long-term effects of the phenothiazines. Interestingly, administration of doses of pimozide which produce only minor changes in [^{3}H]nimodipine binding (Fig. 8) has been reported to decrease [^{3}H]spiperone binding to D_2 receptors in the brain [34]. Pimozide, a neuroleptic of the diphenylbutylpiperidine class, is a potent antagonist at D_2 receptors [34] and calcium channels [6] and, unlike the phenothiazines, can alleviate both the "negative" and "positive" symptoms of schizophrenia [21]. In addition, while phenothiazines are known to produce tardive dyskinesia, pimozide has been used successfully to treat this syndrome [21]. It has therefore been speculated that the unique therapeutic benefits of pimozide result from antagonism of both D_2 receptors and Ca^{2+} channels [33]. Thus, it appears that the changes observed at the 1,4-dihydropyridine sites differ depending on the class of neuroleptics employed, implying that these changes (or lack of them) might relate to the unique therapeutic and/or side effects accompanying treatment with these drugs. It is unlikely that the changes produced by the phenothiazines can account for the persistence of tardive dyskinesia, since they are rapidly reversible upon discontinuation of these drugs. The biochemical basis of these changes at the 1,4-dihydropyridine binding site has not yet been determined.

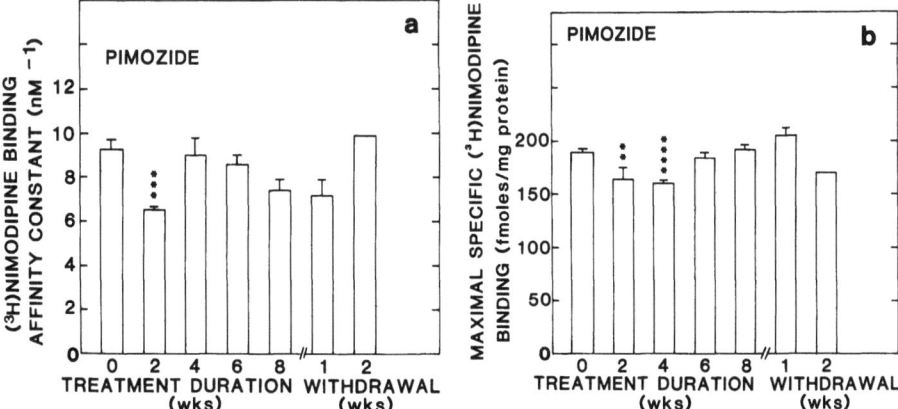

Fig. 8 a, b. Effect of in vivo administration of pimozide to mice on 1,4-dihydropyridine binding sites in the brain. Mice were administered pimozide (0.0025 mg/g diet). [^{3}H]Nimodipine binding was determined at the indicated time and the results are presented as the affinity (**a**) and the B_{max} (**b**) of the radioligand. The data are presented as the mean ± SEM of 3–11 brains. See Fig. 6 for explanation of asterisks

Effects of Morphine on 1,4-Dihydropyridine Binding Sites

The involvement of Ca^{2+} in the actions of opiates is supported by several pieces of evidence. For example, the administration of morphine in rats decreases the Ca^{2+} content of several brain regions [22], and specifically in brain cellular fractions rich in nerve endings [1, 10]. In addition, opiates inhibit depolarization-dependent Ca^{2+} uptake into synaptosomal preparations [9, 14]. Interestingly, EGTA, a Ca^{2+} chelator [13], and trivalent cation Ca^{2+} channel blockers [26] produce analgesia when administered centrally, suggesting an involvement of Ca^{2+} in the normal regulation of pain pathways.

Chronic opiate administration produces effects considerably different from those described above. For example, chronic morphine administration to animals increases vesicular Ca^{2+} content in the brain [10]. In addition, K^+-stimulated Ca^{2+} uptake in synaptosomes is significantly increased by chronic morphine [9] and levorphanol [30] treatment in rats. These findings prompted us to measure the density of brain 1,4-dihydropyridine Ca^{2+} channels consequent to morphine treatment.

Initially, the in vitro effects of morphine at 1,4-dihydropyridine binding sites in the brain was determined. Morphine at micromolar concentrations had no effect on these sites (not shown). However, prolonged in vivo administration of this drug resulted in upregulation of 1,4-dihydropyridine sites in the brain [23, 25]. For example, morphine administration by pellet implantation increased the density of [^{3}H]nitrendipine binding sites in the brains of tolerant dependent mice without changing the ligand affinity [23]. Similar findings were observed in brains obtained from morphine-tolerant-dependent rats (Fig. 9). The development of tolerance to morphine was evinced by the gradual loss of the analgesic effects of this drug (Table 1). As can be seen, morphine produced a significant degree of analgesia 2 and 26 h following pellet

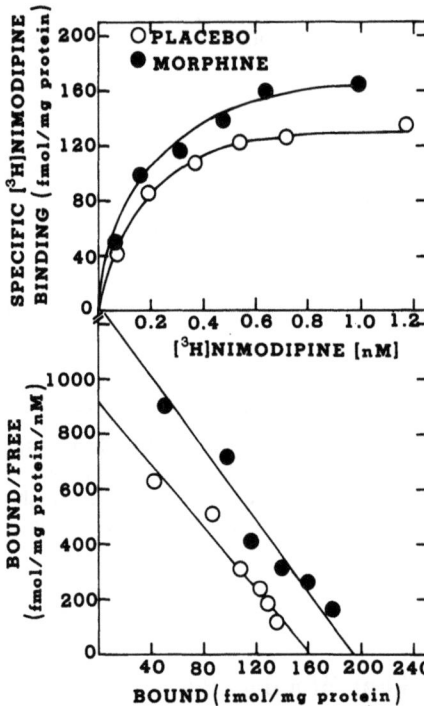

Fig. 9. Upregulation of [^{3}H]nimodipine binding sites in rat brain following implantation of morphine pellets. Rats were implanted with three placebo pellets or three morphine pellets for 3 days. [^{3}H]Nimodipine binding was performed in crude brain homogenates obtained from these rats. The figure is a representative plot of 11 and 4 experiments with similar results obtained from placebo- and morphine-treated rats respectively

implantation. However, by 50 h post implantation, the analgesic effects had largely dissipated. Figure 10 shows brain regional localization of the incrases in [^{3}H]nimodipine binding sites in the rat. Significant elevations in binding were observed in the cortex, hippocampus, hypothalamus, and brainstem but not in the striatum or cerebellum. Furthermore, the increase in the number of 1,4-dihydropyridine sites was dose- and time-dependent (Fig. 11). Optimal increases were observed following implantation of three morphine pellets (75 mg morphine each) for 3 days. This period, as described above, coincides with the emergence of tolerance to morphine and the development of physical dependence (Table 1 and data not shown).

Table 1. Time course of morphine analgesia in the rat following morphine administration

Treatment	Latency time (s)		
	Treatment for 2 h	Treatment for 26 h	Treatment for 50 h
Placebo	12.0 ± 1.7 (5)	12.7 ± 1.7 (5)	10.9 ± 0.8 (5)
Morphine	42.8 ± 4.4 (6)*	42.4 ± 3.3 (6)*	16.7 ± 2.9 (6)

Rats were implanted with three placebo pellets or three morphine pellets (75 mg each), following which analgesia was determined using a hotplate at 55 °C. Hindpaw lick was taken as the appropriate response to thermal nociceptive stimuli. Each rat was tested twice at an interval of 15 min at each of the designated time points. Latency for naive controls was 12.7 ± 0.6 s. Figures in parentheses indicate numbers of animals tested. Data expressed as mean ± SEM.

*Statistically different from control ($P < 0.05$).

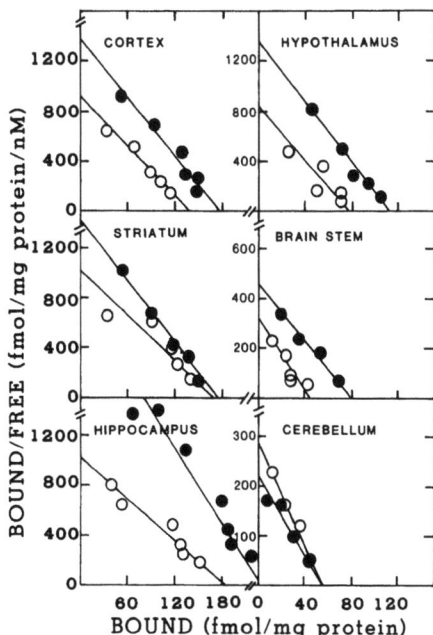

Fig. 10. Regional distribution of morphine-induced increases in [³H]nimodipine binding sites in the brain. [³H]Nimodipine binding studies were performed in brain membranes of controls and rats treated with three morphine pellets for 3 days. The data for each brain region are from an experiment representative of four to eight experiments, each performed in triplicate

Thus, it is suggested that increases in 1,4-dihydropyridine binding sites in the brain might contribute, at least in part, to the manifestation of tolerance to and/or physical dependence on morphine. Interestingly, chronic exposure of cell culture to opioids has been shown to increase both basal and stimulated cyclic AMP formation [8], probably resulting from a loss of tonic inhibition of adenylate cyclase mediated by the guanine nucleotide regulatory protein (G_i). Since cyclic AMP levels appear to modulate the expression of 1,4-dihydropyridine binding sites [18, 32], it is likely that the

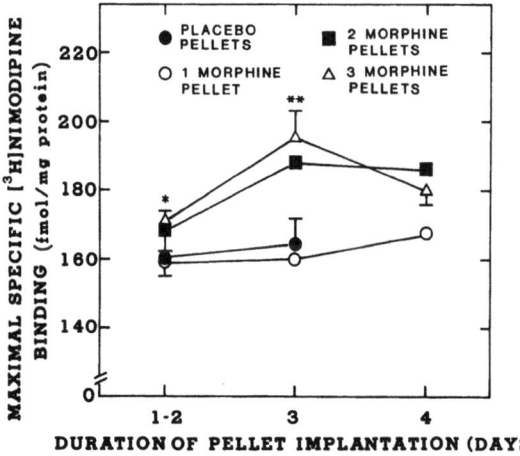

Fig. 11. Dose and time dependence of morphine-induced increase in [³H]nimodipine binding sites in rat brain. Rats were treated with placebo pellets or the indicated number of morphine pellets. [³H]Nimodipine binding assays were performed in triplicate, using six to eight concentrations of the radioligand. Each point represents the mean ± SEM of two to nine rats. * Statistically different from rats treated with placebo pellets for 1–2 days ($P < 0.05$); ** statistically different from rats treated with placebo pellets for 3 days ($P < 0.05$)

increase in these sites results from the increase in cyclic AMP associated with morphine tolerance and dependence.

Indirect support for the hypothesis that the increases in 1,4-dihydropyridine binding sites have a role in physical dependence derives from the finding that various dihydropyridine Ca^{2+} channel antagonists attenuate naloxone-precipitated withdrawal signs in morphine-dependent rats and mice [25]. For example, nimodipine (10 mg/kg, i.p.) decreases the incidence of withdrawal signs elicited by naloxone in morphine-dependent rats such as "wet-dog" shakes, ptosis, teeth chattering, abdominal stretching, and diarrhea. In this respect, the drug was as effective as clonidine, a drug known to be effective in blocking morphine withdrawal [4] (Table 2). Furthermore, the dihydropyridine Ca^{2+} channel antagonists also attenuate naloxone-precipitated

Table 2. Inhibition of naloxone-precipitated withdrawal signs in the rat by Ca^{2+} antagonists and clonidine

Withdrawal Signs	Vehicle (n = 9)	Nimodipine 10 mg/kg, i.p. (n = 9)	Clonidine 100 µg/kg, i.p. (n = 6)
Wet-dog shakes	3.4	1.3*	1.0*
Ptosis	100	55 *	0 *
Teeth chattering	7.0	3.0*	3.8*
Abdominal stretching	0.9	0.5*	0 *
Diarrhea	2.7	0.1*	0 *

Rats were made dependent on morphine by i.p. injections over a 6-day period as described [25]. Morphine-dependent rats were pretreated 20 min prior to naloxone challenge with vehicle, Ca^{2+} channel antagonists, or clonidine. The incidence or severity of withdrawal was determined for 30 min following naloxone injections.
Statistical comparisons between vehicle- and drug-treated groups were performed using the Mann-Whitney U test for all signs except ptosis, which was analyzed by the Chi-Square test.
Wet dog shakes, teeth chattering and abdominal stretching are presented as the average number of occurrences. The percent of animals showing ptosis and the mean severity score for diarrhea (O = no diarrhea, 3 = severe diarrhea) are presented accordingly.
* Statistically different from vehicle-pretreated rats administered naloxone ($P < 0.05$).

Table 3. Inhibition of naloxone-precipitated withdrawal jumping in the mouse by pretreatment with 1,4-dihydropyridine Ca^{2+} channel antagonists

Treamtent	Percent of Vehicle	
Vehicle	100	(n = 16)
Nisoldipine (16.6 mg/kg)	35 ± 35*	(n = 9)
Nimodipine (5 mg/kg)	56 ± 43*	(n = 7)
Nimodipine (10 mg/kg)	35 ± 40*, **	(n = 13)

Morphine-dependent mice were pretreated with i.p. injections of vehicle or Ca^{2+} antagonists at the indicated doses 20 min prior to naloxone challenge. Jumping was determined over the first 5 min following naloxone injections. Values are expressed as percentages (mean ± standard deviation) of vehicle-pretreated mice.
* Statistically different from vehicle-pretreated mice administrered naloxone ($P < 0.05$), using the Mann-Whitney U test. Clonidine (10 µg/kg) completely abolished jumping in this experimental paradigm.
** Statistically different from mice treated with 5 mg/kg nimodipine ($P < 0.05$).

jumping (a classic opioid withdrawal sign in mice) in morphine-dependent mice (Table 3). These latter findings suggest that blockade of dihydropyridine-sensitive Ca^{2+} channels might provide a useful means of treating opioid withdrawal.

Acknowledgement. The authors acknowledge the help of Ms. Donna Bethea in typing the manuscript.

References

1. Cardenas HL, Ross DH (1976) Calcium depletion of synaptosomes after morphine treatment. Br J Pharmacol 57:521–526
2. Davis JM, Erickson S, Dekirmenjian H (1978) In: Psychopharmacology: a generation of progress. Lipton MA, DiMascio A, Killam KF (eds). Raven Press, New York, pp 445–459
3. Glossmann H, Ferry DR, Goll A, Striessnig I, Schober, M (1985) Calcium channels: basic properties as revealed by radioligand binding studies. J Cardiovas Pharmacol 7:S20–S30
4. Gold MS, Redmond DE, Kleber HD (1978) Clonidine blocks acute opiate withdrawal symptoms. Lancet 2:599–602
5. Gould RJ, Murphy KMM, Reynolds IJ, Snyder SH (1984) Calcium channel blockade; possible explanation for thioridazine peripheral side effects. Am J Psychiatry 141:352–357
6. Gould RJ, Murphy KMM, Reynolds IJ, Snyder SH (1983) Antischizophrenic drugs of the diphenylbutylpiperidine type act as calcium channel antagonists. Proc Natl Acad Sci USA 80:5122–5125
7. Gould RJ, Murphy KMM, Snyder SH (1985) Autoradiographic localization of calcium channel antagonist receptors in rat brain with [^{3}H]nitrendipine. Brain Res. 330:217–223
8. Griffin MT, Law PY, Loh HH (1985) Involvement of both inhibitory and stimulatory guanine nucleotide binding proteins in the expression of chronic opiate regulation of adenylate cyclase activity in NG-108 cells. J. Neurochem. 45:1585–1591
9. Guerrero-Munoz F, Cerreta KV, Guerrero ML, Way EL (1979) Effect of morphine on synaptosomal Ca^{++} uptake. J Pharmacol Exp Ther 209:132–136
10. Harris RA, Yamamoto H, Loh HH, Way EL (1977) Discrete changes in brain calcium with morphine analgesia, tolerance-dependence, and abstinence. Life Sci 20:501–506
11. Higiwara S, Byerley L (1983) The calcium channel. Trend Neurosci. 6:189–193
12. Higiwara S, Byerley L (1981) Calcium channel. Ann Rev Neurosci 4:69–125
13. Kakunaga T, Kaneto H, Hano K (1966) Pharmacological studies on analgesia. VII. Significance of the calcium ion in morphine analgesia. J Pharmacol Exp Ther 153:134–141
14. Kamikubo K, Niwa M, Fujimura H, Miura K (1983) Morphine inhibits depolarization-dependent calcium uptake by synaptosomes. Eur J Pharmacol 95:149–150
15. Kane JM, Smith JM (1982) Tardive dyskinesia: prevalence and risk factors, 1959–1979. Arch Gen Psychiatry 39:473–481
16. Leslie SW, Elrod SV, Coleman R, Belknap JK (1979) Tolerance to barbiturate and chlorpromazine-induced central nervous system sedation. Involvement of calcium-mediated stimulus-secretion coupling. Biochem Pharmacol 28:1437–1440
17. Miller RJ (1987) Multiple calcium channel and neuronal function. Science 235:46–52
18. Nirenberg M, Wilson S, Higashida H, Rotter A, Krueger K, Busis N, Ray R, Kenimer JG, Adler M (1983) Modulation of synapse formation by cyclic adenosine monophosphate. Science 235:794–799
19. Nowycky MC, Fox AP, Tsien RW (1985) Three types of neuronal calcium channels with different calcium agonist sensitivity. Nature 316:440–443
20. Panza G, Grebb JA, Sanna E, Wright AC, Hanbauer I (1985) Evidence for down-regulation of [^{3}H]nitrendipine recognition sites in mouse brain after long-term treatment with nifedipine or verapamil. Neuropharmacology 24:1113–1117
21. Pinder RM, Brogden RN, Sawyer PR, Speight TM, Spencer R, Avery GS (1976) Pimozide: a review of its pharmacological properties and therapeutic uses in psychiatry. Drug 12:1–40

22. Quirion R, Lafaille F, Nair NPV (1985) Comparative potencies of calcium channel antagonists and antischizophrenic drugs on central and peripheral calcium channel binding sites. J Pharm Pharmacol 37:437–440
23. Ramkumar V, El-Fakahany EE (1984) Increase in [^{3}H]nitrendipine binding in the brain in morphine-tolerant mice. Eur J Pharmacol 102:371–372
24. Ramkumar V, El-Fakahany EE (1985) Changes in the affinity of [^{3}H]nimodipine binding sites in the brain upon chlorpromazine treatment and subsequent withdrawal. Res Commun Chem Pathol Pharmacol 48:463–466
25. Ramkumar V, El-Fakahany EE (1988) Prolonged morphine treatment increase rat brain dihydropyridine binding sites: possible involvement in the development of morphine dependence. Eur J Pharmacol 146:73–83
26. Reddy SVR, Yaksh TL (1980) Antinociceptive effects of lanthanum, neodymium and europium following intracranial administration. Neuropharmacology 19:181–185
27. Richelson E (1984) Neuroleptic affinities for human brain receptors and their use in predicting adverse effects. J Clin Psychiatry 45:331–336
28. Rosenthal HE (1967) A graphic method for the determination and presentation of binding parameters in complex systems. Anal Biochem 20:525–532
29. Ross DH, Cardenas HL (1979) Nerve cell calcium as a messenger for opiate and endorphin actions. Adv Biochem Psychopharmacol 20:301–311
30. Ross DH, Lynn SC, Cardenas HL (1977) Ions, opiate and cellular adaptation. In: Blume K (ed) Alcohol and opiates: neurochemical and behavioral mechanisms. Academic Press, New York, pp 265–279
31. Rupniak NMJ, Jenner P, Marsden CD (1983) The effect of chronic neuroleptic administration of cerebral dopamine receptor function. Life Sci 32:2289–2311
32. Schmidt A, Renaud JF, Lazdunski M (1985) Short term and long term effects of beta-adrenergic effectors and cyclic AMP on nitrendipine-sensitive voltage-dependent Ca^{2+} channel. J Biol Chem 260:13041–13049
33. Snyder SH (1984) Drug and neurotransmitter receptors in the brain. Science 224:22–31
34. Tecott LH, Kwong LL, Uhr S, Peroutka SJ (1986) Differential modulation of dopamine D_2 receptors by chronic haloperidol, nitrendipine, and pimozide. Biol Psychiatry 21:1114–1122
35. Tsien RW (1983) Calcium channel in excitable membranes. Ann Rev Physiol 45:341–358
36. Waddington JL, Cross AJ, Gamble SJ, Bourne RC (1985) Spontaneous orofacial dyskinesia and dopaminergic function in rats after 6 months of neuroleptic treatment. Science 220:530–532

Nimodipine and Neural Plasticity in the Peripheral Nervous System of Adult and Aged Rats

W. H. Gispen[1], T. Schuurman[2], and J. Traber[2]

[1] Rudolf Magnus Institute for Pharmacology and Institute of Molecular Biology and Medical Biotechnology, University of Utrecht, Padualaan 8, 3584 CH Utrecht, the Netherlands and
[2] Neurobiology Department, Troponwerke, Neurather Ring 1, 5000 Köln, FRG

Introduction

Neural plasticity as defined in this contribution is the capability of the nervous system to adapt to a changing internal or external environment, to previous experience, or to trauma. It is becoming increasingly clear that the nervous system is not a static, but rather a dynamic network of cells allowing adaptive changes at all levels of complexity. These can be studied at the molecular, morphological, neurophysiological, and behavioral level. Neural plasticity is an essential and central feature of adaptation. Nervous system plasticity is of great significance in relation to a number of important health-related problems, such as peripheral nerve, spinal cord, and brain injury, developmental disorders, learning disabilities, and dementia. Profound insight into the mechanism of neural plasticity is a prerequisite for advances in the therapy of these pathologies (Gelijns et al. 1987).

It is evidedent that neural Ca^{2+} homeostasis is a key factor in the control of neuronal plasticity. For instance, changes in synaptic plasticity as brought about by long-term potentiation in rat hippocampal synapses neurons involve Ca^{2+} entry (Lynch et al. 1983) and activation of Ca^{2+}-sensitive processes such as phosphatidylinositol 4,5-bisphosphate breakdown (Bär et al. 1984), protein kinase C-mediated phosphorylation of specific substrate proteins (B-50/F1, Akers et al. 1986; De Graan et al. 1986), or proteolysis by calpain (Lynch and Baudry 1984). Moreover in the aged nervous system, where in general there is a loss of neuronal plasticity, a severe dysregulation of neural Ca^{2+} homeostasis exists (Landfield, this volume). In fact, there are indications that age-, lesion-, or disease-related neuronal cell loss in the brain may in part be attributed to a cytotoxic increase in intracellular Ca^{2+}.

Evidence is accumulating which suggests that Ca^{2+} entry blockers of the dihydropyridine type have profound neuro- and psychopharmacological effects in animals and man (Betz et al. 1985). The mechanism by which such effects are brought about is still unknown. It may be that the known vascular effects are responsible for the observed neural activity, whereas an alternative mechanism involves the activation of the dihydropyridine receptors present in neural tissue (Belleman et al. 1983; Betz et al. 1985).

In view of the above, it was decided to test the effect of nimodipine treatment on different levels of neural plasticity. We report here on nimodipine-enhanced repair of damaged peripheral nerves and the amelioration of age-related deficits in motor performance of aged rats.

Postlesion Plasticity in Peripheral Nervous System

The neuron is an extremely specialized and differentiated cell and has proven to be the most vulnerable cell in the mammalian central and peripheral nervous system. In general, it is assumed that damage to cell bodies of neurons results in an irreversible degeneration and cell death. On the other hand, if the damage is restricted to the neuronal processes (dendrites and axons), regeneration with resulting reinnervation of the target muscle is, in principle, possible. For reasons still not completely understood, it appears that neurons in the peripheral nervous system show better axonal regeneration than those in the central nervous system. The milieu surrounding the damaged axon is important in this respect, for if a motor axon is damaged within the vertebral column, hardly any outgrowth of newly formed sprouts is seen, as is typical of central nervous system neurons. If the same sort of lesion is placed distally outside the vertebral column, axonal regeneration and eventual target muscle reinnervation is evident.

Ramon y Cajal demonstrated in 1928 that postlesion brain axonal regeneration could be facilitated by the implantation of pieces of sciatic nerve. It is well known that the outgrowth and elongation of regenerating axons is guided by a variety of humoral and structural factors, which are of neuronal, glial, and target cell origin (Varon 1985). In the case of central sciatic nerve implantation, it most certainly involves the neurite-promoting activity of laminin, an extracellular matrix protein from Schwann cells (Varon 1985).

Similarities in many respects make postlesion neuronal plasticity a fast replay of processes that take place during neuronal development. In other words, cellular or network repair is very much determined by factors that also govern the development and maturation of the cell or network.

Recently, a small family of proteins was discovered of which the synthesis increases dramatically (10- to 100-fold) following axotomy (Benowitz et al. 1981; Skene and Willard 1981; Willard and Skene 1982). These growth-associated proteins (GAPs, designated GAP50, GAP43, and GAP24 to indicate their respective apparent molecular weights in kilodaltons) have been considered to play a crucial role in the capacity of axons to grow since their genes are expressed in injured nerves that do regenerate (e.g., toad optic nerve, rabbit hypoglossal nerve), and not in nerves that do not successfully re-form their axons (e.g., rabbit optic nerves).

How GAPs contribute to axonal regeneration is still unknown. The best characterized GAP, i.e., GAP 43 (pp 46, F1) or B-50, is a substrate protein to protein kinase C. It has been postulated that the phsophorylation state of B-50/GAP may determine the degree of sensitivity of the regenerating axon to extrinsic neurotrophic or neurite-promoting factors (De Koning and Gispen 1988). As protein kinase C is a Ca^{2+}-sensitive enzyme, Ca^{2+} homeostasis is again of importance in this respect.

Nimodipine and Plasticity in Rat Sciatic Nerve

Nontransection nerve damage, such as a crush lesion, results in Wallerian degeneration accompanied by a rapid and appropriate reinnervation along the intact endoneural tubes. The speed at which the regenerating sprouts grow into the distal portion of the damaged nerve is approximately that of the slow axonal transport. However, among the various species there is a great diversity in regenerative capacity. Certainly, regeneration of the sciatic nerve is slower in humans than in rodents, cats, and dogs.

Recovery of sensory function following a crush of the sciatic nerve of the rat can be accurately followed by applying a small electric current locally to the foot sole (De Koning et al. 1986). A normal rat invariably retracts its paw instantaneously when the skin of the foot sole closes the electric current between the stimulation poles. Reinnervation following crush is evidenced by return of the foot withdrawal reflex elicited locally at the foot sole. We are aware that the sciatic nerve is a mixed sensorimotor nerve, and that in functional tests sensory modalities are usually measured as well as motor modalities. In keeping with the literature, the foot shock foot withdrawal test is taken to describe the sensory modality. To monitor return of motor function, we used the elegant walking pattern test reported by De Medinacelli et al. (1982). The principle of this test is that after dipping the hindfeet in photographic developer, rats that walk over photographic paper in an alley leave behind a pattern of footprints that can be analyzed in detail. From a variety of parameters such as print length, toe spreading, step distance, etc., an index of motor function or functional sciatic index (FSI) can be derived (De Medicanelli et al. 1982). Since the walking pattern is the result of the coordinated use of different muscle groups, it gives a good impression of the qualitative aspects of sciatic nerve regeneration as well.

In the first experiment, 20 rats received a crush lesion in the right sciatic nerve. Nimodipine was administered via food pellets (225 ppm) to ten rats, whereas the other ten received control food pellets not containing nimodipine. Return of sensory function was measured every other day from postoperation day 11 onwards, and daily from day 17 through day 21. The first signs of recovery of function were observed between days 15 and 17, and functional recovery as assessed by the foot withdrawal reflex was complete by day 20. However, no effect of oral administration of nimodipine (225 ppm) was detected. Analysis of motor function of these rats was performed by calculating the FSI from their walking pattern at postoperation days 6, 10, 14, 16, 18, and 20. The FSI of the experimental paw in the control group reached normal levels from day 18 onwards. Furthermore, at all days tested, the FSI of nimodipine-treated rats was smaller than that observed in the untreated rats. Analysis of variance revealed significant enhancement of recovery of function in the nimodipine-treated group (Van der Zee et al. 1987).

In the second experiment, a larger dose of nimodipine was tested (860 ppm). Again, ten rats received control food pellets and ten rats nimodipine-containing pellets. Sensory function was tested at postoperation days 2, 6, 12, 14, 16, 18, and 20. As can be seen in Fig. 1A, first signs of recovery of sensory function were observed in untreated rats at day 15. Nearly total recovery was observed from day 18 onwards. As illustrated in Fig. 1A, nimodipine at this higher dose level significantly enhances the return of sensory function following sciatic nerve damage. In the motor function test, the effect of a higher dose of nimodipine was also more pronounced (Fig. 1B). Nondrug-treated rats showed a normal FSI from postoperation day 18 onwards, whereas the nimodipine-treated group showed smaller FSI values at all days tested (Van der Zee et al. 1987).

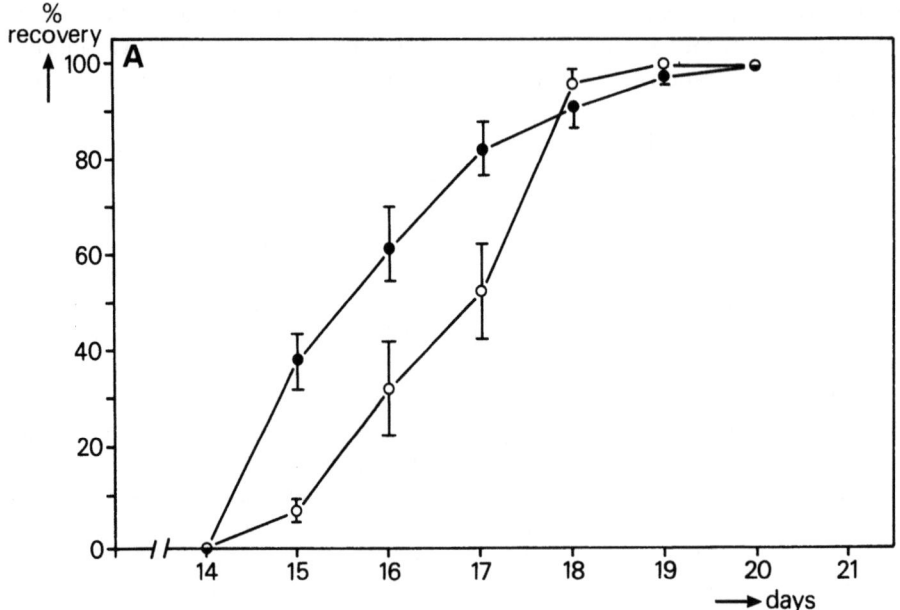

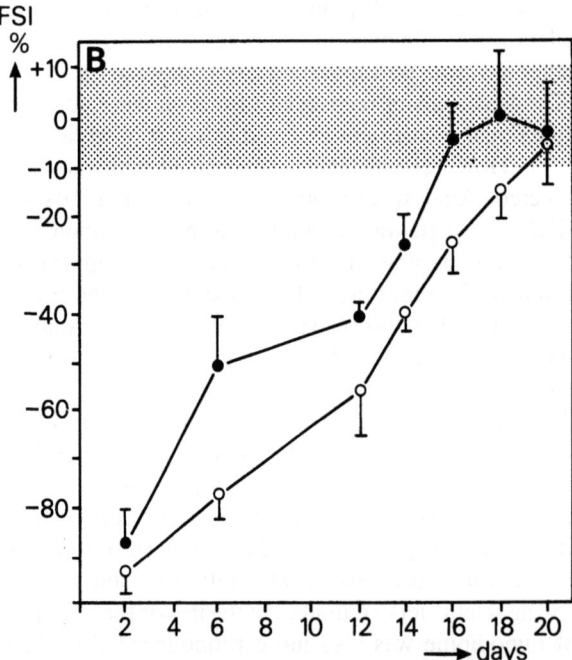

Fig. 1A, B. Return of sensorimotor function following a crush lesion of the sciatic nerve in the rat: effect of oral nimodipine treatment. ●–● 860 ppm, ($n = 10$), o–o control, ($n = 10$). **A:** Foot-reflex withdrawal test. **B:** Functional sciatic nerve index *(FSI)* calculated from the walking pattern. Taken from Van der Zee et al. (1987)

Nimodipine and Neural Plasticity in Aged Rats

It is a generally accepted notion that with increasing age neural plasticity diminishes. The observations that support this hypothesis include the enhanced gliosis and loss of synapses in the aged hippocampus, along with the difficulty of demonstrating synaptic plasticity by means of long-term potentiation in aged rats (Landfield 1983; Landfield et al. 1978; Tielen et al. 1983). Reintroduction of fetal cells in specific age-affected regions results in the counteraction of the functional deficits related to dysfunction of those regions (Gage and Björklung 1983). As nimodipine was able to affect post-lesions plasticity in the peripheral nervous system (see above) and the central nervous system (see Betz et al. 1985), it was decided to study the efficacy of chronic treatment with nimodipine on age-related deterioration of motor performance in the rat.

Age-related deficits in motor function in rats have been noted by many researchers and are discussed in detail by Coper et al. (1986). The decline in motor performance is neither abrupt nor uniform and probably not random; rather it occurs in steps and is almost systematic. Although there are certainly age-related deficits in neuromuscular function (peripheral nerve function, muscle strength, etc.), Coper et al. (1986) suggest that the decline of motor performance with age is most likely primarily due to a loss of precision and a slowing-down of central neural mechanisms.

Motor performance was tested in a battery of motor tests that varied in complexity, measuring locomotion and balance, suspended hanging, and climbing into a pole in essence as described by Gage et al. (1984). The tests are described in detail in Schuurman et al. (1987). Time taken to reach safety platforms via various suspended bridges or the latency to falling off a suspended wire or from a vertically placed pole were scored.

In addition, a detailed analysis of the walking pattern of rats was performed using the test apparatus described by De Medinacelli et al. (1982). Attention was focussed on the appearance of abnormal footprints. These abnormal prints appeared as (a) additional small prints per footprint, (b) fuzzy footprints as a result of exorotation of the foot following placement on the photographic paper, and (c) fuzzy footprints due to lack of elevation of the paw at the onset of a new step (see Fig. 2). If an animal showed any of these abnormalities, it was scored as displaying abnormal locomotion.

In the first experiment, 52 male rats from an inbred Wistar strain (Winkelmann, Borchen, FRG) were used. At 24 months of age, the rats were given either control food pellets or food pellets that contained nimodipine (860 ppm; Bayer, Leverkusen, FRG). In order to obtain two groups with comparable motor coordination ability, the 52 rats were subjected to a simple bridge performance test at the onset of the experimental period and matched according to their individual scores.

As also reported by others (Gage et al. 1984), performance on a rather simple broad bridge was less affected by age than that in the more complex motor coordination test in the vertical plane. The effect of ingestion of nimodipine on motor coordination was tested at weeks 0, 4, 10, and 16 following the onset of treatment. While the most pronounced beneficial effect of nimodipine treatment was observed in the easier bridge tests (Fig. 3), data analysis of the motor coordination tests indicated that nimodipine improved motor function in all tests; however the horizontal and vertical tests presumably measured different aspects of motor coordination in terms of both complexity and neural substrate (Schuurman et al. 1987).

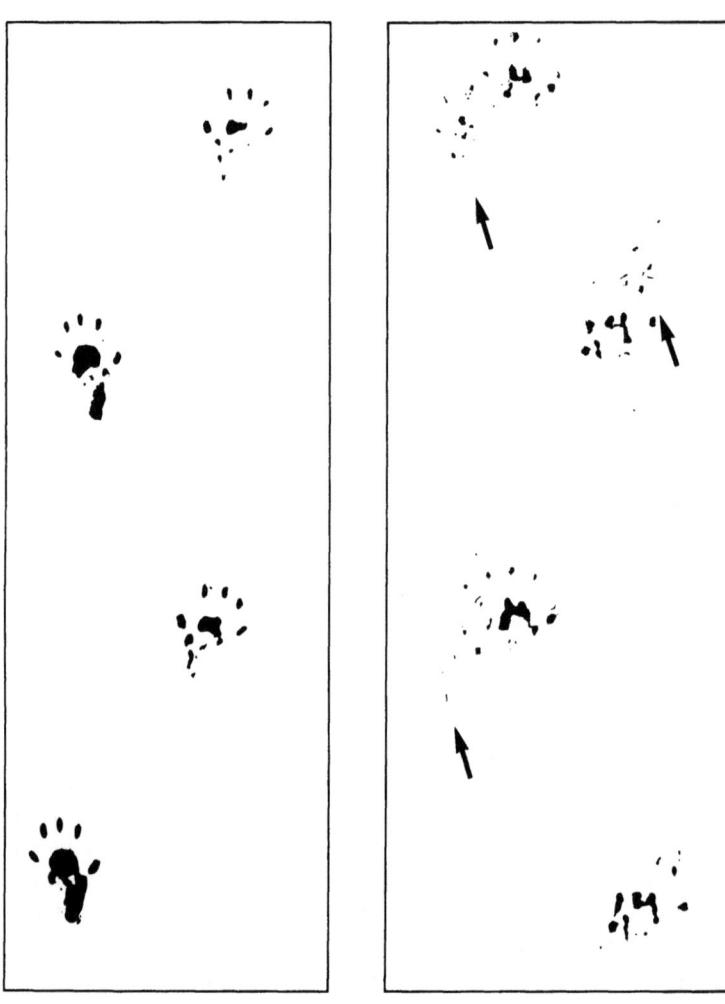

Fig. 2. Age-related deficits in walking pattern of the rat.

As shown in Fig. 4, from the onset of the experiment the incidence of abnormal footprints increased tremendously with increasing age in control food-eating rats. In contrast, nimodipine-fed rats showed no increase in abnormal footprint frequency during the first 2 months of the treatment period. Only at 26 months did nimodipine-treated rats begin to display more abnormal footprints at rate of increase similar to that seen in control rats at 24 months of age. Thus, nimodipine clearly delayed and/or suppressed the occurrence of these motor deficits (Schuurman et al. 1987).

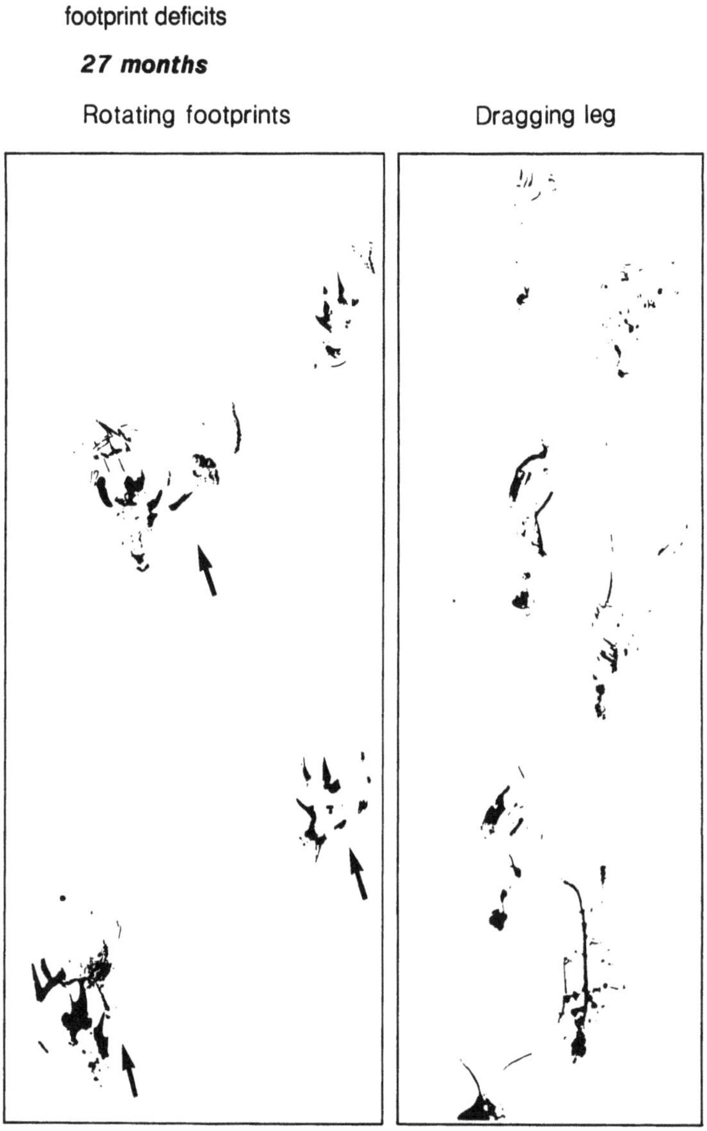

footprint deficits

27 months

Rotating footprints   Dragging leg

Fig. 2

Whatever the cause of the abnormal locomotion (central vs. peripheral), the incidence of footprint abnormality appeared to be a simple and reliable measure of motor dysfunction in the aged rat. Therefore, in a recent study we used this parameter to assess the neurotrophic, therapeutic efficacy of nimodipine ingestion in aged rats already displaying appreciable amounts of motor deterioration.

A group of male rats (Wistar, Winkelmann, Borchen, FRG) 24 months of age were subjected to the footprint test and selected for already existing footprint abnormalities. Depending on the frequency of abnormality, the animals were matched pairwise to form two groups. These groups were then randomly allocated to

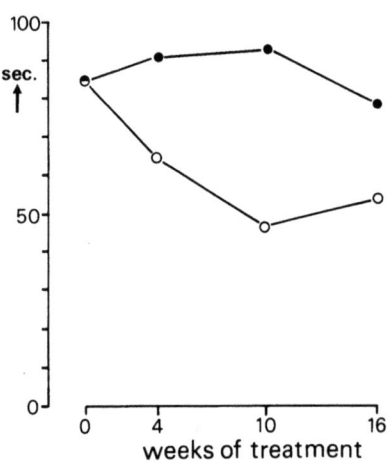

Fig. 3. Effect of nimodipine (860 ppm in food pellets, treatment period in weeks) on motor coordination of aged rats on a broadsuspended bridge (●–● nimodipine, n = 26; O–O control, n = 26). Taken from Schuurman et al. (1987)

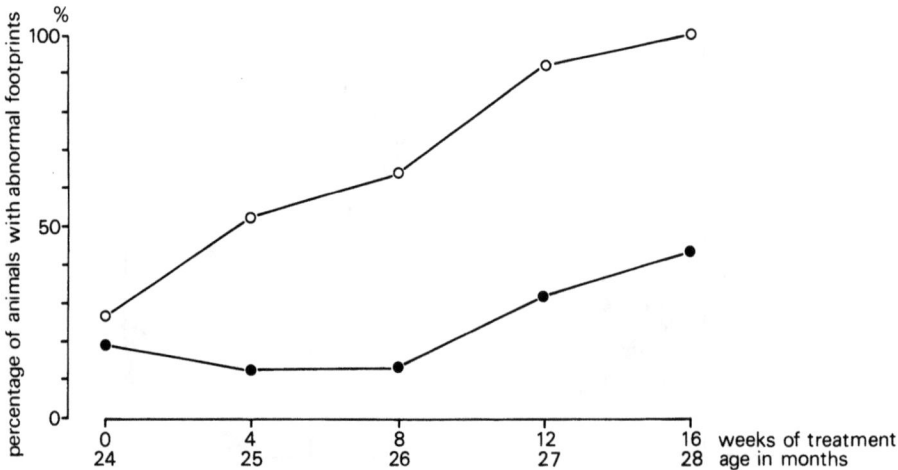

Fig. 4. The effect of oral treatment with nimodipine on the occurrence of abnormal foot prints in aging rats (●–● nimodipine, n = 26; o–o control, n = 26)

nimodipine (860 ppm) or placebo food. Subsequently, the rats were tested after 6, 10, 16, and 20 weeks of treatment. As can be seen in Fig. 5, at the onset of the experiment the percentage of animals showing abnormal footprints of both paws was approximately similar in the two groups of rats. In placebo-fed animals, this percentage gradually increased, whereas in nimodipine-treated rats the percentage markedly decreased, albeit that during the last weeks of the experimental period this index also increased sharply in the nimodipine group. Nonetheless, the data serve as a first

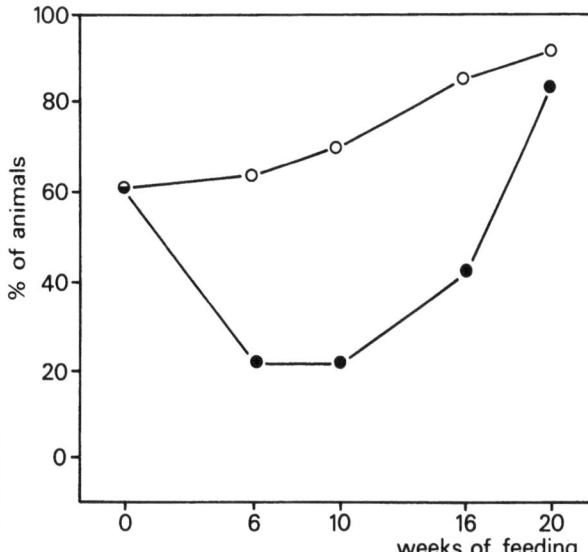

Fig. 5. The number of aging rats that showed abnormal foot prints. Effect of oral nimodipine [860 ppm, (●–●) 860 ppm, ($n = 24$), o–o control, ($n = 24$)]

indication that under certain conditions nimodipine not only delays the onset of age-related motor deficits but may also counteract these deficits once already present.

Following completion of the behavioral tests, 12 rats were selected randomly from both the nimodipine- and the placebo-fed group and subjected to measurement of sciatic nerve and caudal nerve conduction velocity, as described by De Koning and Gispen (1988). Six of each group of 12 were then killed, and their sciatic nerves excised and processed for histology, as described by Bijlsma et al. (1983). In cross sections of the nerve taken 1 cm distal to the sciatic notch, the number of myelinated axons was determined by means of a computer-aided image analysis system. The motor and H-related (Hofman reflex related) sensory nerve conduction velocities in the sciatic nerve were determined by one experimenter, the caudal nerve sensory conduction velocity by another. Both experimenters were unaware of the treatment a given rat had received and of the result obtained in the parallel nerve conduction velocity measurement. It was found that nimodipine treatment of aged rats already displaying appreciable amounts of motor deterioration resulted in an enhancement of all three nerve conduction velocities measured (Van der Zee, manuscript in preparation).

Thus, nimodipine affected not only the sciatic nerve, whose function presumably might affect the footprint appearance, but also the caudal nerve. As nimodipine-treated rats are also more active in the open field test (Schuurman, manuscript in preparation), an exclusive effect on the sciatic nerve could have been due to an indirect mechanism, presumably via a central input in locomotion, resulting in more and better ambulation with improved nerve function as a consequence. Although more data are necessary to warrant firm conclusions, the present data suggest that there is a general improvement of nerve conduction velocity in peripheral nerves of aged rats treated chronically with nimodipine.

Little is known about the fiber density in sciatic nerve of aged rats. The cross section of nerves obtained from placebo-fed rats showed a relatively low fiber density when

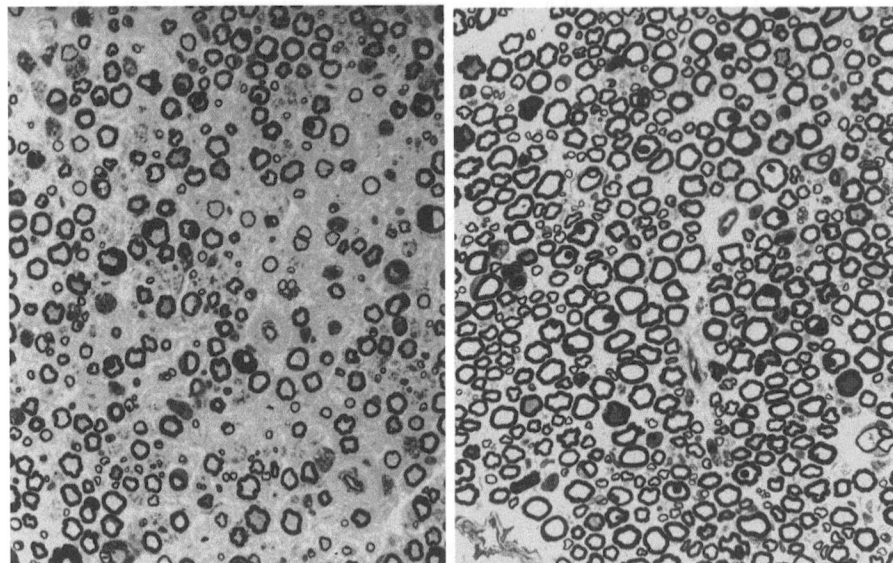

Fig. 6. Cross section of the sciatic nerve taken 1 cm distal from the sciatic notch stained for myelin. *Left panel:* example taken from a control fed rat. *Right panel:* example taken from a nimodipine fed rat

compared with data obtained from young adult rats (Bijlsma et al. 1983). In addition, there were signs of excessive amyloid deposits around blood vessels similar to those reported for blood vessels in the aged rat brain (Luiten, personal communication). Interestingly, sections obtained from nimodipine-fed rats showed little vascular pathology of this kind, and the fiber density was significantly higher than that seen in controls. Typical examples are shown in Fig. 6. Note that in the section of the sciatic nerve of the nimodipine-fed rat there are numerous small myelinated axons reminiscent of a nerve regenerating following crush damage (see for instance Bijlsma et al. 1983). Additional histological work is in preparation to analyze in more detail this effect of nimodipine on the sciatic axons.

Currently, it would appear that nimodipine treatment of aged rats has a marked effect on sciatic nerve conduction velocity and histology. These data may further support the notion that aspects of the peripheral nervous system are of great importance to motor function, and that age-related deficiencies in these may be ameliorated by nimodipine.

Concluding Remarks

In the present study, we examined the effect of the Ca^{2+}-entry blocker nimodipine on neuronal plasticity in the peripheral nervous system of the rat. Both enhancement of postlesion plasticity in rat sciatic nerve and counteraction of an age-related loss of plasticity in this nerve were observed. The data are in good agreement with the results obtained by C.S. Isaacson (personal communication); who observed that systemic nimodipine treatment protects from retrograde septal cell loss following unilateral section of the fornix in adult rats.

Before speculating on the significance of these findings in the context of therapy of plasticity-related pathology, a number of issues should be addressed. Is the nimodipine effect shared by other Ca^{2+}-entry blockers, and can treatment with Ca^{2+} channel activators such as Bay K8644 delay repair mechanisms even further? What are the optimal dose, route, and treatment schedule, etc.? Nevertheless, these data suffice to underscore that nimodipine exerts a neurotrophic action, which may result from a direct interaction between nimodipine and the nervous system. Further research is in progress to unravel the nature of this interaction at the level of both the central and the peripheral nervous system.

Acknowledgements. The authors gratefully acknowledge the cooperation of H. R. Klein, J. H. Brakkee, C. E. E. M. Van der Zee, and R. Gerritsen van der Hoop.

References

Akers RF, Lovinger DM, Colley PA, Linden DJ, Routtenberg A (1986) Translocation of protein kinase C activity may mediate hippocampal long-term potentiation. Science 231:587–589
Bär PR, Wiegant F, Lopes da Silva FH, Gispen WH (1984) Tetanic stimulation affects the metabolism of phosphoinositides in hippocampal slices. Brain Res 321:381–385
Belleman P, Schade A, Towart R (1983) Dihydropyridine receptor in rat brain labelled with [^{3}H]-nimodipine. Proc Natl Acad Sci USA 80:2356
Benowitz L, Shaskoua V, Yoon MG (1981) Specific changes in rapidly transported proteins during regeneration of the gold fish optic nerve. J Neurosci 1:300–307
Betz E, Deck K, Hoffmeister F (1985) Nimodipine: pharmacological and clinical properties. Schattauer, Stuttgart
Bijlsma WA, Jennekens FGI, Schotman P, Gispen WH (1983) Stimulation by ACTH (4–10) of nerve fiber regeneration following sciatic nerve crush. Muscle Nerve 6:104–112
Coper H, Jänicke B, Schulze G (1986) Biopsychological research on adaptivity across the life-span of animals. In: PD Baltes, Featherman DL, Lerner RM (eds) Life-span development and behavior. Erlbaum, Hillsdale pp 207–232
De Graan PNE, Oestreicher AB, Schrama LH, Gispen WH (1986) Phosphoprotein B-50: localization and function. Progr Brain Res 69:37–50
De Koning P, Brakkee JH, Gispen WH (1986) Methods for producing a reproducible crush in the sciatic and tibial nerve of the rat and rapid and precise testing of return of sensory function. J Neurol Sci 74:237–241
De Koning P, Gispen WH (1988) A rationale for the use of melanocortins in the treatment of nervous tissue damage. In: Stein DG, Sabel B (eds) Pharmacological approaches to the treatment of brain and spinal cord injuries. Plenum New York (in press)
De Medinacelli L, Freed WJ, Wyatt RJ (1982) An index of the functional condition of rat sciatic nerve based on measurements made from walking trades. Exp Neurol 77:634–643
Gage FH, Björklund A, Stenevi U, Dunnett SB (1983) Intracerebral grafting in the aging brain. In: Gispen WH, Traber J (eds) Aging of the brain. Elsevier, Amsterdam
Gage FH, Dunnett StB, Björklund A (1984) Spatial learning and motor deficits in aged rats. Neurobiol Age 5:43–48
Gelijns AC, Graaff PJ, Lopes da Silva FH, Gispen WH (1987) Future health care applications resulting from progress in the neurosciences: the significance of neural plasticity research. Health Policy 8:265–276.
Landfield PW (1983) Mechanisms of altered neural function during aging. Dev Neurol 7:51–71
Landfield PW, McGaugh JL, Lynch G (1978) Impaired synaptic potentiation processes in the hippocampus of aged, memory-deficient rats. Brain Res 150:85

Landfield PW, Baskin RK, Pitler TA (1981) Brain aging correlates: retardation by hormonal pharmacological treatments. Science 214:581–584

Lynch G, Baudry M (1984) The biochemistry of memory: a new and specific hypothesis. Science 224:1057–1063

Lynch G, Larson J, Kelso S, Barrionuevo G, Schottler F (1983) Intracellular injections of EGTA block induction of hippocampal long-term potentiation. Nature 305:719–721

Ramon y Cajal S (1928) Degeneration and regeneration of the nervous system. Hafner, New York

Schuurman T, Klein H, Beneke M, Traber J (1987) Nimodipine and motor deficits in the aged rats. Neurosci Res Commun 1:9–15

Skene JHP, Willard M (1981) Changes in axonally transported proteins during axon regeneration in toad retinal ganglion cells. J Cell Biol 89:86–95

Tielen AM, Mollevanger WJ, Lopes da Silva FH, Hollander CF (1983) Neuronal plasticity in hippocampal slices of extremely old rats. Dev Neurol 7:73–84

Varon S (1985) Factors promoting the growth of the nervous system. Neurosciences 3:62

Van der Zee CEEM, Schuurman T, Traber J, Gispen WH (1987) Oral administration of nimodipine accelerates functional recovery following peripheral nerve damage in the rat. Neurosci Lett 83:143–148

Willard M, Skene JHP (1982) Molecular events in axonal regeneration. In: A Nicholls (ed) Repair and regeneration of the nervous system. Springer, Heidelberg, pp 71–89

Anticonvulsant Properties of Dihydropyridine Calcium Antagonists*

F. B. Meyer[1], R. E. Anderson[2], and T. M. Sundt Jr.[1]

[1] Cerebrovascular Research Laboratories, Department of Neurosurgery, Mayo Clinic,
  200 First Street Southwest, Rochester, MN 55905, USA
[2] Mayo Graduate School, Rochester, Minnesota, USA

Introduction

In the past several years a primary thrust of our research effort has been directed toward evaluating the anticonvulsant properties of dihydropyridine Ca^{2+} antagonists.

Hoffmeister first demonstrated that nimodipine was effective in altering pentylenetetrazole-induced seizures in mice [40]. In our work, the modulatory role of this class of agents in experimental epilepsy was first noted in a decreased frequency of ischemic seizures in an animal model of focal cerebral ischemia [52]. Since then, these observations have been extended by demonstrating that dihydropyridines like nimodipine have anticonvulsant properties in seizures induced by electroconvulsive shock (ECS), bicuculline, and pentylenetetrazole [53, 55]. Other teams of investigators have demonstrated this effect in seizures induced in DBA/2 mice and through pentylenetetrazole, cefazolin, or nitrous oxide withdrawal [1, 21, 59].

From a second perspective, administration of the dihydropyridine Ca^{2+} agonist Bay K8644 induces behavior in mice including limb clonus/tonus activity [10]. Intravenous infusion of this agonist in dogs causes generalized convulsions [67]. Cats subjected to chronic ECS develop a tolerance to this stimulus which is associated with an increase in H^3-nitrendipine binding sites in the cerebral cortex [9].

Therefore, there is a growing body of literature which indicates that Ca^{2+} channels modulated by dihydropyridines have a facilitatory role in experimental epilepsy. This article will first briefly review both the role of Ca^{2+} in neuronal excitability and neuronal Ca^{2+} channels. Subsequently, experimental evidence demonstrating the anticonvulsant properties of dihydropyridines will be analyzed. Based on this information an integrative hypothesis incorporating knowledge on Ca^{2+} channels, hippocampal ionic currents, and seizure discharge will be developed.

Calcium and Neuronal Hyperexcitability

There are several lines of research which indicate that Ca^{2+} is essential for seizure discharge. First, in clinical epilepsy, excessive discharge of cortical neurons in a repetitive fashion is the basic dysfunction at the cellular level. The interictal "spike" on electroencephalography is a large extracellular field potential which is associated

---

* This work was supported in part by the New York Academy of Medicine and NIH RO1 NS24329–02

with paroxysmal depolarizing shifts in neuronal membrane potentials triggering action potentials [42]. These paroxysmal depolarizing shifts require disinhibition of a subpopulation of neurons with intrinsic burst-generating potential. Intrinsic bursting appears to be Ca^{2+}-dependent [68].

Brain slice preparations of hippocampal CA3 and CA1 neurons along with voltage-clamp studies have yielded the following information about Ca^{2+} channels [71, 72, 82–84]. There are at least two independent Ca^{2+} currents: transient fast Ca^{2+} spikes and slow Ca^{2+} depolarizations [35, 43, 13]. In addition, there are at least two Ca^{2+}-activated ionic currents: K^+ currents and Cl^- currents [12, 64]. Although the interplay between these currents is unclear, endogenous burst firing is thought to reflect a sequential volley of inward Ca^{2+} currents and outwart Ca^{2+}-activated K^+ currents. A speculative sequence to explain intrinsic bursting would be that inward Ca^{2+} currents generate a slow depolarization which triggers low-threshold Na^+ spikes followed by high-threshold Ca^{2+} spikes. The Ca^{2+}-activated K^+ and Cl^- currents, along with the Na^+/K^+ pump, then cause rapid repolarization [71, 84]. Presumably the ionic fluxes responsible for endogenous bursting of hippocampal neurons are similar to the repetitive firing of neurons during a seizure discharge. It should be noted that the paroxysmal depolarizing shift mentioned above can be attenuated by intracellular injection of the Ca^{2+} blocker D-890 [20, 81].

Second, epileptic discharge requires synchronization of a population of neurons which have the propensity to discharge. This synchronization may be synapsis-mediated. From the classic studies of Katz and Miledi, it is apparent that Ca^{2+} entry into the presynaptic terminal is essential for neurotransmitter release [44–45]. Subsequent investigations suggest that the electron-dense particles in the active zones of the presynaptic terminal are actually voltage-dependent Ca^{2+} channels [2, 70]. This Ca^{2+} channel may be the N channel described below. Ca^{2+}-mediated exocytosis may proceed through phosphorylation of synapsin I, a protein which surrounds vesicles, by a Ca^{2+}/calmodulin-dependent protein kinase [11, 18, 47]. It is interesting that some anticonvulsants like carbamazepine, benzodiazepines, and phenytoin inhibit Ca^{2+}/calmodulin protein phosphorylation, offering a potential alternative explanation for the anticonvulsant effects of these agents [19].

Third, extracellular Ca^{2+} decreases during epileptic activity in vivo induced by ECS, pentylenetetrazole, and penicillin [36, 37]. This decrease in $[Ca^{2+}]_e$ occurs just prior to the onset of burst firing and reaches a stable decline within the first few seconds of seizure discharge [69]. During status epilepticus, there is accumulation of Ca^{2+} in the mitochondria of bursting neurons, predominantly in hippocampal CA3 and CA1 neurons when seizures are induced through bicuculline, allylglycine, or kainic acid [31]. Furthermore, during progressive seizures there is accumulation of fatty acids thought to be related to activation of phospholipase A_2 by Ca^{2+} influx [14].

Fourth, some current anticonvulsants directly effect neuronal Ca^{2+} channels. Phenytoin interferes with synaptosome Ca^{2+} uptake and also inhibits the binding of dihydropyridines to neuronal membranes [29, 65, 66] Furthermore, barbiturates have been demonstrated to block both neuronal L and N channels in dorsal root ganglion neurons [33].

Fifth, punitive neurotransmitters which gate a receptor-operated Ca^{2+} channel are both neurotoxic and epileptogenic. When used to elicit seizure activity there is accumulation of Ca^{2+} in mitochondria located in the dendrites and soma of CA3 and

CA1 pyramidal neurons [26, 31]. Antagonists to these excitatory amino acids are anticonvulsants in seizure-prone DBA/2 mice [16]. In vitro hippocampal slices suggest that these excitatory amino acids alter both extracellular Na^+ and Ca^{2+} concentrations and increase Ca^{2+} uptake in brain slice preparations. The NMDA receptor for these excitatory neurotransmitters gates a receptor-operated Ca^{2+} channel. In addition, it is possible that the kainic acid receptor gates an Na^+ channel, thereby causing membrane depolarization, which in turn opens up voltage-dependent Ca^{2+} channels. The resulting influx of Ca^{2+} then leads to neuronal excitability [50, 74, 80, 85].

Sixth, in addition to the work originating from this laboratory and others pertaining to the anticonvulsant properties of dihydropyridine Ca^{2+} antagonists (reviewed below), two other Ca^{2+} antagonists have anticonvulsant effects: verapamil and flunarizine. Intraventricular infusion of verapamil, a diphenylalkamine Ca^{2+} antagonist, has been found to reduce paroxysmal depolarizing shifts induced by penicillin and pentylenetetrazole [7, 79]. The diphenylpiperazine Ca^{2+} antagonist flunarizine has been used as an add-on therapy in patients with medically resistant seizures with beneficial effects [8, 22, 63]. It is important to note that flunarizine binds to Na^+ channels and may be modifying seizure activity through this mechanism as opposed to Ca^{2+} antagonism [32].

Neuronal Dihydropyridine Calcium Channels

Since dihydropyridine Ca^{2+} antagonists have anticonvulsant effects in experimental epilepsy, it is important to briefly review information pertaining to neuronal Ca^{2+} channels [28]. Authoritative discussions can be found in other chapters of this book. Generally, extracellular Ca^{2+} may enter the cell through channels classified into two major types: voltage-dependent channels sensitive to membrane potential alterations, and receptor-operated channels. Of specific interest for dihydropyridines are the voltage-dependent channels.

Voltage-dependent Ca^{2+} channels are activated by neuronal membrane depolarizations which facilitate transmembrane flux of Ca^{2+} due to concentration gradients. The Ca^{2+} channel is not exclusive in that other cations, including Na^+, Ba^{2+}, La^{3+}, Cd^{2+}, and Mg^{2+}, can traverse the channel [34]. Inactivation of the Ca^{2+} channel occurs through at least two mechanisms: inactivation through prior depolarization and inactivation due to accumulation of Ca^{2+} at the cytoplasmic end of the channel [25, 30]. There are some neurotransmitters which will inhibit voltage-dependent Ca^{2+} channels, including dopamine, serotonin, norepinephrine, somatostatin, adenosine, prostaglandin E_1, and GABA [23, 48, 49]. Three subtypes of voltage-dependent Ca^{2+} channels have been identified in dorsal root ganglion neurons: T, N, and L channels [46, 57, 61]. The N channel is activated by strong depolarizations located in the presynaptic terminal. Therefore, it might be pivotal for neurotransmitter release. The T channel gives rise to small transient Ca^{2+} currents activated by weak depolarizations and subsequently quickly inactivated. Its function is speculative but may be linked to repetitive neuronal bursting. The L channel is a long-lasting current activated by strong membrane depolarizations and is located in the cell soma [73]. Of these three different Ca^{2+} channels, only the L channel is modulated by dihydropyridine agonists and antagonists [60].

[³H]Nimodipine crosses the blood-brain barrier and binds to neuronal dihydropyridine receptors. The density of labeled receptor sites is greatest in the molecular layer of the dentate gyrus, followed by the anterior amygdala, hippocampal CA3 and CA1 neurons, and the pyriform cortex. There are also abundant receptor sites in the olfactory bulb, cerebral cortex lamina I–IV, and the thalamus [4, 15, 62]. It is also apparent that there is widespread distribution of functional dihydropyridine-sensitive Ca^{2+} channels in cultured neurons from various brain regions, with hippocampal voltage-dependent Ca^{2+} channels demonstrating the greatest sensitivity to dihydropyridines [77].

Evidence for the Anticonvulsant Effects of Dihydropyridines

Thus far we have found that nimodipine has anticonvulsant properties in experimental seizures induced through ischemia, post-ischemic reperfusion, bicuculline, pentylenetetrazole, and kainic acid. This agent was effective both when given parenterally, and orally. Furthermore, there appears to be differences in potency between dihydropyridines: in ECS-induced seizures, nimodipine > PN200–110 >> nicardipine. The ineffectiveness of nicardipine may be due to poor blood-brain barrier penetration of this agent, or to poor incorporation into the lipid bilayer where the dihydropyridine receptor is thought to exist. Nimodipine is more lipid-soluble than nicardipine [3, 39, 41]. The difference in potency between nimodipine and PN200–110 is more difficult to explain, since both are lipid-soluble and penetrate the blood-brain barrier, binding to neuronal tissue. One potential explanation is that nimodipine has centrally active breakdown products. Summarized below are some of our results in these various seizure models. All these experiments were performed in the white New Zealand rabbit under 1.0% halothane anesthesia.

Ischemic Seizures

Thirty animals underwent occlusion of the middle cerebral artery for 4 h with continuous measurements of intracellular brain pH, cerebral blood flow, and electroencephalography. This preparation requires a craniectomy to expose the cortex and the circle of Willis. Prior experiments with this model have demonstrated that occlusion of the middle cerebral artery for this duration produces several zones of focal ischemia – a region of severe ischemia which evolves into cortical infarction and a second region of moderate ischemia which has a tendency to deteriorate with time. Measurements of intracellular brain pH and cortical blood flow were performed with the use of a pH-sensitive fluorophor [55, 76]. The electrocorticograms were recorded by placement of gold-plated electrodes around the margins of the craniectomy. Fifteen animals received a continuous infusion of nimodipine 0.5 µg/kg/min commencing 15 min after vessel occlusion, and fifteen received equivalent volumes of the vehicle. As illustrated in Fig. 1, infusion of this Ca^{2+} blocker improved both cortical blood flow and intracellular brain pH in both regions of moderate and severe ischemia ($p < 0.001$) [51]. Most interestingly, nine of 15 control animals had ischemic seizures, compared to only one of 15 nimodipine-treated animals ($p < 0.001$). The decreased incidence of

Fig. 1. Effect of nimodipine in focal cerebral ischemia. Shown are serial measurements of brain pH_i and cortical blood flow in 20 animals which underwent 4 h of middle cerebral artery occlusion: 10 controls and 10 treated with nimodipine 0.5 µg kg^{-1} min^{-1} commencing 20 min after vessel occlusion. Preocclusion brain pH_i was 7.01 ± 0.04 and CBF was 51.8 ± 4.6 ml 100 g^{-1} min^{-1}. Fifteen minutes after vessel occlusion in regions of severe ischemia brain pH_i was 6.64 ± 0.06 in controls and 6.57 ± 0.03 in the animals for treatment, while CBF was 12.7 ± 2.3 and 10.5 ± 1.3 ml 100 g^{-1} min^{-1} respectively. Infusion of nimodipine caused significant improvement in both of these parameters. Four hours after vessel occlusion, brain pH was 6.08 ± 0.15 in controls versus 6.91 ± 0.06 in the nimodipine group ($p < 0.001$). CBF was 5.2 ± 1.5 in controls versus 18.8 ± 3.0 ml 100 g^{-1} min^{-1} in the treated group ($p < 0.001$). Differences from the mean are standard error. [Adapted from 51]

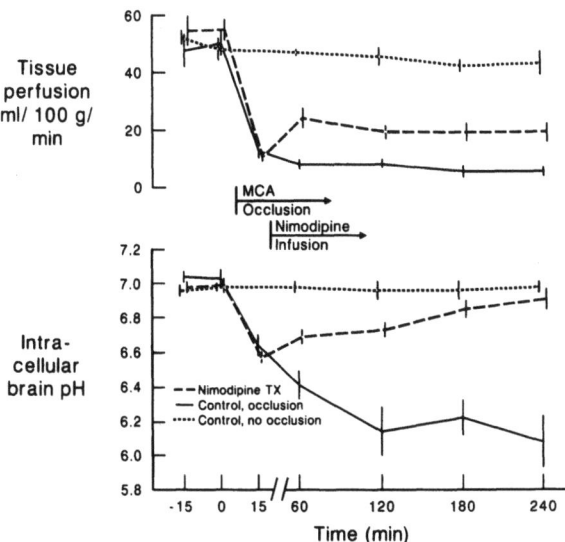

ischemic seizures may have been a secondary reflection of the improvement in blood flow and metabolic parameters. Alternatively, nimodipine might have had a direct anticonvulsant effect.

Reperfusion Seizures

We had previously noted that upon restoration of flow after middle cerebral artery occlusion for 45 min, there is a high incidence of reperfusion seizures. In five animals the middle cerebral artery was temporarily occluded with a miniature microvascular clip. Upon restoration of flow by removal of the clip, four animals developed reperfusion seizures noted on continuous electrocorticographic monitoring. Through a PE-50 catheter previously placed into the ligated ipsilateral external carotid artery, 1 ml containing 40 µg of the vehicle polyethylene glycol was injected over 15 s. The vehicle had no effect on the reperfusion seizures. Subsequently, 40 µg nimodipine was injected. In all four animals, nimodipine immediately arrested seizure activity ($p < 0.05$). Observation for an additional 30 min demonstrated no return of epileptiform activity (Fig. 2).

Topical Bicuculline and Pentylenetetrazole

Ten animals underwent bilateral frontoparietotemporal craniectomies. Gelfoam pledgets soaked in either bicuculline (5 µg) or pentylenetetrazole (5 mg) were applied

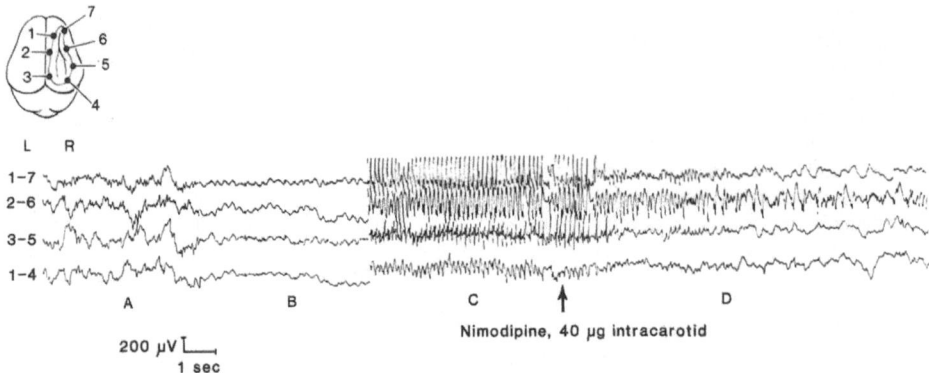

Fig. 2. Effect of nimodipine on reperfusion seizure. *A* Preocclusion tracing was 6–10 Hz. *B* Thirty seconds after vessel occlusion there is EEG attenuation. *C* Fifteen minutes after flow restoration there was seizure discharge in frontotemporal cortex. *D* One second after intracarotid injection of nimodipine 40 µg/kg, seizure activity was suppressed except for a small focus which persisted for approximately 12 s. The remaining cortex demonstrated postictal attenuation. Seizure activity did not return during 30 min observation. [Adapted from 52]

simultaneously to both cortices. Chemically induced seizures occurred within 60 s of application of either convulsant and persisted for a minimum of 2 h thereafter. Each animal also had placement of a PE-50 catheter into one ligated external carotid artery. Although there is a patent circle of Willis in these animals, injection of a test agent through this catheter will deliver the greatest concentration to the ipsilateral hemisphere, with only minimal delivery to the contralateral hemisphere. Therefore, with simultaneous application of convulsants to both hemispheres, one hemisphere serves as a control while the other hemisphere, ipsilateral to the catheter, may be used to test the anticonvulsant properties of an agent. In this experiment, 50 test injections were made in a total of 10 animals. Ten test injections were for the vehicle polyethylene glycol, 10 were for verapamil 0.5 mg/kg, and 30 were for injection of nimodipine 40 µg/kg. In 28 of 30 injections of nimodipine, there was arrest of seizure activity, whereas both the vehicle and verapamil were ineffective ($p < 0.001$). Nimodipine was equally effective against bicuculline- and pentylenetetrazole-induced seizures. Presumably, parenteral injection of verapamil was ineffective in part because of its poor blood-brain barrier penetration (Figs. 3, 4).

Intravenous Pentylenetetrazole

Pentylenetretrazole was administered intravenously to 20 animals with monitoring of hemispheric electrocorticograms by stereotactic placement of electrodes. In 10 control animals, pentylenetetrazole at 60 mg/kg produced seizures lasting for a minimum of 30 min. In 10 test animals, immediately after induction of seizures through injection of an identical dose of pentylenetetrazole, nimodipine was given as a slow intravenous

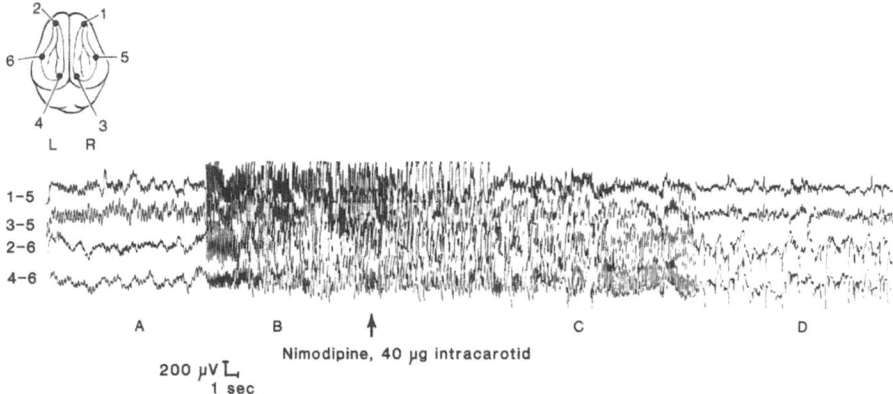

Fig. 3. Effect of nimodipine on bicuculline seizure. *A* Baseline tracing was 6–10 Hz. *B* Thirty seconds after topical application of bicuculline 5 µg/kg to each cerebral hemisphere, bilateral seizure discharges have developed. *C* Within 2 s after injection of nimodipine 40 µg/kg into the right internal carotid artery, the right hemisphere (*top two tracings*) showed attenuation followed by arrest of seizure activity. The control left hemisphere (*bottom two tracings*) continued to have seizure activity. *D* Five minutes after unilateral treatment, seizure activity persisted in the untreated left hemisphere. [Adapted from 52]

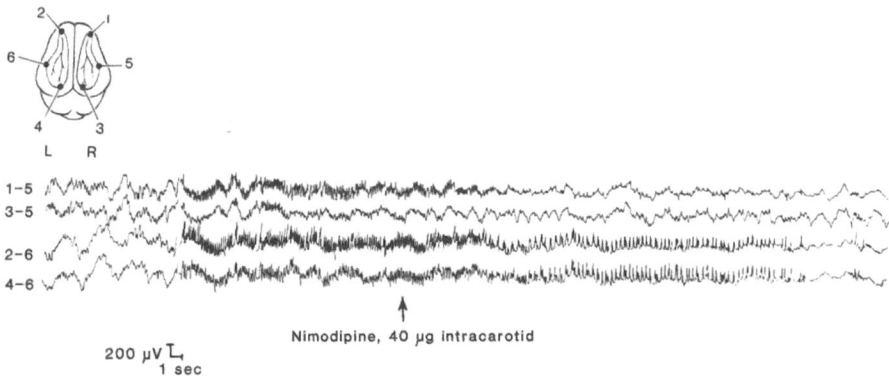

Fig. 4. Effect of nimodipine on pentylenetetrazole seizure. Baseline tracing was 6–10 Hz. Forty-five seconds after topical application of 5 mg pentylenetetrazole to each hemisphere, bilateral seizures have developed. There is some asymmetry in parietotemporal leads (3–5 and 4–6), but frontotemporal tracings show equally severe seizure discharge (1–5 and 2–6). Within 5 s of injecting nimodipine 40 µg/kg into the right internal carotid artery, the right hemisphere (*top two tracings*) showed arrest of seizure activity. The control left hemisphere (*bottom two tracings*) continued to demonstrate epileptiform activity. Ten minutes after unilateral treatment, seizure activity persisted in the untreated left hemisphere. [Adapted from 52]

infusion up to 200 µg/kg. Larger doses caused moderate hypotension. In nine of 10 animals, seizures were abolished ($p < 0.001$). Observation for an additional 30 min demonstrated only occasional spikes or bursts up to 5 s long of epileptiform activity (Fig. 5).

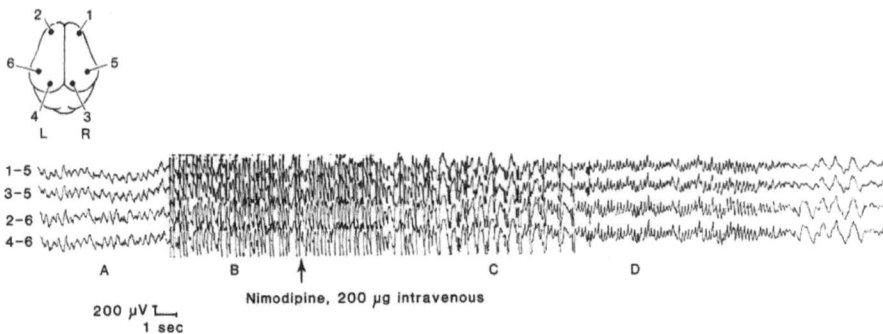

Fig. 5. Effect of intravenous nimodipine on pentylenetetrazole seizure. *A* Baseline recording was 6–10 Hz. *B* Five minutes after intravenous injection of 60 mg/kg pentylenetetrazole, bilateral seizure activity developed. *C* Five seconds after slow intravenous infusion of 200 µg nimodipine, there was seizure attenuation. *D* Twenty-five seconds later seizure activity has ceased. There was no return of epileptiform activity during 30 min subsequent observation. [Adapted from 52]

Electroconvulsive Shock Seizures

Forty-five animals underwent right frontoparietotemporal craniectomy. A pair of 1-mm-diameter silver ball electrodes were manipulated onto the cortex 5 mm apart. Each animal also had placement of electrodes around the margins of the craniectomy for continuous electrocorticographic recordings. Seizures were induced by a 5-s bipolar current of 10 V, 100 Hz, 0.1 ms pulses. The cortical impedance of the stimulating electrodes ranged from 6 to 10 ohms. In 10 control animals, this stimulus induced a seizure lasting 50–75 s with a mean of 64.0 ± 6.2 s. In these controls, the stimulus was delivered every 15 min to determine whether there would be progressive electrophysiological deterioration from repetitive seizures with 15 min recovery. After 75 min or six stimulations, the seizure duration was 63.2 ± 4.5 s. Therefore, in this experimental design, if seizure suppression occurred with repeat administration of a test agent it was not due to cortical fatigue but rather to the cumulative anticonvulsant properties of the test agent. In all test animals, the agent was delivered via intracarotid injection 2 min prior to stimulation. Therefore, each test agent was given in a preload fashion. In addition to the 10 control animals which received no infusion to test cortical reserve, 10 animals were treated with nimodipine given sequentially at doses of 10 µg/kg, 30 µg/kg, and 100 µg/kg; 10 received repeated injections of equivalent concentrations of the vehicle; 10 received sequential doses of phenytoin at 10 µg/kg, 30 µg/kg, 100 µg/kg, 300 µg/kg, 1 mg/kg, and 3 mg/kg; and five received sequential doses of verapamil at 10 µg/kg, 30 µg/kg, 100 µg/kg, 300 µg/kg, and 1 mg/kg. Cumulative doses of verapamil greater than 1.5 mg/kg were cardiotoxic.

A cumulative nimodipine dose of 140 µg/kg administered by intracarotid injection prevented seizure induction whereas the vehicle at equivalent concentrations was ineffective. Verapamil was also ineffective. Most importantly, phenytoin, which is considered to be extremely effective in preventing ECS seizures, was effective only at a cumulative dose of 4440 µg/kg. Therefore, in this experimental paradigm nimodipine was 30 times more effective than phenytoin (Fig. 6).

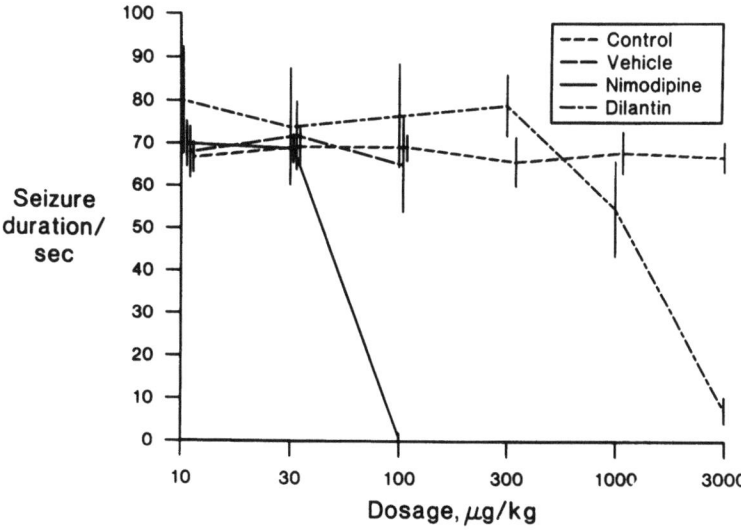

Fig. 6. Effect of nimodipine on ECS seizure. Graph of seizure duration verses log dose in 45 animals which underwent ECS of 10 V, 100 Hz, 0.1 ms pulses for 5 s. Intracarotid nimodipine at 140 µg/kg prevented seizure induction when given 2 min prior to stimulus ($p < 0.001$). The PEG vehicle was ineffective. To obtain a similar anticonvulsant effect 4440 µg/kg phenytoin was required. Not depicted are 10 animals in which verapamil at 1.5 mg/kg total dose was ineffective. In 10 control animals, there was no cortical fatigue during repetitive seizures with 15 min recovery. [Adapted from 53]

Comparison of Nimodipine, PN200-110, Nicardipine, and Phenytoin in Electroconvulsive Shock Seizures

In a paradigm similar to that described above, the anticonvulsant effects of various dihydropyridines were compared with phenytoin in ECS seizures induced with a stimulus of 10 V, 100 Hz, 0.1 ms pulses for 5 s. Each drug was administered by intracarotid injection 2 min prior to shock. Prior to treatment, seizure duration ranged from 43.8 ± 5.1 to 49.6 ± 5.2 s. A cumulative nimodipine dose of 440 µg/kg decreased seizure discharge to 6.6 ± 5.0 s ($p < 0.001$). A total dose of 1.0 mg/kg PN200-110 was required to achieve a similar effect. Nicardipine was ineffective up to doses which caused hypotension. A cumulative phenytoin dose of 7.0 mg/kg was required to suppress seizure activity (Fig. 7). Therefore, in this experiment nimodipine > PN200-110 >> phenytoin in preventing acute ECS seizure discharge [56].

Oral Administration of Nimodipine

In the preceding experiments, nimodipine was administered parenterally. Based on the above results, we elected to test the effects of this agent against pentylenetetrazole seizures when given orally. Twenty animals were assigned to one of two groups: 10 controls, and 10 animals treated with nimodipine 5 mg/kg/day orally. On the 6th day,

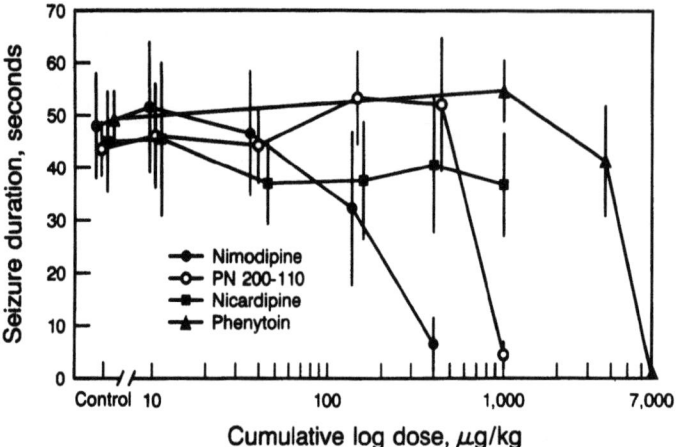

Fig. 7. Comparison of dihydropyridines in ECS seizures. In a paradigm similar to that depicted in Fig. 6, intracarotid nimodipine achieved an anticonvulsant effect at 440 µg/kg. PN200–110 at 1 mg/kg demonstrated a similar seizure reduction. Nicardipine was ineffective up to doses which caused hypotension. Both nimodipine and PN200–110 were significantly more effective than phenytoin ($p < 0.001$). [Adapted from 56]

in a blinded fashion, each animal was tested by administration of increasing doses of pentylenetetrazole intravenously. Electrical activity of each hemisphere was assessed by stereotactic placement of electrodes through small burr holes. The epileptogenicity of pentylenetetrazole was measured in all animals by four electrocorticographic criteria: (1) seizure lasting longer than 5 s, (2) two seizures occurring within 5 min, (3) epileptiform activity for 1 h, and (4) status epilepticus. After baseline electrocorticographic recordings for 30 min to ensure normal electrical activity, pentylenetetrazole was administered intravenously in 10mg/kg aliquots. The EEG was then ovserved for 5 min to assess for any effect. When a cortical response conforming to the above criteria was met, no additional convulsant was given until those criteria were satisfied. Hence the thresholds for the various responses described above could be assessed. The dose of pentylenetetrazole in mg/kg was calculated and noted prior to giving additional convulsant to satisfy the next criterion.

Nimodipine had impressive anticonvulsant effects for each of the four EEG criteria: the dose of pentylenetetrazole required to produce a seizure lasting longer than 5 s was 27.0 ± 5.4 mg/kg in controls as compared to 49.6 ± 9.9 mg/kg in the treated group ($p < 0.001$). The dose of convulsant required to induce status epilepticus was 68.6 ± 6.3 mg/kg in controls versus 105.8 ± 14.1 mg/kg in the treated group ($p < 0.001$). In all four categories, nimodipine increased the seizure threshold by 50%–60%. During the 5 days of oral administration there were no differences between the two groups in systemic parameters, including arterial blood presure, arterial blood gases, or baseline electrical activity (Fig. 8).

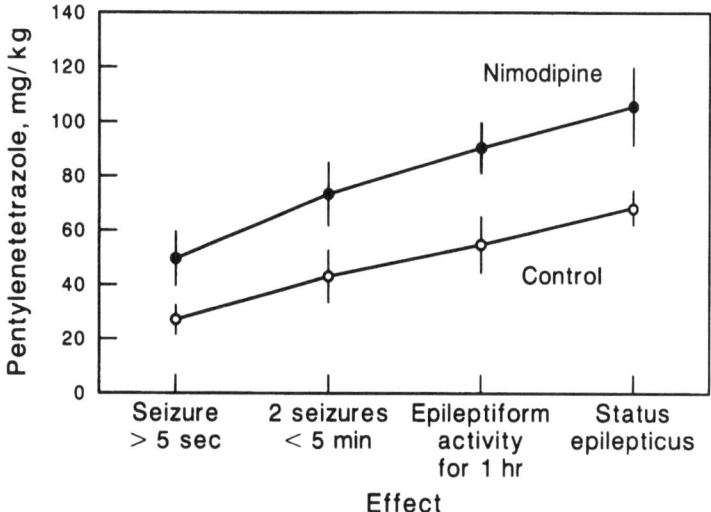

Fig. 8. Effect of oral nimodipine. Graph of stimulus versus effect in 10 controls and 10 animals treated with nimodipine 5 mg kg^{-1} day^{-1} orally for 5 days. Increasing doses of pentylenetetrazole were given intravenously and assessed by four EEG criteria: total dose required to produce first seizure greater than 5 s, two seizures occurring within 5 min, epileptiform activity for 1 h, and status epilepticus. In all categories, nimodipine increased the threshold by 50%–60% ($p < 0.001$). [Adapted from 55]

Other Laboratories

As indicated in the introduction, independent researchers have also demonstrated anticonvulsant properties of dihydropyridines. Hoffmeister et al. [40] tested nimodipine in a more standard fashion, finding that the ED$_{50}$ was 46 mg/kg p.o. against seizures induced in mice by intravenous administration of 50 mg/kg pentylenetetrazole. Nimodipine was ineffective in ECS shock (ED$_{50}$ > 700), contrary to our findings above.

Morocutti and colleagues [59] induced epileptic activity in rabbits by intraventricular infusion of cefazolin. Administration of nimodipine at 0.1 mg/kg intravenously first attenuated seizure activity in the contralateral cortex, followed by arrest of the primary focus ipsilateral to the ventricular infusion. One difficulty with this study is that the dose of parenteral nimodipine used did cause a moderate reduction in arterial blood pressure by 25 mmHg. This reduction might have impaired cerebral perfusion, leading to a decrease in substrate delivery at a time of increased metabolic demand. However, the known cerebral vasodilatory effects of nimodipine makes it difficult to estimate the net result of this drop in blood pressure.

Ascioti et al. [1] analyzed the anticonvulsant effects of various Ca^{2+} antagonists in sound-induced seizures in DBA/2 mice. They found that the potency in order of effectiveness was flunarizine > nimodipine > nifedipine > diltiazem. Nitrendipine and verapamil were ineffective. The ineffectiveness of nitrendipine and verapamil could be explained in part by their poor blood-brain barrier penetration. Finally,

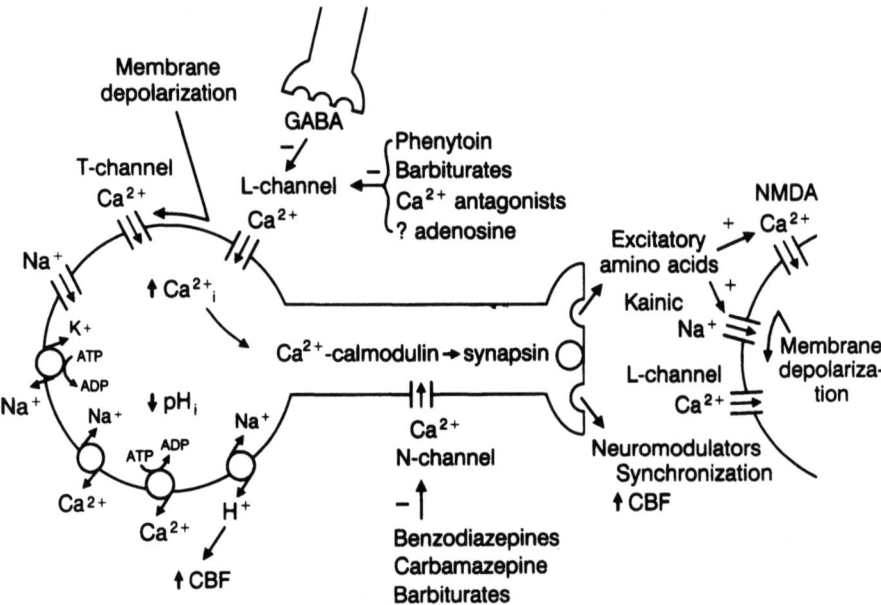

Fig. 9. Hypothetical model of the role of Ca^{2+} in seizure discharge

Dolin et al. [21] demonstrated that pretreatment with nitrendipine prevented nitrous oxide withdrawal seizures in mice. They also found that both nifedipine and nimodipine were effective against seizures induced through ethanol withdrawal in rats.

Calcium and Epilepsy: An Integrative Approach

A hypothetical schema for the role of Ca^{2+} in seizure initiation and propagation is depicted in Fig. 9. This proposal attempts to integrate current information on Ca^{2+} channels, data obtained from hippocampal brain slice preparations, and evidence that Ca^{2+} is essential for epileptic phenomena. In this speculative mechanism, the initiating event is Ca^{2+} influx through either voltage-dependent L channels or receptor-operated channels. This is based on the experimental evidence, reviewed above, that seizure initiation can be attenuated by preadministration of dihydropyridine Ca^{2+}-channel blockers which are known to block L channels. However, preadministration of dihydropyridines has variable effects on seizure discharge evoked by parenteral punitive excitatory amino acids (Meyer, unpublished observations). With the initial influx of Ca^{2+} there is membrane depolarization which then leads to opening of both Ca^{2+} T channels and Na^+ channels. This would conform to brain slice preparations in which a long-lasting Ca^{2+} current (L channel) leads to rapid transient Ca^{2+} currents (T Channel) and Na^+ currents. Rapid repolarization would occur through the outward Ca^{2+}-driven K^+ current and inward Cl^- current and through the Na^+/K^+ pump. This initiating mechanism would also explain why GABA neurons are inhibitory through modulation of the L channel and also provide a site at which the endogenous Ca^{2+} antagonist adenosine could exert its anticonvulsant effect [24].

Synaptic mediation of a discharge would follow by opening of voltage-dependent N channels due to a propagating action potential. Ca^{2+} influx would then bind to calmodulin, leading to exocytosis of neurotransmitter with recruitment of additional neurons. Exocytosis of excitatory amino acids would therefore be dependent on Ca^{2+} influx. In this schema, the punitive neurotransmitters would facilitate Ca^{2+} entry through receptor-operated channels, providing a route through which they would cause hyperexcitability.

To continue a seizure discharge, in addition to rapid membrane repolarization, neurons would be forced to regulate intracellular calcium through rapid Na^+/Ca^{2+} exchange, the Ca^{2+} ATPase, endoplasmic reticulum, and perhaps through mitochondrial sequestration. We might anticipate that since intracellular Na^+ would be elevated, the Na^+/Ca^{2+} antiport pump might be of secondary importance as compared to the latter mechanisms.

This model offers sites at which anticonvulsants may modulate seizure discharge. Phenytoin and some organic Ca^{2+}-channel blockers would primarily block Ca^{2+} influx through L channels, carbamazepine and benzodiazepines would block N channels, and barbiturates would block both channels. This approach also provides two mechanisms by which cerebral blood flow would increase during epileptic discharge: extracellular acidosis and release of vasodilators at the synapse [38, 78].

Potential Therapeutic Anticonvulsant Role

In summary, dihydropyridine Ca^{2+} antagonists like nimodipine appear to have impressive anticonvulsant effects in a variety of models of experimental epilepsy. The fact that this is the case is not completely surprising, since in vitro studies with hippocampal neurons indicate that bursting is Ca^{2+}-dependent. Furthermore, at least one type of neuronal Ca^{2+} channel – the L channel – is modulated by dihydropyridine agonists and antagonists. Given the fact that dihydropyridine receptor sites are located in regions, like the hippocampus, where epileptogenesis is liable to occur, and that agents like nimodipine do cross the blood-brain barrier, it is reasonable to speculate that these agents may prove to be useful in the treatment of epilepsy.

Some words of caution are necessary to maintain a correct perspective on this possible therapeutic role. First, Ca^{2+} regulation and metabolism is extremely complex, including the fact that Ca^{2+} enters neurons through many channels, most of which are dihydropyridine-insensitive.

Second, the role of excitatory amino acids needs clarification. There is strong evidence that these neurotransmitters, including glutamate and aspartate, are intrinsically involved in seizure phenomena [5–6, 17, 26, 31]. In fact, excitatory amino acid antagonists are potent anticonvulsants in pentylenetetrazole and DBA/2 mice [16]. Three major receptors have been identified for the excitatory aminoacids on the basis of selectivity by kainic acid, quisqualate, and N-methyl-D-aspartate (NMDA) [27]. The NMDA receptor has its highest concentration in the hippocampus [58]. When glutamate interacts at either the quisqualate or the kainic acid receptor it causes excitation through opening of Na^+ channels. It is possible that this resulting membrane depolarization leads to opening of voltage-sensitive Ca^{2+} channels. Accordingly, seizures induced by kainic acid can be modified by dihydropyridines (Meyer, unpublished observations). When glutamate binds to the NMDA receptor a Ca^{2+} channel is opened. There is no evidence which suggests that this agonist-operated Ca^{2+} channel is blocked by Ca^{2+} agonists. Therefore, if indeed this NMDA-gated Ca^{2+} channel is critical for seizure discharge, then dihydropyridines may not be effective in clinical epilepsy.

What is apparent is that Ca^{2+}, and Ca^{2+} regulation, are central to the evolution of seizure discharge. Therefore, one can anticipate that a better understanding of the characteristics and interaction of both voltage-sensitive and receptor-operated Ca^{2+} channels will yield more specific Ca^{2+}-channel antagonists that will find a therapeutic role in the treatment of epilepsy.

Finally, the relationship between Ca^{2+} metabolism, seizures, and ischemia is important to emphasize. As elegantly discussed by Siesjö, seizures form a part of the ischemic spectrum [75]. The clinical correlate of this relationship is that hippocampal pyramidal neurons are exquisitely susceptible to ischemia and are also the primary source of adult epileptic activity (temporal lobe epilepsy). Is it possible that the link between these two pathophysiological states is Ca^{2+} homeostasis? It is quite interesting that in our models of focal ischemia, seizures occur at a high frequency and are quickly attenuated by dihydropyridine Ca^{2+} antagonists. As authoritatively discussed by others in this volume, Ca^{2+} influx during ischemia is thought to precipitate a cascade of events, including the activation of phospholipase A_2 with the liberation of free fatty acids and the sequestration of Ca^{2+} in mitochondria, paralyzing function. Perhaps another deleterious effect of Ca^{2+} influx during ischemia is to create a state of hyperexcitability (seizures) with increased metabolic demands at a time when substrate delivery is already compromised. If this postulate is correct, then Ca^{2+}-channel blockers may attenuate ischemic neuronal damage through inhibition of neuronal excitability irrespective of their effects on cerebral blood flow or inhibition of the well-documented Ca^{2+} pathways of neuronal injury.

References

1. Ascioti C, De Sarro GB, Meldrum BS, Nistico G (1986) Calcium entry blockers as anticonvulsants in DBA/2 mice. Br J Pharmacol 88:379
2. Atwood HL, Lnenicka GA (1986) Structure and function in synapses: emerging concepts. Trend Neurosci 9:248–250
3. Baky S (1985) Nicardipine hydrochloride. In: Scriabine A (ed) New drugs annual: cardiovascular drugs. Raven Press, New York pp 153–172
4. Bellemann P, Schade A, Towart R (1983) Dihydropyridine receptor in rat brain labelled with H^3-nimodipine. Proc Natl Acad Sci USA 80:2356–2360
5. Ben-Ari Y, Lagowska J, Tremblay E, Le Gal La Salle G (1979) A new model of focal status epilepticus: intra-amygdaloid application of kainic acid elicits repetitive secondarily generalized convulsive seizures. Brain Res 163:176–179
6. Ben-Ari Y, Tremblay E, Ottersen OP, Meldrum BS (1980) The role of epileptic activity in hippocampal and remote cerebral lesions induced by kainic acid. Brain Res 191:79–97
7. Bingmann D, Speckmann E-J (1985) Calcium antagonists flunarizine and verapamil depress ictal activity in neurons of hippocampal slices. J Neurol 232:259
8. Binnie CD, de Beukelaar F, Meijer JWA, Overweg MJ, Wauquier A, van Wieringen A (1985) Open dose ranging trial of flunarizine as add-on therapy in epilepsy. Epilepsia 26: 424–428

9. Bolger GT, Weissman BA, Bacher J, et al. (In Press) Calcium antagonist binding in cat brain tolerant to electroconvulsive shock. Pharmacol Biochem Behav
10. Bolger GT, Weissman BA, Skolnick P (1984) The behavioral effects of the calcium channel agonist Bay K8644 in the mouse: antagonism by the calcium antagonist nifedipine. Naunyn-Schmiedeberg Arch Pharmacol 328:373–377
11. Browing MD, Huganir R, Greengard P (1985) Protein phosphorylation and neuronal function. J Neurochem 45:11–23
12. Brown DA, Griffith WH (1983) Calcium activated outward current in voltage clamped hippocampal neurons of the guinea pig. J Physiol (Lond) 337:287–301
13. Brown DA, Griffith WH (1983) Persistent slow inward current in voltage clamped hippocampal neurons of the guinea pig. J Physiol (Lond) 337:303–320
14. Chapman AG (1981) Free fatty acid release and metabolism of adenosine and cyclic nucleotides during prolonged seizures. In: Morselli PL, Lloyd KG, Loscher W, Meldrum BS (eds) Neurotransmitters, Seizures, and Epilepsy, Raven Press, New York pp 165–173
15. Cortes R, Supavilai P, Karobath M, Palacios JM (1984) Calcium antagonist binding sites in the rat brain: quantiative autoradiographic mapping using the 1,4-dihydropyridines H^3-PN200–110 and H^3-PY 108–068. J Neural Transmission 60:169–197
16. Croucher MJ, Collins JF, Meldrum BS (1982) Anticonvulsant action of excitatory amino acid antagonists. Science 216:899–901
17. Curtis DR, Watjkins JC (1963) Acidic amino acids with strong excitatory actions on mammalian neurons. J Physiol 160:1–14
18. DeLorenzo RJ (1982) Calmodulin in neurotransmitter release and synaptic function. Fed Proc 41:2265–2272
19. DeLorenzo RJ (1984) Calmodulin systems in neuronal excitability: a molecular approach to epilepsy. Ann Neurol 16 Suppl S104–114
20. Deisz RA, Prince DA (1987) Effect of D890 on membrane properties of neocortical neurons. Brain Res 422:63–73
21. Dolin SJ, Little HJ (1986) The dihydropyridine nitrendipine prevents nitrous oxide withdrawal seizures in mice. Br J Addiction 81:708
22. Desmedt CKL, Niemegeers CJE, Janssen PAJ (1975) Anticonvulsant properties of cinnarizine and flunarizine in rats and mice. Arzneimittelforsch 25:1408–1413
23. Dunlap K, Fischbach GD (1981) Neurotransmitters decrease the calcium conductance activated by depolarizations of embryonic chick sensory neurons. J Physiol (Lond) 317:519–535
24. Dunwiddie TV, Werth J (1982) Sedative and anticonvulsant effects of adenosine analogs in mouse and rat. J Pharmacol Exp Ther 220:70–76
25. Eckert R, Chad JE (1984) Inactivation of Ca channel. Prog Biophys Biol 44:215–267
26. Evans MC, Griffiths T, Meldrum BS (1984) Kainic acid seizures and the reversibility of calcium loading in vulnerable neurons in the hippocampus. Neuropath and Appl Neurobiol 10:285–302
27. Foster AC, Fagg GE (1984) Acidic amino acid binding sites in mammalian neuronal membranes: their characteristics and relationship to synaptic receptors. Brain Res Rev 7:103–164
28. Godfraind T, Miller R, Wibo M (1986) Calcium antagonism and calcium entry blockade. Pharmacological Reviews 38:321–416
29. Greenberg DA, Cooper EC, Carpenter CC (1984) Phenytoin interacts with calcium channels in brain membranes. Ann Neurol 16:616–617
30. Greenberg DA, Carpenter CL, Messing RO (1985) Inactivation of $^{45}Ca^{2+}$ uptake by prior depolarization of PC12 cells. Neuroscience Letters 62:377–381
31. Griffiths T, Evans MC, Meldrum BS (1983) Intracellular calcium accumulation of rat hippocampus during seizures induced by bicuculline or L-allylglycine. Neuroscience 10:385–395
32. Grima M, Schwartz J, Spach MO, Velly J (1986) Anti-anginal arylalkylamines and sodium channels: ^{3}H-batrachotoxinin-A 20-x-benzoate and ^{3}H-tetracaine binding. Br J Pharmacol 89:641–646
33. Gross RA, MacDonald RL (1988) Barbiturates and nifedipine have different and selective effects on calcium currents of mouse DRG neurons in culture. Neurology 38:443–451
34. Hagiwara S, Byerly L (1981) Calcium channel. Annu Rev Neurosci 4:69–125
35. Halliwell JV (1983) Caesium loading reveals two distinct Ca-currents in voltage-clamped guinea pig hippocampal neurons in vitro. J Physiol (Lond) 341:10–11

36. Heinemann U, Louvel J (1983) Changes in Ca^{2+} and K^+ during repetitive electrical stimulation and during pentylenetetrazole induced seizure activity in the sensorimotor cortex of cats. Pflugers Arch 398:310–317
37. Heinemann U, Lux HD, Gutnick MJ (1977) Extracellular free calcium and potassium during paroxysmal activity in cerebral cortex of the cat. Exp Brain Res 27:237–243
38. Heuser D (1978) The significance of cortical extracellular H^+, K^+, and Ca^{2+} activities for regulation of local cerebral blood flow under conditions of enhanced neuronal activity. CIBA Found Sym 56:339–353
39. Higuchi S, Sasaki H, Shiobara Y, Sado T (1977) Absorption, excretion and metabolism of a new dihydropyridine diester cerebral vasodilator in rats and dogs. Xenobiotica 7:469–479
40. Hoffmeister F, Benz U, Heise A, Krause HP, Neuser V (1982) Behavioral effects of nimodipine in animals. Arzneim Forsch 32:347–360
41. Janis RA, Siver J, Triggle DJ (1987) Drug action and cellular calcium regulation. Adv Drug Research 16:309–586
42. Johnston D, Brown TH (1984) Mechanisms of neuronal burst generation. In: Schwartzkroin PA, Wheal HV (eds) Electrophysiology of epilepsy. Academic Press, New York, pp 277–301
43. Johnston D, Hablitz JJ, Wilson WA (1980) Voltage clamp discloses slow inward current in hippocampal burst firing neurons. Nature (Lond) 286:391–393
44. Katz B, Miledi R (1966) Spontaneous and evoked acitivity of motor nerve endings in calcium Ringer. J Physiol (Lond) 203:689–706
45. Katz B, Miledi R (1970) Further study of the role of calcium in synaptic transmission. J Physiol (Lond) 207:789–801
46. Llinas R, Yarom Y (1981) Electrophysiology of mammalian inferior olivary neurons in vitro. Different types of voltage dependent conductances. J Physiol (Lond) 315:549–567
47. Llinas R, McGuinness TL, Leonard CS, Sugimori M, Greengard P (1985) Intraterminal injection of synapsin I or calcium/calmodulin-dependent protein kinase II alters neurotransmitter release at the squid giant synapse. Proc Natl Acad Sci USA 82:3035–3039
48. MacDonaold RL, Skerritt JH, Werz MA (1986) Adenosine agonists reduce voltage-dependent calcium conductances of mouse sensory neurons in cell cultures. J Physiol (Lond) 370:70–90
49. Marchetti C, Carbone E, Lux HD (1986) Effects of dopamine and noradrenaline on Ca channels of cultured sensory and sympathetic neurons of chick. Pfluegers Arch 406:104–111
50. Meldrum BS (1983) Metabolic factors during prolonged seizures and their relation to nerve cell death. Adv Neurobiol 34:261–275
51. Meyer FB, Anderson RE, Yaksh TL, Sundt, TM Jr (1986) Effect of nimodipine on intracellular brain pH, cortical blood flow, and EEG in experimental focal cerebral ischemia. J Neurosurg 64:617–626
52. Meyer FB, Anderson RE, Sundt TM Jr, Sharbrough FW (1986) Selective CNS calcium channel blockers – a new class of anticonvulsants. Mayo Clin Proc 61:239–247
53. Meyer FB, Tally PW, Anderson RE, Sundt TM Jr, Yaksh TL (1986) Inhibition of electrically induced seizures by a dihydropyridine calcium channel blocker. Brain Research 384:180–183
54. Meyer FB, Anderson RE, Sundt TM Jr, Yaksh TL (1986) Intracellular brain pH, indicator tissue perfusion, electroencephalography and histology in severe and moderate focal cortical ischemia in the rabbit. J Cereb Blood Flow Metab 6:71–78
55. Meyer FB, Anderson RE, Sundt TM Jr, Yaksh TL, Sharbrough FW (1987) Suppression of pentylenetetrazole seizures by oral administration of a dihydropyridine calcium antagonist. Epilepsia 28:409–414
56. Meyer FB, Anderson RE, Sundt TM Jr (1988) Anticonvulsant effects of dihydropyridine calcium antagonists on electrically induced seizures. (Submitted to Neuroscience)
57. Miller RJ (1987) Multiple calcium channels and neuronal function. Science 46–52
58. Monaghan DT, Holets VR, Toy DW, Cotman CW (1983) Anatomical distribution of four pharmacologically distinct H^3-L-glutamate binding sites. Nature 306:176–179
59. Morocutti C, Pierelli F, Sanarelli L, Stefano E, Peppe A, Mattioli GL (1986) Antiepileptic effects of a calcium antagonist (nimodipine) on cefazolin induced epileptogenic foci in rabbits. Epilepsia 27:498–503
60. Nowycky MC, Fox AP, Tsien RW (1985) Long-opening mode of gating of neuronal calcium channel and its promotion by the dihydropyridine calcium agonist Bay K8644. Proc Natl Acad Sci USA 82:2178–2182

61. Nowycky MC, Fox AP, Tsien RW (1985) Three types of neuronal calcium channels with different calcium agonist sensitivity. Nature 316:440–443
62. Peroutka SJ, Allen GS (1983) Calcium channel antagonist binding sites labelled by H^3-nimodipine in human brain. J Neurosurg 59:933–937
63. Overweg J, Binnie CD, Meijer JWA, Meinardi H, Nuijten STM, Schmaltz S, Wauquier A (1984) Double blind placebo controlled trial of flunarizine as add-on therapy in epilepsy. Epilepsia 25:217–222
64. Owen DG, Segal M, Barker JL (1984) A Ca^{2+} dependent Cl^- conductance is present in cultured mouse spinal neurons. Nature (Lond) 311:567–570
65. Pincus JH (1973) Diphenylhydantoin and calcium. Arch Neurol 36:239–244
66. Pincus JH, Hsiao K (1981) Phenytoin inhibits both synaptosomal Ca^{2+} uptake and efflux. Exp Neurol 74:293–298
67. Preuss KC, Gross GH, Brooks HL, Warltier DC (1985) Slow channel calcium channel activators, a new group of pharmacologic agents. Life Sci 37:1271–1278
68. Prince DA (1985) Physiological mechanisms of focal epileptogenesis. Epilepsia 26 Suppl S3–S14
69. Pumain R, Kurcewicz I, Louvel J (1983) Fast extracellular calcium transients: involvement in epileptic processes. Science 222:177–179
70. Pumplin DW, Reese TW, Llinas R (1981) Are the presynaptic membrane particles the calcium channels? Proc Natl Acad Sci USA 78: 7210–7213
71. Schwartzkroin PA (1980) Ionic and synaptic determinants of burst generation. In: Lockard JS, Ward AA Jr (eds) Epilepsy: a window to brain mechanisms. Raven Press, New York, pp 83–95
72. Schwartzkroin PA, Slawsky M (1977) Probable calcium spikes in hippocampal neurons. Brain Res 135:157–161
73. Siekevitz P, Carlin RK, Wu K (1985) Soc Neurosc Abstract 11:646
74. Siesjö BK (1984) Cerebral circulation and metabolism. J Neurosurg 60:883–908
75. Siesjö BK (In Press) Historical overview: calcium, ischemia, and death of brain cells. Ann NY Acad Sci
76. Sundt TM Jr, Anderson RE (1980) Intracellular brain pH and the pathway of a fat soluble pH indicator across the blood-brain barrier. Brain Res 186:355–364
77. Thayer SA, Murphy SN, Miller RJ (1986) Widespread distribution of dihydropyridine-sensitive calcium channels in the central nervous system. Mol Pharmacology 30:505–509
78. Wahl M (1985) Local chemical, neural and humoral regulation of crebrovascular resistance vessels. J Cardiovasc Pharmacol 536–546
79. Walden J, Speckmann E-J, Witte OW (1985) Suppression of focal epileptiform discharges by intraventricular infusion of a calcium antagonist. Electro Clin Neurophysiol 61:299–309
80. Wieloch T (1985) Neurochemical correlates to regional selective neuronal vulnerability. Prog Brain Res 63:69–85
81. Witte OW, Speckmann E-J, Walden J (1987) Motor cortical epileptic foci in vivo: actions of a calcium channel blocker on paroxysmal neuronal depolarizations. Electro Clin Neurophysiol 6:43–55
82. Wong RKS, Prince DA (1978) Participation of calcium spikes during intrinsic burst firing in hippocampal neurons. Brain Res 159:385–390
83. Wong RKS, Prince DA, Basbaum AL (1979) Intradendritic recordings from hippocampal neurons. Proc Natl Acad Sci 76:986–990
84. Wong RKS, Traub RD, Miles R (1984) Epileptogenic mechanisms as revealed by studies of the hippocampal slice. In: Schwartzkroin PA, Wheal HV (eds) Electrophysiology of Epilepsy. Academic Press, London pp 254–275
85. Zanotto L, Heinemann U (1983) Aspartate and glutamate induced reductions in extracellular free calcium and sodium concentration in area of CA1 of in vitro hippocampal slices of rats. Neuroscience Letter 35:79–84

Interaction Between Certain Antipsychotic Drugs and Dihydropyridine Receptor Sites*

R. Quirion**, D. Bloom, and N. P. V. Nair

Douglas Hospital Research Centre and Department of Psychiatry, Faculty of Medicine, McGill University, 6875 Boulevard LaSalle, Verdun, Québec, Canada H4H 1R3

Introduction

The existence of highly specific, selective and high-affinity 1,4-dihydropyridine (DHP) binding sites in brain is now well established (recent reviews: Gould, 1987; Miller 1987; Snyder and Reynolds, 1985; Triggle, 1987). It appears that DHP binding sites are allosterically coupled to phenylalkylamine and diltiazem sites, modulating each other's action on the relevant type of Ca^{2+} channels (Gould, 1987; Miller, 1987; Triggle, 1987). Recent data have clearly shown that these organic Ca^{2+} blockers selectively act on one subtype of channels, the L type, in various tissues (Kamp and Miller, 1987; Miller 1987; Nowycky et al. 1985). Other Ca^{2+} channels (including the N and T types) are insensitive to these organic Ca^{2+} antagonists (Nowycky et al., 1985) Miller, 1987).

It is of interest that certain "atypical" neuroleptics such as pimozide and fluspirilene interact (with high affinity) with the DHP binding sites. This is not the case for classical neuroleptics of the butyrophenone family such as haloperidol (Gould et al., 1983). It has been suggested that these interactions with the DHP binding sites could be related to the clinical profile of pimozide and fluspirilene (Gould et al., 1983; Snyder and Reynolds, 1985). We review here the comparative characteristics of DHP binding sites found in the brain of various mammalian species, including man, and summarize available data concerning the DHP-atypical neuroleptic drug interactions.

Localization of DHP Sites in Animal Brain

Autoradiographic studies have clearly shown that [^{3}H]DHP binding sites are discretely distributed in rat and guinea pig brain tissues (Cortes et al., 1983; Cortes et al., 1984; Ferry et al., 1984; Murphy et al., 1982; Quirion, 1983; Quirion et al., 1985a). The highest densities of specific [^{3}H]DHP sites are found in areas enriched in synaptic contacts such as the hippocampus and the cortex (see Fig. 1). It has been shown that lesions of the granule cell layer induce a major depletion of [^{3}H]DHP binding sites in the dentate gyrus, suggesting that these sites are most likely associated with dendritic

---

* This research project has been supported by the Douglas Hospital Research Centre and by Miles/Bayer Laboratories
** «Chercheur-Boursier» of the «Fonds de la Recherche en Santé du Québec».

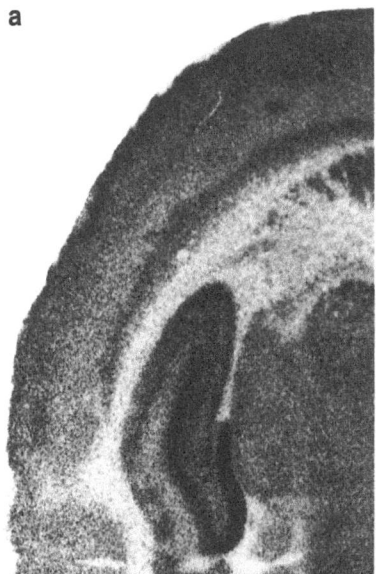

Fig. 1a, b. Photomicrographs of the autoradiographic distribution of [^{3}H]PN200-110 binding sites in rat (a) and human (b) brain tissues at the level of the hippocampus (a) or striatum (b). Note the dense laminar distribution of binding in the rat hippocampal formation and the high amounts of binding in the human brain cortex. See Quirion (1985) and Quirion et al. (1985a) for details

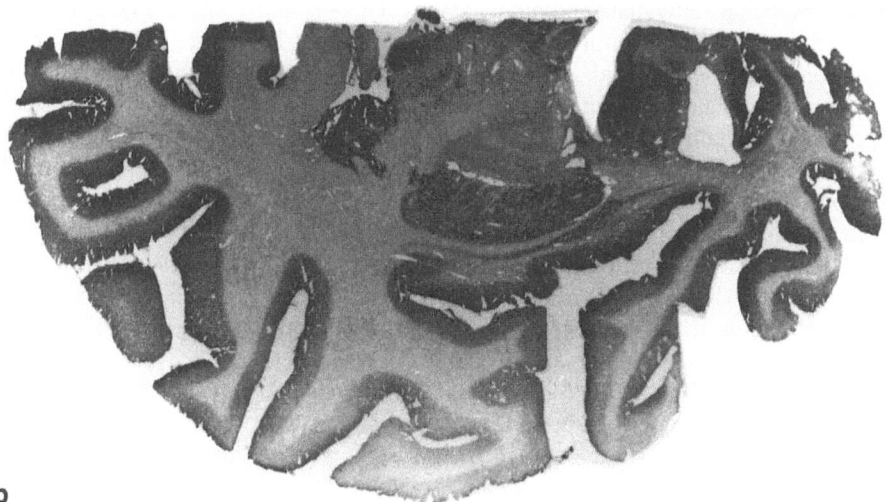

fields in the rat hippocampal formation (Cortes et al., 1983). Other areas enriched with [^{3}H]DHP binding sites included the external plexiform layer of the olfactory bulb, the ventral, lateral and posterior thalamic nuclei and the substantia nigra. Moderate densities of sites were seen in the caudate-putamen and the claustrum. Low levels of binding were generally observed in most remaining regions, including the hypothalamus, midbrain and pons. White matter areas were generally devoid of significant [^{3}H]DHP binding (Quirion et al., 1985a).

The autoradiographic distribution of brain phenylalkylamine binding sites, labelled using a tritiated analogue of verapamil, is similar (if not identical) to that observed with various DHP ligands (Ferry et al., 1984). This further supports the evidence that both classes of Ca^{2+} blockers act on the same type (L) of Ca^{2+} channels in brain tissues (Gould, 1987; Miller, 1987). However, the distribution of putative N-type Ca^{2+} channels labelled with ω-conotoxin is strikingly different from that of DHP/phenylalkylamine sites with high densities, seen in most gray matter areas of the brain (Kerr et al., 1988).

Localization of DHP Sites in Human Brain

Few groups have reported on the presence of high-affinity DHP binding sites in human brain using postmortem tissues (Hanada and Tanaka, 1985; Peroutka and Allen, 1983; Quirion, 1985; Quirion et al., 1985a) as well as freshly biopsied tissues (Quirion et al., 1988). Peroutka and Allen (1983) have shown that [^{3}H]nimodipine binds to an apparent single class of sites (K_d = 0.27 nM) in human frontal cortex. The binding was Ca^{2+}-dependent and competitively inhibited by various DHP-related drugs. Diltiazem potentiated [^{3}H]nimodipine binding, as seen in rat brain membrane preparations. Hanada and Tanaka (1985) reported similar results using [^{3}H]nitrendipine. Additionally, they showed that human brain DHP binding sites were discretely distributed by determining the density of sites in membranes prepared from multiple brain regions. The highest levels of binding were found in the cerebral cortex, thalamus, amygdala and hippocampus (Hanada and Tanaka, 1985).

We have studied the autoradiographic distribution of DHP sites in human brain (Quirion, 1985; Quirion et al., 1985a). DHP binding sites labelled by [^{3}H]PN200–110 are discretely localized in human brain (see Fig. 1). High densities of sites are present in cortex (especially in laminae I and III) and in the granule cell layer of the dentate gyrus. Moderate densities are seen in the caudate, putamen, claustrum and ventral and posterior thalamic nuclei. Other thalamic nuclei, as well as most hypothalamic, pons and midbrain structures, generally contained only limited amounts of specific DHP binding sites. Additionally, the distribution of phenylalkylamine sites is identical to that of DHP binding in human brain (Quirion, 1985). Thus, the characteristics and distribution of DHP receptor sites appear to be similar in the brain of various mammalian species, including man.

More recently, we compared DHP binding parameters in postmortem and freshly biopsied tissues obtained from patients undergoing surgery for removal of brain tissues of epileptic foci (Quirion et al., 1988). Using a broad range of concentrations (0.01–75 nM) of radioligands, we obtained evidence for the existence of low-affinity (25–40 nM) DHP sites in postmortem tissues (postmortem delay of 18 ± 6 h, n = 6). No evidence for the presence of such sites was found in freshly biopsied tissues. This suggests that human brain DHP binding sites may be sensitive to long postmortem delays.

Interaction with Antipsychotic Drugs

Binding Assays

Gould et al. (1983) first reported that antischizophrenic drugs of the diphenylbutylpiperidine (DPBP) type were potent Ca^{2+}-channel antagonists. They demonstrated that pimozide, fluspirilene, penfluridol and clopimozide compete, with nanomolar affinities, for brain [^{3}H]nitrendipine sites, in addition to inhibiting K$^+$-induced Ca^{2+}-dependent contractions of the rat vas deferens. The effect of DPBP drugs on DHP binding was most likely associated with the allosterically coupled phenylalkylamine site, since it was reversed by methoxyverapamil (Gould et al., 1983). More "classical" neuroleptics of the phenothiazine and butyrophenone families apparently lack significant affinity for the DHP/phenylalkylamine receptor complex (Gould et al., 1983). Thus, Gould et al. (1983) suggested that the ability of the DPBP drugs to relieve certain negative symptoms of schizophrenia could be related to their blockade of the L class of Ca^{2+} channels.

We (Quirion et al., 1985b) and others (Flaim et al., 1985; Galizzi et al., 1986; Qar et al., 1987) have confirmed and extended these findings. We found that DPBP drugs like pimozide and fluspirilene were potent competitors for both brain and heart [^{3}H]nitrendipine sites in the rat.

Interestingly, the affinities of the DPBP drugs for the brain DHP sites were between 5 and 10 times greater than for the heart DHP sites, suggesting possible heterogeneity of the DHP site in various tissues (Quirion et al., 1985; Bolger et al., 1987). Haloperidol inhibited [^{3}H]nitrendipine binding only weakly in these preparations. As shown in Table 1, human brain DHP sites are also highly sensitive to DPBP-

Table 1. Comparative affinities of Ca^{2+}-channel antagonists and various antischizophrenic drugs on [^{3}H]nitrendipine binding in human brain (postmortem)

Drug	K$_i$ (nM)
Ca^{2+}-channel antagonists	
Dihydropyridines	
Nimodipine	0.41 ± 0.10
Nitrendipine	1.02 ± 0.30
Phenylalkylamine	
Verapamil	> 20000
Diphenylbutylpiperazines	
Lidoflazine	975 ± 126
Cinnarizine	1200 ± 250
Diphenylalkylamine	
Prenylamine	790 ± 157
Antischizophrenic-neuroleptics	
Butyrophenone	
Haloperidol	2700 ± 300
Diphenylbutylpiperidines	
Fluspirilene	24.6 ± 3.17
Pimozide	65.1 ± 5.47
Penfluridol	210 ± 29.5

K$_i$ is calculated from the formula K$_i$ = IC$_{50}$/1 + F/K$_d$ where IC$_{50}$ is the concentration of drug which inhibited 50% of specifically bound ligand, F represents the free concentration of ligand (0.5 nM) and K$_d$ is the apparent affinity of the sites (0.8 nM for the high-affinity sites). Each value is the mean ± SEM of at least three determinations, each in triplicate. Temporal cortices membrane preparations were used in all experiments (postmortem delays ranging between 5 and 18 h). See Quirion (1985) and Quirion et al. (1988) for experimental details

related drugs such as pimozide and fluspirilene. This demonstrates further the great similarity between rat and human brain DHP receptor sites. It also indicates that DPBP drugs (at therapeutic concentrations) most likely block brain DHP-sensitive Ca^{2+} channels.

Calcium-Channel Blockers and Dopaminergic Systems

The high affinity of certain antischizophrenic drugs for DHP/phenylalkylamine-associated binding sites suggested that the effects of DHP and phenylalkylamine on brain catecholaminergic innervation should be investigated. It has already been shown that DHP and verapamil can block noradrenaline, dopamine and 5-hydroxytryptamine. Bay K8644 evoked release of these transmitters in brain slice preparations or cultured neuronal cells (DiRenzo et al., 1984; Middlemiss, 1985; Middlemiss and Spedding, 1985; Woodward and Leslie, 1986; Perney et al., 1986: but see also Nordstrom et al., 1986). Additionally, Pileblad and Carlsson (1986) have reported that nimodipine reduced the in vivo release and synthesis of dopamine in mouse brain. We obtained similar results in rat brain (decreases in levels of dopamine, homovanillic acid and DOPAC) following an intracerebroventricular injection of 2.5 μg nimodipine/rat (Quirion and Richard, unpublished results). Finally, it has been shown that certain Ca^{2+}-channel blockers prevent haloperidol-induced apomorphine supersensitivity (Grebb et al., 1987) and blocks amphetamine-induced behavioral stimulation in mice (Grebb, 1986). Thus, it appears that organic Ca^{2+}-channel blockers of the DHP and phenylalkylamine series can modulate brain dopaminergic activity.

Possible Clinical Significance

Naturally, the demonstration that certain antischizophrenic drugs (e.g. pimozide, fluspirilene) possess high affinity for DHP/phenylalkylamine sites suggested that Ca^{2+}-channel blockers could be considered as a novel approach for the treatment of schizophrenic patients. This is of special interest since it has been shown that pimozide and penfluridol are effective in alleviating certain symptoms of this disease (Frangos, 1972; Kardo, 1972; Lapierre, 1978).

Verapamil has recently been tried for the treatment of schizophrenia. In an open trial, Bloom et al. (1987) and Tourjman et al. (1987) have reported limited but significant improvement, especially in relation to thought disturbance, activation, irritability and retardation. On the other hand, no beneficial effect was seen in two other studies (Grebb et al., 1986; Schepelern and Koster, 1987). However, it should be noted that all these trials have been conducted in chronic, recalcitrant patients concomitantly receiving maximal doses of neuroleptics. Thus, further assays in less chronic patients or in individuals not previously treated with "classical" neuroleptics may respond better to verapamil therapy. Moreover, we are not aware of any reports using DHP (e.g. nimodipine, nifedipine) as a therapeutic agent in schizophrenia. Such studies appear certainly worthwhile in light of basic research data (see above). Finally, Ca^{2+}-channel blockers might be useful for the treatment of other psychiatric and neurological disorders, including mania (Dubovsky et al., 1982; 1985 Giannini et al., 1984; Dose et al., 1986), bipolar illness (Gitlin and Weiss, 1984), Tourette's disorder (Walsh et al., 1986) and epilepsy (Overweg et al., 1984).

Conclusion

In summary, brain DHP binding sites appear to possess similar characteristics and distribution in all species studied thus far. Moreover, antischizophrenic drugs of the DPBP family, like pimozide and fluspirilene, demonstrate high affinity for the DHP/phenylalkylamine receptor complex associated with the L type of Ca^{2+} channels. This may suggest that Ca^{2+} blockers of the DHP and phenylalkylamine types could act as "atypical" neuroleptics. This possibility should be carefully investigated in various subgroups of schizophrenic patients.

Acknowledgement. The expert secretarial assistance of Mrs. J. Currie is acknowledged.

References

Bloom DM, Tourjman SV, Nair NPV (1987) Verapamil in refractory schizophrenia: a case report. Prog Neuro-Psychopharmacol & Biol Psychiat 11:185–188

Bolger GT, Marcus KA, Daly JW, Skolnick P (1987) Local anesthetics differentiate dihydropyridine calcium antagonist binding sites in rat brain and cardiac membranes. J Pharmacol & Exp Ther 247:922–930

Cortes R, Supavilai P, Karobath M, Palacios JM (1983) The effects of lesions in the rat hippocampus suggest the association of calcium channel blocker binding sites with specific neuronal population. Neurosci Lett 42:249–254

Cortes R, Supavilai P, Karobath M, Palacios JM (1984) Calcium antagonist binding sites in the rat brain: quantitative autoradiographic mapping using the 1,4-dihydrophyridines [^{3}H]PN-200-110 and [^{3}H]PY-108-068. J Neural Trans 60:169–197

Di Renzo G, Amoroso S, Taglialatela M, Annunziato L (1984) Dual effect of verapamil on K^+-evoked release of endogenous dopamine from arcuate nucleus-median eminence complex. Neurosci Lett 50:269–272

Dose M, Emrich HM, Cording-Tommel C, von Zerssen D (1986) Use of calcium antagonists in mania. Psychoneuroendocrinology 11:241–243

Dubovsky SL, Franks RD, Lifschitz M, Coen P (1982) Effectiveness of verapamil in the treatment of a manic patient. Am J Psychiatry 139:502–504

Dubovsky SL, Franks RD, Schrier D (1985) Phenelzine-induced hypomania: effect of verapamil. Biol Psychiatry 20:1009–1014

Ferry DR, Goll A, Gadow C, Glossmann H (1984) (–)-^{3}H-Desmethoxyverapamil labelling of putative calcium channels in brain: autoradiographic distribution and allosteric coupling to 1,4-dihydropyridine and diltiazem binding sites. Naunyn-Schmiedeberg's Arch Pharmacol 327:183–187

Flaim SF, Brannan MD, Swigart SC, Gleason MM, Muschek LD (1985) Neuroleptic drugs attenuate calcium influx and tension development in rabbit thoracic aorta: effects of pimozide, penfluridol, chlorpromazine, and haloperidol. Proc Natl Acad Sci USA 82:1237–1241

Frangos E (1971) Clinical observations on the use of pimozide in the treatment of chronic schizophrenics. Acta Psychiatr Belg 71:698–707

Galizzi JP, Fosset M, Romey G, Laduron P,, Lazdunski M (1986) Neuroleptics of the diphenylbutylpiperidine series are potent calcium channel inhibitors. Proc Natl Acad Sci USA 83:7513–7517

Gianni AJ, Houser WL, Loiselle RH (1984) Antimanic effects of verapamil. Am J Psychiatry 141:1602–1603
Gitlin MJ, Weiss J (1984) Verapamil as maintenance treatment in bipolar illness: a case report. J Clin Psychopharmacol 4:341–343
Gould RJ (1987) Calcium channel antagonists. In: Williams M, Malick JB (eds) Drug discovery and development. The Humana Press, pp 409–442
Gould RJ, Murphy KMM, Reynolds IJ, Snyder SH (1983) Antischizophrenic drugs of the diphenylbutylpiperidine type act as calcium channel inhibitors. Proc Natl Acad Sci USA 80:5122–5125
Grebb JA (1986) Nifedipine and flunarizine block amphetamine-induced behavioral stimulation in mice. Life Sciences 38:1275–1281
Grebb JA, Shelton RC, Taylor EH, Bigelow LB (1986) A negative double-blind, placebo-controlled, clinical trial of verapamil in chronic schizophrenia. Biol Psychiatry 21:691–694
Grebb JA, Shelton RC, Freed WJ (1987) Diltiazem or verapamil prevents haloperidol-induced apomorphine supersensitivity in mice. J Neural Transm 68:241–255
Hanada S, Tanaka C (1985) Regional distribution of [^{3}H]nitrendipine binding in human brain. Neurosci Lett 58:375–380
Kamp TJ, Miller RJ (1987) Voltage-sensitive calcium channels and calcium antagonists. ISI atlas of science: pharmacology 1:133–138
Kerr LM, Filloux F, Olivera BM, Jackson H, Wamsley JK (1988) Autoradiographic localization of calcium channels with [^{125}I]ω-conotoxin in rat brain. Eur J Pharmacol 146:181–183
Kudo Y (1972) A double-blind comparison of pimozide with carpipramine in schizophrenic patients. Acta Psychiatr Belg 72:685–697
Lapierre YD (1978) A controlled study of penfluridol in the treatment of chronic schizophrenia. Am J Psychiatry 135:–959
Middlemiss DN (1985) The calcium channel activator, Bay K 8644, enhances K$^+$-evoked efflux of acetylcholine and noradrenaline from rat brain slices. Naunyn-Schmiedelberg's Arch Pharmacol 331:114–116
Middlemiss DN, Spedding M (1985) A functional correlate for the dihydropyridine binding site in rat brain. Nature 314:94–96
Miller RJ (1987) Multiple calcium channels and neuronal function. Science 235:46–62
Murphy KMM, Gould RJ, Snyder SH (1982) Autoradiographic visualization of [^{3}H]nitrendipine binding sites in rat brain: localization to synaptic zones. Eur J Pharmacol 81:517–519
Nordstrom O, Braesch-Andersen S, Bartfai T (1986) Dopamine release is enhanced while acetylcholine release is inhibited by nimodipine (Bay e 9736). Acta Physiol Scand 126:115–119
Nowycky MC, Fox AP, Tsien RW (1985) Three types of neuronal calcium channel with different calcium agonist sensitivity. Nature 316:440–443
Overweg J, Binnie CD, Meijer JWA, Meinardi H, Nuijfen STM, Schmaltz S, Wauquier A (1984) Double-blind placebo-controlled trial of flunarizine as add-on therapy in epilepsy. Epilepsia 25:217–222
Perney TM, Hirning LD, Leeman SE, Miller R (1986) Multiple calcium channels mediate neurotransmitter release from peripheral neurons. Proc Natl Acad Sci USA 483:6656–6659
Peroutka SJ, Allen GS (1983) Calcium channel antagonist binding sites labeled by ^{3}H-nimodipine in human brain. J Neurosurg 59:933–937
Pileblad E, Carlsson A (1986) In vivo effects of the Ca^{2+}-antagonist nimodipine on dopamine metabolism in mouse brain. J Neural Transm 66:171–187
Qar J, Galizzi J-P, Fosset M, Lazdunski M (1987) Receptors for diphenylbutylpiperidine neuroleptics in brain, cardiac, and smooth muscle membranes. Relationship with receptors for 1,4-dihydropyridines and phenylalkylamines and with Ca^{2+} channel blockade. Eur J Pharmacol 141:261–268
Quirion R (1983) Autoradiographic localization of a calcium channel antagonist, [^{3}H]nitrendipine, binding site in rat brain. Neurosci Lett 36:267–271
Quirion R (1985) Characterization of binding sites for two classes of calcium channel antagonists in human forebrain. Eur J Pharmacol 117:139–142
Quirion R, Lal S, Nair NPV, Stratford JG, Ford RM, Oliver A (1985a) Comparative autoradiographic distribution of calcium channel antagonist binding sites for 1,4-dihydropyridine and phenylalkylamine in rat, guinea pig and human brain. Prog Neuro-Psychopharmacol & Biol Psychiat 9:643–649

Quirion R, Lafaille F, Nair NPV (1985b) Comparative potencies of calcium channel antagonists and antischizophrenic drugs on central and peripheral calcium channel binding sites. J Pharm Pharmacol 37:437–440

Quirion R, Lal S, Olivier A, Robitaille Y, Nair NPV, Ford RM, Stratford JG (1988) Calcium channel binding sites in human brain. Ann NY Acad Sci in press

Schepelern S, Koster A (1987) Verapamil in treatment of severe schizophrenia. Acta Psychiatr Scand 75:557–558

Snyder SH, Reynolds IJ (1985) Calcium-antagonist-drugs: receptor interactions that clarify therapeutic effects. New England Journal of Medicine 313:995–1002

Tourjam SV, Bloom DM, Nair NPV (1987) Verapamil in the treatment of chronic schizophrenia. Psychopharmacol Bull 23:227–229

Triggle DJ (1987) Calcium channel drugs: antagonists and activators. ISI atlas of science: pharmacology 1:329–324

Walsh TL, Lavenstein B, Licamele WL, Bronheim S, O'Leary J (1986) Calcium antagonists in the treatment of Tourette's disorder. Am J Psychiatry 143:1467–1468

Woodward JJ, Leslie SW (1986) Bay K 8644 stimulation of calcium entry and endogenous dopamine release in raat striatal synaptosomes antagonized by nimodipine. Brain Research 370:397–400

Effects of Organic Calcium-Entry Blockers on Stimulus-Induced Changes in Extracellular Calcium Concentration in Area CA1 of Rat Hippocampal Slices*

U. Heinemann, P. Igelmund, R. S. G. Jones, G. Köhr, and H. Walther

Institut für normale und pathologische Physiologie, Universität zu Köln, Robert Koch Str. 39, D-5000 Köln 41, FRG

Introduction

Increases in intracellular Ca^{2+} concentration, $[Ca^{2+}]_i$, serve a number of functions in nerve cells, ranging from regulation of cell morphology [37] to regulation of excitability [36, 48, 60]. In addition, Ca^{2+} seems to be involved in translating short-term excitability changes into long-term alterations of transfer functions in neuronal information processing. Thus, it has been shown that various measures which facilitate Ca^{2+} entry into nerve cells can produce long-term enhancement of synaptic transmission (LTP) [1, 3, 57, 61]. Experimentally, such alterations may even result in pathological states such as epilepsy [24], which in turn are often associated with an enhanced capability for Ca^{2+} uptake into nerve cells [for review: 38, 31]. Under physiological conditions nerve cells appear to tolerate rather big loads of Ca^{2+}; however, when electrolyte regulation is impaired, such Ca^{2+} loading may lead to cell degeneration [e.g., 11, 19, 73].

These facts have led to the question of which mechanisms are involved in the uptake of Ca^{2+} into nerve cells and how such Ca^{2+} movements can be affected. A related problem concerns the finding that many diseases of the CNS affect certain cells with a high priority (selective vulnerability). Indeed, it is now common knowledge that states such as hypoxia, hypoglycemia, and epilepsy lead to degeneration preferentially of cerebellar Purkinje cells, pyramidal cells of neocortex and hippocampus, and to a circumscribed cell loss deep in entorhinal cortex. These cells are characterized by the fact that they either express Ca^{2+} action potentials easily [12, 23, 42, 52] or that they are rich in NMDA receptors, a subtype of glutamate receptors [77] which permit easy access of Ca^{2+} to nerve cells [14, 51, 58].

Stimulus-Induced Changes in Extracellular Calcium Concentration

One way to get insight into some of the areas of uncertainty is to measure changes in extracellular Ca^{2+} concentration, $[Ca^{2+}]_o$, with ion-selective microelectrodes [33, 49]. Indeed, it can be shown that repetitive electrical stimulation causes rather large reductions in $[Ca^{2+}]_o$ of up to 0.6 mM. When the volume fraction of the extracellular space (ES), of glial, and of nerve cells are taken into account this implies rather large

---
* This work was supported by DFG grant He 1128/2−4

cellular Ca^{2+} loads. Excitatory amino acids [34], spreading depression [47], and hypoxia [26, 27], as well as hypoglycemia, also cause decreases in $[Ca^{2+}]_o$, often by more than 1 mM. Such large reductions in $[Ca^{2+}]_o$ are also observed in the photosensitive epilepsy of the baboon [69].

Unlike in hypoxia, however, stimulus- and excitatory amino acid -induced, as well as seizure-dependent changes in $[Ca^{2+}]_o$ show a clear laminar dependence. Thus stimulus-induced decreases in $[Ca^{2+}]_o$ ($\triangle Ca$) in the spinal cord are largest in the dorsal horn [38, 74]. Parallel fiber-induced $\triangle Ca$ in cerebellum are largest in stratum moleculare [66, 67], and cortical surface or thalamic stimulation-induced $\triangle Ca$ are largest in layer II and, in the rat also in layer V [30, 68]. In the entorhinal cortex, the largest $\triangle Ca$ are noted in layer IV/V [76], and in the hippocampal formation the greatest $\triangle Ca$ are seen in ascending order, in stratum granulare (SG) of the dentate gyrus [62] and in stratum pyramidale (SP) of areas CA3 and CA1 [2, 49, 59]. Typical hippocampal stimulus – induced $\triangle Ca$ are illustrated in Fig. 1. The plot of the $\triangle Ca$ against recording depth reveals that half-maximal $\triangle Ca$ are noted in stratum radiatum (SR) at a distance of roughly 170 μm from SP (Fig. 2).

Presynaptic and Postsynaptic Uptake of Calcium

The question arose of into which cellular elements Ca^{2+} flows. This question was addressed by experiments in vitro with hippocampal slices. There it is well

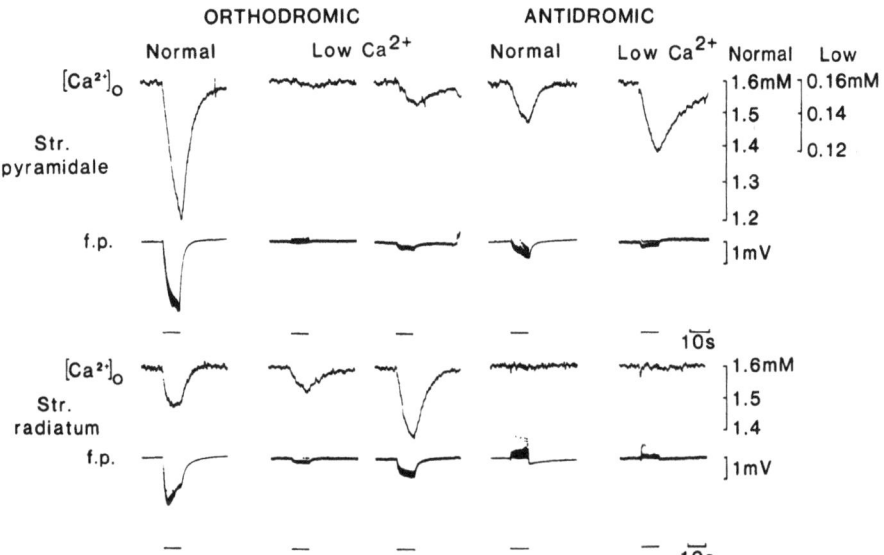

Fig. 1. Stimulus-induced changes in $[Ca^{2+}]_o$ and in extracellular field potentials (f.p.) recorded in area CA1 of in vitro hippocampal slices. Recordings in the *upper two rows* were done in stratum pyramidale, the layer of somata of hippocampal pyramidal cells, recordings in the *lower two rows* in stratum radiatum, the synaptic input layer for Schaffer collaterals. Orthodromic refers to stimulation of the Schaffer collaterals, antidromic to stimulation of the axons of hippocampal pyramidal cells

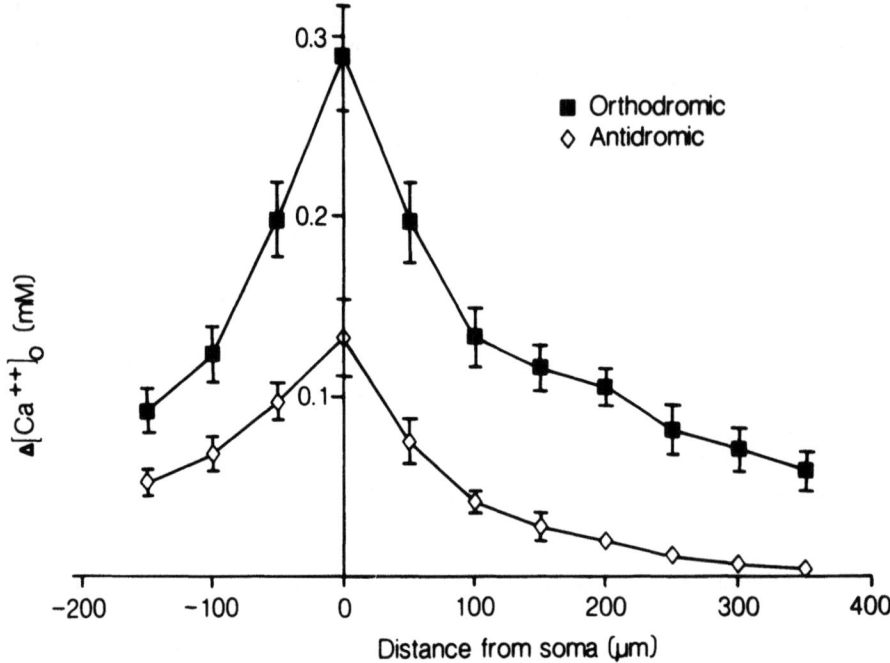

Fig. 2. Laminar profiles of stimulus-induced changes in $[Ca^{2+}]_o$ (stimulus intensity just above threshold for induction of population spikes). Zero refers to recordings in stratum pyramidale, negative distances to recordings in the area of the basal dendritic tree (stratum oriens) and positive distances to soma in the areas of the apical dendritic tree (stratum radiatum)

documented that postsynaptic elements possess Ca^{2+} conductances [5, 78]. Hence, Ca^{2+} uptake from the RS is not restricted to presynaptic elements alone. In order to learn about the relative contributions of presynaptic and postsynaptic Ca^{2+} uptake to decreases in $[Ca^{2+}]_o$, it is necessary to suppress synaptic transmission.

A crude but effective approach is to treat hippocampal slices with kainic acid in toxic concentrations (Walther and Heinemann in prep.). If one applies 500 μM kainate in the presence of 2 mM Ca^{2+}, postsynaptic responses disappear irreversibly while afferent volleys (action potentials in afferent fibers) recover. Cellular responses toward antidromic stimulation do not return within 8 h after application. Nevertheless, repetitive stimulation still causes reductions of $[Ca^{2+}]_o$. These are maximal in SR and amount there to about 60 μM. This compares to $[Ca^{2+}]_o$ decreases of maximally 0.3 mM in intact preparations. Hence about 20% of extracellular Ca^{2+} loss can be ascribed to presynaptic Ca^{2+} entry.

A similar conclusion is reached when synaptic transmission is blocked by lowering $[Ca^{2+}]_o$ [44, 45]. Synaptic responses upon repetitive electrical stimulation are blocked when $[Ca^{2+}]_o$ is lower than 0.22 ± 0.02 mM in the presence of 2 mM Mg^{2+} and 5 mM K^+ [15, 71]. The Ca^{2+} level at which synaptic transmission is blocked is higher in the presence of lower $[K^+]_o$. Hence we worked with a solution containing 0.2 mM Ca^{2+}, 3 mM K^+ and 4 mM Mg^{2+}. Antidromic stimulation under this condition still elicits maximal decreases in $[Ca^{2+}]_o$ in SP, while orthodromically induced Ca^{2+} concentration changes are largest in SR (Figs. 1, 3). These reflect presynaptic Ca^{2+} uptake. Interestingly, small ΔCa are still observed under these conditions in SP. However, when a cut through stratum oriens (SO), SP and half of SR is made between the stimulation and recording electrode, orthodromically evoked ΔCa disappear. [72]. Hence the ΔCa in SP do not reflect postsynaptic Ca^{2+} entry. Under conditions of orthodromic stimulation the use of pharmacological tools now permits augmentation of presynaptic Ca^{2+} entry and thereby eventual recovery of synaptic transmission. One way of doing this is to apply K^+ channel-blocking agents such as a 4-aminopyridine (4-AP) or tetraethylammonium (TEA).

Both, 4-AP and TEA affect Ca^{2+} signals [40]. However, while 4-AP augments predominantly presynaptic Ca^{2+} entry, TEA is more effective on Ca^{2+} decreases evoked by repetitive antidromic stimulation (e.g., Fig. 3). However, such treatment alone recovers synaptic transmission only when the Ca^{2+} level is just below the level of block of synaptic transmission. It was therefore a surprise to us that antagonists of the action of adenosine such as adenosine deaminase or theophylline could restore synaptic transmission with little effect on presynaptic Ca^{2+} entry. When 4-AP was administered in combination with such antagonists in doses which restore synaptic transmission, Ca^{2+} signals in SR were strongly enhanced (often by more than 100%). In SP they reached amplitudes during orthodromic stimulation otherwise only observed during antidromic stimulation [72]. These and other findings suggest that 50%–80% of Ca^{2+} loss in SR can be ascribed to postsynaptic Ca^{2+} entry. The proportion may even be larger in SP.

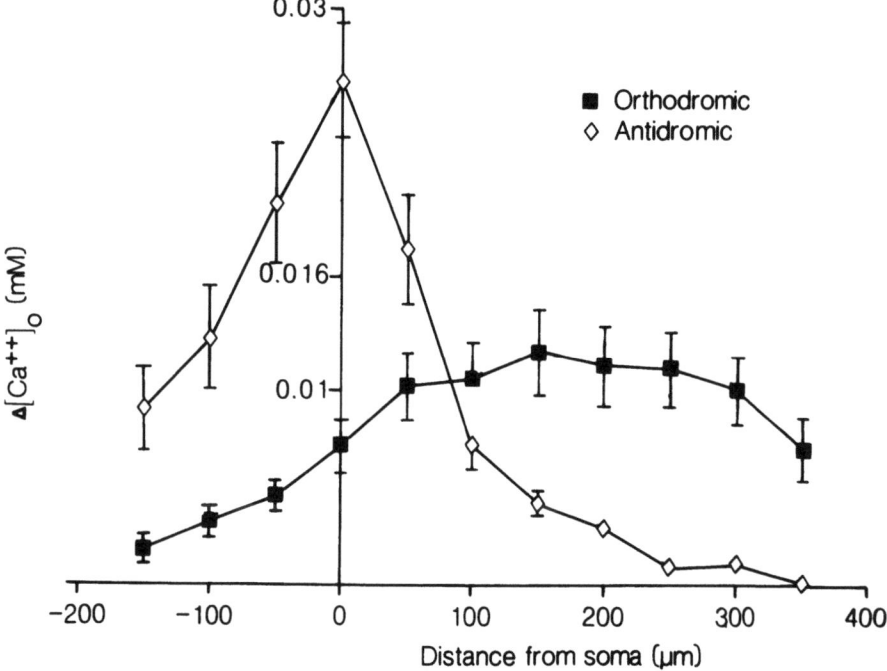

Fig. 3. Laminar profiles of orthodromically and antidromically induced changes in $[Ca^{2+}]_o$ recorded under conditions of blocked synaptic transmission

Role of NMDA Receptors in Extracellular Calcium Loss

The next problem of interest is the pathways by which Ca^{2+} leaves the extracellular space. Since the work of Takeuchi it is clear that nicotinic acetylcholine (ACH) channels possess some degree of Ca^{2+} permeability. In the CNS, nicotinic ACH channels are rare and ACH actions are mostly through K^+ channels. Consequently, ACH application has no direct effect on $[Ca^{2+}]_o$, in contrast to the excitatory amino acids glutamate (glu), aspartate (asp) and DL-homocysteinic acid (DLH) who all readily induce decreases in $[Ca^{2+}]_o$ [34, 68, 79]. While DLH appears to prefer NMDA receptors, glu and asp activate quisqualate (quis) and kainate (kain) receptors as well. Applications of these glutamate receptor-classifying agonists to hippocampal neurons, while measuring E_M and extracellular ionic changes, revealed that quis, kain, and NMDA all induce dose-dependent decreases in $[Na^+]_o$ and rises in $[K^+]_o$, in line with suggestions that the three agonists activate cation-permeable channels [51, 70]. However, the actions of the three agonists differ with respect to $[Ca^{2+}]_o$ changes. While NMDA at any dose induces decreases in $[Ca^{2+}]_o$ which are sufficient to depolarize hippocampal pyramidal cells and to cause reductions in $[Na^+]_o$, quisqualate causes rises in $[Ca^{2+}]_o$ at low doses but decreases in $[Ca^{2+}]_o$ at doses which depolarize hippocampal neurons by more than 30–40 mV (Fig. 4). While NMDA-

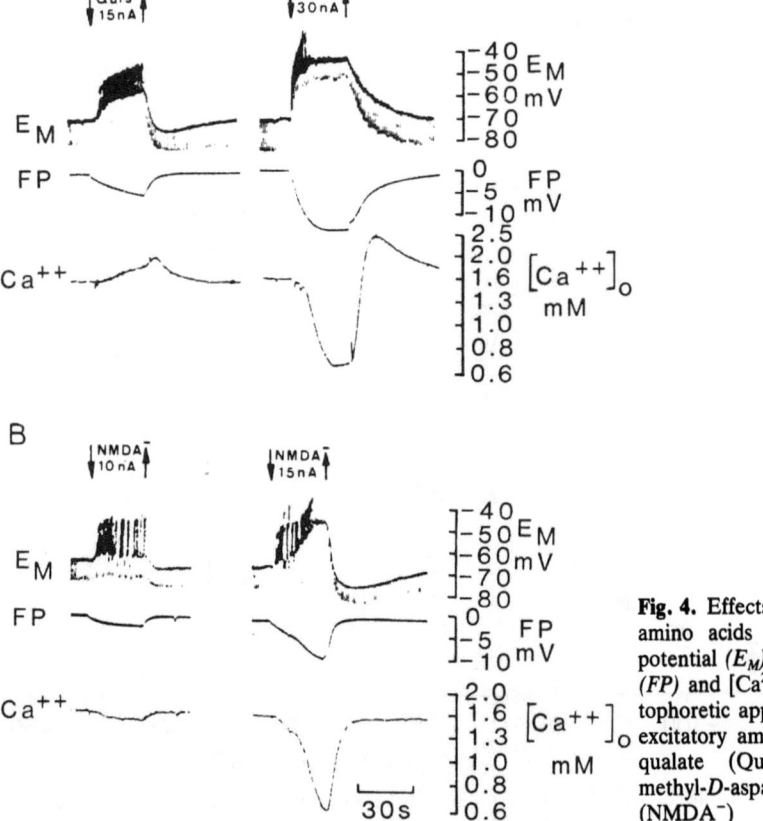

Fig. 4. Effects of excitatory amino acids on membrane potential (E_M) field potential (FP) and $[Ca^{2+}]_o$ during iontophoretic application of the excitatory amino acids quisqualate (Quis⁻) and N-methyl-D-aspartate (NMDA⁻)

induced ΔCa are little affected by low doses of Cd^{2+}, quis- and kain-induced ΔCa are strongly depressed. Similarly, verapamil affects quis- and kain-induced ΔCa much more than NMDA-induced ΔCa. On the other hand, NMDA receptor antagonists such as 2-aminophosphonovalerate (2-APV) and ketamine have no effect on quis- and kain-induced ΔCa, but strongly depress NMDA-induced ΔCa.

Interestingly, both quis- and kain-induced ΔCa have a different time course of recovery than NMDA-induced ΔCa. While Ca^{2+} returns rather rapidly to baseline after termination of NMDA application, quis and kain produce rather large overshoots. These overshoots become smaller when Na-K ATPase activity is blocked due to lowering of $[K^+]_o$ or due to application of ouabain and also when $[Na^+]_o$ is lowered. It is suggested that a sodium calcium exchange process is involved in the generation of rises in $[Ca^{2+}]_o$ and in the generation of post application overshoots.

A larger Ca^{2+} permeability for NMDA-operated channels is also suggested by findings in which extracellular Na^+ was replaced by Tris or choline. Moderate lowering of $[Na^+]_o$ leads to a reduction of quis- and kain-induced ΔCa, while NMDA-induced ΔCa are augmented. However, replacement of more than 50% Na by Tris also causes an enhancement of quis- and kain-induced ΔCa compared to signals recorded in moderately lowered $[Na^+]_o$, indicating that the other glutamate-operated channels also possess some degree of Ca^{2+} permeability.

These findings provide us with tools to analyze stimulus-induced changes in $[Ca^{2+}]_o$. Under normal conditions decreases in $[Ca^{2+}]_o$ are little affected by NMDA-receptor antagonists in SP of hippocampal cells, although a somewhat larger effect is seen in SR, the synaptic input layer [43, 64]. There, ΔCa can be reduced by 10%–40%. However, when a slice is convulsant or when it has been repetitively stimulated with a subsequent augmentation of Ca^{2+} signals in SR as well as in SP, Ca signals in SR can be blocked by up to 80% while those in SP are still little affected. This leads to the conclusion that most of the $[Ca^{2+}]_o$ decrease can be ascribed to Ca^{2+} movements through voltage-activated channels.

Voltage-Activated Calcium Channels and Loss in Extracellular Calcium

Voltage-activated Ca^{2+} channels can be divided into two or three classes [7, 8, 9, 10, 13, 21, 46, 53, 54]. The classical high-threshold voltage-activated (L) channels appear to be sensitive to organic Ca^{2+}-entry blockers, although verapamil in high concentrations appears also to have effects on hippocampal transient Ca^{2+} currents [4, 20, 21, 22, 78]. Low-threshold voltage-activated transient (T) currents appear to be activatable only when the cell membrane potential is below -60 mV and these Ca^{2+} currents appear to be particularly sensitive to Ni^{2+} in low concentrations [22]. The still disputed third type of Ca^{2+} channels (N channels) seems to be sensitive to ω-conotoxin [21]. All Ca^{2+} channels appear to be blocked by the classical Ca^{2+} blockers.

Organic Ca^{2+}-entry blockers have a much stronger effect on stimulus-induced $[Ca^{2+}]_o$ changes in normal medium. They reduce ΔCa by up to 60% [55]. This reduction of Ca^{2+} signals appears, from studies in low-Ca^{2+} medium, to be restricted to postsynaptic Ca^{2+} entry. We investigated three different types of Ca^{2+} antagonists for their ability to reduce ΔCa. Neither fendilline nor nifedipine nor verapamil had any effect on Ca^{2+} signals attributed to presynaptic Ca^{2+} entry, while those evoked by

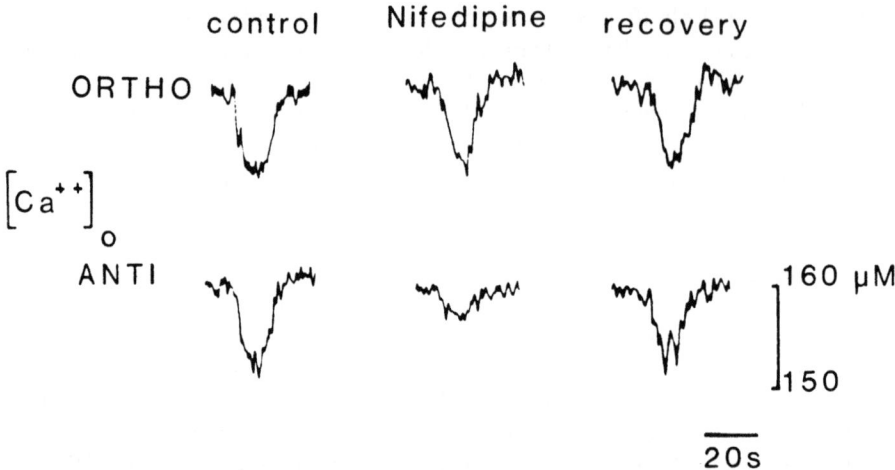

Fig. 5. Effects of nifedipine (10 μM for 10 min) on changes in $[Ca^{2+}]_o$ induced under conditions of blocked synaptic transmission. *Ortho* refers to stimulation of afferent fibers and reflects presynaptic Ca^{2+} entry, and *ANTI* refers to stimulation of fibers from pyramidal cells and reflects predominantly postsynaptic Ca^{2+} entry

antidromic stimulation were depressed by up to 50% (Fig. 5) [41]. Organic Ca^{2+} agonists had no clear effects under our experimental conditions. Inorganic Ca^{2+}-entry blockers such as Mn^{2+} (4mM), Ni^{2+} (2–4 mM, Fig. 6), Co^{2+} (2mM), and Cd^{2+} (0.5 mM) all blocked $[Ca^{2+}]_o$ decreases completely. This concerns both orthodromically and antidromically induced changes in $[Ca^{2+}]_o$. Of interest were studies on the effect of low concentrations of Ni^{2+} on antidromically as well as orthodromically induced Ca^{2+} signals, because of the potency of Ni^{2+} against low-threshold voltage-activated Ca^{2+} currents. Preliminary data suggest that presynaptic Ca^{2+} uptake is little affected by Ni^{2+} at concentrations below 0.1 mM, while antidromically induced Ca^{2+} signals are reduced by 30%. Not yet tested is the effect of ω-conotoxin, but considerable evidence now suggests that this drug reduces transmitter release in various structures, suggesting that the presynaptic Ca^{2+} entry would be mediated by N channels [17, 35]. These data also suggest that antidromically induced postsynaptic Ca^{2+} entry involves all three classes of Ca^{2+} channels and that the experimental conditions determine which Ca^{2+} channel is mostly involved in Ca^{2+} uptake into cells. However, until we do not know whether all high-threshold Ca^{2+} channels possess binding sites for the organic Ca^{2+}-entry blockers these data will remain difficult to interprete. In view of the high concentrations often used in investigations of organic Ca^{2+}-entry blockers, nonspecific effects of local anesthesia cannot be excluded. Interpretation in intact systems is further hampered by the fact that blockade of Ca^{2+} entry also reduces Ca^{2+}-dependent K^+ currents. This sometimes leaves a window for activation of Ca^{2+}-mediated depolarizing responses. In this particular experiment verapamil was bath-applied and intracellular responses were monitored throughout the application. It is noted that a Ca^{2+}-dependent after-hyperpolarization decreases and finally disappears.

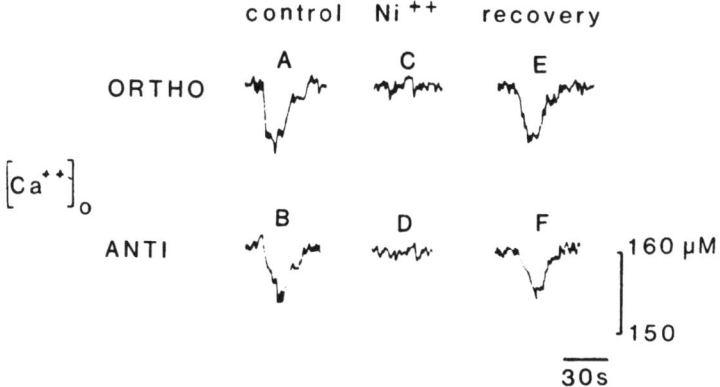

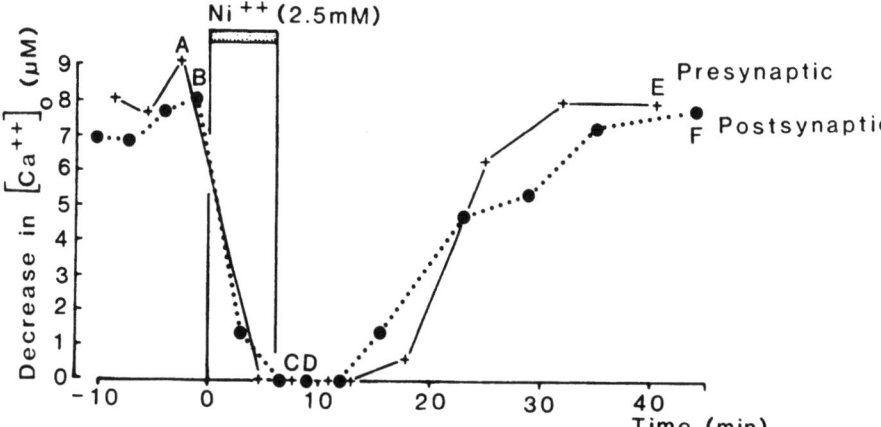

Fig. 6. Effects of Ni^{2+} on pre- and postsynaptic Ca^{2+} entry. *Above:* original tracings; *below:* protocol of changes in the responses during wash-in and wash-out of Ni^{2+}. The Ni^{2+} concentration at the time of blocking of synaptic transmission was less than 0.5 mM

Mechanisms Which Control Cellular Calcium Uptake

In view of the relatively modest actions of Ca^{2+}-entry blockers the question arises which mechanisms normally control Ca^{2+} entry. Two principal protective actions have been defined. One is related to the effects of repolarizing K^+ currents. It has been shown that any drug which blocks K^+ currents enhances Ca^{2+} entry into cells [32, 56]. This applies to drugs such as oenanthotoxin and to K^+ channel-blocking agents such as 4-AP and TEA [31, 40]. Conversely, it can be expected that drugs which enhance K^+ currents also reduce transmembrane Ca^{2+} fluxes.

It is of interest that a large body of transmitters and neuromodulators also act on K^+ currents. This applies often to β-agonists of norepinephrine (NE) and to muscarinic ACH agonists. These all amplify Ca^{2+} currents, while drugs which augment K^+ currents, such as baclofen, reduce stimulus-induced Ca^{2+} signals [29]. On the other

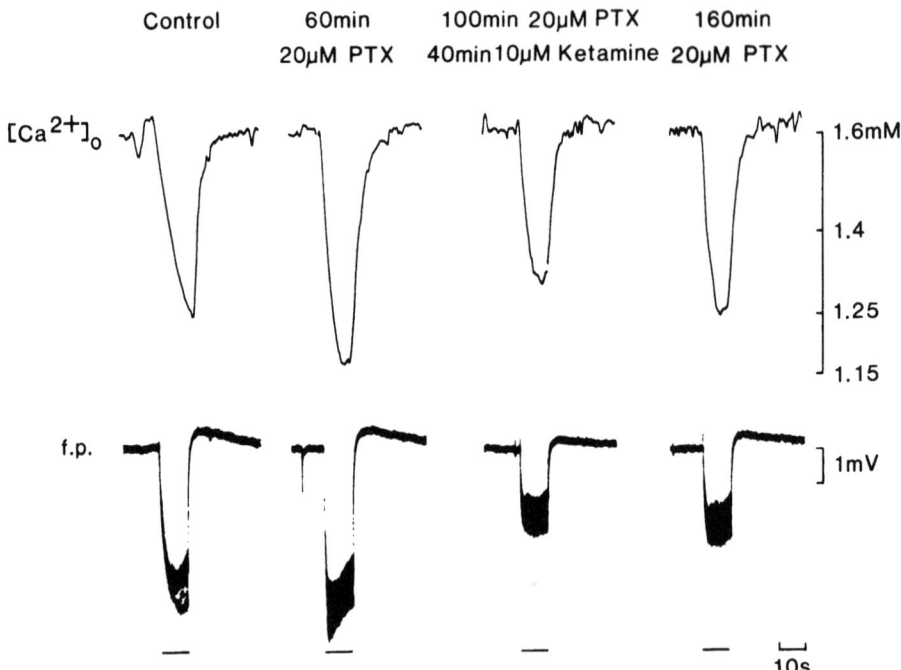

Fig. 7. Effects of picrotoxin *(PTX)*, an antagonist of the action of GABA, and of ketamine, and NMDA antagonist, on stimulus-induced changes in $[Ca^{2+}]_o$ in normal medium

hand a number of neurotransmitters are able to reduce Ca^{2+} currents. This applies to GABA via GABA β-receptors and to NE via α-receptors [18]. Interestingly, NE and its α-agonists have no effects on presynaptic Ca^{2+} entry in area CA1 and the dentate gyrus, while GABA and baclofen depress these Ca^{2+} fluxes. Postsynaptically, all three classes of drugs depress Ca^{2+} currents and Ca^{2+} fluxes. Consequently, there is a considerable enhancement of Ca^{2+} fluxes when slices are treated with antagonists of the action of GABA such as pentylenetetrazole, picrotoxin, and bicuculline (Fig. 7) [25]. Any of these conditions, however, involve also indirect activation of NMDA receptors, since such enhancement is usually antagonized by 2-APV and ketamine.

Changes in Extracellular Calcium Concentration in Chronically Altered Nervous Tissue

From the findings above it is expected that long-lasting alterations in nervous tissue lead to enhanced Ca^{2+} fluxes. This applies to all chronic epilepsies. Indeed, enhanced Ca^{2+} signals were seen in the inborn epilepsy of the photosensitive baboon, in chronic cobalt epilepsy in alumina cream focal epilepsy, and, finally, in kindling epilepsy (a chronic epilepsy induced by daily repeated 1s/50 Hz stimulations). In the latter case the enhanced Ca^{2+} fluxes result from an augmented utilitization of NMDA receptors in synaptic transmission, from a reduction in K^+ currents, and from a loss of noradrenergic receptors [63, 65]. Depending on structure, there may also be an augmentation or a depression of GABAergic inhibitory processes. It is expected that similar alterations occur in any chronic degenerative disease. This would increase vulnerability considerably. Since organic Ca^{2+}-entry blockers have a beneficial effect on acute epilepsies, it may one day be possible to use Ca^{2+}-entry blockers in the therapy of epilepsies [75].

Summary and Conclusions

We have shown in this minireview that neuronal hyperactivity is associated with considerable changes in $[Ca^{2+}]_o$. These changes stem mostly from Ca^{2+} fluxes through voltage-operated channels. The largest component of extracellular Ca^{2+} loss appears to stem from Ca^{2+} fluxes through high-threshold voltage-activated channels. Presynaptic Ca^{2+} uptake and Ca^{2+} fluxes through NMDA-operated channels appear to contribute less to the decreases in $[Ca^{2+}]_o$. This may, however, vary depending on the functional state in the tissue. Thus NMDA-operated Ca^{2+} fluxes appear to contribute a larger component of the change in $[Ca^{2+}]_o$ after induction of LTP and in kindling epilepsy.

Acknowledgements. The expert technical assistance of H. Meyes and M. Groenenwald in the experiments and the secretarial assistance of G. Heske are gratefully acknowledged.

References

1. Baimbridge KG, Miller JJ (1980) Calcium uptake and retention during long term potentiation of neuronal activity in the rat hippocampal slice preparation. Brain Res 221:299–305
2. Benninger C, Kadis J and Prince DA (1980) Extracellular calcium and potassium changes in hippocampal slices. Brain Res 187:165–182
3. Bliss TUP, Gardner-Medwin AR (1973) Longlasting potentiation of synaptic transmission of in the dentate area of unanaesthezied rabbit following stimulation of the perforant path. J Physiol Lond 232:357–379
4. Boll W, Lux HD (1985) Action of organic antagonists on neuronal calcium currents. Neurosci Lett 56:335–339
5. Brown DA, Griffith WH (1983) Calcium-activated outward current in voltage-clamped hippocampal neurones of the guinea pig. J Physiol (Lond) 337:287–301
6. Brown DA, Griffith WH (1983) Persistent slow inward calcium current in voltage-clamped hippocampal neurones of the guinea pig. J Physiol (Lond) 337:303–320
7. Carbone E and Lux HD (1984) A low voltage activated, fully inactivating Ca channel in vertebrate sensory neurones. Nature 310:501–502
8. Carbone E and Lux HD (1984) A low voltage-activated calcium conductance in embryonic chick sensory neurons. Biophys J 46:413–418
9. Carbone E and Lux HD (1987) Kinetics and selectivity of a low-voltage-activated calcium current in chick and rat sensory neurones. J Physiol (Lond) 386:547–570
10. Carbone E and Lux HD (1987) Single low-voltage-activated calcium channels in chick and rat sensory neurones. J Physiol 386:571–601
11. Choi DW (1985) Glutamate neurotoxicity in cortical cell culture is calcium dependent. Neurosci Lett 58:293–297

12. Constanti AL, Galvan M, Franz P and Sim JA (1985) Calcium-dependent inward currents in voltage-clamped guinea pig olfactory cortex neurones. Pflügers Arch 404:259–265
13. Deitmar JW (1984) Evidence for two voltage-dependent calcium currents in the membrane of the ciliate *Stylonychia*. J Physiol (Lond) 355:137–155
14. Dingledine R (1983) N-Methyl-aspertate activates voltage-dependent calcium conductance in rat hippocampal pyramidal cells. J Physiol London 343:385–405
15. Dingledine R and Somjen G (1981) Calcium dependance of synaptic transmission in the hippocampal slice. Brain Res 207:218–222
16. Dingledine R (1986) Involvements of N-methyl-D-aspartate receptors in epileptiform bursting in rat hippocampal slice. Trends Neurosci 9:47–49
17. Dooley DJ, Lupp A, Hertting G (1987) Inhibition of central transmitter release by ω-conotoxin GVIA, a peptide modulator of the N-type voltage sensitive calcium channel. Naunyn-Schmiedeberg's Arch Pharmacol 336:467–470
18. Dunlap K and Fischbach G (1981) Neurotransmitters decrease the calcium conductance activated by depolarization of chick sensory neurones. J Physiol (Lond) 317:519–535
19. Farber JL (1981) The role of calcium in cell death. Life Science 29:1289–1295
20. Fleckenstein A (1977) Specific pharmacology of calcium in myocardium, cardiac pacemakers and vascular smooth muscle. Ann Rev Pharmacol 17:149–166
21. Fox AP. Nowycky MC and Tsien RW (1987). Kinetic and pharmacological properties distinguishing three types of calcium currents in chick sensory neurones. J Physiol (Lond) 394:149–172
22. Fox AP. Nowycky MC and Tsien RW (1987) Single-channel recordings of three types of calcium currents. J Physiol (Lond) 394:173–200
23. Galvan M, Constanti A, Franz P, Sedlmeir C, Sim JA and Endres WJ (1986) Calcium spikes and inward currents in mammalian peripheral and central neurons. In: Calcium electrogenesis and neuronal functioning (eds) Heinemann U, Klee M, Neher E and Singer W. Springer Verlag, Berlin Heidelberg Expl Brain Res Series 14, pp 61–70
24. Goddard GU, McIntyre DC, Leech CU (1969) A permanent change in brain function resulting from daily electrical stimulation. Exp Neurol 25:295–330
25. Hamon B, Heinemann U (1986) Effects of GABA and bicuculline on N-methyl-D-aspartate and quisqualate-induced reductions in extracellular free calcium in area CA1 of the hippocampal slice. Exp Brain Res 64:27–36
26. Hansen AJ (1985) Effect of anoxia on ion distribution in the brain. Physiological Rev 65:101–148
27. Harris RJ and Symon LJ (1984) Extracellular pH, potassium and calcium activities in progressive ischemia of rat cortex. J Cereb Blood Flow Metab 4:178–186
28. Heinemann U and Hamon B (1986) Calcium and epileptogenesis. Exp Brain Res 65:1–10
29. Heinemann U, Hamon B and Konnerth A (1984) GABA and baclofen reduce changes in extracellular free calcium in area CA1 of rat hippocampal slices. Neurosci lett 47:295–300
30. Heinemann U, Konnerth A, Lux HD (1981) Stimulation induced changes in extracellular free calcium in normal cortex and chronic aluminia cream foci of cats. Brain Res 213:246–250
31. Heinemann U, Konnerth A, Pumain R and Wadman W (1986) Extracellular calcium and potassium concentration changes in chronic epileptic tissue. Advances of Neurology 44:641–661
32. Heinemann U and Louvel J (1983) Changes in $[Ca^{2+}]_o$ and $[K^+]_o$ during repetitive electrical stimulation and during pentetrazol induced seizure activity in the sensorimotor cortex of cats. Pflueg Arch 398:310–317
33. Heinemann U, Lux HD and Gutnick MJ (1988) Extracellular free calcium and potassium during paroxysmal activity in the cerebral cortex of the cat. Exp Brain Res 27:237–243
34. Heinemann U and Pumain R (1980) Extracellular calcium activity in cat sensorimotor cortex induced by iontophoretic application of amino acids. Exp Brain Res 40:247–250
35. Hirning LD, Fox AP, McCleskey EW, Olivera BM, Thayer SA, Miller RJ and Tsien RW (1988) Dominant role of N-type Ca^{2+} channels in evoked release of norepinephrine from sympathetic neurons. Science 239:57–61
36. Hofmeier G, Lux HD (1981) The time courses of intracellular free calcium and related electrical effects after injection of $CaCl_2$ into neurons of the snail, *Helix pomatia*. Pflueg Arch 391:242–251
37. Isenberg G (1986) Ca-dependent and Ca-independent F-actin capping proteins determine microfilament assembly in neuronal cells. In: Calcium electrogenesis and neuronal functioning, edited by Heinemann U, Klee M, Neher E and Singer W. Berlin Heidelberg New York London Paris. Springer, pp 271–278 37

38. Janus J, Speckmann EJ and Lehmenkühler A (1981) Relations between extracellular K^+ and Ca^{2+} activities and local field potentials in the spinal cord of the rat during focal and generalized seizure discharges. In: Ion-selective microelectrodes and their use in excitable tissues (ed) Sykova E, Hnik P and Viklicky L, New York London: Plenum Press, p 181-185
39. Jones RSG, Heinemann U (1987) Abolition of the orthodromically evoked IPSP of CA1 pyramidal cells before the EPSP during washout of calcium from the hippocampal slices. Exp Brain Res 65:676-680
40. Jones RSG, Heinemann U (1987) Pre- and postsynaptic K^+- and Ca^{2+}-fluxes in area CA1 of the rat hippocampus in vitro: effects of Ni^{2+}, TEA and 4-AP. Exp Brain Res 68:205-209
41. Jones RSG, Heinemann U (1987) Differential effects of calcium entry blockers on stimulus evoked pre- and postsynaptic influx of calcium in the rat hippocampus in vitro. Brain Res 416:257-266
42. Jones RSG, Heinemann U (1988) Synaptic and intrinsic responses of medial entorhinal cortical cells in normal and magnesium-free medium in vitro. J Neurophysiol in press
43. Köhr G, Heinemann U (1987) Anticonvulsive properties of ketamin and 2-aminophosphono valerate in the low magnesium epilepsy in rat hippocampal slices. Neurosci Res Com 1:17-21
44. Konnerth A, Heinemann U (1983) Effects of GABA on presumed presynaptic Ca^{2+} entry in "in vitro" hippocampal slices. Brain Res 270:185-189
45. Konnerth A, Heinemann U (1983) Presynaptic involvement in frequency facilitation in the hippocampal slice. Neurosci Lett 42:255-260
46. Kostyuk PG, Shuba MF and Savchenko AN (1987) Three types of calcium channels in the membrane of mouse sensory neurons. Biol Membr 4:366-373
47. Kraig RP, Nicholson C (1978) Extracellular ionic variations during spreading depressions. Neurosci 3:1045-1059
48. Krnjevic K and Lisiewicz A (1972) Injections of calcium ions into spinal motoneurones. J Physiol London 255:363-390
49. Krnjevic K, Morris ME and Reiffenstein RF (1982) Stimulation-evoked changes in extracellular K^+ and Ca^{2+} in pyramidal layers of the rat's hippocampus. Can J Physiol Pharmacol 60:1643-1657
50. Krnjevic K, Morris ME, Reiffenstein RF and Ropert N (1982) Depth distribution and mechanism of changes in extracellular K^+ and Ca^{2+} concentrations in the hippocampus. Can J Physiol pharmacol 60:1958-1971
51. Lambert JDC and Heinemann U (1986) Extracellular calcium changes accompanying the action of excitatory amino acids in area CA1 of the hippocampus. Possible implications for the initiation and spread of epileptic discharges. In: Epilepsy and calcium, edited by EJ Speckmann et al., Urban and Schwarzenberg München pp 35-62
52. Llinas R and Sugimori M (1980) Electrophysiological properties of in vitro purkinje cell dendrites in mammalian cerebellar slices. J Physiol London 305:197-213
53. Llinas R and Jahnsen H (1982) Electrophysiology of mammalian thalamic neurones in vitro. Nature 297:406-408
54. Llinas R and Yaarom Y (1981) Electrophysiology of mammalian inferior olivary neurones "in vitro". Different types of voltage-dependent ionic conductances. J Physiol London 315:549-567
55. Louvel J, Abbes S, Godfrain JM (1986) Effect of organic calicum channel blockers on neuronal calcium-dependent processes. In: Calcium electrogenesis and neuronal functioning. Ed by Heinemann U, Klee M, Neher E and Singer W. Heidelberg Tokyo New York, Springer pp 375-385
56. Louvel J, Heinemann U (1983) Changes in $[Ca^{2+}]$, $[K^+]$ and neuronal activity during oenanthotoxin induced epilepsy. EEG clin Neurophysiol 56:457-466
57. Lynch G, Larson J, Kelso S, Barrionuevo G, Schottler F (1983) Intracellular injections of EGTA block induction of hippocampal long term potentiation. Nature 305:719-721
58. MacDermott AB, Mayer ML, Westbrook GL, Smith S and Barker JL (1986) NMDA-receptor activation increases cytoplasmic calcium concentration in cultured spinal cord neurones. Nature 321:519-522
59. Marciani MG, Louvel J, Heinemann U (1982) Aspartate induced changes in extracellular free calcium in "in vitro" hippocampal slices of rats. Brain Res 238:272-277
60. Mayer ML (1985) A calcium-activated chloride current generates the after-depolarization of rat sensory neurones in culture. J Physiol Lond 364:217-239

61. Melchers BPC, Pennartz CMA and Lopes da Silva FH (1987) Differential effects of elevated extracellular calcium concentrations on field potentials in dentate gyrus and CA1 of the rat hippocampal slice preparation. Neurosci Lett 77:37–42
62. Mody I and Heinemann U (1986) Laminar profile of the changes in extracellular calcium concentration induced by repetitive stimulation and excitatory amino acids in the rat dentate gyrus. Neurosci Lett 69:137–142
63. Mody I and Heinemann U (1987) NMDA receptors of dentate gyrus granule cells participate in synaptic transmission following kindling. Nature 326:701–704
64. Mody I, Lambert JDC and Heinemann U (1987) Low extracellular magnesium induces epileptoform activity and spreading depression in rat hippocampal slices. J Neurophysiol 57:869–888
65. Mody I, Stanton P, Heinemann U (1988) Activation of N-methyl-D-aspartate receptors parallels changes in cellular and synaptic properties of dentate gyrus granule cells after kindling. J Neurophys 59:1033–1054
66. Nicholson C, ten Bruggencate G, Steinberg R and Stöckle H (1977) Calcium modulation in brain extracellular microenvironment demonstrated with ion-selective micropipette. Proc Nat Acad Sci 74:1287–1290
67. Nicholson C, ten Bruggencate G, Stöckle H and Steinberg R (1978) Calcium and potassium changes in extracellular microenvironment of cat cerebellar cortex. J Neurophysiol 41:1026–1039
68. Pumain R and Heinemann U (1985) Stimulus and amino-acid induced calcium and potassium changes in the rat neocortex. J Neurophys 53:1–16
69. Pumain R, Menini C, Heinemann U, Louvel J, Silva-Barrat C (1985) Chemical synaptic transmission is not necessary for epileptic seizures to persist in the baboon *Papio papio*. Exp Neurol 89:250–258
70. Pumain R, Kurwecis R, Louvel J (1987) Ionic changes induces by excitatory amino acids in the rat cerebral cortex. Can J Physiol Pharmacol 65:1077–1097
71. Rausche G, Sarvey JM, Heinemann U (1988) Slow synaptic inhibition in relation to frequency habituation in dentate granule cells of rat hippocampal slices (in prep)
72. Schubert P. Heinemann U (1988) Adenosine antagonists combined with 4-aminopheridine causes particular recovery in low Ca medium. Exp Brain Res (in prep)
73. Siesjö BK (1981) Cell damage in the brain: a speculative synthesis. J Cerebral Blood Flow and Metabolism 1:155–185
74. Somjen GG (1980) Stimulus evoked and seizure related responses of extracellular calcium activity in spinal cord compared to those in cerebral cortex. J Neurophysiol 44:617–632
75. Speckmann EJ, Witte OW, Walden J (1986) Involvement of calcium ions in focal epileptic activity of the neocortex. Exp Brain Res 27:385–396
76. Stanton PK, Jones RSG, Mody I, Heinemann U (1987) Epileptiform activity induced by lowering extracellular $[Mg^{2+}]$ in combined hippocampal-entorhinal cortex slices: modulation by receptors for norepinephrine and N-methyl-D-aspartate. Epilepsy Res 1:53–62
77. Watkins JC, Evans RH (1981) Excitatory amino acid transmitters. Ann Rev Pharmacol Toxicol 21:165–204
78. Yaari Y, Hamon B and Lux HD (1987) Development of two type of calcium channels in cultured mammalian hippocampal neurons. Science 235:680–682
79. Zanotto L, Heinemann U (1983) Aspartate and glutamate induced reductions in extracellular free calcium and sodium concentration in area CA1 of "in vitro" hippocampal slices of rats. Neurosci Lett 35:79–84

Alcohol, Neurodegenerative Disorders, and Calcium Channel Antagonist Receptors

D. A. Greenberg, R. O. Messing, S. S. Marks,
C. L. Carpenter, and D. L. Watson

Department of Neurology, University of California, and Ernest Gallo Clinic and Research Center, San Francisco General Hospital, San Francisco, CA 94110, USA

Our current understanding of the physiology and pharmacology of voltage-dependent Ca^{2+} channels owes considerable debt to the use of Ca^{2+}-channel antagonist drugs as probes of channel structure and function. The availability of these probes and the pivotal role of Ca^{2+} channels in normal cellular activity have led workers in diverse fields, including neurology, psychiatry, cardiology, and endocrinology, to search for changes in Ca^{2+}-channel function or number that may be important in the pathophysiology of clinical diseases. This search has gained impetus from studies of genetically determined diseases in animals, such as congenital cardiomyopathy in hamsters (Wagner et al. 1986) and muscular dysgenesis in mice (Pincon-Raymond et al. 1985), in which changes in the expression of Ca^{2+}-channel antagonist receptors have been detected.

Among the most compelling problems in clinical neurology are two groups of disorders that are widely encountered, but for which there is no satisfactory therapeutic approach – alcohol-related disturbances such as physical dependence and withdrawal seizures, and hereditary nervous-system degenerations, including Huntington's disease and Alzheimer's disease. We have used a cultured neural cell line and postmortem human brain specimens to investigate the possible role of voltage-dependent Ca^{2+} channels in the pathogenesis or clinical expression of these conditions.

Alcohol and Calcium Channels

Several investigators have shown that ethanol acutely inhibits voltage-dependent neuronal Ca^{2+} currents and that chronic administration of ethanol to experimental animals can attenuate this acute inhibitory effect (Harris and Hood 1980, Leslie et al. 1983). We used the PC12 cell line to examine the acute and chronic modulation of Ca^{2+} channels by ethanol. In these cells, voltage-dependent Ca^{2+} influx corresponds to the uptake of $^{45}Ca^{2+}$ induced by depolarization with elevated extracellular K^+ (Stallcup 1979), and the channels through which $^{45}Ca^{2+}$ enters the cells are coupled to receptors for dihydropyridine Ca^{2+} channel antagonists (Toll 1982), indicating that they are L-type Ca^{2+} channels.

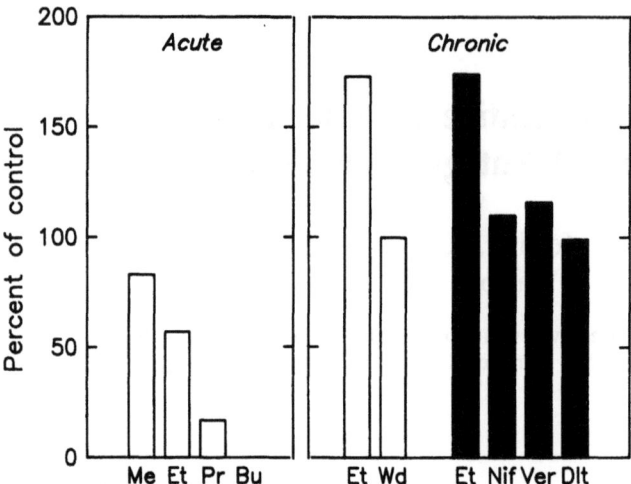

Fig. 1. Effects of acute and chronic ethanol exposure on Ca^{2+} channels in cultured PC12 cells. *Left panel:* Effect of acute (25 min) exposure to 200 mM methanol (*Me*), ethanol (*Et*), propanol (*Pr*), and butanol (*Bu*) on $^{45}Ca^{2+}$ uptake. *Right panel, open bars:* Effect of chronic (6 days) exposure to 200 mM ethanol and of ethanol withdrawal following chronic exposure (*Wd*) on $^{45}Ca^{2+}$ uptake. *Right panel, filled bars:* Effect of chronic (6 days) exposure to 200 mM ethanol, 4 nM nifedipine (*Nif*) 900 nM verapamil (*Ver*), and 1.9 μM diltiazem (*Dlt*) on density of Ca^{2+}-channel antagonist receptors measured with [^{3}H]nitrendipine or (+)-[^{3}H]PN200–110

Acute In Vitro Effects

In PC12 cultures, acute exposure to ethanol for 25 min reduced depolarization-evoked $^{45}Ca^{2+}$ uptake in a concentration-dependent manner (Messing et al. 1986). Statistically significant inhibition occurred with 50 mM ethanol, and inhibition was half-maximal at approximately 200 mM (Fig. 1). A series of straight-chain alcohols reproduced the effect of ethanol with the potency order butanol > propanol > ethanol > methanol, which is consistent with relative potencies for alcohol interactions with membrane proteins and lipids. Taken together with the failure of 200 mM sucrose to alter $^{45}Ca^{2+}$ uptake, these findings indicate that the acute effect of ethanol on Ca^{2+} flux is not due to increased osmolality of the extracellular medium. Ethanol did not alter resting membrane potential measured by uptake of [^{3}H]tetraphenylphosphonium, suggesting a direct interaction with Ca^{2+} channels or their lipid microenvironment.

Chronic in Vitro Effects

To determine whether acute inhibition of Ca^{2+} flux by ethanol is associated with a long-term – and possibly adaptive – cellular response, cells were cultured for up to 10 days in the presence of 200 mM ethanol.

Figure 1 illustrates that in cells exposed to ethanol for 6 days, $^{45}Ca^{2+}$ uptake, measured after rinsing cultures to remove ethanol, increased by about 75%; no further increase in uptake was observed at 8 or 10 days. Shorter exposures or lower doses of ethanol produced correspondingly smaller effects. One interpretation of this finding is that acute inhibition of Ca^{2+} flux by ethanol evokes a compensatory increase in Ca^{2+}-channel activity that offsets the acute effect and allows Ca^{2+}- and depolarization-dependent cellular responses to occur despite continuing acute inhibition of Ca^{2+} flux by ethanol.

The occurrence of neurologic abnormalities such as encephalopathy and seizures upon cessation of alcohol intake is viewed as evidence for physical dependence on the drug. At the cellular level, physical dependence is thought to involve the unmasking of physiologic alterations that restore normal function in the presence of ethanol, but which are dysregulatory when ethanol is removed. Alcohol withdrawal symptoms are generally self-limiting, indicating that this disordered physiologic state is transient. To examine whether the ethanol-induced increase in Ca^{2+} flux is similarly reversible upon withdrawal of ethanol, some cells were grown for 6 days in the presence of 200 mM ethanol and then for additional periods without ethanol. Figure 1 shows that under these conditions, $^{45}Ca^{2+}$ uptake returned to pre-exposure levels within about 16 h.

If long-term exposure to ethanol enhances Ca^{2+} currents by increasing Ca^{2+} flux through L-type channels, then dihydropyridine Ca^{2+}-channel antagonists might be capable of attenuating alcohol withdrawal symptoms. Alternatively, ethanol could induce the expression of another Ca^{2+}-channel subtype that is dihydropyridine-insensitive. In support of the former possibility, we found that in cells treated with 200 mM ethanol for 6 days, as in untreated cells, $^{45}Ca^{2+}$ uptake could be completely abolished by all major classes of Ca^{2+}-channel antagonists, and that the potencies of these drugs were at least as great in ethanol-treated as in control cells (Greenberg et al. 1987).

We next examined whether chronic exposure to ethanol enhanced Ca^{2+} flux by increasing the number of Ca^{2+} channels, as opposed to increasing unitary channel conductance or the probability of channel opening. Under the former circumstance, we expected that the number of binding sites for Ca^{2+}-channel radioligands such as [^{3}H]nitrendipine or (+)-[^{3}H]PN200–110 might be increased after ethanol treatment. Therefore, we measured the binding of [^{3}H]nitrendipine to PC12 membranes (Messing et al. 1986) and of (+)-[^{3}H]PN-200–110 to intact PC12 cells (Marks et al., submitted for publication) following growth for 6 days with and without 200 mM ethanol. As shown in Fig. 1, the density of Ca^{2+}-channel antagonist receptors was increased by about 75%, whereas binding affinity at these sites was unchanged. Thus, the chronic effect of ethanol on voltage-dependent Ca^{2+} flux may result from an ethanol-induced increase in the number of functional Ca^{2+} channels expressed by neural cells. Other investigators have reported similar findings in both PC12 cells and rat brain preparations (Dolin et al. 1987, Skattebøl and Rabin 1987). Ethanol-induced enhancement of Ca^{2+} channel expression may be important in the pathogenesis of alcohol withdrawal syndromes, since Ca^{2+}-channel antagonists have been shown to suppress withdrawal symptoms in both ethanol-treated animals (Little et al. 1986) and human alcoholic patients (Koppi et al. 1987).

The nature of the signal that causes enhanced expression of Ca^{2+} channels on chronic exposure to ethanol is uncertain. One possibility is that acute inhibition of voltage-dependent Ca^{2+} flux is what triggers the long-term response. If this were the case, then the effect of ethanol should be mimicked by chronic treatment with Ca^{2+}-channel antagonist drugs, which also acutely inhibit Ca^{2+} flux. However, when PC12 cells were grown for 6 days with concentrations of various Ca^{2+}-channel antagonists that acutely inhibited $^{45}Ca^{2+}$ uptake to the same extent as 200 mM ethanol, no significant increase in (+)-[^{3}H]PN 200–110 binding was observed (Marks et al., submitted for publication) (Fig. 1). Neither was the Ca^{2+}-channel agonist Bay K8644 able to prevent ethanol-induced up-regulation of (+)-[^{3}H]PN200–110 binding. Therefore, acute inhibition of Ca^{2+} flux by ethanol is unlikely to cause enhanced expression of voltage-dependent Ca^{2+} channels with chronic ethanol exposure. An alternative hypothesis is that enhanced expression of Ca^{2+} channels is triggered by another acute effect of ethanol on cellular Ca^{2+} homeostasis, such as elevation of free intracellular Ca^{2+} levels by mobilization of Ca^{2+} from intracellular sequestration sites (Daniell et al. 1987, Rabe and Weight 1988).

Calcium Channels in Human Alcoholic Brain

A final, related question concerns whether changes in Ca^{2+}-channel activity such as are observed in PC12 cells and experimental animals during long-term administration of ethanol also occur in the brains of human alcoholics. We measured the binding of (+)-[^{3}H]PN200–110 to cerebral cortex samples obtained at autopsy from five nonalcoholic and six abstinent alcoholic patients (Marks et al., submitted for publication). Levels of binding were not significantly increased in brain samples from the alcoholic patients (B_{max} = 59 ± 11 fmol/mg protein) compared with the nonalcoholic controls (47 ± 8 fmol/mg protein). It therefore appears that ethanol does not induce persistent changes in Ca^{2+}-channel expression in brains of alcoholic patients, but the possibility cannot be excluded that transient changes, such as those detected in cultured cells and rat brain synaptosomes, characterize the alcohol withdrawal state in humans. This issue is difficult to study because of the relative brevity of the withdrawal syndrome in relation to the amount of time that typically elapses between death and autopsy.

Calcium Channels in Huntington's Disease and Parkinson's Disease

Ca^{2+} channel antagonist drugs have been reported to produce extrapyramidal side effects, such as parkinsonism, akathisia, and tardive dyskinesia (Jacobs 1983, Chouza et al. 1986), and to be useful in the symptomatic treatment of extrapyramidal movement disorders, including tardive dyskinesia and Gilles de la Tourette syndrome (Ross et al. 1987, Walsh et al. 1986). The most common degenerative disorders involving the extrapyramidal system are Huntington's disease, which produces loss of neurons with cell bodies in the striatum, and Parkinson's disease, which is associated with degeneration of neurons with cell bodies in substantia nigra and nerve terminals in striatum. Injection of the excitotoxin kainic acid into rat striatum reproduces the neuronal loss and many of the biochemical changes found in Huntington's disease (Coyle and Schwarcz 1976, McGeer and McGeer 1976). Intrastriatal kainic acid also reduces Ca^{2+}-channel antagonist receptor binding in striatum by about 90% (Sanna et al. 1986), whereas 6-hydroxydopamine, which destroys dopaminergic neurons pro-

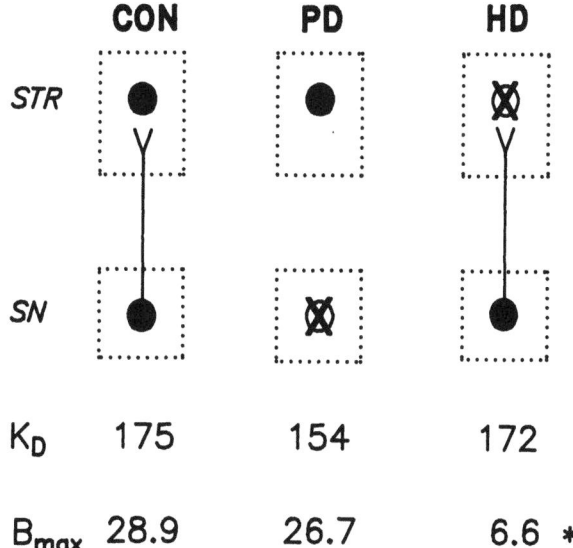

Fig. 2. (+)-[^{3}H]PN200–110 receptor binding in striata from control patients without neurologic disease (*CON*), patients with Parkinson's disease (*PD*), and patients with Huntington's disease (*HD*). Sites of pathology in striatum (*STR*) and substantia nigra (*SN*) are indicated diagrammatically. Values for K$_D$ (equilibrium binding dissociation constant, pM) and B$_{max}$ (binding site density, fmol/mg protein) are mean from five to seven patients. Tissue was provided by Dr. W.W. Tourtellote, National Neurological Research Bank, VA Wadsworth Medical Center, Los Angeles Calif. * $p < 0.05$ relative to controls

jecting from substantia nigra to striatum, has no such effect. We therefore wondered whether corresponding changes in Ca^{2+}-channel antagonist receptors might occur in clinical extrapyramidal disorders.

Binding of the Ca^{2+}-channel antagonist (+)-[^{3}H]PN 200–110 was measured in striatal tissue obtained at autopsy from five patients without neurologic disease, seven with Parkinson's disease, and five with Huntington's dieseas (Watson et al. 1988). As shown in Fig. 2, the affinity of (+)-[^{3}H]PN200–110 for striatal Ca^{2+}-channel antagonist receptors was similar in all three groups. The density of receptors in control and Parkinson's disease specimens was also similar. However, Huntington's disease produced a 75% reduction in receptor density compared to controls, indicating that L-type voltage-dependent Ca^{2+} channels are depleted in the striatum of patients with this disease. It is therefore conceivable that loss of voltage-dependent Ca^{2+} channels contributes to the clinical expression of Huntington's disease and, by implication, that Ca^{2+} channels may be important in normal extrapyramidal function.

References

Chouza C, Scaramelli A, Caamano JL, DeMedina O, Aljanati R, Romero S (1986) Parkinsonism, tardive dyskinesia, akathisia, and depression induced by flunarizine. Lancet 1:1303–1304
Coyle JT, Schwarcz R, (1976) Lesion of striatal neurones with kainic acid provides a model for Huntington's chorea. Nature 263:244–246
Daniell LC, Brass EP, Harris RA (1987) Effect of ethanol on intracellular ionized calcium concentrations in synaptosomes and hepatocytes. Molec Pharmacol 32:831–837
Dolin S, Little H, Hudspith M, Pagonis C, Littleton J (1987) Increased dihydropyridine-sensitive calcium channels in rat brain may underlie ethanol physical dependence. Neuropharmacology 26:275–279
Greenberg DA, Carpenter CL, Messing RO (1987) Ethanol-induced component of ^{45}Ca^{2+} uptake in PC12 cells is sensitive to Ca^{2+} channel modulating drugs. Brain Res 410:143–146

Harris RA, Hood WF (1980) Inhibition of synaptosomal calcium uptake by ethanol. J. Pharmacol Exp Ther 213:562–568

Jacobs MB (1983) Diltiazem and akathisia. Ann Int Med 99:794–795

Koppi S, Eberhardt G, Haller R, Konig P (1987) Calcium-channel-blocking agent in the treatment of acute alcohol withdrawal – caroverine versus meprobamate in a randomized double-blind study. Neuropsychobiology 17:49–52

Leslie SW, Barr E, Chandler J, Farrar RP (1983) Inhibition of fast- and slow- phase depolarization-dependent synaptosomal calcium uptake by ethanol. J Pharmacol Exp Ther 225:571–575

Little HJ, Dolin SJ, Halsey MJ (1986) Calcium channel antagonists decrease the ethanol withdrawal syndrome. Life Sci 39:2059–2065

McGeer EG, McGeer PL (1976) Duplication of biochemical changes of Huntington's chorea by intrastriatal injections of glutamic and kainic acids. Nature 263:517–519

Messing RO, Carpenter CL, Diamond I, Greenberg DA (1986) Ethanol regulates Ca^{2+} channels in clonal neural cells. Proc Natl Acad Sci USA 83:6213–6215

Pincon-Raymond M, Rieger F, Fosset M, Lazdunski M (1985) Abnormal transverse tubule system and abnormal amount of receptors for Ca^{2+} channel inhibitors of the dihydropyridine family in skeletal muscle from mice with embryonic muscular dysgenesis. Dev Biol 112:458–466

Rabe CS, Weight FF (1988) Effects of ethanol on neurotransmitter release and intracellular free calcium in PC12 cells. J Pharmacol Exp Ther 244:417–422

Ross JL, Mackenzie TB, Hanson DR, Charles CR (1987) Diltiazem for tardive dyskinesia. Lancet 1:268

Sanna E, Head GA, Hanbauer I (1986) Evidence for a selective localization of voltage-sensitive Ca^{2+} channels in nerve cell bodies of corpus striatum. J Neurochem 47:1552–1557

Skattebøl A, Rabin RA (1987) Effects of ethanol on $^{45}Ca^{2+}$ uptake in synaptosomes and in PC12 cells. Biochem. Pharmacol 36:2227–2229

Stallcup WB (1979) Sodium and calcium fluxes in a clonal nerve cell line. J Physiol (Lond) 286:525–540

Toll L (1982) Calcium antagonists: High-affinity binding and inhibition of calcium transport in a clonal cell line. J Biol Chem 257:13189–13192

Wagner JA, Reynolds IJ, Weisman HF, Dudeck P, Weisfeldt ML, Snyder SH (1986) Calcium antagonist receptors in cardiomyopathic hamster: selective increases in heart, muscle, brain. Science 232:515–518

Walsh TL, Lavenstein B, Licamele WL, Bronheim S, O'Leary J (1986) Calcium antagonists in the treatment of Tourette's disorder, Am J Psychiatry 143:1467–1468

Watson DL, Carpenter CL, Marks SS, Greenberg DA (1988) Striatal calcium channel antagonist receptors in Huntington's disease and Parkinson's disease. Ann Neurol 23:303–305

5 Endogenous Ligands and Antibodies

Endogenous Ligands for the Calcium Channel: Myths and Realities

D.J. Triggle

126 Cooke, School of Pharmacy, State University of New York, Buffalo, NY 14260, USA

Introduction

The Ca^{2+}-channel antagonists, including the clinically available nifedipine, verapamil, and diltiazem (Fig. 1), represent a clinically and chemically heterogeneous group of drugs. In the clinical setting these agents are all effective, differing only quantitatively, against angina in its several forms. However, verapamil, and to a lesser extent diltiazem, are effective against some cardiac arrhythmias, notably supraventricular tachycardia, while nifedipine and other 1,4-dihydropyridines are completely ineffective (Fleckenstein, 1983; Opie, 1984, 1988). In general, verapamil and diltiazem have both vasodilating and cardiodepressant properties, whilst nifedipine is prominently vasodilatory in its actions. Despite this heterogeneity of action it is clear that there exists a major common locus of action for these drugs – the voltage-dependent Ca^{2+} channel (for reviews see Triggle and Janis, 1984a, 1987; Triggle and Swamy, 1983). In particular, the L-type of Ca^{2+} channel, which appears to be of major significance in the cardiovascular system, represents the primary site of action of the currently available therapeutically employed Ca^{2+}-channel antagonists. It is likely that agents will be developed that will be effective against other classes of Ca^{2+} channels, including the transient T type and the neuronal N type (Janis et al., 1987; Miller, 1987; Narahashi, 1987).

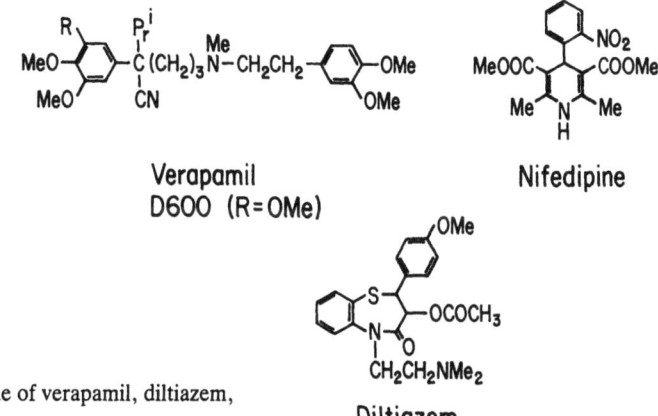

Fig. 1. Structural formulae of verapamil, diltiazem, and nifedipine

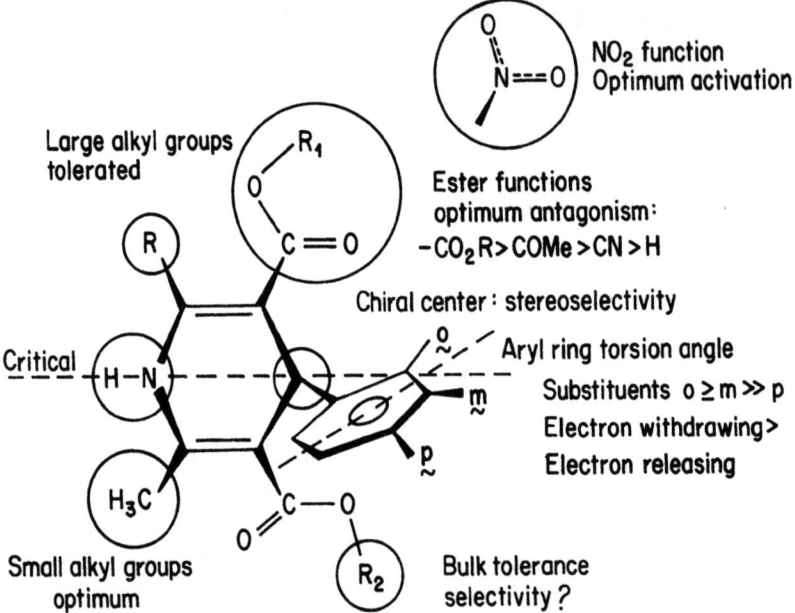

Fig. 2. The structural requirements for activator and antagonist activity in 1,4-dihydropyridines

Calcium Channels as Drug Receptors

It has been apparent for a number of years that there is no single all-encompassing structure-activity relationship for drugs acting at the Ca^{2+} channel and that separate relationships can be described for each structural category. This is particularly true for the 1,4-dihydropyridine group of drugs, where the availability of a relatively large number of agents led to the early generation of the basic outline of a structure-activity relationship derived from both in vivo and in vitro studies. The basic molecular requirements for Ca^{2+} channel activity in the 1,4-dihydropyridines are depicted in Fig. 2. Such studies indicate an organization for the Ca^{2+} channel in which there exist separate, but specific, binding sites for each of the several ligand classes which modulate Ca^{2+}-channel function. These conclusions have been confirmed by subsequent and more direct radioligand binding and by biochemical experiments establishing the allosteric interactions between three major categories of binding site which are carried on a major single protein, molecular weight (M_r) 175000, of the L-type Ca^{2+} channel (Catterall et al., 1988; Froehner, 1988; Leung et al., 1987). These interactions are depicted schematically in Fig. 3.

These binding sites can accommodate 1,4-dihydropyridine activators, including Bay K8644 (Schramm et al., 1983; Takena and Maeno, 1982; Truog et al., 1985) (Fig. 4). Potent and readily demonstrable activator properties have not been reported for agents of the benzothiazepine (diltiazem) or phenylalkylamine (verapamil) classes. However, both diltiazem and D600 have been shown to potentiate Ca^{2+} currents

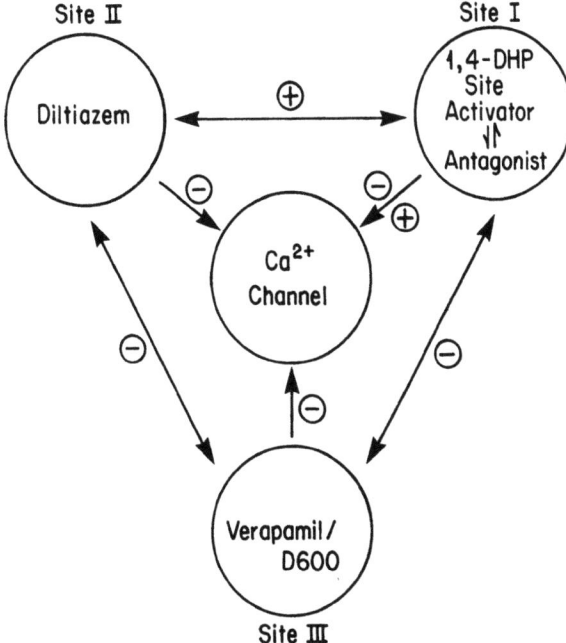

Fig. 3. A schematic representation of the binding site organization at the voltage-dependent Ca^{2+} channel depicting the allosteric linkages between the sites

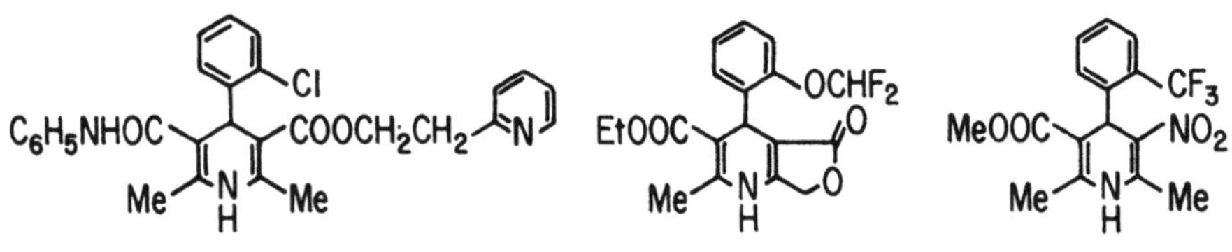

Fig. 4. Structural formulae of 1,4-dihydropyridine activators

in dorsal root ganglion cells under conditions of persistent guanine nucleotide (G) protein activation (Scott and Dolphin 1987). The possession of activator and antagonist properties within a particular ligand class may thus depend both on the presence of the appropriate molecular features in the ligand and upon the availability of the appropriate state of the Ca^{2+} channel. This dualism expands considerably the potential scope of the structural and mechanistic bases of endogenous ligand action.

Major clinical and basic emphasis has centered around the 1,4-dihydropyridine, phenylalkylamine, and benzothiazepine structures (Fig. 1). These are not the only agents that interact at Ca^{2+} channels and have Ca^{2+}-channel modulating properties; structures as diverse as α-adrenoceptor antagonists, general anesthetics, and toxin peptides have been reported active at Ca^{2+} channels (Janis et al., 1987). In most instances this activity is secondary to other, more potent and more specific effects.

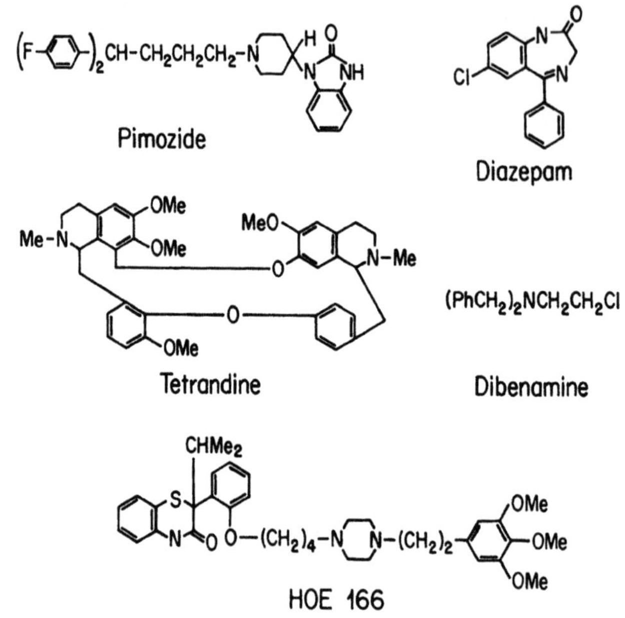

Fig. 5. Representative molecules showing activity at Ca^{2+} channels

However, particular interest attaches to the diphenylbutylpiperidines, which appear to define a fourth specific ligand binding site associated with the Ca^{2+} channel (Gould et al., 1983; Galizzi et al., 1986), the benzodiazepines, both central and peripheral (Rampe and Triggle, 1986), and new structures, including tetrandine (Qian et al., 1984) and HOE 166 (Qar et al., 1988) (Fig. 5). Similarly, Ca^{2+}-channel-activator properties have been described for non-1,4-dihydropyridine structures (Fig. 6), including praziquantel (Chubb et al., 1978; Jim and Triggle, 1979), 1-methyladenine [Jim and Triggle, 1979], and abscissic acid (Huddart et al., 1986).

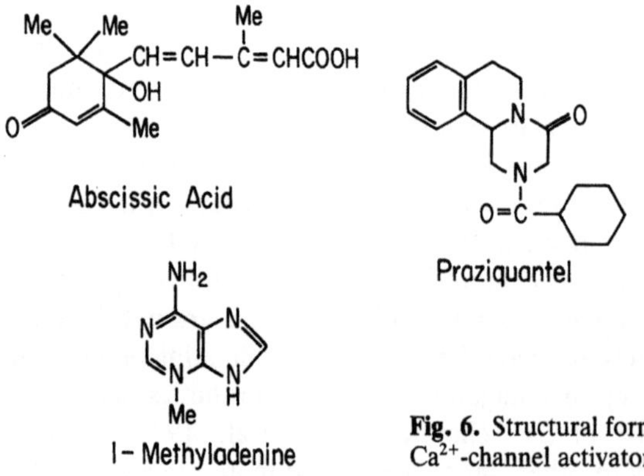

Fig. 6. Structural formulae of some non-1,4-dihydropyridine Ca^{2+}-channel activators

Table 1. Regulation of calcium channels

Stimulus	Tissue	Effect (^{3}H)1,4-DHP B_{max}	K_d	Response	Reference(s)
Chronic nifedipine	Brain Heart	↓ ↓	nc	nd	1,2
Depolarization	PC12	↓	nc	$^{45}Ca^{2+}$ ↓	3
6-OH-dopamine	Heart	↑ (β-adrenoceptor ↑)	nc	nd	4
Kainic acid	Brain	↓ (dopamine recetor ↓)	nc	nd	5
Phorbol ester	Skeletal muscle	↑	nc	$^{45}Ca^{2+}$ ↑	6
Phenylephrine	Heart	↓ (α-, β-adrenoceptor ↓)	nc	nc	7
Alcohol	PC12 synaptosomes	↑ ↑	nc nc	$^{45}Ca^{2+}$ ↑ $^{45}Ca^{2+}$ ↑	8, 9

nc, no change; nd, not determined
1, Panza et al. 1985
2, Gengo et al. 1988a
3, DeLorme et al. 1988
4, Skattebøl and Triggle 1986
5, Skattebøl et al. 1988
6, Navarro 1987
7, Gengo et al. 1988b
8, Messing et al. 1986
9, Skattebøl and Rabin 1987

The existence of both activators and antagonists interacting at a specific channel site, conforming to specific structural requirements, and serving to promote and inhibit channel activity respectively serves to satisfy substantially the definition of a receptor site at which drugs of the 1,4-dihydropyridine and other ligand classes act (Glossmann et al., 1985; Triggle and Janis 1984b).

Additional support for this concept of the Ca^{2+} channel as a receptor derives from several studies describing channel regulation through persistent activation, chronic drug administration, and neuronal lesioning (Table 1) (Ferrante and Triggle 1988). Thus, chronic depolarization results in down-regulation of channel numbers and densities (DeLorme et al., 1988), as does chronic antagonist administration in vivo (Gengo et al., 1988a; Panza et al., 1985). In contrast, peripheral sympathectomy by 6-hydroxydopamine results in the simultaneous up-regulation of both cardiac β-adrenoceptors and Ca^{2+} channels (Skattebøl and Triggle, 1987). There are close parallels between Ca^{2+}-channel regulation and Na^+-channel regulation by electrical and chemical activators (Bar-Sagi and Prives, 1985; Sherman and Catterall, 1984).

Ligand Regulation of Calcium Channels

The studies of drug action at the Ca^{2+} channel led to the early implicit assumption that endogenous factors or ligands might exist that would serve as physiological regulators of channel function and whose activities were mimicked, in whole or in part, by the Ca^{2+}-channel antagonists. The subsequent discovery of the activator ligands of the 1,4-dihydropyridine series expanded further the mechanistic bases of potential endogenous ligand action. Furthermore, the apparent molecular tolerance of drug interaction with the Ca^{2+} channel may translate into a corresponding flexibility of endogenous ligand structure and function.

It is important to note that Ca^{2+} channels are regulated by a variety of chemical factors under physiological conditions. However, this regulation is, for the most part, mediated indirectly via second messengers or coupling proteins. Thus, β-adrenoceptor activation of cardiac Ca^{2+} channels is achieved through the intermediacy of cAMP and channel phosphorylation (Reuter et al., 1986). Similarly, protein kinase C activation has been shown to regulate Ca^{2+}-channel function in a number of preparations, including neurons (Kaczmarek, 1986; Rane and Dunlap, 1986): in skeletal muscle, protein kinase C activation increases the number of 1,4-dihydropyridine-sensitive Ca^{2+} channels (Navarro, 1987).

A number of transmitters serve to regulate channel function through G protein-coupled processes (Brown and Birnbaumer, 1988). Thus dynorphin A acting through ϰ-opiate receptors inhibits N-type Ca^{2+} channels (Gross and Macdonald, 1987), as do somatostatin and baclofen at somatostatin and GABA β-receptors respectively (Lewis et al., 1986; Dolphin and Scott 1987). In principle, such receptor-mediated regulation could be indirect, with G protein-regulated receptor activation generating a second messenger species that modulates channel activity. However, it is clear that direct regulation of voltage-dependent Ca^{2+} channels by G proteins does occur. In cardiac membranes, direct activation of channels by activated Gs or its α-subunit can be measured independent of cAMP (Yatani et al., 1987) and in dorsal root ganglion cells activated G proteins (G_i/G_o) potentiate the activator properties of Ca^{2+}-channel ligands. These observations suggest that activated G proteins can stabilize the activated state of L-type Ca^{2+} channels (Scott and Dolphin, 1987). It remains to be determined whether different classes of Ca^{2+} channel are regulated by different G proteins. However, it is possible that endogenous ligands could regulate Ca^{2+}-channel function indirectly through G proteins.

Thus, Ca^{2+} channels may be regulated indirectly via several distinct processes which produce second messengers, either diffusible or membrane-bound, or which act at associated or associatable coupling processes. These could be termed "endogenous ligands" for the Ca^{2+} channel. However, it appears more useful to restrict the term to those agents that may act directly at the Ca^{2+}-channel protein(s) and mimic the effects of the known Ca^{2+}-channel ligands. No agent acting in this fashion has yet been identified. It may, however, be useful to compare the situation at the Ca^{2+} channel with several other receptor systems for which the existence of endogenous ligand has also been proposed on grounds essentially identical to those already outlined for the Ca^{2+} channel.

Endogenous Ligands and Receptor Systems

Drug action at Ca^{2+} channels presents clear parallels to other receptor systems that have also yielded to the search for endogenous ligands. The most prominent example is clearly that of the opiate receptor, defined originally by the actions of morphine and its potent synthetic analogs, both antagonists and agonists, but without a corresponding physiological substrate(s). Subsequent work defined the series of peptide ligands, including the endorphins and enkephalins (Hughes et al., 1975; Simantov and Snyder, 1976), derived from the processing of proopiomelanocortin and proenkephalin A and B, which define the several subclasses of opiate receptor (reviewed in Hollt, 1986).

The successful discovery of the endogenous opiate ligands and the realization that the benzodiazepines also define a specific receptor (Squires and Braestrup, 1977; Mohler and Okada, 1977), linked to the GABA-modulated Cl-channel (Tallman and Gallager, 1985), has been a major impetus in the search for endogenous valiums. Low-affinity interactions of a number of agents, including hypoxanthine, inosine, and nicotinamide, with the benzodiazepine receptor have been reported, but it is generally agreed that these agents are not likely candidates for endogenous ligand status. In contrast, diazepam binding inhibitor (DBI) a 105 amino acid brain peptide which interacts at benzodiazepine binding sites, possesses proconflict activity and appears to serve as a precursor for the biologically active octadecaneuropeptide (ODN; Gln-Ala-Thr-Val-Gly-Asp-Val-Asn-Thr-Asp-Arg-Pro-Gly-Leu-Asp-Leu-Lys), may be one of a family of peptides, including the endozepines (Marquardt et al., 1986), that serve to modulate the anxiogenic and anxiolytic functions of the benzodiazepine receptor (Costa and Guidotti, 1985; Gray et al., 1986). However, since it is also found in peripheral tissues it may be involved in modulating more than central benzodiazepine receptor function.

Peptides are very attractive candidates for endogenous ligands. It is, therefore, of particular interest that DeBlas and his colleagues, using monoclonal antibodies reactive against the benzodiazepine epitope, have purified *N*-desmethyldiazepam from rat brain (Sangameswaran et al., 1986; Stephenson, 1987). It is argued that this is not a contaminant, but rather may, given the ability of fungi to synthesize benzodiazepines (Luckner, 1984), represent a dietary introduction. Dietary sources are known for other receptor-directed ligands (Hazum, 1983). Morphine is detectable in human milk together with the peptides morphiceptin and β-casomorphin, which are degradation products of casein (Chang et al., 1981; Hazum et al., 1981). Thus endogenous ligands may be dietary factors.

Not all endogenous ligand searches have been as successful as that for the opiates or as putatively successful as that for the benzodiazepines. However, factors have been described that may be endogenous regulators of Na^+, K^+ ATPase (LaBella, 1985; Haber and Haupert, 1987), serotonin uptake (Barbaccia et al., 1983), the phencyclidine receptor (Quirion et al., 1984; Su et al., 1986), presynaptic neurotransmitter release (Benishin et al., 1986), and Na^+ (Lombet et al., 1987) and K^+ (Fosset et al., 1984) channels.

The search for the endogenous regulator of plasmalemmal Na^+-K^+ ATPase is of particular interest, since it is based both upon the existence of a specific target protein with defined specificity for cardiac glycosides and upon observations that volume expansion yields a humoral natriuretic factor of hypothalamic origin that inhibits Na^+-K^+-ATPase. More than one factor may exist; a solely hypothalamic origin is not established, and no purified factors of known structure are yet available. However, it is interesting to note that digoxin-like immunoreactivity has been reported for some fractions, thus raising the possibility that one of the factors may resemble the cardiac glycoside structure.

Endogenous Ligands for Calcium Channels

During the past 5 years a number of reports have described putative endogenous agents, known and unknown in structure, that modulate Ca^{2+}-channel function (Table 2). These findings should be viewed in the perspective of a set of as yet unanswered questions, some general and some specific, concerning endogenous ligand actions at Ca^{2+} channels. Among these questions are:

1. Why should an endogenous ligand(s) exist? Are the observed Ca^{2+}-channel ligand binding sites byproducts of the creation of the peptide organization that comprises the necessary permeation and gating functions of the channel, or are they sites designed as components of the physiological functions of the channel?
2. Will the endogenous ligand be a known or a novel structure? Will the ligand have specificity for a known receptor?
3. Will there be multiple endogenous ligands? Will there be an endogenous ligand for each discrete binding site?
4. Will there be ligands for each channel class, or will there be ligands for only some channel classes?

Table 2. Putative endogenous ligands for calcium channels

Source	M_r (x 1000)	(^{3}H)1,4-DHP binding	$^{45}Ca^{2+}$ Tissue p'col	Heat	Protein	Reference(s)
Brain	5–10	nc[a]	sm ↓	Labile	–	1
Brain	1.2	nc[a,b]	$^{45}Ca^{2+}$ neuronal	Stable	–	2–4
Brain	> 66 (LF)	nc[a,c]	–	Labile	Trypsin-sensitive	5
	1.3 (SF)	nc Irrev.		Stable	Pronase-sensitive	
Stomach	16 (anthralin)	K_d ↓	–	Labile	Pronase-, trypsin-sensitive	6
Blood (erythrocytes)	< 6	–	$^{45}Ca^{2+}$ ↑ aorta	Stable	Pronase-sensitive	7–10
Blood	–	–	$^{45}Ca^{2+}$ ↑ RBC	–	–	11
Blood (human plasma)	–	–	$^{45}Ca^{2+}$ ↑	–	–	12

nc noncompetitive
LF large molecular weight fraction
SF small molecular weight fraction
sm smooth muscle
p'col pharmacology
Irrev irreversible

[a] Not due to Ca^{2+} chelation.
[b] Does not inhibit spiroperidol, imipramine, or muscimol binding
[c] Does not inhibit nitrendipine binding to heart

1, Thayer et al 1984
2, Sanna et al. 1986a
3, Sanna et al. 1987
4, Sanna and Hanbauer 1987
5, Ebersole et al. 1988
6, Mantione et al. 1988
7, Wright and Bookout 1982
8, Wright and McCumbee 1984
9, McCumbee and Wright 1985
10, Huang et al. 1988
11, Zidek et al. 1985
12, Lindner et al. 1987

5. Will the ligand be extracellular or intracellular?
   If extracellular, will it act locally or systemically?
6. Will the endogenous ligand(s) be an activator or antagonist?
7. Will the ligand(s) be active in phasic or tonic fashion?
   Will the ligand be produced and present transiently or on demand (phasic), or be present constantly (tonic)?
8. Will the ligand be produced under physiological or pathological conditions?
9. Will the ligand be a trophic factor?
10. Will the endogenous ligand be engaged in preactivation or postactivation phenomena?
    Could the ligand be produced or generated as a consequence of channel activation?

Little direct evidence exists that close structural analogs of existing ligands occur naturally and may constitute endogenous counterparts. However, it should be noted that diltiazem is structurally analogous to benzodiazepines, and that benzodiazepines and Ca^{2+}-channel ligands show interactions (Rampe and Triggle, 1986). The possible interactions of endogenous benzodiazepine ligands at Ca^{2+} channels may thus be of interest. Malondialdehyde is a highly reactive intermediate formed as a byproduct in prostaglandin and thromboxane biosynthesis; spontaneous condensation with amino acids leads to the generation of 1,4-dihydropyridines (Nair et al., 1986) (Fig. 7). The structures depicted will not be very active at the 1,4-dihydropyridine-sensitive Ca^{2+} channels, but it is possible to envisage other structures that may be formed in vivo from aromatic aldehydes and which may have greater activities.

More plausibly, the existing synthetic ligands may mimic, in whole or in part, the structure of an endogenous molecule or family of molecules. Although it is clear that there exist several discrete binding sites at the L-type Ca^{2+} channel, each capable of accommodating a specific structural category, the topographic relationship between these sites remains to be established. It is thus possible that a single endogenous species could occupy all of these sites simultaneously and thus effectively reduce the number of choices to be made.

Because of the existence of 1,4-dihydropyridines with potent activator and antagonist properties the 1,4-dihydropyridine nucleus has been suggested as a particularly plausible candidate for the status of a molecule that maps an endogenous ligand binding site. Although the 1,4-dihydropyridines are usually regarded as the most potent and selective of the Ca^{2+}-channel ligands, they actually interact at a

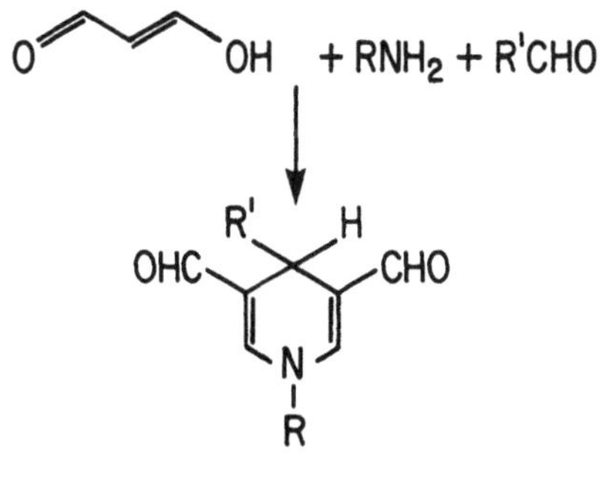

Fig. 7. Possible formation of "endogenous 1,4-dihydropyridine"

variety of receptor systems (Janis et al., 1987). These include interactions at other ion channels, including Na$^+$ channels (Yatani et al., 1988), transport systems, including Na$^+$-K$^+$ ATPase and the adenosine transporter (Janis et al., 1987), interactions at adenosine receptors (Hu et al., 1987), thromboxane A2 receptors (Johnson et al., 1988), PAF receptors (Wade et al., 1986), and phospholipase A2 (Chang et al. 1987). A more detailed review of the multiple effects of Ca^{2+} channel drugs is given by Janis, Silver and Triggle (1987). These observations suggest that the 1,4-dihydropyridine nucleus may serve as a template for structures, presumably peptides, that have the potential to interact at diverse receptor systems, possibly through common organizational features.

A number of peptide toxins are known that act at Ca^{2+} channels (Hamilton and Perez, 1987). Of particular interest are the ω-conotoxins, a group of basic peptides (26–29 residues), with selectivity for neuronal L and N channels (Rivier et al., 1987; Olivera et al., 1987). A larger toxin, taicatoxin, M$_r$ 8000, has different specificity and blocks cardiac L channels (Hamilton and Perez, 1987).

Both high-M$_r$ and low-M$_r$ factors have been isolated of peptide and nonpeptide character. Although the factor isolated from smooth muscle (Mantione et al., 1988) inhibits binding to both peripheral benzodiazepine and 1,4-dihydropyridine receptors and is likely to be a PLA2 isozyme, other factors may have greater specificity. Thus, the low-M$_r$ fraction isolated from rat brain (Sanna et al., 1986a, 1987; Sanna and Hanbauer, 1987) blocks both 1,4-dihydropyridine binding and depolarization-induced Ca^{2+} uptake in cerebellar granule cells, and both high-M$_r$ and low-M$_r$ factors isolated from rat brain (Ebersole et al., 1988) inhibit 1,4-dihydropyridine binding in brain but not in heart.

Factors have been isolated from the erythrocytes and blood of spontaneously hypertensive rats (Huang et al., 1988; Wright and Bookout 1982; Wright and McCumbee 1984; McCumbee and Wright 1985; Zidek et al., 1985) that are hypertensive and increase the Ca^{2+} content of vascular smooth muscle. Similarly, a factor from hypertensive humans increases the cytosolic Ca^{2+} concentration in platelets (Lindner et al., 1987). Little is known about these factors, and neither red blood cells nor platelets are believed to contain voltage-dependent Ca^{2+} channels. No evidence is available as to whether these agents affect ligand binding or channel function at any class of Ca^{2+} channel.

Palmitoyl carnitine represents an agent of known structure proposed as a putative endogenous ligand (Spedding and Mir, 1987; Duncan et al., 1986, 1987). Palmitoyl carnitine levels rise significantly during ischemia, and its polypharmacological properties, including inhibition of Na$^+$-K$^+$ ATPase, Ca^{2+} ATPase, and C kinase and stimulation of Ca^{2+} current, have been attributed to detergent and surface charge effects (Idell-Wenger et al., 1978; Inoue and Pappano, 1983). Binding of (^{3}H)nitrendipine, (^{3}H)verapamil and (^{3}H)diltiazem is inhibited by palmitoyl carnitine with IC$_{50}$ values of approximately $10^{-4}M$. These concentrations were similar to those required to activate Ca^{2+}-dependent responses and block Ca^{2+}-channel antagonist activity. Significantly higher concentrations were required to inhibit ligand binding to dopamine and α- and β-adrenoceptors. While it is possible that palmitoyl carnitine and other fatty acids produced during ischemic cell damage may contribute to the Ca^{2+} overload seen under these conditions or during reperfusion, it is unlikely, given the effective concentrations, that palmitoyl carnitine serves as a physiological regulator.

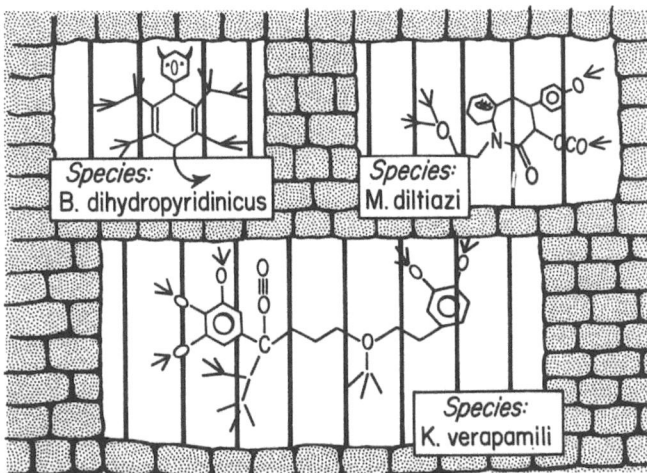

Fig. 8. The endogenous ligand family

The most recent candidate of known structure is endothelin, a 21 amino acid peptide derived from vascular endothelium which is a potent vasoconstrictor whose actions are Ca^{2+}-dependent and sensitive to 1,4-dihydropyridines. The structure of endothelin, with two disulfide bridges, resembles that of other channel-active toxins (Yanagisawa et al., 1988).

Conclusions and Caveats

The search for endogenous ligands for the Ca^{2+} channel is in its infancy. Few would argue that the present data constitute particularly powerful or persuasive evidence in favor of the existence of endogenous ligands. However, few would wish to deny the possibility of the existence of such factors. A major problem in the search for such factors is the multiplicity of choices previously outlined. Each of these choices may force selection of a particular set of experimental procedures that serve to exclude alternatives. Thus, the use of the (^{3}H)1,4-dihydropyridine membrane binding assay as a screening protocol may eliminate factors that act at other classes of Ca^{2+} channels, act at other drug sites or show preferential binding to noninactivated channel states. Similarly, the assumption as to whether an endogenous ligand is likely to be an activator or antagonist of Ca^{2+} channels affects the choice of experimental methods. These difficulties, once appreciated, can be overcome, and the search for Ca^{2+}-channel regulatory factors offers both philosophical and pragmatic satisfaciton, leading as it may to the discovery of new species (Fig. 8). "Seek and ye shall find." (Matthew VII, 7).

References

Bar-Sagi D, Prives J (1985) Negative modulation of sodium channels in cultured chick muscle cells by the channel activator batrachotoxin. J Biol Chem 260:4740–4744

Barbaccia ML, Gandolfi O, Chuang DM, Costa E (1983) Modulation of neuronal serotonin uptake by a putative endogenous ligand of imipramine recognition sites. Proc Natl Acad Sci USA 80:5134–5138

Benishin CG, Pearce LB, Cooper JR (1986) Isolation of a factor (substance B) that antagonizes presynaptic modulation: pharmacological properties. J Pharmacol Exp Therap 239:185–191

Brown AM, Birnbaumer L (1988) Direct G protein gating of ion channels. Amer J Physiol 254:H401–410

Casy AF, Parfitt RT (1987) Opioid analgesics: chemistry and receptors. Plenum, New York

Catterall WA, Seagar MJ, Takahashi M (1988) Molecular properties of dihydropyridine sensitive calcium channels in skeletal muscle. J Biol Chem 263:3535–3538

Chang KJ, Killam A, Hazum E, Cuatrecasas P, Chang JK (1981) Morphiceptin (NH_4-Tyr-Pro-Phe-Pro-$CONH_2$): a potent and specific agonist for morphine (μ) receptors. Science 212:75–77

Chang J, Blazek E, Carlson RP (1987) Inhibition of phospholipase A2 (PLA2) activity by nifedipine and nisoldipine is independent of their calcium-channel-blocking activity. Inflammation 11:353–364

Chub JM, Bennett JL, Akera T, Brody TM (1978) Effects of praziquantel, a new anthelmintic, on electromechanical properties of isolated rat atria. J Pharmacol Exp Therap 207:284–293

Costa E, Guidotti A (1985) Endogenous ligands for benzodiazepine recognition sites. Biochem Pharmacol 34:3399–3403

DeLorme E, Rabe CS, McGee R (1988) Regulation of the number of functional voltage-sensitive Ca^{++} channels on PC12 cells by chronic changes in membrane potential. J Pharmacol Exp Therap 244:838–843

Dolphin AC, Scott RH (1987b) Calcium channel currents and their inhibition by ($-$)-baclofen in rat sensory neurones: modulation by guanine nucleotides. J Physiol 386:1–17.

Ebersole BJ, Gajary ZL, Molinoff PB (1988) Endogenous modulators of binding of (^{3}H)nitrendipine in extracts of rat brain. J Pharmacol Exp Therap 244:971–976

Ferrante J, Triggle DJ (1988) Regulation of Ca^{2+} channels. In preparation

Fleckenstein A (1983) Calcium antagonism in heart and smooth muscle. Wiley, New York

Fosset M, Schmid-Antomarchi H, Hugues M, Romey G, Lazdunski M (1984) The presence in pig brain of an endogenous equivalent of apamin, the bee venom peptide that specifically blocks Ca^{2+}-dependent K^+ channels. Proc Natl Acad Sci USA 81:7228–7232

Froehner SC (1988) New insights into the molecular structure of the dihydropyridine-sensitive calcium channel. Trends Neuro Sci 11:90–92

Galizzi J-P, Fosset M, Romey G, Laduron P, Lazdunski M (1986) Neuroleptics of the diphenylbutyl-piperidine series are potent calcium channel inhibitors. Proc Natl Acad Sci USA 83:7513–7517.

Gengo P, Bowlin N, Wyss VL, Hayes JS (1988b) Effects of prolonged phenylephrine infusion on cardiac adrenoceptors and calcium channels. J Pharmacol Exp Therap 244:100–105

Gengo P, Skattebøl A, Moran JF, Gallant S, Hawthorn M, Triggle DJ (1988a) Regulation by chronic drug administration of neuronal and cardiac calcium channel, beta-adrenoceptor and muscarinic receptor levels. Biochem Pharmacol 37:627–633

Glossmann H, Ferry DR (1985) Assay for calcium channels. Methods Enzymol 109:513–551

Glossmann H, Ferry DR, Goll A, Striessnig J, Zernig G (1985) Calcium channels and calcium channel drugs: recent biochemical and biophysical findings. Arzneim Forsch 35:1917–1935

Gould RJ, Murphy KMM, Reynolds IJ, Snyder SH (1983) Antischizophrenic drugs of the diphenyl-butylpiperidine type act as calcium channel antagonists. Proc Natl Acad Sci USA 80:5122–5125

Gross RA, Macdonald RL (1987) Dynorphin A selectively reduces a large transient (N-type) calcium current of mouse dorsal root ganglion neurons in cell culture. Proc Natl Acad Sci USA 84:5649–5673

Gray PW, Glaister D, Seeburg PH, Guidotti A, Costa E (1986) Cloning and expression of cDNA for human diazepam binding inhibitor, a natural ligand of an allosteric regulatory site of the γ-aminobutyric acid type A receptor. Proc Natl Acad Sci USA 83:7547–7551

Haber E, Haupert GT (1987) The search for a hypothalamic Na^+, K^+-ATPase inhibitor. Hypertension 9:315–324

Hamilton SL, Perez M (1987) Toxins that affect voltage-dependent calcium channels. Biochem Pharmacol 36:3325–3329

Hazum E (1983) Hormones and neurotransmitters in milk. Trends Pharmacol Sci 8:454–456

Hazum E, Sabatka JJ, Chang KJ, Brent DA, Findlay JWA, Cuatrecasas P (1981) Morphine in cow and human milk: could dietary morphine constitute a ligand for specific morphine (μ) receptors? Science 213:1010–1012

Hollt V (1986) Opioid peptide processing and receptor selectivity. Ann Rev Pharmacol Toxicol 26:59–77

Hu P-S, Lindgren E, Jacobson KA, Fredholm BB (1987) Interaction of dihydropyridine calcium channel agonists and antagonists with adenosine receptors. Pharmacol Toxicol 61:121–125

Huang B, McCumbee WB, Wright GL (1988) Calcium channel modulation by a peptide isolated from the blood of spontaneously hypertensive rats. FASEB J 2:3899 abs

Huddart H, Smith RJ, Langston PD, Hetherington AM, Mansfield TA (1986) Is abscissic acid a universally active calcium agonist? New Phytol 104:161–173

Hughes J, Smith TW, Kosterlitz HW, Fothergill LA, Morgan BA, Morris HR (1975) Identification of two related pentapeptides from the brain with potent opiate activity. Nature 258:577–579

Janis RA, Silver P, Triggle DJ (1987) Drug action and cellular calcium regulation. Adv Drug Res 16:311–592

Jim K, Triggle DJ (1979) The action of praziquantel and 1-methyladenine in guinea pig ileal longitudinal muscle. Can J Physiol Pharmacol 57:1460–1463

Johnson GJ, Dunlop PC, Leis LA, From AHL (1988) Dihydropyridine agonist Bay K 8644 inhibits platelet activation by competitive antagonism of thromboxane A2-prostaglandin H2 receptor. Circ Res 62:494–505

Kaczmarek LK (1986) Phorbol esters, protein phosphorylation and the regulation of neuronal ion channels. J Exp Biol 124:375–392

LaBella FS (1985) Endogenous digitalis-like factors. Fed Proc 44:2780–2811

Leung AT, Imagawa T, Campbell K (1987) Structural characterization of the 1,4-dihydropyridine receptor of the voltage-dependent Ca^{2+} channel from rabbit skeletal muscle. J Biol Chem 262:7943–7946

Lewis DL, Weight FF, Luini A (1986) A guanine nucleotide binding protein mediates the inhibition of voltage-dependent calcium current by somatostatin in a pituitary cell line. Proc Natl Acad Sci USA 83:9035–9039

Lindner A, Kenny M, Meacham AJ (1987) Effects of a circulating factor in patients with essential hypertension on intracellular free calcium in normal platelets. New Engl J Med 316:509–513

Lombet A, Fosset M, Romey G, Jacomet Y, Lazdunski M (1987) Identification in mammalian brain of an endogenous substance with Na^+ channel blocking properties similar to those of tetrodotoxin. Brain Res 417:327–334

Luckner M (1984) Secondary metabolism in microorganisms, plants and animals, 2nd edn Springer-Verlag, Berlin pp 272–276

Mantione CL, Goldman ME, Martin B, Bolger GT, Lueddens HWM, Paul SM, Skolnick P (1988) Purification and characterization of an endogenous protein modulator of radioligand binding to "peripheral-type" benzodiazepine receptors and dihydropyridine Ca^{2+} channel antagonist binding sites. Biochem Pharmacol 37:339–347

Marquardt H, Todaro GJ, Shoyab M (1986) Complete amino acid sequences of bovine and human endozepines. Homology with rat diazepam binding inhibitor. J Biol Chem 261:9727–9731

McCleskey EW, Fox AP, Feldman D, Tsien RW (1986) Different types of calcium channels. J Exp Biol 124:177–190

McCumbee WD, Wright GL (1985) Partial purification of a hypertensive substance from rat erythrocytes. Can J Physiol Pharmacol 63:1321–1326

Messing RO, Carpenter CL, Diamond I, Greenberg DA (1986) Ethanol regulates calcium channels in clonal neural cells. Proc Nat Acad Sci USA 83:6213–6215

Miller RJ (1987) Multiple calcium channels and neuronal function. Science 235:46–52

Mohler H, Okada T (1977) Benzodiazepine receptor: demonstration in central nervous system. Science 198:849–851

Nair V, Offerman RJ, Turner GA (1986) Novel fluorescent 1,4-dihydropyridines. J Amer Chem Soc 108:8283–8285

Narahashi T (1987) Drugs acting on calcium channels. In: Calcium and drug actions. Ed PF Baker, pp 255–274, Springer-Verlag Berlin

Navarro J (1987) Modulation of (^{3}H)dihydropyridine receptors by activation of protein kinase C in muscle cells. J Biol Chem 262:4649–4652

Olivera BM, Cruz LJ, de Santos V, LeCheminant GW, Griffin D, Zeikus R, McIntosh JM, Galyean R, Varga J, Gray WR, Rivier J (1987) Neuronal calcium channel antagonists. Discrimination between calcium channel subtypes using ω-conotoxin from *Conus magus* venom. Biochemistry 26:2086–2090

Opie LH (ed) (1984) Calcium antagonists and cardiovascular disease. Raven Press, New York NY

Opie LH (1988) Calcium channel antagonists (part II): use and comparative properties of the three prototypical calcium antagonists in ischemic heart disease, including recommendations based on an analysis of 41 trials. Cardiovasc Drug Therap 1:461–492

Panza G, Grebb JA, Sanna E, Wright AG, Hanbauer I (1985) Evidence for down-regulation of (^{3}H)nitrendipine recognition sites in mouse brain after long term treatment with nifedipine or verapamil. Neuropharmacol 24:1113–1117

Qar J, Barhanin J, Romey G, Henning R, Lerch U, Oekonomopulos R, Urbach H, Lazdunski M (1988) A novel high affinity class of Ca^{2+} channel blockers. Mol Pharmacol 33:363–369

Qian JQ, Thoolan MJMC, van Meel JCA, Timmermans PBMWM, van Zwieten PA (1983) Hypotensive activity of tetrandine in rats. Pharmacology 26:187–197

Quirion R, Dimaggio DA, French ED, Contreras PC, Shiloach J, Pert CB, Everist H, Pert A, O'Donohue TL (1984) Evidence for an endogenous peptide ligand of the phencyclidine receptor. Regul Peptides 5:967–973

Rampe D, Triggle DJ (1986) Benzodiazepines and calcium channel function. Trends Pharmacol Science 7:461–464

Rane SG, Dunlap K (1986) Kinase C activator 1,3-oleoylacetylglycerol attenuates voltage-dependent calcium current in sensory neurons. Proc Natl Acad Sci USA 83:184–188

Reuter H, Kokubun S, Prod'hom B (1986) Properties and modulation of cardiac calcium channels. J Exp Biol 124:191–201

Rivier J, Galyean R, Gray WR, Azimi-Zonooz A, McIntosh JM, Cruz LJ, Olivera BM (1987) Neuronal calcium channel inhibitors. J Biol Chem 262:1194–1198

Sangamenswaran L, Fales HM, Friedrich P, De Blas AL (1986) Purification of a benzodiazepine from bovine brain and detection of benzodiazepine-like immunoreactivity in human brain. Proc Natl Acad Sci USA 83:9236–9240

Sanna E, Hanbauer I (1987) Isolation from rat brain tissue of an inhibiting activity for dihydropyridine binding sites and voltage-dependent Ca^{2+} uptake. Neuropharmacol 26:1811–1814

Sanna E, Wright AG, Hanbauer I (1986a) Modulation of (^{3}H)nitrendipine binding sites and $^{45}Ca^{2+}$ uptake by an endogenous ligand (EL) present in rat brain. Pharmacologist 28:238 abs

Sanna E, Wright AG, Hanbauer I (1987) An endogenous modulator of Ca^{2+} channels in rat brain tissue: isolation and characterization. Soc Neurosci 13:791 abs

Schramm M, Thomas G, Towart R, Franckowiak G (1984) Novel dihydropyridines with positive inotropic action through activation of Ca^{2+} channels. Nature 303:535–537

Scott RH, Dolphin AC (1987) Activation of a G protein promotes agonist responses to calcium channel ligands. Nature 330:760–762

Sherman SJ, Catterall WA (1984) Electrical activity and cytosolic calcium regulate levels of tetrodotoxin-sensitive sodium channels in cultured rat muscle cells. Proc Natl Acad Sci USA 81:262–266

Simantov R, Snyder SH (1976) Morphine-like peptides in mammalian brain: isolation, structure, elucidation and interactions with the opiate receptor. Proc Natl Acad Sci USA 73:2515–2519

Skattebøl A, Hruska RE, Hawthorn M, Triggle DJ (1988) Kainic acid lesions decrease striatal dopamine receptors and 1,4-dihydropyridine sites. Neurosci Lett (in press)

Skattebøl A, Triggle DJ (1987) Effects of ethanol on Ca^{2+} uptake in synaptosomes and PC 12 cells. Biochem Pharmacol 36:2227–2229

Skattebøl A, Triggle DJ (1986) 6-Hydroxydopamine treatment increases beta-adrenoceptors and Ca^{2+} channels in rat heart. Eur J Pharmacol 127:287–289

Squires RF, Braestrup C (1977) Benzodiazepine receptors in rat brain. Nature 266:732–734

Stephenson FA (1987) Benzodiazepines in the brain. Trends Neurosci 10:185–186

Su T-P, Weissman AD, Yeh C-Y (1986) Endogenous ligands for sigma opioid receptors in the whole brain ("sigmaphin"): evidence from binding assays. Life Sciences 38:2199–2210

Takenaka T, Maeno H (1982) A vasoconstrictive compound 1,4-dihydropyridine derivative. Jap J Pharmacol 32:139

Tallman JF, Gallager DW (1985) The Gaba-ergic system: a locus of benzodiazepine action. Ann Rev Neurosci 8:21–44

Thayer SA, Stein L, Fairhurst AS (1984) Endogenous modulator of Ca^{2+} channels. Internat Union Pure Appl Pharmacol London 892P abs

Triggle DJ, Janis RA (1984a) Calcium channel antagonists: new perspectives from the radioligand binding assay. In: Modern methods in pharmacology, ed N Back and S Spector, 2:1–28, New York, Alan R Liss Inc

Triggle DJ, Janis RA (1984b) The 1,4-dihydropyridine receptor: a regulatory component of the Ca^{2+} channel. J Cardiovasc Pharmacol 6: S949–955

Triggle DJ, Janis RA (1987) Calcium channel ligands. Ann Rev Pharmacol Toxicol 27:347–369

Triggle DJ, Swamy VC (1983) Calcium antagonists: some chemical-pharmacological aspects. Circ Res 52 (Suppl I):17–28

Truog RA, Brunner H, Criscione L, Fallert M, Kuhnis H, Meier M, Rogg H (1985) CGP 28 392, a dihydropyridine Ca^{2+} entry stimulator. In: Calcium in biological systems, eds RP Rubin, GB Weiss, JW Putney, pp 441–452. Plenum Press, New York, NY.

Wade PJ, Lad N, Tuffin DP (1986) Interaction of the human platelet PAF acether binding site with calcium and calcium channel antagonists. Prog Lip Res 25:163–165

Wright GL, Bookout C (1982) A hypertensive substance in erythrocytes of spontaneously hypertensive rats. Can J Physiol Pharmacol 60:622–627

Wright GL, McCumbee WD (1984) A hypertensive substance found in the blood of spontaneously hypertensive rats. Life Sciences 34:1521–1528

Yanagisawa M, Kurihara H, Kimura S, Tomobe Y, Kobayahi M, Mitsui Y, Yazaki Y, Goto K, Masaki T (1988) A novel potent vasoconstrictor peptide produced by vascular endothelial cells. Nature 332:411–415

Yatani A, Codina J, Imoto Y, Reeves JP, Birnbaumer L, Brown AM (1987) A G protein directly regulates mammalian cardiac calcium channels. Science 238:1288–1292

Yatani A, Kunze DL, Brown AM (1988) Effects of dihydropyridine calcium channel modulators on cardiac sodium channels. Amer J Physiol 254:H140–147

Zidek W, Hecjkmann U, Vetter H (1985) A circulatory hypertensive factor in spontaneously hypertensive rats and its effects on intracellular free calcium. Regul Peptides Suppl 4:152–155

Endogenous 1,4-Dihydropyridine-Displacing Substances Acting on L-Type Ca^{2+} Channels: Isolation and Characterization of Fractions from Brain and Stomach

R.A. Janis, D.E. Johnson, A.V. Shrikhande, R.T. McCarthy, A.D. Howard, R. Greguski, and A. Scriabine

Institute for Preclinical Pharmacology, Miles Inc., 400 Morgan Lane, West Haven, Connecticut 06516, USA

Introduction

Is It Likely That an Endogenous 1,4-Dihydropyridine-like Ligand Exists?

Several lines of evidence suggest that endogenous 1,4-dihydropyridine (DHP) ligands that act on DHP binding sites may exist. First, the binding of these dihydropyridines is of very high affinity (typically 0.01–1 nM), and is highly stereoselective (Triggle and Janis 1987; Janis et al. 1987). Second, the density of DHP binding sites can be modulated by treating animals chronically with various Ca^{2+}-channel ligands (Panza et al. 1985; Gengo et al. 1988; J. Ferrante and D.J. Triggle, in preparation). Third, synthetic pathways for DHPs that may occur spontaneously in vivo have been proposed (Nair et al. 1986). Finally, the discovery of Bay K8644 and related drugs that activate Ca^{2+} channels supports the idea that analogous endogenous substances exist (Schramm et al. 1983). The observation that Bay K8644 stabilizes a spontaneously occurring mode of L-type calcium-channel gating suggested that binding to the DHP receptor could alter physiological function (Hess et al. 1984).

Endogenous ligands may also exist for high-affinity binding sites allosterically linked to the DHP site, such as those at which fluspirilene, HOE 166 (see Lazdunski, this volume), or gallopamil bind. Synthetic activator or agonist ligands are known for the DHP binding site but not for the allosterically-linked sites. However, one report suggests that Ca^{2+} channel antagonists may act like agonists depending on the state of the guanosine 5-triphosphate (GTP) binding protein (Scott and Dolphin 1987).

There have been several preliminary reports suggesting the existence of endogenous substances that modify DHP binding. Janis et al. (1983) found that serum potently decreases [^{3}H]DHP binding to cardiac membranes. Thayer et al. (1984) reported the isolation of a fraction that both inhibited DHP binding and modified ileal contraction. Further reports on the latter fraction have not appeared, but two other groups have reported on brain fractions that inhibit [^{3}H]DHP bindig and ^{45}Ca influx (see Hanbauer and Sanna 1986, Ebersole et al. 1988; and corresponding chapters in this volume). An endogenous substance previously reported to act on DHP binding sites has now been identified as a phospholipase (Mantione et al. 1988). Substance P has been reported to stimulate both D888 and nitrendipine binding, but only in hippocampal membranes (Battaini et al. 1986). However, in our studies we have not observed effects of substance P on [^{3}H]DHP binding to hippocampal membranes (unpublished observations). For further discussion of the parallels between the DHP binding site and known receptors, and of putative endogenous ligands for Ca^{2+} channels as well as other receptor systems, see Triggle (this volume).

Criteria for Demonstrating the Existence of Endogenous DHP-like Ligands

An endogenous DHP-like ligand should exert reversible, specific, and reasonably high affinity of inhibition of [^{3}H]DHP binding to membranes from excitable cells. It should produce potent time- and voltage-dependent block of L-type, but not T-type Ca^{2+} channels, nor should it block other types of ion channels. It should be demonstrated that the putative ligand is synthesized endogenously and is capable of modulating Ca^{2+} channel activity. An exception to this would be ingested ligands that are subsequently stored and used as modulators. At the present time, no pure endogenous substance has been shown to exert potent effects on both [^{3}H]DHP binding and Ca^{2+} channel current over the same concentration range.

Methods

Purification procedures

Fresh bovine brains were obtained from a local slaughterhouse. Lyophilized bovine brain and lamb stomach were obtained from Burlington Bio-Medical Corp. (New York). Typically, 10–15 kg fresh or 0.1–1 kg lyophilized calf brain or lamb stomach was homogenized (Polytron, Brinkman) in acid by the method of Bennett (1979) as modified by Quirion et al. (1984). The homogenate was clarified by centrifugation (16000 g for 45 min) and the supernatant was extracted with 2 volumes of petroleum ether. The aqueous phase was adjusted to pH 3 with NaOH and subjected to ultrafiltration on a Pellicon apparatus (Millipore) equipped with a $\cong$ 30 kDa cut-off membrane. The ultrafiltrate was then subjected to preparative reverse phase (RP) HPLC on a Vydac octadecylsilane column (ODS; 5.7 × 30 cm) employing a gradient of 0%–80% acetonitrile in 0.1% trifluoroacetic acid (TFA) over 2 h at a flow rate of 50 ml/min. Fractions (150 ml) were collected, aliquots were dried (Speed Vac, Savant) and tested for their ability to inhibit the specific binding of [^{3}H]DHP to rat cardiac membranes. Fractions were considered "active" if they inhibited [^{3}H]DHP binding by at least two fold over a background inhibition of $\cong$ 15%, which was observed in many fractions. Fractions of interest corresponding to 30%–35% acetonitrile for bovine brain (or 45%–50% acetonitrile for lamb stomach), were pooled and lyophilized or further purified before lyophilization.

Fractionation of the active region from preparative RP-HPLC by size exclusion HPLC (SEC) was performed on a BioSil TSK-250 column (7.5 × 300 mm; BioRad) utilizing a mobile phase of 40% acetonitrile/0.1% TFA at a flow rate of 0.5 ml/min. The column was calibrated using molecular weight standards dissolved in the above buffer from 66 kDa (bovine serum albumin; Ve, 4.2 ml) to 0.06 kDA (NaCl; Ve, 11.2 ml). Aliquots were dried and assayed for [^{3}H]DHP binding inhibitory activity. The activity that elutes with a molecular weight of < 1 dKa is herein denoted as FBI (brain) or SSi (stomach).

Analytical Procedures

Inhibition of binding of [^{3}H]DHPs (nimodipine, nitrendipine, nisoldipine, Bay K8644, or PN200-110) was carried out at 25°C in 0.25-2.5 ml assay volumes as previously described (Janis et al. 1983). The percent inhibition of binding was independent of the ligand used. Protease inhibitors were added in some assays (see "Results and Discussion"). [^{3}H]Ouabain and [^{3}H]saxitoxin binding were determined as described by Colvin et al. (1985) and Weigele and Barchi (1978). The binding of [^{3}H]nitrendipine to anti-DHP antibodies was studied as described by Campbell et al. (1986). The antibodies were supplied by Dr. K. Campbell (University of Iowa).

Electrophysiological effects of fractions were determined by previously published methods used to study the effect of nimodipine (Cohen and McCarthy 1987). Fractions were added either extracellularly or into the dialysis solution of the recording pipette. The whole-cell variant of the patch-electrode voltage clamp technique was used. Tail current analysis with curve peeling was used to separate the contributions of the two types of Ca^{2+} channels.

Inorganic phosphate was measured by the method of Bartlett (1959) and carbohydrate by that of Dubois et al. (1956). Amino acid analysis was performed on an Applied Biosystems instrument (420A/130A) using postcolumn PITC chemistry. The high-performance thin-layer chromatography (HPTLC) method of Macala et al. (1983) was used for lipid identification. HPTLC plates were charred in 3% cupric acetate in 8% (v/v) phosphoric acid for lipid detection. Paired plates were developed and areas eluted to determine the location of active fractions.

As an aid in the quantitation of the amount of activity present in column fractions, the following definition will be used: 1 pU activity is defined as that amount of activity which inhibits the specific binding of [^{3}H]DHP (0.2 nM [^{3}H]PN200-110, 1 nM [^{3}H]nitrendipine, of 1 nM [^{3}H]nimodipine) to rat cardiac membranes by 50%. Unlabeled nimodipine was the standard displacer.

Results and Discussion

Acid Extracts of Bovine Brain

Our standard purification protocol consisted of homogenization in acid, centrifugation, petroleum ether extraction, ultrafiltration, and preparative RP-HPLC on ODS. As shown in Fig. 1, fractions corresponding to 30%-35% acetonitrile inhibited [^{3}H]DHP binding. Electrophysiological measurements were subsequently performed to more stringently address the question of whether the activity was DHP-like in nature.

Active fractions upon intracellular dialysis in the pipette solution produced time- and voltage-dependent block of L-type Ca^{2+} channel current in GH$_3$ cells. The characteristics of this inhibition were consistent with those seen for DHP as determined by tail current analysis. However, the fraction was inactive when added to the extracellular bathing solution.

The fraction produced a concentration-dependent inhibition of [^{3}H]DHP binding to cardiac and brain membranes, but stimulated DHP binding to skeletal muscle

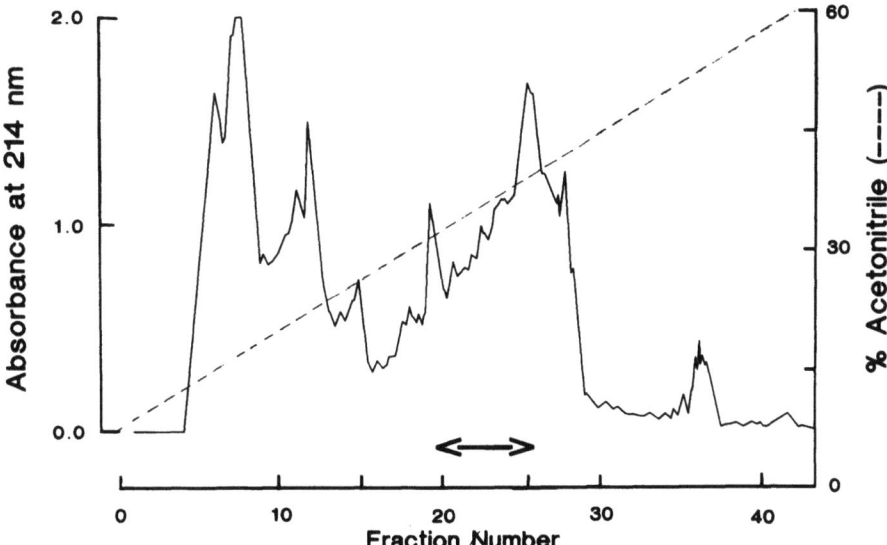

Fig. 1. Preparative reverse phase chromatography of acid-extracted fresh calf brains. The < 30 kDa ultrafiltrate from acid-extracted calf brains was loaded onto a Waters PrepPAK-500 column containing Vydac C18 packing preequilibrated in 0.1% trifluoroacetic acid (TFA). Nonabsorbed material was removed by extensive washing with 0.1% TFA. The column was eluted with a 124-min linear gradient of 0%–80% acetonitrile in 0.1% TFA. *Arrows* denote the region containing FBI and myelin basic protein as determined by [^{3}H]DHP binding, following a subsequent size exclusion HPLC step

membranes. The stimulation suggested that the fraction did not inhibit binding to cardiac and brain membranes by reducing free [^{3}H]DHP. Dialysis experiments confirmed that the fraction, in contrast to albumin, did not bind [^{3}H]DHP. The inhibition of DHP binding produced by this fraction was reversed in the presence of 10 mM Ca^{2+}. The inhibition observed was not affected by boiling or by the inclusion of protease inhibitors during binding, but was destroyed by ashing. Polyacrylamide electrophoresis in the presence of sodium dodecyl sulfate showed the presence of two bands with relative molecular weights of 18 and 20 kDa. Amino acid analysis and tryptic maps of the 18-kDa protein matched that of myelin basic protein (MBP). The IC$_{50}$ value for inhibition of [^{3}H]DHP binding to cardiac membranes of the 18-kDa protein was $\cong$ 4 μM, the same as that observed for commercially available preparations of MBP.

We attempted to obtain fragments of this peptide that would be more potent in the [^{3}H]DHP binding assay. Fragmentation and loss of activity were produced by trypsin, V8 protease, Lys-C and thrombin, but not by aminopeptidase M, carboxypeptidase Y, or formic acid treatment. Fragments more potent than MBP were not obtained.

Of considerable interest was the observation that commercially available MBP inhibited [3H]DHP binding but did not block L-type Ca^{2+} channel current. The above finding raised the question as to whether some substance(s), copurifying with MBP on preparative RP-HPLC, was responsible for the observed activity. Indeed, fractionation of the preparative RP-HPLC-purified activity by SEC in the presence of 40%

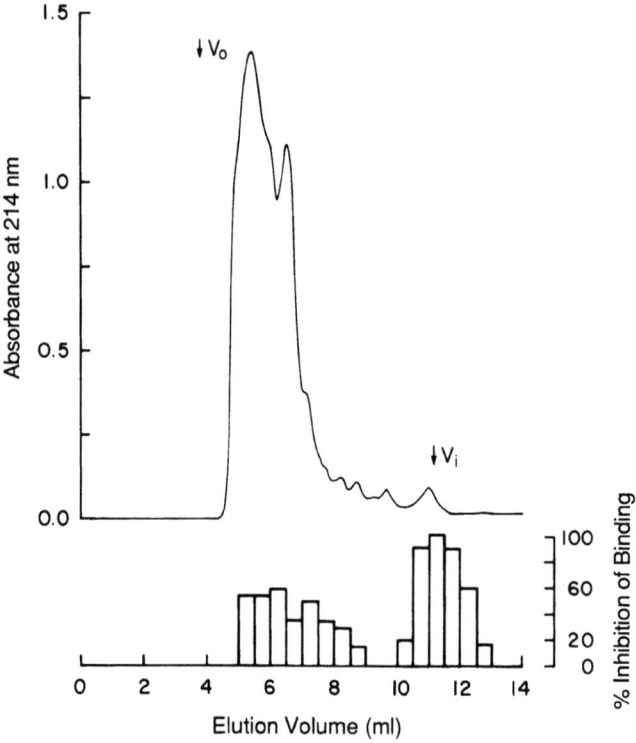

Fig. 2. Separation of myelin basic protein from FBI by size exclusion chromatography. Of the active fraction shown in Fig. 1, 2 mg (by weight) were purified on a Bio-Sil TSK-250 column using 0.1% TFA, 40% acetonitrile as the mobile phase

acetonitrile revealed the presence of two inhibitors of DHP binding: a large molecular weight fraction corresponding to MBP (MBP-containing fraction) and a small molecular weight fraction with an apparent size of < 1 kDa (Fig. 2). This low molecular weight fraction, designated FBI, differed from MBP in that its inhibitory activity could reach as high as 100% (Fig. 3), whereas the MBP-containing fraction produced a maximal inhibition of $\cong$ 50% even when present at very high concentrations (> 2mg/ml). Furthermore, the activity of FBI in the binding assay was not reversed by 10 mM Ca^{2+}, whereas that of the MBP-containing fraction was completely reversed. Based on our unit definition (1 pU = 1 pmol of substance having the same affinity as nimodipine), the application of 5 mg (by weight) preparative RP-HPLC-purified activity on SEC resulted in a yield of $\cong$ 50 pU FBI. Thus, the yield of FBI was 50 ng/kg brain (w/w) assuming a molecular weight for FBI equal to that of nimodipine (0.4 kDa). FBI obtained from the SEC was found to inhibit the binding of [^{3}H]nitrendipine to anti-DHP antibodies. However, after further purification, this inhibitory activity was lost, but activity in the [^{3}H]DHP binding assay and on calcium-channel current were not.

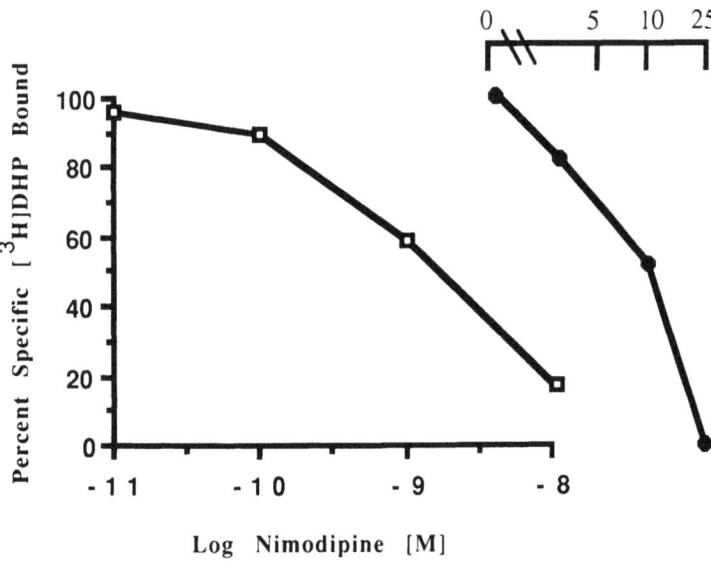

Fig. 3. Effect of increasing amounts of FBI from fresh brain purified by size exlusion chromatography. The fraction was dissolved in 10% dimethylsulfoxide and incubated for 45 min with 0.2 nM [^{3}H]DHP in 250 µl assay volume at 25°C. A concentration of 1% dimethylsulfoxide present in 25-µl aliquots of the fraction did not effect control [^{3}H]DHP binding. Increasing concentrations of unlabeled nimodipine were used for the standard curve

Subsequent electrophysiological experiments revealed that following SEC only FBI, but not the MBP-containing fraction, was capable of blocking L-type Ca^{2+} channel current. Thus, MBP dissociated from FBI by SEC, was now inactive on Ca^{2+} channel current, as previously noted for commercial preparations of MBP. In addition, SEC-purified FBI was capable of blocking L-type Ca^{2+} channel current when applied extracellularly (Fig. 4). This finding is consistent with the interpretation that FBI complexed with MBP was inactive extracellularly. The lack of effect by the MBP/FBI complex on L-type Ca^{2+} channel current from outside may reflect an inability of the complex to cross the membrane. On the other hand, intracellular activity of the complex may reflect dissociation of FBI from MBP in the intracellular but not the extracellular solutions used in electrophysiology.

The chemical nature of FBI was explored by analysis for amino acids, carbohydrates, fatty acids, and certain other lipids. High-sensitivity amino acid analysis failed to detect amino acids in the FBI fraction. Based on the amount of activity used, it can be concluded that either the sample was free of amino acids, or, if the activity is due to a peptide, it must be considerably more potent than nimodipine. Neither carbohydrate nor fatty acids were detected in the samples analyzed. Thus, the known fatty acids that we have studied that inhibit [^{3}H]DHP binding (see "Activities of Some Known Substances") were not present in the FBI fraction; they would have been easily detectable by thin-layer chromatography. In addition, a major fatty acid-rich fraction that inhibits [^{3}H]DHP binding that we have isolated by hexane-extraction was

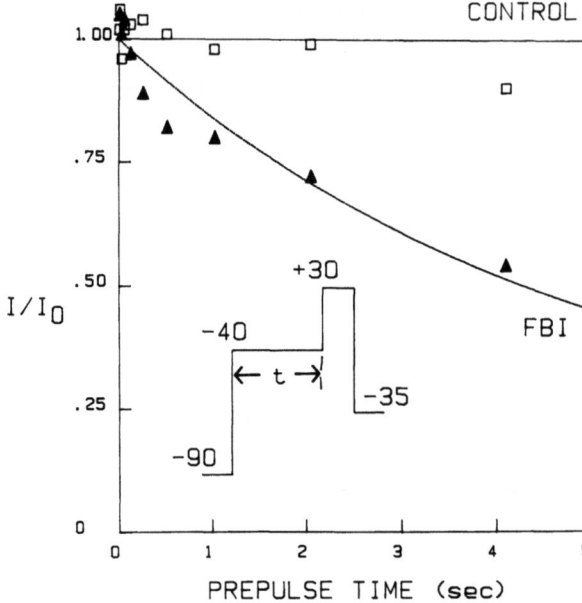

Fig. 4. Measurement of L-type calcium channel tail currents demonstrate time- and voltage-dependent onset of FBI block at −40 mV from a hyperpolarized potential (−90 mV). A pre-pulse of variable length at −40 mV is followed by a strong depolarization to assay for current through available L-type channels upon repolarization; ≅ 10 pU FBI activity was applied extracellularly

found to be inactive on Ca^{2+} channel currents. However, the concentration-inhibition curve for [^{3}H]DHP binding by FBI was steep (Fig. 3), similar to that produced by high concentrations of fatty acids, acyl carnitine and cholines (see "Activities of Some Known Substances").

FBI did not inhibit [^{3}H]ouabain or [^{3}H]saxitoxin binding to brain membranes in amounts that inhibit [^{3}H]DHP binding. The lack of effect on [^{3}H]ouabain binding is of particular interest because it indicates that certain nonesterified fatty acids and lysophospholipid, which are known to inhibit ouabain binding and Na, K-ATPase, are not present in significant amounts (Tamura et al. 1985; Kelly et al. 1986; Kelly, 1987).

FBI appears to be different from the other brain fractions that have been reported to inhibit [^{3}H]DHP binding. The small fraction of Ebersole et al. (1988) is inactive on heart membranes, the fraction of Hanbauer et al. (this volume) inhibits T-type Ca^{2+} channels, and that of Thayer et al. (1984) was reported to be of 5–10 kDa. These characteristics differentiate FBI from the other factors.

Acid Extracts of Lamb Stomach

The procedures used for the acid extraction of lamb stomach were the same as those for lyophilized bovine brain. Major fractions active in the ligand binding assay were eluted from preparative ODS HPLC columns at 43%–50% acetonitrile. This fraction produced a time- and voltage-dependent inhibition of L-type (but not T-type) Ca^{2+} channel current when added to the inside of the recording pipette.

Further purification yielded a < 1 kDa fraction (SSI) and a 8-kDa basic peptide, both of which inhibited DHP binding. The IC_{50} for inhibition of [^{3}H]DHP binding for the 8-kDa peptide was $\cong$ 10 µM, and, like other basic peptides studied, this inhibition was reversed by Ca^{2+}. Preliminary studies with proteolytic enzymes did not yield fragments more active in the DHP binding assay. The addition of the 8-kDa peptide to the patch pipette suggested that it produced only a very weak inhibition of Ca^{2+} channel current, in contrast to the apparently potent inhibition by the complex. The 8-kDa peptide, like MBP, inhibited the binding of other ^{3}H-labeled ligands, e.g., [^{3}H]D-ala-D-leu-enkephalin (µ opioid) more than it did [^{3}H]DHPs. The sequence of the 8-kDa peptide was determined; little homology to any known peptides was found.

The activity of the small stomach inhibitory fraction (SSI) did not bind to C18 columns in the presence of 0.1% TFA. SSI inhibited L-type Ca^{2+}-channel current when added extracellularly. Like FBI, the inhibition curve for DHP binding was steeper than that expected for 1 : 1 mass action at the DHP recognition site. Lipid analysis by thin-layer chromatography demonstrated that SSI did not contain detectable amounts of fatty acids. Therefore, it is unlikely that inhibitory activity in the ligand binding assay is due to the fatty acids that we have tested.

SSI, in contrast to the FBI fraction obtained from sizing columns, did not inhibit the binding of [^{3}H]nitrendipine to anti-DHP antibodies. The characteristics of some of the endogenous inhibitors of [^{3}H]DHP binding are compared in Table 1.

Table 1. Effects of fractions and substances that inhibit [^{3}H]DHP binding

Source of fraction or substance	Ca^{2+} channel current		[^{3}H]DHP binding to	
	L-Type	T-Type	Membranes	Antibodies
Bovine brain, acid extract (FBI)	Inhibit	0	Inhibit	Inhibit[a]
Myelin basic protein	0	0	Inhibit	0
Lamb stomach, acid extract (SSI)	Inhibit	0	Inhibit	0
Lamb stomach, acid extract (8 kDa)	Inhibit[b]	0	Inhibit	0

[a] Fraction from sizing column but not from cyano or C4 columns.
[b] A weak inhibitory activity was seen.

Rough estimations were made of the amounts of lyophilized lamb stomach or brain needed to isolate 1 µg SSI or FBI, arbitrarily assuming a molecular weight and affinity the same as nimodipine. The values for stomach and brain were 60 and 20 kg, respectively.

Activities of Some Known Substances

Having found that MBP inhibited [^{3}H]DHP binding, we tested a variety of basic peptides for binding inhibition (Table 2). The most potent basic peptide tested, protamine, had an IC_{50} value of more than 100 times that of nimodipine.

Several reports have demonstrated effects of basic peptides on other systems. For example, low concentrations (< 10 µg/ml) of protamine, histones H1, 2A, 2B, 3, and 4, MBP, and polylysines of > 2 kDa have also been reported to inhibit calmodulin-

Table 2. Inhibition of [³H]DHP binding by basic peptides[a]

	IC$_{50}$, µM
Myelin basic protein	3.6
Histone H2A	3.1
Poly-L-lysine (3.4 kDa)	1.3
Poly-L-lysine (8.4 kDa)	0.9
Protamine	0.26

a Inactive basic substances included aprotinin, Lys-bradykinin, spermine, and putrescine.

stimulated cyclic nucleotide phosphodiesterase (Itano et al. 1980). Protamine, histone, and MBP inhibit myosin light-chain kinase at 2–100 µg/ml (Iwasa et al. 1981). The apparent Hill slopes and potency for this inhibition is similar to that for inhibition of [³H]DHP binding. Similarly, MBP and histone H1 inhibit protein kinase C with IC$_{50}$ values of 1 and 3 µM respectively (Su et al. 1986). MBP, protamine, and histones all inhibit the binding of other ³H-labeled ligands. For example, MBP (50 µg/ml) inhibited [³H]Tyr-D-Ala-Gly-Me-Phe-Gly-ol δ opiate) binding while histone H2A (50 µg/ml) and protamine (20 µg/ml) inhibited [³H]ouabain binding to a greater extent than that of [³H]nitrendipine (G. Bolger, unpublished studies).

Fatty acids in the C$_{16}$–C$_{20}$ range had IC$_{50}$ values of 50–180 µM. Acyl carnitines and choline (C$_{16}$–C$_{18}$) and L-α-lysophosphatidyl choline-palmitoyl, had, at best, IC$_{50}$ values of ≅ 25 µM. In all cases the apparent Hill slope values were > 2. It was previously proposed that palmitoyl carnitine directly activated Ca^{2+} channels based on its ability to increase established Ca^{2+}-induced contractions of guinea pig ileum and to inhibit the binding of [³H]DHP and [³H]diltiazem (IC$_{50}$ for both, 120 µM) and [³H]D888 (IC$_{50}$, 80 µM; Spedding and Mir 1987).

A large number of endogenous substances (e. g., norepinephrine, adenosine, acetylcholine, ATP, γ-aminobutyric acid) are known to modulate L-type Ca^{2+} channels. However, there is no evidence that any of these substances act directly on the channel or DHP binding site protein (Janis et al. 1987). The effects of these substances on L-type Ca^{2+} channels are most likely mediated by GTP-binding proteins and/or by protein kinases. For example, the inhibitory effects of somatostatin (but not nifedipine) on Ca^{2+} current in AtT-20 cells is blocked by pertussis toxin (Lewis et al. 1986). Biogenic amines and peptide hormones are not thought to act directly on the DHP receptor/Ca^{2+} channel complex. These findings are consistent with their inactivity in our [³H]DHP binding assay (unpublished studies).

Summary

Two fractions have been isolated that inhibit [³H]DHP binding and L-type Ca^{2+} channel current. One of these (FBI) was from an acid extract of bovine brain, one (SSI) from a similar extract of lamb stomach. Both low molecular weight substances, FBI and SSI, coeluted with basic proteins from preparative C18 reverse phase columns. FBI was associated with MBP, and SSI with an 8-kDa peptide the amino acid sequence of which has not been previously reported. Both fractions inhibit

[^{3}H]DHP binding and produce a time- and voltage-dependent block of Ca^{2+} channel current upon intracellular dialysis. In the absence of the < 1 kDa fraction, MBP and the 8-kDa protein were inactive or poorly active (respectively) in the electrophysiological assay. Studies are underway to determine the chemical structure of the active substances in FBI and SSI.

Most commercially available lipids and all peptide hormones tested in [^{3}H]DHP binding assay were inactive. Active basic peptides included MBP, the 8-kDa peptide from stomach, certain histones, and protamine. Protamine was the most active commercially available substance, with an IC$_{50}$ value of 0.26 μM, about one-tenth that of the other basic peptides. Acyl carnitines and cholines with side chains of C$_{14}$-C$_{20}$ inhibited [^{3}H]DHP binding with IC$_{50}$ values of 25-80 μM.

Acknowledgements. We thank Donna Winters for technical assistance in ligand binding and isolation procedures and Gary Davis for amino acid analysis and for sequencing of the 8-kDa peptide from stomach. We also thank Dr. Jeff Conn for assistance in lipid analysis.

References

Bartlett GR (1959) Phosphorous assay in column chromatography. J Biol Chem 234:466-468

Battaini F, Govoni S, Del Vesco R, Magnoni MSC, Trabucchi M (1986) Interaction between substance P and calcium ion homeostasis in rat hippocampus. Soc Neurosci Abstr 12:176a

Bennett HPJ, Browne CA, Goltzman D, Solomon S (1979) Isolation of peptide hormones by reversed-phase high pressure liquid chromatography. In: Gross E, Meienhoger J (eds) Peptides, structure and biological function, Pierce Chem Co, Rockville, IL, pp 121-125

Campbell KP, Sharp AH, Kahl SD (1987) Anti-dihydropyridine antibodies exhibit [^{3}H]nitrendipine binding properties similar to the membrane receptor for the 1,4-dihydropyridine Ca^{2+} channel antagonists. J Cardiovasc Pharmacol 9(Suppl 4): S113-S121

Cohen CJ, McCarthy RT (1987) Nimodipine block of calcium channels in rat anterior pituitary cells. J Physiol 387:195-225

Colvin RA, Ashavaid TF, Herbette LG (1985) Strucutre-function studies of canine cardiac sarcolemmal membranes. I. Estimation of receptor site densities. Biochim Biophys Acta 12:601-608

Dubois M, Gilles KA, Hamilton JK, Rebers RA, Smith F (1956) Colorimetric method for determination of sugars and related substances. Anal Chem 28:350-356

Ebersole BJ, Gajary ZL, Molinoff PB (1988) Endogenous modulators of binding of [^{3}H]nitrendipine in extracts of rat brain. J Pharmacol Exp Ther 244: 971-976

Gengo P, Skattebol A, Moran JF, Gallant S, Hawthorn M, Triggle DJ (1988) Regulation by chronic drug administration of neuronal and cardiac calcium channel, γ-adrenoceptor and muscarinic receptor levels. Biochem Pharmacol 37:627-633

Hanbauer I, Sanna E (1986) Endogenous modulator for nitrendipine binding sites. Clin Neuropharm 9, S4:220-222

Hess P, Lansmann JB, Tsien RW (1984) Different modes of calcium channel gating behavior favoured by dihydropyridine Ca^{2+} agonists. Nature 311:538-544

Itano T, Itano R, Penniston JT (1980) Interactions of basic polypeptides and proteins with calmodulin. Biochem J 189:455-459

Iwasa Y, Iwasa T, Matsui K, Higashi K, Miyamoto E (1981) Interaction of calmodulin with chromatin associated proteins and myelin basic protein. Life Sci 29:1369-1377

Janis RA, Krol GJ, Noe AJ, Pan M (1983) Radioreceptor and high-performance liquid chromatographic assays for the calcium channel antagonist nitrendipine in serum. J Clin Pharmacol 23:266-273

Janis RA, Silver PJ, Triggle DJ (1987) Drug action and cellular calcium regulation. Adv Drug Res 16:309-591

Kelly RA, O'Hara DS, Mitch WE, Smith TW (1986) Identification of NaK-ATPase inhibitors in human plasma as nonesterified fatty acids and lysophospholipids. J Biol Chem 261:11704–11711

Kelly RA (1987) Endogenous cardiac glycosidelike compounds. Hypertension 10 (Suppl):I87–I92

Lewis DL, Luini A, Weight FF (1986) A guanine nucleotide-binding protein mediates inhibition of voltage-dependent calcium current by somatostatin in a pituitary cell line. Proc Natl Acad Sci USA 83:9035–9039

Macala LJ, Yu RK, Ando S (1983) Analysis of brain lipids by high performance thin-layer chromatography and densitometry. J Lipid Res 24:1243–1250

Mantione CR, Goldman ME, Martin B, Bolger GT, Lueddens HWM, Paul SM, Skolnick P (1988) Purification and characterization of an endogenous protein modulator of radioligand binding to "peripheral-type" benzodiazepine receptors and dihydropyridine Ca^{2+} channel antagonist binding sites. Biochem Pharmacol 37:339–347

Nair V, Offerman RJ, Turner GA (1986) Novel fluorescent 1,4-dihydropyridines. J Amer Chem Soc 108:8283–8285

Panza G, Grebb JA, Sanna E, Wright AG, Hanbauer I (1985) Evidence for down regulation of [^{3}H]nitrendipine recognition sites in mouse brain after long term treatment with nifedipine or verapamil. Neuropharmacology 24:1113–1117

Quirion R, DiMaggio DA, French ED, Contreras PC, Shiloach J, Pert CB, Everist H, Pert A, O'Donohue TL (1984) Evidence for an endogenous peptide ligand for the phencyclidine receptor. Peptides 5:967–973

Schramm M, Thomas G, Towart R, Franchowiak G (1983) Novel dihydropyridines with positive inotropic action through activation of Ca^{2+} channels. Nature 303:535–537

Scott RH, Dolphin AC (1987) Activation of a G protein promotes agonist responses to calcium channel ligands. Nature 330:760–762

Spedding A, Mir AK (1987) Direct activation of Ca^{2+} channels by palmitoyl carnitine, a putative endogenous ligand. Br J Pharmac 92:457–468

Su H-D, Kemp, BE, Turner RS, Kuo JF (1986) Synthetic myelin basic protein peptide analogs are specific inhibitors of phospholipid/calcium-dependent protein kinase (protein kinase C). Biochem Biophys Res Commun 134:78–84

Tamura M, Kuwano H, Kinoshita T, Inagami T (1985) Identification of linoleic and oleic acids as endogenous Na^+, K^+-ATPase inhibitors from acute volume-expanded hog plasma. J Biol Chem 260:9672–9677

Thayer SA, Stein L, Fairhurst AS (1984) Endogenous modulator of calcium channels. IUPHAR 9th Intl Congress of Pharmacol 892p (abst)

Triggle DJ, Janis RA (1987) Calcium channel ligands. Annu Rev Pharmacol Toxicol 27:347–369

Weigele JB, Barchi RL (1978) Analysis of saxitoxin binding in isolated rat synaptosomes using a rapid filtration assay. FEBS Lett 91:310–314

Endothelium-Derived Novel Vasoconstrictor Peptide Endothelin: A Possible Endogenous Agonist for Voltage-Dependent Ca^{2+} Channels*

M. Yanagisawa, H. Kurihara, S. Kimura, K. Goto, and T. Masaki

Institute of Basic Medical Sciences, University of Tsukuba, Tsukuba, Ibaraki 305, Japan

Introduction

The tonus of blood vessels is regulated by various neural and hormonal signals together with the local regulatory mechanisms intrinsic to the vessel wall. The discovery by Furchgott and Zawadzki [1] that the acetylcholine-induced vasodilatation is dependent on the presence of an intact endothelium has stimulated intense interest in the role of the endothelium in modulating vascular responsiveness. It has been hypothesized that, when stimulated by a variety of vasoactive agents, including acetylcholine, bradykinin, and Ca^{2+} ionophore A23187, endothelial cells (EC) release short-lived endothelium-derived relaxing factor(s) (EDRF) causing relaxation of the underlying smooth muscle [2, 3]. One EDRF has recently been identified as nitric oxide or a closely related nitrosyl compound [4]. Substantial data are now available which demonstrate that, in addition to mediating vasodilatation, EC can also facilitate contractile responses of the vascular smooth muscle (see [5] for a review). Various chemical and mechanical factors, including noradrenaline, thrombin, neuropeptide Y, A23187, arachidonic acid, hypoxia/anoxia, stretch, and increased transmural pressure, have been found to cause vasoconstriction dependent on and/or enhanced by endothelium. These responses are thought to be mediated by at least several different classes of diffusible substances collectively termed endothelium-derived constricting factors (EDCFs). Although two of these EDCFs are believed to be peptide(s) and eicosanoid(s), none of them has been chemically identified yet.

Recent reports have also described a protease-sensitive vasoconstrictor activity with a relatively low molecular weight ($M_r < 8000$) in the supernatant of cultured EC [6, 7]. We have now identified and molecularly cloned a novel potent vasoconstrictor peptide from porcine [8], rat [9], and human [10] EC. This peptide, endothelin (ET), exhibits a characteristically long-lasting activity both in vivo and in vitro and seems to be the most potent of known vasoconstrictors. Biosynthesis of ET in cultured EC is apparently regulated at the transcriptional level in response to various chemical and mechanical stimuli, indicating the existence in mammals of a novel endothelium-mediated cardiovascular control system. Further, although belonging to an entirely new family of bioactive peptide, the structure of ET bears a resemblance to several

---

* This study was supported in part by research grants from the Ministry of Education, Science, and Culture of Japan.

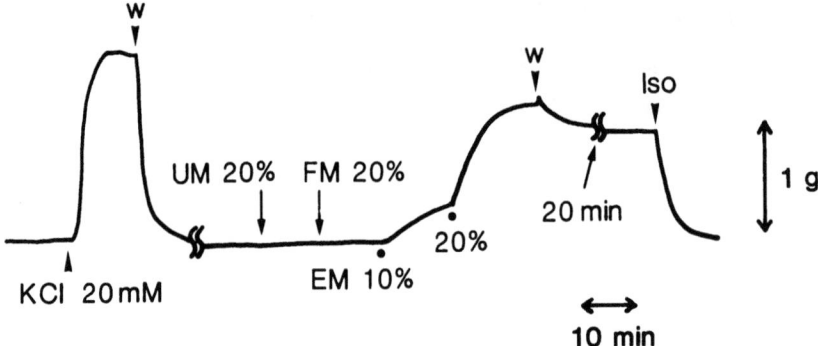

Fig. 1. Constrictive response of a porcine coronary artery strip to culture supernatant from porcine aortic EC. *UM*, unconditioned medium (Eagle's MEM plus 10% horse serum); *FM*, medium conditioned with fibroblasts (human, IMR-90); *EM*, medium conditioned with EC; *Iso*, 10^{-7} M isoproterenol; *W*, wash with Krebs-Ringer solution

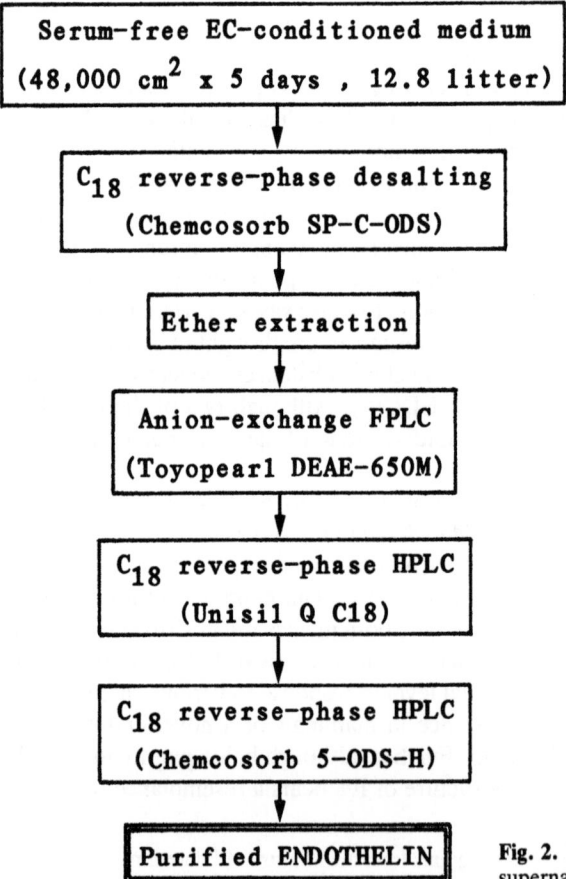

Fig. 2. Purification of ET from culture supernatant of porcine aortic EC

groups of peptide neurotoxins that act on membrane ion channels. This finding, in conjunction with the pharmacological properties of the action of ET, may raise the possibility that ET acts as an endogenous activator of voltage-dependent, dihydropyridine (DHP) sensitive Ca^{2+} channels in vascular smooth muscle cells.

Isolation and Structure Determination of Endothelin

We cultured EC obtained from porcine aortas in Eagle's minimum essential medium supplemented with 10% horse serum. The supernatant from confluent as well as preconfluent monolayer cultures of the EC caused an endothelium-independent, slow-onset contraction when added to porcine coronary artery strips in a concentration-dependent manner (Fig. 1). The supernatant constricted a variety of arteries and veins from various species, including rat, guinea pig, rabbit, dog, cat and human. The constrictive activity was extremely long-lasting and difficult to wash out (lasted more than 3 h despite repeated washes with Krebs-Ringer solution), although rapidly and completely reversed by the addition to the bath of isoproterenol or glyceryl trinitrate. The activity was dependent on extracellular Ca^{2+}; the constrictive response was abolished in a Ca^{2+}-free, EGTA-containing solution. Pretreatment of the conditioned medium with trypsin abolished the activity. The activity seemed to be gradually accumulated in the supernatant during the culture period of 1–7 days. The addition of serum was not essential for the production of the activity by the EC. Further, similar level of the activity was found even in the supernatant from the EC monolayer that had been maintained in serum-free medium for up to 4–5 weeks. These observations suggest that ET is not a derivative of a serum component.

We succeeded in purifying the peptidergic vasoconstrictor from the supernatant of a large-scale, serum-free monolayer culture of the EC [8]. The purification strategy is summarized in Fig. 2. Briefly, the conditioned medium was first desalted and concentrated by a preparative reversed-phase column and the organic solvent removed by ether extraction. The concentrated medium was then purified by three steps of HPLC, the constrictor activity contained in the fractions from each step being monitored by the bioassay with porcine coronary artery strips. The final ET fraction, which is confirmed to be eluted as a single peak on analytical anion-exchange and reverse-phase HPLC, was subjected to amino acid analysis, gas-phase peptide sequencing and carboxy-terminal analysis with carboxypeptidase Y. Porcine ET was shown to be an acidic 21-residue (2492 M_r) peptide with free amino- and carboxy-termini containing two sets of intrachain disulfide bridges (Fig. 3). The structure was confirmed by preparing synthetic ET by a solid-phase chemistry with selective cross-linking of the four Cys residues. The synthetic peptide exhibited a complete biological activity, and retention times identical to those of natural peptide on a C_{18} reverse-phase HPLC (Fig. 4) and a DEAE anion-exchange HPLC.

Cloning and Structure of cDNA for Endothelin Precursor

We subsequently cloned the mRNA encoding for ET by screening, with a synthetic DNA probe, the cDNA libraries constructed for EC from porcine aortas and human

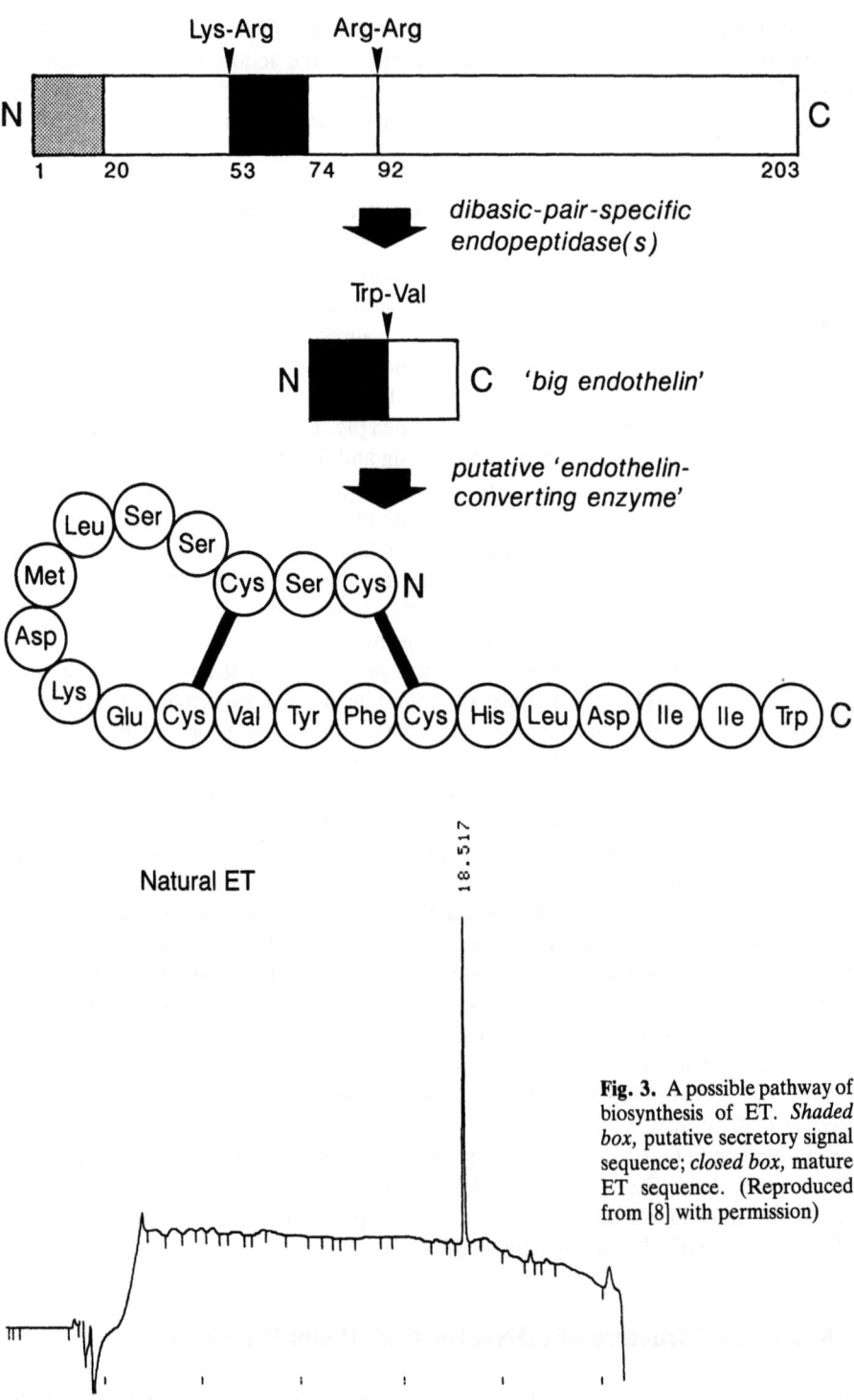

Fig. 3. A possible pathway of biosynthesis of ET. *Shaded box*, putative secretory signal sequence; *closed box*, mature ET sequence. (Reproduced from [8] with permission)

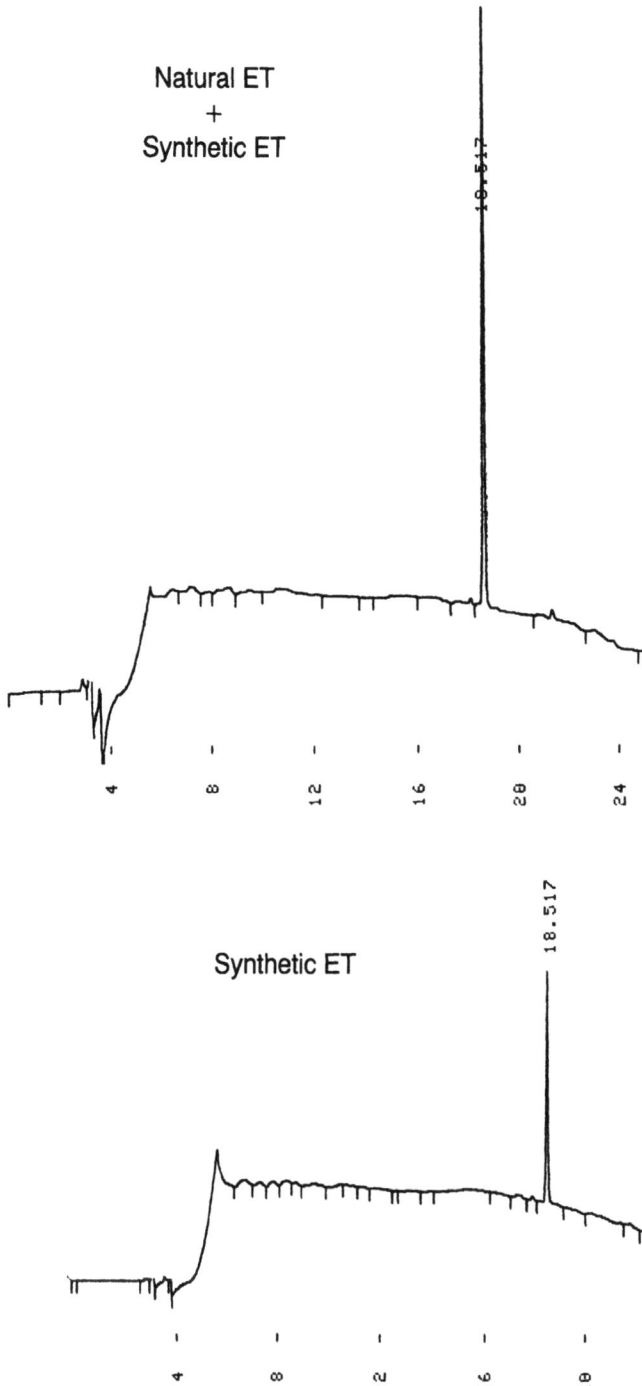

Fig. 4. Comparison of natural and synthetic ET on C_{18} reverse-phase HPLC. *Column,* Chemcosorb 5-ODS-H, 4.6 × 250 mm (Chemco). A linear gradient (10%–50% v/v) of CH_3CN in 0.1% trifluoroacetic acid was applied at a flow rate of 1 ml/min. The eluate absorbance at 210 nm was recorded

umbilical veins [8, 10]. The nucleotide and deduced amino acid sequence of the cDNA for porcine and human ET precursors are shown in Fig. 5. The amino-terminal 18 residues of both of the deduced amino acid sequences are characteristic of secretory signal (pre-) sequences with a hydrophobic core followed by residues with small polar side chains. The presence in vascular EC of mRNA encoding the prepro- form of ET indicates that this peptide is produced by de novo synthesis and processing in a manner similar to that of many peptide hormones. The porcine and human precursor

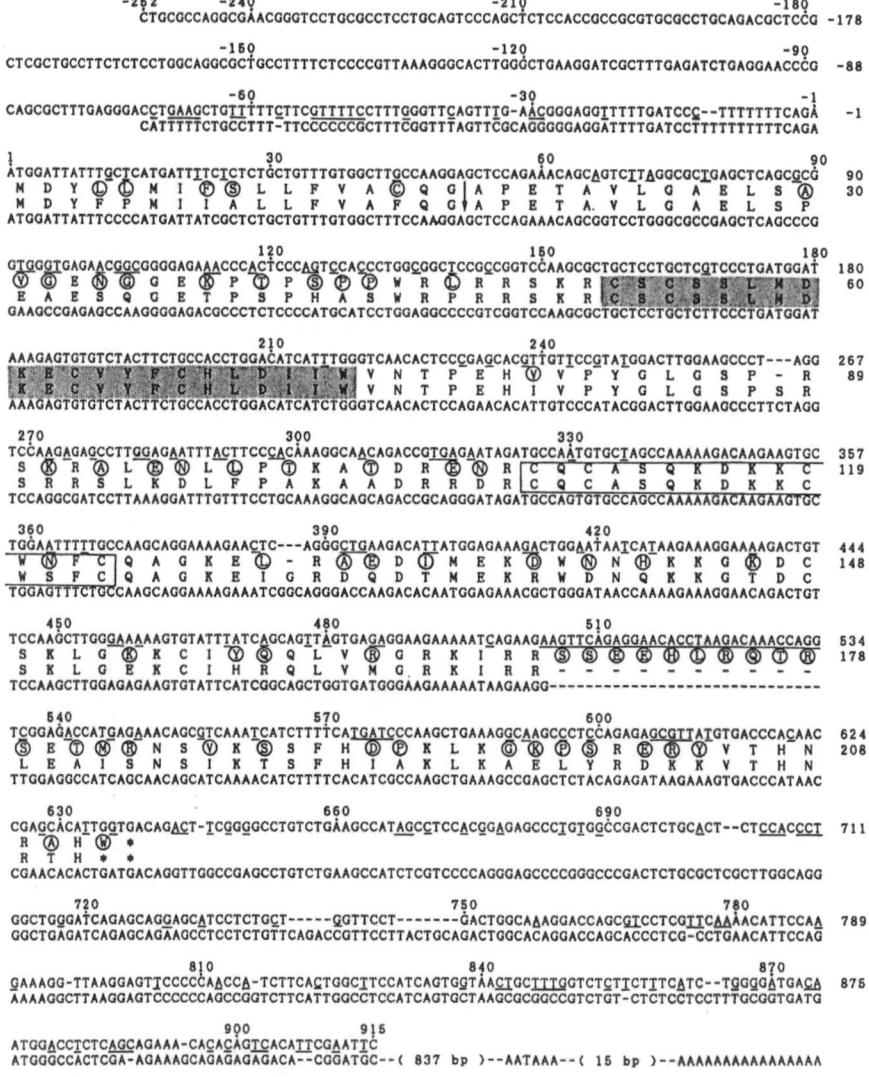

Fig. 5. Nucleotide and deduced amino acid sequences of porcine *(p)* and human *(h)* preproET cDNA. Nucleotide and amino acid substitutions are *underlined* and *circled*, respectively. *Arrow*, a putative cleavage site of the signal peptide. Sequences corresponding to mature ET and the "ET-like peptides" are indicated by *shaded* and *open boxes*, respectively. (Reproduced from [10] with permission)

shows a strong homology in both the nucleotide (79% identities) and the amino acid sequences (69% identities; 80% identities plus conservative substitutions). Especially the sequences flanking the mature ET and the ET-like peptide (boxed in Fig. 5) are nearly perfectly conserved. Although we have not yet determined the structure of human ET at the peptide level, the cDNA sequence shows that mature human ET is very likely identical to porcine ET.

As anticipated, paired basic residues Lys-Arg, which are known to be recognized by the processing endopeptidases [11], directly precede the ET sequences. Surprisingly, however, the carboxy-termini of both porcine and mature human ET lacks the dibasic pairs. This indicates that the mature ET found in the EC-conditioned medium is generated through a very unusual proteolytic processing of the putative 38-residue (human) or 39-residue (porcine) "big endothelin" between Trp and Val residues, presumably involving an endopeptidase with a chymotrypsin-like specificity (Fig. 3). The localization and regulation of this putative "endothelin-converting enzyme" may have an important implication on the control of the biosynthesis of ET.

Expression of Preproendothelin mRNA

The expression of preproET mRNA was examined in porcine tissues and cultured aortic EC by Northern blot analysis with the preproET cDNA as a probe (Fig. 6). The blots showed that the mRNA was basally expressed not only in the cultured EC but also in the aortic intima in situ. Porcine brain, atrium, lung, or kidney did not contain a detectable amount of preproET mRNA. This suggest that little ET is produced by the EC of the microvasculature within these tissues. However, ET showed a potent in

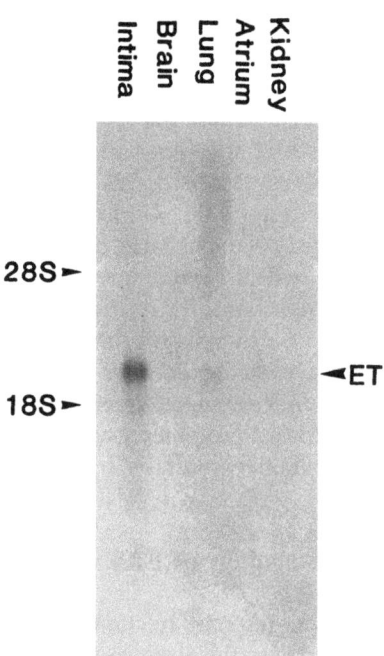

Fig. 6. Northern blots of RNA from porcine tissues probed with porcine preproET cDNA. Intimal cells were directly scraped off from the fresh thoracic aortas; 10 μg total RNA per lane. Exposure (7 days) was much longer than in Fig. 7

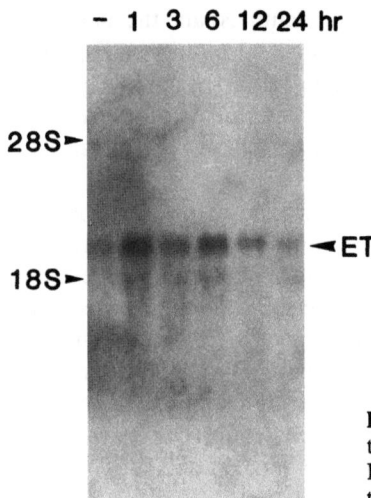

Fig. 7. Induction of preproET mRNA by thrombin in cultured porcine aortic EC. RNA from confluent EC (5 µg total RNA per lane) was extracted immediately before (−) and at the designated number of hours after the addition of bovine thrombin (2 NIH U/ml) to the medium. Autoradiography was exposed for 10 h

vivo pressor activity (Fig. 9; see below). ET might be produced chiefly in larger vessels and carried to precapillary resistance vessels as a circulating hormone. It is likely that the basal tonus of vascular beds in vivo can actually be influenced by the constitutive level of ET production since preproET mRNA is expressed in EC in situ. However, the amount of the mRNA in the aortic intima in situ was much less (approximately one-tenth) than that in the cultured EC. According to the well-accepted hypothesis that cultured EC approximate to a state of "injury" in vivo, it is speculated that ET might be produced more actively by the injured EC, like those in atherosclerotic lesions, and could contribute to the pathogenesis of the vasospasm that is preferentially seen in these lesions.

PreproET mRNA is not only constitutively expressed but also markedly induced by various vasoactive agents including thrombin (Fig. 7), adrenaline, transforming growth factor β (H. Kurihara, unpublished data) ionomycin, and phorbol esters (M. Yanagisawa, unpublished data). The mRNA seems to be transiently induced also by a fluid-mechanical shear stress (M. Yoshizumi, personal communication). Therefore, the production of ET is regulated in response to a wide variety of chemical and mechanical stimuli probably at the level of mRNA transcription. There seems to be an extensive overlap between the stimuli that induce ET and those that induce EDRF. These findings suggest possible involvement of ET in a variety of physiological and pathological processes such as the regulation of blood pressure and/or local blood distribution, hemostasis, hypertension, vasospasm, and atherosclerosis.

Pharmacology of Endothelin: An Endogenous Ca^{2+}-Channel Agonist?

The dose-response relationship of the vasoconstrictor effect of ET on porcine coronary artery strips is shown in Fig. 8. The maximum tensions developed are compar-

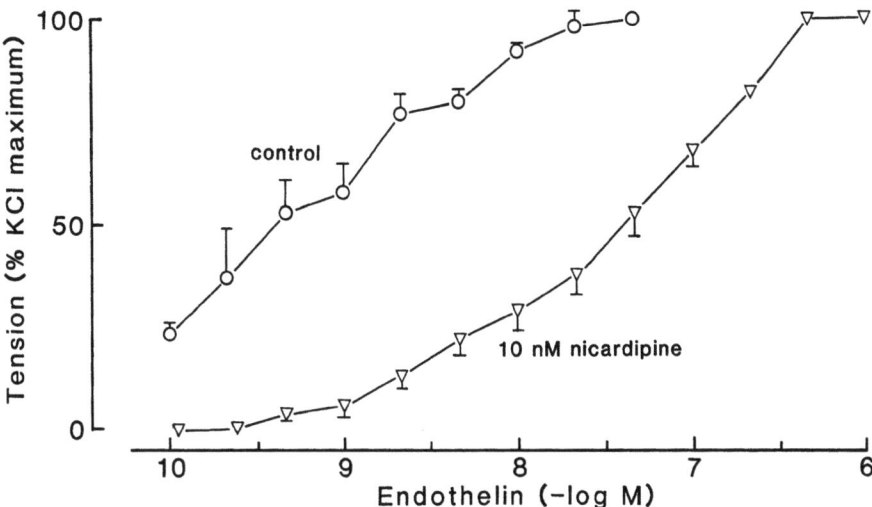

Fig. 8. Effect of nicardipine on dose-response relationship for constrictive response of deendothelialized porcine coronary artery strips to cumulatively applied ET ($n = 5$). The contraction was measured as increases in isometric tension and expressed as percentages of the maximum KCl-induced tension increment. Nicardipine was applied to the bath 15 min before the addition of ET

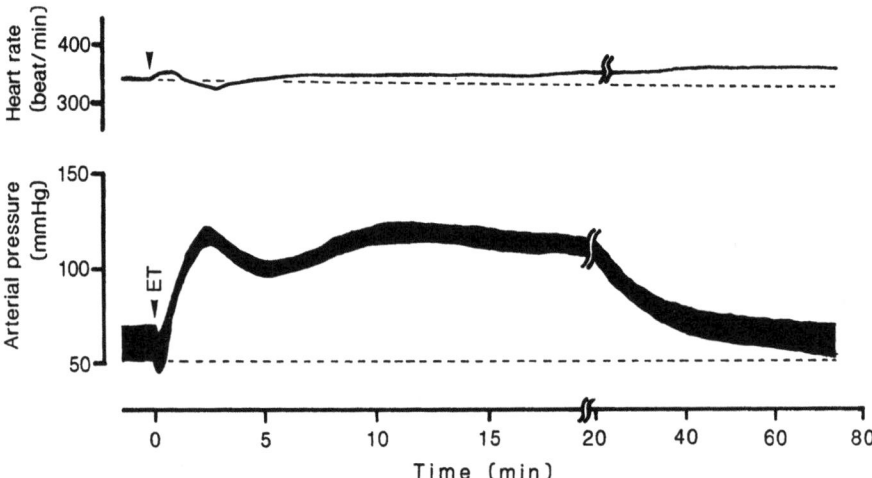

Fig. 9. In vivo pressor effect of ET (1 nmol/kg i.v., bolus in an anesthetized rat pretreated with atropine (0.25 mg/kg i.v), propranolol (1 mg/kg i.v.) and bunazosin (1 mg/kg i.v.). Blood pressure was measured directly from a cannula inserted into the right carotid artery.

able to those of KCl-induced contraction. The EC_{50} value in this assay was $4-5 \times 10^{-10}$ M, which is at least two orders of magnitude lower than any other known vasoconstrictors effective on this artery. This indicates that ET is the most potent mammalian vasoconstrictor known to date. In vivo, intravenous bolus injection of ET (0.1–3 nmol/kg) caused a transient fall followed by a sustained rise in arterial pressure in anesthetized, chemically denervated rats (Fig. 9). The mechanism of the initial

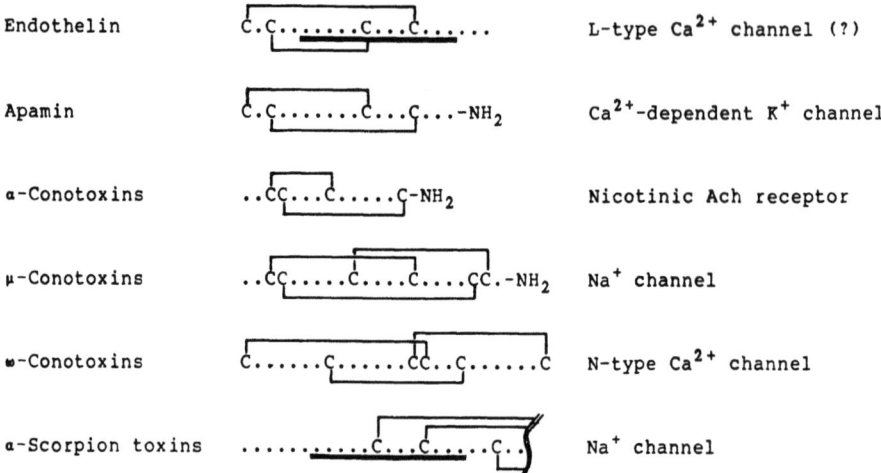

Fig. 10. Structural comparison of ET and several peptide neurotoxins acting on membrane ion channels. Only the half-cystine residues are shown. *Lines,* disulfide bridges; *solid bars,* regional homologies between ET and α-scorpion toxins [8]

transient hypotensive response is currently unknown. However, preliminary observations suggest that ET may induce the production of prostacylin and/or EDRF from endothelium (J. Vane, personal communication). The most prominent feature of the pressor response is its time course; typically, more than 40–60 min is required for return of arterial pressure to the baseline levels. These findings imply that ET may be involved in a long-term (hour- to day-basis) regulation of blood pressure.

The general structure of ET with multiple disulfide bonds within a single, relatively short peptide chain is previously unknown among bioactive peptides of mammalian origin. This type of configuration is rather common to groups of peptide neurotoxins from lower animals that act on membrane channels (Fig. 10). Further, the sequence of ET showed a significant regional homology to α-scorpion toxins, which bind to and inhibit the inactivation of tetrodotoxin-sensitive neuronal Na^+ channels [8] (underlinings in Fig. 10). These findings raise the possibility that ET, like these neurotoxins, acts on membrane ion channels.

In fact, the action of ET is, as mentioned above, dependent on the presence of extracellular Ca^{2+}. Further, Fig. 8 shows that the ET-induced contraction is very effectively inhibited by the preadministration of low doses of DHP Ca^{2+}-channel blocker nicardipine. The dose-response curve was shifted to the right approximately 100-fold in the presence of 10^{-8} M nicardipine. However, the inhibition was completely overcome by sufficiently higher doses of ET. Thus, nicardipine antagonizes the effect of ET in an apparently competitive manner, suggesting that the DHP Ca^{2+}-channel antagonist and ET share the cellular target(s) for their action. Therefore, ET appears to act principally by activating the DHP-sensitive Ca^{2+} channels. It is consistent with this hypothesis that ET actually exerts a positive inotropic effect on isolated rat atrial muscles, which is also efficiently antagonized by nicardipine (T. Ishikawa, personal communication). Although further studies are required to prove that ET acts on the channel directly, we propose that endothelin is an endogenous agonist for the DHP-sensitive Ca^{2+} channels.

What are the implications of these considerations? The potent antagonism against the action of endogenous ET by DHP Ca^{2+} channel blockers may partly explain the powerful and clinically important antihypertensive and spasmolytic effect of these drugs. The DHP-sensitive Ca^{2+} channels may function not only as voltage-gated channels but also as receptors for the endogenous messenger molecule. ET would potentially be a useful tool in dissecting the structure, function, and diversity of the Ca^{2+} channels.

References

1. Furchgott RF, Zawadzki JV (1980) The obligatory role of endothelial cells in the relaxation of arterial smooth muscle by acetylcholine. Nature 288:373–376
2. Furchgott RF (1984) The role of endothelium in the responses of vascular smooth muscle to drugs. Annu Rev Pharmacol Toxicol 24:175–197
3. Vanhoutte PM, Rubanyi GM, Miller VM, Houston DS (1986) Modulation of vascular smooth muscle contraction by the endothelium. Annu Rev Physiol 48:307–320
4. Palmer RMJ, Ferrige AG, Moncada S (1987) Nitric oxide release accounts for the biological activity of endothelium-derived relaxing factor. Nature 327:524–526
5. Rubanyi GM (1988) Endothelium-derived vasoconstrictor factors. In: Ryan US (ed) Endothelial cells, vol 3. CRC Press, Boca Raton, Florida, pp 61–74
6. Hickey KA, Rubanyi G, Paul RJ, Highsmith RF (1985)Characterization of a coronary vasoconstrictor produced by cultured endothelial cells. Am J Physiol 248:C550–C556
7. O'Brien RF, Robbins RJ, McMurtry IF (1987) Endothelial cells in culture produce a vasoconstrictor substance. J Cell Physiol 132:263–270
8. Yanagisawa M, Kurihara H, Kimura S, Tomobe Y, Kobayashi M, Mitsui Y, Yazaki Y, Goto K, Masaki T (1988) A novel potent vasoconstrictor peptide produced by vascular endothelial cells. Nature 332:411–415
9. Yanagisawa M, Inoue A, Ishikawa T, Kasuya Y, Kimura S, Kumagaye S, Nakajima K, Watanabe TX, Sakakibara S, Goto K, Masaki T (1988) Primary structure, synthesis and biological activity of rat endothelin, an endothelium-derived vasoconstrictor peptide. Proc Natl Acad Sci USA In press
10. Itoh Y, Yanagisawa M, Ohkubo S, Kimura C, Kosaka T, Inoue A, Ishida N, Mitsui Y, Onda H, Fujino M, Masaki T (1988) Cloning and sequence analysis of cDNA encoding the precursor of a human endothelium-derived vasoconstrictor peptide, endothelin: identity of human and porcine endothelin. FEBS Lett 231:440–444
11. Turner AJ (1986) Processing and metabolism of neuropeptides. Essays Biochem 22:69–119

Ca²⁺ Channel Antibodies: Subunit-Specific Antibodies as Probes for Structure and Function

K. P. Campbell, A. T. Leung, A. H. Sharp, T. Imagawa, and S. D. Kahl

Department of Physiology and Biophysics, University of Iowa College of Medicine, Iowa City, Iowa 52242, USA

Introduction

Voltage-dependent Ca^{2+} channels are known to exist in cardiac, skeletal, and smooth muscle cells as well as in neuronal and secretory cells [1, 2]. 1,4-Dihydropyridines are potent blockers of voltage-dependent Ca^{2+} channels [3], and the receptor for 1,4-dihydropyridines has been found to be highly enriched in the transverse tubular system of skeletal muscle [4]. Curtis and Catterall [5] were the first to purify the dihydropyridine receptor from rabbit skeletal muscle T-system membranes. Analysis of their preparation of receptor by sodium dodecyl sulfate polyacrylamide gel electrophoresis (SDS-PAGE) suggested that the dihydropyridine receptor consisted of three subunits: an α subunit of 160000 Da, a β subunit of 50000 Da, and a γ subunit of 32000 Da. The apparent molecular weight of the α subunit in their preparation shifted from 160000 to 130000 upon reduction, whereas the molecular weight of the β and γ subunits did not change upon reduction. The dihydropyridine receptor has also been purified from skeletal muscle membranes by Borsotto et al. [6] and Flockerzi et al. [7]. These groups also identified three subunits in their preparations of dihydropyridine receptor but the exact composition of subunits and molecular weight of the subunits differ from the original report of Curtis and Catterall [5]. Our laboratory has shown that the purified 1,4-dihydropyridine receptor from rabbit skeletal muscle triads contains four protein components of 175000 Da ($α_2$), 170000 Da ($α_1$), 52000 Da (β) and 32000 Da (γ) and that the 170000 Da and 175000 Da components are distinct polypeptides [8]. The 170000 Da polypeptide (α subunit) has been shown by photoaffinity labeling with [³H]azidopine and [³H]PN200–110 to contain the dihydropyridine binding site of the receptor [9, 10], and the 170000 Da ($α_1$ subunit) polypeptide and 52000 Da polypeptide (β subunit) have been shown to be substrates for various protein kinases [11–15]. Finally, the primary structure of the $α_1$ subunit shows considerable sequence and structural similarities to the α subunit of the sodium channel [16].

The structure or function of the lower molecular weight subunits of the dihydropyridine receptor has yet to be clearly identified. The smaller polypeptides (32000 and 52000 Da polypeptides) have been associated with the dihydropyridine receptor only by their presence on SDS polyacrylamide gels in the purified receptor preparations, and some of these smaller polypeptides have not been observed in certain preparations of purified receptor. In this report, we describe our work with monoclonal antibodies to the dihydropyridine receptor of the voltage-dependent Ca^{2+} channel.

Our results demonstrate that the 175000 Da (α_2), 170000 Da (α_1), and 52000 Da (β) polypeptides are distinct and integral subunits of the dihydropyridine receptor of the voltage-dependent Ca^{2+} channel.

Experimental Procedures

Heavy microsomes or isolated triads were purified from adult rabbit skeletal muscle in the presence of protease inhibitors as described by Sharp et al. [9]. Transverse tubular vesicles were isolated from rabbit skeletal muscle according to Rosemblatt et al. [17] in the presence of protease inhibitors. Light sarcoplasmic reticulum vesicles were isolated from rabbit skeletal muscle in the presence of protease inhibitors by the method of Campbell et al. [18]. Protein was quantitated by the method of Lowry et al. [19] as modified by Peterson [20]. [^{3}H]PN200–110 binding to isolated membranes was determined as previously described by Leung et al. [8].

Dihydropyridine receptor was purified from triads using wheat-germ agglutinin (WGA) Sepharose affinity chromatography and diethylaminoethanol (DEAE) cellulose ion-exchange chromatography as described by Leung et al. [8, 15]. All buffers used in the preparation contained 0.5 M sucrose, and the solubilization buffer contained the following protease inhibitors: pepstatin A (0.6 μg/ml), aprotinin (0.5 μg/ml), iodoacetamide (18.5 μg/ml), leupeptin (0.5 μg/ml), benzamidine (0.75 mM), and phenylmethylsulfonyl fluoride (PMSF; 0.1 mM). All other buffers contained 0.1 mM PMSF and 0.75 mM benzamidine. The partially purified dihydropyridine receptor from the initial WGA-Sepharose affinity column was referred to as the NAG N-acetyl-D-glucosamine) eluted dihydropyridine receptor, or NAG-eluate. Detergent-solubilized proteins were quantitated by the method of Lowry et al. [19] as modified by Peterson [20] after the proteins were precipitated with 5% trichloroacetic acid in the presence of 0.5 mg sodium deoxycholate. The purified dihydropyridine receptor was analyzed by SDS-PAGE on 5%–16% gradient gels according to the method of Laemmli [21] under both nonreducing (10 mM N-ethylmaleimide in sample buffer) and reducing (5mM dithiothreitol in sample buffer) conditions. Subunit stoichiometry of the purified dihydropyridine receptor was determined by densitometric scanning of 10 μg of the purified dihydropyridine receptor from nine preparations, isolated using either digitonin or 3-[(3-cholamidopropyl) dimethylammonio]-1-propanesulfonate (CHAPS) for solubilization. The Coomassie Blue stained gel was scanned with a Hoefer Model GS300 scanning densitometer. The relative densities of the various bands were determined using the GS350H Data System software from Hoefer.

Hybridoma cells lines were prepared from mice which were initially immunized with isolated triads which are enriched in the dihydropyridine receptor. Tail bleeds from the immunized mice were screened using an immunoblot assay with purified dihydropyridine receptor, and positive mice were then boosted with two intraperitoneal injections of NAG-eluted dihydropyridine receptor followed by an intravenous injection of purified dihydropyridine receptor 2 days before fusion. Spleen cells from the mice were fused with NS-1 myeloma cells [22]. Hybrid cells were grown and passaged in RPMI-1640 medium supplemented with 10% fetal bovine serum.

Hybridoma supernatants were screened by an immunodot assay [23] against light sarcoplasmic reticulum vesicles, skeletal muscle triads, NAG-eluted dihydropyridine receptor, and the void from the WGA-Sepharose (which is depleted of dihydropyridine receptor). The immunodot assay positive monoclonal antibodies were further screened for their ability to immunoprecipitate the [^{3}H[PN200–110-labeled digitonin-solubilized receptor. Monoclonal antibody beads were prepared by incubating 15 bed volumes of hybridoma supernatants with goat anti-mouse IgG (GAM-IgG) Sepharose (Cooper, diluted to an IgG binding capacity of 1 mg/ml with Sepharose CL 4B) to form MAb-GAM-IgG beads. Triad vesicles were labeled with 10 nM [^{3}H]PN200–110 and solubilized with 1% digitonin. The solubilized membranes were then diluted 1:10 with 50 mM Tris-HCl (pH 7.4). Five hundred μl of this mixture was incubated with 50 μl MAb-GAM-IgG Sepharose at 4°C for 2 h with gentle mixing. The mixture was then centrifuged in an Eppendorf centrifuge and the supernatants were removed and assayed for dihydropyridine receptor activity using the PEG precipitation assay (8). The beads were washed twice with 1 ml buffer (100 mM NaCl, 50 mM Tris-HCl pH 7.4) containing 0.1% digitonin and then counted in a scintillation counter.

Skeletal muscle membranes or purified dihydropyridine receptor were separated by SDS-PAGE on 5%–16% gradient gels and transferred to nitrocellulose membranes using a modification of the procedure of Towbin et al. [24]. BLOTTO (Bovine Lacto Transfer Technique Optimizer; 50 mM NaH$_2$PO$_4$, 0.9% NaCl, pH 7.4, 5% nonfat dry milk) [25] was used for blocking of the nitrocellulose transfers and dilution of the antibodies. Nitrocellulose transfers were first incubated with hybridoma supernatants (1:10 or 1:20 dilution) and then with peroxidase-conjugated GAM-IgG secondary antibody (Cooper, 1:1000 dilution). WGA-peroxidase (Sigma) was used to stain WGA-positive glycoproteins on nitrocellulose blots. The nitrocellulose blots were blocked with 0.05% Tween-PBS (50 mM NaH$_2$PO$_4$, 0.9% NaCl, pH 7.4) and incubated with WGA-peroxidase (1:2000) in 0.05% Tween-PBS. The color was developed in both cases using 4-chloro-1-naphthol as the substrate.

Materials

[^{3}H]PN200–110 was obtained from Amersham. Electrophoretic reagents were obtained from Bio-Rad and molecular weight standards from Bethesda Research Laboratories. Protease inhibitors and peroxidase-conjugated WGA were otained from Sigma. Digitonin was from Fisher and Sigma and prepared as previously described [9]. All other reagents were of reagent-grade quality.

Results

The purified dihydropyridine receptor has been shown by Coomassie Blue staining of SDS-polyacrylamide gels to contain four polypeptide components of molecular masses 175 000, 170 000, 52 000 and 32 000 under nonreducing conditions and molecular masses 170 000, 150 000, 52 000 and 32 000 under reducing conditions (Fig. 1). The four polypeptide components of the dihydropyridine receptor have been referred to as the α_1 subunit (170 000 Da polypeptide), the α_2 subunit, formerly called α (175 000/150 000

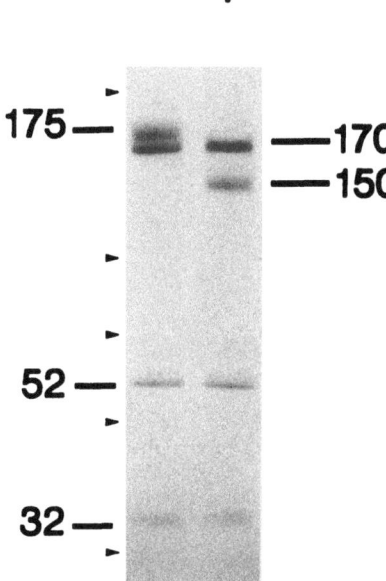

Fig. 1. SDS-polyacrylamide gel electrophoresis of the purified dihydropyridine receptor of the voltage-dependent Ca^{2+} channel. The purified dihydropyridine receptor (10 µg per lane) was separated on a 5%–16% SDS-polyacrylamide gel under nonreducing (−) and reducing (+) conditions. The gels were stained with Coomassie blue and destained. The α_1 subunit (170), α_2 subunit (175; 150), β subunit (52), and γ subunit (32) of the dihydropyridine receptor are indicated. *Arrowheads* indicate the positions of the molecular weight standards (from left to right): 200 000, 97 400, 68 000, 43 000, 25 700, 18 400, and 14 300 Da

Da polypeptide), the β subunit (52 000 Da polypeptide), and the γ subunit (32 000 Da polypeptide). In our previous publications [9, 11, 15] the 170 000 Da polypeptide was referred to as the δ subunit. Coomassie Blue stained polyacrylamide gels of the dihydropyridine receptor purified from either digitonin or CHAPS-solubilized triads were scanned with a densitometer to determine the relative quantities of the four polypeptides (Fig. 2). The absorbances of the various bands were integrated and then divided by the apparent molecular mass of the respective band to yield a relative ratio of the polypeptides. The purified dihydropyridine receptors from nine different preparations were analyzed; the results are summarized in Table 1. The 175 000/150 000 Da, 170 000 Da, 52 000 Da and 32 000 Da polypeptides exhibited a stoichiometric ratio of 1.0:0.79:1.0:1.0. No difference in the stoichiometric ratio of the polypeptides was seen between the preparations using CHAPS and those using

Table 1. Subunit stoichiometry of the purified dihydropyridine receptor of the voltage-dependent Ca^{2+} channel

Subunit	M_r	Relative Intensity % ± SE[a]	Stoichiometric ratio
α_1	170 000	28.1 ± 1.0	1.00
α_2	175 000 150 000	19.6 ± 0.8	0.79
β	52 000	8.8 ± 0.3	1.03
γ	32 000	5.5 ± 0.3	1.04

[a] Standard error of the mean, $n = 18$
The purified dihydropyridine receptor was subjected to SDS-PAGE on 5%–16% gradient gels (10 µg per lane), stained with Coomassie Blue and scanned with a Hoefer Model GS 300 scanning densitometer. The data are compiled from two scans of each of nine different preparations

digitonin for the solubilization of the dihydropyridine receptor. Therefore, it appears from SDS-PAGE analysis of the purified dihydropyridine receptor that the receptor consists of four subunits.

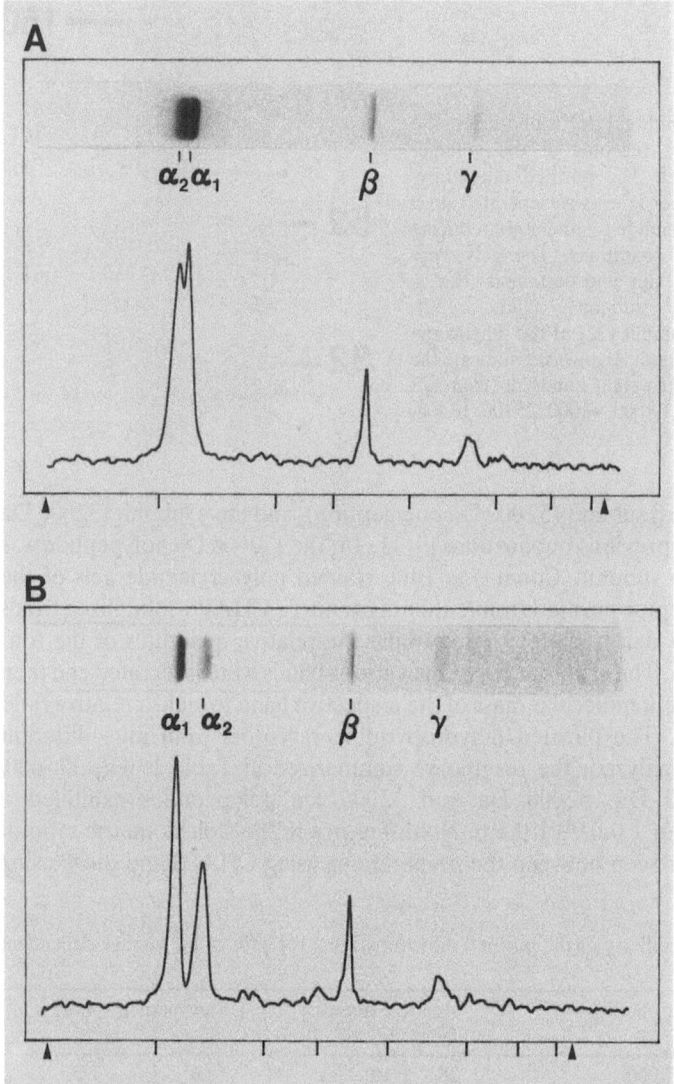

Fig. 2A, B. Densitometric scans of SDS-polyacrylamide gels of the purified dihydropyridine receptor of the voltage-dependent Ca^{2+} channel. The purified dihydropyridine receptor was separated on a 5%–16% SDS-polyacrylamide gel under nonreducing (**A**) and reducing (**B**) conditions. The gels were stained with Coomassie blue, and following destaining the densitometric scans were obtained with a Hoefer GS300 scanning densitometer. The top and dye front of the gels are indicated by the *left* and *right arrowheads*, respectively. *Hash marks* indicate the positions of the molecular weight standards (from left to right) M_r of 200000, 97400, 68000, 43000, 25700, 18400, and 14300

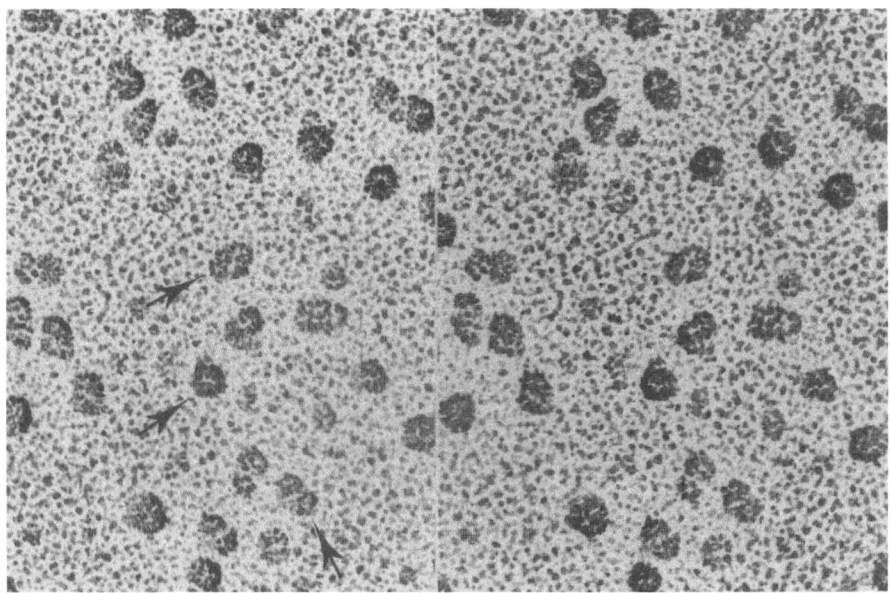

Fig. 3. Stereomicrographs of freeze-dried, rotary-shadowed dihydropyridine receptor of the voltage-dependent Ca^{2+} channel. The purified dihydropyridine receptor was freeze-dried, rotary-shadowed with carbon-platinum, and imaged in an electron microscope. Note variations in shape from round to elongated of the globular molecule and separation into two halves *(arrows)*. A stereo viewer with a magnification of two to three fold should be used to fuse the micrographs. ($\times$ 300000) (With permission of the Journal of Biological Chemistry)

The purified dihydropyridine receptors are globular, with a round or slightly elongated profile, depending on the orientation of the receptor on the mica (Fig. 3). The round profiles have a heavier platinum shadow, indicating that they are taller. The average diameter of the receptor is 16 ± 0.9 nm. The elongated profiles are less heavily shadowed and have a length of up to 22 nm. The receptor also appears to be primarily composed of two components of similar size, separated by a small central gap (Fig. 3). Thus, the dihydropyridine receptor has an ovoidal shape with long and short diameters of 16 and 22 nm.

Monoclonal antibodies against the 1,4-dihydropyridine receptor of the voltage-dependent Ca^{2+} channel were produced by immunizing mice with rabbit skeletal muscle triads followed by booster immunizations with purified dihydropyridine receptor. An immunodot assay was used for screening of the hybridoma supernatants, and antiserum from the mouse used for the fusion was used as a control in each screening. Preparations containing different amounts of dihydropyridine receptor were used to differentiate among the antibodies against the dihydropyridine receptor and those that react with other proteins in triads and sarcoplasmic reticulum (Fig. 4). A hybridoma supernatant was considered positive in the immunodot assay if it reacted with dihydropyridine receptor and/or membranes enriched in the dihydropyridine receptor but showed no reactivity with preparations that are devoid of the dihydropyridine receptor.

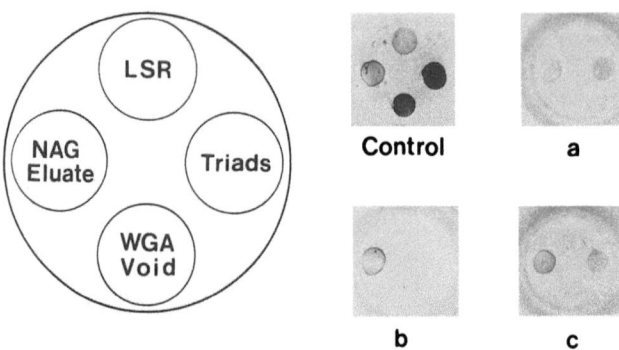

Fig. 4. Immunodot assay for anti-dihydropyridine receptor monoclonal antibodies. Light sarcoplasmic reticulum vesicles *(LSR)*; rabbit skeletal muscle triads *(Triads)*; the void of the WGA-Sepharose column after incubation with digitonin-solubilized triads *(WGA-void)*; dihydropyridine receptor eluted from the WGA-Sepharose column with N-acetylglucosamine *(NAG-eluate)* were dotted (0.5 µl) onto the nitrocellulose at the four quadrants of each well of a millititer plate (Millipore) and allowed to dry as diagrammed. Specific [^{3}H]PN200-110 binding activity for the preparations are: LSR, 0.5 fmol/µl; Triads, 21.7 fmol/µl; WGA-void, 0.1 fmol/µl; and NAG eluate, 21.8 fmol/µl. The plates were blocked with 3% BSA-TBS (20 mM Tris-HCl, 200 mM NaCl, pH 7.5) and allowed to react with hybridoma supernatants. A peroxidase-conjugated goat anti-mouse IgG secondary antibody (Cooper) at 1:1000 dilution in 3% BSA-TBS was used, and the plates were developed using 4-chloro-1-naphthol as the substrate. *Control,* Serum from an immunized mouse used for the fusion, diluted 1:500 in 3% BSA-TBS; *a, b, and c,* results for the immunodot assay using 50 µl positive hybridoma supernatants IIC12, IIF7, and IIID5, respectively (With permission of the Journal of Biological Chemistry)

Immunodot assay positive antibodies were next tested for their ability to immunoprecipitate the [^{3}H]PN200-110-labeled receptor from solubilized triads. Monoclonal antibodies from hybridoma supernatants were preincubated with GAM-IgG Sepharose beads to form MAb-GAM-IgG beads which were then used to immunoprecipitate the digitonin-solubilized [^{3}H]PN200-110-labeled dihydropyridine receptor. The radioactivity on the beads was counted to determine directly the amount of labeled dihydropyridine receptor bound by the antibody. Figure 5 shows the results of an immunoprecipitation assay with a monoclonal antibody (MAb VD2$_1$) to the β subunit (52000 Da polypeptide) of the receptor in comparison with an unrelated T-system monoclonal antibody (MAb IXE12$_2$) and WGA-Sepharose. Monoclonal antibody to the β subunit was found to specifically immunoprecipitate the [^{3}H]PN200-110-labeled dihydropyridine receptor from the assay mixture and was equally efficient as WGA Sepharose in binding the receptor. Similar results were obtained with monoclonal antibodies to the α$_1$ subunit (170000 Da polypeptide). The anti-dihydropyridine receptor antibodies were also shown to bind saturably to the [^{3}H]PN200-110-labeled dihydropyridine receptor, and a close inverse correlation was found between the amount of dihydropyridine receptor immunoprecipitated by the antibody and the amount of [^{3}H]PN200-110-labeled dihydropyridine receptor remaining in the supernatant (Fig. 6). The highest level of dihydropyridine receptor immunoprecipitated by a monoclonal antibody to the α$_1$ subunit (170000 Da polypeptide) ranged from 80% to 95% of the total amount present in the assay mixture. The results show that this assay was able to select those antibodies that bind to the digitonin-solubilized

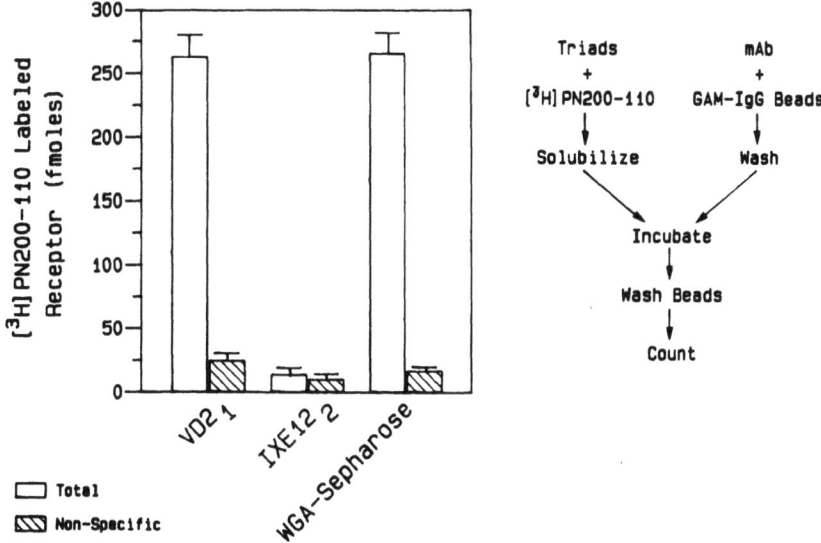

Fig. 5. Immunoprecipitation of the dihydropyridine receptor with monoclonal antibody to the β subunit. Hybridoma supernatants were tested for their ability to immunoprecipitate the [^{3}H]PN200–110-labeled receptor from digitonin-solubilized membranes as shown at right. IXE12$_2$ is an unrelated antibody used as a control and WGA-Sepharose is used as a positive control. The nonspecifically labeled receptor was determined in the presence of 10 μM nitrendipine. *Error bars* represent the standard error of the mean from three independent repeats of the experiments

Fig. 6. Immunoprecipitation of [^{3}H]PN200–110-labeled dihydropyridine receptor with monoclonal antibody to the α$_1$ subunit. Immunoprecipitation assays were carried out as described in "Experimental Procedures." Various amounts of MAb IIC12-GAM-IgG Sepharose were used with the volume of the beads kept constant using Sepharose CL 4B. The maximum amount of [^{3}H]PN200–110-labeled receptor immunoprecipitated in this experiment corresponds to 29.8. ± 1.9 fmol (83.9 ± 5.2%). This correlates with the amount remaining in the supernatant of 5.7 ± 0.5 fmol (16.1 ± 1.4%). The immunoprecipitation of

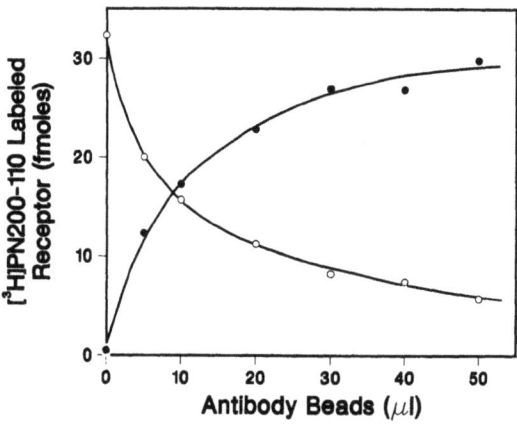

nonspecific binding activity determined in the presence of 10 μM nitrendipine was less than 6% of the total binding (With permission of the Journal of Biological Chemistry

[^{3}H]PN200–110-labeled dihydropyridine receptor and do not compete directly with the dihydropyridine binding site on the receptor.

The molecular components of the skeletal muscle dihydropyridine receptor were characterized by immunoblot assays using monoclonal antibodies capable of immunoprecipitating the dihydropyridine receptor from digitonin-solubilized mem-

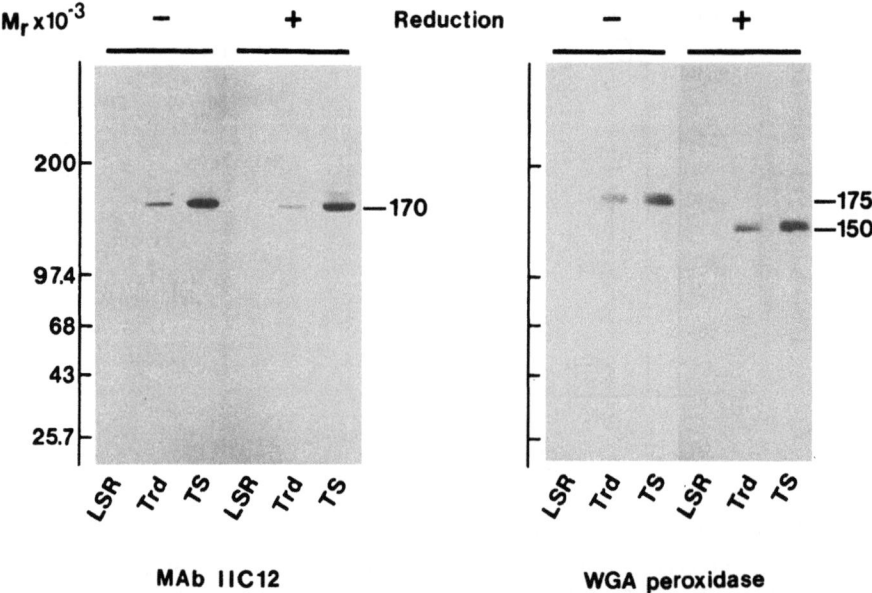

Fig. 7. Immunoblot staining of rabbit skeletal muscle membrane fractions with monoclonal antibodies to the dihydropyridine receptor. Indirect immunoperoxidase staining of membrane fractions with anti-dihydropyridine receptor antibodies was performed as described in the "Experimental Procedures." Light sarcoplasmic reticulum *(LSR)*, triads *(Trd)*, and transverse tubular *(TS)* membranes were subjected to SDS-PAGE on a 3%–12% gradient gel under nonreducing (+ 10 mM N-ethylmaleimide) and reducing (+ 5 mM dithiothreitol) conditions. *Left panel* shows nitrocellulose transfers of the gel stained with monoclonal antibody IIC12 (anti-α_1); *Right panel* shows nitrocellulose transfers of the gel stained with WGA-peroxidase. The α_1 subunit (170 kd) and the α_2 subunit (175, 150 kd) are indicated (With permission of the Journal of Biological Chemistry)

branes. A monoclonal antibody to the α_1 subunit stained a polypeptide of 170000 Da on nitrocellulose transfers of transverse tubular membranes, and isolated triads separated on SDS-PAGE (Fig. 7). Identical results were obtained for two other monoclonal antibodies to the α_1 subunit. WGA-peroxidase staining of the same membrane fractions has demonstrated that the 175000 Da polypeptide (α_2 subunit) is the major WGA-positive glycoprotein in these membranes. The apparent molecular weight of the α_1 subunit remained unchanged with reduction, while the apparent molecular weight of the α_2 subunit shifted from 175000 to 150000 upon reduction. The 170000 Da polypeptide (α_1 subunit) and the 175000 Da glycoprotein (α_2 subunit) were not detected in light sarcoplasmic reticulum membranes, a preparation devoid of dihydropyridine receptor.

Figure 8 shows the immunoblot staining of the various fractions from the purification of the dihydropyridine receptor. The α_1 and β subunits of the dihydropyridine receptor were detected by immunoblot staining with mAb IIC12 (Fig. 8A) and mAb VD2$_1$ (Fig. 8B), respectively. The β subunit copurified with the α_1 subunit at all steps of the purification. These two proteins are present in triads and solubilized triads, absent in the void of the WGA-Sepharose column, enriched in the peak fractions from the

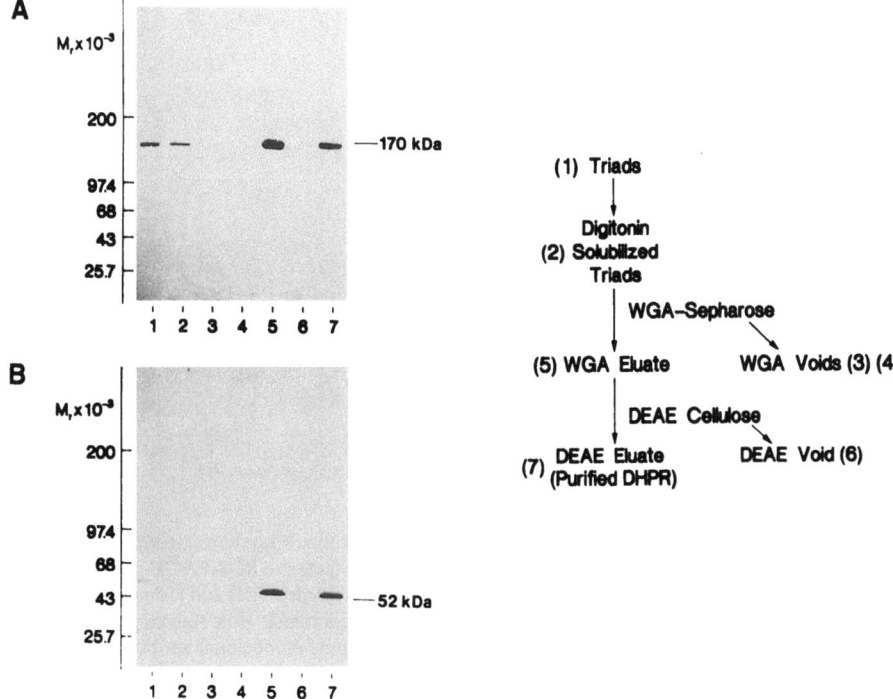

Fig. 8A, B. Immunoblot staining of fractions in purification of the dihydropyridine receptor from skeletal muscle triads. The various fractions from the purification of the dihydropyridine receptor from triads were subjected to SDS-PAGE on a 5%–16% gradient gel and transferred to nitrocellulose. The immunoblots were stained with mAb IIC12, anti-α_1 (A) and mAb VD2$_1$, anti-β (B). The samples on the transfers are: triads, 150 µg *(lane 1)*; digitonin-solubilized triads, 150 µg *(lane 2)*; first void from WGA-Sepharose column, 150 µg *(lane 3)*; second void from WGA Sepharose column, 150 µg *(lane 4)*; eluate from WGA-Sepharose, 10 µg *(lane 5)*; void from DEAE-cellulose, 10 µg *(lane 6)*; peak fractions from DEAE-cellulose, 7.5 µg *(lane 7)*. The α_1 subunit (170 kd) and β subunit (52 kda) are indicated

WGA-Sepharose column, absent in the void of the DEAE-cellulose column, and once again enriched in the peak fractions of the DEAE-cellulose column. The intensity of staining of each subunit in the various fractions also appear to parallel each other. We have also used CHAPS for the solubilization and purification of the dihydropyridine receptor from triads and have shown the presence of the β subunit in the CHAPS-purified dihydropyridine receptor by immunoblot staining with mAb VD2$_1$. Figure 9 shows Coomassie Blue staining and immunoblot staining of the purified dihydropyridine receptor. The α_1 subunit was the only protein stained by anti-α_1 monoclonal antibodies, and its molecular weight remains unchanged with reduction. The α_2 subunit was the only protein stained by WGA-peroxidase, and its molecular weight shifted upon reduction. Finally, the β subunit antibody stained a 52-kDa polypeptide under both reducing and nonreducing conditions and did not stain the α_1, α_2, or γ subunits of the receptor.

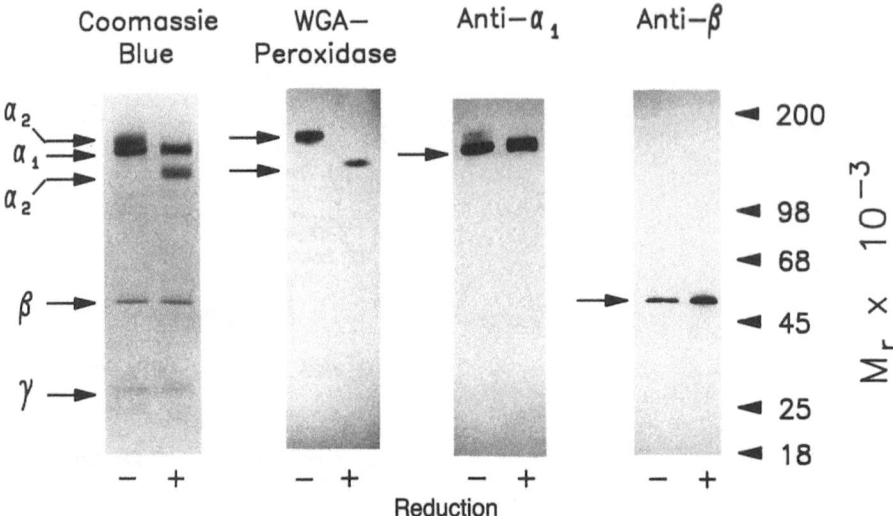

Fig. 9. Coomassie blue and immunoblot staining of the purified dihydropyridine receptor from rabbit skeletal muscle. The purified dihydropyridine receptor was subjected to SDS-PAGE on a 5%–16% gradient gel under nonreducing (−)conditions (+ 20 mM n-ethylmaleimide) and reducing (+)conditions (+ 10 mM dithiothreitol). *Left panel* is a photograph of a Coomassie Blue stained gel; *right three panels* are immunoblots with WGA peroxidase to the α_2 subunit, monoclonal antibodies to the α_1 subunit, and monoclonal antibodies to the β subunit. The α_1, α_2 and β subunits are indicated by arrows. Molecular weight standards are indicated on the right

Discussion

The 1,4-dihydropyridine receptor of the voltage-dependent Ca^{2+} channel has been purified in our laboratory from rabbit skeletal muscle triads. We have found skeletal muscle triads to be the best starting material for the purification of the 1,4-dihydropyridine receptor because skeletal muscle triads contain one major WGA-positive glycoprotein (see Fig. 7). In addition, skeletal muscle triads are enriched in [^{3}H]PN200–110 binding activity (10–40 pmol/mg) and the yield of skeletal muscle triads ranges from 600 to 1000 mg/kg tissue as compared to 10–30 mg/kg for transverse tubular membranes.

The purified dihydropyridine receptor was characterized by SDS-PAGE with Coomassie Blue staining and immunoblotting with monoclonal antibodies and WGA peroxidase. Densitometric scanning of the Coomassie Blue stained gels of the purified dihydropyridine receptor provided evidence that each of the four polypeptides are indeed integral subunits of the dihydropyridine receptor (Fig. 3). Under reducing conditions, in which the α_1 and α_2 subunits were well resolved on the gel, and the α_2 subunit migrated with an apparent molecular mass of 150000, a stoichiometric ratio of 1.0:0.79:1.0:1.0 was obtained for the α_1, α_2, β, and γ subunits, respectively (Table 1). The stoichiometric ratio of 1:1:1 among the α_1, β, γ subunits strongly suggests that the β subunit is an integral component of the dihydropyridine receptor. The anomalous ratio of 0.79:1 between the α_2 subunit and the other subunits of the dihydropyridine receptor can be explained by its glycoprotein nature. The α_2 subunit has been shown to be heavily glycosylated, and if one were to assume that the protein component of the

α_2 subunit has a molecular mass of 120 000 Da then the stoichiometric ratio would approach 1:1:1:1 for the four subunits of the dihydropyridine receptor.

Electron-microscopic characterization of the purified dihydropyridine receptor provided additional structural information on the dihydropyridine receptor. Rotary-shadowed stereographs of the freeze-dried dihydropyridine receptor revealed a homogeneous preparation of ovoidal particles 16×22 nm in size, demonstrating that the protein components of the purified dihydropyridine receptor exist as a single complex with two symmetrical halves. The two halves of the complex may represent the two larger components in association with the two smaller polypeptides. Our data are consistent with the hydrodynamic studies of the cardiac dihydropyridine receptor reported by Horne and Oswald [26], in which the dihydropyridine receptor was reported to be a large ellipsoidal transmembrane protein with a molecular weight of approximately 370 000.

Immunoblot staining of the purified dihydropyridine receptor on SDS-PAGE under reducing and nonreducing conditions with the monoclonal antibodies and WGA-peroxidase shows that the α_1 and α_2 subunits are distinct proteins, and the α_2 subunit is the WGA-positive glycoprotein component of the purified dihydropyridine receptor. The β and γ subunits are not related to the α_1 or α_2 subunits since they are not stained by the monoclonal antibodies to the α_1 subunit or WGA-peroxidase. The α_2 subunit (175 000 Da glycoprotein) appears to be equivalent to the 160 000 Da subunit of the dihydropyridine receptor described by Curtis and Catterall [5] and Borsotto et al. [6]. It undergoes a decrease in its apparent molecular mass upon reduction and is stained by WGA-peroxidase on nitrocellulose blots. The α_1 subunit of the dihydropyridine receptor appears to be equivalent to the 142 000 Da subunit described by Flockerzi et al. [7] since it has been found to be a phosphoprotein [11].

The association of the four subunits of the dihydropyridine receptor had been shown previously only by protein staining of polyacrylamide gels of the purified dihydropyridine receptor. Work from our laboratory has further demonstrated their association by other approaches. A monoclonal antibody specific for the β subunit of the dihydropyridine receptor from rabbit skeletal muscle is capable of immuno-precipitating the [^{3}H]PN200-110-labeled dihydropyridine receptor from digitonin-solubilized triads. Since the 170 000 Da subunit of the dihydropyridine receptor contains the dihydropyridine binding site [9, 10], the immunoprecipitation of dihydropyridine binding activity by an antibody specific to the 52 000 Da polypeptide demonstrates that the 52 000 Da and 170 000 Da polypeptides are associated in digitonin-solubilized triads. It also shows that the 52 000 polypeptide is not an unrelated polypeptide that copurifies nonspecifically with the dihydropyridine receptor. Immunoblot staining of the various fractions from the chromatographic procedures used in the purification of the dihydropyridine receptor revealed that this 52 000 Da polypeptide copurifies with the 170 000 Da subunit throughout the entire purification process (Fig. 3), demonstrating the close association between these two proteins. Immunoprecipitation of the α_1 subunit also resulted in the coprecipitation of the α_2 subunit [8], and when the digitonin-solubilized dihydropyridine receptor was bound to WGA Sepharose, a buffer containing 1% SDS was required to separate the α_1 subunit from the α_2 subunit [9]. Considered together, these results demonstrate the close association between the α_2 subunit and the α_1 subunit (dihydropyridine binding subunit) of the receptor.

The 52000 Da polypeptide had not been reported to copurify with the dihydropyridine receptor when certain procedures were used for the purification. Lazdunski and coworkers, using CHAPS to solubilize the receptor, have reported a single polypeptide of 170000 Da under nonreducing conditions that is converted to a polypeptide of 140000 Da and several small polypeptides upon reduction [6]. We have shown that with our purification procedure, the substitution of digitonin with CHAPS for the solubilization of the receptor produced the same four protein components in the purified receptor, and the 52000 Da polypeptide was present in the purified receptor, as determined by staining with mAb $VD2_1$.

In summary, work from our laboratory has demonstrated that the 1,4-dihydropyridine receptor of the voltage-dependent Ca^{2+} channel purified from rabbit skeletal muscle contains four distinct subunits that are closely associated in a 1:1:1:1 stoichiometric ratio. The α_1 subunit is the dihydropyridine binding subunit of the receptor [9, 10] and probably also contains the ion-conducting pore of the Ca^{2+} channel complex [16]. The function of the α_2 subunit and lower molecular weight subunits (β, γ) remains to be elucidated. Preliminary results from Coronado's laboratory [27] have shown that the monoclonal antibody ($VD2_1$) to the 52000 Da subunit (β) described in this report is capable of activating the Ca^{2+} channel whereas polyclonal antibodies to the 32000 Da subunit (γ) are capable of inhibiting the Ca^{2+} channel, suggesting that these lower molecular weight subunits might have regulatory roles in the function of the dihydropyridine-receptor-Ca^{2+} complex.

Summary

Monoclonal antibodies to the 1,4-dihydropyridine receptor of the voltage-dependent Ca^{2+} channel have been produced and used in immunoprecipitation and immunoblotting experiments to probe the structure and function of the Ca^{2+} channel. The purified 1,4-dihydropyridine receptor from rabbit skeletal muscle contains four polypeptide components of 175000 Da (α_2), 170000 Da (α_1), 52000 Da (β), and 32000 Da (γ) when analyzed by SDS-PAGE under nonreducing conditions. Densitometric scanning of Coomassie Blue stained SDS-polyacrylamide gels of the purified dihydropyridine receptor has shown that the four polypeptide components exist in a 1:1:1:1 stoichiometric ratio. Electron microscopy of the freeze-dried, rotary-shadowed dihydropyridine receptor has shown that the preparation contains a homogeneous population of 16 × 22 nm ovoidal particles large enough to contain all four polypeptides of the receptor.

Monoclonal antibodies to the 170000 Da polypeptide (α_1 subunit) and monoclonal antibodies to the 52000 Da polypeptide (β subunit) were able to specifically imunoprecipitate the [^{3}H]PN200–110-labeled dihydropyridine receptor from digitonin-solubilized membranes. Immunoblot staining of the purified dihydropyridine receptor using monoclonal antibodies to the α_1 subunit and WGA peroxidase have demonstrated that the α_1 and the β_2 subunits are distinct proteins, and that the α_2 subunit is the glycoprotein component of the dihydropyridine receptor. The apparent molecular weight of the α_1 subunit on SDS-PAGE remained unchanged with reduction while the apparent molecular weight of the α_2 subunit shifted from 175000 to 150000 upon reduction. Immunoblotting experiments with a monoclonal antibody to the β subunit have shown that the 52000 Da polypeptide copurifies with the 175000 Da and 170000 Da polypeptides at all stages of the purification, and that the higher molecular weight subunits of the receptor were not labeled by the monoclonal antibody to the β subunit. In conclusion, we have demonstrated that the 175000 Da ($α_2$), 170000 Da ($α_1$) and 52000 Da (β) polypeptides are integral and distinct subunits of the purified dihydropyridine receptor of the voltage-dependent Ca^{2+} channel.

Literatur

1. Tsien RW (1983) Calcium channels in excitable cell membranes. Annu Rev Physiol 45:341–358
2. Reuter H (1983) Calcium channel modulation by neurotransmitters, enzymes and drugs. Nature (London) 301:569–574
3. Janis RA, Triggle DJ (1984) 1,4-Dihydropyridine Ca^{2+} channel antagonists and activators: a comparison of binding characteristics with pharmacology. Drug Dev Res 4:257–274
4. Fosset M, Jaimovich E, Delpont E, Lazdunski M (1983) [^{3}H]Nitrendipine receptors in skeletal muscle: properties and preferential localization in transverse tubules. J Biol Chem 258:6086–6092
5. Curtis BM, Catterall WA (1984) Purification of the calcium antagonist receptor of the voltage-sensitive calcium channel from skeletal muscle transverse tubules. Biochem 23:2113–2118
6. Borsotto M, Barhanin J, Fosset M, Lazdunski M (1985) The 1,4-dihydropyridine receptor associated with the skeletal muscle voltage-dependent Ca^{2+} channel: purification and subunit composition. J Biol Chem 290:14255–14263
7. Flockerzi V, Oeken HJ, Hofmann F, Pelzer D, Cavalie A, Trautwein W (1986) Purified dihydropyridine-binding site from skeletal muscle t-tubules is a functional calcium channel. Nature (London) l323:66–68
8. Leung AT, Imagawa T, Campbell KP (1987) Structural characterization of the 1,4 dihydropyridine receptor of the voltage-dependent Ca^{2+} channel from rabbit skeletal muscle: evidence for two distinct high molecular weight subunits. J Biol Chem 262:7943–7946
9. Sharp AH, Imagawa T, Leung AT, Campbell KP (1987) Identification and characterization of the dihydropyridine-binding subunit of the skeletal muscle dihydropyridine receptor. J Biol Chem 262:12309–12315
10. Ferry DR, Kampf K, Goll A, Glossman H (1985) Subunit composition of skeletal muscle transverse tubular calcium channels evaluated with the 1,4-dihydropyridine photoaffinity probe, [^{3}H]azidopine. EMBO J 4:1933–1940
11. Imagawa T, Leung AT, Campbell KP (1987) Phosphorylation of the 1,4-dihydropyridine receptor of the voltage-dependent Ca^{2+} channel by an intrinsic protein kinase in isolated triads from rabbit skeletal muscle. J Biol Chem 262:8333–8339
12. Curtis BM, Catterall WA (1985) Phosphorylation of the calcium antagonist receptor of the voltage-sensitive calcium channel by cAMP-dependent protein kinase. Proc Natl Acad Sci USA 82:2518–2532
13. Hosey MM, Bosortto M, Lazdunski M (1986) Phosphorylation and dephosphorylation of dihydropyridine-sensitive voltage-dependent Ca^{2+} channel in skeletal muscle membranes by cAMP and Ca^{2+}-dependent processes. Proc Natl Acad Sci USA 83:3733–3737
14. Flockerzi V, Oeken HJ, Hofmann F (1986) Purification of a functional receptor for calcium-channel blockers from rabbit skeletal-muscle microsomes. Eur J Biochem 161:217–224
15. Leung AT, Imagawa T, Block B, Franzini-Armstrong C, Campbell KP (1988) Biochemical and ultrastructural characterization of the 1,4-dihydropyridine receptor from rabbit skeletal muscle: evidence for a 52000-Da subunit. J Biol Chem 263:994–1001
16. Tanabe T, Takeshima H, Mikami A, Flockerzi V, Takahasi H, Kangawa K, Kojima M, Matsuo H, Hirose T, Numa S (1987) Primary structure of the receptor for calcium channel blockers from skeletal muscle. Nature 328:313–328
17. Rosemblatt M, Hidago C, Vergara C, Ikemoto N (1981) Immunological and biochemical properties of transverse tubule membranes isolated from rabbit skeletal muscle. J Biol Chem 256:8140–8148

18. Campbell KP, Franzini-Armstrong C, Shamoo AE (1980) Further characterization of light and heavy sarcoplasmic reticulum vesicles. Identification of the "sarcoplasmic reticulum feet" associated with heavy sarcoplasmic reticulum vesicles. Biochem Biophys Acta 602:97–116
19. Lowry OH, Rosebrough NJ, Farr AL, Randall RJ (1951) Measurement with the folin phenol reagent. J Biol Chem 193:265–275
20. Peterson GL (1977) A simplification of the protein assay method of Lowry et al. which is more generally applicable. Anal Biochem 83:346–356
21. Laemmli UK (1970) Cleavage of structural proteins during the assembly of the head of bacteriophage T4. Nature 227:680–685
22. Kennett RH (1980) Fusion by centrifugation of cells suspended in polyethylene glycol. In: Kennett RH, McKearn TJ, Bechtol KG (eds) Monoclonal antibodies, Plenum, New York, pp 365–367
23. Hawkes R, Niday E, Gordon J (1982) A dot-immunobinding assay for monoclonal and other antibodies. Anal Biochem 119:142–147
24. Towbin H, Staehelin T, Bordon J (1979) Electrophoretic transfer of proteins from polyacrylamide gels to nitrocellulose sheets: procedure and some applications. Proc Natl Acad Sci USA 76:4350–4354
25. Johnson DA, Gautsch JW, Sportsman JR, Elder JH (1984) Improved technique utilizing nonfat dry milk for analysis of proteins and nucleic acids transferred to nitrocellulose. Gene Anal Techn 1:3–8
26. Horne WA, Weiland GA, Oswald RE (1986) Solubilization and hydrodynamic characterization of the dihydropyridine receptor from rat ventricular muscle. J Biol Chem 261:3588–3594
27. Vilven J, Leung AT, Imagawa T, Sharp AH, Campbell KP and Coronado R (1988) Interaction of calcium channels of skeletal muscle with monoclonal antibodies specific for its dihydropyridine receptor. Biophysical J 53:556a

Endogenous Ligands for Voltage-Sensitive Calcium Channels in Extracts of Rat and Bovine Brain*

B. J. Ebersole and P. B. Molinoff

[1] Department of Pharmacology, University of Pennsylvania School of Medicine, Philadelphia, PA 19104–6084, USA

Introduction

The role of calcium as a second messenger has been well established. A change in the intracellular concentration of calcium as a result of activation of receptors or ion channels can initiate many cellular responses, including secretion, release of neurotransmitters, and modulation of enzyme activities (Rubin et al., 1985). Changes in the concentration of intracellular free calcium can result from the agonist-stimulated release of calcium from intracellular stores (Berridge and Irvine 1984; Spat et al. 1986). Intracellular calcium levels may also be increased by an influx of calcium from the extracellular environment, and several types of channels that permit the entry of calcium into the cell have been described. In some cells, calcium may enter through "receptor-operated" channels that are opened in response to stimulation of a receptor by agonists (Bolton 1979). Another class of calcium channels, designated "voltage-sensitive calcium channels" (VSCC), open to permit the influx of calcium in response to depolarization of the cell membrane (Tsien 1983). The VSCC have been further classified on the basis of conductance, activation and inactivation characteristics, and pharmacological profile. One type of VSCC, the "L" channel, is sensitive to blockade by calcium-channel blockers of the 1,4-dihydropyridine class and is found in neurons (Nowycky et al. 1985) as well as other types of tissues (e.g., Nilius et al. 1985; McClesky et al. 1987). Two other types of VSCC, the T and N channels, have been described; they are insensitive to blockade by 1,4-dihydropyridines (McClesky et al. 1987; Nowycky et al. 1985).

Ligand-binding studies with labeled 1,4-dihydropyridines, including [^{3}H]NT and [^{3}H]PN200–110, have been performed to characterize VSCC (Triggle and Janis 1987). The interaction of other classes of organic calcium-channel blockers with binding sites for 1,4-dihydropyridines appears to involve allosteric mechanisms. For

---

Abbrevations:
DMEM, Dulbecco's modified Eagle's medium;
HPLC, high-performance liquid chromatography;
LF, SF, large-molecular-weight, small-molecular-weight inhibitory factor;
[^{3}H]NT, [^{3}H]nitrendipine;
VSCC, voltage-sensitive calcium channels

* This work was supported by the U. S. Public Health Service (Grants GM 34781 and NS 18479) and by funds provided by Miles Laboratories, Inc.

example, the benzothiazepine diltiazem increases the binding of [^{3}H]1,4-dihydropyridines, while the phenylalkylamine verapamil noncompetively inhibits binding of [^{3}H]1,4-dihydropyridines (Ehlert et al. 1987; Boles et al. 1984; Reynolds et al. 1986). These findings suggest that there are several overlapping or interactive sites that may mediate the regulation of calcium channels.

In several systems, the presence of high-affinity binding sites for drugs has suggested the existence of endogenous ligands for these sites. The partial purification of heat- and acid-stable compounds in rat brain that inhibit the binding of [^{3}H]NT to membranes prepared from rat hippocampus (Sanna and Hanbauer 1987) and rat brain, but not rat heart (Ebersole et al. 1988), has been described. Gel-filtration studies revealed the presence of both large-molecular-weight (MW > 50000) and small-molecular-weight (MW ≈ 1300) factors that inhibit the binding of [^{3}H]NT (Ebersole et al. 1988). Further characterization of SF revealed that it was sensitive to inactivation by pronase, suggesting that it may be a peptide or contain peptide bonds. The inhibition of the binding of [^{3}H]NT by SF was irreversible or only slowly reversible, and when assays were carried out in the presence of SF the B_{max} for binding of [^{3}H]NT was reduced with no change in the affinity of the binding sites for [^{3}H]NT.

To determine the possible functional significance of SF, the effects of SF prepared from rat brain on voltage-dependent influx of calcium into cultured PC12 cells have been investigated. PC12 cells, derived from a rat adrenal medullary tumor, have been shown to posses voltage-dependent calcium channels that can be measured by electrophysiological techniques (Brown et al. 1982) and in studies of the uptake of $^{45}Ca^{2+}$ (Stallcup 1979). Binding sites for [^{3}H]1,4-dihydropyridines have also been described (Greenberg et al. 1986); Kunze et al. 1987; Toll 1982). Preliminary studies with extracts of bovine brain revealed the presence of SF similar to those present in extracts of rat brain. Because of the need for significant amounts of purified SF for determining the molecular basis of inhibition, we have initiated studies designed to purify SF from bovine brain.

Methods

Preparation of Brain Extracts

All procedures were carried out at 4°C unless otherwise stated.

Frozen, unstripped bovine calf brains were obtained from Pel-Freez (Rogers, Arkansas) and were thawed at 4°C overnight. Brains obtained from male Sprague-Dawley rats were used fresh. Tissues were weighed and homogenized in 2 volumes of 50 mM Tris-HCl, pH = 7.5 at 25°C (Tris buffer) with a Brinkmann Polytron at setting 6. The homogenate was centrifuged at 24000 g for 15 min. For preparation of rat cytosol, the supernatant resulting from this initial centrifugation was centrifuged at 150000 g for 30 min. The supernatant resulting from this second centrifugation was heated to 95°C for 15 min and centrifuged at 24000 g for 15 min. For rat brain, the supernatant resulting from this third centrifugation was designated "rat SF". For bovine brain, heat-stable large-MW components were removed from the supernatant by ultrafiltration through an Amicon YM-10 membrane. The ultrafiltrate resulting from this step was designated "bovine SF".

[³H]Dihydropyridine Binding Assay

The binding of [^{3}H]NT to membranes prepared from rat brain was measured as previously described (Ebersole et al. 1988). An assay volume of 0.5 ml was used unless otherwise indicated. The binding of [^{3}H]PN200-110 was measured under similar conditions, except that the assay volume was increased to 1.0 ml. For both radioligands, specific binding was defined as binding that was inhibited when assays were carried out in the presence of 1 µM unlabeled nifedipine.

To allow quantitation of the specific activity of SF in a given sample, dose-response curves for the inhibition of binding of [^{3}H]NT by SF were constructed. A "unit" of SF activity was defined as the IC$_{50}$ (in µl) for inhibition by a preparation of SF. Estimates of the extent of purification were obtained by comparing the number of activity units of SF relative to the content of Lowry-reactive substance (Lowry et al. 1951).

Chromatography of SF

Gel-Filtration Chromatography

Gel-filtration chromatography of both rat and bovine SF was carried out as previously described (Ebersole et al. 1988). Samples of rat cytosol or bovine SF (2.0 ml) were applied to a 1 × 48 cm column of Sephadex G-25 Superfine that had been equilibrated in 50 mM Tris buffer containing 200 mM NaCl, and 2.0-ml fractions were collected at a flow rate of 0.5 ml/min.

Chromatography of Bovine SF on Q-Sepharose

The chromatography procedure used for rat SF (Ebersole et al. 1988) was modified for chromatography of bovine SF. Bovine SF was diluted 1:1.5 (vol:vol) with distilled water to reduce the ionic strength of the extract and was applied to a column of Q-Sepharose (acetate form) that had been equilibrated in distilled water. The sample was applied at a flow rate of 0.2 ml/min. The column was then washed with distilled water until the optical density at 280 nm was less than 0.2. SF was eluted with 100 mM ammonium acetate at pH = 6.9. Fractions containing inhibitory activity were pooled and dried by Speed-Vac centrifugation. Dried samples were stored at $-20°C$.

DEAE-HPLC of Bovine SF

Bovine SF that had been eluted from Q-Sepharose and dried was reconstituted in 5 mM ammonium acetate and applied to a Beckman Spherogel DEAE-HPLC column (acetate form) that had been equilibrated in 9 mM ammonium acetate. SF activity was eluted with a gradient of ammonium acetate (9–200 mM, pH = 6.9). Fractions were dried and resuspended in a minimal volume of 5 mM ammonium acetate.

$^{45}Ca^{2+}$ Influx in PC 12 Cells

PC 12 cells were grown in DMEM (Flow Labs) at 37°C in a humidified atmosphere of 5% Co_2/95% air. For studies of $^{45}Ca^{2+}$ influx, cells were subcultured into 35-mm dishes or six-well cluster dishes at a density of 70000 cells/cm^2 and were grown for 2 days before use. Voltage-dependent influx of $^{45}Ca^{2+}$ was measured as described by Greenberg et al. (1985). Briefly, plates were rinsed with physiological salt solution containing 5 mM KCl and then preincubated in the same buffer for 15 min at room temperature. Following aspiration of the preincubation buffer, cells were exposed to buffer containing 1 µCi/ml $^{45}CaCl_2$ for 2 min in the presence of either 5 or 50 mM KCl. The incubation buffer was removed and the plates were washed three times with cold buffer containing 5 mM KCl and drained. Cells were solubilized in 0.1 N NaOH and aliquots removed for determination of protein and for quantitation of $^{45}Ca^{2+}$ uptake by liquid scintillation counting. The difference in $^{45}Ca^{2+}$ uptake between resting and depolarizing conditions represents the influx of Ca^{2+} resulting from depolarization of the cell membrane. The dihydropyridine sensitivy of the influx of $^{45}Ca^{2+}$ was confirmed by performing experiments in the presence of 1 µM nifedipine. In experiments designed to evaluate the effect of SF, preparations of SF were present during the preincubation period and during the incubation with $^{45}Ca^{2+}$.

Results and Discussion

Purification of SF from Rat and Bovine Brain

Gel-Filtration Chromatography

Gel filtration chromatography of cytosol from rat brain revealed the presence of both LF and SF that could inhibit the binding of [^{3}H]NT to membranes prepared from rat brain (Fig. 1). When fractions were boiled and reassayed, only the SF (MW ≈ 1300) remained active. A comparison of dose-response curves constructed for cytosol, for the pooled large-MW fractions, and for the pooled small-MW fractions revealed that all of the inhibitory activity in the cytosol was accounted for by the contents of these two peaks, and that each peak contributed approximately half of the inhibitory activity present in cytosol. Because heat treatment was adequate to remove LF, boiled cytosol from rat brain was used for subsequent purification of SF.

Chromatography of cytosol prepared from bovine brain also revealed the presence of LF and SF (data not shown). Gel-filtration chromatography of boiled bovine brain extract revealed an SF (MW ≈ 1300) similar to that prepared from rat brain (Fig. 2). In contrast to the results obtained with cytosol from rat brain, however, a substantial amount of LF remained after heat treatment of extract from bovine brain (Fig. 2). To remove large-MW material from preparations of bovine SF, boiled extract was filtered through a membrane with an MW cut-off of 10000. The ultrafiltrate resulting from this treatment (Fig. 3) was used for further purification of SF from bovine brain.

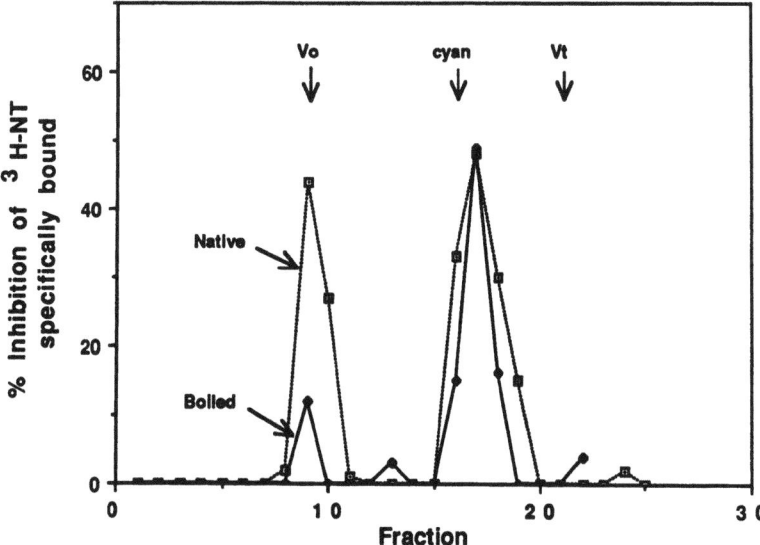

Fig. 1. Gel filtration of rat SF on Sephadex G-25. V_0 and V_t, the void volume and total volume as determined by the elution profiles of blue dextran 2000 and tyrosine respectively; *cyan* cyanocobalamin (MW 1355). Aliquots of 350 µl of each fraction were assayed for the ability to inhibit the binding of [^{3}H]NT (0.6 nM) to membranes prepared from rat brain

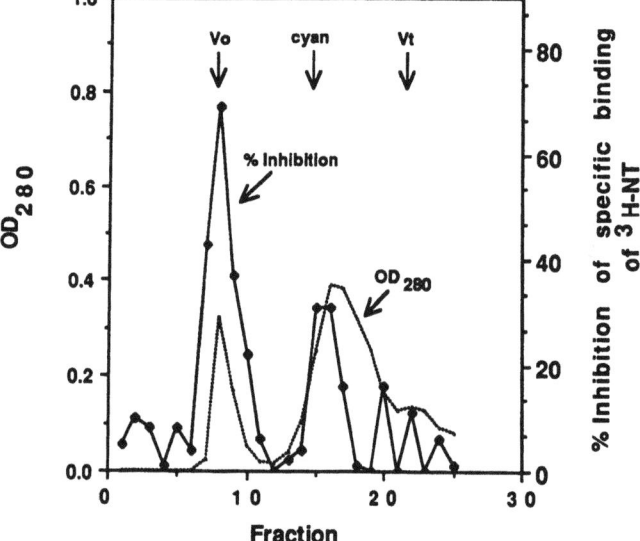

Fig. 2. Gel filtration of bovine extract on Sephadex G-25. OD_{280}, optical density at 280 nm. Column calibration and assay of activity in fractions as described in legend to Fig. 1

Ion-Exchange Chromatography of SF

The activity of SF from both rat and bovine brain (Fig. 4) was retained by Q-Sepharose and was eluted in the presence of 0.1 M ammonium acetate. Approximately two-thirds of the UV-absorbing material, and less than 10% of the SF, was

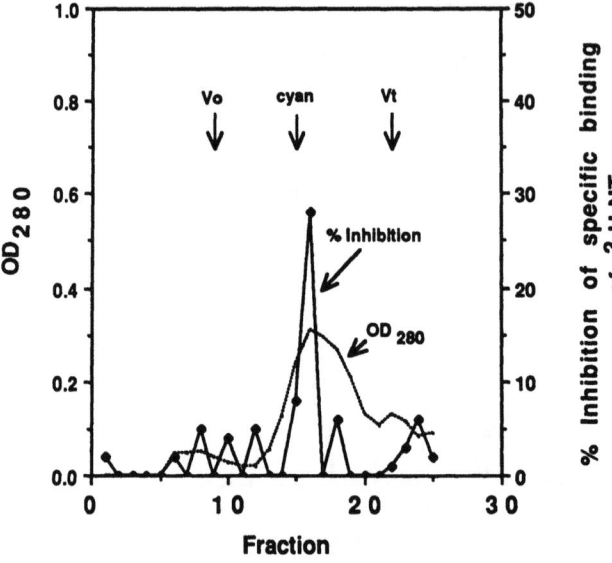

Fig. 3. Gel filtration of ultrafiltered bovine extract on Sephadex G-25. Column calibration and assay of activity in fractions as described in legend to Fig. 1

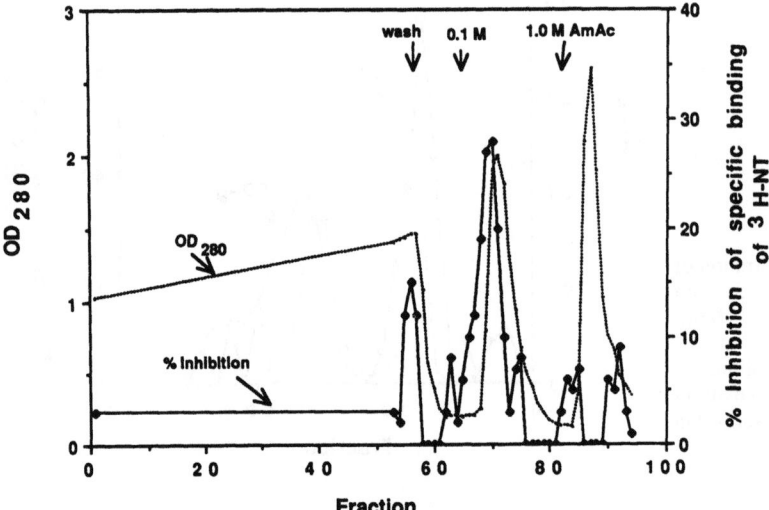

Fig. 4. Ion-exchange chromatography of ultrafiltered bovine SF on Q-Sepharose. The column dimension were 2.5 × 40 cm, and 7.0 ml fractions were collected. A total of 400 ml diluted SF was applied at a flow rate of 0.2 ml/min to a column of Q-Sepharose (acetate form) that had been equilibrated in distilled water. *Wash,* wash with distilled water; *0.1 M,* elution with 0.1 *M* ammonium acetate; *1.0 M,* elution with 1.0 *M* ammonium acetate. Aliquots of 350 µl of each fraction were assayed for the ability to inhibit the binding of [^{3}H]NT (0.6 nM) to membranes prepared from rat brain

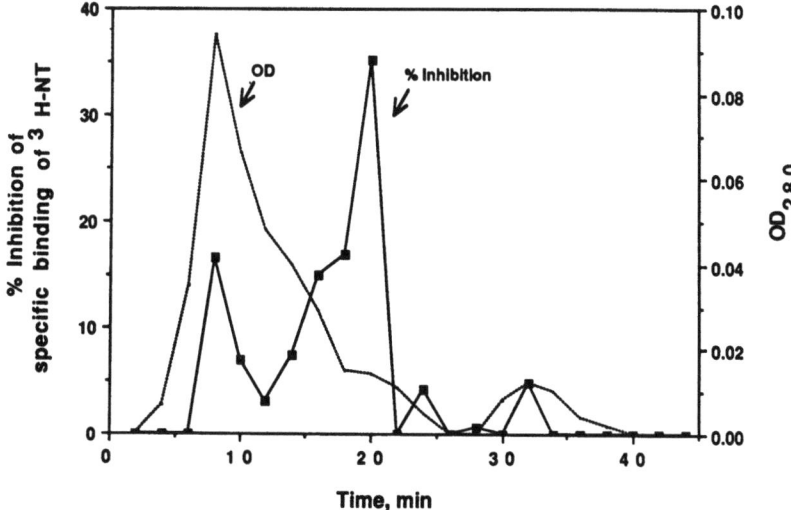

Fig. 5. DEAE-HPLC of purified bovine SF. Bovine SF that had been purified on Q-Sepharose was resuspended in 5 mM ammonium acetate, and a sample corresponding to 48 μg Lowry-reactive material was applied to a Spherogel DEAE-HPLC column. The column was eluted at a flow rate of 1.0 ml/min with a linear gradient of 9-200 mM ammonium acetate at pH = 6.9. Fraction volume was 2.0 ml. Aliquots of 350 μl of each fraction were assayed for the ability to inhibit the binding of [^{3}H]NT (0.6nM) to membranes prepared from rat brain

recovered in the effluent and wash fractions. The remainder of the SF recovered from the column was present in fractions that were eluted with 0.1 M ammonium acetate.

Bovine SF that had been purified by chromatography on Q-Sepharose was further purified by HPLC on a Spherogel DEAE column (Fig. 5). The SF inhibitory activity was retained by the matrix and was eluted with a relatively low concentration of ammonium acetate. Because the content of Lowry-active material was below the limits of reliable detection, it was not possible to assess the extent of purification resulting from this step. However, SF was separated from a large peak of UV-positive material that was not retained by the column.

Effect of Bovine SF on the Binding of [^{3}H]NT

The inhibitory activity of bovine SF was reduced by treatment with pronase, similar to previous findings obtained in studies with rat SF (data not shown), suggesting that bovine SF is a peptide or contains peptide bonds. Inhibition of the binding of [^{3}H]NT by bovine SF was found to be irreversible. Saturation studies of the binding of [^{3}H]NT in the presence of bovine SF purified by chromatography on Q-Sepharose revealed that the inhibition was due to a decrease in the density of binding sites for [^{3}H]NT, with no change in the affinity of the sites for [^{3}H]NT (Fig. 6).

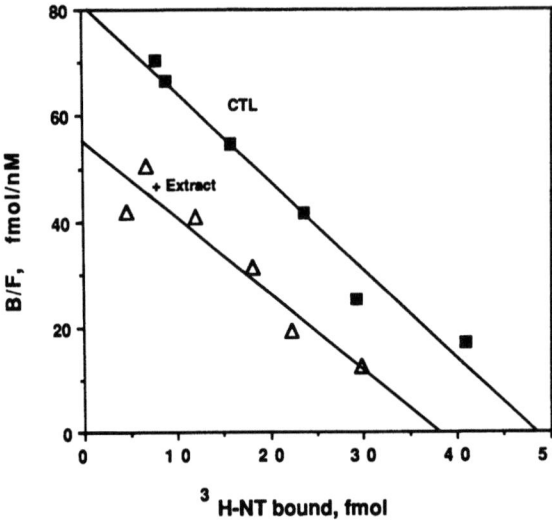

Fig. 6. Scatchard analysis of the binding of [^{3}H]NT in the presence and absence of bovine SF purified by chromatography on Q-Sepharose. Each assay contained 175 µg brain membrane protein, 1 mM CaCl$_2$, varying concentrations of [^{3}H]NT, 1 µM nifedipine or vehicle, and Tris buffer *(CTL)* or purified SF *(+ Extract)* in a final volume of 1.0 ml. Control: B$_{max}$ = 244 fmol/mg protein, K$_d$ = 0.5 nM; + SF, B$_{max}$ = 205 fmol/mg protein, K$_d$ = 0.6 nM

Effect of Rat SF on Voltage-Dependent Influx of ^{45}Ca^{2+} in PC 12 Cells

SF prepared from rat brain inhibited the binding of ^{3}H-PN200–110 to membranes prepared from PC 12 cells. The effect of a dose of rat SF that caused inhibition of 50% of the binding of [^{3}H]PN200–110 to membranes prepared from PC 12 cells on depolarization-induced influx of ^{45}Ca^{2+} in PC 12 cells is shown in Table 1. As expected for influx of ^{45}Ca^{2+} via voltage-dependent calcium channels, depolarization-induced uptake of ^{45}Ca^{2+} was blocked in the presence of nifedipine. When cells were preincubated and assayed in the presence of an amount of rat SF sufficient to cause 50% of the maximal inhibition of binding of [^{3}H]PN200–110 to PC 12 membranes, the depolarization-induced uptake of ^{45}Ca^{2+} was greatly attenuated. The effect of nifedipine was not blocked by SF, and the effects of SF and nifedipine were not greater than the inhibition seen with nifedipine alone. These results suggest that rat SF is capable of causing a functional blockade of voltage-dependent calcium channels as well as inhibition of the binding of [^{3}H]1,4-dihydropyridines.

Table 1. Voltage-dependent influx of ^{45}Ca^{2+} in PC 12 cells

	5 mM K$^+$	50 mM K$^+$ (cpm ^{45}Ca^{2+}/mg protein)	Net influx
Control	1279 ± 104	3491 ± 121	2212 ± 160
+ Nifedipine (2 µM)	1134 ± 44	1395 ± 78	261 ± 90
+ SF	862 ± 77	2124 ± 161	1262 ± 178
+ Nifedipine + SF	835 ± 37	1166 ± 118	331 ± 124

Cells were incubated for 2 min in buffer containing 1 µCi/ml ^{45}CaCl$_2$ and either 5 mM (resting) or 50 mM (depolarizing) KCl. The difference in ^{45}Ca^{2+} uptake between resting and depolarizing conditions represents the net influx of Ca^{2+} resulting from depolarization of the cell membrane.

These studies do not address the question of the mechanism of inhibition of channel function, i. e., whether the inhibition of flux is due to direct interactions of SF with the channel protein or whether SF interacts with some other cell component which in turn interacts with the channel protein. It is also possible that exposure of cells to SF leads to a covalent modification of the channel, either directly or by the action of an enzyme on the channel protein, which would be consistent with the irreversible inhibition of binding of [^{3}H]NT observed in membrane preparations. Establishing the molecular mechanisms responsible for the inhibition will require purified preparations of SF.

The relation of the SF described in these studies to those reported by other investigators is not clear. The SF described by Sanna and Hanbauer (1987) is heat-stable, acid-stable, inhibits the binding of [^{3}H]NT by reducing the density of binding sites for [^{3}H]NT, and inhibits veratridine-stimulated uptake of ^{45}CaCl$_2$ into cultured cerebellar granule cells. The similarity of the properties of the inhibitory factor described by Sanna and Hanbauer (1987) to those of SF suggests that these factors may be identical or closely related. However, a side-by-side comparison of the properties of these factors has not yet been carried out and will be necessary to establish the relationship between them.

References

Berridge MJ, Irvine RF (1984) Inositol trisphosphate, a novel second messenger in cellular signal transduction. Nature (London) 312:315–321

Boles RG, Yamamura HI, Schoemaker H, Roeske WR (1984) Temperature-dependent modulation of ^{3}H-nitrendipine binding by the calcium channel antagonists verapamil and diltiazem in rat brain synaptosomes. J Pharmacol Exp Ther 229:333–339

Bolton TB (1979) Mechanisms of action of transmitters and other substances on smooth muscle. Physiol Rev 59:606–718

Brown AM, Camerer H, Kunze DL, Lux DH (1982) Similarity of unitary Ca^{2+} currents in three different species. Nature (London) 299:156–158

Ebersole BJ, Gajary ZL, Molinoff PB (1988) Endogenous modulators of binding of ^{3}H-nitrendipine in extracts of rat brain. J Pharmacol Exp Ther 244:971–976

Ehlert FJ, Itoga E, Roeske WR, Yamamura HI (1982) The interaction of ^{3}H-nitrendipine with receptors for calcium antagonists in the cerebral cortex and heart of rats. Biochem Biophys Res Comm 104:937–943

Greenberg DA, Carpenter CL, Cooper EC (1985) Stimulation of calcium uptake in PC12 cells by the dihydropyridine agonist BAY K8644. J Neurochem 45:990–993

Greenberg DA, Carpenter CL, Messing RO (1986) Depolarization-dependent binding of the calcium channel antagonist (+)-[^{3}H]PN200–110 to intact cultured PC12 cells. J Pharmacol Exp Ther 238:1021–1027

Kunze DL, Hamilton SL, Hawkes MJ, Brown AM (1987) Dihydropyridine binding and calcium channel function in clonal rat adrenal medullary tumor cells. Mol Pharmacol 31:401–409

Lowry OH, Rosebrough NJ, Farr AL, Randall RJ (1951) Protein measurement with the Folin phenol reagent. J Biol Chem 193:265–275

McClesky EW, Fox AP, Feldman DH, Cruz LJ, Olivera BM, Tsien RW, Yoshikami D (1987) ω-Conotoxin: direct and persistent blockade of specific types of calcium channels in neurons but not muscle. Proc Natl Acad Sci USA 84:4327–4331

Nilius B, Hess P, Lansman JB, Tsien RW (1985) A novel type of cardiac calcium channel in ventricular cells. Nature (London) 316:443–446

Nowycky MC, Fox AP, Tsien RW (1985) Three types of neuronal calcium channel with different calcium agonist sensitivity. Nature (London) 316:440–443

Reynolds IJ, Snowman AM, Snyder SH (1986) (−)-^{3}H-Desmethoxyverapamil labels multiple calcium channel modulator receptor in brain and skeletal muscle membranes: differentiation by temperature and dihydropyridines. J Pharmacol Exp Ther 237:731–738

Rubin RP, Weiss GB, Putney JW Jr (eds) (1985) Calcium in biological systems. Plenum, New York

Sanna E, Hanbauer I (1987) Isolation from rat brain tissue of an inhibiting activity for dihydropyridine binding sites and voltage-dependent Ca^{2+} uptake. Neuropharmacology 26:1811–1814

Spat A, Bradford PG, McKinney JS, Rubin RP, Putney JW Jr (1986) A saturable receptor for ^{32}P-inositol-1,4,5-trisphosphate in hepatocytes and neutrophils. Nature (London) 319:514–516

Stallcup WB (1979) Sodium and calcium fluxes in a clonal nerve cell line. J Physiol (London) 286:525–540

Toll L (1982) Calcium antagonists: high-affinity binding and inhibition of calcium transport in a clonal cell line. J Biol Chem 257:13189–13192

Triggle DJ, Janis RA (1987) Calcium channel ligands. Annu Rev Pharmacol Toxicol 27:347–369

Tsien RW (1983) Calcium channels in excitable cell membranes. Annu Rev Physiol 45:341–358

An Endogenous Purified Peptide Modulates Ca^{2+} Channels in Neurons and Cardiac Myocytes

I. Hanbauer[1], E. Sanna[1], G. Callewaert[2], and M. Morad[2]

[1] Hypertension Endocrine Branch, National Heart, Lung, and Blood Institute, National Institutes of Health, Bethesda, MD 20892, USA
[2] Department of Physiology, University of Pennsylvania, Philadelphia, PA 19104, USA

Introduction

Transport of Ca^{2+} through membrane channels plays an important role in excitation-contraction coupling of cardiac and smooth muscle, in neurosecretion, and in neuronal signaling. The discovery that dihydropyridines can regulate voltage-activated Ca^{2+} channels (Fleckenstein et al. 1972) set the stage for studies on the structure and function of these channels and provided a biochemical probe useful in the search for possible endogenous modulators (EMs). Studies on the existence of endogenous Ca^{2+}-channel modulators were triggered by various reports providing electrophysiological and pharmacological evidence for the involvement of organic Ca^{2+}-channel antagonists in Ca^{2+}-channel regulation (Fleckenstein 1977; Janis and Diamond 1981; Tsien 1984). Supporting the idea of the possible existence of EMs were reports showing that sympathetic denervation of the heart up-regulated the dihydropyridine binding sites in this tissue (Skattebol 1986) and that up-regulation of ^{3}H-nitrendipine binding sites occurred in mouse brain after chronic treatment with morphine (Ramkumar and El-Fakahany1984). Janis et al. (1988) have reported that a number of endogenous substances alter Ca^{2+}-channel activity by either inhibiting ^{3}H-dihydropyridine binding or modifying potential-dependent Ca^{2+} currents. For example, dynorphine A inhibited Ca^{2+}-channel activity, while calcitoninlike peptides enhanced the Ca^{2+} channel without altering ^{3}H-dihydropyridine binding (Nohmi et al. 1985; Tsunoo et al. 1986; MacDonald and Merz 1987). Mir and Spedding (1986) reported that palmitoyl carnitine (IC$_{50}$ = 1.2 × 10^{-4} M) decreased ^{3}H-dihydropyridine binding and reversed the inhibitory action of nifedipine or verapamil on the heart. Other endogenous substances, including catecholamines, biogenic amines, amino acids, and adenosine, are also shown to inhibit Ca^{2+} influx in neurons but do not alter the specific binding of ^{3}H-dihydropyridines.

In this report, we shall summarize a four-step procedure which we have used to isolate and identify a novel EM of Ca^{2+} channels. We show that our endogenous ligand has a molecular weight of about 1000 daltons, displaces nitrendipine in neuronal cultures, and regulates Ca^{2+} current in neuronal or myocardial cells.

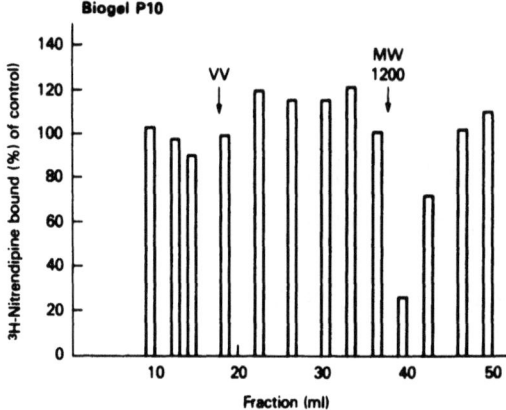
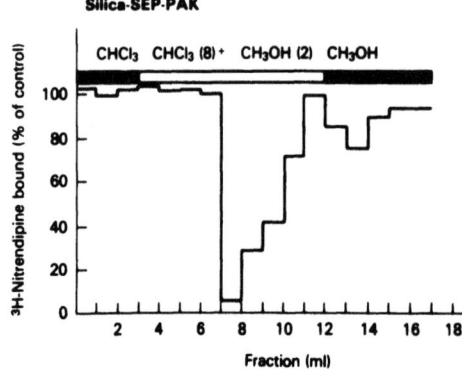
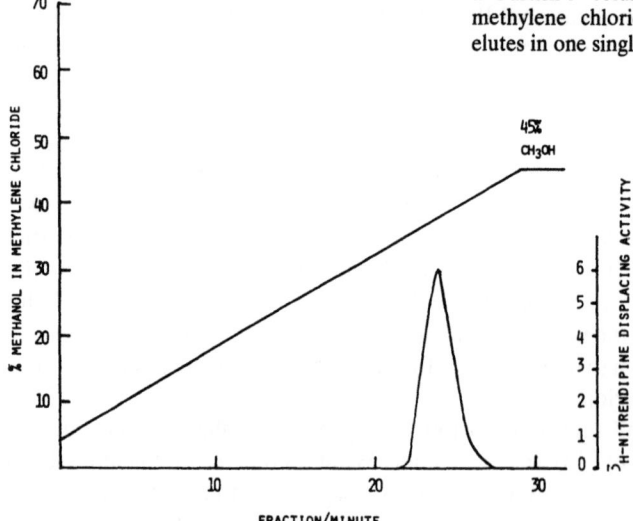

Fig. 1. Purification of an endogenous modulator (EM) for nitrendipine binding sites. *Top* Chromatography on Biogel-P10 column equilibrated and eluted with 0.1 N acetic acid. *Middle* Separation on silica Sep-pak. The sample was loaded on the cartridge (Waters Assoc.) and washed with chloroform. The major amount of EM was eluted with chloroform: methanol = 4:1. *Bottom* HPLC on a Partisil-5 column: initial condition 95% methylene chloride + 5% methanol. EM elutes in one single peak with 33% methanol

Extraction of Nitrendipine-Inhibitory Activity

Aqueous extracts of brains or hearts of 40 rats were filtered through a Biogel P10 column and the fractions eluted from the column were tested for ^{3}H-nitrendipine-displacing activity. Brain extracts contained nitrendipine-displacing material that coeluted with a molecular weight marker of 1200 daltons (Fig. 1). This material was also soluble in methanol or ethanol and insoluble in chloroform. In contrast, when heart extracts were passed over a Biogel P10 column, none of the eluted fractions contained nitrendipine-displacing activity. To design a procedure for the further purification of the nitrendipine-displacing material present in brain extract, we examined its chromatographic behavior on different Sep-pak cartridges (Waters Assoc.). The data in Table 1 show that the nitrendipine-displacing activity was present in the effluent of C_{18} Sep-pak cartridges in the absence or presence of different ion-pairing reagents. This result suggests that the nitrendipine-displacing material may contain highly polar groups that prevent it from being retained on C_{18}. Interestingly, comparative studies on the chromatographic characteristics of the tetrapeptide Glu-Ala-Glu-Asp (Sigma) showed that it also was not retained on C_{18} Sep-pak cartridge under similar chromographic conditions. In contrast, the nitrendipine-displacing material did bind to silica Sep-pak cartridge (Waters Assoc.) and was eluted by a methylenechloride: methanol gradient (Fig. 1). Based on this information, the active material was further purified by HPLC technique using a Partisil-10 or Partisil-5 silica column (Whatman) and a linear gradient from 5% to 45% methanol in methylenechloride for 40 min. Figure 1 shows that nitrendipine-displacing activity eluted in one single peak from a Partisil-5 column with 33% methanol 67% methylenechloride. The active material eluting in this peak was resolved into two mass fragments using californium-252 plasma desorption mass spectrometry. The major fragment had a molecular mass of 948 daltons while that of the smaller fragment (40% of counts/channel) was 1021 daltons.

To establish specificity for the nitrendipine-displacing activity purified by Partisil-10 HPLC, we tested its effect on the specific binding of radioligands known to have high affinity for neurotransmitter receptors also present in synaptosomal brain membranes. The data in Table 2 demonstrate that nitrendipine-displacing material failed to alter the radioligand binding to dopamine, serotonin, or glutamate receptor sites in striatal membranes. It showed displacing activity for ^{3}H-flunitrazepam binding sites in cerebellar membranes. Although reports in the literature demonstrated that various benzodiazepines inhibited different types of Ca^{2+}-channel currents (Taft and DeLorenzo 1984; Rampe and Triggle 1987), we do not know at the present time

Table 1. Behavior of nitrendipine-displacing activity on different types of Sep-pak cartridges

Type of cartridge	Initial condition	Eluting condition
C_{18}	0.1% TFA	Same as initial
C_{18}	0.0023 M HFBA pH 1.9	Same as initial
C_{18}	0.0023 M HFBA + TEA pH 3.0	Same as initial
C_{18}	0.0023 M HFBA + TEA pH 4.0	Same as initial
Silica	100% methylenechloride	80% methylenechloride + 20% methanol

TFA, trifluoroacetic acid; HFBA, heptafluorobutyric acid; TEA, triethylamine

Table 2. Displacing activity of the Ca^{2+}-channel modulator for radioligands of various membrane receptors

Radioligand	Tissue	Displacing activity[a]
^{3}H-Diltiazem (5 nM)	Heart	6.7
^{3}H-Nitrendipine (200 pM)	Heart	7.5
^{3}H-Nitrendipine (100 pM)	Brain	22
^{3}H-Flunitrazepam (4 nM)	Brain	24
^{3}H-MK 801 (10 nM)	Brain	1.0
^{3}H-Imipramine (4 nM)	Brain	0.75
^{3}H-Spiroperidol (400 pM)	Brain	0.85

[a] (Ligand bound in absence of endogenous ligand) / (ligand bound in presence of endogenous ligand)

whether the decrease of ^{3}H-flunitrazepam and ^{3}H-nitrendipine binding was caused by the same substance or by different coeluting substances. The nitrendipine-displacing material extracted from brain also decreased ^{3}H-nitrendipine or ^{3}H-diltiazem binding to heart membranes but was less potent than in brain membranes (Table 2).

Effect of Nitrendipine-Displacing Material on Veratridine-Stimulated Ca^{2+} Influx in Primary Cultures of Cerebellar Granule Cells

Several reports suggest that ^{3}H-nitrendipine binding sites constitute a component of the supramolecular entity of the voltage-dependent Ca^{2+} channel (Curtis and Catterall 1983; Glossmann and Ferry 1983; Borsotto et al. 1984; Tanabe et al. 1987). Moreover, it was shown that the increase of $^{45}Ca^{2+}$ flux elicited by K^+ depolarization or veratridine was attenuated by dihydropyridines (Carboni et al. 1985). The data in Table 3 show that in primary cultures of cerebellar cells, veratridine (2×10^{-5} M)

Table 3. Effect of an endogenous modulator (EM) purified from brain on the veratridine-stimulated $^{45}Ca^{2+}$ uptake and cGMP accumulation in cerebellar granule cells

Addition to incubation medium			$^{45}Ca^{2+}$ uptake (nmol mg protein^{-1} 20 s^{-1})	cGMP accumulation (pmol mg protein^{-1} 2 min^{-1})
EM Active	Acid-hydrolyzed	Veratridine[a]		
None	None	No	4.3	0.11
None	60 µl	No	3.9	n.d.
None	None	Yes	11.2	17.25
30 µl	None	Yes	8.1	11.04
60 µl[b]	None	Yes	5.3	5.12
None	60 µl	Yes	11.0	n.d.

Values were obtained by two different measurements in triplicate. n.d., not determined
[a] For $^{45}Ca^{2+}$ uptake and cGMP accumulation studies, 2×10^{-5} M and 5×10^{-5} M veratridine respectively was used
[b] 60 µl EM caused 90% inhibition of ^{3}H-nitrendipine binding

elicits a two- to threefold increase of $^{45}Ca^{2+}$ influx and a 15-fold increase of cGMP content. Addition of nitrendipine-displacing material isolated from brain tissue to the incubation medium 1 min prior to the addition of $^{45}Ca^{2+}$ caused a dose-dependent decrease of $^{45}Ca^{2+}$ influx and cGMP accumulation in cerebellar granule cells (Table 3). Ca^{2+} influx and cGMP formation in granule cells was not altered when acid-hydrolyzed material was tested.

Chemical Properties of Nitrendipine-Displacing Material

We have studied the behavior of nitrendipine-displacing material during high-voltage electrophoresis on a 0.2% agarose column equilibrated in 25 mM ammonium formate buffer, at pH 3.0 or 4.0, at 1100 V and 4.5 mA. Under these conditions the nitrendipine-displacing material showed no mobility, which suggests that in these pH conditions the material is at its isoelectric point (Sanna et al. 1988). Chromatography of the HPLC-purified material on silica TLC plates showed that the material with nitrendipine-displacing activity was contained in a single spot that stained with iodine and ninhydrin. In an attempt to determine whether the active material is of peptidergic nature, we studied the effect of digestion with pronase, proteinase K, or acid hydrolysis. The results of these experiments, summarized in Table 4, demonstrate that the nitrendipine-displacing material is resistant to both proteases but loses its displacing activity after acid hydrolysis (5.7 N HCl for 30 min at 155 °C). We have examined the hydrolysate for its amino acid composition and found that it consisted of 55% Asp, 25% Glu, 5% Gly, 5% Thr, and two unidentified peaks. The attempt to determine the amino acid sequence of the material was unsuccessful. We are presently studying the possibility of a blockade of free N-terminal by carbohydrates or pyrrolidine ring formation.

Table 4. Effect of various proteases and acid hydrolysis on the nitrendipine-displacing activity of an endogenous ligand isolated from rat brain

| Endogenous ligand | | ^{3}H-Nitrendipine bound |
Addition	Pretreatment	(fmol/mg protein)
None	None	80
60 µl	None	39
60 µl	Pronase	40
60 µl	Proteinase K	47
60 µl	Acid hydrolysis	81
60 µl	Xanthine oxidase	42

The values represent the mean of two different assays in triplicate. Aliquots containing ^{3}H-nitrendipine-displacing material eluted from HPLC partisil-10 SCX column were preincubated in the presence of either pronase (1 mg/ml, 7 h at 37 °C) or proteinase K (50 µg/ml, 7 h at 37 °C in 50 mM Tris buffer, pH 7.8, and xanthine oxidase (1 U/ml, 30 min at 37 °C) in 50 mM Tris buffer, pH 7.5

Ca^{2+}-Channel Current Measurements in Heart and Neuroblastoma Cells

The whole-cell patch-clamp technique was used to measure the Ca^{2+}-channel current (I_{Ca}) in isolated guinea pig ventricular cells (Mitra and Morad 1986) and mouse neuroblastoma cells. The internal and external solutions were designed to maximize the measurements of I_{Ca} and minimize currents associated with K^+ and Na^+ channels. In neuroblastoma cells, addition of 1–2 units of EM caused a marked suppression of both the low- and high-threshold Ca^{2+} current (Fig. 2B) within a few milliseconds of addition of EM. I_{Ca} was generally suppressed by about 70% and the effect was slowly reversible. In guinea pig ventricular myocytes, on the other hand, the high-threshold Ca^{2+} current was strongly enhanced (100%–400%) by the EM but required about 100 s to achieve maximal activation (Fig. 2A). The enhancing effect could be recorded at all membrane potentials tested and was very slowly reversible on wash-out of EM.

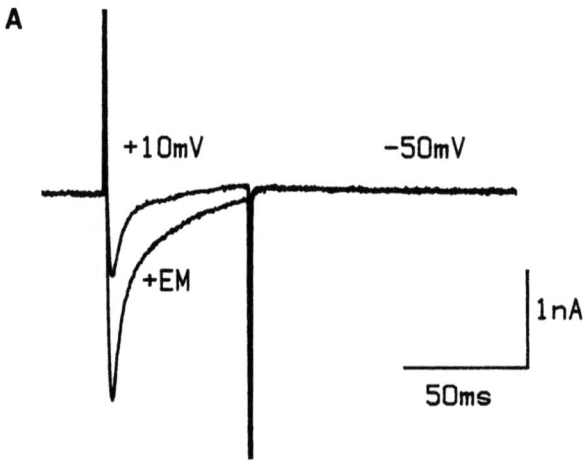

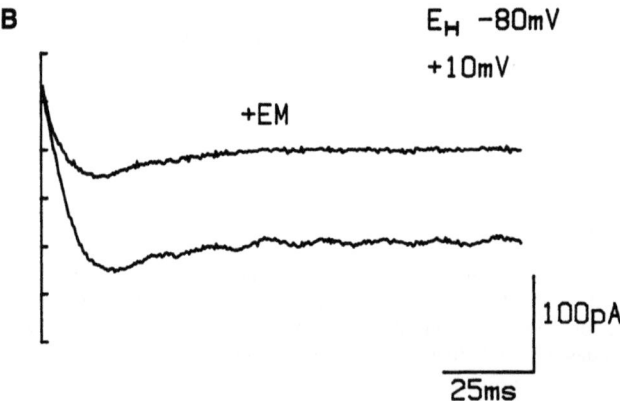

Fig. 2A, B. Effect of EM (1 unit) on the high-threshold Ca^{2+} current in guinea pig ventricular myocytes (**A**) and in neuroblastoma cells (**B**). Calcium currents were measured under voltage-clamp conditions that minimized other ionic currents. In **B**, Ni^{2+}-sensitive currents are shown (difference between the currents before and after 5 mM Ni^{2+})

Table 5. Effect of various compounds on the increase of I_{Ca} in cardiac cells elicited by an endogenous modulator for nitrendipine binding sites

Compounds tested	Increase of I_{Ca}
Propranolol (5 µM)	Not changed
WB-4101 (200 µM)	Not changed
Nifedipine (10 µM)	Inhibited
Glycine (100 µM)	Not mimicked
L-Glutamate (100 µM)	Not mimicked
GABA (10 µM)	Not mimicked

The enhancement of Ca^{2+} current by EM was not mediated through either the β- or the α-adrenergic receptors and was independent of the presence and absence of cAMP in the internal dialyzing solution. Table 5 illustrates the effect of a number of agents tested on I_{Ca} in cardiac myocytes. These results suggest that EM interacts directly with the Ca^{2+} channel, serving as an agonist on the cardiac myocytes and as an antagonist on the neuroblastoma cells. Based on the onset of EM action on neuronal and cardiac Ca^{2+} channels, we speculate that the antagonistic effect of EM in the neurons occurs via a direct interaction with the channel at its external side, while the agonistic effect on the myocytes may be mediated by the activation of an as yet unidentified intracellular mediator.

References

Borsotto M, Norman RI, Fosset M, Lazdunski M (1984) Solubilization of the nitrendipine receptor from skeletal muscle transverse tubule membranes. Interactions with specific inhibitors of the voltage-dependent Ca^{2+} channel. Eur J Biochem 142:449–455

Carboni E, Wojcik WJ, Costa E (1985) Dihydropyridine changes the uptake of Ca^{2+} induced by depolarization in primary cultures of cerebellar granule cells. Neuropharmacology 24:1123–1126

Curtis BM, Catterall WA (1983) Solubilization of the calcium antagonist receptor from rat brain. J Biol Chem 258:7280–7283

Fleckenstein A, Tritthart H, Doring H-J, Byon YK (1972) Bay a 1040 – ein hochaktiver Ca^{2+}-antagonistischer Inhibitor der elektro-mechanischen Koppelungsprozesse im Warmblüter-Myokard. Arnzeimittelforsch 22:22–33

Fleckenstein A (1977) Specific pharmacology of calcium in myocardium, cardiac pacemakers, and vascular smooth muscle. Ann Rev Pharmacol Toxicol 17:149–166

Glossmann H, Ferry DR (1983) Solubilization and partial purification of putative calcium channels labelled with [^{3}H]-nitrendipine. Naunyn Schmiedeberg's Arch Pharmacol 323:279–291

Hanbauer I, Sanna E (1986) Endogenous modulator for nitrendipine binding sites. Clinical Neuropharmacology 9 Suppl 4:220–222

Hanbauer I, Sanna E (1988) Presence in brain of an endogenous ligand for nitrendipine binding sites that modulates Ca^{2+} channel activity. Ann NY Acad Sci 522:96–105

MacDonald RL, Werz MA (1987) Dynorphin A decreases voltage-dependent calcium conductance of mouse dorsal root ganglion neurones. J Physiol (Lond) 377:237–249

Mir AK, Spedding M (1986) Proc Brit Pharmacol Soc 88:381 P

Mitra R, Morad M (1986) Two types of calcium channels in guinea pig ventricular myocytes. Proc Natl Acad Sci USA 83:5340–5344

Nohmi M, Shinnick-Gallagher P, Green PW, Gallagher JP, Cooper CW (1985) Soc Neurosci Abstr 11:708

Ramkumar V, El-Fakahany EE (1984) Increase in [^{3}H]-nitrendipine binding site in the brain in morphine-tolerant mice. Eur J Pharmacol 102:371–2

Rampe D, Triggle DJ (1987) Benzodiazepine interactions at neuronal and smooth muscle Ca^{2+} channels. Eur J Pharmacol 134:189–197

Sanna E, Hanbauer I (1987) Isolation from rat brain tissue of an inhibiting activity for dihydropyridine binding sites and voltage-dependent Ca^{2+} uptake. Neuropharmacology 26:1811–1814

Sanna E, Wright AG Jr, Daly JW, Hanbauer I (1988) Dihydropyridine-sensitive Ca^{2+} channels in rat brain: modulation by an endogenous ligand. In: Fidia Research Series, Symposia in Neuroscience VI Liviana Press, Padova, Italy, pp 123–132

Skattebol A, Triggle DJ (1986) 6-Hydroxydopamine treatment increases B-adrenoceptors and Ca^{2+} channels in rat heart. Eur J Pharmacol 127:287–289

Taft WC, Delorenzo RJ (1984) Micromolar-affinity benzodiazepine receptors regulate voltage-sensitive calcium channels in nerve terminal preparations. Proc Natl Acad Sci USA 81:3118–3122

Tanabe T, Takeshima H, Mikami A, Flockerzi V, Takahashi H, Kangawa K, Kojima M, Matsuo H, Hirose T, Numa S (1987) Primary structure of the receptor for Ca^{2+} channel blockers from skeletal muscle. Nature (Lond) 328:313–318

Tsien RW (1983) Calcium channels in excitable cell membranes. Annu Rev Physiol 45:341–358

Tsunoo A, Yoshii M, Narahashi T (1986) Block of calcium channels by enkephalin and somatostatin in neuroblastoma-glioma hybrid NG 108–15 cells. Proc Natl Acad Sci USA 83:9832–9836

Antibodies Against the ADP/ATP Carrier Interact with the Calcium Channel and Induce Cytotoxicity by Enhancement of Calcium Permeability

H.-P. Schultheiss[1], I. Janda[1], U. Kühl[1], G. Ulrich[1], and M. Morad[2]

[1] Department of Internal Medicine, Klinikum Großhadern, University of Munich, D-8000 Munich 70, FRG
[2] Department of Physiology, University of Pennsylvania, Philadelphia, PA 19104, USA

Introduction

Current concepts of autoimmunity emphasize the role of antibodies in the development and maintenance of autoimmune disorders. Recently we showed that the sera of patients with dilated cardiomyopathy – a suspected virus-induced autoimmune disease – contain circulating autoantibodies directed against the ADP/ATP carrier [1, 2]. The adenine nucleotide translocator is an intrinsic hydrophobic protein located in the inner mitochondrial membrane. The protein has a relative molecular weight of around 30000 and exists in the native form as a dimer [3]. Since the inner mitochondrial membrane is a priori impermeant to hydrophilic metabolites, the transfer of ATP to the cytosol with its energy-consuming processes and the return of ADP to the inner mitochondrial space for regeneration by oxidative phosphorylation requires a particular transport catalysis. The ADP/ATP shuttle, the only active nucleotide transport system in mitochondria, is highly specific and corresponds exactly to the requirements of the ATP production in aerobic cells [4]. In this study we investigated the effects of the antibodies against the ADP/ATP carrier on isolated adult rat myocytes. Immunobinding and immunofluorescence studies showed a specific binding of the antibodies to a cell-surface protein. Exposure of the myocytes to the antibodies resulted in enhanced calcium current and cell damage, which was prevented by the addition of calcium-channel blockers. These studies provide evidence for an interaction of the anti-ADP/ATP carrier antibodies with the calcium channel and suggest a new mechanism for cytotoxicity.

Immunological Characterization of the Antibodies Against the ADP/ATP Carrier

To characterize the targets of the antibodies specific for the carboxyatractylate protein complex, different methods such as radioimmunoassay, immunoprecipitation, and Western blotting were used. To test the possible effect of the antibodies on the nucleotide transport in vitro, the exchange rate of mitochondria was determined by the inhibitor-stop method combined with the back exchange [3, 5].

The specificity of antibodies raised against the heart ADP/ATP carrier was tested by the immunoblot technique. In Western blot of heart mitochondrial proteins exposed to the anti-ADP/ATP carrier antibodies, the antibody binds only to the 30000

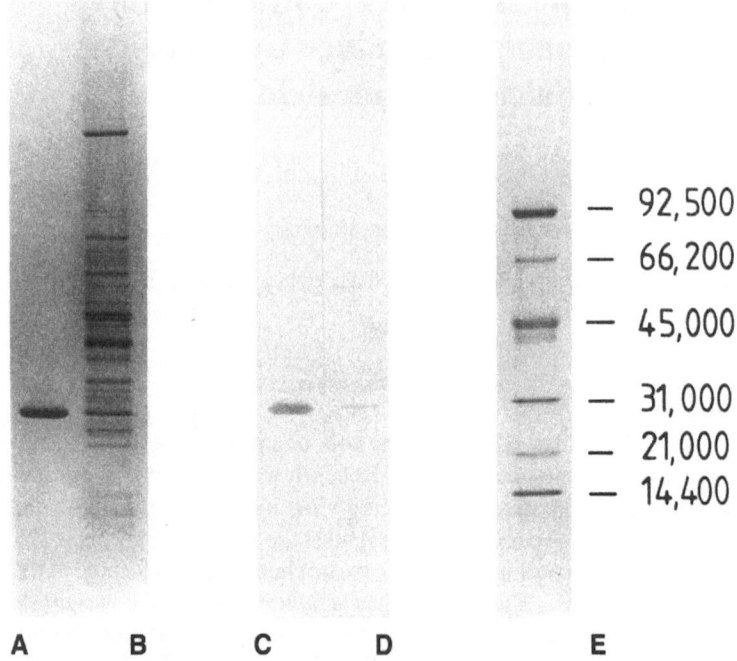

Fig. 1A–E. Detection of antibodies against the ADP/ATP carrier by the immunoblot technique. The isolated ADP/ATP carrier (**A**) and total mitochondrial proteins from heart (**B**) as well as marker proteins (**E**) were separated on sodium dodecyl sulfate (SDS) polyacrylamide slab gels. One section of the gel was stained with Coomassie blue (**A, B, E**), the other section was electrophoretically blotted on a nitrocellulose sheet and incubated with anti-ADP/ATP carrier antiserum and then with horseradish peroxidase conjugated anti-IgG (**C, D**)

MW band, which corresponds to the isolated ADP/ATP carrier (Fig. 1). For a further immunochemical characterization of ADP/ATP carrier antibodies, we elaborated a solid-phase double-antibody immunoradiometric assay (IRMA). First, the surface of the plastic microtiter well was coated with the isolated ADP/ATP carrier protein from heart, liver, and kidney in the form of the CAT-protein complex. Residual binding sites on the plastic were blocked with a nonspecific protein such as albumin. The plate was then washed and incubated with antiserum to allow the antibodies to bind to the antigen. After further washing, a "second antibody", ^{125}I-labeled protein A, was added to the wells, which binds to the Fc part of the first antibody. The amount of bound ^{125}I-labeled protein A was determined by a γ-spectrometer.

Figure 2 gives the titration of the antiserum against the heart protein. The antiserum showed cross-reactivity between the three proteins from heart, kidney, and liver although the antiserum against heart protein had the highest binding activity with the heart protein. Maximum binding to kidney protein was about 50% and to liver protein about 20% of the binding to heart protein.

When the antiserum against kidney protein was tested, the highest titer was found against the kidney protein (Fig. 2). It also contained cross-reacting antibodies, as

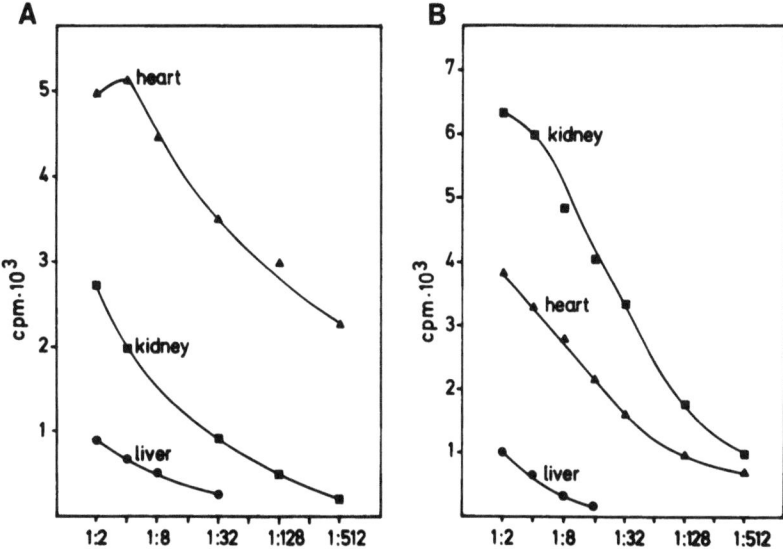

Fig. 2. Specificity and cross-reactivity of heart carboxyatractylate protein antiserum (**A**) and kidney CAT protein antiserum (**B**). Antisera isolated from heart (▲-▲), kidney (■-■), and liver (●-●) were incubated in microtiter plates, and binding of the two antisera was evaluated by solid phase radioimmunoassay. After the incubation of the antigen coated wells with 90 µl of the antiserum and washing three times, unlabeled protein A in varying concentrations (1–128 ng or 2.5–640 ng, respectively) was added. Afterwards, specific-bound antibodies, which were not blocked by cold protein A, were detected by ^{125}I-labeled protein A. The degree of inhibition of binding by ^{125}I-labeled protein A was a function of bound nonradioactive protein A and served as the basis for a quantitative assay

shown by the binding on the translocator protein isolated from heart. The binding of antikidney antiserum on the liver protein, again, was significantly lower. Controls, in which the specific antiserum was omitted or substituted by control sera, routinely gave 2%–6% of the activity obtained when specific antisera were used.

The organ specificity of antiheart antiserum was also confirmed by immunoadsorption studies. The maximum amount of antibody bound by a given immunoabsorbant (plateau binding value) was determined by titrating a fixed amount of antiserum with increasing amounts of antigen. By means of this assay the interaction of antibodies with the isolated carrier protein from heart, kidney, and liver was examined. While the residual activity of the antiheart antibody was only about 10% after immunoadsorption of heart protein, the residual activity after immunoadsorption on kidney (approximately 45%) and on liver protein (80%) was significantly higher (Fig. 3). As neither proteins from kidney nor liver were able to displace antiheart antibodies completely, organ-specific antibodies must be postulated, although an incomplete cross-reaction between the three organs exists. This partial cross-reactivity could also be demonstrated by immunoadsorption studies on mitochondria or by crossed immunoelectrophoresis (data not shown). In an attempt to gain further evidence for organ specificity of antibodies against the ADP/ATP carrier from heart, we determined whether the antiserum inhibits mitochondrial nucleotide transport in heart,

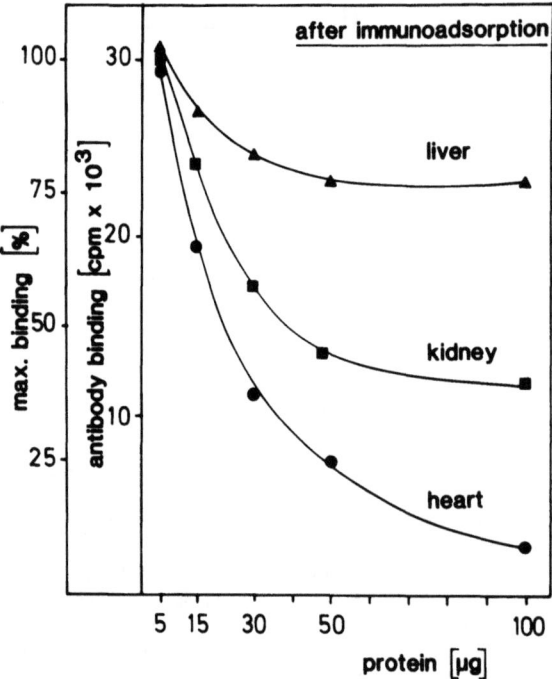

Fig. 3. Binding of heart carboxyatractylate protein antiserum to the ADP/ATP carrier isolated from heart after immunoadsorption on isolated carrier proteins from heart, kidney, and liver. Increasing amounts of isolated carboxyatractylate-protein complexes (5–100 µg) were incubated with a fixed amount of isolated IgG of antiheart protein antiserum. After "preincubation" of the antigen-antibody complexes with protein A bearing *Staphylococcus aureus* (Cowan I), residual antibody activity of the supernatant was determined in an immunoradiometric assay with carboxyatractylate protein heart precoated wells (see legend to Fig. 2)

liver, and kidney. The ADP/ATP exchange rate was measured by the inhibitor-stop method combined with back exchange. Figure 4 shows immunoinactivation of the adenine nucleotide translocator of heart mitochondria by the anti-heart-protein antiserum. Serial dilution of antiserum resulted in a progressive decrease of the blocking effect of the antibody. However, further raising of antibody concentration per milligram mitochondrial protein did not result in total inhibition of the translocation process. A possible explanation for this phenomenon is that the high carrier density on the inner mitochondrial membrane (5 pmol/cm^2) hinders the binding of the antibodies sterically. This suggestion is supported by results observed in the reconstituted system, where the density of active carrier molecules is only 2–5 per milligram liposome [6]. When antibodies were applied to purified carrier protein in the reconstituted system, inhibition of the reconstituted transport was nearly complete, depending on the antibody concentration.

In support of organ-specific binding of antibodies to the carrier protein, it is of interest to note that the anti-heart-protein antiserum showed no inhibition of the adenine nucleotide transport in kidney and liver mitochondria (Fig. 5). Furthermore, the inhibitory activity of anti-heart-protein antiserum was not abolished by adsorption to kidney or liver mitochondria while preadsorption of the antiserum to heart mitochondria did abolish its inhibitory activity.

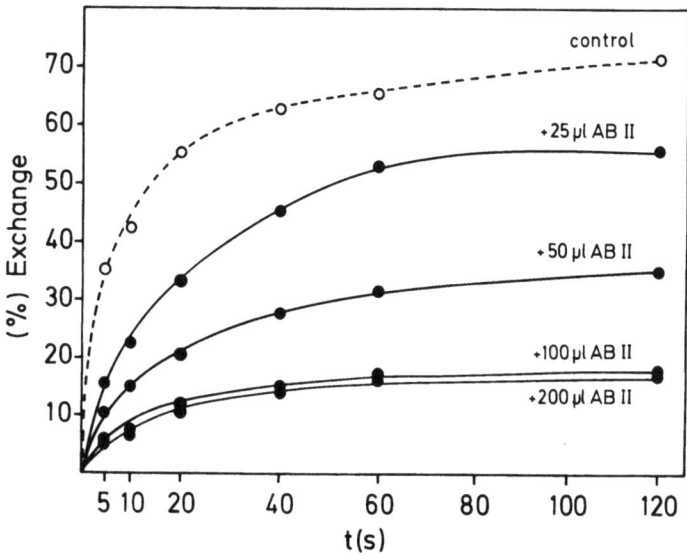

Fig. 4. Influence of anti-heart carboxyatractylate CAT antiserum *(AB)* on nucleotide transport from heart mitochondria. Heart mitochondria – loaded with ^{14}C-labeled ADP – were incubated with antiserum at 4°C for 30 min. The ADP/ATP exchange rate was determined by back exchange combined with the inhibitor-stop method. Exchange was performed at 2°C after serial dilutions of the blocking antibodies *(AB II)* (isolated IgG, corresponding to 25, 50, 100, 200 µl). ●–●, Antiserum; O–O, control

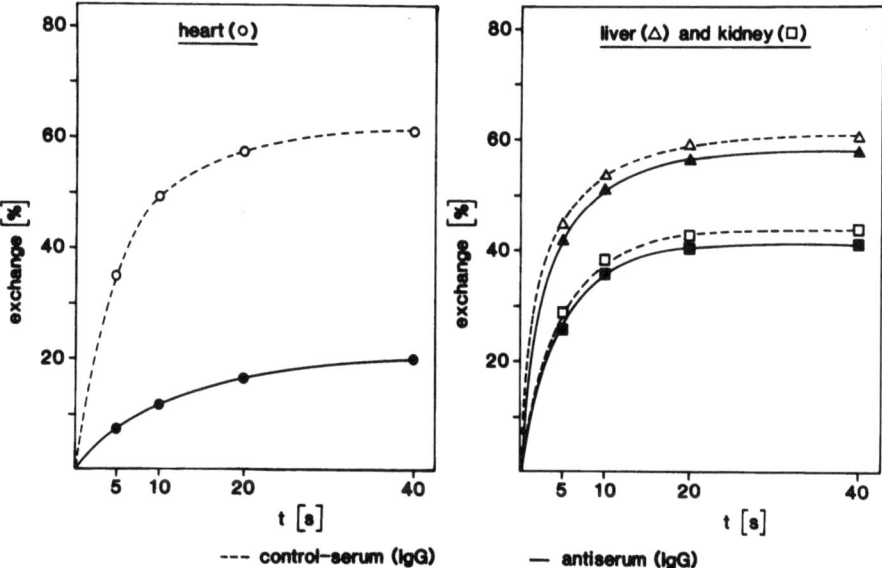

Fig. 5. Influence of antiheart carboxyatractylate protein antiserum on nucleotide transport from heart (O), liver (△), and kidney (□) mitochondria. Isolated and preloaded (^{14}C-labeled ADP) mitochondria were incubated with equilibrated antiserum (isolated IgG) for 30 min at 4°C. Afterwards the ADP/ATP exchange rate was determined by the inhibitor-stop method combined with back exchange. Exchange time was 5, 10, 20, and 40 s at 4°C for heart and kidney mitochondria and at 8°C for liver mitochondria

Specific Binding of Antibodies Against the ADP/ATP Carrier to the Cell Surface of Cardiac Myocytes

Evidence for cross-reactivity between the ADP/ATP carrier and a cell surface protein was first obtained by indirect immunofluorescence. Incubation of frozen sections of heart tissue with the above-characterized anti-ADP/ATP carrier antibodies showed, besides intracellular staining, antibody binding to the plasma membrane. This was confirmed by positive staining of isolated cardiac myocytes showing a sarcolemmal immunofluorescence (Fig. 6). After neutralization of anti-ADP/ATP carrier antibodies by preadsorption with the isolated ADP/ATP carrier, intracellular staining and staining of the cell surface disappeared (Fig. 6). Preimmune sera did not react with the cell surface.

This cross-reactivity between antibodies and the cell surface was also demonstrated by radioimmunobinding assay. Figure 7 shows a time-dependent binding of anti-

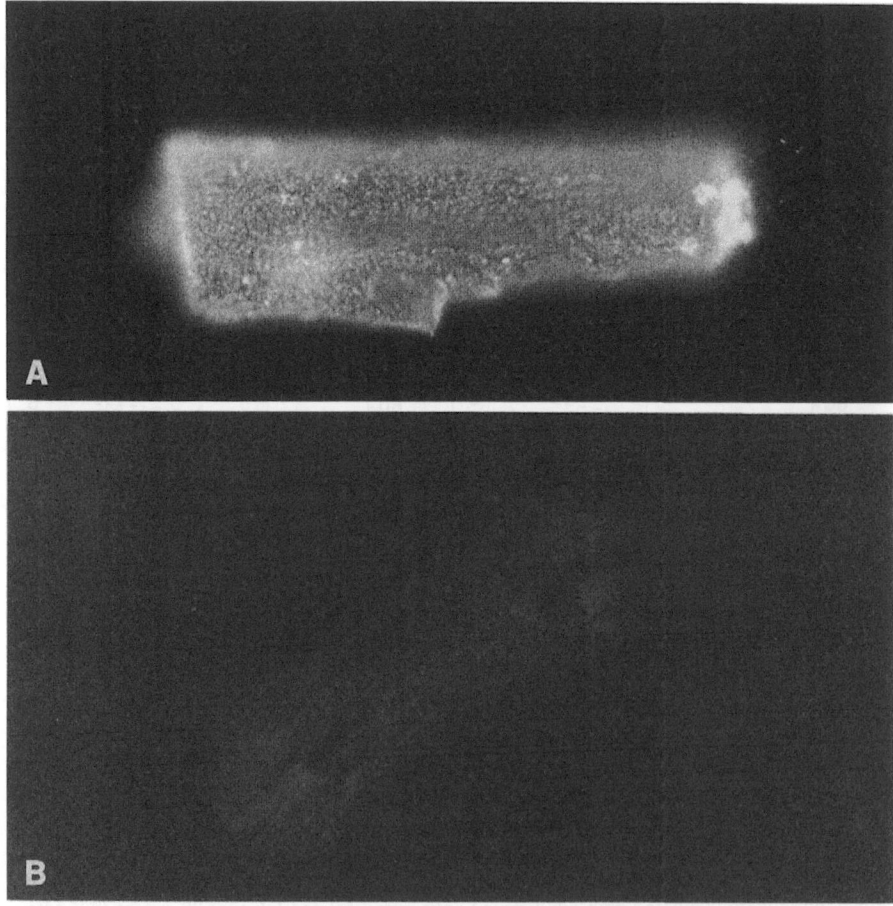

Fig. 6A, B. Immunofluorescence on isolated living cardiac myocytes. **A** Cell-surface staining of isolated adult rat cardiac myocytes with anti-ADP/ATP carrier antibodies. **B** Disappearance of cell-surface reaction after neutralization of antibodies with purified ADP/ATP carrier

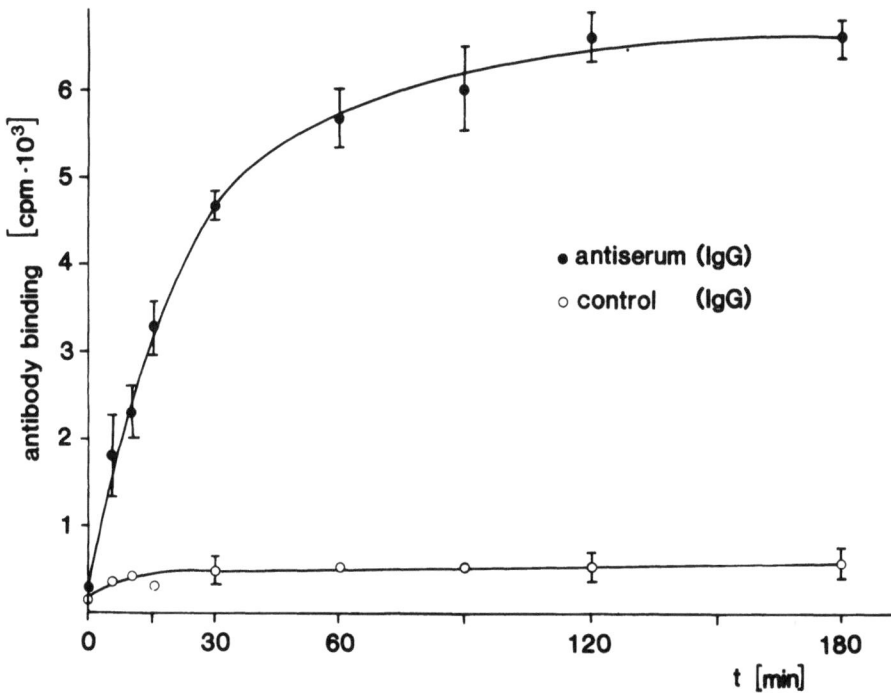

Fig. 7. Time-dependent binding of anti-ADP/ATP carrier antibodies to isolated cardiac myocytes measured with radioimmunobinding assay (antibody dilution 1:2000). Cardiac myocytes were prepared according to Hohl et al. [8]. After isolation 2×10^5 myocytes in 400 µl washing buffer containing 0.5% bovine serum albumin and 1 mM Ca^{2+} were incubated with antibodies (IgG$_5$). ^{125}I-labeled protein A was used to label the bound antibodies. For comparison all data were calculated for 1×10^6 cells

bodies to myocytes. After 60 min most of the antibodies were bound to the cell. Prolonged incubation for up to 180 min could not increase antibody binding significantly. Preimmune serum IgG (control) did not bind to myocytes. Antibody dilution from 1:1000 to 1:6000 resulted in a reduction of antibody binding until 1:5000, where no significant binding could be detected. The specificity of the antigen-antibody reaction was demonstrated by suppression of the binding to myocytes by immunoadsorption. Figure 8 shows that preincubation of the antibody with increasing amounts of the isolated ADP/ATP carrier (1 to 1500 ng) resulted in a concentration-dependent decrease of antibody binding.

Incubation of cardiac myocytes with concentrations of anti-ADP/ATP carrier antibodies (affinity-purified IgG) higher than 1:1000 resulted in a concentration-dependent decrease of cell viability. Deterioration of myocytes, monitored visually, involved in sequence: rhythmic contraction, bleb formation, contracture to an almost cuboid shape, cell rounding, and finally cell death (Fig. 9). These cells took up trypan blue very rapidly.

This antibody-mediated cytotoxic effect was strictly calcium dependent. Using an antibody concentration of 1:100 and a calcium concentration of 1 mM, 20% of

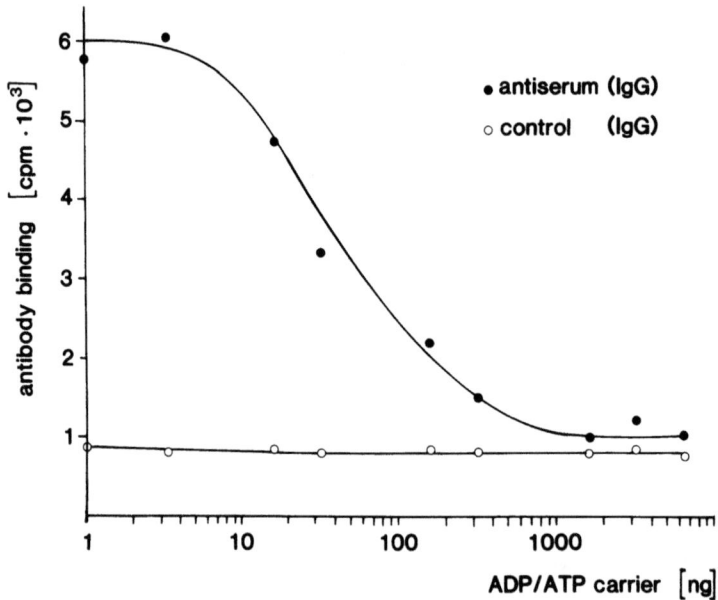

Fig. 8. Inhibition of antibody binding to isolated cardiac myocytes after preincubation with purified ADP/ATP carrier. Antiserum in a dilution of 1:2000 was incubated with increasing amounts of purified ADP/ATP carrier and then measured for residual binding in radioimmunobinding assay

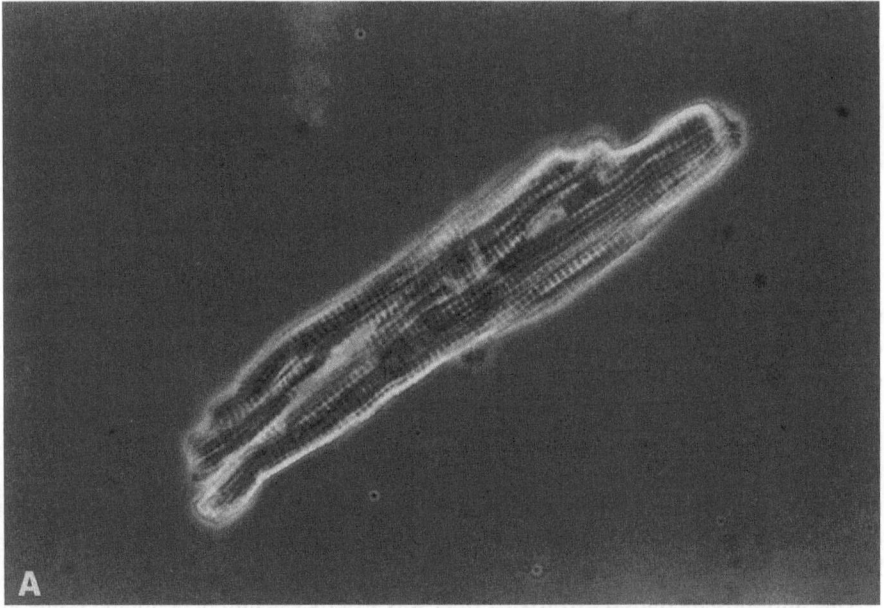

Fig. 9

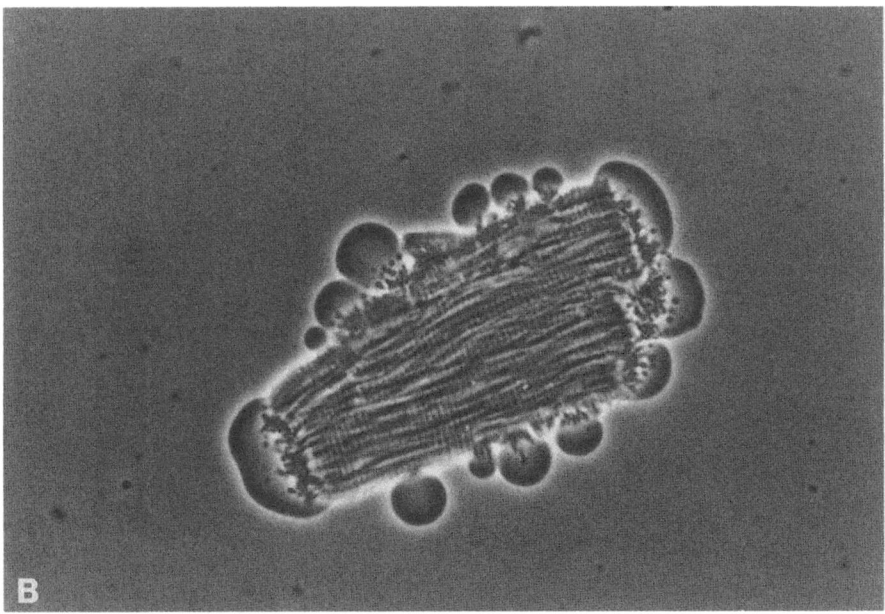

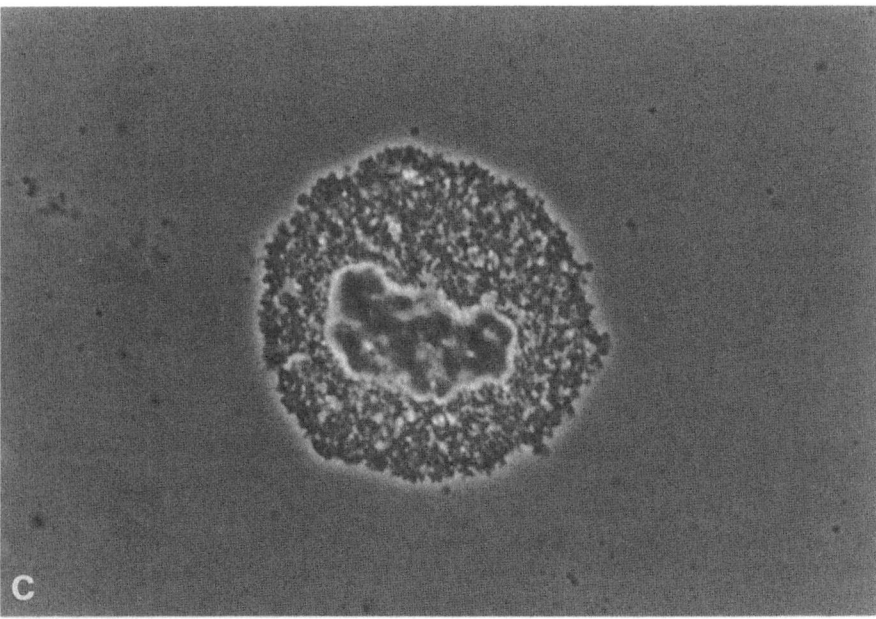

Fig. 9A–C. Appearance of isolated cardiac myocytes during incubation with anti-ADP/ATP carrier antibodies (dilution 1:100, 1 mM Ca^{2+}). **A** Normal rod-shaped cell. **B** Bleb formation. **C** Final cell death

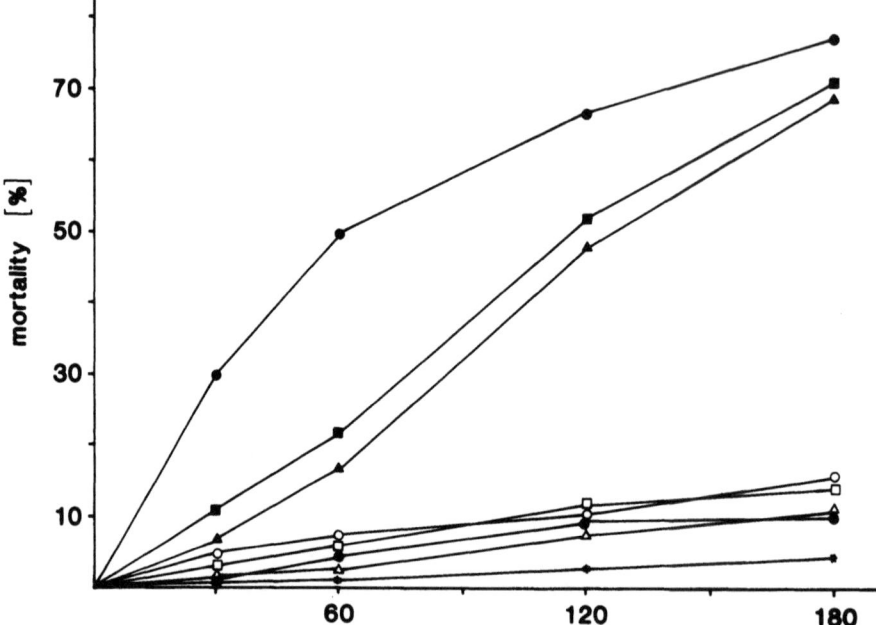

Fig. 10. Calcium concentration dependent mortality of isolated cardiac myocytes incubated with anti-ADP/ATP carrier antibodies: Ca^{2+} concentration 1.0 mM: ● + antiserum, ○ control; Ca^{2+} concentration 0.5 mM: ■ + antiserum, □ control; Ca^{2+} concentration 0.1 mM: ▲ + antiserum, △ control; Ca^{2+} concentration 0: ◒ + antiserum, ✱ control. Antibody (IgG) dilution, 1:100

myocytes died after 30 min, 40% after 1 h, and about 80% after 3 h. In contrast, only 10% of myocytes incubated with preimmune serum IgG (control) did not survive 3 h (Fig. 10). With decreasing amounts of calcium the effect of antibody-mediated cytotoxicity is gradually diminished. With an antibody concentration of 1:100 and nominally no calcium only 10% of cells died during 3 h. The corresponding control shows a mortality of 5% during 3 h.

The use of calcium-channel antagonist together with anti-ADP/ATP carrier antibodies (IgG 1:100) in myocyte suspension (1 mM Ca^{2+}) reduced dramatically the cytotoxic effect of antibodies. Nifedipine and nitrendipine (both 10^{-6} M) seemed to be more potent in protecting the cells than did verapamil (10^{-6} M). Without antibodies none of the calcium-channel antagonists had a significant influence on cell viability. The use of the β-receptor antagonist propranolol (10^{-7} M) together with anti-ADP/ATP carrier antibodies had no effect on cell viability.

The results thus far suggest that antibodies against the ADP/ATP carrier bind specifically to the cell surface of cardiac myocytes and enhance their calcium permeability, leading to cell death. To obtain further support for this idea, we measured the effect of antibodies on the calcium current of isolated rat myocytes, using previously described enzymatic isolation technique [8, 9].

Rat myocytes were used immediately following enzymatic isolation and were stable for at least 8–10 h. In some experiments rat myocytes were kept in culture for 2–3 days prior to use. No significant differences in I_{Ca} were observed in fresh or cultured rat myocytes. Cells were voltage clamped using the whole-cell patch clamp method [10]. Experimental conditions and the intracellular dializing solutions were similar to those previously described [9]. Calcium current was measured in solutions containing $10^{-6}-10^{-5}$ M tetrodotoxin (TTX) which suppressed the fast sodium current. In rat myocytes, in addition, the holding potential was set at -50 to -45 mV to inactivate any remaining TTX-insensitive I_{Na}. The antibody was applied with a fast perfusion system which enabled a rapid exchange (50–100 ms) of the solution around the cell [11]. The addition of antibody markedly enhanced I_{Ca} and slowed its inactivation. The effect of antisera or affinity-purified IgG on the inactivation time course was often more marked in rat myocytes than in frog cells (data not shown). Figure 11 shows the time course of potentiation of I_{Ca} (induced by pulsing from -50 to $+10$ mV) in a rat ventricular myocyte, which was dialized with a solution containing high concentrations of CsCl and TEA to block outward currents (see legend, Fig. 11). The onset of enhancement of I_{Ca} was rapid, and generally 20–30 s was sufficient for the effect of antibody to take place. The enhancement of I_{Ca} was only slowly reversible (Fig. 11). The potentiating effect of the antibody on peak I_{Ca} in rat ventricular cells was variable. The variability of the enhancing effect of the antibody could in part be attributed to the rapid run-down of I_{Ca} in rat myocytes. Consistent with this hypothesis, washout of the antibody was also accompanied by a decrease of I_{Ca} significantly below the control preantibody levels. Irrespective of the effect of antibody on peak I_{Ca} there was always a marked slowing of the inactivation time course. The potentiating effect of antibody on I_{Ca} was often so strong as to make effective voltage control of the membrane

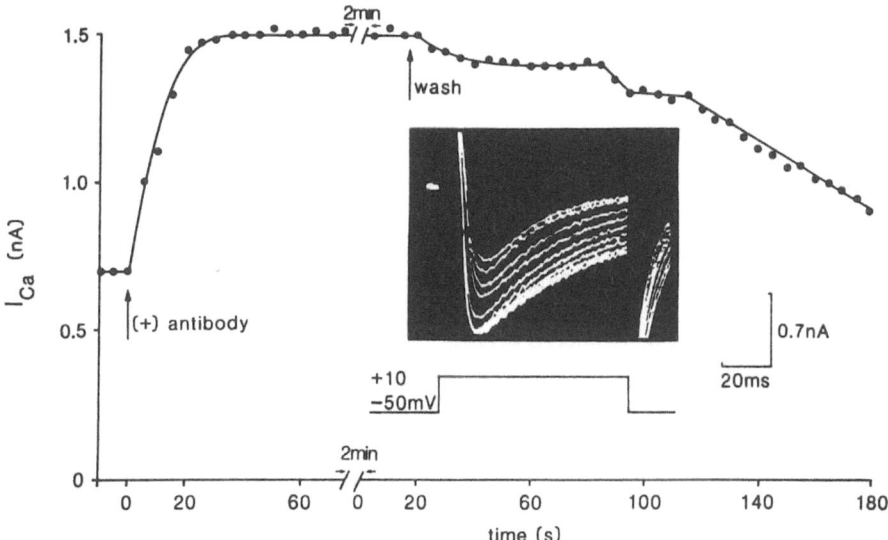

Fig. 11. Effect of antibody on I_{Ca} of rat ventricular myocytes. Time course of the effect of antibodies and its reversal upon removal of antibodies. *Inset*, original superimposed traces of I_{Ca}, activated at $+10$ mV from a holding potential of -50 mV, upon addition of antibodies (IgG, T52). External solutions contained 2 mM Ca^{2+} and 10^{-5} M TTX. Internal solutions contained 100 mM CsCl, 20 mM TEA, 20mM glucose, 5 mg ATP, 10 mM EGTA, 1 mM Ca^{2+}, 10 mM Hepes, at pH 2.3 buffered with CsOH at room temperature

difficult at potentials near -10 mV. The enhancing effect of antibody on I_{Ca} could be blocked by addition of nifedipine (10^{-6} to 5×10^{-5} M). β-Blockers (e. g., propranolol at 10^{-6} M), however, had little or no effect on the enhancement of I_{Ca} by the antibody. Irrespective of the mechanism of antibody action, the strong enhancement of I_{Ca} by the antibody suggests that the antibody binds directly to a channel moiety and modifies its kinetics, leading to enhanced calcium permeability of the cell. This finding is consistent with the observation that exposure of isolated myocytes to high concentrations of antibodies leads to cell deterioration and death. This antibody-induced cytotoxicity thus appears to occur secondary to calcium overload.

Conclusions

The experiments described here provide evidence for a cross-reactivity of antibodies against the ADP/ATP carrier of the inner mitochondrial membrane with the calcium channel – a cell surface protein. Our results demonstrate that these antibodies specifically enhance calcium transport through the calcium channel by increasing the magnitude of calcium current and slowing its inactivation. Consistent with this finding, isolated intact rat myocytes became unstable within 60 min exposure to antibody, leading to cell death. In contrast, control cells plus preimmune serum remained stable for at least 10 h. Cytotoxicity was delayed or completely suppressed when organic calcium-channel blockers (verapamil, nitrendipine, and nifedipine) were used in addition to the antibody. Further support for the idea that the antibody – primarily directed against the ADP/ATP carrier – binds to the calcium channel comes from our observation that these antibodies bind to the 175-kDa and the 55-kDmoiety of the isolated calcium channel (data not shown). The cross-reactivity between ADP/ATP carrier and calcium channel might reflect common structural and perhaps functional similarities between these two transmembrane proteins, as was also shown for other membrane channels.

We have recently shown the existence of autoantibodies to the ADP/ATP carrier in a significant proportion of patients suffering from virus myocarditis or dilated cardiomyopathy – suspected autoimmune diseases [1, 2]. These autoantibodies interact with isolated cardiac myocytes and have a cytotoxic effect. Further data indicate that these autoantibodies also modulate the function of the calcium channel and induce cytotoxicity secondary to calcium overload. These findings may be of great clinical relevance in the understanding of these diseases.

As abnormalities in Ca^{2+} metabolism of myocardium are thought to be involved in genesis of cardiac myopathies [12, 13], the results described here suggest that the etiology of dilated cardiomyopathy may be related to an autoantibody-mediated enhancement of calcium current and subsequent Ca^{2+} overload of myocytes.

References

1. Schultheiss H-P, Bolte H-D (1985) Immunological analysis of autoantibodies against the adenine nucleotide translocator in dilated cardiomyopathy. J Mol Cell Cardiol 17:603–617
2. Schultheiss H-P (1987) The mitochondrium as antigen in inflammatory heart disease. Eur Heart J 8, Suppl J:203–210

3. Klingenberg M (1985) The ADP/ATP carrier in mitochondrial membranes. In: Martonosi A (ed) The enzymes of biological membranes, vol 4. Plenum, New York, pp 511–553
4. Klingenberg M, Heldt HW (1982) The ADP/ATP translocation in mitochondria and its role in intracellular compartmentation. In: Sies H (ed) Metabolic compartmentation. Academic, London, pp 101–122
5. Schultheiss H-P, Klingenberg M (1984) Immunochemical characterization of the adenine nucleotide translocator – organ specificity and conformation specificity. Eur J Biochem 143:599–605
6. Krämer R, Klingenberg M (1979) Reconstitution of adenine nucleotide transport from beef heart mitochondria. Biochemistry 18:4209–4215
7. Klingenberg M, Appel M (1980) Is there a common binding center in the ADP, ATP carrier for substrate and inhibitors? Amino acid reagents and the mechanism of the ADP, ATP translocator. FEBS Lett 119:195–199
8. Hohl CM, Altschuld RA, Brierley GP (1983) Effects of calcium on the permeability of isolated adult rat heart cells to sodium. Arch Biochem Biophys 221:197–205
9. Mitra R, Morad M (1986) Two types of calcium channels in guinea pig ventricular myocytes. Proc Natl Acad Sci USA 83:5340–5344
10. Hamill OP, Marty A, Neher E, Sakmann B, Sigworth F (1981) Improved patch clamp technique for high resolution current recording from cells and cell free membrane patches. Pflügers Arch Gesamte Physiol 391:85–100
11. Konnerth A, Lux HD, Morad M (1987) Proton-induced transformation of calcium channel in chick dorsal root ganglion cells. J Physiol 386:603–633
12. Weisman HF, Weisfeldt ML (1987) Toward an understanding of the molecular basis of cardiomyopathies. J Am Coll Cardiol 10:1135–1142
13. Limas CJ, Olivari MT, Goldenberg IF, Levine TB, Benditt DG, Simon A (1987) Calcium uptake by cardiac sarcoplasmic reticulum in human dilated cardiomyopathy. Cardiovasc Res 21:601–605

Calcium, Aging and Disease

Calcium Aging, and Diseases

T. Fujita

Third Division, Department of Medicine, Kobe University School of Medicine, 7–5–2 Kusunoki-cho, Chuo-ku, Kobe 650, Japan

Introduction

In addition to being an essential nutrient, calcium is the major component of hard tissues such as bone and teeth, where 99% of the total body calcium is located to provide strength and resistance to physical stress. The concentration of calcium in the extracellular compartment, mainly in circulating blood plasma, is much lower than that in the hard tissues – ratio ca. 1:10000 – and is maintained at a constant level with surprising accuracy. Calcium-regulating hormones such as parathyroid hormone, calcitonin, and 1,25 $(OH)_2$ vitamin D play important roles in the homeostasis of plasma calcium. A constant plasma calcium level is necessary for normal functioning of the neuromuscular system and the heart.

Calcium is also found intracellularly. Cytosolic free calcium is also maintained at a constant level, 1/10000 that in the plasma, although much higher calcium concentrations are found in organelles, e. g., in mitochondria and endoplasmic reticulum, and in cytoskeleton, these sites acting as reservoirs for cellular calcium. Intracellular calcium is essential for signal transduction across the cell membrane and within the cell. Besides acting as a messenger by itself, calcium is essential in the function of calcium-binding proteins such as calmodulin and of calcium-activated enzymes such as calpain and protein kinase C.

In addition to the actions of calcium in each of these three compartments – skeletal, extracellular, and intracellular – the vast difference in calcium concentration among the compartments is of prime importance. When the marked concentration gradient across an intercompartmental border is attenuated, the cell or organism inevitably loses part of its function. This occurs in aging and in diseases characterized by functional impairment of each cell or the whole organism. Calcium is thus unique in its nonuniform distribution with enormous concentration gradients in cells and organisms.

Calcium Environment

In view of the close similarity of the constituents of the human body to those of seawater, but not to those of the land, life may have come from the sea (Fujita, 1987). Some primitive form of life must have developed somewhere in or near seawater during chaotic changes on the surface of the Earth. One of the important differences between creatures living in seawater and those living on land, like human beings, is in the supply of calcium from the environment. Fish living in seawater, for example, obtains as much calcium as required from the seawater constantly streaming through the gills. Life on land, on the contrary, is a constant struggle against calcium deficiency, since no calcium is found in the air. The only source of calcium in landdwelling animals is food, which never supplies as much calcium as seawater. The longer we live in a calcium-deficient environment, the more calcium-deficient we inevitably become and the harder the struggle naturally gets (Fujita, 1986).

Calcium-Regulating Hormones

Parathyroid Hormone

The key to the understanding of calcium metabolism is parathyroid hormone (PTH). Parathyroid glands were discovered in 1850 in the rhinoceros and later described in all kinds of mammals and birds, but never in fish. Parathyroid hormone, a single-chain peptide consisting of 84 amino acids, was identified and sequenced in man as well as in cow, pig, rat, and chicken. The presence of peptides with PTH-like immunoreactivity in fish plasma has been claimed but never confirmed. The possible origin of the PTH-like material is the corpuscle of Stannius. No evidence, however, has yet been presented as to the presence of authentic parathyroid glands or PTH in fish.

PTH is important in maintaining serum calcium at a constant level. Secreted in response to hypocalcemia, it normalizes serum calcium by mobilizing calcium from bone through increased resorption, increasing reabsorption of filtered calcium from renal tubules, and augmenting calcium absorption from the gut through increased $1,25(OH)_2$ vitamin D synthesis in the proximal tubule. Even in the absence of marked and sustained hypocalcemia, PTH may be secreted in response to a constantly negative calcium balance, possibly through a minor and transient fall of serum calcium, or mild $1,25(OH)_2$ vitamin D deficiency due to aging of the kidney. PTH increases in chronic renal failure as the glomerular filtration rate falls below 40 ml/min. Age-linked increase of serum PTH is also well established, but the part played by the progressive loss of renal function in old age has not been defined. The terminal rise of PTH in rats predicting renal failure and death may be prevented, with considerable prolongation of life span, by administering a low-protein diet containing adequate calcium. In addition to a marked increase of parathyroid gland weight in chronic renal failure, age-linked increases and increases in groups of lower social standing, presumably with low dietary calcium intake, have been reported. PTH thus appears to be nature's way of compensating calcium deficiency.

The action of PTH is mediated by activation of adenylate cyclase and an increase of cytosolic free calcium. Thus PTH may be called one of the natural calcium ionophores. The profound calcium concentration gradient between extracellular and intracellular compartment may be attenuated by the action of PTH.

Calcitonin

Calcitonin is an antagonist of PTH in many respects. Secreted in response to hypercalcemia, it directly inhibits osteoclastic bone resorption, decreasing the release of calcium from bone into the bloodstream. It is surprising that the physiological action of calcitonin in humans has not yet been established, almost 30 years after its discovery (Copp et al., 1962). In contrast to PTH, calcitonin may be called the hormone of calcium abundance, required only when too much calcium is coming out of bone. Since humans are always deficient in calcium, especially towards old age, normal individuals may not require calcitonin. Nevertheless, calcitonin is a useful therapeutic agent for conditions associated with abnormally increased bone resorption, such as hypercalcemia and osteoporosis.

Calcium and Cells

The sharp calcium concentration gradient across the cell membrane is necessary for all cell activity. A small rise in intracellular calcium causes physiological excitation of nerve cells, contraction of muscle cells, and secretion from endocrine and exocrine cells. Further increase with time, due to simple wear and tear, peroxide accumulation, cross-linkage, or other factors leading to functional incompetence of cellular membranes, may blunt the intercompartmental calcium concentration gradients.

At least some of the functional disturbances of aging may be due to such blunting of the calcium gradients. Aging is defined as an increased risk of diseases. When, as in many diseases, cell membranes are weakened or damaged because of genetic predisposition, physical causes, viral or bacterial infection, toxic substances, or autoantibodies, cell membranes are no longer capable of maintaining the sharp concentration gradients between the compartments, causing further blunting. This results in disturbance of cell motility, abnormal contraction, disturbance of secretion, and uncontrollable proliferation. The final consequence would be a complete loss of the calcium concentration gradient between compartments, or cell death.

All diseases are thus associated with cellular dysfunction. All cellular dysfunctions are associated with abnormal signal transduction and abnormal calcium distribution and dynamics. In other words, abnormal calcium distribution, which may be caused by PTH secreted in response to calcium deficiency, may provide a backdrop for aging and diseases (Fig. 1).

Calcium and Aging

Calcium deficiency in aging is accelerated by progressive loss of kidney tissue, leading to decreased biosynthesis of 1,25 $(OH)_2$ vitamin D_3, tendency of less exposure to sunshine, resulting in decreased conversion of previtamin D to vitamin D in the skin and poor dietary intake of vitamin D itself.

As shown in Fig. 2, calcium deficiency progresses with age. Serum calcium remains constant, with minor fluctuations, but serum PTH rises with age. Bone calcium is lost in response to calcium deficiency, estrogen lack, and secondary hyperparathyroidism.

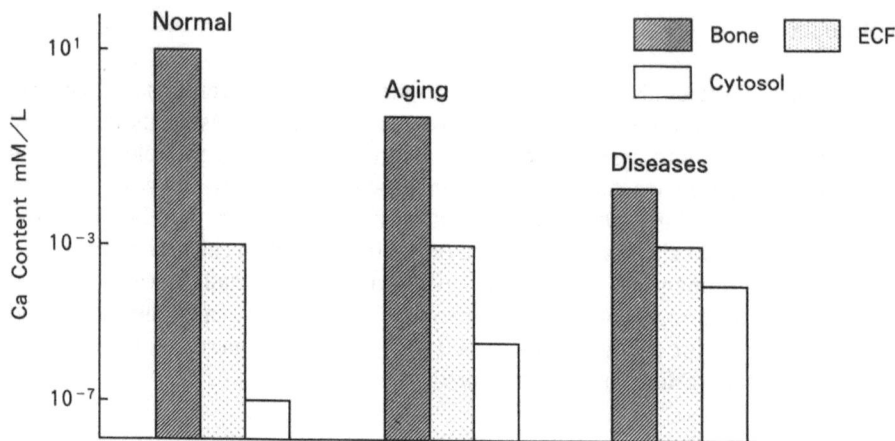

Fig. 1. The normal sharp calcium concentration gradients among the bone, extracellular and intracellular compartments are blunted in aging and, much more so, in disease

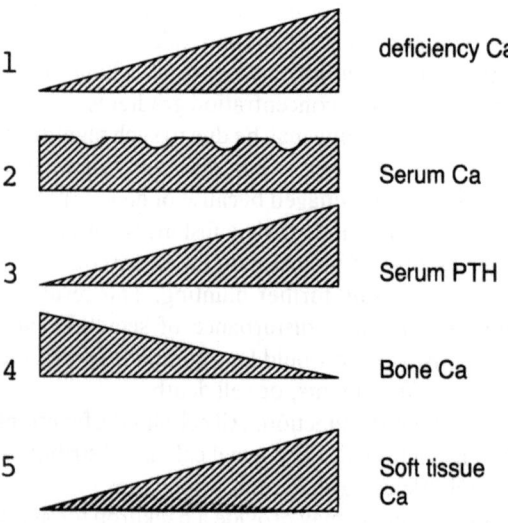

Fig. 2. *1* Calcium deficiency augments with advancing age. *2* Serum calcium remains constant with minor fluctuations. *3* Serum PTH rises progressively with age in response to calcium deficiency. *4* Bone calcium is gradually lost in response to calcium deficiency and secondary hyperparathyroidism. *5* Soft tissue calcium increases with age ("calcium shift")

Release of even a small proportion of bone calcium may represent calcium flooding for soft tissues, where little calcium is normally found. This increase of calcium content in soft tissue along with loss of bone calcium, or "calcium shift" from bone to soft tissue, was originally suggested by Elkeles (1957). The fact that calcium deficiency results in calcium excess in soft tissues is the "calcium paradox of aging." The list of diseases which may be caused by calcium deficiency and aging is quite long.

Typical examples are osteoporosis, or age-dependent bone loss (Nordin, 1962), and hypertension, which may be caused by smooth muscle contraction due to increased intracellular calcium induced by calcium deficiency and excess of PTH (McCarron et al., 1987). Arteriosclerosis begins earlier than macroscopic atherosclerosis and calcification, as elastin undergoes degeneration on calcium deposition (Lansing, 1950). Low calcium intake predisposes to myocardial infarction (Knox, 1973). The progress of nephritis is mediated by mesangial contraction due to increased intracellular calcium. Dialysis encephaloneuropathy, senile dementia, and even amyotrophic lateral sclerosis may result from deposition in the nervous system of calcium released from bone in response to calcium deficiency and secondary hyperparathyroidism.

References

Copp DH, Cameron EC, Cheney BA, Davidson AGF and Henze KG (1962) Evidence for calcitonin – a new hormone from the parathyroid that lowers blood calcium. Endocrinology 70:636–649
Elkeles A (1957) A comparative radiological study of calcified atheroma in males and females over 50 years of age. Lancet 2:714–715
Fujita T (1986) Aging and calcium. Mineral and Electrolyte Metabolism 12:149–156
Fujita T (1987) Calcium and your health (in English). Japan Publications, Tokyo
Knox EG (1973) Ischemic heart disease and dietary intake of calcium. Lancet 1:1465–1467
Lansing AI, Alex M, Rosenthal TB (1950) Calcium and elastin in human arteriosclerosis. J Gerontol 5:112–119
McCarron DA, Morris CD, Cole C (1987) Dietary calcium in human hypertension. Science 217:267–269
Nordin BEC (1962) Calcium balance and calcium requirement in spinal osteoporosis. Amer J Clin Nutr 10:384–390

Subject Index

abscissic acid 552
acetylcholine 242, 348
adenosine 384, 413
adenylate cyclase 366, 419
adenylimidodiphosphate 52
ADP/ATP carrier antibodies 619–631
adrenaline 79
adrenomedullary secretion 347–362
afterhyperpolarization 473
aging 465–475, 491–501, 635–639
alcohol 541–548, 553
Alzheimer's disease 474, 475
amiloride 76, 117, 120, 121
amiodarone 240
amlodipine 82–90, 240, 354
anaesthetics 317–325
angiotensin 284, 382
antiarrhythmics 317–333
antidromic stimulation 529–537
antiheart antiserum 621
antipsychotic drugs 520–525
arrhythmias 310–316
ATP 52, 79
azidopamil 159
azidopine 159, 168, 169–189, 202

basolateral membrane 410
BAY K 8644 56–60, 63–69, 109, 110, 112, 117, 185, 206, 213, 220, 224, 232, 240, 255, 257, 262, 275, 287, 329, 333, 338, 340, 351, 353, 381, 423, 444, 445, 453, 456–463, 524, 544, 550, 551, 564, 566
BAY P 8857 240
benzothiazepines 163, 168–189, 174, 194, 354
bepridil 58, 159, 163
bilayer pathway model 235
bradykinin 436, 437
buffer barrier hypothesis 287

cadmium 343, 356
caffeine 287
calbindin 22, 24, 26
calcineurin 25, 29, 52, 111

calcitonin 637
calcium agonists 54–61, 63–69, 94, 109, 110, 150, 213, 335–346, 550–559
calcium antagonists 5, 81–91, 115–125, 150, 159–166, 257, 333, 335, 354, 355, 378, 466, 550–559
– inorganic 72, 117, 128–136, 213, 349, 353, 358, 438, 534, 535
– receptor 187, 193, 194, 196, 201, 206, 207, 293, 301, 333, 541–545, 564
calcium ATPase 377
calcium-calmodulin system 412
calcium channel
– activation 45
– antibody 178, 200–210, 232–237, 423, 586–600
– blocker, see calcium antagonists
– brain 182, 205, 206
– closed times 44, 67–68, 133, 447–462
– conductance 184, 233, 257, 445, 453, 456
– density 97, 98
– dephosphorylation 50
– endogenous modulators 611
– gating mechanisms 459–463
– heart 39–52, 293–309
– inactivation 45, 103–112, 125, 153, 456, 461, 629
– lipid bilayer incorporation 60
– L-type 40–52, 95–99, 116–125, 159, 164, 168, 182, 187, 193, 212, 442–463, 466, 467, 522, 533, 554, 564
– mRNA 272–280
– Markov model 56
– model 134, 152, 207, 443, 448, 450, 460
– modes 54, 56, 125, 456
– N-type 57, 116–117, 168, 182, 442–463, 522, 533, 534
– open probability 43, 47, 49, 66, 94, 257, 457
– open times 44, 52, 54, 67, 68, 133–135, 233, 447–462
– phosphorylation 50–52, 100–112, 182, 193, 204, 554
– photoaffinity labeling 159, 168–169, 194

- proton-induced transformation 71–79
- rate constants 451, 458
- reconstitution 183, 209, 213, 217–230, 252–271
- regulatory mechanisms 187
- selectivity 455
- skeletal muscle 159, 193, 200–209, 213
- states 44, 68, 69, 89, 187, 197, 356, 447–463
- structure 160, 168–192, 200–210, 211–216, 233, 586–599
- tissue distribution 165, 168–189
- T-type 57, 95–99, 116–125, 159, 347–362, 442–463, 533

calcium chelators 377
calcium current
- activation 45, 64, 125, 451
- β-adrenergic modulation 39–52
- deactivation 64, 460
- inactivation 45, 47, 466–468
- first latency 45, 49, 54, 56
- kinetics 39–52, 56, 67, 68, 117, 118, 349, 442–463
- Purkinje fibers 82–90
- recovery after inactivation 149
- run down 104–111, 466
- single channel measurements 40–52, 54–61, 93–99, 104–112, 132–134, 219–226, 259, 275, 444–463, 616
- tail current 149, 460, 461, 468
- waiting time 449
- whole cell measurements 40, 41, 49, 50, 51, 54–61, 63–69, 71–79, 81–90, 93, 104–112, 116–125, 130–135, 434, 435, 444

calcium flux 206, 284, 348, 378, 541–545
calcium ionophore 411
calcium modulated proteins 16–32
calcium overload 311, 312, 327, 630
calcium regulating hormones 635
calcium release 138–154, 286
calcium shift 638
calcium transient 139, 141, 142, 147, 150
calmidazolium 342
calmodulin 16–32, 342, 386, 412
calpain 26
calretinin 22, 26
caltractin 29
cAMP 50, 52, 108, 109, 287, 394–403, 412, 554
cardiac hypertrophy 329
cardiomyopathy 294
catecholamine release 347–362
cells
- cardiac 39–52, 64, 81–91, 317
- chromaffine 71, 77, 78, 348–362
- endothelial 288, 289, 575
- fibroblasts 92–99
- juxtaglomerular 372, 404

- neuronal 71–79, 115–125, 128–136, 433–440, 442–463, 541
- pancreatic 363–371
- parathyroid 452
- PC 12 604
- pituitary 103–112, 335–345

cellular damage 333
cerebral ischemia 506, 507
cesium 339
CGP 28-392 63–69, 351, 551
charge movement 138–154
chelators 376
chloride currents 276
chlorpromazine 242, 479–483
cholecystokinin 365
cholera toxin 365, 436
cinnarizine 356, 523
clonidine 488
clopimozide 164
con A 202, 208
ω conotoxin 76, 115–125, 129, 161, 168, 172, 522
cytosolic calcium 337, 349, 410, 419
cytotoxicity 619–634

D 575 257, 262
D 600 140, 142–146, 150, 213, 220, 224, 348, 355, 444, 445, 550, 551
D 890 186, 260, 262
Dahl rats 327
diacylglycerol 437
diazepam 552
dibenamine 552
dihydropyridine 5, 54–61, 63–70, 81–90, 104–112, 140–142, 159, 168–192, 194, 206, 231–251, 252–271, 327–334
- anticonvulsant properties 503–516
- binding site see receptor
- charged 83–90
- enantiomers 54, 57–59, 63, 269, 423, 550
- G-protein 54–62
- heart 81–91, 293–309
- receptor 54, 55, 56, 69, 89, 161, 231–251, 293–309, 478–489, 520–525, 565–574
- vascular smooth muscle 283–292

digitonin 201, 204, 352
diltiazem 72, 150, 159, 201, 206, 262, 310, 354, 381, 481, 482, 542, 550, 551, 558, 602, 614
diphenylbutylpiperidines 163, 165, 168
diphenylhydantoin 58
dodecanol 317
dopamine 524, 553
drosophila 163, 168
drug design 231

EDRF 283–292, 575
EF-hand 17–32

endogenous ligands 549–559, 564, 565, 566, 570, 571, 572, 601–610, 611–618
endoproteinase 195
endothelin 286, 288, 575–585
endothelium 283–292
enkephalin 571
epilepsy 514, 515, 536, 537
EPSP 469–472
ethacizin 317, 320
ethanol 317, 541–545
ethmozin 317, 320
excitation-contraction coupling 138–154
excitatory amino acids 437, 532
exocytosis 77, 78

fendiline 533
flecainide 317, 318
flunarizine 310, 354
flunitrazepam 613
fluphenazine 482–484
fluspirilone 164, 523, 564
forskolin 287, 398
frequency potentiation 469–472
fura-2 289

GABA 537
gallopamil 564
glutamate 437
glycine 437–439
G-proteins 54–61, 363–371, 419–421, 425, 434, 436, 551, 554
growth hormone 336
GTP 58, 59
GTP binding protein 564
guarded receptor hypothesis 323

H 160/51 63
haloperidol 523
heart contractility 9–15
hexanol 317
hippocampal neurons 465–477
hippocampal slices 528–540
histamine 284
HOE 166 552, 564
Huntington's disease 544, 545
5-hydroxytryptamine 284
hyperparathyroidism 637
hypertension 293, 327, 639

inositol triphosphate 286, 344, 365, 436, 437
ionomycin 351, 411
ionophore A 23187 350
ischaemia 293, 506, 507
isoproterenol 59

kainate 439, 440
kainic acid 553
ketamine 536

learning 473
lectinus 201
lidoflazine 523
liposomes 219
local anesthetic 241
LU 49888 168

macula densa 372, 404
magnesium 128–136, 438, 469
menthol 117
methacholine 352
methoxyverapamil 381
methyladenine 552
MK 801 438, 439
modulated receptor hypothesis 89, 323
molecular modelling 239
morphine 294, 485–489
muscular dysgenesis 159
myosin light chain 16, 21, 22

Na-Ca exchange 352, 374
Na,K-ATPase 374
naloxone 488, 489
natriuresis 389
nerve conduction velocity 499
neural plasticity 491
neuroleptics 164
neuropeptides 79, 433–436
neuropharmacology 433–440
neurotransmitter release 77, 78, 351, 434
neutron diffraction 237, 238, 244
nicardipine 83–85, 189, 354, 506, 511, 512, 584
nicotine 352
nifedipine 5, 74, 76, 81, 117, 140–142, 164, 261, 275, 284, 310, 328, 353, 354, 378, 533, 534, 542, 549, 553, 604, 628
niludipine 354
nimodipine 5, 104, 236, 240, 257, 262, 330, 343, 354, 466, 468, 479–488, 491–501, 523, 524, 566
nisoldipine 5, 81–90, 239, 339, 354, 488, 566
nitrendipine 5, 54, 55, 58, 81, 164, 187, 201, 206, 253, 257, 262, 328, 338, 353, 354, 438, 439, 440, 523, 542, 543, 558, 566, 601, 607, 613, 615, 628
nitric oxide 288, 575
NMDA 437–440, 532, 533
norepinephrine 284
nucleoside 169
nucleotides 169, 272, 621

octanol 317
oocyte 273
orthodromic stimulation 529–537

osteoporosis 637
ouabain 352, 376, 566
oxygen deprivation 293

palmitoyl carnitine 558
parathyroid hormone 327, 333, 394–403, 418–431
Parkinson's disease 544, 545
parvalbumin 16, 20, 23, 26
penfluridol 164, 523
pertussis toxin 420, 434–437
phencyclidine 438
phenylalkylamines 161–165, 168–189, 194, 241, 269, 354, 521
phenylephrine 553
phenytoin 510, 511, 512
phorbol ester 386, 553
phosphatidylinositol 286, 384
phospholipase C 366–371
phosphorylation 184, 197
photoaffinity labeling 168–192, 246
picrotoxin 536
pimozide 164, 340, 482–485, 523, 552
plant membranes 163
PN 200-110 57, 58, 159, 164, 185, 202, 206, 213, 254, 257, 262, 298, 348, 354, 506, 511, 512, 521, 522, 542–545, 566, 588, 601
potassium channel 208
praziquantel 552
prenylamine 523
prolactin 335, 343
propranolol 240, 628
protein kinase C 52, 58, 184, 193, 194, 344, 383, 436, 554
pyrophosphatase 52

quin-2 289, 349
quinidine 317, 318
quisqualate 532

receptor-operated channels 384
renal hemodynamics 414
reperfusion arrthythmias 311
renin secretion 372–393
Ringer, Sidney 9
ryanodine 138, 312

sarcoplasmic reticulum 286, 311, 312
saxitoxin binding 566
S 207–180 257, 262
schizophrenia 523–525
seizures 506–514
sensorimotor function 494
sinoatrial nodes 310
skeletal muscle 138–154
slice neurons 466–469
sodium channels 58, 207, 208
sodium current 71–79, 115–125, 129–136, 186, 277, 319, 339
spontaneously hypertensive rats 326
stroke prone rats 330–333
strontium fluxes 348
substance P 434
synaptic transmission 469–470, 531

tetrandine 552
tetrodotoxin 339
thrombin 288
thyrotropin releasing hormone 344
troponin C 16–32
trypsin 195
tryptophan 248
tubuloglomerular feedback 404

vanadate 376
vascular injury 326–333
vascular smooth muscle 282–289
vasoactive intestinal peptide 344
vasopressin 382
verapamil 72, 117, 120, 121, 129, 201, 206, 213, 262, 310, 354, 381, 481–483, 510, 523, 533, 542, 549–551, 602, 628
veratridine 342, 614
vitamin D_3 637
voltage-sensor 138–154, 213

walking pattern 495–498

xenopus oocytes 272–279

YC 170 63, 551
202-791 63, 64, 94, 140–147, 351–354, 422

MIX
Papier aus verantwortungsvollen Quellen
Paper from responsible sources
FSC® C105338

If you have any concerns about our products,
you can contact us on
ProductSafety@springernature.com

In case Publisher is established outside the EU,
the EU authorized representative is:
**Springer Nature Customer Service Center GmbH
Europaplatz 3, 69115 Heidelberg, Germany**

Printed by Libri Plureos GmbH
in Hamburg, Germany